Universitext

Universitext

Universitext is a series of textbooks that presents material from a wide variety of mathematical disciplines at master's level and beyond. The books, often well class-tested by their author, may have an informal, personal even experimental approach to their subject matter. Some of the most successful and established books in the series have evolved through several editions, always following the evolution of teaching curricula, into very polished texts.

Thus as research topics trickle down into graduate-level teaching, first textbooks written for new, cutting-edge courses may make their way into *Universitext.*

More information about this series at http://www.springer.com/series/223

Tobias Mühlenbruch • Wissam Raji

On the Theory of Maass Wave Forms

Tobias Mühlenbruch
Department of Mathematics
and Computer Science
FernUniversität in Hagen
Hagen, Germany

Wissam Raji
Department of Mathematics
American University of Beirut
Beirut, Lebanon

ISSN 0172-5939 ISSN 2191-6675 (electronic)
Universitext
ISBN 978-3-030-40477-2 ISBN 978-3-030-40475-8 (eBook)
https://doi.org/10.1007/978-3-030-40475-8

Mathematics Subject Classification (2010): 11F11, 11F12, 11F99

This Springer imprint is published by the registered company Springer Nature Switzerland AG.
The registered company address is: Gewerbestrasse 11, 6330 Cham, Switzerland

Preface

"Why write a textbook on Maass wave forms?" This question was the usual reaction to our plan for writing this textbook. There are several textbooks on Maass wave forms already available. What distinguishes our textbook from others?

We believe that most of the available textbooks on Maass wave forms and related objects fall into two categories. Textbooks in the first category deal with classical modular forms (what we mean by "classical" is explained in Remark 1.63) and their arithmetic properties. That is, the focus in these books is on the connection between holomorphic modular forms and related objects in arithmetic geometry (e.g., elliptic curves and Galois representations). Some good examples of textbooks here are those of Apostol [12], Serre [182], Shimura [184], Lang [116], Schoenberg [176], Koblitz [109], and Diamond and Shurman [66]. Books in this category, by and large, do not tackle the subject of Maass wave forms, but some do briefly discuss it.

The second category of textbooks consists of those that discuss $\mathrm{GL}_2(\cdot)$ or $\mathrm{SL}_2(\cdot)$ and their associated Lie algebras. The main focus of books here is usually on the representations of Lie algebras, where holomorphic modular forms often appear. Maass wave forms are natural counterparts to those representations. Examples of textbooks in this category are those of Borel [17], Bump [52], Iwaniec [101], and Kubota [111].

Our textbook takes a different approach. We focus on Maass wave forms as the main objects of interest and strive to give Maass wave forms a firm analytical treatment, only briefly pointing out connections with other fields in mathematics like representation theory.

Maass wave forms are introduced as functions on the complex upper half-plane that are eigenfunctions of the hyperbolic Laplace operator, have a prescribed behavior under transformations by elements of the full modular group $\mathrm{SL}_2(\mathbb{Z})$, and admit an expansion of a certain form that contains exponentials and Whittaker functions. Instead of first placing Maass wave forms in the context of representation theory, ideas from geometric analysis and the theory of special functions are used to obtain results.

The textbook is structured as follows. We start with an informal exploratory Introductory Roadmap that serves as a guide to the topics presented.

Chapters 1 and 2 give a brief introduction to the theory of classical modular forms and related objects. The goal of these chapters is to briefly discuss modular forms and to introduce the necessary related objects and results (e.g., L-series and period polynomials) that are needed in later chapters. The selection of the presented results is biased to suit our needs, and we also omit several proofs that can be easily found in the references mentioned above. In many cases, we skip proofs in this chapter. However, we prove the analogous results for the Maass wave forms case, which usually come with greater generality.

Chapter 3 introduces Maass wave forms of real weight. After defining multiplier systems, we discuss three differential operators: the hyperbolic Laplace operator and the two Maass operators. Then, we introduce Maass wave forms of real weight and give some examples. We conclude this chapter by discussing Hilbert spaces of Maass forms, Hecke operators, and the Friedrichs extension of the hyperbolic Laplace operator, ending with a few remarks on the associated Selberg's conjecture.

Chapters 4 and 5 are the main chapters of this textbook. We introduce the concept of families of Maass wave forms and discuss the associated L-series in Chapter 4. Chapter 5 introduces period functions associated with (families of) Maass wave forms. The above objects are presented in analogy with classical modular forms and their associated L-series and period polynomials.

Chapter 6 connects Maass cusp forms (of weight 0) with discrete dynamical systems associated with the Artin billiard.

The last chapter, Chapter 7, introduces weak harmonic Maass wave forms and their associated objects like mock modular period functions and regularized L-series.

Readers of this textbook are required to have basic knowledge in complex analysis, real analysis, and abstract algebra. Although elementary concepts are being introduced in every chapter, some special knowledge of certain topics in analysis is assumed. Readers may refer to Appendix A, where some theorems and results are briefly presented with helpful references indicated in case more details are desired.

Some exercises are included. Detailed solutions to those exercises are also available online. These solutions can be found at our personal homepages and our ResearchGate pages.

Acknowledgements

We would like to express our sincere appreciation to all who have participated in improving this textbook. We are particularly grateful to our colleagues who encouraged us to pursue this project. We thank our reviewers for giving us numerous suggestions on how to improve the manuscript and for the encouraging and insightful comments. We also say a big thank you to Ken Ono for his support.

Another big thank you goes to Jeff Breeding-Allison, Lloyd Kilford, Larry Rolen, and M. Shane Tutwiler who assisted in proofreading and in the linguistic editing. We also thank our families for their understanding and support.

Hagen, Germany
Beirut, Lebanon
May 2020

Tobias Mühlenbruch
Wissam Raji

Contents

List of Symbols

$\mathbb{N}$	Set of natural numbers $\mathbb{N} = \{1, 2, 3, \ldots\}$
$\mathbb{Z}$	Set of integers $\mathbb{Z} = \{\ldots, -2, -1, 0, 1, 2, \ldots\}$
$\mathbb{Q}$	Set of rational numbers
$\mathbb{R}$	Set of real numbers
$\mathbb{C}$	Set of complex numbers
$\mathbb{H}$	Set of complex upper half-plane $\mathbb{H} = \{z \in \mathbb{C} \colon \ \Im(z) > 0\}$
$\mathbb{H}^-$	Set of complex lower half-plane $\mathbb{H}^- = \{z \in \mathbb{C} \colon \ \Im(z) < 0\}$
$\mathbb{C}'$	Set of complex cut-plane $\mathbb{C}' = \mathbb{C} \setminus (-\infty, 0]$
$\mathbb{C}[X]_k$	Set of complex polynomials in formal variable X of degree $\leq k$
$\mathrm{SL}_2(\cdot)$	Set/group of 2×2 matrices with determinant 1 and entries in "$\cdot$"
$\mathrm{GL}_2(\cdot)$	Set/group of 2×2 invertible matrices with entries in "$\cdot$"
$\mathrm{Mat}_2(\cdot)$	Group of 2×2 matrices with entries in "$\cdot$"
$\mathrm{trace}(V)$	The trace of matrix V
x^t	Transpose of a matrix or vector x
$\Re(z)$	Real part of z
$\Im(z)$	Imaginary part of z
$\mathrm{Im}(g)$	Image of the map g
$\mathrm{Ker}(g)$	Kernel of the map g
$\mathrm{sign}(a)$	Sign function of real argument a
$\lvert a \rvert$	Absolute value of complex argument a
$\gcd(a, b)$	Greatest common divisor of integers a and b
$n \mid c$	n divides c

$f(x) = O(g(x))$ as $x \to \infty$	"Big O notation": $\limsup_{x\to a} \left\lvert \frac{f(x)}{g(x)} \right\rvert < \infty$ if $g(x) \neq 0$
$f(x) \ll g(x)$	$f(x) = O(g(x))$
$f(x) \sim g(x)$ as $x \to \infty$	Asymptotic behavior: $\lim_{x\to\infty} \frac{f(x)}{g(x)} = 1$
$z \searrow 0$	One-sided limit: z tends to 0 and $z > 0$

Introductory Roadmap

This section is intended to serve as a roadmap to the content of this textbook. We sketch the content of this textbook through simple examples and brief explanations of certain concepts.

Chapter 1 – A Short Introduction to Modular Forms

In this chapter, we introduce classical modular forms. Since there are already many good textbooks on modular forms, we keep this section as short as possible by presenting only the necessary concepts and objects for the subsequent chapters on Maass wave forms and often leaving out certain details. In many cases, we omit the proofs mainly because we prove their analogues in the Maass forms case. The reader proficient with classical modular forms can easily skip this chapter.

In this chapter, we start by defining the notion of a modular form. For simplicity, we consider in this informal introduction the defining properties of a modular cusp form F. Let $k \in 2\mathbb{N}$ be an even natural number. A function $F : \mathbb{H} \to \mathbb{C}$ on the complex upper half-plane is called modular cusp form of weight k if

1. $F : \mathbb{H} \to \mathbb{C}$ is a holomorphic function
2. $F\left(\frac{az+b}{cz+d}\right) = (cz+d)^k\, F(z)$ for all $z \in \mathbb{H}$ and $a, b, c, d \in \mathbb{Z}$ with $ad - bc = 1$
3. $F(z) \to 0$ for $\Re(z) \to \infty$ and $\Im(z)$ bounded (i.e., $z \to i\infty$ on a vertical strip)
4. F admits a Fourier expansion of the form $F(z) = \sum_{n=1}^{\infty} a(n)\, e^{2\pi i n z}$ with suitable coefficients $a(n)$

A typical example is the discriminant function

$$\Delta(z) = e^{2\pi i z} \prod_{n=1}^{\infty} \left(1 - e^{2\pi i n z}\right)^{24} = \sum_{n=1}^{\infty} \tau(n)\, e^{2\pi i n z} \qquad (z \in \mathbb{H}). \tag{0.1}$$

We also give other important examples of modular cusp forms: Eisenstein series (§1.3.4), holomorphic Poincaré series (§1.4), and theta series (§1.5).

Having defined the notion of modular (cusp) forms of weight k, we introduce Hecke operators T_n ($n \in \mathbb{N}$). We discuss these operators in Section 1.6 and give their basic properties.

In Section 1.7, we introduce L-series associated with a modular cusp form. We expand on the theory of L-series and their properties because of their important role in the later chapters. The associated L-series $L(s)$ and L-function $L^\star$ of a modular cusp form $F(z) = \sum_{n=1}^{\infty} a(n)\, e^{2\pi i n z}$ are defined as

$$L(s) = \sum_{n \in \mathbb{N}} a(n)\, n^{-s} \qquad (\Re(s) > \frac{k}{2} + 1) \quad \text{and}$$

$$L^\star(s) = \int_0^\infty F(iy)\, y^{s-1}\, dy \qquad (s \in \mathbb{C}).$$

One interesting result is Hecke's converse theorem, which states that the following two spaces are equivalent:

1. the space of modular cusp forms of weight k
2. the space of L-series $L(s)$ that converges absolutely on the half-plane $\Re(s) > \frac{k}{2} + 1$ such that its associated L-function

$$L^\star(s) := (2\pi)^{-s}\Gamma(s)\, L(s)$$

extends to a holomorphic function on $\mathbb{C}$ and satisfies the functional equation

$$L^\star(s) = (-1)^{\frac{k}{2}}\, L^\star(k-s).$$

This converse result will have an analog in the context of Maass cusp forms that will be presented in later chapters.

Chapter 2 – Period Polynomials

We introduce the concept of period polynomials associated with modular cusp forms in this chapter. These polynomials are discussed in detail and several related representations are given. The theory of period polynomials and subsequently period functions will be discussed again in the context of Maass cusp forms in Section 5.1.

To illustrate, the period polynomial of a modular cusp form F of weight $k \in 2\mathbb{N}$ (for the full modular group) is given by:

$$P(X) = \int_0^{i\infty} (z - X)^{k-2}\, F(z)\, dz \qquad (X \in \mathbb{C}).$$

Obviously, $P(X)$ is a polynomial of degree at most $k-2$. In Section 2, we give other representations via Eichler integrals, the $(k-1)$-fold antiderivative, an additional integral representation, and the Niebur integral.

The period polynomials lead naturally to a cohomological description of the space of cusp forms, where the period polynomials are representatives of elements in such a group. This is known as the Eichler cohomology group, introduced in Section 2.2. The associated result, mapping the Eichler cohomology group to the space of modular cusp forms, is known as the Eichler cohomology theorem, discussed in Section 2.3.

We conclude this chapter with Section 2.4, where we construct a representation of the Hecke operators that act on period polynomials, following an approach by Choie and Zagier [57] and some remarks in Section 2.5.

Chapter 3 – Maass Wave Forms of Real Weight

In this chapter, we introduce Maass wave forms and give their properties.

Consider the second-order differential operator

$$\Delta_k := -y^2\left(\frac{\partial^2}{\partial x^2}+\frac{\partial^2}{\partial y^2}\right)+iky\frac{\partial}{\partial x}$$

for some real weight k. We call Δ_k the hyperbolic Laplace operator of weight k. (For simplicity, we will not explain the weight factor in detail in this introduction.) Following the definition of modular cusp forms, we define Maass cusp forms of real weight k (for the full modular group). We restrict ourselves here to the case $k \in 2\mathbb{Z}$.

A function $u : \mathbb{H} \to \mathbb{C}$ on the complex upper half-plane is called the Maass cusp form of weight k if

1. u is a real-analytic function satisfying $\Delta_k\, u = \left(\frac{1}{4}-\nu^2\right) u$ for some spectral parameter ν
2. $u\left(\frac{az+b}{cz+d}\right) = e^{-ik\arg(cz+d)}\, u(z)$ for all $z \in \mathbb{H}$ and $a, b, c, d \in \mathbb{Z}$ with $ad-bc=1$
3. $u(z) \to 0$ for $\Re(z) \to \infty$ and $\Im(z)$ bounded (i.e., $z \to i\infty$ on a vertical strip)
4. u admits a Fourier-Whittaker expansion of the form

$$u(x+iy) = \sum_{\substack{n\in\mathbb{Z}\\ n+\kappa\neq 0}} A(n)\, \mathbb{W}_{\operatorname{sign}(n)\frac{k}{2},\nu}\left(4\pi\,|n+\kappa|\,\Im(z)\right) e^{2\pi i(n+\kappa)\Re(z)}$$

where $\mathbb{W}_{k,\nu}(y)$ denotes the W-Whittaker function with parameters k and ν. (Some details about Whittaker functions are found in Section 3.2.1.) The parameter κ is suitably chosen, satisfying $0 \le \kappa < 1$ and $\kappa - \frac{k}{12} \in \mathbb{Z}$.

We then give some basic properties of Maass cusp forms in Section 3.3. For example, there exist first-order differential operators $\mathrm{E}_k^{\pm}$, known as Maass operators, that raise or lower the weight by 2. If u is a Maass cusp form of weight k, then $\mathrm{E}_k^{\pm}\, u$ is a Maass cusp form of weight $k \pm 2$ (or vanishes).

In Section 3.4, we give a few examples for Maass forms: the embedding of holomorphic modular forms, non-holomorphic Eisenstein series, and a Maass cusp form derived from the Dedekind η-function. We continue by discussing the Hilbert space spanned by Maass cusp forms in Section 3.5 and introduce Hecke operators in Section 3.6.

Section 3.7 discusses the Friedrichs extension, a self-adjoint extension of the differential operator Δ_k: First, we define a suitable L^2-Hilbert space, where the space C_0^2 of two times continuously differential functions with compact support is a dense subset. Then, we use standard arguments from operator theory to show that there exists a self-adjoint operator A_k which coincides on the dense subset C_0^2.

We conclude Chapter 3 with some remarks about Selberg's conjecture which is a conjecture about the actual range of the spectral parameter ν of Maass cusp forms.

Chapter 4 – Families of Maass Cusp Forms, L-Series, and Eichler Integrals

In the previous chapter, we discussed Maass cusp forms u of weight k. In Chapter 4, instead of discussing an individual Maass form u of weight k, we consider sequences of Maass forms $\big(u_l\big)_{l\in 2\mathbb{Z}+k}$. The individual entries u_l are Maass forms of weight l which are coupled via the Maass operators. The Maass cusp forms u_l satisfy

$$\mathrm{E}_l^+ \, u_l = -\big(1+2\nu+l\big)\, u_{l+2} \quad \text{and} \quad \mathrm{E}_l^- \, u_l = -\big(1+2\nu-l\big)\, u_{l-2}$$

for all $l \in 2\mathbb{Z}+k$. We call the sequence $\big(u_l\big)_l$ of Maass cusp forms a family of Maass cusp forms.

In Section 4.1, we introduce the concept of families of Maass cusp forms and their properties. We also show that the suitable scaled Fourier-Whittaker expansions of the forms u_l admit the same coefficients.

This change of focus allows us to treat the weight l as a parameter. In some sense, we might say that the individual Maass cusp form u_l in a family $\big(u_l\big)_{l\in 2\mathbb{Z}+k}$ is not so important and that it can be easily interchanged with another Maass cusp form $u_{l'}$ of the same family.

Section 4.2 extends the concept of L-series and L-functions to families of Maass cusp forms. For example, we define the $L^\star$-function again by an integral transformation similar to the modular cusp form case. We write

$$L_l^\star(s) = \mathcal{M}\Big(u_l(i\cdot)\Big)(s) \overset{(1.154)}{=} \int_0^\infty u_l(iy)\, y^{s-1}\, \mathrm{d}y$$

for each weight $l \in 2\mathbb{Z}+k$. These L-function $L_l^\star$ satisfy a functional equation of the form

$$L_l^\star(-s) = (-1)^{\frac{k-l}{2}}\, L_l^\star(s).$$

We present and discuss several converse theorems in analogy to Hecke's converse theorem stated in Corollary 1.217. We conclude this section with the introduction of D-functions and the following result:

Theorem *Let $k \in \mathbb{R}$, $0 \leq \kappa < 1$, and $|\Re(\nu)| < \frac{1}{2}$ such that $k \in 12\kappa + 12\mathbb{Z}$ and if $2\nu + k$ or $2\nu - k \in 1 + 2\mathbb{Z}$, then $k \notin \mathbb{Z}$.*

There is a canonical correspondence between the following objects:

1. *an admissible family $\left(u_l\right)_{l \in k+2\mathbb{Z}}$ of Maass cusp forms*
2. *a pair of functions $D_0(s)$ and $D_1(s)$ such that*
   a. *D_δ, $\delta = 0, 1$, have a series representation of the form*

$$D_0(s) = \sum_{\substack{n \in \mathbb{Z} \\ n+\kappa \neq 0}} b_n \big(i(n+\kappa)\big)^{-s}, \quad \text{and} \quad D_1(s) = \sum_{\substack{n \in \mathbb{Z} \\ n+\kappa \neq 0}} b_n \big(-i(n+\kappa)\big)^{-s}$$

   that converge absolutely on a right half-plane

   b. *for both choices of $\pm$, the function*

$$s \mapsto \frac{1}{\sin(\pi s)} \Big(D_0(s) - e^{\pm \pi i s} D_1(s) \Big)$$

   extends to an entire function of finite order

   c. *the two functions*

$$D_\delta^\star(s) = (2\pi)^{-s}\, \Gamma(s)\, D_\delta(s) \qquad (\delta = 0, 1)$$

   are entire and satisfy the functional equation

$$D_0^\star(s) = e^{-\frac{\pi i k}{2}}\, D_1^\star(1 - 2\nu - s)$$

   for all $s \in \mathbb{C}$

In Section 4.3, we introduce the nearly periodic functions associated with families of Maass cusp forms.

In particular, we consider a holomorphic function $f : \mathbb{C} \smallsetminus \mathbb{R} \to \mathbb{C}$ that satisfies $f(z+1) = e^{2\pi i \kappa} f(z)$, $z \in \mathbb{C} \smallsetminus \mathbb{R}$, and is bounded by a multiple of $|\Im(z)|^{-C}$ for some $C > 0$. Moreover, the function $f(z) - e^{\frac{\pi i k}{2}} z^{2\nu - 1} f\left(-\frac{1}{z}\right)$ extends holomorphically across the right half-plane and is bounded by a multiple of $\min\left\{1, |z|^{2\Re(\nu)-1}\right\}$ in the right half-plane.

The main result in Section 4.3 establishes the equivalence between nearly period functions f satisfying the above conditions and families of Maass cusp forms. We also show their relation to the L- and the D-functions. An additional representation of nearly periodic functions as integral transformation of a Maass cusp form u is also given.

Chapter 5 – Period Functions

In Section 5.1, we introduce the concept of period functions associated with families of Maass cusp forms. We present a slightly simplified definition; the full definition is given in Section 5.1.

For $\nu \in \mathbb{C}$, we call a holomorphic function $P : \mathbb{C} \smallsetminus (-\infty, 0] \to \mathbb{C}$ a period function if P satisfies the three-term equation

$$P(\zeta) = P(\zeta + 1) + (\zeta + 1)^{2\nu - 1}\, P\left(\frac{\zeta}{\zeta + 1}\right)$$

on $\mathbb{C}'$ and satisfies the growth condition

$$P(\zeta) = \begin{cases} O\left(z^{\max\{0, 2\Re(\nu) - 1}\right) & \text{as } \Im(z) = 0,\ \zeta \searrow 0 \text{ and} \\ O\left(z^{\min\{0, 2\Re(\nu) - 1}\right) & \text{as } \Im(z) = 0,\ \zeta \to \infty. \end{cases}$$

We show in §5.1.1 the following equivalence: There are natural bijective correspondences between

1. nearly periodic functions f on $\mathbb{C} \smallsetminus \mathbb{R}$ as introduced informally above
2. period functions P in the sense above.

Additionally, we study in §5.1.4 the connection of period functions to families of Maass cusp forms.

Since we visited the Eichler cohomology group and the Eichler cohomology theorem briefly in §2.2 and §2.3, it is natural to present the recently discovered analogous results for Maass wave forms in Section 5.2. We keep this section on an introductory level since the underlying techniques are rather complex.

We conclude Chapter 5 with Section 5.3 discussing a representation of Hecke operators that act on period functions. This extends the representation of Hecke operators on Maass cusp forms of weight 0 (discussed in §3.6).

Chapter 6 – Continued Fractions and the Transfer Operator Approach

We change the focus in this chapter. Instead of discussing details of Maass cusp forms or their period functions and their relations, we discuss a particular dynamical system, the Artin billiard. It appears later that the discrete dynamical system characterizing the Artin billiard is related to period functions (and hence to Maass cusp forms).

To imagine the Artin billiard system, first consider a usual billiard table like one used for pool or snooker. The basic idea behind such a billiard game is to play the balls along straight lines (if we disregard things like spin of a ball and ball pockets of the table) and reflections on the boundary of the table. The straight lines are the (segments of) geodesics on the Euclidean two-dimensional plane. Moreover, the billiard table can be modeled by a rectangular set in the Euclidean plane. The reflections at the boundary of the billiard table are reflections at the boundary of the rectangular set. The dynamical system is the movement of the ball. This translates, in our model, to the orbits of freely moving particles (of unit speed) on the rectangular set.

Now, we describe the difference between a usual billiard and the Artin billiard. We first exchange the model of the billiard table with the set

$$\left\{z \in \mathbb{H};\, 0 < \Re(z) < \frac{1}{2},\ |z| > 1\right\},$$

which is as a subset of the upper complex plane and replace the Euclidean metric with the hyperbolic metric. This means that the billiard table is a triangle in the Poincaré model (the upper complex half-plane) of the hyperbolic plane. The vertices of the triangle are i, $\frac{1+\sqrt{3}i}{2}$, and $i\infty$. Again, the dynamical system is the free movement of a ball on (segments of) geodesics with reflections at the boundary of such a theoretical billiard table.

In Section 6.1, we recall continued fractions and describe the connection to a discredited version of the Artin billiard as a dynamical system. We explain how to describe the orbits in the dynamical system using continued fractions.

In the next section, §6.2, we motivate transfer operators. This concept was introduced by David Ruelle to describe and understand the associated dynamical zeta functions of certain discrete dynamical systems. In our case, the main focus is on the Gauss map

$$[0,1) \to [0,1); \quad x \mapsto \begin{cases} \frac{1}{x} - \left\lfloor \frac{1}{x} \right\rfloor & \text{for } x \neq 0 \text{ and} \\ 0 & \text{for } x = 0 \end{cases}$$

and its associated discrete dynamical system given by iterative use of the Gauss map on the unit interval $[0, 1)$. The construction of the transfer operator $\mathcal{L}_s$ associated with the Gauss map is presented. We end the section with a description of Mayer's transfer operator $\tilde{\mathcal{L}}_s$, which is a simple extension of the transfer operator for the Gauss map.

In Section 6.3, we deal with eigenfunctions of the Mayer transfer operator $\tilde{\mathcal{L}}_s$ with eigenvalue 1 (or equivalently with eigenfunctions of $\mathcal{L}_s$ with eigenvalues ± 1). We show that these eigenfunctions are in fact related to period functions. This means of course that we just found a connection between Maass cusp forms, which are eigenfunctions of the hyperbolic Laplace operator, and eigenfunctions of Mayer's transfer operator (with eigenvalue 1), which are derived using methods from dynamical systems and mathematical physics.

We conclude Chapter 6 with a short introduction to the Selberg zeta function and discuss the relation between the Selberg zeta function and period functions. One highlight here is the following result.

Let $s \neq \frac{1}{2}$ be a complex number with $\Re(s) > 0$. Then

1. there exists a nonzero function h with $\mathcal{L}_s h = -h$ if and only if s is the spectral parameter corresponding to an odd Maass wave form
2. there exists a nonzero function h with $\mathcal{L}_s h = h$ if and only if s is either the spectral parameter corresponding to an even Maass form, $2s$ is a zero of the Riemann zeta function, or $s = 1$.

In fact, the existence of such an eigenfunction h of the transfer operator $\mathcal{L}_s$ is directly linked to the vanishing of the Selberg zeta function $Z(s)$. Of course, the stated result in Section 6.4 also defines the function spaces in which h lives. The above result is interesting since it gives a direct relation between spectral parameters of Maass cusp forms and the zeros of the Selberg zeta function.

Chapter 7 – Weak Harmonic Maass Wave Forms

The last chapter is devoted to the introduction of the interesting concepts of weak harmonic Maass forms and mock modular forms. We explain one of the key differences between them. To do so, we first look at one of the key differences between classical modular forms F and Maass wave forms.

One of the defining properties of a modular form F is that F is a holomorphic function, as we already mentioned in section "Chapter 1 – A Short Introduction to Modular Forms" of this introductory chapter. For a Maass form u (of real weight k), we replace this fundamental condition by u being an eigenfunction of the hyperbolic Laplace operator Δ_k, as described in section "Chapter 3 – Maass Wave Forms of Real Weight". Other properties, like the transformation behavior, are adapted accordingly to fit into this underlying requirement.

The concept of weak harmonic Maass forms takes a different direction by defining a different hyperbolic Laplace operator $\tilde{\Delta}_k$ by

$$\tilde{\Delta}_k = -y^2 \left(\frac{\partial^2}{\partial x^2} + \frac{\partial^2}{\partial y^2} \right) + iky \left(\frac{\partial}{\partial x} + i \frac{\partial}{\partial y} \right).$$

Comparing this second-order differential operator with the definition of Δ_k above, we see that the difference is in how the part depending on the weight k is defined. For example, using the identity $\frac{\partial}{\partial \bar{z}} = \frac{1}{2} \frac{\partial}{\partial x} + \frac{i}{2} \frac{\partial}{\partial y}$ for the decomposition of $z = x + iy$ into real and imaginary part, we see that $\tilde{\Delta}_k$ can be expressed as

$$\tilde{\Delta}_k = -y^2 \left(\frac{\partial^2}{\partial x^2} + \frac{\partial^2}{\partial y^2} \right) + 2iky \left(\frac{\partial}{\partial \bar{z}} \right).$$

This motivates the natural decomposition of an eigenfunction of $\tilde{\Delta}_k$ into a holomorphic and non-holomorphic parts.

In Section 7.1, we introduce weak harmonic Maass forms as follows: for $k \in 2\mathbb{Z}$, a two times continuously differentiable function $f : \mathbb{H} \to \mathbb{C}$ is called a weak harmonic Maass wave form of weight k if it satisfies the following three conditions:

1. $\tilde{\Delta}_k f = 0$
2. f satisfies $f\left(\frac{az+b}{cz+d}\right) = (cz+d)^k f(z)$ for all $z \in \mathbb{H}$ and $M \in \Gamma(1)$
3. $f(z) = O\left(e^{Cy}\right)$ as $\Im(z) = y \to \infty$

We show that any weak harmonic Maass wave form f can be decomposed into two parts $f = f_+ + f_-$, where f_+ is the holomorphic part of f and f_- the non-holomorphic part. The holomorphic part f_+ is also known as a mock modular form.

Section 7.1 also gives basic properties of weak harmonic Maass forms.

The next section, Section 7.2, is an introductory section on mock modular forms. We introduce the concept of mock modular forms and the associated mock Jacobi forms without going deeply into details.

In Section 7.3, we continue the discussion of weak harmonic Maass cusp forms by looking at their associated L-series and L-functions. Again, this extends the classical concept of L-functions associated with modular cusp forms, discussed in Section 1.7.

The main problem for this extension is that the integral transformation

$$\int_0^{i\infty} F(z)\, z^{s-1}\, \mathrm{d}z$$

used in the setting of modular cusp forms F is not well defined anymore in the context of weak harmonic Maass cusp forms. A new approach on how to define the above integral is needed. This leads to the concept of the regularized integral

$$R.\int_0^{i\infty} F(w)\, w^{s-1}\, \mathrm{d}w := \left[\int_0^{i\infty} e^{uiw}\, F(w)\, w^{s-1}\, \mathrm{d}w\right]_{u=0}.$$

The precise meaning of this regularization is given in Section 7.3.

The regularized integral is used as a technical tool to introduce L-functions associated with weakly holomorphic Maass cusp forms.

In Section 7.4, we introduce the concepts of period functions associated with mock modular functions, i.e., the holomorphic part of weakly harmonic Maass wave forms. The main focus is to point out relations to the classical theory of L-functions and period polynomials presented in §1.7 and §2.

Chapter 1
A Brief Introduction to Modular Forms

In this chapter, we give an overview on the theory of "classical" modular forms. What we mean by classical here are holomorphic modular forms on the full modular group or on its congruence subgroups. We give the definition of modular forms for real weights and give properties of the multiplier system. A few explicit examples of modular forms (Eisenstein series, discriminant function, holomorphic Poincaré series, and theta series) are also presented. Then, we introduce Hecke operators and L-functions and give their properties that will be needed in later chapters.

Also part of our brief introduction to topics of modular forms is the next chapter, where we focus on period polynomials and the Eichler-Shimura isomorphism theorem.

1.1 Notations and Some Simple Results

We denote the sets of natural numbers and the set of integers by

$$\mathbb{N} = \{1, 2, 3, \ldots\} \quad \text{and} \quad \mathbb{Z} = \{\ldots, -2, -1, 0, 1, 2, \ldots\}.$$

The set of rational numbers, real numbers, and complex numbers are denoted by $\mathbb{Q}$, $\mathbb{R}$, and $\mathbb{C}$, respectively. Let $z = x + iy \in \mathbb{C}$ denote a complex number with $x, y \in \mathbb{R}$. Its real and imaginary parts are denoted by $\Re(z) = x$ and $\Im(z) = y$, respectively.

Let

$$\mathbb{H} = \big\{z \in \mathbb{C}\colon\ \Im(z) > 0\big\} \quad \text{and} \quad \mathbb{H}^- = \big\{z \in \mathbb{C}\colon\ \Im(z) < 0\big\} \tag{1.1}$$

denote the *upper half-plane* and the *lower half-plane*, respectively.

T. Mühlenbruch, W. Raji, *On the Theory of Maass Wave Forms*, Universitext,
https://doi.org/10.1007/978-3-030-40475-8_1

In addition, we define the *group of linear fractional transformations* or *general linear group*, denoted by $\mathrm{GL}_2(\mathbb{R})$, to be the set of all 2×2 invertible matrices with real entries:

$$\mathrm{GL}_2(\mathbb{R}) := \left\{ \begin{pmatrix} a & b \\ c & d \end{pmatrix} : a, b, c, d \in \mathbb{R} \text{ and } ad - bc \neq 0 \right\}. \tag{1.2}$$

The subset $\mathrm{GL}_2(\mathbb{R})^+$ denotes the subset of $\mathrm{GL}_2(\mathbb{R})$ with positive determinant:

$$\mathrm{GL}_2(\mathbb{R})^+ := \left\{ \begin{pmatrix} a & b \\ c & d \end{pmatrix} : a, b, c, d \in \mathbb{R} \text{ and } ad - bc > 0 \right\}. \tag{1.3}$$

A subgroup of $\mathrm{GL}_2(\mathbb{R})$, known as the *special linear group* $\mathrm{SL}_2(\mathbb{R})$, is given by:

$$\mathrm{SL}_2(\mathbb{R}) := \left\{ \begin{pmatrix} a & b \\ c & d \end{pmatrix} : a, b, c, d \in \mathbb{R} \text{ and } ad - bc = 1 \right\}. \tag{1.4}$$

The next subgroup, a special subgroup with nice properties, is $\mathrm{SL}_2(\mathbb{Z})$, the set of all 2×2 matrices in $\mathrm{SL}_2(\mathbb{R})$ with integer entries:

$$\mathrm{SL}_2(\mathbb{Z}) := \left\{ \begin{pmatrix} a & b \\ c & d \end{pmatrix} \in \mathrm{SL}_2(\mathbb{R}) : a, b, c, d \in \mathbb{Z} \right\}. \tag{1.5}$$

We call this group the *full modular group*.

Interesting normal subgroups of $\mathrm{SL}_2(\mathbb{Z})$ are known as the principal congruence subgroups. For $N \in \mathbb{N}$, the *principal congruence subgroup of level N* given by

$$\Gamma(N) := \left\{ \begin{pmatrix} a & b \\ c & d \end{pmatrix} \in \mathrm{SL}_2(\mathbb{Z}) : \begin{pmatrix} a & b \\ c & d \end{pmatrix} \equiv \begin{pmatrix} 1 & 0 \\ 0 & 1 \end{pmatrix} \pmod{N} \right\}. \tag{1.6}$$

Notice that the definition of the principal congruence subgroups implies directly the inclusion

$$\Gamma(M) \subset \Gamma(N) \subset \Gamma(1) = \mathrm{SL}_2(\mathbb{Z}) \tag{1.7}$$

for $M, N \in \mathbb{N}$ with $M \mid N$. Moreover, $\Gamma(N)$ is of finite index in $\Gamma(1)$ since $\Gamma(1)/\Gamma(N) \cong \mathrm{SL}_2(\mathbb{Z}/N\mathbb{Z})$. We refer to [142, Theorem 4.2.5] for an explicit calculation of the index.

Remark 1.1 The notation $\Gamma(1)$ is also commonly used to denote the full modular group.

Definition 1.2 A *congruence subgroup* is a subgroup of $\Gamma(1)$ that contains $\Gamma(N)$ for some $N \in \mathbb{N}$.

Remark 1.3 There exist also finite-index subgroups that are not a congruence subgroup. For example, Wohlfahrt already shows in [208, Theorem 5] that non-congruence subgroups of $\mathrm{SL}_2(\mathbb{Z})$ of finite index have at least index 7. Moreover, he mentions the existence of an example of index 7 [208, p. 531]. The exploratory preprint [125] gives an overview on non-congruence subgroups of the full modular group and its effect on modular forms.

Two very useful congruence subgroups of finite index in $\Gamma(1)$ are

$$\begin{aligned} \Gamma_0(N) :&= \left\{ \begin{pmatrix} a & b \\ c & d \end{pmatrix} \in \mathrm{SL}_2(\mathbb{Z}) \colon\ c \equiv 0 \pmod N \right\} \\ &= \left\{ V \in \mathrm{SL}_2(\mathbb{Z}) \colon\ V \equiv \begin{pmatrix} \star & \star \\ 0 & \star \end{pmatrix} \pmod N \right\} \qquad \text{and} \end{aligned} \tag{1.8}$$

$$\begin{aligned} \Gamma_1(N) :&= \left\{ \begin{pmatrix} a & b \\ c & d \end{pmatrix} \in \mathrm{SL}_2(\mathbb{Z}) \colon\ c \equiv 0 \text{ and } a \equiv d \equiv 1 \pmod N \right\} \\ &= \left\{ V \in \mathrm{SL}_2(\mathbb{Z}) \colon\ V \equiv \begin{pmatrix} 1 & \star \\ 0 & 1 \end{pmatrix} \pmod N \right\}. \end{aligned} \tag{1.9}$$

The first one is sometimes called the *Hecke congruence subgroup of level N*.

The companion groups to $\Gamma_0(N)$ and $\Gamma_1(N)$ are the groups

$$\begin{aligned} \Gamma^0(N) :&= \left\{ \begin{pmatrix} a & b \\ c & d \end{pmatrix} \in \mathrm{SL}_2(\mathbb{Z}) \colon\ b \equiv 0 \pmod N \right\} \\ &= \left\{ V \in \mathrm{SL}_2(\mathbb{Z}) \colon\ V \equiv \begin{pmatrix} \star & 0 \\ \star & \star \end{pmatrix} \pmod N \right\} \qquad \text{and} \end{aligned} \tag{1.10}$$

$$\begin{aligned} \Gamma^1(N) :&= \left\{ \begin{pmatrix} a & b \\ c & d \end{pmatrix} \in \mathrm{SL}_2(\mathbb{Z}) \colon\ b \equiv 0 \text{ and } a \equiv d \equiv 1 \pmod N \right\} \\ &= \left\{ V \in \mathrm{SL}_2(\mathbb{Z}) \colon\ V \equiv \begin{pmatrix} 1 & 0 \\ \star & 1 \end{pmatrix} \pmod N \right\}. \end{aligned} \tag{1.11}$$

Notice that the inclusions

$$\begin{aligned} &\Gamma(N) \subset \Gamma_1(N) \subset \Gamma_0(N) \subset \Gamma(1) \quad \text{and} \\ &\Gamma(N) \subset \Gamma^1(N) \subset \Gamma^0(N) \subset \Gamma(1) \end{aligned} \tag{1.12}$$

follow directly from the definitions. We refer to [142, Chapter IV] for more information about congruence subgroups.

Two important elements of each of the above matrix groups are the matrices:

$$\mathbf{1} := \begin{pmatrix} 1 & 0 \\ 0 & 1 \end{pmatrix} \quad \text{and} \quad -\mathbf{1} := \begin{pmatrix} -1 & 0 \\ 0 & -1 \end{pmatrix}, \tag{1.13}$$

where $\mathbf{1}$ is called the identity matrix.

We will also need sometimes the set of 2×2 matrices with integer, and real entries given respectively:

$$\begin{aligned} \mathrm{Mat}_2(\mathbb{Z}) : &= \left\{ \begin{pmatrix} a & b \\ c & d \end{pmatrix} : \; a, b, c, d \in \mathbb{Z} \right\} \quad \text{and} \\ \mathrm{Mat}_2(\mathbb{R}) : &= \left\{ \begin{pmatrix} a & b \\ c & d \end{pmatrix} : \; a, b, c, d \in \mathbb{R} \right\}. \end{aligned} \tag{1.14}$$

Definition 1.4 Let $V = \begin{pmatrix} a & b \\ c & d \end{pmatrix}$ be a matrix with real entries. We define the *determinant* of V by

$$\det V = ad - bc$$

and its *trace* by

$$\mathrm{trace}\,(V) = a + d.$$

Define the Riemann sphere as

$$\mathbb{S} := \mathbb{C} \cup \{\infty\}. \tag{1.15}$$

Definition 1.5 We define the group action

$$\begin{aligned} \mathrm{GL}_2(\mathbb{R}) \times \mathbb{S} &\to \mathbb{S} \\ (M, z) &\mapsto M\,z \end{aligned}$$

by

$$\begin{pmatrix} a & b \\ c & d \end{pmatrix} z := \begin{cases} \frac{az+b}{cz+d} & \text{if } z \in \mathbb{C},\, z \neq -\frac{d}{c} \\ \frac{a}{c} & \text{if } z = \infty \text{ and} \\ \infty & \text{if } z = -\frac{d}{c} \end{cases} \tag{1.16}$$

for $M = \begin{pmatrix} a & b \\ c & d \end{pmatrix} \in \mathrm{GL}_2(\mathbb{R})$ and $z \in \mathbb{S} = \mathbb{C} \cup \{\infty\}$ where $M\,z$ is called *linear fractional transformation* or *Möbius transformation* of z under M.

Remark 1.6 The Möbius transformation is indeed a well-defined group action. We actually have

$$(MV)z = M(Vz) \tag{1.17}$$

for all $z \in \mathbb{S}$ and $M, V \in \mathrm{GL}_2(\mathbb{R})$.

The next lemma answers the question about the kernel of the Möbius transformations, that is the set of matrices that *act trivially* on $\mathbb{S}$.

Lemma 1.7 *The kernel of the group* $\mathrm{GL}_2(\mathbb{R})$ *acting on* $\mathbb{S}$ *by Möbius transformation is given by*

$$\left\{ \begin{pmatrix} \lambda & 0 \\ 0 & \lambda \end{pmatrix} : \lambda \in \mathbb{R}_{\neq 0} \right\}.$$

Proof The proof is left for the reader as an exercise. □

The statement of the lemma means that each element of the kernel set acts as the identity operation:

$$\begin{pmatrix} \lambda & 0 \\ 0 & \lambda \end{pmatrix} z = z \qquad \text{for all } z \in \mathbb{S} \text{ and } \lambda \in \mathbb{R}_{\neq 0}.$$

Multiplying this equation with a matrix $\begin{pmatrix} a & b \\ c & d \end{pmatrix} \in \mathrm{GL}_2(\mathbb{R})$ gives

$$\begin{pmatrix} \lambda a & \lambda b \\ \lambda c & \lambda d \end{pmatrix} z = \begin{pmatrix} a & b \\ c & d \end{pmatrix} z.$$

This allows us to identify the matrix element $\begin{pmatrix} a & b \\ c & d \end{pmatrix} \in \mathrm{SL}_2(\mathbb{R})$ with the fractional linear transformation $\frac{az+b}{cz+d}$ up to multiplication of a common unit in the entries. In particular, the kernel of the $\mathrm{SL}_2(\mathbb{R})$ action on $\mathbb{S}$ by Möbius transformation is given by the matrices $\{\mathbf{1}, -\mathbf{1}\}$.

Now, we consider the group $\mathrm{SL}_2(\mathbb{R})$ and its action on $\mathbb{C} \cup \{\infty\}$. It is easy to see that $\mathrm{SL}_2(\mathbb{R})$ preserves the upper half-plane $\mathbb{H}$, the compactified real line $\mathbb{R} \cup \{\infty\}$, and the lower half-plane $\mathbb{H}^-$. This is due to the fact that

$$\Im(Vz) = \Im\left(\frac{az+b}{cz+d}\right) = \det V \, \frac{\Im(z)}{|cz+d|^2} \tag{1.18}$$

for all $V = \begin{pmatrix} a & b \\ c & d \end{pmatrix} \in \mathrm{GL}_2(\mathbb{R})$. Summarizing, we get the following result:

Lemma 1.8 *We have*

$$\mathrm{SL}_2(\mathbb{R})\,\mathbb{H} \subset \mathbb{H}, \quad \mathrm{SL}_2(\mathbb{R})\,\mathbb{H}^- \subset \mathbb{H}^- \quad \textit{and} \quad \mathrm{SL}_2(\mathbb{R})\left(\mathbb{R}\cup\{\infty\}\right) \subset \mathbb{R}\cup\{\infty\}.$$

Exercise 1.9 Let $V = \begin{pmatrix} a & b \\ c & d \end{pmatrix} \in \mathrm{SL}_2(\mathbb{R})$. Show that the differential satisfies

$$\mathrm{d}(Vz) = \frac{\mathrm{d}z}{(cz+d)^2}. \tag{1.19}$$

Remark 1.10 The letter "*d*" appears here in two different settings. It often appears as an entry in a matrix $\begin{pmatrix} a & b \\ c & d \end{pmatrix}$ where usually d denotes a real or integer coefficient. A typical example for this usage is (1.18). Also the letter "*d*" appears denoting the differential in $\mathrm{d}z$, for example in (1.19). However, this different usages of the letter "*d*" is unlikely to lead to any ambiguity.

We will also need the notion of the greatest common divisor.

Definition 1.11 Let $(c, d) \in \mathbb{Z}^2_{\neq(0,0)}$ be a pair of integers that are not both zero. The *greatest common divisor* $\gcd(c, d)$ of the pair is defined as

$$\gcd(c, d) = \max\left\{n \in \mathbb{N}\colon\ n \mid c \text{ and } n \mid d\right\}.$$

Here, the notation $n \mid c$ means that n divides c, i.e., there exists a $j \in \mathbb{Z}$ such that $nj = c$. We call c and d *coprime* if $\gcd(c, d) = 1$.

1.2 The Full Modular Group and Its Subgroups

In this section, we discuss the needed basic properties of the full modular group and its subgroup. At many instances, we omit the proofs of certain theorems or leave them as exercises.

1.2.1 The Full Modular Group

Recall the full modular group

$$\Gamma(1) = \mathrm{SL}_2(\mathbb{Z}) = \left\{ \begin{pmatrix} a & b \\ c & d \end{pmatrix} : \ a, b, c, d \in \mathbb{Z},\ ad - bc = 1 \right\}$$

defined in (1.5). Two important elements of $\mathrm{SL}_2(\mathbb{Z})$ are the matrices

$$S = \begin{pmatrix} 1 & 1 \\ 0 & 1 \end{pmatrix} \quad \text{and} \quad T = \begin{pmatrix} 0 & -1 \\ 1 & 0 \end{pmatrix}. \tag{1.20}$$

In fact, these two elements are the generators of the full modular group in the sense that every matrix $M \in \mathrm{SL}_2(\mathbb{Z})$ can be written as a word in S and T.

Remark 1.12 One can easily see that the generators S and T of the full modular group satisfy the identities

$$T^2 = -\mathbf{1} = (TS)^3. \tag{1.21}$$

Remark 1.13 The above choice of generators of the full modular group is not uniform throughout the literature. Sometimes, books or articles simply reverse the meaning of S and T. Other texts may use different generators and/or different letters to denote these generators. For example, [100, p. 15, Theorem 1.1] reverses the letters S and T and [55, (2.5)] denotes the same generators by T and Q. It is recommended that the reader always checks which generators of the full modular group are used and how they are denoted.

Theorem 1.14 *Every $A \in \Gamma(1)$ can be expressed in the form*

$$A = T^{\epsilon_1} S^{n_1} T S^{n_2} T \ldots T S^{n_k},$$

where ϵ_1 and n_i's are integers with $n_1, n_k \in \mathbb{Z}$ and $n_i \in \mathbb{Z}_{\neq 0}$ for all $i \in \{2, \ldots, k-1\}$ and $\epsilon_1 \in \{0, 2\}$.

Notice that this representation is not unique.

Proof A matrix $A \in \Gamma(1)$ has a representation

$$A = \pm S^{n_1} T S^{n_2} T \ldots T S^{n_k},$$

where the n_i's are integers with $n_1, n_k \in \mathbb{Z}$ and $n_i \in \mathbb{Z}_{\neq 0}$ for all $i \in \{2, \ldots, k-1\}$. This is a direct application of the Euclidean algorithm. More details on this representation can be found in [12, Chapter II]. Now, we only need to encode the leading scalar multiple ± 1.

Recall that the scalar multiplication can be encoded as multiplication by a diagonal matrix:

$$\pm 1\, M = \begin{pmatrix} \pm 1 & 0 \\ 0 & \pm 1 \end{pmatrix} M \qquad \text{for every matrix } M \in \mathrm{GL}_2(\mathbb{C}).$$

Moreover, the matrix T in (1.20) satisfies

$$T^0 = \mathbf{1} \quad \text{and} \quad T^2 = -\mathbf{1} = \begin{pmatrix} -1 & 0 \\ 0 & 1 \end{pmatrix}.$$

Combining these arguments, we see that each $A \in \Gamma(1)$ can be written as a word of the form

$$A = T^{\epsilon_1} S^{n_1} T S^{n_2} T \ldots T S^{n_k},$$

where ϵ_1 and n_i's are integers with $n_1, n_k \in \mathbb{Z}$ and $n_i \in \mathbb{Z}_{\neq 0}$ for all $i \in \{2, \ldots, k-1\}$ and $\epsilon_1 \in \{0, 2\}$. □

Definition 1.15 For each $A \in \Gamma(1)$, we define the *Eichler length* $l(M)$ of the matrix A by

$$l(A) := \min \left\{ \epsilon_1/2 + |n_1| + \ldots + |n_k| + k : \; M = T^{\epsilon_1} S^{n_1} T S^{n_2} T \ldots T S^{n_k} \right\}.$$

The minimum is taken over the sum of the absolute values of the exponents of each word in S and T that represents A. In addition, we put $l(\mathbf{1}) := 0$.

1.2.2 Fundamental Domains for the Full Modular Group

We first define the notion of a fundamental domain.

Definition 1.16 Let $\Gamma \subset \mathrm{SL}_2(\mathbb{Z})$ be a congruence subgroup. A *fundamental domain* $\mathcal{F} = \mathcal{F}_\Gamma$ of Γ is an open subset $\mathcal{F} = \mathcal{F}_\Gamma \subset \mathbb{H}$ such that:

1. No two distinct points of $\mathcal{F}$ are equivalent under Γ. In other words, for two distinct points $z_1, z_2 \in \mathcal{F}$, $z_1 \neq z_2$, there is no matrix $V \in \Gamma$ such that $V z_1 = z_2$.
2. Every point in $\mathbb{H}$ is equivalent to some point in the closure $\overline{\mathcal{F}}$ of $\mathcal{F}$ under Γ: for every point $z \in \mathbb{H}$ there exists a $V \in \Gamma$ such that $Vz \in \overline{\mathcal{F}}$.

(This means that Γ acts discontinuously.)

It is important to bear in mind that the fundamental domain is not unique as the remark below illustrates.

Remark 1.17 Let $\mathcal{F}$ be a fundamental domain for the group $\Gamma \subset \mathrm{SL}_2(\mathbb{R})$ in the sense of Definition 1.16. Then, for each $V \in \Gamma$ the set

$$V\mathcal{F} := \left\{ V z : \; z \in \mathcal{F} \right\}$$

is also a fundamental domain of Γ. It satisfies

$$V\mathcal{F} \cap \mathcal{F} = \emptyset \iff V \neq \pm\mathbf{1}.$$

Definition 1.18 The *standard fundamental domain* $\mathcal{F}$ of the full modular group $\Gamma(1)$ is given by the open set

$$\mathcal{F}_{\Gamma(1)} = \left\{ z \in \mathbb{H}: \ |z| > 1, |\Re(z)| < \frac{1}{2} \right\}. \tag{1.22}$$

Exercise 1.19 Let $\partial\mathcal{F}_{\Gamma(1)}$ denote the set-boundary of the fundamental domain $\mathcal{F}_{\Gamma(1)}$. Show that $\partial\mathcal{F}_{\Gamma(1)}$ can be decomposed into 4 sets $C_1, \ldots, C_4$ such that $C_i \cap C_j$ contains at most one point and $C_1 \cup \cdots \cup C_4 = \partial\mathcal{F}_{\Gamma(1)}$. Moreover, $C_1, \ldots, C_4$ can be chosen such that

$$S\, C_1 = C_2 \qquad \text{and} \qquad T\, C_3 = C_4.$$

1.2.3 Fundamental Domains for Subgroups of $\Gamma(1)$

Let Γ to be a congruence subgroup of $\Gamma(1)$. This implies, in particular, that the index of Γ in $\Gamma(1)$ is finite, i.e., $[\Gamma(1) : \Gamma] < \infty$.

The fundamental domain of the group Γ is denoted by $\mathcal{F}_\Gamma$. Also $\overline{D}$ means the closure and D° means the interior of a set D. We now give the fundamental domains for congruence subgroups in terms of the standard fundamental domain of $\Gamma(1)$.

Theorem 1.20 *Let* Γ *be a subgroup of* $\Gamma(1)$ *with finite index* μ. *Let* $A_1, \ldots, A_\mu$ *denote the right coset representatives of* $\Gamma(1)$ *in* Γ *with (i.e.,* $\Gamma(1) = \bigcup_{i=1}^{\mu} \Gamma A_i$*). Then*

$$\mathcal{F}_\Gamma = \bigcup_{i=1}^{\mu} A_i \mathcal{F}_{\Gamma(1)} \tag{1.23}$$

is a fundamental domain for Γ.

We sometimes call the above fundamental domain $\mathcal{F}_\Gamma$ the *standard fundamental domain* for Γ. The reason for this name is that $\mathcal{F}_\Gamma$ is derived from the standard fundamental domain $\mathcal{F}_{\Gamma(1)}$ for the full modular group given in Definition 1.18.

Remark 1.21 The standard fundamental domain, as defined above, depends on the coset representation of the subgroup in the full modular group. If another set of representatives of the right cosets of Γ in $\Gamma(1)$ is taken, say $A'_1, \ldots A'_\mu$, then the standard fundamental domain becomes

$$\mathcal{F}'_\Gamma = \bigcup_{i=1}^{\mu} A'_i \mathcal{F}_{\Gamma(1)}.$$

The word "standard" here does not imply uniqueness of the domain. It only means that $\mathcal{F}_\Gamma$ is derived from the "standard" fundamental domain $\mathcal{F}_{\Gamma(1)}$ of the full modular group.

We now give an example of a subgroup of finite index in $\Gamma(1)$ and determine its cosets. We then use the cosets to determine the fundamental domain of the subgroup using Theorem 1.20.

Example 1.22 Let p be a prime. Then, for every $V \in \Gamma(1)$, $V \notin \Gamma_0(p)$, there exists $P \in \Gamma_0(p)$ and an integer k with $0 \leq k < p$ satisfying

$$V = P\,T\,S^k. \tag{1.24}$$

The fundamental domain of $\Gamma_0(p)$ is given by the set

$$\mathcal{F}_{\Gamma_0(p)} = \left[\bigcup_{k=0}^{p-1} T\,S^k\,\mathcal{F}_{\Gamma(1)}\right] \cup \mathcal{F}_{\Gamma(1)}. \tag{1.25}$$

Remark 1.23 An illustration of the standard fundamental domain $\mathcal{F}_{\Gamma_0(2)}$ is given in Figure 1.1. Some recent algorithms to compute fundamental domains of subgroups of $\Gamma(1)$ of finite index can be found in [112] and [200] (Figure 1.2).

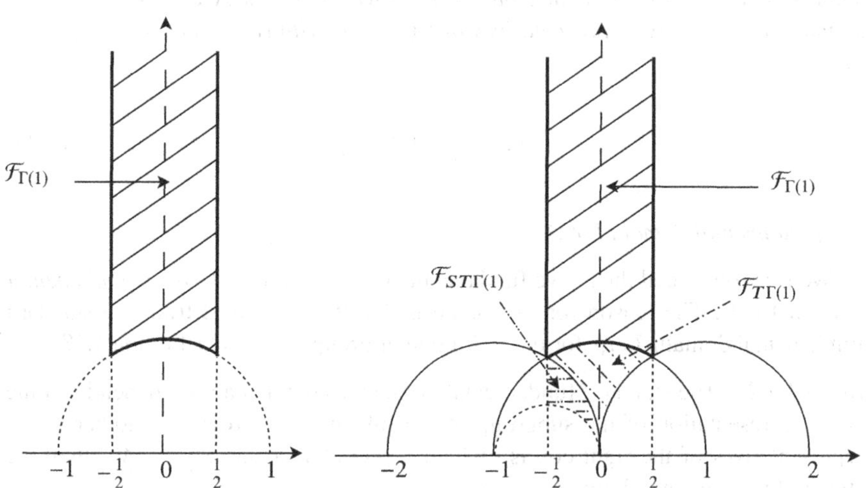

Fig. 1.1 The left diagram illustrates the standard fundamental domain $\mathcal{F}_{\Gamma(1)}$ of the full modular group $\Gamma(1)$, the right diagram illustrates the standard fundamental domain $\mathcal{F}_{\Gamma_0(2)}$ of the Hecke congruence subgroup $\Gamma_0(2)$ with coset representatives **1**, T, and $T\,S$.

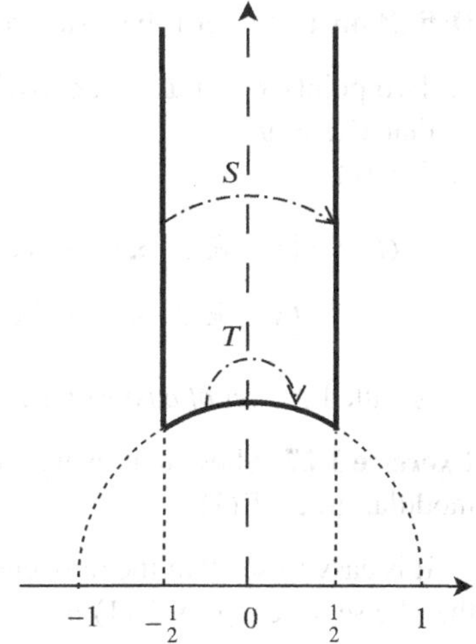

Fig. 1.2 The diagram illustrates the standard fundamental domain $\mathcal{F}_{\Gamma(1)}$ of the full modular group $\Gamma(1)$, indicating which part of the boundary gets mapped onto each other by S and T, respectively.

1.2.4 Cuspidal Points

We discuss the notion of cuspidal points or cusps. Simply said, cusps are all elements of $\mathbb{Q} \cup \{\infty\}$ the set. One could also say, cusps are all "rational numbers" in the projective real line. (Here we also understand $\infty = \frac{1}{0}$ as rational number.)

In preparation of a formal definition of cups, recall the Riemann sphere $\mathbb{S} = \mathbb{C} \cup \{\infty\}$ as introduced in (1.15). We denote the closure of a set $D \subset \mathbb{C} \cup \{\infty\}$ in $\mathbb{S}$ by $\overline{D}^{\mathbb{S}}$.

Definition 1.24 Let Γ be a subgroup of $\mathrm{SL}_2(\mathbb{R})$ and $\mathcal{F}_\Gamma$ a fundamental domain. The elements of the set

$$C_{\Gamma,\mathcal{F}_\Gamma} := \overline{\mathcal{F}_\Gamma}^{\mathbb{S}} \cap \left(\mathbb{R} \cup \{\infty\}\right)$$

are called the *cusps of* Γ *in* $\mathcal{F}_\Gamma$.

Remark 1.25 We point out again that the above closure of the fundamental domain $\mathcal{F}_\Gamma$ is taken in the Riemann sphere $\mathbb{S}$ and not in $\mathbb{H}$. To illustrate the point, we give both closures for the standard fundamental domain $\mathcal{F}_{\Gamma(1)}$:

$$\overline{\mathcal{F}_{\Gamma(1)}} = \left\{z \in \mathbb{H} : \ |\Re(z)| \le \frac{1}{2}, |z| \ge 1\right\} \qquad \text{and}$$

$$\overline{\mathcal{F}_{\Gamma(1)}}^{\mathbb{S}} = \left\{z \in \mathbb{H} : \ |\Re(z)| \le \frac{1}{2}, |z| \ge 1\right\} \cup \{\infty\} = \overline{\mathcal{F}_{\Gamma(1)}} \cup \{\infty\}.$$

The only cusp of $\Gamma(1)$ in $\mathcal{F}_{\Gamma(1)}$ is ∞.

Definition 1.26 Let Γ be a subgroup of $\mathrm{SL}_2(\mathbb{R})$.

1. Two points x and w are *equivalent* under the group Γ if there exists $V \in \Gamma$ such that $V x = w$.
2. The set

$$\begin{aligned} C_\Gamma :&= \{x \in \mathbb{R} \cup \{\infty\} : x \text{ is equivalent to a cusp of } \Gamma \text{ in } \mathcal{F}_\Gamma\} \\ &= \{x \in \mathbb{R} \cup \{\infty\} : \exists V \in \Gamma \text{ and } w \text{ a cusp of } \Gamma \text{ in } \mathcal{F}_\Gamma \text{ satisfying } x = V w\} \end{aligned} \tag{1.26}$$

is called the *set of cusps of* Γ.

Exercise 1.27 Show that every rational number is equivalent to ∞ under the full modular group $\Gamma(1)$.

It is easy to see that the only cusp of $\Gamma(1)$ in $\mathcal{F}_{\Gamma(1)}$ is ∞. Using 1.27, one can see that the set of cusps of $\Gamma(1)$ is

$$\begin{aligned} C_{\Gamma(1)} &= \{x \in \mathbb{R} \cup \{\infty\} : \exists V \in \Gamma(1) \text{ such that } x = V \infty\} \\ &= \mathbb{Q} \cup \{\infty\}. \end{aligned}$$

Lemma 1.28 *Let $\Gamma \subset \Gamma(1)$ be a subgroup with finite index μ and representatives $A_1, \ldots, A_\mu$ of the right cosets. Its standard fundamental domain is denoted by $\mathcal{F}_\Gamma$. The number of cusps of Γ in $\mathcal{F}_\Gamma$ satisfies*

$$\sharp C_{\Gamma,\mathcal{F}_\Gamma} \leq \mu,$$

and we have

$$C_{\Gamma,\mathcal{F}_\Gamma} = \{A_1 \infty, \ldots, A_\mu \infty\}.$$

Remark 1.29 The matrices $A_i, i \in \{1, \ldots, \mu\}$, in Lemma 1.28 are sometimes called the *scaling matrices* of the cusp q_i for the group $\Gamma \subset \Gamma(1)$.

Recall that an element $\begin{pmatrix} a & b \\ c & d \end{pmatrix}$ of $\mathrm{SL}_2(\mathbb{R})$ is called *parabolic* if $|a + d| = 2$.

Definition 1.30 Let $\Gamma \subset \mathrm{SL}_2(\mathbb{R})$ and $x \in \mathbb{R} \cup \{\infty\}$. The point x is called a *parabolic point of* Γ, if there exists a parabolic matrix $M \in \Gamma$ satisfying $M x = x$.

Lemma 1.31 *Let $\frac{a}{b} \in \mathbb{Q} \cup \{\infty\}$ with $\gcd(a, b) = 1$ where we understand $\frac{1}{0} := \infty$. Put*

$$M_{\frac{a}{b}} := \begin{pmatrix} 1 - ab & a^2 \\ -b^2 & 1 + ab \end{pmatrix}. \tag{1.27}$$

We have:

1. $M_{\frac{a}{b}} \in \Gamma(1)$.
2. $M_{\frac{a}{b}}$ *is parabolic and* $M_{\frac{a}{b}} \frac{a}{b} = \frac{a}{b}$.
3. *There exists an* $A \in \Gamma(1)$ *such that* $A\,\infty = \frac{a}{b}$. *We have* $M^n_{\frac{a}{b}} = AS^nA^{-1}$ *for every* $n \in \mathbb{Z}$.
4. *Suppose* $V \in \Gamma(1)$ *and* V *fixes* $\frac{a}{b}$. *Then* $V = \pm M^n_{\frac{a}{b}}$ *for some* $n \in \mathbb{Z}$.

In particular, we may choose $A = \begin{pmatrix} a & \star \\ b & \star \end{pmatrix}$ *and have*

$$M^n_{\frac{a}{b}} = AS^nA^{-1} = \begin{pmatrix} 1 - nab & na^2 \\ -nb^2 & 1 + nab \end{pmatrix}.$$

Proof The proof is done by direct calculation and will be left as an exercise. □

Definition 1.32 Let $z \in \mathbb{S}$ and Γ a group. The *stabilizer subgroup* Γ_z of z in Γ is $\{V \in \Gamma : Vz = z\}$.

Remark 1.33 Let $\Gamma \subset \mathrm{SL}_2(\mathbb{R})$ and $A \in \mathrm{SL}_2(\mathbb{R})$. It is straightforward to check that $\Gamma_{Az} = A\Gamma_z A^{-1}$.

Proposition 1.34 *Let* $\Gamma \subset \Gamma(1)$ *be a subgroup of finite index* μ, *let* $q \in \mathbb{Q} \cup \{\infty\}$, *and let* Γ_q *be the stabilizer of* q *in* Γ. *Then, there exists* $M \in \Gamma$ *such that* M *is parabolic and* $\Gamma_q = \langle M, -\mathbf{1}\rangle$. *Also, every element in* $\Gamma_q \setminus \{\pm \mathbf{1}\}$ *is parabolic.*

(We may take $M = M_q$ *defined in Lemma 1.31.)*

Remark 1.35 Notice that we already used the following fact several times. Assume that two elements $A, A' \in \Gamma(1)$ satisfy $A\,\infty = \frac{a}{b} = A'\,\infty$. Then, $A = A'S^t$ for some integer t.

Definition 1.36 Let $\Gamma \subset \Gamma(1)$ such that $\Gamma(1) = \bigcup_{i=1}^{\mu} A_i\Gamma$ with $A_1 = \mathbf{1}$.

The *width of the cusp* ∞ is the smallest $\lambda_1 \in \mathbb{N}$ satisfying $S^{\lambda_1} \in \Gamma$.

For the other cusps $A_j\,\infty$, we define the *width of the cusp* $A_j\,\infty$ to be the smallest $\lambda_j \in \mathbb{N}$ satisfying $M^{\lambda_j}_{\frac{a_j}{b_j}} \in \Gamma$, where $A_j\,\infty =: \frac{a_j}{b_j}$ such that $\gcd(a_j, b_j) = 1$ and $b_j \neq 0$.

Example 1.37 We know that $C_{\Gamma(1),\mathcal{F}_{\Gamma(1)}} = \{\infty\}$. Since $S\infty = \infty$ and $S \in \Gamma(1)$, we see immediately that the width of the cusp ∞ is $\lambda_\infty = 1$.

We give now a more involved example.

Example 1.38 We consider the *theta group* Γ_θ. Its index is $[\Gamma(1) : \Gamma_\theta] = 3$ and the group is generated by

$$\Gamma_\theta = \left\langle \begin{pmatrix} 1 & 2 \\ 0 & 1 \end{pmatrix}, \begin{pmatrix} 0 & -1 \\ 1 & 0 \end{pmatrix} \right\rangle = \langle S^2, T \rangle.$$

The representatives of the right cosets of Γ_θ in $\Gamma(1)$ are $\mathbf{1}$, S, and ST given by:

$$\mathbf{1} = \begin{pmatrix} 1 & 0 \\ 0 & 1 \end{pmatrix}, \quad S = \begin{pmatrix} 1 & 1 \\ 0 & 1 \end{pmatrix} \quad \text{and} \quad ST = \begin{pmatrix} 1 & -1 \\ 1 & 0 \end{pmatrix}.$$

As a result, we can see that the inequivalent cusps are ∞ and 1.

For the cusp $1 = \frac{1}{1} = ST\,\infty$, we find

$$M_{\frac{1}{1}} \overset{(1.27)}{=} \begin{pmatrix} 0 & 1 \\ -1 & 2 \end{pmatrix} = \begin{pmatrix} 0 & 1 \\ -1 & 0 \end{pmatrix} \begin{pmatrix} 1 & -2 \\ 0 & 1 \end{pmatrix} = T^3\, S^{-2}.$$

Hence, $M_{\frac{1}{1}}$ is already in Γ_θ. This implies that the width of the cusp 1 is 1.

The cusp ∞ has width 2, since it is already clear from the generators of Γ_θ that $S \notin \Gamma_\theta$ but $S^2 \in \Gamma_\theta$.

A fundamental domain F_{Γ_θ} of the theta group Γ_θ is given by

$$\mathcal{F}_{\Gamma_\theta} = \mathcal{F}_{\Gamma(1)} \cup S\,\mathcal{F}_{\Gamma(1)} \cup ST\,\mathcal{F}_{\Gamma(1)} = \mathcal{F}_{\Gamma(1)} \cup \begin{pmatrix} 1 & 1 \\ 0 & 1 \end{pmatrix} \mathcal{F}_{\Gamma(1)} \cup \begin{pmatrix} 1 & -1 \\ 1 & 0 \end{pmatrix} \mathcal{F}_{\Gamma(1)},$$

which is illustrated in Figure 1.3.

The next lemma presents a consistency condition for the width of cusps. What happens if you take two sets of representatives for the right cosets? Obviously, the

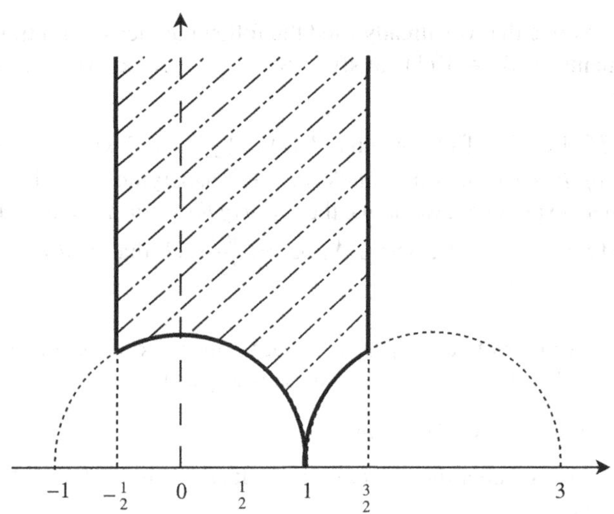

Fig. 1.3 The figure illustrates the fundamental domain $\mathcal{F}_{\Gamma_\theta}$ of the theta group Γ_θ discussed in Exercise 1.38. Clearly visible are the two cusps: ∞ of length 2 and 1 of length 1.

standard fundamental domain changes as well as the location of the cusps. But will the width of the cusps remain the same?

Lemma 1.39 *Let Γ be a subgroup of $\Gamma(1)$ of finite index μ with two sets $\{A_i\}$ and $\{A_i'\}$ of representatives of the right cosets of Γ in $\Gamma(1)$. Suppose that the representatives are arranged so that*

$$A_i' \in A_i\Gamma \qquad \textit{for all } 1 \le i \le \mu.$$

Assume that each cusp $q_i := A_i\,\infty$ has width λ_i and each cusp $q_i' := A_i'\,\infty$ has width λ_i'.

Then, $\lambda_i = \lambda_i'$ for all i.

Proof The proof is left as an exercise. □

1.3 Modular Forms – Properties and Examples

1.3.1 Introduction – Modular Forms

Let $\Gamma \subset \Gamma(1)$ be a subgroup of finite index in the full modular group. In principle, we would like to study invariant functions under the group action. However, the space of invariant holomorphic and bounded functions is not that interesting. For example, the space of holomorphic and bounded functions that are invariant under the full modular group $\Gamma(1)$ consists only of the constant functions.

Informally speaking: In order to get a more interesting space of functions, we consider meromorphic functions F on $\mathbb{H}$ that satisfy a slightly different transformation law

$$F(Mz) = v(M)\, j(M,z)^k\, F(z) \qquad \text{for all } z \in \mathbb{H} \text{ and for all } M \in \Gamma. \tag{1.28}$$

The terms $v(M)$ and $j(M,z)$ denote a suitable multiplier system and the automorphic factor introduced below. Functions F satisfying (1.28) are called *modular forms*.

The *automorphic factor* is defined by

$$j : \mathrm{Mat}_2(\mathbb{R}) \times \mathbb{C} \to \mathbb{C} \quad (M,z) \mapsto j(M,z) := cz + d, \tag{1.29}$$

where $M = \begin{pmatrix} a & b \\ c & d \end{pmatrix}$ and the real parameter $k \in \mathbb{R}$ in (1.28) is the *weight* of F. The *multiplier system*, or *multiplier*, is a function $v : \Gamma \to \mathbb{C}$ satisfying $|v(M)| = 1$ for all $M \in \Gamma$ such that (1.28) allows nontrivial solutions. The rigorous definition of multiplier systems and modular forms will be given below in §1.3.2 and §1.3.3, respectively.

We now present some facts.

1. It can be easily shown that the multiplier v does not depend on z using the open mapping theorem.
2. For a real weight k, we need to specify a branch cut to make this notion precise. For a nonvanishing complex number $w \in \mathbb{C}_{\neq 0}$, we put

$$w^k := |w|^k \, e^{ik \arg(w)} \qquad \text{with } -\pi \leq \arg(w) < \pi. \tag{1.30}$$

Notice that we adhere to the convention that $\arg(-\mathbf{1}) = -\pi$.

3. Now we assume that there exists a nonzero function F satisfying the transformation law (1.28). Given two matrices $M_1, M_2 \in \Gamma$ and $z \in \mathbb{H}$, we consider

$$F\big(M_1 M_2 z\big).$$

Using (1.17) we see that

$$\begin{aligned} v(M_1)v(M_2)\, j(M_1, M_2 z)^k j(M_2, z)^k \, F(z) &= F\big(M_1 M_2 z\big) \\ &= v(M_1 M_2)\, j(M_1 M_2, z)^k \, F(z). \end{aligned}$$

Since F does not vanish everywhere, we just showed the following:

Lemma 1.40

1. *Let $k \in \mathbb{R}$ be a real weight. If v is a multiplier associated with nonconstant functions respecting the transformation law* (1.28)*, then v satisfies*

$$v(M_1 M_2)\, j(M_1 M_2, z)^k = v(M_1)v(M_2)\, j(M_1, M_2 z)^k j(M_2, z)^k \tag{1.31}$$

for all $M_1, M_2 \in \Gamma$ and $z \in \mathbb{H}$.

2. *Let $k \in \mathbb{Z}$ be an integer. Then we have no branching problems when we take powers of the automorphic factors. We then have*

$$j(M_1 \, M_2, z)^k = j(M_1, M_2 z)^k j(M_2, z)^k \tag{1.32}$$

for all $M_1, M_2 \in \mathrm{Mat}_2\,(\mathbb{R})$ and $z \in \mathbb{H}$. Moreover, the multiplier v is, in this case, multiplicative:

$$v(M_1 \, M_2) = v(M_1)\, v(M_2) \tag{1.33}$$

for all $M_1, M_2 \in \Gamma$.

Remark 1.41 Property (1.31) is known as the *consistency condition* of the multiplier v.

Exercise 1.42 Consider $k \in \mathbb{R}$. Show that

$$\left|j(M_1 M_2, z)^k\right| = \left|j(M_1, M_2 z)^k\right| \left|j(M_2, z)^k\right| \tag{1.34}$$

for all $z \in \mathbb{H}$ and $M_1, M_2 \in \mathrm{Mat}_2(\mathbb{R})$.

Exercise 1.43 Let $q = \frac{a}{b} \in \mathbb{Q}$ with coprime integers $a \in \mathbb{Z}$ and $b \in \mathbb{Z}_{\neq 0}$. Recall the matrix M_q in (1.27) of Lemma 1.31. We assume $M_q^t \in \Gamma(1)$ for some $t \in \mathbb{Z}$.

For every $k \in \mathbb{R}$, we have

$$\left(M_q^t z - q\right)^k = j\left(M_q^t, z\right)^{-k} (z-q)^k. \tag{1.35}$$

1.3.2 Multiplier Systems

We now give the rigorous definition of the multiplier system.

Definition 1.44 Let $\Gamma \subset \Gamma(1)$ be a subgroup with finite index and $k \in \mathbb{R}$. A function

$$v : \Gamma \to \mathbb{C}_{\neq 0}$$

is called a *multiplier system* or a *multiplier* with respect to (Γ, k) provided:

1. $|v(M)| = 1$ for all $M \in \Gamma$ and
2. v satisfies the consistency condition (1.31), for every M_1, M_2 in Γ.

Remark 1.45 If $k \in \mathbb{Z}$, then (1.31) reduces to (1.33).

In the following example, we determine the value of the multiplier system at $\pm\mathbf{1}$.

Example 1.46 Take $M_1 = M_2 = \mathbf{1} \in \Gamma$ and use (1.31). Since $\mathbf{1}^2 = \mathbf{1}$, we have

$$v(\mathbf{1}^2)\,(1)^k = v(\mathbf{1})v(\mathbf{1})\,1^{2k},$$

which implies

$$v(\mathbf{1}) = v(\mathbf{1})^2.$$

Since v is a multiplier system and takes only values with $|v(\cdot)| = 1$, we get

$$v(\mathbf{1}) = 1.$$

Now, let $M_1 = M_2 = -\mathbf{1}$. The same calculation with $(-\mathbf{1})^2 = \mathbf{1}$ leads to

$$v(\mathbf{1})\,(1)^k = v(-\mathbf{1})^2\,(-1)^k(-1)^k.$$

Since we already know $v(\mathbf{1}) = 1$, we have

$$1 = v(-\mathbf{1})^2\,(-1)^{2k} \iff v(-\mathbf{1})^2 = (-1)^{-2k} \overset{(1.30)}{=} e^{2\pi ik}.$$

Hence,

$$v(-\mathbf{1}) \in \left\{\pm e^{\pi ik}\right\}.$$

Now, consider a nonzero function F satisfying (1.28). Using the transformation law (1.28) for a function F, we get

$$F(z) = F\big((-\mathbf{1})\,z\big) \overset{(1.28)}{=} v(-\mathbf{1})\,(-1)^k\,F(z)$$

for all $z \in \mathbb{H}$. Dividing $F(z)$ on both sides gives

$$1 = v(-\mathbf{1})\,(-1)^k = v(-\mathbf{1})\,e^{-\pi ik}.$$

As a result,

$$v(-\mathbf{1}) = e^{\pi ik}. \tag{1.36}$$

Definition 1.47 We assume $-\mathbf{1} \in \Gamma$. Condition (1.36) is called the *nontriviality condition* for multipliers.

Exercise 1.48 Let $\Gamma \subset \Gamma(1)$ be a subgroup with finite index, $k \in \mathbb{R}$ and $v : \Gamma \to \mathbb{C}_{\neq 0}$ a multiplier that satisfies the nontriviality condition (1.36). Assume $T \in \Gamma$ with $T = \begin{pmatrix} 0 & -1 \\ 1 & 0 \end{pmatrix}$ as defined in (1.20). Show that the multiplier v satisfies

$$v(T) = e^{-\frac{\pi i}{2}k}, \tag{1.37}$$

respecting the argument convention (1.30).

Exercise 1.49 Consider the full modular group $\Gamma(1)$, a $k \in \mathbb{R}$, and a multiplier $v : \Gamma \to \mathbb{C}_{\neq 0}$ which satisfies the nontriviality condition (1.36). Recall that $S, T \in \Gamma(1)$ with $S = \begin{pmatrix} 1 & 1 \\ 0 & 1 \end{pmatrix}$ and $T = \begin{pmatrix} 0 & -1 \\ 1 & 0 \end{pmatrix}$ as defined in (1.20).

Using $(ST)^3 = -\mathbf{1}$ show that the multiplier v satisfies

$$\begin{aligned} v(S) &= e^{\frac{\pi i}{6}k} \quad \text{and} \\ v(T) &= e^{-\frac{\pi i}{2}k}, \end{aligned} \tag{1.38}$$

respecting the argument convention (1.30).

Example 1.50 Recall the function

$$\Delta : \mathbb{H} \to \mathbb{C}; \quad z \mapsto \Delta(z) := e^{2\pi i z} \prod_{n\in\mathbb{N}} \left(1 - e^{2\pi i n z}\right)^{24},$$

that was already introduced in (0.1). This function is also known as the *discriminant function*. It satisfies the transformation law (1.28) with trivial multiplier $\upsilon \equiv 1$, weight 12, and the full modular group $\Gamma(1)$.

The *Dedekind η-function*, given by

$$\eta : \mathbb{H} \to \mathbb{C}; \quad z \mapsto \eta(z) := e^{\frac{2\pi i z}{24}} \prod_{n\in\mathbb{N}} \left(1 - e^{2\pi i n z}\right),$$

satisfies

$$\Delta(z) = (\eta(z))^{24} \qquad (z \in \mathbb{H}).$$

Moreover, η satisfies the transformation laws

$$\eta(Tz) = e^{-\frac{\pi i}{4}} z^{\frac{1}{2}} \eta(z) \quad \text{and} \quad \eta(Sz) = e^{\frac{\pi i}{12}} \eta(z).$$

More details about the Dedekind η-function can be found in [12, Chapter III] or in [116, Chapter IX, §1].

Let $r \in \mathbb{R}_{>0}$ be a positive real number. Then, the $(24r)^{\text{th}}$ power of the η-function, η^{24r}, satisfies

$$\left(\eta(Tz)\right)^{24r} = e^{-6\pi i r} z^{12r} \left(\eta(z)\right)^{24r} \quad \text{and} \quad \left(\eta(Sz)\right)^{24r} = e^{2\pi i r} \left(\eta(z)\right)^{24r} \tag{1.39}$$

for all $z \in \mathbb{H}$. Comparing these transformation laws to (1.28), we are inspired to define the missing factor as a value of a suitable multiplier υ_{24r}.

Let $k > 0$ be positive real weight. We define the *η-multiplier* υ_k by its values on the generators S and T of $\Gamma(1)$:

$$\upsilon_k(T) := e^{-\frac{\pi i k}{2}} \quad \text{and} \quad \upsilon_k(S) := e^{\frac{\pi i k}{6}}. \tag{1.40}$$

Then, we extend υ_k to Γ using the consistency condition (1.31) mentioned in Remark (1.41). In particular, υ_k is a multiplier system with respect to $(\Gamma(1), k)$, $k > 0$.

Example 1.51 Another typical example of a multiplier system is the Dirichlet character χ that is defined as follows.

We call a function $\chi : \mathbb{Z} \to \mathbb{C}_{\neq 0}$ a *Dirichlet character* of *modulus* $n \in \mathbb{N}$ if it satisfies the following conditions:

1. $\chi(l) = \chi(l+n)$ for all $l \in \mathbb{Z}$,
2. if $\gcd(l, n) > 1$ then $\chi(l) = 0$ and if $\gcd(l, n) = 1$ then $\chi(l) \neq 0$,
3. $\chi(lm) = \chi(l)\,\chi(m)$ for all $l, m \in \mathbb{Z}$.

We refer to [144] and [76, Dirichlet Character] for more details about Dirichlet characters.

Let $k \in 2\mathbb{N}$ and consider the congruence subgroup $\Gamma_0(N)$ and χ a Dirichlet character of modulus N. We construct a multiplier v on $\Gamma_0(N)$ compatible with k by

$$v\left(\begin{pmatrix} a & b \\ c & d \end{pmatrix}\right) := \chi(d)$$

for all $\begin{pmatrix} a & b \\ c & d \end{pmatrix} \in \Gamma_0(N)$. The above properties of a Dirichlet character ensure that v is a multiplier system for $\big(\Gamma_0(N), k\big)$.

We need another technical definition relating cusps and values of multipliers.

Definition 1.52 Let $\Gamma \subset \Gamma(1)$ be of finite index, v be a multiplier system of weight k on Γ, and let q be a cusp of Γ. Denote by λ_q the width of the cusp q in Γ introduced in Definition 1.36.

We define a real number κ_q such that

$$v\big(M_q^{\lambda_q}\big) = e^{2\pi i \kappa_q}, \tag{1.41}$$

where κ_q can be chosen satisfying $0 \leq \kappa_q < 1$. Here, the matrix M_q is given in (1.27).

Example 1.53 Recall the η-multiplier v_k, $k > 0$, defined in Example 1.50. It is a multiplier with respect to $\big(\Gamma(1), k\big)$. In particular, we stated in (1.40) that $v_k(S)$ has the value

$$v_k(S) = e^{\frac{\pi i k}{6}}.$$

Recall also that M_∞ defined in (1.27) satisfies $M_\infty = \begin{pmatrix} 1 & 1 \\ 0 & 1 \end{pmatrix} = S$, that the cusp ∞ of the full modular group $\Gamma(1)$, and that the width of the cusp ∞ is $\lambda_\infty = 1$, see Example 1.37.

Then, κ_∞ introduced in Definition 1.52 above satisfies

$$e^{2\pi i \kappa_\infty} \overset{(1.41)}{=} v\big(M_\infty^{\lambda_\infty}\big) \overset{E\,1.37}{=} v\big(S\big) \overset{(1.40)}{=} e^{\frac{\pi i k}{6}}.$$

Hence,

$$\kappa_\infty \equiv \frac{k}{12} \mod 1.$$

1.3.3 Modular Forms

In the following section, we give the definition of a modular form on a subgroup of finite index in the full modular group. We state that such functions have Fourier expansions at all the cusps of the fundamental domain of the subgroup and recall several important properties. We conclude this section with interesting examples.

However, we need another result before we give the definition of a modular form. The following theorem illustrates an argument, which will be used later. The theorem simply states that if a function satisfies a specific transformation law on the generators of the full modular group, then it satisfies the same transformation law on every other element of that group.

Theorem 1.54 *Let $k \in 2\mathbb{Z}$ be an even integer and let $F : \mathbb{H} \to \mathbb{C}$ be a function satisfying the two relations*

$$F(z+1) = F(z) \quad \text{and} \quad F\left(\frac{-1}{z}\right) = z^k\, F(z) \tag{1.42}$$

for all $z \in \mathbb{H}$. Then, F satisfies

$$F\big(M\,z\big) = j(M, z)^k\, F(z) \tag{1.43}$$

for all $z \in \mathbb{H}$ and all $M \in \Gamma(1)$.

Remark 1.55 Theorem 1.54 shows that if F satisfies (1.42), then F satisfies the transformation law (1.28) for even integral weight $k \in 2\mathbb{Z}$, trivial multiplier $\upsilon \equiv 1$, and the full modular group $\Gamma(1)$.

Proof (of Theorem 1.54) Each element $M \in \Gamma(1)$ can be expressed as a word in the generators S and T of $\Gamma(1)$, see Corollary 1.14, and has an Eichler length $l(M)$, see Definition 1.15. We make an induction argument in the Eichler length to prove the theorem.

Consider the case $l(M) = 0$ as a starting point. This implies immediately $M = \mathbf{1}$ and (1.43) simplifies to the trivial identity

$$F(z) = F\big(\mathbf{1}\,z\big) \overset{!}{=} j(\mathbf{1}, z)^k\, F(z) = 1^k\, F(z).$$

Suppose that (1.43) holds for all elements $M' \in \Gamma(1)$ with Eichler length $l(M') < m$ for some $m \in \mathbb{N}$. We consider an element $M \in \Gamma(1)$ with Eichler length $l(M) = m$. Since $\Gamma(1)$ is generated by S and T, we know that M can be written as (at least) one of the three cases

$$M = M'\,S, \quad M = M'\,S^{-1}, \quad \text{or} \quad M = M'\,T,$$

where M' has a smaller Eichler length ($l(M') < m$). Assume $M = M'S$ for some $M' \in \Gamma(1)$ with $l(M') \leq m - 1$. Hence,

$$
\begin{aligned}
F(M\,z) = F(M'\,S\,z) &= j(M', S\,z)^k\, F(S\,z) \\
&= j(M', S\,z)^k j(S, z)^k\, F(z) \\
&\overset{(1.32)}{=} j(M'S, z)^k\, F(z) \\
&= j(M, z)^k\, F(z).
\end{aligned}
$$

The other two cases $M = M'\,S^{-1}$ and $M = M'\,T$ follow by the same argument, based on (1.32). □

We now present the Fourier expansion of functions satisfying (1.28) in addition to other conditions. If we impose certain growth conditions on those functions, those functions will be modular forms. We do not give the proof but it can be found in [106, Chapter II].

Theorem 1.56 *Let $\Gamma \subset \Gamma(1)$ be a subgroup of finite index, $k \in \mathbb{R}$ a real weight, and $v : \Gamma \to \mathbb{C}_{\neq 0}$ be a multiplier with respect to (Γ, k). Suppose*

$$F : \mathbb{H} \to \mathbb{C}; \quad z \mapsto F(z)$$

is a meromorphic function that satisfies the transformation law (1.28). *Moreover, F has at most finitely many poles in the closure of the standard fundamental domain $\overline{\mathcal{F}_\Gamma} \cap \mathbb{H}$ of Γ. Let $q_1 = \infty, q_2, \ldots, q_\mu \in \mathbb{Q} \cup \{\infty\}$ be the cusps of Γ in the standard fundamental domain $\mathcal{F}_\Gamma$ and $A_1, A_2, \ldots, A_\mu$ be representatives of the associated right cosets of Γ in $\Gamma(1)$ satisfying $q_j = A_j\,\infty$, as introduced in Lemma 1.28. For each $j \in \{1, \ldots, mu\}$ let λ_j be the width of the cusp q_j and κ_j be the corresponding real numbers associated with q_j by* (1.41).

Then for each $j \in \{1, \ldots, \mu\}$, there exists a nonnegative real y_j such that F has an expansion at the cusp q_j given by

$$F(z) = \sigma_j(z) \sum_{n=-\infty}^{\infty} a_j(n)\, e^{2\pi i (n+\kappa_j)\, A_j^{-1} \frac{z}{\lambda_j}} \qquad \textit{for all } z \in \mathbb{H}, \Im\left(A_j^{-1} z\right) > y_j, \tag{1.44}$$

where $\sigma_j(z)$ is given by

$$\sigma_{k,j}(z) = \begin{cases} 1 & \textit{if } q_j = \infty, \\ (z - q_j)^{-k} & \textit{if } q_j \neq \infty. \end{cases} \tag{1.45}$$

The above result implies the following fact, stated as an exercise.

Exercise 1.57 Assume that $\mathcal{F}_\Gamma$ and $\mathcal{F}'_\Gamma$ are two standard fundamental domains of Γ (i.e., both are generated using different sets of representatives for the right cosets of Γ in $\Gamma(1)$) and assume that $F(z)$ is a meromorphic function on $\mathbb{H}$ satisfying (1.28). $F(z)$ has a finite number of poles in $\overline{\mathcal{F}_\Gamma} \cap \mathbb{H}$ if and only if $F(z)$ has a finite number of poles in $\overline{\mathcal{F}'_\gamma} \cap \mathbb{H}$.

Definition 1.58 Let F be as in Theorem 1.56 and consider the expansions of F at the cusp's $q_j \in C_{\Gamma,\mathcal{F}_\Gamma}$.

1. If only a finite number of terms with $n < 0$ appear in (1.44) at every cusp q_j, we say that F *is meromorphic at* q_j.
2. If the first nonzero $a_j(n)$ occurs for $n = -n_j < 0$ for some $n_j \in \mathbb{N}$, we say F has a *pole at* q_j *of order* $n_j - \kappa_j$.
3. If the first nonzero $a_j(n)$ occurs for $n = n_0 \geq 0$ for some $n_j \in \mathbb{N}$, we say F is *regular at* q_j *with a zero at* q_j *of order* $n_0 + \kappa_j$.

We now give the formal definition of modular forms.

Definition 1.59 Let $k \in \mathbb{R}$ and let v be a multiplier system with respect to (Γ, k). A meromorphic function $F : \mathbb{H} \to \mathbb{C}$ is called a *weakly holomorphic modular form of weight k and multiplier system v on* Γ provided:

1. $F(z)$ satisfies (1.28) for all elements of Γ.
2. $F(z)$ has a finite number of poles in $\overline{\mathcal{F}_\Gamma} \cap \mathbb{H}$.
3. $F(z)$ is meromorphic at each cusp $q_j \in C_{\Gamma,\mathcal{F}_\Gamma}$ which is a parabolic point.

We will denote the vector space of modular forms of weight k and multiplier system v on a group Γ by $M^!_{k,v}(\Gamma)$. A useful subspace is the space $S^!_{k,v}(\Gamma) \subset M^!_{k,v}(\Gamma)$ where the Fourier expansion (1.44) of each weakly holomorphic modular form $F \in S^!_{k,v}(\Gamma)$ has vanishing zero-terms: $a_j(0) = 0$.

We call $F(z)$ an *entire modular form* of weight k with multiplier v on Γ if $F \in M^!_{k,v}(\Gamma)$ is holomorphic in $\mathbb{H}$ and $F(z)$ is regular at each cusp q_j. The vector space of such entire modular forms will be denoted $M_{k,v}(\Gamma)$.

We call $F \in M_{k,v}(\Gamma)$ a *cusp form* if $F(z)$ is a modular form with a zero of positive order at each cusp q_j. The vector space of such cusp forms will be denoted by $S_{k,v}(\Gamma)$.

In addition, we call $F(z)$ a *modular function* if $F(z)$ is a modular form of weight 0 and trivial multiplier, i.e., $v(M) = 1$ for all $M \in \Gamma$.

Remark 1.60 The above definition of weakly modular forms, modular forms, or cusp forms is commonly used in the literature. For example, see [126, Definition 4.2.1].

Remark 1.61 If the multiplier is trivial, i.e., $v \equiv 1$, we use the notation $M^!_k(\Gamma)$, $M_k(\Gamma)$, and $S_k(\Gamma)$ instead of $M^!_{k,1}(\Gamma)$, $M_{k,1}(\Gamma)$, and $M_{k,1}(\Gamma)$.

Remark 1.62 The notations $M_{k,v}(\Gamma)$ and $S_{k,v}(\Gamma)$ for entire modular and cusp forms are now standard (up to small variations). Also, elements of the space $M^!_{k,v}(\Gamma)$ are called weakly holomorphic modular forms in modern literature. However, you can sometimes find different notations, mostly in some older sources like [106]. For example, the space $M^!_{k,v}(\Gamma)$ is sometimes denoted by $\{\Gamma, k, v\}$, the space of entire modular forms $M_{k,v}(\Gamma)$ is sometimes denoted by $C^+(\Gamma, k, v)$, and the space of cusp forms $S_{k,v}(\Gamma)$ is sometimes denoted by $C^0(\Gamma, k, v)$.

Remark 1.63 We refer to modular forms or cusp forms sometimes as "classical" modular forms or "classical" cusp forms in the remaining chapters. This is to highlight that the theory of modular forms has been a topic of the mathematical community for some time already (and therefore deserving the word "classical"). However, the older literature usually restricts the study to even, integral, or half-integral weights, since these forms with these weights often display special arithmetic properties. We use the word "classical" a bit more loosely to also refer to the real weight case.

We list some facts about modular forms. More details and proofs can be found in [66, 106, 142].

Theorem 1.64 *Let* Γ *and* $f \in M_0(\Gamma)$ *be a modular function on* Γ*. (So* $f(z)$ *is holomorphic in* $\mathbb{H}$ *and regular at all cusps of the standard fundamental domain* $\mathcal{F}_\Gamma$*.) Then* f *is constant.*

Corollary 1.65 *Let* Γ *be a subgroup of* $\Gamma(1)$ *with finite index. If* f *is a modular function on* Γ *and* f *is bounded in* $\mathbb{H}$*, then* f *is entire and hence constant.*

The rate of growth of a cusp form is given by the following theorem.

Lemma 1.66 *Let* Γ *be a subgroup of* $\Gamma(1)$ *with finite index and let* $F \in S_{k,v}(\Gamma)$*. There exists a positive constant* $K > 0$ *such that*

$$|F(z)| \leq K\, \Im(z)^{-\frac{k}{2}} \qquad \textit{for all } z \in \mathbb{H}.$$

Exercise 1.67 Let Γ be a subgroup of $\Gamma(1)$ with finite index and let $F \in S_{k,v}(\Gamma)$. Show that there exists $\delta > 0$ such that F satisfies the growth estimate

$$F(z) = O\left(e^{-2\pi\delta\Im(z)}\right) \qquad \text{as } \Im(z) \to \infty.$$

We now show that the Fourier coefficients of the expansion of a cusp form at infinity have polynomial growth.

Theorem 1.68 *Let* $F \in S_{k,v}(\Gamma)$ *with* $\Gamma \subset \Gamma(1)$ *being of finite index. Assume that* F *has an expansion at* ∞ *given by*

$$F(z) = \sum_{n=n_0}^{\infty} a(n)\, e^{2\pi i (n+\kappa)\frac{z}{\lambda}}$$

with $n_0 + \kappa > 0$*. Then, the Fourier coefficients* $a(n)$ *satisfy*

$$a(n) = O\left(n^{\frac{k}{2}}\right) \qquad \textit{as } n \to \infty. \tag{1.46}$$

Remark 1.69 The bound on Fourier coefficients presented in (1.46) is not optimal. The Ramanujan-Petersson conjecture sharpens the above bound conjecturally to $a(n) = O\left(n^{\frac{k-1}{2}}\right)$ as $n \to \infty$. The proof of this conjecture goes back to Deligne for weight $k = 12$ [64].

A reference to more modern publications is for example [178]. Schulze-Pillot and Yenirce [178] investigate the Ramanujan-Petersson conjecture for arbitrary level and arbitrary character.

It is worth mentioning that when $k < 0$, we can do much better. This is illustrated in the following corollary.

Corollary 1.70 *Let $k < 0$ be a negative weight and Γ be a subgroup of $\Gamma(1)$ with finite index. If $F \in S_{k,v}(\Gamma)$, then $F \equiv 0$.*

We now determine the growth of an entire modular form.

Lemma 1.71 *Let Γ be a subgroup of $\Gamma(1)$ with finite index. If $F \in M_{k,v}(\Gamma)$ is an entire modular form, then*

$$|F(z)| \leq K\left(1 + \Im(z)^{-k}\right) \qquad (z \in \mathbb{H})$$

for some $K > 0$.

This lemma can be used to determine the growth of the Fourier coefficients of entire modular forms at the infinite cusps. The following result is widely known as the *Hecke bound* for the Fourier coefficients of modular cusp forms of a given weight k.

Theorem 1.72 *Let Γ be a subgroup of $\Gamma(1)$ with finite index and let $F \in M_{k,v}(\Gamma)$ be a modular form of weight $k \geq 0$. Assume that F has an expansion at infinity given by*

$$F(z) = \sum_{n=n_0}^{\infty} a(n)\, e^{2\pi i (n+\kappa)\frac{z}{\lambda}}$$

with $n_0 + \kappa \geq 0$. Here, $\lambda \in \mathbb{N}$ and $0 \leq \kappa < 1$ denote the width of the cusp ∞ and its associated κ-value, see Definitions 1.36 and 1.52, respectively.

Then $a(n) = O\left(n^k\right)$ as $n \to \infty$.

We recall that there are no nonzero entire forms of negative weight on a finite-index subgroup of $\mathrm{SL}_2(\mathbb{Z})$.

Lemma 1.73 *If $F \in M_{k,v}(\Gamma)$ with negative weight $k < 0$, then there exists a positive constant $K > 0$ such that*

$$|F(z)| \leq K\, \Im(z)^{-k/2} \qquad \text{for all } z \in \mathbb{H}.$$

Proposition 1.74 *Let $F(z) \in M_{k,\upsilon}(\Gamma)$ with negative weight $k < 0$. Then, the modular form $F \equiv 0$ vanishes everywhere.*

Exercise 1.75 Let $F \in M^!_{k,1}(\Gamma)$ and $G \in M^!_{l,1}(\Gamma)$. Then $F \cdot G \in M^!_{k+l,1}(\Gamma)$, where the function product is defined as $\big(F \cdot G\big)(z) := F(z) \cdot G(z)$.

Exercise 1.76 Show that every holomorphic function F on $\mathbb{H}$ which is invariant under $\Gamma(1)$ and bounded at a neighborhood of ∞ has to be a constant function.

1.3.4 Eisenstein Series

Here, we present examples of entire modular forms known as Eisenstein series. Almost every textbook on modular forms mentions Eisenstein series and proves their basic properties. So we will not give proofs of the theorems presented but one can refer to [12, Chapter VI], [142, Chapter VII] for more details. Eisenstein series have been generalized in many directions. There are Eisenstein series of half-integer weight, on subgroups of the full group, on higher level groups, etc. The importance of Eisenstein series, besides forming the basis of the complex vector space of modular forms, is due to their interesting applications in number theory. In fact, their Fourier coefficients are divisor functions as will be seen later. Moreover, and because of the finite dimensionality of the space of modular forms, many arithmetic identities arise naturally.

Definition 1.77 We define the *prime summation* $\sum'_{(m,n)}$ as follows:

$$\sum_{(m,n)}{}' := \sum_{\substack{(m,n)\in\mathbb{Z}^2 \\ (m,n)\neq(0,0)}} := \sum_{m\in\mathbb{Z}} \sum_{\substack{n\in\mathbb{Z} \\ (m,n)\neq(0,0)}}, \tag{1.47}$$

which is a double summation over the nonvanishing pairs $(m, n) \in \mathbb{Z}^2_{\neq(0,0)}$.

Remark 1.78 The prime notation is a typical notation for Eisenstein series generated by summing a lattice. Typically, we sum over some expressions which are not defined for a finite number of lattice points. The prime notation indicates that those lattice points are omitted in the summation.

We now define Eisenstein series on the full modular group for weights $k > 2$.

Definition 1.79 Let $2 < k \in 2\mathbb{N}$. We define the *Eisenstein series G_k of weight k* by

$$G_k : \mathbb{H} \to \mathbb{C}; \quad z \mapsto G_k(z) := \sum_{(m,n)}{}' (mz+n)^{-k}. \tag{1.48}$$

Remark 1.80

1. The series above converges absolutely for every fixed z and uniformly on all compact subsets of $\mathbb{H}$.
2. It is important to mention that $G_k \equiv 0$ for all odd k since the sum runs over all integers. (The terms $mz + n$ and $(-m)z + (-n)$ appear in the series and cancel each other).
3. Also,

$$\lim_{z \to i\infty} G_k(z) = 2\zeta(k) \neq 0$$

holds for even integer weights $k > 2$. Here, the limit $z \to i\infty$ is understood as $\Im(z) \to \infty$ and $\Re(z)$ is bounded. Also, the only terms in the series that survive the limit $z \to \infty$ are the ones with $m = 0$.

Exercise 1.81 Prove Remark 1.80, Item 3: For $2 < k \in 2\mathbb{N}$ a positive even integer weight shows

$$\lim_{z \to i\infty} G_k(z) = 2\zeta(k),$$

where the limit $z \to i\infty$ is understood as $\Im(z) \to \infty$ and $\Re(z)$ is bounded.

The Fourier expansion of Eisenstein series at the cusp ∞ is given in the following theorem.

Theorem 1.82 *Let $k \in 2\mathbb{Z}$, $k > 2$.*

1. *We have $G_k \in M_{k,1}\big(\Gamma(1)\big)$, where the multiplier* 1 *is the trivial multiplier.*
2. *The Fourier expansion of $G_k(z)$ is given by*

$$G_k(z) = 2\zeta(k) + 2\frac{(2\pi i)^k}{(k-1)!}\sum_{n=1}^{\infty}\sigma_{k-1}(n)e^{2\pi i n z} \qquad (z \in \mathbb{H}), \tag{1.49}$$

where

$$\sigma_k(n) = \sum_{\substack{d|n \\ d>0}} d^k \tag{1.50}$$

is the kth arithmetic divisor function.

Remark 1.83 It is important to note that, being inspired by Theorem 1.82, we can define a weight 2 Eisenstein series by

$$G_2(z) = 2\zeta(2) + 2(2\pi i)^2\sum_{n=1}^{\infty}\sigma_1(n)e^{2\pi i n z} \qquad (z \in \mathbb{H}). \tag{1.51}$$

Note that for $k = 2$, the series (1.48) will not converge and one has to define G_2 in terms of its Fourier series. G_2 is well defined on $\mathbb{H}$, converges absolutely for all $z \in \mathbb{H}$, and is a holomorphic function on $\mathbb{H}$.

Note that $G_2(z)$ is not a modular form but still satisfies an interesting transformation law under the inversion $z \mapsto Tz = \frac{-1}{z}$, keeping in mind that it is periodic of period 1. Thus,

$$z^{-2} G_2\left(-\frac{1}{z}\right) = G_2(z) - \frac{2\pi i}{z}. \tag{1.52}$$

$G_2(z)$ is a good example that motivates the theory of rational period functions that we will mention briefly in later sections.

We conclude the section on Eisenstein series by defining the normalized version of Eisenstein series. We would like to normalize the Eisenstein series G_k so that the constant term in the Fourier expansion of the normalized series is 1. Recalling the expansion of G_k in (1.49) we have the following definition of the normalized Eisenstein series.

Definition 1.84 Let $k \in 2\mathbb{Z}$, $k > 2$ be a positive even weight. We define the *normalized Eisenstein series*

$$E_k(z) = \frac{1}{2\zeta(k)} G_k(z). \tag{1.53}$$

for all $z \in \mathbb{H}$.

Exercise 1.85 Let $k \in 2\mathbb{Z}$, $k > 2$ be a positive even weight. Show

$$E_k(z) = \sideset{}{'}\sum_{\substack{c \in \mathbb{Z}_{\geq 0},\, d \in \mathbb{Z} \\ \gcd(c,d)=1 \\ c=0 \Rightarrow d=1}} (cz+d)^{-k} \tag{1.54}$$

holds for all $z \in \mathbb{H}$.

1.3.5 The Discriminant Function $\Delta(z)$

In this section, we give an example of a cusp form on the full modular group. We recall a well-known function called the discriminant function Δ, a cusp form of weight 12 for the full modular group, which we already mentioned in Equation (0.1) and in Exercise 1.50. (To be more precise, we already mentioned the discriminant function Δ in section "Chapter 1 – A Short Introduction to Modular Forms" of the Introductory Roadmap. Since the discussion there is very brief, we continue to refer to Equation (0.1) and Exercise 1.50.) Δ plays an important role in the theory of elliptic curves and modular forms. In fact, it is the discriminant of the curve whose variables are the Weierstress-$\wp$ function and its derivative and whose coefficients are Eisenstein series.

We finally introduce and define the discriminant function properly.

Definition 1.86 The *discriminant function* $\Delta : \mathbb{H} \to \mathbb{C}$ is the Δ-function first stated in (0.1):

$$\Delta(z) = e^{2\pi i z} \prod_{n=1}^{\infty} \left(1 - e^{2\pi i n z}\right)^{24} \qquad (z \in \mathbb{H}). \tag{1.55}$$

Remark 1.87 We see that the term $e^{2\pi i z}$ appears quite often in (1.55). Indeed this term also appears in the Fourier expansions for example in (1.49) of G_k. To shorten the notation and to increase the readability of the expansions it is common to use $q = e^{2\pi i z}$. It is then easy to write Fourier expansions as power series in q:

$$\sum_{n=0}^{\infty} a_n \, e^{2\pi i n z} = \sum_{n=0}^{\infty} a_n \, q^n .$$

However, we decided that we will not use the q-notation in this book. We believe that the use of q as notation for $e^{2\pi i z}$ implies often implicitly that n is an integer power. However, we use later on expressions $e^{2\pi i n z}$ with real n, see for example (3.47) in Chapter 3. We will avoid this possible confusion by not using the q-notation.

It is clear that this product converges uniformly on compact subsets of $\mathbb{H}$. The product expansion also implies immediately that Δ is translation invariant:

$$\Delta(z+1) = \Delta(z) \qquad \text{for all } z \in \mathbb{H}.$$

Theorem 1.88 $\Delta(z) \in S_{12,1}\big(\Gamma(1)\big)$.

Remark 1.89 The coefficients $\tau(n)$ of the Fourier expansion of the discriminant function Δ form a function

$$\tau : \mathbb{N} \to \mathbb{Z}; \quad n \mapsto \tau(n).$$

This function is known as the *Ramanujan τ function*, which is an important number-theoretic function. By construction, $\tau(1) = 1$ since only one term $e^{2\pi i z}$ appears in the product definition of Δ in (1.55).

Example 1.90 The first few values of the Ramanujan τ function are for example listed in [196, A000594]. Combining these values of the Ramanujan τ function with the Fourier expansion of the discriminant function Δ gives

$$\Delta(z) = q - 24q^2 + 252q^3 - 1472q^4 + 4830q^5 - 6048q^6 - 16744q^7 + \dots \quad (1.56)$$

for all $z \in \mathbb{H}$ and using the notation $q = e^{2\pi i z}$.

Exercise 1.91 Calculate the leading coefficients 1, −24, and 252 appearing in the Fourier expansion of the discriminant function Δ in (1.56) of Example 1.90.

1.3.6 Petersson Scalar Product and the Hilbert Space of Cusp Forms

Let $\Gamma \subset \Gamma(1)$ be a subgroup of finite index μ and let $k \in \mathbb{R}$ be a weight and v a multiplier with respect to (Γ, k). We define the Petersson scalar product by

$$\begin{aligned} \langle \cdot, \cdot \rangle_{\Gamma,k} :\quad & M_{k,v}(\Gamma) \times S_{k,v}(\Gamma) \to \mathbb{C} \\ & (f, g) \mapsto \langle f, g \rangle_{\Gamma,k} := \frac{1}{\mu} \int_{\mathcal{F}} f(z)\, \overline{g(z)}\, \Im(z)^k\, \Im(z)^{-2} \mathrm{d}\lambda_2(z), \end{aligned} \quad (1.57)$$

where $\mathcal{F}$ is a standard fundamental domain of the group Γ, see §1.2.3. A typical choice of a fundamental domain of Γ is derived in Theorem 1.20.

The integration measure $\lambda_2(z)$ denotes the usual Euclidean surface Lebesgue measure given by

$$\mathrm{d}\lambda_2(z) = \mathrm{d}x\, \mathrm{d}y, \quad (1.58)$$

if $z = x + iy \in \mathbb{C}$ with $x, y \in \mathbb{R}$. The derived measure

$$\Im(z)^{-2}\, \mathrm{d}\lambda_2(z) := y^{-2}\, \mathrm{d}x\, \mathrm{d}y \quad (1.59)$$

for $z = x + iy \in \mathbb{H}$ is called the *hyperbolic surface measure.*

Remark 1.92 The *hyperbolic distance function* $d : \mathbb{H} \times \mathbb{H} \to \mathbb{R}_{\geq 0}$ is given by

$$\cosh d(z, z') = 1 + \frac{1}{2} \frac{|z - z'|}{\Im(z)\, \Im(z')} \qquad (z, z' \in \mathbb{H}). \quad (1.60)$$

The induced metric $\mathrm{d}s$ satisfies

$$(\mathrm{d}s)^2 = \frac{\mathrm{d}x^2 + \mathrm{d}y^2}{y^2} \qquad (z = x + iy \in \mathbb{H}). \tag{1.61}$$

Lemma 1.93 *The hyperbolic surface measure $\Im(z)^{-2}\mathrm{d}\lambda_2(z)$ is invariant under coordinate changes by Möbius transformations:*

$$\Im(M\,z)^{-2}\mathrm{d}\lambda_2(M\,z) = \Im(z)^{-2}\mathrm{d}\lambda_2(z) \tag{1.62}$$

for all $z \in \mathbb{H}$ and $M \in \mathrm{SL}_2(\mathbb{R})$.

Proof The proof can be done by direct calculation. □

Exercise 1.94 Show $\int_{\mathcal{F}_{\Gamma(1)}} \Im(z)^{-2}\,\mathrm{d}\lambda_2(z) = \frac{\pi}{3}$.

Lemma 1.95 *For $(f, g) \in M_{k,v}(\Gamma) \times S_{k,v}(\Gamma)$, the evaluation of $\langle\cdot,\cdot\rangle_{\Gamma,k}$ in (f, g) is finite: $\langle f, g\rangle_{\Gamma,k} \in \mathbb{C}$.*

Lemma 1.96 *For $(f, g) \in M_{k,v}(\Gamma) \times S_{k,v}(\Gamma)$ and $M \in \Gamma$ we have*

$$\begin{aligned}&\frac{1}{\mu}\int_{M\mathcal{F}} f(z)\,\overline{g(z)}\,\Im(z)^k\,\Im(z)^{-2}\mathrm{d}\lambda_2(z)\\ &\quad= \frac{1}{\mu}\int_{\mathcal{F}} f(z)\,\overline{g(z)}\,\Im(z)^k\,\Im(z)^{-2}\mathrm{d}\lambda_2(z).\end{aligned} \tag{1.63}$$

Proof Using substitution, one can easily get the result. □

Remark 1.97 Let $\mathcal{F}_\Gamma$ be a fundamental domain of Γ of the form given in (1.23):

$$\mathcal{F}_\Gamma = \bigcup_{i=1}^{\mu} A_i\,\mathcal{F}_{\Gamma(1)}$$

with $A_1, \ldots, A_\mu$ a right coset representation of Γ in $\Gamma(1)$. Let $M \in \Gamma$. The fundamental domain $\mathcal{F}'_\Gamma$ associated with the new right coset representation is

$$\mathcal{F}'_\Gamma = \bigcup_{i=1}^{\mu} A_i M\,\mathcal{F}_{\Gamma(1)}.$$

Lemma 1.96 shows that the Petersson scalar product $\langle\cdot,\cdot\rangle_{\Gamma,k}$ defined in (1.57) does not depend on the choice of the standard fundamental domain $\mathcal{F}_\Gamma$ as derived in (1.23).

Theorem 1.98 *Let $\Gamma \subset \Gamma(1)$ be a subgroup of finite index μ and let $k \in \mathbb{R}$ be a weight and v a multiplier with respect to (Γ, k). The Petersson scalar product $\langle\cdot,\cdot\rangle_{\Gamma,k}$ satisfies*

1. *$\langle\cdot,\cdot\rangle_{\Gamma,k}$ is a scalar product: it is well defined and satisfies*

$$\langle \alpha f, g\rangle_{\Gamma,k} = \alpha\langle f, g\rangle_{\Gamma,k} = \langle f, \overline{\alpha} g\rangle_{\Gamma,k} \qquad \big(f \in M_{k,v}(\Gamma), S_{k,v}(\Gamma), \alpha \in \mathbb{C}\big),$$
$$\langle f_1 + f_2, g\rangle_{\Gamma,k} = \langle f_1, g\rangle_{\Gamma,k} + \langle f_2, g\rangle_{\Gamma,k}$$

and

$$\langle f, g_1 + g_2\rangle_{\Gamma,k} = \langle f, g_1\rangle_{\Gamma,k} + \langle f, g_2\rangle_{\Gamma,k}$$

for $f, f_1, f_2 \in M_{k,v}(\Gamma)$, $g, g_1, g_2 \in S_{k,v}(\Gamma)$ and $\alpha \in \mathbb{C}$; and

2. *We have*

$$\langle f, g\rangle_{\Gamma,k} = \overline{\langle g, f\rangle_{\Gamma,k}}$$

for $f, g \in S_{k,v}(\Gamma)$ both cusp forms.

Proof That the Petersson scalar product is well defined follows directly from Lemma 1.95. It is independent of the choice of the fundamental domain, as can be seen from Remark 1.97. The linearity in the first component and anti-linearity in the second follows directly from the linearity of integrals. The Hermitian property also follows directly from the definition in (1.57), assuming that both forms f, g are cusp forms. □

Definition 1.99 A complex *Hilbert space* $\mathcal{H}$ is a vector space with a scalar product $\langle\cdot,\cdot\rangle : \mathcal{H} \times \mathcal{H} \to \mathbb{C}$.

Corollary 1.100 *Let $\Gamma \subset \Gamma(1)$ be a subgroup of finite index μ and let $k \in \mathbb{R}$ be a weight and v a multiplier with respect to (Γ, k). The space $S_{k,v}(\Gamma)$ is a Hilbert space with the Petersson scalar product $\langle\cdot,\cdot\rangle_{\Gamma,k}$ as its scalar product.*

Proof The proof of Corollary 1.100 is left to the reader as an exercise. □

1.4 Holomorphic Poincaré Series

In this section, we introduce holomorphic Poincaré series and state some relevant results. Poincaré series are important objects in the theory of modular forms. The space of cusp forms of a given weight k forms a Hilbert space and, using Petersson inner product, one can show that Poincaré series generate the space of cusp forms. Moreover, their Fourier coefficients are combinations of Kloosterman sums that give Poincaré series their arithmetic importance. We refer to [100, Chapters III and IV] for more details.

One can easily show that we can write the sum over all coprime pairs as a sum over the quotient set $\Gamma_\infty\backslash\Gamma(1)$. Here Γ_∞ denotes the subset of $\Gamma(1)$ that stabilizes the cusp ∞ in the sense of Definition 1.32:

$$\Gamma_\infty := \left\{V \in \Gamma(1) : V\infty = \infty\right\} = \left\{\begin{pmatrix} \star & \star \\ 0 & 1 \end{pmatrix}\right\}.$$

This leads to the following different representations of the coset $\Gamma_\infty\backslash\Gamma(1)$.

Lemma 1.101

1. *Let $c \in \mathbb{Z}_{\geq 0}$, $d \in \mathbb{Z}$ such that* $\gcd(c,d) = 1$. *Assume that the two matrices $M_{(c,d)}, M'_{(c,d)} \in \mathrm{SL}_2(\mathbb{Z})$ have the same given lower row (c,d). Then both matrices differ by a power of S: $M'_{(c,d)} = S^n\, M_{(c,d)}$ for some $n \in \mathbb{Z}$.*
2. *A complete set of representatives of the right cosets*

$$\Gamma_\infty\backslash\Gamma(1) = \left\{\left[\begin{pmatrix} \star & \star \\ c & d \end{pmatrix}\right]_{\Gamma(1)_\infty} : c \in \mathbb{Z}_{\geq 0}, d \in \mathbb{Z}, (c,d) \neq (0,0) \text{ and } \gcd(c,d) = 1\right\}$$

is given by

$$\left\{\begin{pmatrix} \star & \star \\ 0 & 1 \end{pmatrix}, \begin{pmatrix} \star & \star \\ c & d \end{pmatrix} : c \in \mathbb{N},\ d \in \mathbb{Z} \text{ and } \gcd(c,d) = 1\right\}. \tag{1.64}$$

3. *Another complete set of representatives of the right cosets $\Gamma_\infty\backslash\Gamma(1)$ is given by*

$$\left\{\begin{pmatrix} \star & \star \\ c & d \end{pmatrix} \in \Gamma(1) :\ c \in \mathbb{Z},\ -d \in \mathbb{N}, \text{ and } \gcd(b,d) = 1\right\} \cup \left\{\begin{pmatrix} \star & \star \\ 1 & 0 \end{pmatrix}\right\}. \tag{1.65}$$

Proof

1. First, we show that for each coprime $(c,d) \in \mathbb{Z}_{\neq(0,0)}$ there exists $a, b \in \mathbb{Z}$ such that

$$\begin{pmatrix} a & b \\ c & d \end{pmatrix} \in \Gamma(1).$$

This follows immediately from the Chinese remainder theorem, see, e.g., Appendix A.1. Since left-multiplication by S given in (1.20) does not change the lower row,

$$S\begin{pmatrix} a & b \\ c & d \end{pmatrix} = \begin{pmatrix} a+c & b+d \\ c & d \end{pmatrix},$$

we have

$$\left\{S^n \begin{pmatrix} a & b \\ c & d \end{pmatrix} : \ n \in \mathbb{Z}\right\} \subset \Gamma(1).$$

Moreover, the values a and b of the upper row are only determined modulo c and d, respectively. This shows the first part of the lemma.

2. Using Proposition 1.34 to calculate the stabilizer $\Gamma_\infty = \langle S, -\mathbf{1}\rangle$ of the cusp ∞ of $\Gamma(1)$, we find

$$\Gamma_\infty \begin{pmatrix} \star & \star \\ c & d \end{pmatrix} \subset \Gamma(1)$$

and

$$\left(\Gamma_\infty \begin{pmatrix} \star & \star \\ c & d \end{pmatrix}\right) \cap \left(\Gamma_\infty \begin{pmatrix} \star & \star \\ c' & d' \end{pmatrix}\right) = \emptyset$$

for all $(c, d), (c', d') \in \mathbb{Z}_{\neq(0,0)}$, $(c, d) \neq \pm(c', d')$.

On the other hand, each element of $\Gamma(1)$ has coprime entries (c, d) in its lower row. Hence, it belongs to the left coset $\Gamma_\infty \begin{pmatrix} \star & \star \\ c & d \end{pmatrix}$ up to a common sign.

This shows that $\Gamma(1)$ can be written as the disjoint union

$$\Gamma(1) = \Gamma_\infty \begin{pmatrix} \star & \star \\ 0 & 1 \end{pmatrix} \cup \bigcup_{\substack{(c,d) \\ c\in\mathbb{N},\, d\in\mathbb{Z} \\ \gcd(c,d)=1}} \Gamma_\infty \begin{pmatrix} \star & \star \\ c & d \end{pmatrix}$$

of left cosets and proves the second part of the lemma.

3. The third part of the lemma is left to the reader as an exercise.

□

We define Poincaré series as follows:

Definition 1.102 Let $m \in \mathbb{Z}_{\geq 0}$ and $k \in \mathbb{N}$ even, $k > 2$. The series

$$P_{m,k}(z) := \sum_{M\in\Gamma_\infty\backslash\Gamma(1)} j(M, z)^{-k}\, e^{2\pi i m Mz} \qquad (z \in \mathbb{H}) \tag{1.66}$$

is called the mth *Poincaré series of weight* k.

Note that for $m = 0$, one gets the Eisenstein series of weight k.

Lemma 1.103 *Let $m \in \mathbb{Z}_{\geq 0}$ and $k \in 2\mathbb{N}$, $k \geq 4$. The Poincaré series $P_{m,k}(z)$ satisfies the modular transformation law for weight k and trivial multiplier $v = 1$:*

$$P_{m,k}(Mz) = j(M, z)^k P_{m,k}(z) \qquad \textit{for all } M \in \Gamma(1).$$

Moreover, if $m \in \mathbb{N}$, *then* $P_{m,k} \in S_k\big(\Gamma(1)\big)$ *is a cusp form.*

What happens if we take the Petersson scalar product between Poincaré series and a cusp form?

Theorem 1.104 *Let* $m \in \mathbb{N}$, $k \in 2\mathbb{N}$, $k \geq 4$ *a weight, and* $f \in S_k\big(\Gamma(1)\big)$ *a cusp form whose Fourier expansion at the cusp* ∞ *is given by* $f(z) = \sum_{n>0}^{\infty} a(n)\, e^{2\pi i n z}$. *The Petersson scalar product between the Poincaré series* $P_{m,k}$ *and the cusp form* f *is given by*

$$\big\langle P_{m,k}, f\big\rangle = \frac{\overline{a(m)}}{\mu}\, \frac{\Gamma(k-1)}{(4\pi m)^{k-1}}. \tag{1.67}$$

Here, μ *denotes the hyperbolic surface area of the fundamental domain* $\mathcal{F}_{\Gamma(1)}$, *see, e.g., Exercise 1.94.*

Corollary 1.105 *Let* $k \in 2\mathbb{N}$, $k \geq 4$ *be a weight. The space of cusp forms* $S_k\big(\Gamma(1)\big)$ *is spanned by Poincaré series* $P_{m,k}$, $m \geq 0$.

Proof To show that $S_k\big(\Gamma(1)\big)$ is spanned by Poincaré series, we show that each cusp form can be expressed as a linear combination of Poincaré series.

Assume that $f \in S_k\big(\Gamma(1)\big)$ is orthogonal to all Poincaré series:

$$\langle P_{m,k}, f\rangle = 0 \qquad \text{for all } m \in \mathbb{Z}_{\geq 0}.$$

Theorem 1.104 implies that all Fourier coefficients of f vanish. Hence, $f(z) = 0$ for all $z \in \mathbb{H}$. This means that all cusp forms lie in the span of the Poincaré series. □

We now present the Fourier expansion of the Poincaré series. However, we first need to introduce the Kloosterman sums $S(m, n; c)$ and the J-Bessel functions.

Definition 1.106 For $m, n, c \in \mathbb{N}$ the *Kloosterman sum* $K(m, n; c)$ is given by

$$S(m, n; c) := \sum_{\substack{0 \leq l < c \\ \gcd(l,c)=1}} e^{\frac{2\pi i (ml + nl^\star)}{c}},$$

where $l^\star$ is defined as a solution of $l\, l^\star \equiv 1 \bmod c$.

Remark 1.107 The Kloosterman sums are usually denoted by the letter S. This notation goes back to Kloosterman's original paper [105].

The following result was already shown in [92, Theorem 292].

Lemma 1.108 *For* $m, s \in \mathbb{N}$ *with* $s > 1$ *we have the identity*

$$\sum_{c=1}^{\infty} c^{-s} \left(\sum_{\substack{0 \le l < c \\ \gcd(l,c)=1}} e^{2\pi i \frac{lm}{c}} \right) = \frac{\sigma_{s-1}(m)}{m^{s-1}\zeta(s)},$$

where $\zeta(s) = \sum_{n=1}^{\infty} n^{-s}$ *denotes the* Riemann ζ-function *(for* $\Re(s) > 1$*) and* $\sigma_{s-1}(m)$ *the arithmetic divisor function introduced in Theorem 1.82.*

Following the standard definitions for Bessel functions, see [76, Bessel functions] or [153, (10.2.2)], we define the J-Bessel function by its series expansion.

Definition 1.109 Let $k \in \mathbb{Z}$ be an integer. The *Bessel function* J_k (of the first kind) is given by

$$J_k(z) = \left(\tfrac{1}{2}z\right)^k \sum_{l=0}^{\infty} (-1)^l \frac{\left(\frac{1}{4}z^2\right)^l}{l!\,\Gamma(k+l+1)},$$

where $\Gamma(x)$ denotes the Γ-function.

Remark 1.110 Let $k \in \mathbb{Z}$ be an integer. The Bessel function J_k is an entire function and solves the Bessel differential equation [153, (10.2.1)]

$$z^2 \frac{\mathrm{d}J(z)}{\mathrm{d}z^2} + z\frac{\mathrm{d}J(z)}{\mathrm{d}z} + \left(z^2 - k^2\right)J(z) = 0 \qquad (z \in \mathbb{C}). \tag{1.68}$$

Definition 1.111 The *Kronecker* δ*-function* $\delta_{n,m}$ is defined by

$$\delta_{n,m} := \begin{cases} 1 & \text{if } n = m \text{ and} \\ 0 & \text{if } n \neq m. \end{cases} \tag{1.69}$$

The Fourier expansion of Poincaré series is given by the following theorem.

Theorem 1.112 *For* $m \in \mathbb{Z}_{\geq 0}$ *and* $k \in 2\mathbb{N}$, $k \geq 4$, *the Poincaré series* $P_{m,k}$ *has the Fourier expansion*

$$P_{m,k}(z) = \sum_{n \in \mathbb{Z}_{\geq 0}} a_n e^{2\pi i n z} \qquad (z \in \mathbb{H})$$

at the cusp ∞. *The coefficients* a_n *are given as follows:*

1. If $m = 0$, *then*

$$a_n = \begin{cases} 1 & \text{if } n = 0 \text{ and} \\ \frac{(2\pi i)^k\, \sigma_{k-1}(n)}{(k-1)!\,\zeta(k)} & \text{if } n > 0. \end{cases}$$

2. *If* $m > 0$, *then*

$$a_n = \delta_{n,m} + \sum_{c \in \mathbb{N}} S(m, n; c) \frac{(2\pi i)^k}{c} \left(\frac{n}{m}\right)^{\frac{k-1}{2}} J_{k-1}\left(\frac{4\pi\sqrt{mn}}{c}\right).$$

Exercise 1.113 For $k \in 2\mathbb{N}$, $k \geq 4$, show that the 0th Poincaré series $P_{0,k}$ and the normalized Eisenstein series E_k satisfy

$$P_{0,k} = E_k. \tag{1.70}$$

1.4.1 Further Remarks on Modular Forms

The theory of modular forms is well understood and quite deep. We have presented only the very basics of the theory. In this section, we present the dimension of the complex vector space of modular forms and we observe its finite dimensionality. Note that the results mentioned below are given in the integer weight case on the full modular group. One can determine dimensions in half-integer weight cases using the famous Riemann-Roch theorem and can also determine easy generalizations on subgroups of finite index of the full modular group. We refer to [66, Chapter III], [142, Chapter II] for proofs and for more information on this topic.

Let $k \in 2\mathbb{Z}_{\geq 0}$. The dimension of the space of entire modular forms for the full modular group and weight k is

$$\dim M_k\big(\Gamma(1)\big) = \begin{cases} \left[\frac{k}{12}\right] & \text{for } k \equiv 2 \bmod 12 \text{ and} \\ \left[\frac{k}{12}\right] + 1 & \text{for } k \not\equiv 2 \bmod 12. \end{cases} \tag{1.71}$$

Moreover,

$$\dim M_k\big(\Gamma(1)\big) = 1 + \dim\ S_k\big(\Gamma(1)\big). \tag{1.72}$$

Using (1.71) and (1.72), one can show the following dimensions. For $k \in \{2, 4, 6, 8, 10\}$ we have

$$\dim\ S_k\big(\Gamma(1)\big) = 0.$$

Moreover,

$$\dim\ S_{12}\big(\Gamma(1)\big) = 1,$$

and

$$\dim\ S_{14}\big(\Gamma(1)\big) = 0.$$

1.5 Theta Series

Let us examine the following old question in number theory: Is it possible to write every natural number as a sum of four squares of integers? A direct follow-up question would be to count the number of ways one can write a natural number as different sums of four squares of integers. The first question was answered by Joseph Louis Lagrange in 1770. The answer to the second question was attributed to Carl Gustav Jacob Jacobi. To formulate the questions more precisely, consider the counting function

$$\begin{aligned} &r_4 : \mathbb{N} \to \mathbb{Z}_{\geq 0}; \\ &\qquad n \mapsto r_4(n) := \sharp\left\{(x_1, x_2, x_3, x_4) \in \mathbb{Z}^4 \colon\ x_1^2 + x_2^2 + x_3^2 + x_4^2 = n\right\}, \end{aligned}$$

where $\mathbb{N} = \mathbb{Z}_{\geq 1} = \{1, 2, 3, \dots\}$ denotes the set of natural numbers.
We have the following results.

Lagrange's four-square theorem: $r_4(n) \geq 1$ for all $n \in \mathbb{N}$.
Jacobi's theorem: For all $n \in \mathbb{N}$,

$$r_4(n) = \begin{cases} 8\sigma_1(n), & \text{if } 4 \nmid n, \text{ and} \\ 8\sigma_1(n) - 32\sigma_1\left(\frac{n}{4}\right), & \text{if } 4 \mid n. \end{cases}$$

Here $\sigma_1(n)$ denotes the "sum of all divisors" function. The notation $4 \nmid n$ means "4 does not divide n" and $4 \mid n$ means "4 does divide n."

Jacobi's proof leads to the concepts of theta-functions and modular forms by forming the generating function of $r_4(n)$. Such generating functions turned out to appear in the Fourier expansions of certain functions that have nice transformation properties under some discrete groups. These functions are modular forms.

1.5.1 Jacobi Theta Function

Definition 1.114 The *Jacobi theta function* $\Theta : \mathbb{C} \times \mathbb{H} \to \mathbb{C}$ is defined by the series

$$\Theta(w, z) := \sum_{n \in \mathbb{Z}} e^{\pi i n^2 z}\, e^{2\pi i n w} \tag{1.73}$$

for all $w \in \mathbb{C}$ and $z \in \mathbb{H}$.

Remark 1.115 We defined the Jacobi theta function as a function on $\mathbb{C} \times \mathbb{H}$ and denoted the variables with $w \in \mathbb{C}$ and $z \in \mathbb{H}$. This notation of the variables and the ordering of the two spaces $\mathbb{C}$ and $\mathbb{H}$ are not standardized throughout the literature.

Sometimes, the spaces are swapped, meaning that the Jacobi theta function is defined on $\mathbb{H} \times \mathbb{C}$. We recommend that one always check the underlying definition of the Jacobi theta function carefully to avoid misunderstandings.

Lemma 1.116 *The Jacobi theta function $\Theta(w, z)$ is an entire function in the first argument for fixed $z \in \mathbb{H}$. It is a holomorphic function in the second argument for fixed $w \in \mathbb{C}$.*

Proof Let $t_0 > 0$ and $M > 0$ be arbitrary positive constants, fixed throughout this proof. We consider the sets

$$H_t := \left\{z \in \mathbb{H} : \Im(z) \geq t\right\} \subset \mathbb{H}$$

and

$$U_M := \left\{w \in \mathbb{C} : |w| \leq M\right\} \subset \mathbb{C}.$$

For all $z \in H_t$ and $w \in U_M$, we have the series estimate

$$\begin{aligned} |\Theta(w, z)| &\leq \sum_{n\in\mathbb{Z}} \left| e^{\pi i n^2 z} e^{2\pi i n w} \right| \\ &\overset{\Im(z)\geq t_0}{\leq} \sum_{n\in\mathbb{Z}} e^{-\pi i n^2 t_0} \left| e^{2\pi i n w} \right| \\ &\overset{|w|\leq M}{\leq} \sum_{n\in\mathbb{Z}} e^{-\pi i n^2 t_0} e^{2\pi |n| M} \\ &< \infty. \end{aligned}$$

Hence, $\Theta(\cdot, z)$ is entire on $\mathbb{C}$ in the first argument for each fixed $z \in \mathbb{H}$. Similarly, $\Theta(w, \cdot)$ is holomorphic in the second argument for each fixed $w \in \mathbb{C}$. □

We have the following transformation behavior of the Jacobi theta function in the first argument:

Proposition 1.117

1. $\Theta(w + 1, z) = \Theta(w, z)$ *for all* $w \in \mathbb{C}$ *and* $z \in \mathbb{H}$.
2. $\Theta(w + z, z) = e^{-\pi i z} e^{-2\pi i w} \Theta(w, z)$ *for all* $w \in \mathbb{C}$ *and* $z \in \mathbb{H}$.
3. *If* $w \in \mathbb{C}$ *and* $z \in \mathbb{H}$ *satisfy* $w = \frac{1+z}{2} + m + nz$ *for some* $m, n \in \mathbb{Z}$*, then* $\Theta(w, z) = 0$.

Proof The first property follows directly from the expansion of Θ. We have

$$\Theta(w + 1, z) \overset{(1.73)}{=} \sum_{n\in\mathbb{Z}} e^{\pi i n^2 z} e^{2\pi i n(w+1)}$$

$$
\begin{aligned}
&= \sum_{n\in\mathbb{Z}} e^{\pi i n^2 z}\, e^{2\pi i n w} \\
&\overset{(1.73)}{=} \Theta(w, z)
\end{aligned}
$$

for all $w \in \mathbb{C}$ and $z \in \mathbb{H}$.

The second property is based on a completion of a square argument:

$$
n^2 + 2n = (n+1)^2 - 1. \tag{1.74}
$$

Applying this to the expansion of Θ, we find

$$
\begin{aligned}
\Theta(w+z, z) &\overset{(1.73)}{=} \sum_{n\in\mathbb{Z}} e^{\pi i n^2 z}\, e^{2\pi i n(w+z)} \\
&= \sum_{n\in\mathbb{Z}} e^{\pi i (n^2+2n) z}\, e^{2\pi i n w} \\
&\overset{(1.74)}{=} e^{-\pi i z} \sum_{n\in\mathbb{Z}} e^{\pi i (n+1) z}\, e^{2\pi i n w} \\
&= e^{-\pi i z}\, e^{-2\pi i w} \sum_{n\in\mathbb{Z}} e^{\pi i (n+1) z}\, e^{2\pi i (n+1) w} \\
&\overset{(1.73)}{=} e^{-\pi i z}\, e^{-2\pi i w}\, \Theta(w, z)
\end{aligned}
$$

for all $w \in \mathbb{C}$ and $z \in \mathbb{H}$.

To prove the last property, we consider $\Theta\left(\frac{1+z}{2}, z\right)$ for some $z \in \mathbb{H}$. We find

$$
\begin{aligned}
\Theta\left(\frac{1+z}{2}, z\right) &\overset{(1.73)}{=} \sum_{n\in\mathbb{Z}} e^{\pi i n^2 z}\, e^{2\pi i n \frac{1+z}{2}} \\
&= \sum_{n\in\mathbb{Z}} e^{\pi i (n^2+n) z} \underbrace{e^{\pi i n}}_{=(-1)^n} \\
&= \sum_{n\in\mathbb{N}} e^{\pi i (n^2+n) z} - \sum_{n\in\mathbb{Z}_{\leq 0}} e^{\pi i (n^2+n) z} \\
&\overset{n=-m-1}{=} \sum_{n\in\mathbb{N}} e^{\pi i (n^2+n) z} - \sum_{m\in\mathbb{N}} e^{\pi i \left((-m-1)^2 + (-m-1)\right) z}.
\end{aligned}
$$

Using

$$
(-m-1)^2 + (-m-1) = m^2 + 2m + 1 - m - 1 = m^2 + m,
$$

we find

$$\begin{aligned}\Theta\left(\frac{1+z}{2}, z\right) &= \sum_{n\in\mathbb{N}} e^{\pi i(n^2+n)z} - \sum_{m\in\mathbb{N}} e^{\pi i\left((-m-1)^2+(-m-1)\right)z} \\ &= \sum_{n\in\mathbb{N}} e^{\pi i(n^2+n)z} - \sum_{m\in\mathbb{N}} e^{\pi i(m^2+m)z} \\ &= \sum_{n\in\mathbb{N}} \underbrace{e^{\pi i(n^2+n)z} - e^{\pi i(n^2+n)z}}_{=0} \\ &= 0.\end{aligned}$$

Using the first two transformation relations, we use the above result to show

$$0 = \Theta\left(\frac{1+z}{2}, z\right) = \Theta\left(\frac{1+z}{2} + n + mz, z\right)$$

for all $m, n \in \mathbb{Z}$. This concludes the proof of Proposition 1.117. □

Now, we consider the transformation behavior in the second argument. A first trivial result is given by the following:

Exercise 1.118 Show that $\Theta(w, z+2) = \Theta(w, z)$ for all $w \in \mathbb{C}$ and $z \in \mathbb{H}$.

To show the transformation behavior under $z \mapsto \frac{-1}{z}$ in the second argument of the Jacobi theta function, first consider the following auxiliary result.

Lemma 1.119 ([190, Chapter IV, Equation (3) on Page 120]) *For fixed $t > 0$ and $a \in \mathbb{R}$ we have*

$$\sum_{n\in\mathbb{Z}} e^{-\pi t(n+a)^2} = \frac{1}{\sqrt{t}} \sum_{n\in\mathbb{Z}} e^{-\pi \frac{n^2}{t}} e^{2\pi i n a}. \tag{1.75}$$

Proof Consider the function $f : \mathbb{R} \to \mathbb{R}$ given by

$$f(x) = e^{-\pi t(x+a)^2}.$$

Its Fourier transform is given by

$$\widehat{f}(k) := \int_{-\infty}^{\infty} f(x)\, e^{-2\pi i k x}\, \mathrm{d}x = \frac{1}{\sqrt{t}} e^{-\pi \frac{k^2}{t}} e^{2\pi i k a}.$$

Using the Poisson summation formula (Appendix A.11), the result is obtained. □

Proposition 1.120 *For $z \in \mathbb{H}$ and $w \in \mathbb{C}$, we have*

$$\Theta\left(w, \frac{-1}{z}\right) = \sqrt{\frac{z}{i}}\, e^{\pi i z w^2}\, \Theta(zw, z). \tag{1.76}$$

Here, $\sqrt{\frac{z}{i}}$ *denotes the branch of the square root defined on* $\mathbb{H}$ *which is positive if* $z \in i\mathbb{R}_{>0}$.

Remark 1.121 The square root in the proposition above can be defined as follows: if $z = r\, e^{i\phi}$ with $r > 0$ and $\phi = \arg(z) \in (0, \pi)$, then $\sqrt{\frac{z}{i}}$ is given by

$$\sqrt{\frac{z}{i}} = \sqrt{r\, e^{i(\phi - \frac{\pi}{2})}} \overset{!}{=} \sqrt{r}\, e^{i\frac{\phi - \frac{\pi}{2}}{2}} = \sqrt{r}\, e^{i\left(\frac{\phi}{2} - \frac{\pi}{4}\right)}.$$

Proof (of Proposition 1.120) Since $\Theta(\cdot, \cdot)$ is entire in the first argument and holomorphic in the second argument, it suffices to check (1.75) for $w = x + i0 \in \mathbb{R}$ and purely imaginary $z = it \in i\mathbb{R}_{>0}$. If (1.75) holds for these values, it holds everywhere, due to being entire in its first argument for a fixed second argument and being holomorphic in its second argument for a fixed first argument.

This leads to the question on whether the first equality in the sequence of equations

$$\begin{aligned}
\Theta\left(x, \frac{-1}{it}\right) = \Theta\left(x, \frac{i}{t}\right) &\overset{(1.73)}{=} \sum_{n \in \mathbb{Z}} e^{-\pi \frac{n^2}{t}}\, e^{2\pi i n x} \\
&= \sqrt{t}\, e^{-\pi t x^2} \sum_{n \in \mathbb{Z}} e^{-\pi n^2 t}\, e^{-2\pi n t x} \\
&\overset{(1.73)}{=} \sqrt{\frac{z}{i}}\, e^{\pi i z w^2}\, \Theta(zw, z)
\end{aligned}$$

holds. Reformulating the identity in question, we show

$$\frac{1}{\sqrt{t}} \sum_{n \in \mathbb{Z}} e^{-\pi \frac{n^2}{t}}\, e^{2\pi i n x} = e^{-\pi t x^2} \sum_{n \in \mathbb{Z}} e^{-\pi n^2 t}\, e^{-2\pi n t x}. \tag{1.77}$$

Using (1.75), we express the left-hand side of (1.77) as

$$\begin{aligned}
\frac{1}{\sqrt{t}} \sum_{n \in \mathbb{Z}} e^{-\pi \frac{n^2}{t}}\, e^{2\pi i n x} &\overset{(1.75)}{=} \sum_{n \in \mathbb{Z}} e^{-\pi t (n + x)^2} \\
&= \sum_{n \in \mathbb{Z}} e^{-\pi t (n^2 + 2nx + x^2)} \\
&= e^{-\pi t x^2} \sum_{n \in \mathbb{Z}} e^{-\pi n^2 t}\, e^{-2\pi n t x}.
\end{aligned}$$

This is exactly the right-hand side of (1.77), proving the identity. □

Summarizing, we show the following transformation behavior of the Jacobi theta function.

Theorem 1.122 *For $z \in \mathbb{H}$ and $w \in \mathbb{C}$, we have*

$$\Theta(w+1, z) = \Theta(w, z) \quad \text{and} \quad \Theta(w+z, z) = e^{-\pi i z}\, e^{-2\pi i w}\, \Theta(w, z) \tag{1.78}$$

in the first coordinate and

$$\Theta(w, z+2) = \Theta(w, z) \quad \text{and} \quad \Theta\left(w, \frac{-1}{z}\right) = \sqrt{\frac{z}{i}}\, e^{\pi i z w^2}\, \Theta(zw, z) \tag{1.79}$$

in the second coordinate. Here $\sqrt{\frac{z}{i}}$ denotes the branch of the square root defined on $\mathbb{H}$ which is positive if $z \in i\mathbb{R}_{>0}$.

Proof The identities in the first coordinate follow from Proposition 1.117. The identities in the second coordinate follow from Exercise 1.118 and Proposition 1.120. □

Corollary 1.123 *The function $\theta : \mathbb{H} \to \mathbb{C}$ given by*

$$\theta(z) := \Theta(0, z) \tag{1.80}$$

satisfies

$$\theta(z+2) = \theta(z) \quad \text{and} \quad \theta\left(\frac{-1}{z}\right) = \sqrt{\frac{z}{i}}\, \theta(z). \tag{1.81}$$

The next corollaries give the growth behavior of $\theta(z)$ as $z \to i\infty$ and as $z \to 1$.

Corollary 1.124 *The function θ satisfies the limit*

$$\lim_{\Im(z)\to\infty} \theta(z) = 1. \tag{1.82}$$

Proof Using the series definition of Θ, we get

$$\begin{aligned} \theta(z) &\overset{(1.80)}{=} \Theta(0, z) \\ &\overset{(1.73)}{=} \sum_{n\in\mathbb{Z}} e^{\pi i n^2 z}. \end{aligned}$$

For $n \in \mathbb{Z}$, each individual summand satisfies

$$\lim_{\Im(z)\to\infty} e^{\pi i n^2 z} = \lim_{\Im(z)\to\infty} \left(e^{\pi i z}\right)^{n^2} = \begin{cases} 0 & \text{if } n^2 \neq 0 \text{ and} \\ 1 & \text{if } n^2 = 0. \end{cases}$$

We used the fact that $0 < \left|e^{\pi i z}\right| < 1$ and $\left(e^{\pi i z}\right)^0 = 1$ for all $z \in \mathbb{H}$ and that

$$\lim_{\Im(z)\to\infty} e^{\pi i z} = 0.$$

Hence,

$$\theta(z) = \sum_{n\in\mathbb{Z}} e^{\pi i n^2 z} = 1 + \sum_{n\in\mathbb{Z}_{\neq 0}} e^{\pi i n^2 z} \longrightarrow 1 \qquad \text{as } \Im(z) \to \infty.$$

□

Corollary 1.125 *The function θ satisfies*

$$\theta\left(1 - \frac{1}{z}\right) = \sqrt{\frac{z}{i}} \sum_{n\in\mathbb{Z}} e^{\pi i \left(n+\frac{1}{2}\right)^2 z} \tag{1.83}$$

for all $z \in \mathbb{H}$.

In particular, we have the growth estimate

$$\theta\left(1 - \frac{1}{z}\right) = O\left(\sqrt{\frac{z}{i}}\, e^{\frac{\pi i z}{2}}\right) \qquad \textit{as } \Im(z) \to \infty. \tag{1.84}$$

Proof Using (1.80) together with the expansion of Θ in (1.73), we find

$$\begin{aligned}
\theta(1+z) &\overset{(1.80)}{=} \Theta(0, 1+z) \\
&\overset{(1.73)}{=} \sum_{n\in\mathbb{Z}} e^{\pi i n^2 (1+z)} \\
&= \sum_{n\in\mathbb{Z}} e^{\pi i n^2}\, e^{\pi i n^2 z} \\
&= \sum_{n\in\mathbb{Z}} (-1)^n\, e^{\pi i n^2 z},
\end{aligned}$$

using the fact that $n \equiv n^2 \bmod 2$ for all $n \in \mathbb{Z}$. Using again that $(-1)^n = e^{\pi i n}$, we find

$$\begin{aligned}
\theta(1+z) &= \sum_{n\in\mathbb{Z}} (-1)^n\, e^{\pi i n^2 z} \\
&= \sum_{n\in\mathbb{Z}} e^{\pi i n^2 z}\, e^{\pi i n}
\end{aligned}$$

$$= \sum_{n\in\mathbb{Z}} e^{\pi i n^2 z} e^{2\pi i n \frac{1}{2}}$$

$$\overset{(1.73)}{=} \Theta\left(\frac{1}{2}, z\right).$$

Now, using the transformation property (1.76) of Θ in Proposition 1.120,

$$\begin{aligned}
\theta\left(1+\frac{1}{z}\right) &= \Theta\left(\frac{1}{2}, \frac{-1}{z}\right) \\
&\overset{(1.76)}{=} \sqrt{\frac{z}{i}}\, e^{\frac{\pi i z}{4}}\, \Theta\left(\frac{z}{2}, z\right) \\
&\overset{(1.73)}{=} \sqrt{\frac{z}{i}}\, e^{\frac{\pi i z}{4}} \sum_{n\in\mathbb{Z}} e^{\pi i n^2 z} e^{\pi i n z} \\
&= \sqrt{\frac{z}{i}} \sum_{n\in\mathbb{Z}} e^{\pi i \left(n^2+n+\frac{1}{4}\right) z} \\
&= \sqrt{\frac{z}{i}} \sum_{n\in\mathbb{Z}} e^{\pi i \left(n+\frac{1}{2}\right)^2 z}.
\end{aligned}$$

This shows (1.83) of the corollary.

The growth estimate now follows from the growth estimate of the individual terms in the series expansion. The dominating terms are $e^{\pi i \left(\frac{1}{2}\right)^2 z}$ and $e^{\pi i \left(-\frac{1}{2}\right)^2 z}$ corresponding to $n = 0$ and $n = -1$ respectively. They both give a growth estimate of $O\left(\sqrt{\frac{z}{i}}\, e^{\frac{\pi i z}{2}}\right)$ as $\Im(z) \to \infty$. The other terms contribution is of lower order. □

Next, we consider a product expansion of the Jacobi theta function, following the arguments presented in [190, Chapter IV, §1].

To do so, we temporarily define a product $\tilde{\Theta}(w, z)$ that has the same transformational properties as the Jacobi theta function Θ. Then, we show that the two functions $\tilde{\Theta}$ and Θ agree.

Definition 1.126 The triple product function $\tilde{\Theta} : \mathbb{C} \times \mathbb{H} \to \mathbb{C}$ is defined as

$$\tilde{\Theta}(w, z) := \prod_{n\in\mathbb{N}} \left(1 - e^{2\pi i n z}\right)\left(1 + e^{\pi i (2n-1) z} e^{2\pi i w}\right)\left(1 + e^{\pi i (2n-1) z} e^{-2\pi i w}\right) \tag{1.85}$$

for all $w \in \mathbb{C}$ and $z \in \mathbb{H}$.

Remark 1.127 Using the notation $q = e^{\pi i z}$, the triple product function can be written as

$$\tilde{\Theta}(w, z) = \prod_{n\in\mathbb{N}} \left(1 - q^{2n}\right)\left(1 + q^{2n-1} e^{2\pi i w}\right)\left(1 + q^{2n-1} e^{-2\pi i w}\right).$$

Proposition 1.128

1. *The Jacobi theta function $\tilde{\Theta}(w, z)$ is an entire function in the first argument for fixed $z \in \mathbb{H}$. It is a holomorphic function in the second argument for fixed $w \in \mathbb{C}$.*
2. $\tilde{\Theta}(w+1, z) = \tilde{\Theta}(w, z)$ *for all $w \in \mathbb{C}$ and $z \in \mathbb{H}$.*
3. $\tilde{\Theta}(w+z, z) = e^{-\pi i z}\, e^{-2\pi i w}\, \tilde{\Theta}(w, z)$ *for all $w \in \mathbb{C}$ and $z \in \mathbb{H}$.*
4. $\tilde{\Theta}(w, z) = 0 \iff w = \frac{1+z}{2} + m + nz$ *for some $m, n \in \mathbb{Z}$.*
5. *The zeros of $\tilde{\Theta}(w, z)$ are simple.*

Proof

1. Let $t_0 > 0$. We have the estimate
$$\left|e^{\pi i z}\right| \leq e^{-\pi t_0} < 1$$
for all $z \in \mathbb{H}$ with $\Im(z) \geq t_0$. This implies that the series
$$\sum_{n \in \mathbb{N}} \left|e^{\pi i (2n-1) z}\right|$$
converges absolutely. Hence, the product $\tilde{\Theta}(w, z)$ in (1.85) of Definition 1.126 defines an entire function in $w \in \mathbb{C}$ for fixed $z \in \mathbb{H}$ and defines a holomorphic function in $z \in \mathbb{H}$ for fixed $w \in \mathbb{C}$, see [190, Chapter V].
2. The product expansion of $\tilde{\Theta}(w, z)$ in (1.85) implies immediately that
$$\begin{aligned}
&\tilde{\Theta}(w+1, z) \\
&\overset{(1.85)}{=} \prod_{n \in \mathbb{N}} \left(1 - e^{2\pi i n z}\right)\left(1 + e^{\pi i (2n-1) z}\, e^{2\pi i (w+1)}\right) \cdot \\
&\qquad\qquad \cdot \left(1 + e^{\pi i (2n-1) z}\, e^{-2\pi i (w+1)}\right) \\
&= \prod_{n \in \mathbb{N}} \left(1 - e^{2\pi i n z}\right)\left(1 + e^{\pi i (2n-1) z}\, e^{2\pi i w}\right)\left(1 + e^{\pi i (2n-1) z}\, e^{-2\pi i w}\right) \\
&\overset{(1.85)}{=} \tilde{\Theta}(w, z).
\end{aligned}$$
3. We have
$$\begin{aligned}
&\tilde{\Theta}(w+z, z) \\
&\overset{(1.85)}{=} \prod_{n \in \mathbb{N}} \left(1 - e^{2\pi i n z}\right)\left(1 + e^{\pi i (2n-1) z}\, e^{2\pi i (w+z)}\right)\left(1 + e^{\pi i (2n-1) z}\, e^{-2\pi i (w+z)}\right) \\
&= \prod_{n \in \mathbb{N}} \left(1 - e^{2\pi i n z}\right)\left(1 + e^{\pi i (2n+1) z}\, e^{2\pi i w}\right)\left(1 + e^{\pi i (2n-3) z}\, e^{-2\pi i w}\right)
\end{aligned}$$

$$= \left(\frac{1+e^{-\pi i z}\, e^{-2\pi i w}}{1+e^{\pi i z}\, e^{2\pi i w}}\right) \cdot$$

$$\cdot \prod_{n\in\mathbb{N}} \left(1-e^{2\pi i n z}\right)\left(1+e^{\pi i(2n-1)z}\, e^{2\pi i w}\right)\left(1+e^{\pi i(2n-1)z}\, e^{-2\pi i w}\right)$$

$$\stackrel{(1.85)}{=} \left(\frac{1+e^{-\pi i z}\, e^{-2\pi i w}}{1+e^{\pi i z}\, e^{2\pi i w}}\right) \tilde{\Theta}(w,z).$$

Using the identity

$$\frac{1+a}{1+\frac{1}{a}} = a$$

with $a = e^{-\pi i z}\, e^{-2\pi i w}$, we find

$$\begin{aligned}\tilde{\Theta}(w+z,z) &= \left(\frac{1+e^{-\pi i z}\, e^{-2\pi i w}}{1+e^{\pi i z}\, e^{2\pi i w}}\right) \tilde{\Theta}(w,z)\\ &= e^{-\pi i z}\, e^{-2\pi i w}\, \tilde{\Theta}(w,z).\end{aligned}$$

4. Recall that a product vanishes if at least one of its factors vanishes. Since $\left|e^{2\pi i n z}\right| < 1$ for all $z \in \mathbb{H}$, we see that the factors $1-e^{2\pi i n z} = 1-\left(e^{2\pi i z}\right)^n \neq 0$ for all $z \in \mathbb{H}$ and $n \in \mathbb{N}$.

   The second and third factors

$$1+e^{\pi i(2n-1)z}\, e^{2\pi i w} \quad \text{and} \quad 1+e^{\pi i(2n-1)z}\, e^{-2\pi i w}$$

in the product representation (1.85) vanish if

$$e^{\pi i(2n-1)z}\, e^{2\pi i w} \stackrel{!}{=} -1 = e^{-\pi i} \quad \text{and} \quad e^{\pi i(2n-1)z}\, e^{-2\pi i w} \stackrel{!}{=} -1 = e^{-\pi i},$$

respectively. This is the case if either

$$(2n-1)z + 2w = -1 \bmod 2$$

or

$$(2n-1)z - 2w = -1 \bmod 2.$$

Equivalently, this is the case if either

$$nz - \frac{z}{2} + \frac{1}{2} = -w \bmod 1$$

or

$$nz - \frac{z}{2} + \frac{1}{2} = w \bmod 2.$$

We can unify the above two conditions to

$$w = \frac{1}{2} + \frac{z}{2} + m + nz$$

for some $m \in \mathbb{Z}$. In other words, we have shown the following equivalence:

$$\tilde{\Theta}(w, z) = 0 \iff w = \frac{1+z}{2} + m + nz \quad \text{for some } m, n \in \mathbb{Z}.$$

5. The previous arguments, where we calculate the location of the zeros of $\tilde{\Theta}(w, z)$, also show the following: for every fixed $w \in \mathbb{C}$ and $z \in \mathbb{H}$ satisfying $w = \frac{1}{2}+\frac{z}{2}+m+nz$ for some $m, n \in \mathbb{Z}$, exactly one factor of either $1+e^{\pi i(2n-1)z}\, e^{2\pi i w}$ or $1+e^{\pi i(2n-1)z}\, e^{-2\pi i w}$ in the product representation (1.85) vanishes. Hence, the zero in w and z is simple.

□

Exercise 1.129 Show that for every $z \in \mathbb{H}$ there exists a constant $c(z)$ such that

$$\Theta(w, z) = c(z)\, \tilde{\Theta}(w, z) \tag{1.86}$$

for all $w \in \mathbb{C}$.

Theorem 1.130 *We have*

$$\Theta(w, z) = \tilde{\Theta}(w, z) \tag{1.87}$$

for all $w \in \mathbb{C}$ and $z \in \mathbb{H}$.

Proof Exercise 1.129 shows that for every $z \in \mathbb{H}$ there exists a constant $c(z)$ such that $\Theta(w, z) = c(z)\, \tilde{\Theta}(w, z)$ holds for all $w \in \mathbb{C}$. To establish the identity, we have to show that $c(z) = 1$ for all $z \in \mathbb{H}$.

Step 1: Calculate $c(z)$ for $w = \frac{1}{2}$.
Put $w = \frac{1}{2}$. Then (1.86) together with the defining equations (1.73) and (1.85) imply

$$\begin{aligned}
&\sum_{n\in\mathbb{Z}} e^{\pi i n^2 z}\, e^{2\pi i n \frac{1}{2}} \\
&\quad = \sum_{n\in\mathbb{Z}} (-1)^n\, e^{\pi i n^2 z} \\
&\quad \overset{(1.86)}{=} c(z) \prod_{n\in\mathbb{N}} \left(1 - e^{2\pi i n z}\right) \left(1 + e^{\pi i(2n-1)z}\, e^{2\pi i \frac{1}{2}}\right) \left(1 + e^{\pi i(2n-1)z}\, e^{-2\pi i \frac{1}{2}}\right)
\end{aligned}$$

$$= c(z) \prod_{n\in\mathbb{N}} \left(1 - e^{2\pi i n z}\right)\left(1 - e^{\pi i(2n-1)z}\right)\left(1 - e^{\pi i(2n-1)z}\right)$$

$$= c(z) \prod_{n\in\mathbb{N}} \left(1 - \left(e^{\pi i z}\right)^n\right)\left(1 - \left(e^{\pi i z}\right)^{2n-1}\right).$$

We used the following fact:

$$\prod_{n\in\mathbb{N}} \left(1 - \left(e^{\pi i z}\right)^{2n}\right)\left(1 - \left(e^{\pi i z}\right)^{2n-1}\right) = \prod_{n\in\mathbb{N}} \left(1 - \left(e^{\pi i z}\right)^n\right).$$

Hence,

$$c(z) = \frac{\sum_{n\in\mathbb{Z}}(-1)^n \left(e^{\pi i z}\right)^{n^2}}{\prod_{n\in\mathbb{N}} \left(1 - \left(e^{\pi i z}\right)^n\right)\left(1 - \left(e^{\pi i z}\right)^{2n-1}\right)} \tag{1.88}$$

for all $z \in \mathbb{H}$.

Step 2: Calculate $c(z)$ for $w = \frac{1}{4}$.

Let us calculate $\Theta\left(\frac{1}{4}, z\right)$ on the left-hand side of (1.86). We have

$$\begin{aligned}
\Theta\left(\frac{1}{4}, z\right) &\overset{(1.73)}{=} \sum_{n\in\mathbb{Z}} e^{\pi i n^2 z}\, e^{2\pi i n \frac{1}{4}} \\
&= \sum_{n\in\mathbb{Z}} i^n\, e^{\pi i n^2 z} \\
&= \sum_{n\in 2\mathbb{Z}} i^n\, e^{\pi i n^2 z} + \underbrace{\sum_{n\in 1+2\mathbb{Z}} i^n\, e^{\pi i n^2 z}}_{=0} \\
&= \sum_{n\in 2\mathbb{Z}} i^n\, e^{\pi i n^2 z} \\
&\overset{2n=m}{=} \sum_{m\in\mathbb{Z}} (-1)^m\, e^{4\pi i m^2 z}.
\end{aligned}$$

We used the fact that the set $1 + 2\mathbb{Z}$ contains all odd integers. In particular, if $n \in 1 + 2\mathbb{Z}$, then $-n \in 1 + 2\mathbb{Z}$, too. Hence, the terms

$$i^n\, e^{\pi i n^2 z} \quad \text{and} \quad i^{-n}\, e^{\pi i(-n)^2 z} = -i^n\, e^{\pi i n^2 z}$$

cancel each other.

Now, we calculate $\tilde{\Theta}\left(\frac{1}{4}, z\right)$ on the right-hand side of (1.86). We have

$$
\begin{aligned}
&\tilde{\Theta}\left(\frac{1}{4}, z\right) \\
&\overset{(1.85)}{=} \prod_{n\in\mathbb{N}} \left(1 - e^{2\pi i n z}\right) \left(1 + e^{\pi i (2n-1)z} \underbrace{e^{2\pi i \frac{1}{4}}}_{=e^{\frac{\pi i}{2}} = i}\right) \left(1 + e^{\pi i (2n-1)z} \underbrace{e^{-2\pi i \frac{1}{4}}}_{=e^{-\frac{\pi i}{2}} = -i}\right) \\
&= \prod_{n\in\mathbb{N}} \left(1 - e^{2\pi i n z}\right) \underbrace{\left(1 + i\, e^{\pi i (2n-1)z}\right) \left(1 - i\, e^{\pi i (2n-1)z}\right)}_{=1+e^{\pi i (4n-2)z}} \\
&= \prod_{n\in\mathbb{N}} \left(1 - e^{2\pi i n z}\right) \left(1 + e^{\pi i (4n-2)z}\right),
\end{aligned}
$$

using $(1-a)(1+a) = 1 - a^2$ with $a = i\, e^{\pi i (2n-1)z}$. Re-indexing the factors, we get

$$
\begin{aligned}
\tilde{\Theta}\left(\frac{1}{4}, z\right) &= \prod_{n\in\mathbb{N}} \left(1 - e^{2\pi i n z}\right) \left(1 + e^{\pi i (4n-2)z}\right) \\
&= \prod_{n\in\mathbb{N}} \left(1 - e^{2\pi i n z}\right) \prod_{n\in\mathbb{N}} \left(1 + e^{\pi i (4n-2)z}\right) \\
&\overset{\substack{2n=4m-2 \\ 2n=4m}}{=} \prod_{m\in\mathbb{N}} \left(1 - e^{\pi i (4m-2)z}\right) \left(1 - e^{\pi i (4m)z}\right) \prod_{n\in\mathbb{N}} \left(1 + e^{\pi i (4n-2)z}\right),
\end{aligned}
$$

using the fact that we can split the set of even natural numbers in two sets consisting of $m = 0 \bmod 4$ and $m = 2 \bmod 4$, respectively. Next, we again apply the binomial formula $(1-a)(1+a) = 1 - a^2$ with $a = e^{\pi i (4m-2)z}$ and get

$$
\begin{aligned}
\tilde{\Theta}\left(\frac{1}{4}, z\right) &= \prod_{m\in\mathbb{N}} \left(1 - e^{\pi i (4m-2)z}\right) \left(1 - e^{\pi i (4m)z}\right) \prod_{n\in\mathbb{N}} \left(1 + e^{\pi i (4n-2)z}\right) \\
&= \prod_{m\in\mathbb{N}} \left(1 - e^{\pi i (4m)z}\right) \left(1 + e^{\pi i (8m-4)z}\right).
\end{aligned}
$$

Summarizing, we find

$$
\begin{aligned}
c(z) &\overset{(1.86)}{=} \frac{\Theta\left(\frac{1}{4}, z\right)}{\tilde{\Theta}\left(\frac{1}{4}, z\right)} \\
&= \frac{\sum_{m\in\mathbb{Z}} (-1)^m \left(e^{\pi i z}\right)^{4m^2}}{\prod_{m\in\mathbb{N}} \left(1 - \left(e^{\pi i z}\right)^{4m}\right) \left(1 + \left(e^{\pi i z}\right)^{8m-4}\right)}
\end{aligned}
\tag{1.89}
$$

for all $z \in \mathbb{H}$.

Step 3: Show $c(4z) = c(z)$.

Comparing the expressions of the right-hand sides in (1.88) and (1.89), we get

$$\begin{aligned} c(4z) &\overset{(1.88)}{=} \frac{\sum_{n\in\mathbb{Z}}(-1)^n \left(e^{\pi i(4z)}\right)^{n^2}}{\prod_{n\in\mathbb{N}}\left(1-\left(e^{\pi i(4z)}\right)^n\right)\left(1-\left(e^{\pi i(4z)}\right)^{2n-1}\right)} \\ &= \frac{\sum_{n\in\mathbb{Z}}(-1)^n \left(e^{\pi iz}\right)^{4n^2}}{\prod_{n\in\mathbb{N}}\left(1-\left(e^{\pi iz}\right)^{4n}\right)\left(1-\left(e^{\pi iz}\right)^{8n-4}\right)} \\ &\overset{(1.89)}{=} c(z) \end{aligned}$$

for all $z \in \mathbb{H}$.

Step 4: Conclude $c(z) = 1$.

Iterating the identity $c(4z) = c(z)$, we get

$$c(z) = c(4z) = c\left(4^2 z\right) = c\left(4^k z\right)$$

for every $k \in \mathbb{Z}_{\geq 0}$ and $z \in \mathbb{H}$.

Now, consider the term $\left(e^{\pi i(4^k z)}\right)^m$ that appears in $c\left(4^k z\right)$. We see that for a given $z \in \mathbb{H}$, we have

$$\lim_{k\to\infty}\left(e^{\pi i(4^k z)}\right)^m = \begin{cases} 1 & \text{if } m = 0 \text{ and} \\ 0 & \text{if } m \neq 0. \end{cases}$$

This allows us to argue as follows:

$$\begin{aligned} c(z) &= \lim_{k\to\infty} c\left(4^k z\right) \\ &= \lim_{k\to\infty} \frac{1}{\prod_{n\in\mathbb{N}}\left(1-\left(e^{\pi i(4^k z)}\right)^n\right)\left(1-\left(e^{\pi i(4^k z)}\right)^{2n-1}\right)} + \\ &+ \frac{\sum_{n\in\mathbb{Z}_{\neq 0}}(-1)^n \left[\lim_{k\to\infty}\left(e^{\pi i(4^k z)}\right)^{n^2}\right]}{\prod_{n\in\mathbb{N}}\left(1-\left[\lim_{k\to\infty}\left(e^{\pi i(4^k z)}\right)^n\right]\right)\left(1-\left(\left[\lim_{k\to\infty} e^{\pi i(4^k z)}\right)^{2n-1}\right]\right)} \\ &= 1 + \frac{\sum_{n\in\mathbb{Z}_{\neq 0}} 0}{\prod_{n\in\mathbb{N}} 1} \\ &= 1. \end{aligned}$$

□

Corollary 1.131 *For $z \in \mathbb{H}$, we have*

$$\theta(z) = \prod_{n \in \mathbb{N}} \left(1 - e^{2\pi i n z}\right)\left(1 + e^{\pi i (2n-1) z}\right)^2. \tag{1.90}$$

Proof To prove the corollary, consider Theorem 1.130 and the product expansion of $\tilde{\Theta}(w,t)$ in Definition 1.126. Using $\theta(z) = \Theta(0,z)$ we immediately get the result. □

1.5.2 Theta Series of Index m

Recall Definition 1.114 of the Jacobi theta function in the previous subsection. We define a variant of the index-m Jacobi theta function as follows.

Definition 1.132 Let $l, m \in \mathbb{N}$. The *Jacobi theta function* of *index* m, $\Theta_{m,l} : \mathbb{C} \times \mathbb{H} \to \mathbb{C}$, is defined as

$$\begin{aligned} \Theta_{m,l}(w,z) :&= \sum_{\substack{r \in \mathbb{Z} \\ r \equiv l \bmod 2m}} e^{\frac{\pi i r^2 z}{4m}} e^{2\pi i r w} \\ &= \sum_{n \in \mathbb{Z}} e^{\frac{\pi i (l+2mn)^2 z}{4m}} e^{2\pi i (l+2mn) w} \end{aligned} \tag{1.91}$$

for $w \in \mathbb{C}$ and $z \in \mathbb{H}$.

Remark 1.133 The Jacobi theta function of index m is a Jacobi form of weight $\frac{1}{2}$ and index m on some subgroup of $\mathrm{SL}_2(\mathbb{Z})$ where we define the Jacobi forms as follows.

A *Jacobi form* of *level* 1, *weight* k, and *index* m is a function $\phi : \mathbb{C} \times \mathbb{H} \to \mathbb{C}$ such that

1. $\phi\left(\frac{w}{cz+d}, \frac{az+b}{cz+d}\right) = (cz+d)^k e^{\frac{2\pi i m c w^2}{cz+d}} \phi(w,z)$ for all $\begin{pmatrix} a & b \\ c & d \end{pmatrix} \in \mathrm{SL}_2(\mathbb{Z})$, $w \in \mathbb{C}$ and $z \in \mathbb{H}$,
2. $\phi(w + \lambda z + \mu, z) = e^{-2\pi i m(\lambda^2 z + 2\lambda w)} \phi(w,z)$ for all integers $\lambda, \mu \in \mathbb{Z}$,
3. ϕ has a Fourier expansion of the form

$$\phi(w,z) = \sum_{n \in \mathbb{Z}_{\geq 0}} \sum_{r^2 \leq 4mn} c(n,r)\, e^{2\pi i (nz + rw)}$$

   with suitable coefficients $c(n,r)$, and
4. the coefficients $c(n,r)$ in the Fourier expansion above satisfy $c(n,r) = 0$ for all $4mn < r^2$.

We call a Jacobi form $\phi(w, z)$ a *Jacobi cusp form* if it satisfies the stronger condition that $c(n, r) = 0$ for all $4mn \leq r^2$.

We refer to [74] for more details on Jacobi forms. Some interesting remarks and properties can also be found in [63, §4.1]. In particular, a Jacobi (cusp) form admits a theta series expansion of the form

$$\phi(w, z) = \sum_{l \in \{0, \dots, 2m-1\}} h_l(z)\, \Theta_{m,l}(w, z),$$

where the coefficients $h_l(z)$ are modular (cusp) forms of weight $k - \frac{1}{2}$. The above decomposition is known as the *theta decomposition* of the Jacobi form $\phi(w, z)$. Note that the theta coefficients can be formed into vector-valued modular forms.

The recent article [205] also studies the theta decomposition of Jacobi forms. It mentions in the introduction some appearances of theta decompositions in the literature.

The above remark implies the following transformation properties for Jacobi theta functions of index m.

Theorem 1.134 *For $l, m \in \mathbb{N}$, let $\Theta_{m,l}$ denote a Jacobi theta function of index m. The function $\Theta_{m,l}$ satisfies the transformation properties*

1. $\Theta_{m,l}\left(\frac{w}{cz+d}, \frac{az+b}{cz+d}\right) = (cz+d)^{\frac{1}{2}}\, e^{\frac{2\pi i m c w^2}{cz+d}}\, \Theta_{m,l}(w, z)$ *for all matrices* $\begin{pmatrix} a & b \\ c & d \end{pmatrix} \in \mathrm{SL}_2(\mathbb{Z})$, $z \in \mathbb{H}$ *and* $w \in \mathbb{C}$, *and*
2. $\Theta_{m,l}(w + \lambda z + \mu, z) = e^{-2\pi i m(\lambda^2 z + 2\lambda w)}\, \Theta_{m,l}(w, z)$ *for all integers* $\lambda, \mu \in \mathbb{Z}$.

1.5.3 Theta Series and Sums of Squares

Recall the introduction of Section 1.5. We are interested in the following question: how often we can write a given integer $n \in \mathbb{Z}_{\geq 0}$ as a sum of four squares?

Let's start first with the question on whether we can write an integer $n \in \mathbb{Z}_{\geq 0}$ as sum of two squares and in how many ways. We are interested in the value of the function

$$r_2 : \mathbb{Z}_{\geq 0} \to \mathbb{Z}_{\geq 0}; \qquad n \mapsto r_2(n) := \sharp\big\{(a, b) \in \mathbb{Z}^2 : a^2 + b^2 = n\big\}. \tag{1.92}$$

Similarly, we define the function r_4 counting the number of ways that an integer $n \in \mathbb{N}$ can be written as a sum of four squares:

$$r_4 : \mathbb{Z}_{\geq 0} \to \mathbb{Z}_{\geq 0}; \qquad n \mapsto r_4(n) := \sharp\big\{(a, b, c, d) \in \mathbb{Z}^4 : a^2 + b^2 + c^2 + d^2 = n\big\}. \tag{1.93}$$

Table 1.1 Some values of the functions $r_2(n)$ and $r_4(n)$ defined in (1.92) and (1.93) respectively.

n	$r_2(n)$	$r_4(n)$		n	$r_2(n)$	$r_4(n)$
0	1	1		11	0	96
1	4	8		12	0	96
2	4	24		13	8	112
3	0	32		14	0	192
4	4	24		15	0	192
5	8	48		16	4	24
6	0	96		17	8	144
7	0	64		18	4	312
8	4	24		19	0	160
9	4	104		20	8	144
10	8	144		21	0	256

Example 1.135 Some values of r_2 are given in Table 1.1. For example, we have $r_2(0) = 1$ since $0^2 = 0^2 + 0^2$ and $r_2(1) = 4$ since $1 = (\pm 1)^2 + 0^2 = 0^2 + (\pm 1)^2$.

In order to describe the behavior of the function r_2, we use the generating function, following the argument presented in [190, Chapter X, §3]. First, we start with a simple exercise, combining the identities in (1.73), (1.80), and (1.92) respectively (1.93), together with some simple combinatorics.

Exercise 1.136 Show the identities

$$\theta(z)^2 = 1 + \sum_{n\in\mathbb{N}} r_2(n)\, e^{\pi i n z} \tag{1.94}$$

and

$$\theta(z)^4 = 1 + \sum_{n\in\mathbb{N}} r_4(n)\, e^{\pi i n z}. \tag{1.95}$$

Consider the function

$$F_1 : \mathbb{H} \to \mathbb{C}; \quad z \mapsto F_1(z) := 2 \sum_{n\in\mathbb{Z}} \frac{1}{e^{\pi i n z} + e^{-\pi i n z}}. \tag{1.96}$$

Of course, we have the trivial identity

$$\begin{aligned} F_1(z) &\overset{(1.96)}{=} 2 \sum_{n\in\mathbb{Z}} \frac{1}{e^{\pi i n z} + e^{-\pi i n z}} \\ &= \sum_{n\in\mathbb{Z}} \frac{1}{\cos(n\pi z)} \end{aligned} \tag{1.97}$$

for all $z \in \mathbb{H}$ using the trigonometric identity

$$\cos(\phi) = \frac{1}{2}\left(e^{i\phi} + e^{-i\phi}\right).$$

As a first step, we show that F_1 in (1.97) is well defined.

Lemma 1.137 *For every $z \in \mathbb{H}$, the series*

$$2\sum_{n\in\mathbb{Z}} \frac{1}{e^{\pi inz} + e^{-\pi inz}} \quad \text{and} \quad 1 + 4\sum_{n\in\mathbb{N}} \frac{e^{\pi inz}}{1 + e^{2\pi inz}}$$

converge absolutely and are equal.

This implies that F_1 is well defined on $\mathbb{H}$ and we have

$$F_1(z) = 2\sum_{n\in\mathbb{Z}} \frac{1}{e^{\pi inz} + e^{-\pi inz}} = 1 + 4\sum_{n\in\mathbb{N}} \frac{e^{\pi inz}}{1 + e^{2\pi inz}} \tag{1.98}$$

for all $z \in \mathbb{H}$. Moreover, F_1 is holomorphic on $\mathbb{H}$.

Proof Let us first show that the two series are equal. Using

$$\frac{1}{e^{\pi inz} + e^{-\pi inz}} = \frac{1}{e^{-\pi inz}} \frac{1}{e^{2\pi inz} + 1} = \frac{e^{\pi inz}}{e^{2\pi inz} + 1},$$

we find

$$\begin{aligned} 2\sum_{n\in\mathbb{Z}} \frac{1}{e^{\pi inz} + e^{-\pi inz}} &= 1 + 2\sum_{n\in\mathbb{N}} \frac{1}{e^{\pi inz} + e^{-\pi inz}} + 2\sum_{n\in\mathbb{N}} \frac{1}{e^{\pi i(-n)z} + e^{-\pi i(-n)z}} \\ &= 1 + 2\sum_{n\in\mathbb{N}} \frac{e^{\pi inz}}{1 + e^{2\pi inz}} + 2\sum_{n\in\mathbb{N}} \frac{e^{\pi inz}}{1 + e^{2\pi inz}} \\ &= 1 + 4\sum_{n\in\mathbb{N}} \frac{e^{\pi inz}}{1 + e^{2\pi inz}}. \end{aligned}$$

This shows the right identity in (1.98).

Now take a fixed $z \in \mathbb{H}$. Since $\left|e^{\pi inz}\right| \le \left|e^{\pi iz}\right| < 1$ for all $n \in \mathbb{N}$, we have

$$\left|1 + e^{2\pi inz}\right| \ge \varepsilon > 0$$

with $\varepsilon := 1 - \left|e^{\pi iz}\right| = 1 - e^{-\pi\Im(z)} > 0$ independent of n. Hence we can estimate

$$\begin{aligned}\left|\sum_{n\in\mathbb{N}}\frac{e^{\pi inz}}{1+e^{2\pi inz}}\right| &\le \sum_{n\in\mathbb{N}}\left|\frac{e^{\pi inz}}{1+e^{2\pi inz}}\right| \\ &= \sum_{n\in\mathbb{N}}\frac{\left|e^{\pi inz}\right|}{\left|1+e^{2\pi inz}\right|} \\ &\le \sum_{n\in\mathbb{N}}\frac{\left|e^{\pi iz}\right|^n}{\varepsilon} \\ &= \frac{1}{\varepsilon}\sum_{n\in\mathbb{N}}\left(e^{-\pi\Im(z)}\right)^n \\ &< \infty,\end{aligned}$$

using that the geometric series $\sum_{n=0}^{\infty} q^n = \frac{1}{1-q}$ converges absolutely for all $|q| < 1$.

The same trick shows that the series $2\sum_{n\in\mathbb{Z}}\frac{1}{e^{\pi inz}+e^{-\pi inz}}$ converges absolutely for all $z \in \mathbb{H}$. This proves that F_1 in (1.96) is well defined and holomorphic on $\mathbb{H}$. □

Before we show the transformation behavior of $F_1(z)$ under $z \mapsto \frac{-1}{z}$, first consider the following auxiliary result.

Lemma 1.138 ([190, Chapter IV, Equation (4) on Page 120]) *For fixed $t > 0$ and $a \in \mathbb{R}$, we have*

$$\sum_{n\in\mathbb{Z}}\frac{e^{-2\pi ian}}{\cosh\left(\frac{\pi n}{t}\right)} = \sum_{n\in\mathbb{Z}}\frac{t}{\cosh\left(\pi(n+a)t\right)}. \tag{1.99}$$

Proof Recall that the trigonometric function $\cosh(\phi)$, the *cosinus hyperbolicus*, is defined by

$$\cosh(\phi) := \frac{1}{2}\left(e^{\phi}+e^{-\phi}\right) \tag{1.100}$$

and satisfies

$$\cosh(i\phi) = \cos(\phi) \tag{1.101}$$

for all $\phi \in \mathbb{C}$.

Consider the function $g : \mathbb{R} \to \mathbb{R}$ given by

$$g(x) = \frac{e^{-\pi iax}}{\cosh\left(\frac{\pi x}{t}\right)}.$$

Its Fourier transform is given by

$$\widehat{g}(k) := \int_{-\infty}^{\infty} g(x)\, e^{-2\pi i k x}\, \mathrm{d}x = \frac{t}{\cosh\big(\pi(k+a)t\big)}.$$

Using the Poisson summation formula (see Appendix A.11), we find immediately Relation 1.99. □

The above auxiliary lemma implies one part of the following transformation behavior of f.

Proposition 1.139 *The function F_1 satisfies*

$$F_1\left(\frac{-1}{z}\right) = \frac{z}{i}\, f(z) \tag{1.102}$$

and

$$F_1(z+2) = F_1(z) \tag{1.103}$$

for all $z \in \mathbb{H}$.

Proof The transformation (1.102) follows for $z = it$, $t > 0$, directly from the representation (1.97) together with (1.100) (with $a = 0$) from Lemma 1.138. By holomorphic continuation, we extend (1.102) to all $z \in \mathbb{H}$.

For the second transformation property, again consider the representation (1.97) of F_1. Since $n \in \mathbb{N}$ and the trigonometric function cos is 2π-periodic, we conclude

$$\begin{aligned} F_1(z+2) &\overset{(1.97)}{=} \sum_{n\in\mathbb{Z}} \frac{1}{\cos(n\pi z + 2n\pi)} \\ &= \sum_{n\in\mathbb{Z}} \frac{1}{\cos(n\pi z)} \\ &\overset{(1.97)}{=} F_1(z). \end{aligned}$$

□

Lemma 1.140 *We have*

$$\lim_{\Im(z)\to\infty} F_1(z) = 1. \tag{1.104}$$

Proof The exponentials appearing in the denominators satisfy

$$\lim_{\Im(z)\to\infty} \left|e^{\pm\pi i n z}\right| = \lim_{\Im(z)\to\infty} e^{\mp\pi n \Im(z)} = \begin{cases} 0 & \text{if } \mp n < 0, \\ 1 & \text{if } n = 0 \text{ and} \\ \infty & \text{if } \mp n > 0. \end{cases}$$

Hence,

$$\lim_{\Im(z)\to\infty} \frac{1}{e^{\pi inz}+e^{-\pi inz}} = \begin{cases} 0 & \text{if } n \neq 0 \text{ and} \\ 2 & \text{if } n = 0. \end{cases}$$

We conclude that

$$\begin{aligned} \lim_{\Im(z)\to\infty} F_1(z) &\overset{(1.96)}{=} \lim_{\Im(z)\to\infty} 2\sum_{n\in\mathbb{Z}} \frac{1}{e^{\pi inz}+e^{-\pi inz}} \\ &= 1 + \sum_{n\in\mathbb{Z}_{\neq 0}} \lim_{\Im(z)\to\infty} \frac{1}{e^{\pi inz}+e^{-\pi inz}} \\ &= 1. \end{aligned}$$

□

Lemma 1.141 *The function* F_1 *defined in* (1.96) *satisfies*

$$F_1\left(1+\frac{1}{z}\right) = 4\frac{z}{i}e^{\frac{\pi iz}{2}} + O\left(|z|\, e^{\frac{-3\pi\Im(z)}{2}}\right), \tag{1.105}$$

as $\Im(z) \to \infty$.

Proof We use Lemma 1.138 with $a = \frac{1}{2}$ and get (for $z = it$, $t > 0$)

$$\begin{aligned} F_1\left(1+\frac{1}{it}\right) &\overset{(1.97)}{=} \sum_{n\in\mathbb{Z}} \frac{1}{\cos\left(\pi n + \frac{\pi n}{it}\right)} \\ &= \sum_{n\in\mathbb{Z}} \frac{1}{(-1)^n \cos\left(i\frac{\pi n}{t}\right)} \\ &\overset{(1.100)}{=} \sum_{n\in\mathbb{Z}} \frac{(-1)^n}{\cosh\left(\frac{\pi n}{t}\right)} \\ &\overset{(1.99)}{=} \sum_{n\in\mathbb{Z}} \frac{t}{\cosh\left(\pi\left(n+\frac{1}{2}\right)t\right)} \\ &= \frac{it}{i}\sum_{n\in\mathbb{Z}} \frac{1}{\cos\left(\pi\left(n+\frac{1}{2}\right)it\right)}, \end{aligned}$$

using $\cos(-x) = \cos(x)$ in the last step. Using analytic continuation in z, we extend the growth estimate from $z = it$ to all $z \in \mathbb{H}$.

Splitting the right series into terms indexed by $n \in \{-1, 0\}$ and $n \notin \{-1, 0\}$ we get

$$F_1\left(1+\frac{1}{z}\right)=\frac{z}{i}\sum_{n\in\mathbb{Z}}\frac{1}{\cos\left(\pi\left(n+\frac{1}{2}\right)z\right)}$$

$$=\frac{z}{i}\frac{2}{\cos\left(\frac{\pi z}{2}\right)}+\frac{z}{i}\sum_{n\in\mathbb{Z}\setminus\{-1,\}}\frac{1}{\cos\left(\pi\left(n+\frac{1}{2}\right)z\right)}.$$

So, we get as an estimate

$$F_1\left(1+\frac{1}{z}\right)=\frac{z}{i}\frac{2}{\cos\left(\frac{\pi z}{2}\right)}+O\left(\frac{z}{i}\sum_{n\in\mathbb{Z}_{\neq 0}}\frac{1}{\cos\left(\pi\left(n+\frac{1}{2}\right)z\right)}\right)$$

$$=\frac{z}{i}\frac{4}{e^{\frac{\pi i z}{2}}+e^{-\frac{\pi i z}{2}}}+O\left(t\,e^{-\frac{3\pi\Im(z)}{2}}\right).$$

Since

$$\frac{4}{e^{\frac{\pi i z}{2}}+e^{-\frac{\pi i z}{2}}}\approx 4\,e^{\frac{\pi i z}{2}}$$

as $\Im(z)\to\infty$, we conclude the growth estimate

$$F_1\left(1+\frac{1}{z}\right)=4\frac{z}{i}e^{\frac{\pi i z}{2}}+O\left(t\,e^{-\frac{3\pi\Im(z)}{2}}\right)$$

as $\Im(z)\to\infty$. □

Comparing F_1 and θ^2, we get the following.

Theorem 1.142 *The functions F_1 in (1.96) and θ^2 are equal:*

$$F_1(z)=\theta^2(z)\quad \textit{for all } z\in\mathbb{H}. \tag{1.106}$$

Proof Collecting the transformation properties of F_1 in Proposition 1.139, together with its holomorphicity (Lemma 1.137) and its growth estimates in Lemmata 1.140 and 1.92 along with similar properties for θ^2 deduced from Corollaries 1.123, 1.124, and 1.125, we see that the quotient function

$$g(z):=\frac{F_1(z)}{\theta^2(z)}\qquad\text{for all } z\in\mathbb{H}$$

is holomorphic on $\mathbb{H}$ (since θ does not vanish – this follows from Corollary 1.131 since none of the factors in the product expansion vanishes on $\mathbb{H}$). The function g satisfies

$$g(z+2) = g(z) \quad \text{and} \quad g\left(\frac{-1}{z}\right) = g(z) \qquad \text{for all } z \in \mathbb{H}.$$

Moreover, the function is bounded at the cusps ∞ and 1 since the growth estimates for both F_1 and θ cancel there. Recalling the Möbius transformation of S^2 and T (see (1.20)) and that the theta group Γ_θ is generated by S^2 and T (see Example 1.38), we conclude that

$$g(Mz) = g(z) \qquad \text{for all } M \in \Gamma_\theta \text{ and } z \in \mathbb{H}.$$

Since $\Gamma_\theta \subset \Gamma(1)$ is a subgroup of finite index – the index is calculated to be 3 in Exercise 1.38 – we may apply Theorem 1.64 and we conclude that g is a constant function. Hence, there exists a $c \in \mathbb{C}$ (independent of $z \in \mathbb{H}$) satisfying

$$F_1(z) = c\,\theta^2(z)$$

for all $z \in \mathbb{H}$.

The constant c has to be 1: using

$$\lim_{\Im(z)\to\infty} \theta^2(z) = 1$$

in (1.82) and

$$\lim_{\Im(z)\to\infty} F_1(z) = 1$$

in (1.104), we conclude

$$\begin{aligned} c &= \frac{F_1(z)}{\theta^2(z)} \\ &= \lim_{\Im(z)\to\infty} \frac{F_1(z)}{\theta^2(z)} \\ &= \frac{\lim_{\Im(z)\to\infty} F_1(z)}{\lim_{\Im(z)\to\infty} \theta^2(z)} \\ &= \frac{1}{1} = 1. \end{aligned}$$

□

One may wonder how the above result may help us with our motivating question of understanding the counting function $r_2(n)$ in (1.92). Since $r_2(n)$ appears in the Fourier expansion of θ^2 as coefficients, see (1.94) in Exercise 1.136, we have the following.

Corollary 1.143 *We have*

$$\sum_{n\in\mathbb{Z}_{\geq 0}} r_2(n)\, e^{\pi inz} = 1 + 4\sum_{n\in\mathbb{N}} \frac{e^{\pi inz}}{1+e^{2\pi inz}} \tag{1.107}$$

for all $z \in \mathbb{H}$.

Proof Using (1.106) of Theorem 1.142 and substituting the expansions of θ^2 in (1.94) and F_1 in (1.98), we immediately get (1.107). □

Corollary 1.144 *We have*

$$r_2(n) = 4\big(d_1(n) - d_3(n)\big) \qquad \text{for all } n \in \mathbb{N}. \tag{1.108}$$

Here, $d_1 \colon \mathbb{N} \to \mathbb{Z}$ *and* $d_3 \colon \mathbb{N} \to \mathbb{Z}$ *denote the counting functions given by*

$$\begin{aligned} d_1(n) :&= \sharp\big\{k \in \mathbb{N}\colon\ 4k+1 \mid n\big\} \quad \text{and} \\ d_3(n) :&= \sharp\big\{k \in \mathbb{N}\colon\ 4k+3 \mid n\big\}. \end{aligned} \tag{1.109}$$

Remark 1.145 The counting functions $d_1(n)$ and $f_3(n)$ in (1.109) count the number of divisors of n of the form $4k+1$ and $4k+3$, respectively.

Before we prove Corollary 1.144, we need a combinatorial result.

Lemma 1.146 *We have*

$$\frac{e^{\pi inz}}{1+e^{2\pi inz}} = \big(d_1(n) - d_3(n)\big)\, e^{\pi inz} \tag{1.110}$$

for all $n \in \mathbb{N}$ *and* $z \in \mathbb{H}$.

Proof Using the identity

$$\frac{1}{1+a^2} = \frac{1-a^2}{1-a^4} = \frac{1}{1+a^4} - \frac{a^2}{1-a^4}$$

with $a = e^{\pi inz}$, we find

$$\frac{e^{\pi inz}}{1+e^{2\pi inz}} = \frac{e^{\pi inz}}{1-e^{4\pi inz}} - \frac{e^{3\pi inz}}{1-e^{4\pi inz}}.$$

Using the geometric series expansions

$$\begin{aligned}\frac{b^n}{1-b^{4n}} &= \sum_{m\in\mathbb{Z}_{\geq 0}} b^{n(4m+1)} \\ &\overset{n(4m+1)=k}{=} \sum_{k\in\mathbb{N}} d_1(k)\, b^k\end{aligned}$$

and

$$\begin{aligned}\frac{b^{3n}}{1-b^{4n}} &= \sum_{m\in\mathbb{Z}_{\geq 0}} b^{n(4m+3)} \\ &\overset{n(4m+3)=k}{=} \sum_{k\in\mathbb{N}} d_3(k)\, b^k\end{aligned}$$

with $b = e^{\pi i z}$, we find

$$\begin{aligned}\frac{e^{\pi i n z}}{1+e^{2\pi i n z}} &= \frac{e^{\pi i n z}}{1-e^{4\pi i n z}} - \frac{e^{3\pi i n z}}{1-e^{4\pi i n z}} \\ &= \sum_{k\in\mathbb{N}} d_1(k)\, e^{\pi i k z} - \sum_{k\in\mathbb{N}} d_3(k)\, e^{\pi} i k z.\end{aligned}$$

The series on the right-hand side above converge since $d_{1/3}(k) \leq k$ holds for all $k \in \mathbb{N}$. □

Proof (of Corollary 1.144) Since (1.107) for all $z \in \mathbb{H}$, we conclude

$$\begin{aligned}\sum_{n\in\mathbb{Z}_{\geq 0}} r_2(n)\, e^{\pi i n z} &\overset{(1.107)}{=} 1 + 4\sum_{n\in\mathbb{N}} \frac{e^{\pi i n z}}{1+e^{2\pi i n z}} \\ &\overset{(1.110)}{=} 1 + 4\sum_{n\in\mathbb{N}} \big(d_1(n) - d_3(n)\big) e^{\pi i n z}\end{aligned}$$

for all $z \in \mathbb{H}$. Hence, the coefficients of the Fourier expansion agree and we get

$$r_2(n) = 4\big(d_1(n) - d_3(n)\big)$$

for all $n \in \mathbb{N}$. □

We considered the question about the behavior of $r_2(n)$, $n \in \mathbb{N}$, that we answered in Corollary 1.144. The key idea was that we consider the function, $\theta^2(z)$, which contains the values $r_2(n)$ as coefficients of its Fourier expansion.

We now discuss $r_4(n)$. First, we recall the relation between $r_4(n)$ and the Fourier expansion of $\theta^4(z)$ in (1.95) of Exercise 1.136. As above for θ^2, see Theorem 1.142, we would like to relate $\theta^4(z)$ to another function that we call F_2.

Definition 1.147 The function $F_2 : \mathbb{H} \to \mathbb{C}$ is defined by the series

$$F_2(z) := \sum_{m\in\mathbb{Z}} \sum_{n\in\mathbb{Z}}{}' \frac{1}{\left(\frac{m}{2}z+n\right)^2} - \sum_{m\in\mathbb{Z}} \sum_{n\in\mathbb{Z}}{}' \frac{1}{\left(mz+\frac{n}{2}\right)^2}. \tag{1.111}$$

Here the notation $\sum_{m\in\mathbb{Z}} \sum_{n\in\mathbb{Z}}{}'$ means that we calculate first the inner series (over n) and then the outer series (over m); the prime $'$ indicates that we skip over the term with $(m, n) = (0, 0)$.

Remark 1.148 Recall that the above series are conditionally convergent and not absolutely convergent. Hence, the order of the summation plays a crucial role.

The next results are presented without proof. We refer to [116, Chapter I] for more details.

Lemma 1.149 *Recall the Eisenstein series G_2 in* (1.51) *of Remark 1.83. We have*

$$F_2(z) = G_2\left(\frac{z}{2}\right) - 4\, G_2(2z) \tag{1.112}$$

for all $z \in \mathbb{H}$.

Remark 1.150 Lemma 1.149 shows in particular that F_2 is well defined since it is represented in (1.112) in terms of the Eisenstein series G_2.

Proposition 1.151 *The function F_2 satisfies the transformation property*

$$-\, z^{-2}\, F_2\left(\frac{-1}{z}\right) = F_2(z) \tag{1.113}$$

for all $z \in \mathbb{H}$.

It satisfies the growth estimates

$$\lim_{\Im(z)\to\infty} F_2(z) = -\pi^2 \tag{1.114}$$

and

$$F_2\left(1-\frac{1}{z}\right) = O\left(|z|^2\, e^{-2\pi\,\delta\,\Im(z)}\right) \qquad \text{as } \Im(z) \to \infty \tag{1.115}$$

for some (small) $\delta > 0$.

Before we continue with the discussion of the function F_2, we need to introduce a modified divisor function, where we skip all summands of σ_1 divisible by 4.

Definition 1.152 We define the *modified divisor function* $\sigma_1^\star : \mathbb{N} \to \mathbb{N}$ by

$$\sigma_1^\star(n) := \sum_{\substack{0<d\mid n \\ 4\nmid d}} d. \tag{1.116}$$

Proposition 1.153 *We have*

$$F_2(z) = -\pi^2 - 8\pi^2 \sum_{n\in\mathbb{N}} \sigma_1^\star(n)\, e^{\pi i n z} \tag{1.117}$$

for all $z \in \mathbb{H}$.

Exercise 1.154 Show that the divisor functions $\sigma_1(n)$ defined in (1.50) and the modified divisor function $\sigma_1^\star(n)$ in (1.116) satisfy

$$\sigma_1^\star(n) = \begin{cases} \sigma_1(n) & \text{if } 4 \nmid n \text{ and} \\ \sigma_1(n) - 4\sigma_1\left(\frac{n}{4}\right) & \text{if } 4 \mid n \end{cases} \tag{1.118}$$

for all $n \in \mathbb{N}$.

Next, we match θ^4 and F_2.

Theorem 1.155 *The function* F_2 *and* θ^4 *satisfy*

$$F_2(z) = -\pi^2\,\theta^4(z) \tag{1.119}$$

for all $z \in \mathbb{H}$.

Corollary 1.156 *We have*

$$\theta^4(z) = 1 + \sum_{n\in\mathbb{N}} 8\sigma_1^\star(n)\, e^{\pi i n z} \tag{1.120}$$

for all $z \in \mathbb{H}$.

Proof Combine (1.117) in Proposition 1.153 with (1.119) in Theorem 1.155. □

Corollary 1.157 *We have*

$$r_4(n) = 8\sigma_1^\star(n) \tag{1.121}$$

for all $n \in \mathbb{N}$.

Proof We use Corollary 1.156 and Exercise 1.136. This gives

$$1 + \sum_{n\in\mathbb{N}} r4(n)\, e^{\pi i n z} \overset{(1.95)}{=} \theta^4(z) \overset{(1.120)}{=} 1 + \sum_{n\in\mathbb{N}} 8\sigma_1^\star(n)\, e^{\pi i n z}$$

for all $z \in \mathbb{H}$. Hence, the coefficients have to agree and we conclude (1.121). □

Remark 1.158 We just proved Jacobi's theorem, which we mentioned in the introduction of Section 1.5. As one can see, the study of the arithmetic properties of the counting functions r_2 and r_4 leads to modular forms via their generating functions θ^2 and θ^4, respectively.

Now, we know that the theta function is associated with the counting functions r_2 and r_4. What about other r_k's? We discuss this only for $k \in \mathbb{N}$ that are divisible by 8.

Definition 1.159 Let $8 \mid k \in \mathbb{N}$. The *counting function* $r_k : \mathbb{Z}_{\geq 0} \to \mathbb{Z}_{\geq 0}$ is defined by

$$r_k(n) = \sharp\{(x_1, \ldots, x_k) \in \mathbb{Z}^k\colon\ x_1^2 + \ldots + x_k^2 = n\}. \tag{1.122}$$

Lemma 1.160 *Let* $8 \mid k \in \mathbb{N}$. *The generating function of* r_k *satisfies*

$$\begin{aligned} \theta^k(z) &= \sum_{n_1,\ldots,n_k \in \mathbb{Z}} e^{\pi i (n_1^2 + \ldots + n_k^2) z} \\ &= 1 + \sum_{n \in \mathbb{N}} r_k(n)\, e^{\pi i n z} \end{aligned} \tag{1.123}$$

for all $z \in \mathbb{H}$.

Proof Similar to Exercise 1.136, we calculate

$$\begin{aligned} \theta(z)^k &\overset{(1.80)}{=} \Theta(0, z)^k \\ &\overset{(1.73)}{=} \left(\sum_{n \in \mathbb{Z}} e^{\pi i n^2 z} \right)^k \\ &= \sum_{n_1,\ldots,n_k \in \mathbb{Z}} e^{\pi i (n_1^2 + \ldots + n_k^2) z} \\ &\overset{(1.122)}{=} \sum_{m \in \mathbb{Z}_{\geq 0}} r_k(m)\, e^{\pi i m z}. \end{aligned}$$

□

Corollary 1.161 *Let* $8 \mid k \in \mathbb{N}$. *We have*

$$\theta^k(z+2) = \theta^k(z) \tag{1.124}$$

for all $z \in \mathbb{H}$.

Proof This follows immediately from $\theta(z+2) \overset{(1.81)}{=} \theta(z)$ in Corollary 1.123. □

To show the transformation property under $z \mapsto \frac{-1}{z}$, first consider the following axillary result.

Lemma 1.162 *Let* $8 \mid k \in \mathbb{N}$. *We have*

$$\sum_{n_1,\ldots,n_k\in\mathbb{Z}} e^{-\pi(n_1^2+\ldots+n_k^2)t} = \frac{1}{y^{\frac{k}{2}}} \sum_{n_1,\ldots,n_k\in\mathbb{Z}} e^{-\pi\frac{n_1^2+\ldots+n_k^2}{t}} \tag{1.125}$$

for all $t > 0$.

Proof Using Lemma 1.119, we find

$$\begin{aligned}\sum_{n_1,\ldots,n_k\in\mathbb{Z}} e^{-\pi(n_1^2+\ldots+n_k^2)t} &= \sum_{n_1,\ldots,n_k\in\mathbb{Z}} e^{-\pi n_1^2 t} \cdots e^{-\pi n_k^2 t} \\ &\overset{(1.75)}{=} \sum_{n_1,\ldots,n_k\in\mathbb{Z}} \frac{1}{\sqrt{t}} e^{-\pi\frac{n_1^2}{t}} \cdots \frac{1}{\sqrt{t}} e^{-\pi n_k^2 t} \\ &= \frac{1}{t^{\frac{k}{2}}} \sum_{n_1,\ldots,n_k\in\mathbb{Z}} e^{-\pi\frac{n_1^2+\ldots+n_k^2}{t}}.\end{aligned}$$

□

Proposition 1.163 *Let* $8 \mid k \in \mathbb{N}$. *We have*

$$z^{-\frac{k}{2}}\, \theta^k\left(\frac{-1}{z}\right) = \theta^k(z) \tag{1.126}$$

for all $z \in \mathbb{H}$.

Proof Consider $z = it$ with $t > 0$. We have

$$\begin{aligned}\theta^k(it) &\overset{(1.123)}{=} \sum_{n_1,\ldots,n_k\in\mathbb{Z}} e^{-\pi(n_1^2+\ldots+n_k^2)t} \\ &\overset{(1.125)}{=} \frac{1}{t^{\frac{k}{2}}} \sum_{n_1,\ldots,n_k\in\mathbb{Z}} e^{-\pi\frac{n_1^2+\ldots+n_k^2}{t}} \\ &\overset{(1.123)}{=} \frac{1}{t^{\frac{k}{2}}}\, \theta^k\left(\frac{-1}{it}\right).\end{aligned}$$

Using the holomorphy of θ allows us to extend the above identity to all $z \in \mathbb{H}$. This proves (1.126). □

Proposition 1.164 *Let* $8 \mid k \in \mathbb{N}$. *The function* θ^k *satisfies*

$$j(M,z)^{-\frac{k}{2}}\, \theta^k\big(M\,z\big) = \theta^k(z) \tag{1.127}$$

for all $z \in \mathbb{H}$ *and* $M \in \Gamma_\theta$.

Remark 1.165 The above proposition shows that θ^k (for $8 \mid k \in \mathbb{N}$) satisfies the transformation behavior of a modular form of weight $\frac{k}{2}$. We have to check the growth in the cusps ∞ and 1 of $\mathcal{F}_{\Gamma_\theta}$ to check whether θ^k is a modular form.

Lemma 1.166 *Let* $8 \mid k \in \mathbb{N}$. *The function* θ^k *satisfies*

$$\lim_{\Im(z)\to\infty} \theta^k(z) = 1 \tag{1.128}$$

and

$$\theta^k\left(1 - \frac{1}{z}\right) = O\left(|z|^{\frac{k}{2}} e^{-\frac{k\pi}{2}\Im(z)}\right) \qquad \text{as } \Im(z) \to \infty. \tag{1.129}$$

Proof The first identity is a direct consequence of Corollary 1.124. The second follows from Corollary 1.125. □

Theorem 1.167 *Let* $8 \mid k \in \mathbb{N}$. *The function* θ^k *is a modular form of weight* $\frac{k}{2}$ *for the theta group* Γ_θ*:*

$$\theta^k \in M_{\frac{k}{2}}(\Gamma_\theta). \tag{1.130}$$

Proof θ^k satisfies the transformation property of modular forms, see Proposition 1.164. The growth conditions in Lemma 1.166 imply that the Fourier expansion of θ^k is regular at the cusps ∞ and 1. We also know that θ^k is holomorphic. Hence, $\theta^k \in M_{\frac{k}{2}}(\Gamma_\theta)$. □

1.6 Hecke Operators

In this section, we define Hecke operators on the space of modular forms of integer weight and give their basic properties. We first give the definition of what is known as the slash operator, a technical tool that usually facilitates calculation. We then define Hecke operators and show that Hecke operators map the space of modular forms to itself. This gives rise to what are known as Hecke eigenforms, which are modular forms whose coefficients are multiplicative.

1.6.1 The Slash Operator

Definition 1.168 Let f be a meromorphic function on $\mathbb{H}$ and k a real number. We define the *slash operator of weight k* by

$$\left(f\big|_k M\right)(z) := j(M,z)^k\, f(Mz) \qquad (z \in \mathbb{H}), \tag{1.131}$$

where $M \in \mathrm{GL}_2(\mathbb{R})^+$ is a 2×2 matrix with positive determinant and the automorphic factor j is defined in (1.29).

The slash operator allows us to write (1.28) as

$$f\big|_k M = v(M)\, f \qquad \text{for } M \in \Gamma.$$

We sometimes extend the slash operator to work with formal sums of matrices:

$$f\big|_k (M+V) := f\big|_k M + f\big|_k V \qquad \text{for } M \in \Gamma. \tag{1.132}$$

Lemma 1.169 *Let $k \in 2\mathbb{Z}$ and $f : \mathbb{H} \to \mathbb{C}$. We have*

$$\left(f\big|_k M\right)\big|_k V = f\big|_k MV \tag{1.133}$$

for all matrices $M, V \in \mathrm{GL}_2(\mathbb{R})$.

Proof We have

$$\begin{aligned}\left(\left(f\big|_k M\right)\big|_k V\right)(z) &= j(V,z)^{-k}\left(f\big|_k M\right)(V z)\\ &= j(V,z)^{-k}\, j(M, V z)^{-k}\, f\big(M(V z)\big)\\ &= j(MV, z)^{-k}\, f(MV z)\\ &= \left(f\big|_k MV\right)(z)\end{aligned}$$

for all $z \in \mathbb{H}$. □

Definition 1.170 Let $n \in \mathbb{N}$. The *set $\mathcal{R}_\star$ of finite linear combinations of elements* $\mathrm{Mat}_2(\mathbb{Z})$ with integer coefficients is defined as

$$\mathcal{R}_\star := \mathbb{Z}\big[\mathrm{Mat}_2(\mathbb{Z})\big]. \tag{1.134}$$

The *set $\mathcal{R}_n$ of finite linear combinations of elements* $\mathrm{Mat}_2(\mathbb{Z})$ *with fixed determinant* n with integer coefficients is defined as

$$\mathcal{R}_n := \mathbb{Z}\big[\{M \in \mathrm{Mat}_2(\mathbb{Z});\, \det M = n\}\big]. \tag{1.135}$$

Obviously, $\mathcal{R}_\star$ (or $\mathcal{R}_n$) is an additive ring: if $\sum_i a_i M_i, \sum_j b_j V_j \in \mathcal{R}_\star$, then $\sum_i a_i M_i + \sum_j b_j V_j \in \mathcal{R}_\star$. The other relations follow analogously.

We can extend the definition of the slash operator to elements in $\mathcal{R}_\star$.

Definition 1.171 Let f be a meromorphic function on $\mathbb{H}$ and k be a real number. We define the *slash operator of weight k* on $\mathcal{R}_\star$ by

$$(f|_k \sum_i a_i M_i)(z) := \sum_i a_i \left(j(M,z)^k f(M_i z) \right) \qquad (z \in \mathbb{H}, \sum_i a_i M_i \in \mathcal{R}_\star). \tag{1.136}$$

The definition on $\mathcal{R}_n$ with (formal) linear combinations of matrices of fixed determinant $n \in \mathbb{N}$ is analogous.

1.6.2 Hecke Operators of Index $n \in \mathbb{N}$

Definition 1.172 Let k be an integer and let $f : \mathbb{H} \to \mathbb{C}$. For $n \in \mathbb{N}$ and $z \in \mathbb{H}$, we define the *Hecke operator of index n* by

$$\begin{aligned} (T_n f)(z) :&= n^{k-1} \sum_{d|n} d^{-k} \sum_{b=0}^{d-1} f\left(\frac{nz + bd}{d^2} \right) \\ &= n^{k-1} f|_k \left(\sum_{d|n} \sum_{b=0}^{d-1} \begin{pmatrix} \frac{n}{d} & b \\ 0 & d \end{pmatrix} \right) (z). \end{aligned} \tag{1.137}$$

Definition 1.173 Let $p \in \mathbb{N}$ be a prime and $k \in 2\mathbb{Z}$. We define the map $\tilde{T}_p : M_k^!(\Gamma(1)) \to M_k^!(\Gamma(1))$ given by

$$\begin{aligned} \tilde{T}_p f :&= f|_k \begin{pmatrix} p & 0 \\ 0 & 1 \end{pmatrix} + \sum_{j=0}^{p-1} f|_k \begin{pmatrix} 1 & j \\ 0 & p \end{pmatrix} \\ &= f|_k \left(\begin{pmatrix} p & 0 \\ 0 & 1 \end{pmatrix} + \sum_{j=0}^{p-1} \begin{pmatrix} 1 & j \\ 0 & p \end{pmatrix} \right). \end{aligned} \tag{1.138}$$

Remark 1.174 We have given a practical definition of Hecke operators above. It is important to mention though that we are actually summing over a complete system of representatives $\mathcal{T}(n)$ of equivalence classes of $\Gamma(1)\backslash H(n)$ where

$$\mathcal{T}(n) := \left\{ \begin{pmatrix} a & b \\ 0 & d \end{pmatrix} : \ 0 < d \mid n, \ ad = n, \ 0 \leq b < d \right\} \tag{1.139}$$

and

$$H(n) := \{ A \in \mathrm{Mat}_2(\mathbb{Z}) : \ \det A = n \}. \tag{1.140}$$

Exercise 1.175 Let $n \in \mathbb{N}$, $A_1 \in H(n)$, and $V_1 \in \Gamma(1)$. Show that there exist a triangular matrix $A_2 \in \mathcal{T}(n)$ and an element $V_2 \in \Gamma(1)$ such that

$$A_1\, V_1 = V_2\, A_2. \tag{1.141}$$

Moreover,

$$j(A_1, V_1\, z)\, j(V_1, z) = j(V_2, A_2\, z)\, j(A_2, z)$$

for all $z \in \mathbb{H}$.

Proposition 1.176 *Let weight k be an integer and $n \in \mathbb{N}$. If $f : \mathbb{H} \to \mathbb{C}$ is periodic, then $T_n f$ is also periodic. In other words,*

$$f(z+1) = f(z) \quad \Longrightarrow \quad \big(T_n f\big)(z+1) = \big(T_n f)\big)(z) \tag{1.142}$$

for all $z \in \mathbb{H}$.

We now describe the effect of Hecke operators on Fourier expansions. A detailed treatment of this expansion can be found in [12, Chapter VI].

Theorem 1.177 *Let $k \in 2\mathbb{Z}$ and $n \in \mathbb{N}$. Suppose that the function $f : \mathbb{H} \to \mathbb{C}$ has a Fourier expansion of the form*

$$f(z) = \sum_{m \in \mathbb{Z}} c(m)\, e^{2\pi i m z} \qquad (z \in \mathbb{H}).$$

Then the function $T_n f$ admits the Fourier expansion

$$\big(T_n f\big)(z) = \sum_{m \in \mathbb{Z}} \gamma_n(m)\, e^{2\pi i m z} \tag{1.143}$$

with

$$\gamma_n(m) := \sum_{0<d \mid \gcd(n,m)} d^{k-1}\, c\left(\frac{mn}{d^2}\right). \tag{1.144}$$

A direct consequence is the following corollary.

Corollary 1.178 *In the setting of Theorem 1.177, consider the coefficients $\gamma_n(m)$ as defined in* (1.144). *These satisfy*

$$\gamma_n(m) = \gamma_m(n) \tag{1.145}$$

for all $m, n \in \mathbb{N}$.

Corollary 1.179 *Suppose $k \in 2\mathbb{Z}$ and $n \in \mathbb{N}$. If $f : \mathbb{H} \to \mathbb{C}$ satisfies*

$$f(V\, z) = j(V, z)^k\, f(z)$$

for all $V \in \Gamma(1)$ *and* $z \in \mathbb{H}$, *then* $T_n\, f$ *satisfies*

$$\big(T_n\, f\big)(V\,z) = j(V,z)^k\,\big(T_n\, f\big)(z)$$

for all $V \in \Gamma(1)$ *and* $z \in \mathbb{H}$.

Remark 1.180 One can also see that Theorem 1.177 gives the relations

$$\gamma_n(0) = \sigma_{k-1}(n)\,c(0) \quad \text{and} \quad \gamma_n(1) = c(n).$$

This simply reflects that γ_n takes Fourier coefficients of cusp forms to cusp forms.

Corollary 1.181 *Let* $n \in \mathbb{N}$ *and* $k \in 2\mathbb{Z}$.

1. *If* $f \in M_k^!\big(\Gamma(1)\big)$, *then* $T_n f \in M_k^!\big(\Gamma(1)\big)$.
2. *If* $f \in M_k\big(\Gamma(1)\big)$, *then* $T_n f \in M_k\big(\Gamma(1)\big)$.
3. *If* $f \in S_k\big(\Gamma(1)\big)$, *then* $T_n f \in S_k\big(\Gamma(1)\big)$.

Theorem 1.182 *Let* $n, m \in \mathbb{N}$ *be coprime and suppose that* $f : \mathbb{H} \to \mathbb{C}$ *is periodic. Then*

$$T_m\,T_n\,f = T_{mn}\,f. \tag{1.146}$$

Remark 1.183 Equation (1.146) follows from the fact that the Hecke operator T_n has the representation

$$T_n\,f = n^{k-1} \sum_{A \in \mathcal{T}(n)} f\big|_k A. \tag{1.147}$$

Calculating $T_m\,T_n\,f$ yields

$$\begin{aligned} T_m\,T_n\,f &= (mn)^{k-1} \sum_{\substack{A \in \mathcal{T}(m)\\ B \in \mathcal{T}(n)}} f\big|_k BA \\ &= (mn)^{k-1} \sum_{C \in \mathcal{T}(mn)} f\big|_k C \\ &= T_{mn}\,f. \end{aligned}$$

One of the main results on Hecke operators is stated below.

Theorem 1.184 *For* $m, n \in \mathbb{N}$ *and* $f : \mathbb{H} \to \mathbb{C}$ *periodic, we have*

$$T_m\,T_n\,f = \sum_{d \mid \gcd(m,n)} d^{k-1}\,T_{\frac{mn}{d^2}}\,f. \tag{1.148}$$

Of course, the roles of n and m above can be reversed. This leads to

Corollary 1.185 *For $m, n \in \mathbb{N}$ and $f : \mathbb{H} \to \mathbb{C}$ being periodic, we have*

$$T_m\, T_n\, f = T_n\, T_m\, f.$$

Proof Using (1.148), we have

$$\begin{aligned} T_m\, T_n\, f &= \sum_{d \mid \gcd(m,n)} d^{k-1}\, T_{\frac{mn}{d^2}}\, f, \\ &= T_n\, T_m\, f \end{aligned}$$

since the expression in the middle is symmetric in n and m. □

1.6.3 Eigenforms

Let $k \in 2\mathbb{N}$ and $f \in M_k^!\big(\Gamma(1)\big)$. Suppose that f admits a Fourier expansion $f(z) = \sum_{m\in\mathbb{Z}} c(m)\, e^{2\pi i m z}$ for all $z \in \mathbb{H}$ and satisfies

$$T_n f = c(n)\, f \qquad \text{for all } n \in \mathbb{N}.$$

Then, the Fourier coefficients satisfy

$$c(m)\, c(n) = \sum_{0<d \mid \gcd(m,n)} d^{k-1}\, c\left(\frac{mn}{d^2}\right) \tag{1.149}$$

using Theorem 1.177. For example, the arithmetic divisor function $\sigma_{k-1}(n)$, defined in (1.50), satisfies this relation.

Definition 1.186 For $k \in 2\mathbb{N}$, let $0 \neq f \in M_k^!\big(\Gamma(1)\big)$ satisfy

$$T_n\, f = \lambda(n)\, f \qquad \text{for some } \lambda(n). \tag{1.150}$$

We then call f an *eigenfunction* of the Hecke operator T_n with *eigenvalue* $\lambda(n)$.

If f is an eigenfunction of T_n for all $n \in \mathbb{N}$, we call f a *simultaneous eigenfunction* of all Hecke operators. We call a simultaneous eigenfunction f *normalized* if the first Fourier coefficient of $f(z) = \sum_n c(n)\, e^{2\pi i n z}$ is 1 ($c(1) = 1$).

Theorem 1.187 *Let $k \in 2\mathbb{N}$. If $f \in M_k\big(\Gamma(1)\big)$ is a simultaneous eigenfunction with nonvanishing eigenvalues $\lambda(n) \neq 0$, $n \in \mathbb{N}$. If f admits the Fourier expansion $f(z) = \sum_{n\in\mathbb{Z}_{\geq 0}} c(n)\, e^{2\pi i n z}$, then its first Fourier coefficient $c(1)$ does not vanish.*

Theorem 1.188 *Let $k \in 2\mathbb{N}$ and let $0 \neq f \in S_k\big(\Gamma(1)\big)$ admit the Fourier expansion*

$$f(z) = \sum_{n\in\mathbb{N}} c(n)\, e^{2\pi i n z}.$$

The following statements are equivalent:

1. *f is a normalized simultaneous Hecke eigenform,*
2. $c(m)\,c(n) = \sum_{0 \le d | \gcd(m,n)} d^{k-1}\,c\left(\frac{mn}{d^2}\right)$ *for all $m, n \in \mathbb{N}$, and*
3. $T_n f = c(n)\,f$ *for all $n \in \mathbb{N}$.*

Recall the normalized Eisenstein series E_k that we introduced in Definition 1.84. For $k \in 2\mathbb{N}$, $k \ge 4$, the Eisenstein series E_k is an entire modular form of weight k and trivial multiplier for the full modular group $\Gamma(1)$.

The action of the Hecke operators on E_k is described in the following theorem.

Theorem 1.189 *Let $k \in 2\mathbb{N}$, $k \ge 4$, and $n \in \mathbb{N}$. We have*

$$T_n\, E_k = \sigma_{k-1}(n)\, E_k \tag{1.151}$$

where $\sigma_k(n)$ is the kth arithmetic divisor function defined in (1.50).

Corollary 1.190 *Let $k \in 2\mathbb{N}$ and suppose $f \in M_k\big(\Gamma(1)\big) \smallsetminus S_k\big(\Gamma(1)\big)$. Then the first of the following statements implies the second.*

1. *f is a normalized simultaneous Hecke eigenform.*
2. *$f = \alpha_k\, E_k$ with $\alpha_k = \frac{(k-1!\,\zeta(k)}{(2\pi i)^k}$ and E_k the normalized Eisenstein series given in Definition 1.84.*

Next, we consider Hecke operators and describe how these operators interact with the Petersson scalar product.

Definition 1.191 Let $\mathcal{H}$ be a complex Hilbert space. We call an operator $T : \mathcal{H} \to \mathcal{H}$ *self-adjoint* if

$$\langle Tf, g\rangle = \langle f, Tg\rangle \tag{1.152}$$

for all $f, g \in \mathcal{H}$.

Proposition 1.192 *For $n \in \mathbb{N}$ and $4 \le k \in 2\mathbb{N}$ a positive even weight, the Hecke operator T_n is a self-adjoint operator on the Hilbert space of cusp forms $S_k\big(\Gamma(1)\big)$.*

Corollary 1.193 *Let $4 \le k \in 2\mathbb{N}$ a positive even weight and $n \in \mathbb{N}$. If $f \in S_k\big(\Gamma(1)\big)$ is an eigenfunction of the Hecke operator T_n, then its eigenvalue is real.*

Now we can state a result by Hecke.

Theorem 1.194 *Let $4 \le k \in 2\mathbb{N}$. The space $S_k\big(\Gamma(1)\big)$ of cusp forms admits an orthonormal basis consisting only of simultaneous eigenfunctions of all Hecke operators.*

1.7 L-Functions Associated with Cusp Forms

This section introduces the notion of L-series and its relation to modular cusp forms, which is also a classical topic in the theory of modular forms. We present it in more detail to serve as a reference section for our later treatment of L-series associated with Maass cusp forms in §4.2.

Definition 1.195 Let $C \in \mathbb{R}$ be a constant and $\left(a(n)\right)_{n\in\mathbb{N}}$ be a sequence of complex numbers satisfying the growth estimate

$$a(n) = O\left(n^C\right) \qquad (n \in \mathbb{Z}_{\geq 0}).$$

For given $0 \leq \kappa < 1$, the function

$$L(s) := \sum_{n\in\mathbb{N}}^{\infty} a(n)\, n^{-s} \qquad (\Re(s) > C + 1) \tag{1.153}$$

is called the *L-series* of $\left(a(n)\right)_n$.

Exercise 1.196 Show that the L-series defined above converges for all $\Re(s) > C + 1$.

Remark 1.197

1. If the sequence $\left(a(n)\right)_n$ contains the Fourier coefficients of a cusp form F for the full modular group, we call the L-series in (1.153) the *L-series of F*.
2. The L-series in (1.153) is also known as *Dirichlet series*.
3. Other areas of interest are the domains of conditional convergence or uniform convergence of Dirichlet series. We refer to [133] and references therein for more information on these topics.

1.7.1 The Mellin Transform

Definition 1.198 Let $f : (0, \infty) \to \mathbb{C}$ be a Lebesgue-integrable function. Its *Mellin transform* is defined as

$$\mathcal{M}(f)(s) := \int_0^\infty f(t)t^{s-1}\, dt \tag{1.154}$$

for all $s \in \mathbb{C}$ where the integral exists.

Remark 1.199 We use Γ to denote two different objects. Depending on the context, it might denote a congruence subgroup or it might denote the Γ-function. However, as in Remark 1.10, we believe that this duality is unlikely to lead to any ambiguity.

Example 1.200 A typical example is the Γ-function and its integral representation

$$\begin{aligned}\Gamma(s) :&= \int_0^\infty e^{-y} y^{s-1}\,\mathrm{d}y \\ &= \mathcal{M}\left(e^{-\cdot}\right)(s) \qquad (\Re(s) > 0).\end{aligned} \tag{1.155}$$

The next lemma summarizes some obvious existence results.

Lemma 1.201 *Let* $f : (0, \infty) \to \mathbb{C}$ *be continuous with asymptotic estimates*

$$f(t) = \begin{cases} O\left(t^{-a}\right) & \text{as } t \to 0 \text{ and} \\ O\left(t^{-b}\right) & \text{as } t \to \infty \end{cases}$$

for some real $a < b$. *Then* $\mathcal{M}(f)(s)$ *exists for every* $s \in \mathbb{C}$ *with* $a < \Re(s) < b$.

If f *satisfies additionally the stronger estimate*

$$f(t) = O\left(e^{-ct}\right) \qquad \text{as } t \to \infty$$

for some $c > 0$, *then* $\mathcal{M}(f)(s)$ *exists for all* $\Re(s) > a$.

Proof Assume that f satisfies the first estimates. Splitting the integral at 1,

$$\begin{aligned}\int_0^\infty f(t)t^{s-1}\,\mathrm{d}t &= \int_0^1 f(t)t^{s-1}\,\mathrm{d}t + \int_1^\infty f(t)t^{s-1}\,\mathrm{d}t \\ &= O\left(\int_0^1 t^{s-a-1}\,\mathrm{d}t\right) + O\left(\int_1^\infty t^{s-b-1}\,\mathrm{d}t\right).\end{aligned}$$

We see from the first and second terms respectively that the integral exists if $\Re(s-a) > 0$ and $\Re(s-b) < 0$. Hence, the Mellin transform is well defined for $a < \Re(s) < b$.

The second part follows from the rough estimate

$$f(t) = O\left(e^{-ct}\right) = O\left(t^{-b}\right) \qquad \text{as } t \to \infty$$

for every $b \in \mathbb{R}$. □

Definition 1.202 For real $a < b$, let $\varphi : \{s \in \mathbb{C} : a < \Re(s) < b\} \to \mathbb{C}$ be a holomorphic function satisfying the asymptotic estimate

$$\varphi(s) \to 0 \qquad \text{uniformly as } |\Im(s)| \to \infty.$$

We also assume that the path-integral $\int_{c-i\infty}^{c+i\infty} |\varphi(s)|\mathrm{d}s$ exists for every c with $a+\delta < c < b-\delta$ (for arbitrary but fixed $\delta > 0$).

The *inverse Mellin transform* is defined by

$$\mathcal{M}^{-1}(\varphi)(x) := \frac{1}{2\pi i}\int_{c-i\infty}^{c+i\infty} \varphi(s)x^{-s}\,\mathrm{d}s \qquad (x>0) \tag{1.156}$$

for every $a < c < b$.

Lemma 1.203 *The inverse Mellin transform $\mathcal{M}^{-1}(\varphi)$ exists for every $a < c < b$ and is independent of c.*

Proof Since by assumption $\varphi(s) \to 0$ uniformly as $|\Im(s)| \to \infty$ on the strip $a+\delta < \Re(s) < b-\delta$, we may move the path of integration in (1.156) to any $a < c < b$ without changing the integral. This shows that $\mathcal{M}(f)$ does not depend on the choice of c. □

Proposition 1.204 (Mellin inversion formula) *For real $a < b$, let $\varphi : \{s \in \mathbb{C}\colon\ a < \Re(s) < b\} \to \mathbb{C}$ be holomorphic such that it satisfies the asymptotic*

$$\varphi(s) \to 0 \qquad \textit{uniformly as } |\Im(s)| \to \infty.$$

We also assume that the path-integral $\int_{c-i\infty}^{c+i\infty} |\varphi(s)|\mathrm{d}s$ exists for every c with $a+\delta < c < b-\delta$ (for arbitrary but fixed $\delta > 0$).

In this situation the function $f := \mathcal{M}^{-1}(\varphi)$ satisfies the inversion formula

$$\varphi(s) = \mathcal{M}(f)(s)$$

for every s in the strip $a < \Re(s) < b$.

Conversely, suppose that $f : (0,\infty) \to \infty$ is piecewise continuous, and suppose that the integral $\int_0^\infty |f(t)| t^{\Re(s)-1}\,\mathrm{d}t$ exists for all $a < \Re(s) < b$. Then,

$$f(x) = \mathcal{M}^{-1}\Big(\mathcal{M}(f)\Big)(x) \qquad (x > 0).$$

Proof We follow the arguments in [61, pp. 103–105]. Fix a c with $a < c < b$, i.e., fix a path of integration in (1.156). Now, choose $a < c_1 < c < c_2 < b$. We have

$$\begin{aligned}\int_0^\infty & f(t)\,t^{s-1}\,\mathrm{d}t \\ &= \int_0^1 f(t)\,t^{s-1}\,\mathrm{d}t + \int_1^\infty f(t)\,t^{s-1}\,\mathrm{d}t \\ &= \frac{1}{2\pi i}\int_0^1\int_{c-i\infty}^{c+i\infty} \varphi(\tilde{s})t^{-\tilde{s}}\,\mathrm{d}\tilde{s}\,t^{s-1}\,\mathrm{d}t + \frac{1}{2\pi i}\int_1^\infty\int_{c-i\infty}^{c+i\infty} \varphi(\tilde{s})t^{-\tilde{s}}\,\mathrm{d}\tilde{s}\,t^{s-1}\,\mathrm{d}t.\end{aligned}$$

The first integral can be estimated by

$$\left|\frac{1}{2\pi i}\int_0^1\int_{c-i\infty}^{c+i\infty} \varphi(\tilde{s})t^{-\tilde{s}}\,\mathrm{d}\tilde{s}\,t^{s-1}\,\mathrm{d}t\right| \le \frac{1}{2\pi}\int_{-\infty}^{\infty} |\varphi(c_1+iy)|\,\mathrm{d}y \int_0^1 t^{c-c_1-1}\,\mathrm{d}t < \infty,$$

and the second by

$$\left|\frac{1}{2\pi i}\int_0^1\int_{c-i\infty}^{c+i\infty}\varphi(\tilde{s})t^{-\tilde{s}}\,\mathrm{d}\tilde{s}\,t^{s-1}\,\mathrm{d}t\right| \leq \frac{1}{2\pi}\int_{-\infty}^{\infty}|\varphi(c_1+iy)|\,\mathrm{d}y\int_1^{\infty}t^{c-c_2-1}\,\mathrm{d}t<\infty,$$

We may apply Fubini's Theorem and interchange the integrals, getting

$$\begin{aligned}
&\int_0^{\infty} f(t)\,t^{s-1}\,\mathrm{d}t\\
&=\frac{1}{2\pi i}\int_0^1\int_{c-i\infty}^{c+i\infty}\varphi(\tilde{s})t^{-\tilde{s}}\,\mathrm{d}\tilde{s}\,t^{s-1}\,\mathrm{d}t+\frac{1}{2\pi i}\int_1^{\infty}\int_{c-i\infty}^{c+i\infty}\varphi(\tilde{s})t^{-\tilde{s}}\,\mathrm{d}\tilde{s}\,t^{s-1}\,\mathrm{d}t\\
&=\frac{1}{2\pi i}\int_{c_2-i\infty}^{c_2+i\infty}\frac{\varphi(\tilde{s})}{\tilde{s}-s}\,\mathrm{d}\tilde{s}-\frac{1}{2\pi i}\int_{c_1-i\infty}^{c_1+i\infty}\frac{\varphi(\tilde{s})}{\tilde{s}-s}\,\mathrm{d}\tilde{s}.
\end{aligned}$$

Applying Cauchy's integral formula, we see that the difference at the right-hand side equals $\varphi(s)$ since the integrand vanishes as $|\Im(\tilde{s})|\to\infty$. For the second part, use the substitution $x=e^u$ and put $g:=\mathcal{M}(f)$. We get

$$\begin{aligned}
\mathcal{M}^{-1}(g)(x) &= \frac{1}{2\pi i}\int_{c-i\infty}^{c+i\infty} g(s)\,x^{-s}\,\mathrm{d}s\\
&\overset{x\mapsto e^u}{=} \frac{1}{2\pi}\int_{-\infty}^{\infty} g(c+it)\,e^{-u(c+it)}\,\mathrm{d}t\\
&= \frac{1}{2\pi}\int_{-\infty}^{\infty}\left(\int_0^{\infty} f(y)\,y^{c+it-1}\,\mathrm{d}y\right)e^{-u(c+iy)}\,\mathrm{d}y\\
&\overset{y\mapsto e^v}{=} \frac{1}{2\pi}\int_{-\infty}^{\infty}\left(\int_{-\infty}^{\infty} f\left(e^v\right)e^{v(c+it-1)}\,e^v\mathrm{d}v\right)e^{-u(c+iy)}\,\mathrm{d}y\\
&= \frac{e^{-uc}}{2\pi}\int_{-\infty}^{\infty}\int_{-\infty}^{\infty} f\left(e^v\right)e^{it(v-u))}\,e^{vc}\,\mathrm{d}v\mathrm{d}y.
\end{aligned}$$

Now, the Fourier integral theorem shows that the right-hand side is equal to

$$e^{-uc}e^{uc}\,f\left(e^u\right)=f(x).$$

We conclude

$$\mathcal{M}^{-1}\big(\mathcal{M}(f)\big)=f$$

to finish the proof. □

1.7.2 L-Series Associated with Cusp Forms on $\Gamma(1)$

Let $k \in 2\mathbb{N}$ and let $F \in S_k(\Gamma(1))$. The cusp form F has a Fourier expansion of the form

$$F(z) = \sum_{n=1}^{\infty} a(n)\, e^{2\pi i n z} \qquad (z \in \mathbb{H}). \tag{1.157}$$

The Fourier coefficients $a(n)$ satisfy the growth estimate

$$a(n) = O\left(n^{\frac{k}{2}}\right) \qquad \text{as } n \to \infty \tag{1.158}$$

(see Theorem 1.68). We associate with F the L-series of F given in (1.153):

$$L(s) = \sum_{n \in \mathbb{N}} a(n)\, n^{-s} \qquad (\Re(s) > \frac{k}{2} + 1).$$

Definition 1.205 Let $k \in 2\mathbb{N}$ and let $F \in S_k(\Gamma(1))$ with Fourier expansion (1.157). Define the *L-function* $L^\star(s)$ by taking the Mellin transform of F along the upper imaginary axis

$$L^\star(s) := \int_0^{\infty} F(iy)\, y^{s-1}\, \mathrm{d}y \tag{1.159}$$

for all complex s.

Remark 1.206 The L-function defined in (1.159) is sometimes called the *completed L-function.*

Lemma 1.207

1. *$L^\star(s)$ is well defined for all $s \in \mathbb{C}$.*
2. *$L(s)$ associated with the cusp form F exists for $\Re(s) \geq \frac{k}{2} + 1 > 0$.*
3. *We have*

$$\int_0^{\infty} F(iy)\, y^{s-1}\, \mathrm{d}y = L^\star(s) = (2\pi)^{-s}\Gamma(s)\, L(s) \tag{1.160}$$

for $\Re(s) \geq \frac{k}{2} + 1 > 0$, where $\Gamma(s)$ denotes the Γ-function, see (1.155) *or Appendix A.7.*

Proof

1. The existence can be shown as follows: Consider the defining integral

$$\int_0^{\infty} y^{s-1}\, F(iy)\, \mathrm{d}y.$$

Splitting the path of integration at 1 and applying the substitution $y \mapsto \frac{1}{y}$, we get

$$\begin{aligned}
\int_0^\infty y^{s-1}\, F(iy)\, \mathrm{d}y &= \int_0^1 y^{s-1}\, F(iy)\, \mathrm{d}y + \int_1^\infty y^{s-1}\, F(iy)\, \mathrm{d}y \\
&\overset{y\mapsto\frac{1}{y}}{=} \int_1^\infty y^{-s-1}\, F\left(\frac{-1}{iy}\right) \mathrm{d}y + \int_1^\infty y^{s-1}\, F(iy)\, \mathrm{d}y \\
&\overset{(1.28)}{=} \int_1^\infty y^{k-s-1}\, F(iy)\, \mathrm{d}y + \int_1^\infty y^{s-1}\, F(iy)\, \mathrm{d}y \\
&= \int_1^\infty \left(y^{k-s-1} + y^{l}\right) F(iy)\, \mathrm{d}y. \qquad (1.161)
\end{aligned}$$

The last integral exists since $F(iy) = O\left(e^{-\delta y}\right)$ as $y \to \infty$ for some $\delta > 0$, see Exercise 1.67. This implies that the L-function

$$L^\star(s) = \int_0^\infty y^{s-1}\, F(iy)\, \mathrm{d}y$$

exists for all s.

2. By estimate (1.158), there exists a constant $C > 0$ such that

$$|a(n)| \leq C\, n^{\frac{k}{2}}$$

for all $n \in \mathbb{N}$. Exercise 1.196 implies this statement.

3. Define $G : (0, \infty) \to \mathbb{R}$ by

$$G(y) := C \sum_{n=1}^{\infty} n^{\frac{k}{2}}\, e^{-2\pi n y} \qquad (y > 0).$$

We can easily see that $G(y)$ dominates $F(x + iy)$ for all $x \in \mathbb{R}$ as the following calculation shows

$$\begin{aligned}
|F(x+iy)| &\leq \sum_{n=1}^{\infty} \left| a(n)\, e^{2\pi i n (x+iy)} \right| \\
&= \sum_{n=1}^{\infty} |a(n)|\, e^{-2\pi n y} \\
&\leq C \sum_{n=1}^{\infty} n^{\frac{k}{2}}\, e^{-2\pi n y} \\
&= G(y)
\end{aligned}$$

for all $y > 0$ and $x \in \mathbb{R}$.

Now, for $\Re(s) > \frac{k}{2}+1$, Fubini's theorem allows us to interchange integration and summation in

$$\begin{aligned}\int_0^\infty G(y)\, y^{s-1}\, dy &= C \int_0^\infty \sum_{n=1}^\infty n^{\frac{k}{2}}\, e^{-2\pi n y}\, y^{s-1}\, dy \\ &= C \sum_{n=1}^\infty n^{\frac{k}{2}} \int_0^\infty e^{-2\pi n y} y^{s-1}\, dy \\ &= C\,(2\pi)^{-s} \sum_{n=1}^\infty n^{\frac{k}{2}-s} \int_0^\infty e^{-y} y^{s-1}\, dy \\ &= C\,(2\pi)^{-s}\Gamma(s) \sum_{n=1}^\infty n^{\frac{k}{2}-s},\end{aligned}$$

where we use the integral representation of the Γ-function

$$\Gamma(s) = \int_0^\infty e^{-y} y^{s-1}\, dy \qquad (\Re(s) > 0),$$

see (1.155) or Appendix A.7. The domain $\Re(s) > \frac{k}{2}+1$ of absolute convergence of the series $\sum_{n=1}^\infty n^{\frac{k}{2}-s}$ ensures that the Mellin transform of G exists.

Now, applying the dominated convergence theorem, we find that $L^\star$

$$\begin{aligned}L^\star(s) &= \int_0^\infty F(y)\, y^{s-1}\, dy \\ &= \int_0^\infty \sum_{n=1}^\infty a(n)\, e^{-2\pi n y}\, y^{s-1}\, dy \\ &= \sum_{n=1}^\infty a(n) \int_0^\infty e^{-2\pi n y} y^{s-1}\, dy \\ &= \sum_{n=1}^\infty a(n)\,(2\pi n)^{-s} \int_0^\infty e^{-y} y^{s-1}\, dy \\ &= (2\pi)^{-s}\Gamma(s) \sum_{n=1}^\infty a(n) n^{-s}\end{aligned}$$

for all s with $\Re(s) > \frac{k}{2}+1$.

□

Lemma 1.208 *The L-function $L^\star$ of the cusp form F extends holomorphically to the complex plane and satisfies the functional relation*

$$L^\star(k-s) = (-1)^{\frac{k}{2}}\, L^\star(s) \qquad (s \in \mathbb{C}). \tag{1.162}$$

Proof Recall that F satisfies (1.28) for the trivial multiplier system. In particular, F satisfies the identity

$$z^k\, F(z) = j(T,z)^k\, F(z) \overset{(1.28)}{=} F\big(Tz\big) = F\left(\frac{-1}{z}\right) \qquad (z \in \mathbb{H}),$$

applying the element $T \in \Gamma(1)$ defined in (1.20). Also notice that the upper imaginary axis is invariant under the map $z \mapsto \frac{-1}{z}$ and acts on the upper imaginary axis $\{iy\colon\ y > 0\}$ as $y \mapsto \frac{1}{y}$. Applying this substitution to the defining integral of $L^\star$ in (1.159), we get

$$\begin{aligned} L^\star(s) &\overset{(1.159)}{=} \int_0^\infty F(iy)\, y^{s-1}\, dy \\ &\overset{y\mapsto\frac{1}{y}}{=} -\int_\infty^0 F\left(\frac{-1}{iy}\right) y^{1-s}\, y^{-2} dy \\ &\overset{(1.28)}{=} \int_0^\infty (iy)^k\, F(iy)\, y^{-1-s} dy \\ &= i^k \int_0^\infty F(iy)\, y^{k-1-s}\, dy \\ &\overset{(1.159)}{=} i^k\, L^\star(k-s). \end{aligned}$$

Since $\frac{1}{i^k} = (-i)^k = (-1)^{\frac{k}{2}}$, this proves (1.162) assuming that $L^\star$ is a holomorphic function on $\mathbb{C}$.

Next, we show that $L^\star$ extends holomorphically to $\mathbb{C}$. Using the first part of Lemma 1.207, we know that $L^\star(s)$ exists for all $s \in \mathbb{C}$. Since $L^\star(s)$ is holomorphic for $\Re(s) > \frac{k}{2} + 1$ (see (1.160)) and $L(s)$ and $\Gamma(s)$ are holomorphic on a suitable right half-plane, the L-function extends holomorphically to $\mathbb{C}$. □

Exercise 1.209 Use the expression of $L^\star(s)$ in (1.161) to show the functional equation (1.162).

Exercise 1.210 For $k \in 2\mathbb{N}$ and $F \in S_k\big(\Gamma(1)\big)$, consider the associated L-series $L(s)$ defined in (1.153). Show the following statements:

1. $L(s)$ extends to a holomorphic function on $\mathbb{C}$, and
2. $L(s)$ satisfies the functional equation

$$\frac{(2\pi)^{k-s}}{\Gamma(k-s)} L(s) = i^k \frac{(2\pi)^s}{\Gamma(s)} L(k-s) \tag{1.163}$$

for all $s \in \mathbb{C}$.

Combining Lemma 1.207 and Lemma 1.208, we proved:

Proposition 1.211 *Let $F \in S_k(\Gamma(1))$ be a cusp form of weight $k \in 2\mathbb{N}$ (and trivial multiplier). Then, the L-series $L(s)$ of F defined in* (1.153) *converges absolutely for $\Re(s) > \frac{k}{2} + 1$. Its L-function $L^\star(s)$ given in* (1.159) *extends to a holomorphic function on $\mathbb{C}$ and satisfies the functional equation* (1.162).

An obvious question is whether the following holds: given L and $L^\star$, does there exist a cusp form F such that L is the L-series of F? Before we answer this question, we recall the inverse Mellin transform (see AppendixA.2).

Definition 1.212 For $a < b$, let $g(s)$ be an analytic function on the strip $a < \Re(s) < b$ such that $g(s) \to 0$ uniformly as $|\Im(s)| \to \infty$. The *inverse Mellin transform is defined by*

$$\mathcal{M}^{-1}(\varphi) = \frac{1}{2\pi i} \int_{c-i\infty}^{c+i\infty} x^{-s} \varphi(s)\, ds \qquad (x > 0). \tag{1.164}$$

Lemma 1.213 *Let $L(s) = \sum_{n \in \mathbb{N}} a_n n^{-s}$ be an L-series that is absolutely convergent on the half-plane $\Re(s) > \frac{k}{2} + 1$. Assume that its associated L-function $L^\star(s) := (2\pi)^{-s}\Gamma(s)\, L(s)$ is a holomorphic function on the same half-plane.*

Then, the function $f : i(0,\infty) \to \mathbb{C}$ on the upper imaginary axis defined by $f := \mathcal{M}^{-1}(L^\star)$ satisfies

$$f(iy) = \sum_{n=1}^{\infty} a_n e^{-2\pi n y} \qquad (y > 0).$$

Proof Using Stirling's estimate of the Γ-function in Appendix A.3, we find the upper bound

$$\begin{aligned} |L^\star(s)| &= |(2\pi)^{-s}\Gamma(s)\, L(s)| \\ &= (2\pi)^{-\sigma}\, |L(s)||\gamma(s)| \\ &\le C_2\, (2\pi)^{-\sigma} \left(\sum_{n=1}^{\infty} |a_n| n^{-\sigma} \right) e^{-\frac{\pi|t|}{2}}\, |t|^{\sigma - \frac{1}{2}} \end{aligned}$$

for $s = \sigma + it$ with $\sigma > \frac{k}{2} + 1$ bounded, $|t| > 2$, and $C_2 > 0$ a constant defined in Appendix A.3. We used $a_n = O\left(n^{\frac{k}{2}}\right)$ that follows from the fact that $L(s)$ converges absolutely for $\Re(s) > \frac{k}{2} + 1$.

This bound allows us to apply the inverse Mellin transform defined in (1.156) to $L^\star$. Taking $c > \frac{k}{2} + 1$, we get

$$\begin{aligned}\mathcal{M}^{-1}(L^\star)(y) &= \frac{1}{2\pi i}\int_{c-i\infty}^{c+i\infty} L^\star(s) y^{-s}\, \mathrm{d}s \\ &= \frac{1}{2\pi i}\int_{c-i\infty}^{c+i\infty} (2\pi)^{-s}\Gamma(s)\, L(s)\, y^{-s}\, \mathrm{d}s\end{aligned}$$

for all $y > 0$. Let us assume that we may interchange the path integration with the summation hidden in the L-series $L(s)$. We then get

$$\begin{aligned}\mathcal{M}^{-1}(L^\star)(y) &= \frac{1}{2\pi i}\int_{c-i\infty}^{c+i\infty} (2\pi)^{-s}\Gamma(s)\, L(s)\, y^{-s}\, \mathrm{d}s \\ &= \sum_{n=1}^{\infty} a_n \frac{1}{2\pi i}\int_{c-i\infty}^{c+i\infty} \Gamma(s)\,(2\pi n y)^{-s}\, \mathrm{d}s \\ &= \sum_{n=1}^{\infty} a_n \frac{1}{2\pi i}\int_{c-i\infty}^{c+i\infty} \Gamma(s)\,(2\pi n y)^{-s}\, \mathrm{d}s.\end{aligned}$$

Using (1.155), together with the first part of Proposition 1.204, we get

$$\frac{1}{2\pi i}\int_{c-i\infty}^{c+i\infty} \Gamma(s)\, y^{-s}\, \mathrm{d}s = e^{-y} \qquad (y > 0).$$

Hence, the above inverse Mellin transform gives

$$\begin{aligned}\mathcal{M}^{-1}(L^\star)(y) &= \sum_{n=1}^{\infty} a_n \frac{1}{2\pi i}\int_{c-i\infty}^{c+i\infty} \Gamma(s)\,(2\pi n y)^{-s}\, \mathrm{d}s \\ &= \sum_{n=1}^{\infty} a_n\, e^{-2\pi n y}.\end{aligned}$$

We still have to show that we may interchange integration and summation. Using the upper bound we calculated for $|L^\star(s)|$ above, we may use the dominated convergence theorem to show this. □

Lemma 1.214 *For $k \in 2\mathbb{N}$, let $L(s) =$ be an L-series that is absolutely convergent on the half-plane $\Re(s) > \frac{k}{2} + 1$. Assume that its associated L-function $L^\star(s) := (2\pi)^{-s}\Gamma(s)\, L(s)$ extends to a holomorphic function on $\mathbb{C}$ and satisfies the functional equation $L^\star(s) = (-1)^{\frac{k}{2}}\, L^\star(k - s)$.*

Then the function $f : i(0,\infty) \to \mathbb{C}$ on the upper imaginary axis defined by $f := \mathcal{M}^{-1}(L^\star)$ satisfies

$$(iy)^k f(iy) = f\left(\frac{-1}{iy}\right) \qquad (y > 0).$$

Proof We already established in the proof of Lemma 1.213 that the inverse Mellin transform $\mathcal{M}^{-1}(L^\star)$ is well defined. For $c > \frac{k}{2} + 1$, we have

$$\begin{aligned} f(iy) &= \mathcal{M}^{-1}(L^\star)(y) \\ &= \frac{1}{2\pi i} \int_{c-i\infty}^{c+i\infty} L^\star(s) y^{-s}\, \mathrm{d}s \end{aligned}$$

for all $y > 0$. Applying the functional equation $L^\star(s) = (-1)^{\frac{k}{2}} L^\star(k-s)$, we get

$$\begin{aligned} f(iy) &= \frac{1}{2\pi i} \int_{c-i\infty}^{c+i\infty} L^\star(s)\, y^{-s}\, \mathrm{d}s \\ &\overset{(1.162)}{=} \frac{(-1)^{\frac{k}{2}}}{2\pi i} \int_{c-i\infty}^{c+i\infty} L^\star(k-s) y^{-s}\, \mathrm{d}s \\ &= \frac{-(-1)^{\frac{k}{2}}}{2\pi i} \int_{k-c+i\infty}^{k-c-i\infty} L^\star(s) y^{s-k}\, \mathrm{d}s \\ &= y^{-k} \frac{(-1)^{\frac{k}{2}}}{2\pi i} \int_{k-c-i\infty}^{k-c+i\infty} L^\star(s) \left(\frac{1}{y}\right)^{-s} \mathrm{d}s \end{aligned}$$

for all $y > 0$. The integral on the right-hand side is again interpreted as the inverse Mellin transform, integrating along the path $(k-c) + i\mathbb{R}$ instead of $c + i\mathbb{R}$. Hence, we find

$$\begin{aligned} f(iy) &= y^{-k} \frac{(-1)^{\frac{k}{2}}}{2\pi i} \int_{k-c-i\infty}^{k-c+i\infty} L^\star(s) \left(\frac{1}{y}\right)^{-s} \mathrm{d}s \\ &= (-1)^{\frac{k}{2}} y^{-k} \mathcal{M}^{-1}(L^\star) \left(\frac{1}{y}\right) \\ &= (-1)^{\frac{k}{2}} y^{-k} f\left(i\frac{1}{y}\right) \end{aligned}$$

for all $y > 0$. Since k is even, we write $-1 = i^{-2}$ and get

$$f(iy) = (iy)^{-k} f\left(\frac{-1}{iy}\right) \qquad (y > 0).$$

□

Lemma 1.215 *Let $f : i(0,\infty) \to \mathbb{C}$ be a function given by the Fourier expansion*

$$f(iy) = \sum_{n=1}^{\infty} a_n\, e^{-2\pi n y} \qquad (y > 0)$$

such that the series $\sum_{n=1}^{\infty} a_n\, e^{-2\pi n y}$ converges absolutely for all $y > 0$.

Then f extends holomorphically to a function $\tilde{f} : \mathbb{H} \to \mathbb{C}$ given by

$$\tilde{f}(z) = \sum_{n=1}^{\infty} a_n\, e^{2\pi i n z} \qquad (z \in \mathbb{H}).$$

The function f satisfies $\tilde{f}(z+1) = \tilde{f}(z)$ for all $z \in \mathbb{H}$.

Proof Consider the following estimate based on the triangular inequality

$$\left|\sum_{n=1}^{\infty} a_n\, e^{2\pi i n z}\right| \le \sum_{n=1}^{\infty} |a_n|\, e^{-2\pi n y},$$

where $z = x + iy \in \mathbb{H}$ with real x and $y > 0$. Since $\sum_{n=1}^{\infty} a_n\, e^{-2\pi n y}$ converges absolutely for every $y > 0$, we see that the right-hand side and, hence, the left-hand side of the above inequality converge. Therefore,

$$\tilde{f} : \mathbb{H} \to \mathbb{C}; \quad z \mapsto \tilde{f}(z) := \sum_{n=1}^{\infty} a_n\, e^{2\pi i n z}$$

is well defined and forms a holomorphic function on $\mathbb{H}$ (since no dependency of $\bar{z}$ appears). The periodicity of $\tilde{f}$ is obvious from the Fourier expansion (using $e^{2\pi i n} = 1$). □

Collecting all results from the previous three lemmas we get the following converse statement.

Proposition 1.216 *For $k \in 2\mathbb{N}$, let $L(s) = \sum_{n\in\mathbb{N}} a_n n^{-s}$ be an L-series that is absolutely convergent on the half-plane $\Re(s) > \frac{k}{2} + 1$. Assume that its associated L-function $L^\star(s) := (2\pi)^{-s}\Gamma(s)\, L(s)$ extends to a holomorphic function on $\mathbb{C}$ and satisfies the functional equation $L^\star(s) = (-1)^{\frac{k}{2}} L^\star(k-s)$.*

Then, the function $f : \mathbb{H} \to \mathbb{C}$ defined by

$$f(z) = \sum_{n=1}^{\infty} a_n\, e^{2\pi i n z}$$

is a cusp form of weight k with trivial multiplier on the full modular group.

Proof Lemma 1.213 shows that $f = \mathcal{M}^{-1}(L^\star)$ is well defined on the upper imaginary axis and has the required Fourier expansion. Lemma 1.215 shows that f extends holomorphically to $\mathbb{H}$ and satisfies $f(z+1) = f(z)$ for all $z \in \mathbb{H}$. Lemma 1.214 shows the relation $j(T,z)^k\, f(z) = f(T\,z)$ for all $z \in i(0,\infty)$ on the upper imaginary axis. This relation holds for all $z \in \mathbb{H}$ since f is holomorphic (by the unique continuation principle, see Appendix A.8). Applying Theorem 1.54, f satisfies the required transformation property (1.43). Hence, $f \in S_k\big(\Gamma(1)\big)$. □

Collecting Propositions 1.211 and 1.216, we immediately have the following:

Corollary 1.217 (Hecke's converse theorem [93]) *Let $k \in 2\mathbb{N}$. The following two spaces are equivalent:*

1. $S_k\big(\Gamma(1)\big)$.
2. *The space of L-series $L(s)$ that converges absolutely on the half-plane $\Re(s) > \frac{k}{2}+1$ such that its associated L-function*

$$L^\star(s) := (2\pi)^{-s}\,\Gamma(s)\,L(s)$$

extends to a holomorphic function on $\mathbb{C}$ and satisfies the functional equation

$$L^\star(s) = (-1)^{\frac{k}{2}}\, L^\star(k-s).$$

The maps given by

$$S_k\big(\Gamma(1)\big) \ni f \mapsto \int_0^\infty F(iy)\, y^{s-1}\, dy \overset{(1.159)}{=} L^\star$$

and

$$L^\star \mapsto \mathcal{M}^{-1}\big(L^\star\big)$$

given by the Mellin transform (1.159) *and inverse Mellin transform* (1.164) *respectively are isomorphisms between the two spaces.*

Remark 1.218 An interesting question is whether a type of Hecke's converse theorem exists for subgroups $\Gamma \subset \Gamma(1)$ of finite index. Only partial results exist on this topic. For example, regarding Hecke congruence subgroups, there exists Weil's converse theorem [52, Section 1.5].

1.7.3 L-Functions of Hecke Eigenforms

We assume the following setting throughout this section. Let $k \in 2\mathbb{N}$ and consider a normalized Hecke eigenform $f \in M_k\big(\Gamma(1)\big)$. This means, in particular, that f satisfies

$$T_n f = a(n) f \qquad (n \in \mathbb{N}),$$

where $a(n)$ is the nth Fourier coefficient in the expansion (1.157). The most important aspect of eigenforms is that its L-series can be written as an Euler product and this is due to the multiplicative property of its Fourier coefficients.

Using Theorem 1.184, we get the following:

Corollary 1.219 *For $k \in 2\mathbb{N}$ and let $f \in M_k(\Gamma(1))$ be a normalized Hecke eigenform with Fourier coefficients $a(n)$. Then, the Fourier coefficients satisfy*

$$a(m)a(n) = \sum_{0<d|\gcd(m,n)} d^{k-1} a\left(\frac{mn}{d^2}\right) \tag{1.165}$$

for all $m, n \in \mathbb{N}$.

Proof We simply apply Theorem 1.184:

$$\begin{aligned} a(m)\, a(n)\, f &= T_m T_n f(z) \\ &\overset{(1.148)}{=} \sum_{0<d|\gcd(m,n)} d^{k-1}\, T_{\frac{mn}{d^2}} f \\ &= \sum_{0<d|\gcd(m,n)} d^{k-1}\, a\left(\frac{mn}{d^2}\right) f. \end{aligned}$$

for all $m, n \in \mathbb{N}$. □

Corollary 1.220 *For $k \in 2\mathbb{N}$, let $f \in M_k(\Gamma(1))$ be a normalized Hecke eigenform with Fourier coefficients $c(n)$. Let p be a prime. Then, the Fourier coefficients satisfy*

$$\begin{aligned} &a(p^n m) = a(p^n)a(m) \qquad \text{for all } m, n \in \mathbb{N} \text{ with } \gcd(m, p) = 1 \text{ and} \\ &a(p^{n+1}) = a(p)\, a(p^n) - p^{k-1}a(p^{n-1}) \qquad \text{for all } n \in \mathbb{N}. \end{aligned} \tag{1.166}$$

Proof Using Corollary 1.219, we calculate the coefficients. For all $m, n \in \mathbb{N}$ with $\gcd(m, p) = 1$, we get

$$a(p^n)a(m) \overset{(1.165)}{=} \sum_{0<d|\gcd(p^n,m)} d^{k-1}\, a\left(\frac{p^n m}{d^2}\right) \overset{p\nmid m}{=} a(p^n m).$$

Next consider $a(p)\, a(p^n)$ for $n \in \mathbb{N}$. We have

$$a(p)\, a(p^n) \overset{(1.165)}{=} \sum_{0<d|\gcd(p,p^n)} d^{k-1}\, a\left(\frac{pp^n}{d^2}\right) = a(pp^n) + p^{k-1}a(p^{n-1}).$$

This shows the second identity in (1.166). □

Lemma 1.221 ([116, p. 21]) *Let $a(n)$, $n \in \mathbb{N}$, be a sequence of numbers and let p be a prime. The power series identity*

$$\frac{1}{1-a(p)X+p^{k-1}X^2}=\sum_{i=0}^{\infty}a\left(p^i\right)X^i \tag{1.167}$$

is true if and only if $a(n)$ satisfies

$$a(1)=1 \qquad \textit{and} \qquad a\left(p^{i+1}\right)=a(p)\,a\left(p^i\right)-p^{k-1}a\left(p^{i-1}\right) \quad \textit{for all } i\in\mathbb{N}. \tag{1.168}$$

Proof Assume that $a(n)$ satisfies (1.168). Then,

$$\begin{aligned}
&\left(1-a(p)X+p^{k-1}X^2\right)\sum_{i=0}^{\infty}a\left(p^i\right)X^i\\
&= \sum_{i=0}^{\infty}a\left(p^i\right)X^i-\sum_{i=0}^{\infty}a(p)a\left(p^i\right)X^{i+1}+\sum_{i=0}^{\infty}p^{k-1}a\left(p^i\right)X^{i+2}\\
&= a(1)+a(p)X+\left(\sum_{i=2}^{\infty}a\left(p^i\right)X^i\right)-a(p)a(1)X-\\
&\quad -\left(\sum_{i=1}^{\infty}a(p)a\left(p^i\right)X^{i+1}\right)+\sum_{i=0}^{\infty}p^{k-1}a\left(p^i\right)X^{i+2}\\
&\overset{a(1)=1}{=} 1+\left(\sum_{i=2}^{\infty}a\left(p^i\right)X^i\right)-\left(\sum_{i=1}^{\infty}\left(a(p)a\left(p^i\right)-p^{k-1}a\left(p^{i-1}\right)\right)X^{i+1}\right)\\
&\overset{(1.167)}{=} 1+\left(\sum_{i=2}^{\infty}a\left(p^i\right)X^i\right)-\left(\sum_{i=1}^{\infty}a\left(p^{i+1}\right)X^{i+1}\right)\\
&= 1.
\end{aligned}$$

Hence, (1.167) holds.

For the converse, assume (1.167) holds. Hence,

$$\begin{aligned}
1&=\left(1-a(p)X+p^{k-1}X^2\right)\sum_{i=0}^{\infty}a\left(p^i\right)X^i\\
&=\sum_{i=0}^{\infty}a\left(p^i\right)X^i-\sum_{i=0}^{\infty}a(p)a\left(p^i\right)X^{i+1}+\sum_{i=0}^{\infty}p^{k-1}a\left(p^i\right)X^{i+2}
\end{aligned}$$

$$= \left(a(0) + a(p)X + \sum_{i=2}^{\infty} a\left(p^i\right) X^i\right) - \left(a(p)a(1)X + \sum_{i=1}^{\infty} a(p)a\left(p^i\right) X^{i+1}\right) +$$
$$+ \sum_{i=0}^{\infty} p^{k-1} a\left(p^i\right) X^{i+2}$$
$$= a(0) + \left(a(p) - a(p)a(1)\right)X + \sum_{i=2}^{\infty} \left(a\left(p^i\right) - a(p)a\left(p^{i-1}\right) + p^{k-1}a\left(p^{i-2}\right)\right) X^i.$$

Comparing the coefficients of the formal power series on the left-hand side and the right-hand side gives $a(0) = 1$ in the constant X^0 term, $a(p) = a(p)a(1)$ in the term X, and $a\left(p^i\right) = a(p)a\left(p^{i-1}\right) + p^{k-1}a\left(p^{i-2}\right)$ for all $i \geq 2$. Hence the $a(n)$'s satisfy (1.168). □

Now consider the L-series associated with a normalized Hecke eigenform f. We have the following result.

Proposition 1.222 *For $k \in 2\mathbb{N}$, let $f \in S_k(\Gamma(1))$ be a cusp form with associated L-series $L(s)$ defined in* (1.153).

If f is a normalized Hecke eigenform, then the associated L-series $L(s)$ admits the Euler product

$$L(s) = \prod_{p \text{ prime}} \left(1 - a(p)p^{-s} + p^{k-1-2s}\right)^{-1} \qquad \left(\Re(s) > \frac{k}{2} + 1\right), \tag{1.169}$$

where the product runs over the set of all prime numbers p.

Proof Each $n \in \mathbb{N}$ can be uniquely written as

$$n = p_1^{i_1} \cdots p_l^{i_l}$$

for some $l \in \mathbb{Z}_{\geq 0}$, $i_1, \ldots i_l \in \mathbb{N}$, and $p_1, \ldots, p_l$ distinct primes, where the case $l = 0$ is understood as the empty product 1.

Let $f \in S_k(\Gamma(1))$ be a normalized Hecke eigenform and consider the series $\sum_{i=0}^{\infty} \frac{a(p^i)}{p^{si}}$.

Using the identities derived in Corollary 1.220, we get

$$\begin{aligned} L(s) &= \sum_{n \in \mathbb{N}} \frac{a(n)}{n^s} \\ &= a(1) + \sum_{l \in \mathbb{N}} \sum_{\substack{i_1, \ldots, i_l \in \mathbb{N} \\ p_1, \ldots, p_l \text{distinct primes}}} \frac{a\left(p_1^{i_1} \cdots p_l^{i_l}\right)}{\left(p_1^{i_1} \cdots p_l^{i_l}\right)^s} \end{aligned}$$

$$\overset{(1.166)}{=} a(1) + \sum_{l\in\mathbb{N}} \sum_{\substack{i_1,\ldots,i_l\in\mathbb{N} \\ p_1,\ldots,p_l \text{ distinct primes}}} \frac{a\left(p_1^{i_1}\right)}{\left(p_1^{i_1}\right)^s} \cdots \frac{a\left(p_l^{i_l}\right)}{\left(p_l^{i_l}\right)^s}$$

$$= \prod_{p \text{ prime}} \sum_{i\in\mathbb{Z}_{\geq 0}} \frac{a\left(p^i\right)}{\left(p^i\right)^s},$$

where we used $a(1) = 1$ for normalized Hecke eigenforms. Applying Lemma 1.221, we find

$$L(s) = \prod_{p \text{ prime}} \sum_{i\in\mathbb{Z}_{\geq 0}} \frac{a\left(p^i\right)}{p^{is}}$$

$$\overset{(1.167)}{=} \prod_{p \text{ prime}} \frac{1}{1 - a(p)p^{-s} + p^{k-1}p^{-2s}}$$

for all s with $\Re(s) > \frac{k}{2} + 1$. This proves (1.169). □

Does the converse of Proposition 1.222 also hold? We will derive a partial result.

Definition 1.223 A sequence $a(n)$, $n \in \mathbb{N}$ of (complex) numbers is called *multiplicative* if $a(mn) = a(m)\,a(n)$ holds for all coprime $m, n \in \mathbb{N}$. It is called *completely multiplicative* if $a(mn) = a(m)\,a(n)$ holds for all $m, n \in \mathbb{N}$.

Proposition 1.224 *Let $k \in 2\mathbb{N}$ and assume that $a(n)$, $n \in \mathbb{N}$, is a nonvanishing multiplicative sequence of complex numbers such that the associated L-series $L(s)$ defined in* (1.153) *belongs to the second space described in Corollary 1.217. (Nonvanishing means that $a(n) \neq 0$ for at least one $n \in \mathbb{N}$.)*

If $L(s)$ additionally admits the Euler product (1.169)*, then*

$$f : \mathbb{H} \to \mathbb{C}; \quad f(z) := \sum_{n\in\mathbb{N}} a(n)\, e^{2\pi n z}$$

is a normalized Hecke eigenform with $f \in S_k\big(\Gamma(1)\big)$.

Proof Corollary 1.217 shows $f \in S_k\big(\Gamma(1)\big)$. It remains to show that f is indeed a normalized Hecke eigenform.

$a(n)$ being multiplicative implies

$$a(n) = a(1\,n) = a(1)\,a(n) \qquad \text{for all } n \in \mathbb{N}.$$

Hence, either $a(1) = 1$ or $a(n) = 0$ for all $n \in \mathbb{N}$. The second option contradicts the nonvanishing condition of $a(n)$.

We may expand the L-series as in the first step of the proof of Proposition 1.222:

$$
\begin{aligned}
L(s) &= \sum_{n\in\mathbb{N}} \frac{a(n)}{n^s} \\
&= a(1) + \sum_{l\in\mathbb{N}} \sum_{\substack{i_1,\dots,i_l\in\mathbb{N} \\ p_1,\dots,p_l \text{distinct primes}}} \frac{a\left(p_1^{i_1}\cdots p_l^{i_l}\right)}{\left(p_1^{i_1}\cdots p_l^{i_l}\right)^s} \\
&= a(1) + \sum_{l\in\mathbb{N}} \sum_{\substack{i_1,\dots,i_l\in\mathbb{N} \\ p_1,\dots,p_l \text{ distinct primes}}} \frac{a\left(p_1^{i_1}\right)}{\left(p_1^{i_1}\right)^s} \cdots \frac{a\left(p_l^{i_l}\right)}{\left(p_l^{i_l}\right)^s}
\end{aligned}
$$

($a(n)$ is multiplicative)

$$
= \prod_{p \text{ prime}} \sum_{i\in\mathbb{Z}_{\geq 0}} \frac{a\left(p^i\right)}{\left(p^i\right)^s},
$$

where we used in the last step that $a(1) = 1$.

On the other hand, we know that the L-series admits an Euler product (1.169). Hence

$$
\prod_{p \text{ prime}} \sum_{i\in\mathbb{Z}_{\geq 0}} \frac{a\left(p^i\right)}{\left(p^i\right)^s} \stackrel{!}{=} \prod_{p \text{ prime}} \left(1 - a(p)p^{-s} + p^{k-1-2s}\right)^{-1} \qquad \left(\Re(s) > \frac{k}{2} + 1\right).
$$

Replacing p^s with the formal variable X shows that the sequence $a(n)$ satisfies the identity (1.167). Applying Lemma 1.221 now implies that $a(n)$ satisfies (1.168). Hence

$$
a(m)\,a(n) = \sum_{0<d|\gcd(m,n)} d^{k-1}\, a\left(\frac{mn}{d^2}\right) \qquad \text{for all } m, n \in \mathbb{N}.
$$

Applying Theorem 1.188 shows that f is indeed a normalized simultaneous Hecke eigenform. □

Remark 1.225 In [60, Lemma 2.4] Conrey, Farmer, and Imamoglu obtained the following estimates.

For $k \in 2\mathbb{N}$, let $F \in S_k(\Gamma(1))$ be a normalized Hecke eigenform and let $L(s)$ be its associated L-series. Then, for $\Re(s) \geq \frac{3k}{4}$ we have

$$
|L(s) - 1| \leq 5^{-\frac{k}{4}}, \tag{1.170}
$$

and for $\Re(s) \geq \frac{k}{2}$ we have

$$
L(s) \leq 1 + 2\sqrt{k}\log(2k). \tag{1.171}
$$

1.8 Concluding Remarks

The above recollection of results related to modular forms might give the impression that the theory of modular forms is some obscure area of complex analysis. William A. Stein made the same observation in [188, §1.5] and states briefly six interesting applications of modular forms. For example, Wiles's proof [204] of Fermat's last theorem uses modular forms extensively and showed new techniques for solving certain diophantine equations.

An useful resource is the *L-functions and Modular Forms Database* [124]. This database contains lots of numerical information about modular forms and other related objects. For example, the discriminant function Δ defined in §1.3.5 is listed in the *L-functions and Modular Forms Database* under [124, Newform orbit 1.12.a.a], which means level 1, weight 12, character orbit 1.a (trivial), see also https://www.lmfdb.org/ModularForm/GL2/Q/holomorphic/1/12/a/a/.

Chapter 2
Period Polynomials

In this chapter, we discuss period polynomials that emerge from Eichler integrals on the full modular group. Period polynomials and their coefficients have significant impact on the theory of modular forms. In fact, the coefficients of these polynomials are certain values of L-functions. Moreover, by defining the cohomology of period polynomials, one can show that there is an isomorphism between two copies of the space of modular forms and the cohomology group. This is known as the Eichler-Shimura isomorphism. We give all the basic properties of period polynomials and prove the Eichler-Shimura isomorphism theorem. We continue by describing the action of Hecke operators on period polynomials, following an approach by Choie and Zagier [57]. We omit some of the proofs due to the abundance of references in the literature or, in many cases, because we prove their analogues in the Maass forms case.

2.1 Various Ways to Period Polynomials

This section introduces different approaches towards period polynomials. The most direct way, in our opinion, is to introduce period polynomials as a special integral transformation of modular cusp forms. We present this approach in §2.1.1. Other approaches use Eichler integrals, the $(k-1)$-fold antiderivative, other integral representations for the Eichler integrals, and the Niebur integral representation in the subsequent subsections §2.1.2–§2.1.6.

T. Mühlenbruch, W. Raji, *On the Theory of Maass Wave Forms*, Universitext,
https://doi.org/10.1007/978-3-030-40475-8_2

2.1.1 Period Polynomials via Integral Transformations

Definition 2.1 For a cusp form $F \in S_k(\Gamma(1))$, we call the function $P = P_F$ defined by the integral transformation

$$P(X) := P_F(X) := \int_0^{i\infty} (z - X)^{k-2} F(z)\, \mathrm{d}z \qquad (X \in \mathbb{C}) \tag{2.1}$$

the *period polynomial* P *of* F (for $\Gamma(1)$). The path of integration is the upper imaginary axis $i\mathbb{R}_{>0}$, connecting the cusp 0 to the cusp ∞.

Let $\mathbb{C}[X]_{k-2}$ denote the set of polynomials in X of degree at most $k-2$.

Lemma 2.2 *The integral transformation in* (2.1) *is well defined and the resulting period polynomial* P *is indeed a polynomial. The degree of* P *is smaller than or equal to* $k-2$. *Thus* $P(X) \in \mathbb{C}[X]_{\leq k-2}$.

Proof Recall that $F \in S_k(\Gamma(1))$. To show that the integral exists, we split the path of integration at i and consider the two integrals separately.

Consider $\int_i^{i\infty} (z-X)^{k-2} F(z)\, \mathrm{d}z$. Since F satisfies the growth condition derived in Exercise 1.67, we see that F decays exponentially at the cusp $i\infty$. Hence the integrand decays exponentially fast and the integral exists.

Since F satisfies the transformation property (1.28), we have in particular

$$z^k F(z) = F\left(\frac{-1}{z}\right).$$

Using the substitution $z \mapsto \frac{-1}{z}$, we get

$$\begin{aligned}
\int_0^{i} (z - X)^{k-2} F(z)\, \mathrm{d}z &\overset{z\mapsto Tz}{=} -\int_{i\infty}^{i} \left(\frac{-1}{z} - X\right)^{k-2} F\left(\frac{-1}{z}\right) z^{-2} \mathrm{d}z \\
&\overset{(1.28)}{=} \int_i^{i\infty} \left(\frac{-1}{z} - X\right)^{k-2} z^{k-2} F(z)\, \mathrm{d}z \\
&= \int_i^{i\infty} (-1 - zX)^{k-2} F(z)\, \mathrm{d}z.
\end{aligned}$$

Again, the growth condition derived in Exercise 1.67 ensures the existence of the integral for all $X \in \mathbb{C}$.

As a result, we see that

$$\int_0^{i\infty} (z - X)^{k-2} F(z)\, \mathrm{d}z = \int_0^{i} (z - X)^{k-2} F(z)\, \mathrm{d}z + \int_i^{i\infty} (z - X)^{k-2} F(z)\, \mathrm{d}z$$

exists for all $X \in \mathbb{C}$.

Since $(z-X)^{k-2}$ is a polynomial in X of degree $k-2$, we also see that

$$\int_0^{i\infty} (z-X)^{k-2}\, F(z)\, \mathrm{d}z \in \mathbb{C}[X]_{k-2}.$$

□

Remark 2.3 Recall the binomial formula

$$(z-X)^{k-2} = \sum_{l=0}^{k-2} \binom{k-2}{l} z^l\, X^{k-l-2}.$$

Applying this identity to the integrand in (2.1) and comparing the result with (1.159), we see directly that period polynomials and L-functions are related. We discuss this connection later.

We used in the proof above that the term $(z-X)^{k-2}$ has certain transformation properties under $z \mapsto Tz$ with T given in (1.20). The next lemma shows that these properties essentially extend to all $M \in \mathrm{SL}_2(\mathbb{R})$.

Lemma 2.4 *Let $k \in 2\mathbb{Z}$ and X denote a complex variable. We have*

$$\big(Mz - MX\big)^{k-2} = \frac{(z-X)^{k-2}}{j(M,z)^{k-2}\, j(M,X)^{k-2}} \tag{2.2}$$

for all $M \in \mathrm{SL}_2(\mathbb{R})$, $z \in \mathbb{H}$, and $X \in \mathbb{C}$ provided that $j(M,X) \neq 0$.

Proof We leave the proof to the reader as an exercise. □

A direct application is the following transformation behavior for period polynomials.

Lemma 2.5 *Let $k \in 2\mathbb{N}$, $F \in S_k\big(\Gamma(1)\big)$, and $M \in \Gamma(1)$. We have*

$$P_F\big|_{2-k}M = \int_{M^{-1}0}^{M^{-1}i\infty} (z-X)^{k-2}\, F(z)\, \mathrm{d}z \qquad (X \in \mathbb{C}). \tag{2.3}$$

The path of integration is the geodesic line $M^{-1}i\mathbb{R}_{>0}$ connecting the cusp $M^{-1}0$ to the cusp $M^{-1}i\infty$.

Proof Consider $F \in M_k\big(\Gamma(1)\big)$ for $k \in 2\mathbb{N}$ and trivial multiplier. Using (2.1) and a suitable substitution we find

$$\begin{aligned} P_F\big|_{2-k}M(X) &\overset{(1.131)}{=} j(M,X)^{k-2}\, P_F(M\,X) \\ &\overset{(2.1)}{=} j(M,X)^{k-2} \int_0^{i\infty} (z - MX)^{k-2}\, F(z)\, \mathrm{d}z \end{aligned}$$

$$\overset{z\mapsto Mz}{=} j(M,X)^{k-2}\int_{M^{-1}0}^{M^{-1}i\infty}(Mz-MX)^{k-2}\,F(Mz)\,j(M,z)^{-2}\,\mathrm{d}z$$

$$\overset{(2.2)}{=}\int_{M^{-1}0}^{M^{-1}i\infty}(z-X)^{k-2}\,j(M,z)^{-k}F(Mz)\,\mathrm{d}z$$

$$\overset{(1.28)}{=}\int_{M^{-1}0}^{M^{-1}i\infty}(z-X)^{k-2}\,F(z)\,\mathrm{d}z$$

for all $X\in\mathbb{C}$. □

We now derive consistency relations of period polynomials and describe their behavior on the generators of the full group.

Lemma 2.6 *For $k\in 2\mathbb{N}$, consider a cusp form $F\in S_k\big(\Gamma(1)\big)$. The associated period polynomial P_F satisfies*

$$P_F+P_F\big|_{2-k}T=0 \quad \textit{and} \quad P_F+P_F\big|_{2-k}(TS)+P_F\big|_{2-k}(TS)^2=0, \tag{2.4}$$

where S and T are given in (1.20).

Proof Recall the relations $T^2=-\mathbf{1}=(TS)^3$ as shown in Remark 1.12. We consider the relation $T^2=-\mathbf{1}$ first. Since $T=\begin{pmatrix}0&-1\\1&0\end{pmatrix}$ and $T^2=\begin{pmatrix}-1&0\\0&-1\end{pmatrix}=-\mathbf{1}$, we find $T^{-1}0=i\infty$ and $T^{-1}i\infty=0$. The cusps 0 and $i\infty$ are mapped onto each other and the path of integration is reversed. Hence,

$$\begin{aligned}P_F(X)&\overset{(2.1)}{=}\int_0^{i\infty}(z-X)^{k-2}\,F(z)\,\mathrm{d}z\\&=-\int_{T^{-1}0}^{T^{-1}i\infty}(z-X)^{k-2}\,F(z)\,\mathrm{d}z\\&\overset{(2.3)}{=}-\big(P_F\big|_{2-k}T\big)(X)\end{aligned}$$

for all $X\in\mathbb{C}$.

Now consider the matrices $TS=\begin{pmatrix}0&-1\\1&1\end{pmatrix}$ and $(TS)^2=\begin{pmatrix}-1&-1\\1&0\end{pmatrix}$. Applying the Möbius transformations gives

$$(TS)^{-1}\,0=i\infty,\quad (TS)^{-2}\,0=-1,\quad (TS)^{-1}\,i\infty=-1\quad\text{and}\quad (TS)^{-2}\,i\infty=0.$$

Hence the path of integration $i\mathbb{R}_{>0}$ connecting the cusp 0 to $i\infty$ can be transformed to

$$i\mathbb{R}_{>0}=(TS)^{-1}\,i\mathbb{R}_{>0}\cup(TS)^{-2}\,i\mathbb{R}_{>0}.$$

Now, we use the fact that the integrand is holomorphic in z and vanishes at the cusps. So the integral can be written as

$$\begin{aligned}
P_F(X) &\overset{(2.1)}{=} \int_0^{i\infty} (z-X)^{k-2}\, F(z)\, dz \\
&= \int_0^{-1} (z-X)^{k-2}\, F(z)\, dz + \int_{-1}^{i\infty} (z-X)^{k-2}\, F(z)\, dz \\
&= -\int_{(TS)^{-2}\,0}^{(TS)^{-2}\,i\infty} (z-X)^{k-2}\, F(z)\, dz - \int_{(TS)^{-1}\,0}^{(TS)^{-1}\,i\infty} (z-X)^{k-2}\, F(z)\, dz \\
&\overset{(2.3)}{=} -\big(P_F\big|_{2-k}(TS)^2\big)(X) - \big(P_F\big|_{2-k}TS\big)(X)
\end{aligned}$$

for all $X \in \mathbb{C}$. □

The above lemma inspires the following definition, extending the notion of period polynomials for $\Gamma(1)$.

Definition 2.7 Let $k \in 2\mathbb{N}$ and $P \in \mathbb{C}[X]_{k-2}$ be a polynomial of degree at most $k-2$. We call P a *period polynomial* P (for $\Gamma(1)$) if it satisfies the two functional equations

$$P + P\big|_{2-k}T = 0 \quad \text{and} \quad P + P\big|_{2-k}(TS) + P\big|_{2-k}(TS)^2 = 0. \tag{2.5}$$

We denote the space of such period polynomials by $\mathbb{P}_{k-2}$.

Exercise 2.8 Let $k \in 2\mathbb{N}$. Show that the polynomial $P(X) := X^{k-2} - 1$ is a period polynomial.

The next proposition explains the relation between L-functions and period polynomials, which we have already hinted at in Remark 2.3.

Proposition 2.9 *Let $k \in 2\mathbb{N}$ and $F \in S_k\big(\Gamma(1)\big)$. The L-function $L^\star$ and the period function P, defined in* (1.159) *and* (2.1), *respectively, associated with F satisfy*

$$\begin{aligned}
P_F(X) &= \sum_{l=0}^{k-2} \binom{k-2}{l} (-X)^{k-l-2}\, i^{l+1}\, L^\star(l+1) \\
&= -\sum_{l=0}^{k-2} \binom{k-2}{l} X^l\, (-i)^{k-l-1}\, L^\star(k-l-1) \qquad (X \in \mathbb{C}).
\end{aligned} \tag{2.6}$$

Proof Using the binomial formula, we have

$$(z-X)^{k-2} = \sum_{l=0}^{k-2} \binom{k-2}{l} z^l\, (-X)^{k-l-2}.$$

Thus, we get

$$\begin{aligned}
P(X) &\overset{(2.1)}{=} \int_0^{i\infty} (z-X)^{k-2}\, F(z)\, \mathrm{d}z \\
&= \int_0^{i\infty} \sum_{l=0}^{k-2} \binom{k-2}{l} z^l (-X)^{k-l-2}\, F(z)\, \mathrm{d}z \\
&= \sum_{l=0}^{k-2} \binom{k-2}{l} (-X)^{k-l-2} \int_0^{i\infty} z^l\, F(z)\, \mathrm{d}z \\
&\overset{z\mapsto iy}{=} \sum_{l=0}^{k-2} \binom{k-2}{l} i^{l+1} (-X)^{k-l-2} \int_0^{\infty} y^l\, F(iy)\, \mathrm{d}y \\
&\overset{(1.159)}{=} \sum_{l=0}^{k-2} \binom{k-2}{l} i^{l+1} (-X)^{k-l-2}\, L^\star(l+1),
\end{aligned}$$

assuming that the integrals on the right-hand sides exist. This was shown in Lemma 1.207.

The second identity in (2.6) follows from a similar calculation. Using the binomial formula,

$$(z-X)^{k-2} = (X-z)^{k-2} = \sum_{l=0}^{k-2} \binom{k-2}{l} X^l\, (-z)^{k-l-2},$$

since $k \in 2\mathbb{N}$. □

Exercise 2.10 Let $k \in 2\mathbb{N}$ and $F \in S_k\big(\Gamma(1)\big)$. Show that the L-series and the period function P, defined in (1.153) and (2.1), respectively, associated with F satisfy

$$P_F(X) = -\sum_{l=0}^{k-2} \frac{(k-2)!}{l!} \frac{L(k-l-1)}{(2\pi i)^{k-l-1}}\, X^l \qquad (X \in \mathbb{C}). \tag{2.7}$$

2.1.2 *Eichler Integrals and Period Polynomials*

There is another connection between period polynomials and Eichler integrals. Eichler integrals can be seen as generalizations of abelian integrals. We introduce abelian integrals to give the reader another perspective on the theory.

Definition 2.11 Let $F : \mathbb{H} \to \mathbb{C}$ be a meromorphic function on the upper half-plane that is meromorphic at the infinite cusp. Then, $F(z)$ is called an *abelian integral* if there exists a sequence of numbers $p_M \in \mathbb{C}$, $M \in \Gamma(1)$, such that

$$F\big(M\,z\big) = F(z) + p_M \tag{2.8}$$

for all $M \in \Gamma(1)$.

As an application, we now show that the antiderivative of a modular form of weight 2 on the full modular group is an abelian integral.

Definition 2.12 Let $f : \mathbb{H} \to \mathbb{C}$ have a Fourier expansion at ∞ of the form

$$f(z) = \sum_{n=-\mu}^{\infty} a(n) e^{2\pi i n z} \qquad (z \in \mathbb{H}).$$

An *antiderivative of* f is defined as

$$F(z) = \frac{a(-\mu)}{(-2\pi i \mu)} e^{-2\pi i \mu z} + \ldots + \frac{a(-1)}{(-2\pi i)} e^{-2\pi i z} + a(0) z + \sum_{n=1}^{\infty} \frac{a(n)}{(2\pi i n)} e^{2\pi i n z}$$

for all $z \in \mathbb{H}$.

Lemma 2.13 *Let $f \in M_2^!\big(\Gamma(1)\big)$ and assume that $f(z)$ is holomorphic on $\mathbb{H}$. Write its Fourier expansion at the cusp ∞ as*

$$f(z) = \sum_{n=-\mu}^{\infty} a(n) e^{2\pi i n z} \qquad (z \in \mathbb{H}).$$

Then the antiderivative F of f given in Definition 2.12 is an abelian integral.

Now we return back to Eichler integrals. One can easily see that Eichler integrals of weight 0 are actually abelian integrals.

Definition 2.14 Let $k \in 2\mathbb{N}$ and let $F : \mathbb{H} \to \mathbb{C}$ be a holomorphic function. The function F is called an *Eichler integral* of weight $2-k$ if there exists a sequence of polynomials $P_M(z) \in \mathbb{C}[z]_{k-2}$ of degree at most $k-2$ such that F satisfies the transformation property

$$j(M, z)^{k-2}\, F\big(M\,z\big) = F(z) + P_M(z) \tag{2.9}$$

for all $M \in \Gamma(1)$ and $z \in \mathbb{H}$.

Lemma 2.15 *Let $k \in 2\mathbb{N}$ and let F be an Eichler integral of weight $2-k$. The associated sequence of polynomials $\big(P_M\big)_{M\in\Gamma(1)}$ (see Definition 2.14) satisfies*

$$P_1 = P_{-1} = 0 \tag{2.10}$$

and

$$P_{MV} = P_M\big|_{2-k} V + P_V \tag{2.11}$$

for all $M, V \in \Gamma(1)$.

Proof Consider the transformation property (2.9) of Eichler integrals. Obviously we have

$$F(z) = j(\mathbf{1}, z)^{2-k}\, F(\mathbf{1}z) \quad \text{and} \quad F(z) = j(-\mathbf{1}, z)^{2-k}\, F\big((-\mathbf{1})z\big)$$

since $\mathbf{1}z = z$, $j(\mathbf{1}, z) = 1$, $(-\mathbf{1})z \overset{(1.13)}{=} \begin{pmatrix} -1 & 0 \\ 0 & -1 \end{pmatrix} z \overset{(1.16)}{=} z$, $j(-\mathbf{1}, z) = -1$, and $j(-\mathbf{1}, z)^{2-k} = 1$, since k is an even integer. Hence the transformation property (2.9) gives the following identities:

$$F(z) = F(z) + P_{\mathbf{1}}(z) \quad \text{and} \quad F(z) = F(z) + P_{-\mathbf{1}}(z).$$

This implies $P_{\mathbf{1}} = P_{-\mathbf{1}} = 0$.

For the last part, we start with the transformation property (1.32) of the automorphic factor j in Lemma 1.40 (since $2 - k \in 2\mathbb{Z}$). Applying (1.32) to $M, V \in \Gamma(1)$, we find

$$j(MV, z)^{2-k} = j(M, vz)^{2-k}\, j(V, z)^{2-k} \qquad (z \in \mathbb{H}).$$

Combining it with the transformation property (2.9) of Eichler integrals, we find

$$\begin{aligned}
F(z) + P_{MV}(z) &\overset{(2.9)}{=} j(MV, z)^{2-k}\, F(MV\, z) \\
&\overset{(1.32)}{=} j(V, z)^{2-k} \Big(j(M, Vz)^{2-k}\, F(M\, Vz) \Big) \\
&\overset{(2.9)}{=} j(V, z)^{2-k} \Big(F(Vz) + P_M(Vz) \Big) \\
&= j(V, z)^{2-k}\, F(Vz) + j(V, z)^{2-k}\, P_M(Vz) \\
&\overset{(2.9)}{=} F(z) + P_V(z) + j(V, z)^{2-k}\, P_M(Vz)
\end{aligned}$$

for all $z \in \mathbb{H}$. Subtracting $F(z)$ on both sides, we get

$$P_{MV}(z) = P_V(z) + j(V, z)^{2-k}\, P_M(Vz) \qquad (z \in \mathbb{H}).$$

□

Proposition 2.16 *Let $k \in 2\mathbb{N}$. There exists a map from the space of Eichler integrals of weight $2 - k$ that vanishes under S to the space $\mathbb{P}_{k-2}$ of period polynomials:*

$$F \mapsto P := P_T \overset{(2.9)}{=} F\big|_{2-k}(\mathbf{1} - T). \tag{2.12}$$

The kernel of the map is in the set of modular forms of weight $2 - k$ *for* $\Gamma(1)$.

Proof We prove the lemma in two steps. First, we show that $F - F\big|_{2-k}T$ is a period polynomial of degree at most $k-2$ provided that F is an Eichler integral of the same weight. Then we conclude by calculating the kernel.

Let F be an Eichler integral of weight $2 - k$. Then the transformation property (2.9) of Eichler integrals implies that $P := P_T$ given by

$$F\big|_{2-k}T = F + P_T$$

is a polynomial of degree at most $k - 2$. Using the matrix identity $T^2 = -\mathbf{1}$ and repeated application of (2.10) of Lemma 2.15, we find

$$0 = P_{-\mathbf{1}} = P_{T^2} \overset{(2.11)}{=} P_T\big|_{2-k}T + P_T.$$

Hence $P = P_T$ satisfies the left relation in (2.5) of Definition 2.7.

Similarly, using $(TS)^3 = -\mathbf{1}$ in (1.21) and repeatedly applying (2.11) of Lemma 2.15, we find

$$\begin{aligned}
0 &= P_{-\mathbf{1}} = P_{(TS)^3} \\
&\overset{(2.11)}{=} P_{TSTST}\big|_{k-2}S + P_S \\
&\overset{(2.11)}{=} \Big(P_{TSTS}\big|_{k-2}T + P_T\Big)\Big|_{k-2}S + P_S \\
&\overset{(1.133)}{=} P_{TSTS}\big|_{k-2}TS + P_T\big|_{k-2}S + P_S \\
&= P_T\big|_{k-2}STSTS + P_S\big|_{k-2}TSTS + P_T\big|_{k-2}STS + P_S\big|_{k-2}TS + P_T\big|_{k-2}S + P_S.
\end{aligned}$$

Since $P_S = F\big|_{2-k}(\mathbf{1} - S)$ vanishes, as we assumed in the proposition, we find

$$\begin{aligned}
0 &= P_T\big|_{k-2}STSTS + P_T\big|_{k-2}STS + P_T\big|_{k-2}S \\
&= P_T\big|_{k-2}S\big|_{2-k}\big((TS)^2 + TS + \mathbf{1}\big).
\end{aligned}$$

Hence $P_T\big|_{k-2}S$ vanishes under $(TS)^2 + TS + \mathbf{1}$. Using the matrix identity $(TS)^3 = -\mathbf{1}$, we see

$$S(TS)^2 = (-\mathbf{1})T, \qquad S(TS) = TS(-\mathbf{1})T \quad \text{and} \quad S = (TS)^2(-\mathbf{1})T.$$

Hence P_T satisfies

$$\begin{aligned}0 &= P_T\big|_{k-2}S\big|_{2-k}\big((TS)^2 + TS + \mathbf{1}\big)\\ &= P_T\big|_{2-k}\big(\mathbf{1} + TS + (TS)^2\big)\big|_{2-k}(-\mathbf{1})T.\end{aligned}$$

Since $(-\mathbf{1})T = \begin{pmatrix} -1 & -1 \\ 0 & -1 \end{pmatrix}$, we conclude that P_T satisfies the identity

$$\left(P_T\big|_{k-2}STST + P_T\big|_{k-2}ST + P_T\right)\Big|_{2-k} T = 0.$$

Using $j(S, X)^{k-2} = 1$ and $P = -P_T$, the above identity reads as

$$\left(P\big|_{k-2}STST + P\big|_{k-2}ST + P\right)(-X - 1) = 0.$$

This proves the right-hand relation in (2.5) of Definition 2.7. Summarizing, we just showed that $P = P_T$ is a period polynomial of degree at most $k - 2$.

Next, we calculate the kernel of the map $F \mapsto P := F - F\big|_{2-k}T$. An Eichler integral F is in the kernel if $F - F\big|_{2-k}M$ vanishes. Since $P_S = 0$ vanishes by assumption, we find that every F in the kernel satisfies the identities

$$F = F\big|_{2-k}T \qquad \text{and} \qquad F = F\big|_{2-k}S.$$

As a result,

$$F = F\big|_{2-k}M \qquad\qquad \text{for all } M \in \Gamma(1).$$

Also, F is holomorphic on $\mathbb{H}$. Hence F satisfies all criteria of a modular form of weight $2 - k$ for $\Gamma(1)$. In other words, if the Eichler integral F is in the kernel of the map (2.12), then $F \in M^!_{2-k}\big(\Gamma(1)\big)$. □

The remaining part of this section introduces different realizations of Eichler integrals, the $(k - 1)$-fold antiderivative, the integral $\int_z^{i\infty} f(w)\,(z - w)^{k-2}\,dw$ for $f \in S_k\big(\Gamma(1)\big)$, and the integral $\int_z^{i} f(w)\,(z - w)^{k-2}\,dw$ for $f \in M_k\big(\Gamma(1)\big)$.

2.1.3 The (k − 1)-Fold Antiderivative

Similar to the notion of the antiderivative, as introduced in Definition 2.12, we define the $(k - 1)$-fold antiderivative as follows:

Remark 2.17 In Definition 2.12, $F(z)$ is also known as the $(k - 1)$-fold antiderivative.

Exercise 2.18 Let $F : \mathbb{H} \to \mathbb{C}$ be a holomorphic function, $k \in 2\mathbb{N}$, and $M \in \Gamma(1)$. Prove *Bol's identity*

$$\frac{d^{k-1}}{dz^{k-1}}\left(j(M,z)^{k-2}\,F\big(M\,z\big)\right) = j(M,z)^{-k}\,F^{(k-1)}\big(M\,z\big) \qquad (z \in \mathbb{H}), \tag{2.13}$$

where $F^{(k-1)}$ denotes the $(k-1)^{\text{th}}$ derivative $F^{(k-1)}(z) = \frac{\mathrm{d}^{k-1}}{\mathrm{d}z^{k-1}}F(z)$ of F.

Remark 2.19 Bol's identity (2.13) goes back to G. Bol's paper [15, p. 28].

Lemma 2.20 *Let $k \in 2\mathbb{N}$ and $f \in M_k^!\big(\Gamma(1)\big)$. We also assume that $f(z)$ is holomorphic on $\mathbb{H}$ and that its Fourier expansion at the cusp ∞ is given by*

$$f(z) = \sum_{n=-\mu}^{\infty} a(n)e^{2\pi i n z} \qquad (z \in \mathbb{H}).$$

The $(k-1)$-fold antiderivative F of f given in Definition 2.12 is an Eichler integral of weight $2-k$.

Proof Since $f \in M_k^!\big(\Gamma(1)\big)$ by assumption, we get

$$j(M,z)^{-k}\,f\big(M\,z\big) - f(z) = 0$$

for all $M \in \Gamma(1)$ and $z \in \mathbb{H}$. Denote by F the $(k-1)$-fold antiderivative of f. The definition directly implies that

$$\frac{\mathrm{d}^{k-1}}{\mathrm{d}z^{k-1}}F(z) = f(z) \tag{2.14}$$

by interchanging summation and differentiation. Using Bol's identity in (2.13) and the chain rule, we find

$$\begin{aligned}\frac{\mathrm{d}^{k-1}}{\mathrm{d}z^{k-1}}\left(j(M,z)^{k-2}\,F\big(M\,z\big) - F(z)\right) &\overset{(2.13)}{=} j(M,z)^{-k}\,F^{(k-1)}\big(M\,z\big) - F^{(k-1)}(z)\\ &= j(M,z)^{-k}\,f\big(M\,z\big) - f(z)\\ &= 0.\end{aligned} \tag{2.15}$$

Integrating $k-1$ times the left-hand and right-hand sides of (2.15), we get

$$j(M,z)^{k-2}\,F\big(M\,z\big) - F(z) = p_M(z) \qquad (z \in \mathbb{H})$$

for suitable polynomials $p_M(z) \in \mathbb{C}_{\le k-2}[z]$ of degree at most $k-2$ and for all $M \in \Gamma(1)$. Hence, F satisfies condition (2.9) of Definition 2.14. This shows that F is an Eichler integral of weight $2-k$. □

Remark 2.21 The authors in [22] considered a generalized definition of the Eichler integral F given as the $(k-1)$-fold derivative in Definition 2.12. Assuming that $f \in M^!_{k,1}(\Gamma(1))$ has the Fourier expansion

$$f(z) = \sum_{n=-\nu}^{\infty} a_n\, e^{2\pi i n z},$$

they associate with f the *formal Eichler integral*

$$\mathcal{E}_f(z) := \sum_{0 \neq n = -\nu}^{\infty} \frac{a_n}{n^{k-1}}\, e^{2\pi i n z} \tag{2.16}$$

and use this definition to develop their theory.

In fact, the above definition of Eichler integral can be easily derived from the integral definition for integer weight k. However, we call ours a generalized definition because in addition one can define Eichler integrals as in (2.21) if one is working with weakly holomorphic modular forms. Another definition used for half-integer weight modular forms with slightly different approach from the classical approach, due to Lawrence and Zagier [119], is to study the non-holomorphic Eichler integral

$$F(z) := \int_{\overline{z}}^{i\infty} f(w)(w-z)^{k-2}\, dw,$$

where z now is in the lower half-plane $\mathbb{H}^-$ and with choosing the appropriate branch.

This diversion from the classical definition is the branching problems that one encounters when working with non-integer weight modular forms.

A vector-valued version of the $(k-1)$-fold derivative and its relation to Eichler integrals was also discussed in [87].

2.1.4 An Integral Representation of the Eichler Integral for Cusp Forms

Definition 2.22 Let $k \in 2\mathbb{N}$ and $f \in S_k(\Gamma(1))$. We define the function $\tilde{F} : \mathbb{H} \to \mathbb{C}$ by the integral transform

$$\tilde{F}(z) := \int_{z}^{i\infty} f(w)\,(w-z)^{k-2}\, dw \qquad (z \in \mathbb{H}). \tag{2.17}$$

The integral in (2.17) is well defined since the cusp form $f \in M_k(\Gamma(1))$ satisfies the exponential growth condition given in Exercise (1.67).

Lemma 2.23 *Let $k \in 2\mathbb{N}$ and $f \in S_k\big(\Gamma(1)\big)$.*

1. *The function $\tilde{F} : \mathbb{H} \to \mathbb{C}$ given in (2.17) is an Eichler integral of weight $2-k$.*
2. *If F denotes the $(k-1)$-fold antiderivative of f defined in Lemma 2.20, then F and $\tilde{F}$ satisfy*

$$-(k-2)!\,F(z) = \tilde{F}(z) \qquad (z \in \mathbb{H}). \tag{2.18}$$

Proof We leave the proof of the first part as an exercise to the reader.

For the second part, we first consider the derivative of $\tilde{F}$.

$$\begin{aligned}
\frac{\mathrm{d}}{\mathrm{d}z}\tilde{F}(z) &= \frac{\mathrm{d}}{\mathrm{d}z}\int_z^{i\infty} f(w)\,(w-z)^{k-2}\,\mathrm{d}w\\
&= \lim_{\varepsilon\searrow 0}\frac{1}{\varepsilon}\left(\int_{z+\varepsilon}^{i\infty} f(w)\,(w-z-\varepsilon)^{k-2}\,\mathrm{d}w - \int_z^{i\infty} f(w)\,(w-z)^{k-2}\,\mathrm{d}w\right)\\
&= \lim_{\varepsilon\searrow 0}\frac{1}{\varepsilon}\left(\int_z^{i\infty} f(w+\varepsilon)\,(w-z)^{k-2}\,\mathrm{d}w - \int_z^{i\infty} f(w)\,(w-z)^{k-2}\,\mathrm{d}w\right)\\
&= \int_z^{i\infty} \lim_{\varepsilon\searrow 0}\frac{1}{\varepsilon}\Big(f(w+\varepsilon) - f(w)\Big)\,(w-z)^{k-2}\,\mathrm{d}w\\
&= \int_z^{i\infty} f'(w)\,(w-z)^{k-2}\,\mathrm{d}w\\
&= f(w)\,(w-z)^{k-2}\Big|_{w=z}^{w=i\infty} - (k-2)\int_z^{i\infty} f(w)\,(w-z)^{k-3}\,\mathrm{d}w.
\end{aligned}$$

Since f, as a cusp form, satisfies the growth condition in Exercise 1.67, we see that the first term vanishes. Hence,

$$\frac{\mathrm{d}}{\mathrm{d}z}\tilde{F}(z) = -(k-2)\int_z^{i\infty} f(w)\,(w-z)^{k-3}\,\mathrm{d}w.$$

Iterating the argument, we find

$$\frac{\mathrm{d}^{k-2}}{\mathrm{d}z^{k-2}}\tilde{F}(z) = (-1)^{k-2}\,(k-2)!\int_z^{i\infty} f(w)\,\mathrm{d}w,$$

and applying it once again, we get

$$\begin{aligned}
\frac{\mathrm{d}^{k-1}}{\mathrm{d}z^{k-1}}\tilde{F}(z) &= (-1)^{k-2}\,(k-2)!\,\frac{\mathrm{d}}{\mathrm{d}z}\int_z^{i\infty} f(w)\,\mathrm{d}w\\
&= -(-1)^{k-2}\,(k-2)!\,f(z).
\end{aligned}$$

The last step uses $\frac{\mathrm{d}}{\mathrm{d}z}\int_z^{i\infty} f(w)\,\mathrm{d}w = -f(z)$. Hence

$$\frac{\mathrm{d}^{k-1}}{\mathrm{d}z^{k-1}}\tilde{F}(z) = -(k-2)!\; f(z)$$

since $k \in 2\mathbb{N}$. This equation and (2.14) together imply that

$$\frac{\mathrm{d}^{k-1}}{\mathrm{d}z^{k-1}}\tilde{F}(z), = -(k-2)!\;\frac{\mathrm{d}^{k-1}}{\mathrm{d}z^{k-1}}F(z),$$

where F is the $(k-1)$-fold antiderivative of f.

It remains to show that (2.18) holds. By $(k-1)$ times indefinite integration, we get

$$\tilde{F}(z) = -(k-2)!\; F(z) + P(z), \tag{2.19}$$

where $P(z) \in \mathbb{C}[z]_{k-2}$ is a polynomial of degree at most $k-2$ coming from the indefinite integration. To calculate the polynomial $P(z)$, we consider the limit as $z \to i\infty$. On the left-hand side, we find

$$\lim_{z\to i\infty}\tilde{F}(z) = \lim_{z\to i\infty}\int_z^{i\infty} f(w)\,(w-z)^{k-2}\,\mathrm{d}w = 0.$$

To calculate $\lim_{z\to i\infty} F(z)$, recall that f, being a cusp form, has a Fourier expansion at the cusp ∞ of the form

$$f(z) = \sum_{n=1}^{\infty} a(n)\,e^{2\pi i n z} \qquad (z \in \mathbb{H}).$$

Then the $(k-1)$-fold antiderivative of f is given by

$$F(z) = \sum_{n=1}^{\infty} \frac{a(n)}{(2\pi i n)^{k-1}}\,e^{2\pi i n z} \qquad (z \in \mathbb{H}).$$

The Fourier coefficients of F satisfy the estimate

$$\left|\frac{a(n)}{(2\pi i n)^{k-1}}\right| \le \left|\frac{a(n)}{n^{k-1}}\right| = O(1)$$

for all $n \in \mathbb{N}$. This implies that F can be estimated by f:

$$|F(z)| \le \sum_{n=1}^{\infty}\left|\frac{a(n)}{(2\pi i n)^{k-1}}\right| e^{-2\pi n \Im(z)} = O\left(e^{-2\pi\delta\Im(z)}\right)$$

as $\Im(z) \to \infty$ for some $\delta > 0$. Hence,

$$\lim_{z \to i\infty} F(z) = 0.$$

Together, both limits show that the additive polynomial $P(z)$ in (2.19) must be identically 0. This shows (2.18) and concludes the proof. □

Exercise 2.24 For $k \in 2\mathbb{N}$, let $f \in S_k(\Gamma(1))$ and $\tilde{F}$ be the Eichler integral defined by the integral (2.17). Show, using only the integral representation (2.17), that the $\tilde{F}$ associated with the period polynomial P in (2.12) admits the integral representation (2.1) and is a period polynomial. In particular, show

$$\tilde{F}\big|_{2-k}(T - \mathbf{1}) = -P_f \quad \text{and} \quad \tilde{F}\big|_{2-k}(S - \mathbf{1}) = 0, \tag{2.20}$$

where P_f is given in (2.1).

Remark 2.25 For $k \in 2\mathbb{N}$, let $f \in S_k(\Gamma(1))$ be a cusp form. Using the inclusions $S_k(\Gamma(1)) \subset S_k^!(\Gamma(1)) \subset M_k^!(\Gamma(1))$ we may associate with f the formal Eichler integral $\mathcal{E}_f(z)$ given in (2.25) of Remark 2.21. Then, $\mathcal{E}_f(z)$ can be interpreted as the Eichler integral $\tilde{F}(z)$ given in (2.17). We have

$$\begin{aligned} \mathcal{E}_f(z) &= -\frac{(2\pi i)^{k-1}}{\Gamma(k-1)} \int_z^{i\infty} f(z)\,(w - z)^{k-2}\,\mathrm{d}w \\ &= -\frac{(2\pi i)^{k-1}}{\Gamma(k-1)}\,\tilde{F}(z). \end{aligned} \tag{2.21}$$

This relation appears in [22, (1.8)].

2.1.5 *An Integral Representation of the Eichler Integral for Entire Forms*

Definition 2.26 Let $k \in 2\mathbb{N}$ and $f \in M_k(\Gamma(1))$. We define the function $\widehat{F} : \mathbb{H} \to \mathbb{C}$ by the integral transform

$$\widehat{F}(z) := \int_z^i f(w)\,(w - z)^{k-2}\,\mathrm{d}w \qquad (z \in \mathbb{H}). \tag{2.22}$$

Obviously the integral in (2.22) is well defined for all $z \in \mathbb{H}$.

Exercise 2.27 Let $k \in 2\mathbb{N}$ and $f \in M_k(\Gamma(1))$. Then the $\widehat{F}$ defined in (2.22) is an Eichler integral of weight $2 - k$.

Lemma 2.28 *Let $k \in 2\mathbb{N}$ and $f \in M_k(\Gamma(1))$. If F denotes the $(k-1)$-fold antiderivative of f defined in Lemma 2.20, then F and $\widehat{F}$ satisfy*

$$\widehat{F}(z) = -(k-2)!\left(F(z) + \frac{k-2}{(k-1)!} f(i) z^{k-1}\right) + P(z) \qquad (z \in \mathbb{H}) \qquad (2.23)$$

for some polynomial $P(z) \in \mathbb{C}[z]_{k-2}$ of degree at most $k-2$ that depends only on f.

Proof Similar to the proof of the second part of Lemma 2.23, we consider the derivative of $\widehat{F}$:

$$\begin{aligned}
\frac{\mathrm{d}}{\mathrm{d}z}\widehat{F}(z) &= \frac{\mathrm{d}}{\mathrm{d}z}\int_z^i f(w)\,(w-z)^{k-2}\,\mathrm{d}w \\
&= \lim_{\varepsilon \searrow 0} \frac{1}{\varepsilon}\left(\int_{z+\varepsilon}^i f(w)\,(w-z-\varepsilon)^{k-2}\,\mathrm{d}w - \int_z^i f(w)\,(w-z)^{k-2}\,\mathrm{d}w\right) \\
&= \lim_{\varepsilon \searrow 0} \frac{1}{\varepsilon}\left(\int_z^i f(w+\varepsilon)\,(w-z)^{k-2}\,\mathrm{d}w - \int_z^i f(w)\,(w-z)^{k-2}\,\mathrm{d}w\right) \\
&= \int_z^i \lim_{\varepsilon \searrow 0} \frac{1}{\varepsilon}\Big(f(w+\varepsilon) - f(w)\Big)\,(w-z)^{k-2}\,\mathrm{d}w \\
&= \int_z^i f'(w)\,(w-z)^{k-2}\,\mathrm{d}w \\
&= f(w)\,(w-z)^{k-2}\Big|_{w=z}^{w=i} - (k-2)\int_z^{i\infty} f(w)\,(w-z)^{k-3}\,\mathrm{d}w \\
&= f(i)\,(i-z)^{k-2} - (k-2)\int_z^{i\infty} f(w)\,(w-z)^{k-3}\,\mathrm{d}w.
\end{aligned}$$

Iterating the argument, we find that

$$\begin{aligned}
&\frac{\mathrm{d}^{k-2}}{\mathrm{d}z^{k-2}}\widehat{F}(z) \\
&\quad = \frac{\mathrm{d}^{k-3}}{\mathrm{d}z^{k-3}}\left(\frac{\mathrm{d}}{\mathrm{d}z}\widehat{F}(z)\right) \\
&\quad = (-1)^{k-3}\,(k-2)!\,f(i)\,(i-z) - \\
&\qquad\qquad - (k-2)\frac{\mathrm{d}^{k-4}}{\mathrm{d}z^{k-4}}\left(\frac{\mathrm{d}}{\mathrm{d}z}\int_z^{i\infty} f(w)\,(w-z)^{k-3}\,\mathrm{d}w\right) \\
&\quad = (-1)^{k-3}\,(k-2)!\,f(i)\,(i-z)
\end{aligned}$$

$$
\begin{aligned}
&\quad - (k-2)\frac{\mathrm{d}^{k-4}}{\mathrm{d}z^{k-4}}\left(f(i)\,(i-z)^{k-3} - (k-3)\int_z^{i\infty} f(w)\,(w-z)^{k-4}\,\mathrm{d}w \right) \\
&= (-1)^{k-3}\,(k-2)!\,f(i)\,(i-z) + (-1)^{k-3}\,(k-2)!f(i)\,(i-z) \\
&\quad + (k-2)(k-3)\frac{\mathrm{d}^{k-4}}{\mathrm{d}z^{k-4}}\int_z^{i\infty} f(w)\,(w-z)^{k-4}\,\mathrm{d}w \\
&= (k-2)\Big((-1)^{k-3}(k-2)!\,f(i)\,(i-z)\Big) + (-1)^{k-2}\,(k-2)!\int_z^{i} f(w)\,\mathrm{d}w \\
&= (-1)^{k-2}\,(k-2)!\left(\int_z^{i} f(w)\,\mathrm{d}w \;-\; (k-2)\,f(i)\,(i-z)\right).
\end{aligned}
$$

Since k is even, we get

$$
\frac{\mathrm{d}^{k-2}}{\mathrm{d}z^{k-2}}\widehat{F}(z) = (k-2)!\left(\int_z^{i} f(w)\,\mathrm{d}w \;-\; (k-2)\,f(i)\,(i-z)\right).
$$

Applying the derivative once again, we find

$$
\begin{aligned}
\frac{\mathrm{d}^{k-1}}{\mathrm{d}z^{k-1}}\widehat{F}(z) &= (k-2)!\,(k-2)\,f(i) \;+\; (k-2)!\,\frac{\mathrm{d}}{\mathrm{d}z}\int_z^{i} f(w)\,\mathrm{d}w \\
&= (k-2)!\,(k-2)\,f(i) \;-\; (k-2)!\,f(z) \\
&= (k-2)!\,(k-1)\,f(i) + (k-2)!f(i) \;-\; (k-2)!\,f(z) \qquad (2.24) \\
&= (k-1)!\,f(i) - (k-2)!\Big(f(z) - f(i)\Big).
\end{aligned}
$$

Equations (2.24) and (2.14) together imply

$$
\frac{\mathrm{d}^{k-1}}{\mathrm{d}z^{k-1}}\widehat{F}(z) = (k-2)!(k-2)\,f(i) \;-(k-2)!\,\frac{\mathrm{d}^{k-1}}{\mathrm{d}z^{k-1}}F(z), \qquad (2.25)
$$

where F is the $(k-1)$-fold antiderivative of f.

It remains to show that (2.23) holds given the above identity for the derivatives. By $(k-1)$ times indefinite integration, we get

$$
\widehat{F}(z) = -(k-2)!\left(F(z) + \frac{k-2}{(k-1)!}\,f(i)z^{k-1}\right) + Q(z),
$$

where $Q(z) \in \mathbb{C}[z]_{k-2}$ is a polynomial of degree at most $k-2$ coming from the indefinite integration. This shows (2.23) and concludes the proof. □

Exercise 2.29 Let $k \in 2\mathbb{N}$ and $f \in S_k(\Gamma(1))$. Show that $\tilde{F}$ and $\widehat{F}$ defined in (2.17) and (2.22) respectively satisfy

$$\tilde{F}(z) = \widehat{F}(z) + P(z), \tag{2.26}$$

where $P \in \mathbb{C}[z]_{k-2}$ is the polynomial of degree at most $k-2$ given by

$$P(z) = \int_i^{i\infty} f(w)\,(w-z)^{k-2}\,\mathrm{d}w \tag{2.27}$$

for all $z \in \mathbb{H}$.

Remark 2.30 Let $z_0 \in \mathbb{H}$ be a fixed reference point on $\mathbb{H}$. Analogous to the above exercise, we can easily see that the integral $\int_z^{z_0} f(w)\,(w-z)^{k-2}\,\mathrm{d}w$ is also an Eichler integral. The difference of the two terms is the polynomial $\int_{z_0}^{i} f(w)\,(w-z)^{k-2}\,\mathrm{d}w \in \mathbb{C}[z]_{k-2}$ of degree at most $k-2$.

2.1.6 The Niebur Integral Representation for Entire Forms

Definition 2.31 Let $k \in 2\mathbb{N}$ and $g \in M_k\big(\Gamma(1)\big)$. We define the function $\widehat{G}^- : \mathbb{H} \to \mathbb{C}$ by the integral transform

$$\widehat{G}^-(z) := \overline{\int_z^i g(w)\,\big(w-\overline{z}\big)^{k-2}\,\mathrm{d}w} \qquad (z \in \mathbb{H}), \tag{2.28}$$

where $\overline{\cdot}$ denotes complex conjugation of the overline expression. We call $\widehat{G}^-$ the *Niebur integral* of g.

Obviously the integral in (2.22) is well defined for all $z \in \mathbb{H}$. One of the most important aspects of this integral is that one can use it to define Eichler integrals for modular forms of real weight. One can notice from the definition that there will be no branching problem simply because we are using $\overline{z}$ for $z \in \mathbb{H}$ inside the integral and thus the factor $(w-\overline{z})^{k-2}$ does not vanish.

Exercise 2.32 Let $k \in 2\mathbb{N}$ and $g \in M_k\big(\Gamma(1)\big)$. Show that $\widehat{G}^-$ defined in (2.28) satisfies

$$\widehat{G}^-\big|_{2-k}M = \widehat{G}^- + \widehat{P}_M \tag{2.29}$$

for all $M \in \Gamma(1)$ with polynomials $\widehat{P}_M \in \mathbb{C}[z]_{k-2}$ given by

$$\widehat{P}_M(z) := \overline{\int_i^{M^{-1}i} g(w)\,\big(w-\overline{z}\big)^{k-2}\,\mathrm{d}w}. \tag{2.30}$$

Lemma 2.33 *Let* $k \in 2\mathbb{N}$ *and* $g \in M_k(\Gamma(1))$. *The family of polynomials* $\widehat{P}_M \in \mathbb{C}[z]_{k-2}$, $M \in \Gamma(1)$ *associated with* f *by the Niebur integral* $\widehat{G}^-$ *and its transformation formula (shown in Exercise 2.32) satisfies* (2.10) *and the transformation relation* (2.11).

Proof This lemma can be proved exactly by the same arguments used in the proof of Lemma 2.15. The only modification necessary is referring to the transformation property shown in Exercise 2.32 instead of the one in Definition 2.14. □

Exercise 2.34 Recall the usual definitions of partial derivatives

$$\frac{\partial}{\partial z} H = \frac{1}{2}\left(\frac{\partial}{\partial x} H - i\frac{\partial}{\partial y} H\right) \qquad \text{and} \qquad \frac{\partial}{\partial \overline{z}} H = \frac{1}{2}\left(\frac{\partial}{\partial x} H + i\frac{\partial}{\partial y} H\right) \tag{2.31}$$

for $z = x + iy$ and $H : \mathbb{H} \to \mathbb{C}$ a function (not necessarily holomorphic) that is derived from the Cauchy-Riemann equations for analytic functions, see Appendix A.4. If $\widehat{G}^-$ is a Niebur integral associated with the modular form $g \in M_k(\Gamma(1))$, then $\widehat{G}^-$ satisfies

$$\frac{\partial}{\partial \overline{z}} \widehat{G}^-(z) = \overline{g(z)}\left((z - \overline{z})^{k-2}\right). \tag{2.32}$$

Exercise 2.35 Let $k \in 2\mathbb{N}$ and $g \in S_k(\Gamma(1))$. We define the function $\tilde{G}^- : \mathbb{H} \to \mathbb{C}$ by the integral transform

$$\tilde{G}^-(z) := \overline{\int_z^{i\infty} g(w)\left(w - \overline{z}\right)^{k-2} dw}. \qquad (z \in \mathbb{H}) \tag{2.33}$$

1. Then $\tilde{G}^-$ satisfies

$$\tilde{G}^-\big|_{2-k} M = \tilde{G}^- + \tilde{P}_M \tag{2.34}$$

   for all $M \in \Gamma(1)$ with polynomials $\tilde{P}_M(z) \in \mathbb{C}[z]_{k-2}$ given by

$$\tilde{P}_M(z) := \overline{\int_{i\infty}^{M^{-1}i\infty} g(w)\left(w - \overline{z}\right)^{k-2} dw}. \tag{2.35}$$

2. We have

$$\widehat{G}^-(z) = \tilde{G}^-(z) + P(z) \tag{2.36}$$

   with $P \in \mathbb{C}[z]_{k-2}$ is the polynomial of degree at most $k - 2$ given by

$$P(z) = \overline{\int_i^{i\infty} g(w)\left(w - \overline{z}\right)^{k-2} \mathrm{d}w} \tag{2.37}$$

for all $z \in \mathbb{H}$.

2.2 The Eichler Cohomology Group

In this section, we introduce the Eichler cohomology group in order to prepare the reader for the Eichler-Shimura isomorphism theorem. To introduce the Eichler cohomology group, we consider a collection of polynomials $\left(P_M\right)_{M\in\Gamma(1)}$.

Definition 2.36 Let $k \in 2\mathbb{N}$ and $\left(P_M\right)_{M\in\Gamma(1)}$ be a collection of polynomials $P_M \in \mathbb{C}[X]_{k-2}$ of degree at most $k-2$.

1. We call the collection $\left(P_M\right)_{M\in\Gamma(1)}$ a *cocycle* if its elements satisfy

$$P_{MV} = P_M\big|_{2-k}V + P_V \qquad \text{for all } M, V \in \Gamma(1). \tag{2.38}$$

2. We call the collection $\left(P_M\right)_{M\in\Gamma(1)}$ a *coboundary* if there exists a fixed polynomial $P \in \mathbb{C}[X]_{k-2}$ such that $P_M = P\big|_{2-k}M - P$ for all $M \in \Gamma(1)$.

The set of cocycles and the set of coboundaries are denoted by $Z^1\left(\Gamma(1), \mathbb{C}[X]_{k-2}\right)$ and $B^1\left(\Gamma(1), \mathbb{C}[X]_{k-2}\right)$, respectively.

It is obvious that the set of cocycles and the set of coboundaries are additive groups.

Exercise 2.37 Show that every coboundary is a cocycle.

Exercise 2.38 Let $k \in 2\mathbb{N}$. Show that every cocycle

$$\left(P_M\right)_{M\in\Gamma(1)} \in Z^1\left(\Gamma(1), \mathbb{C}[X]_{k-2}\right)$$

satisfies

$$P_{\pm 1} = 0 \qquad \text{and} \qquad P_{\pm T}\big|_{2-k}(\pm T) + P_{\pm T} = 0. \tag{2.39}$$

Exercise 2.39 Let $k \in 2\mathbb{N}$. Show that every cocycle

$$\left(P_M\right)_{M\in\Gamma(1)} \in Z^1\left(\Gamma(1), \mathbb{C}[X]_{k-2}\right)$$

satisfies

$$P_M\big|_{2-k}M^{-1} + P_{M^{-1}} = 0 \tag{2.40}$$

for all $M \in \Gamma(1)$.

Definition 2.40 We call the quotient group

$$H^1\big(\Gamma(1), \mathbb{C}[X]_{k-2}\big) := \frac{Z^1\big(\Gamma(1), \mathbb{C}[X]_{k-2}\big)}{B^1\big(\Gamma(1), \mathbb{C}[X]_{k-2}\big)} \tag{2.41}$$

the *Eichler cohomology group*. Elements of $H^1\big(\Gamma(1), \mathbb{C}[X]_{k-2}\big)$ are denoted by $\{P_M\}$, indicating the family of polynomials $\big(P_M\big)_{M\in\Gamma(1)}$ that represent a corresponding cocycle.

We next introduce a natural and important subspace of $H^1\big(\Gamma(1), \mathbb{C}[X]_{k-2}\big)$, the space of parabolic cohomology.

Definition 2.41 Let $k \in 2\mathbb{N}$. A cocycle $\big(P_M\big)_{M\in\Gamma(1)} \in Z^1\big(\Gamma(1), \mathbb{C}[X]_{k-2}\big)$ is *parabolic* if there exists a polynomial $Q \in \mathbb{C}[X]_{k-2}$ such that P_S indexed by the parabolic generator S of $\Gamma(1)$ is given by

$$P_S = Q\big|_{2-k}S - Q. \tag{2.42}$$

The additive subgroup of parabolic cocycles is denoted by $Z^1_{\text{par}}\big(\Gamma(1), \mathbb{C}[X]_{k-2}\big) \subset Z^1\big(\Gamma(1), \mathbb{C}[X]_{k-2}\big)$. We call the quotient group

$$H^1_{\text{par}}\big(\Gamma(1), \mathbb{C}[X]_{k-2}\big) := \frac{Z^1_{\text{par}}\big(\Gamma(1), \mathbb{C}[X]_{k-2}\big)}{B^1\big(\Gamma(1), \mathbb{C}[X]_{k-2}\big)} \tag{2.43}$$

the *parabolic Eichler cohomology group*.

Exercise 2.42 Show that a parabolic cocycle

$$\big(P_M\big)_{M\in\Gamma(1)} \in Z^1_{\text{par}}\big(\Gamma(1), \mathbb{C}[X]_{k-2}\big)$$

satisfies the following: for each parabolic element $V \in \Gamma(1)$, there exists a polynomial $Q_V^\star \in \mathbb{C}[X]_{k-2}$ such that P_V can be expressed as

$$P_V = Q_V^\star\big|_{2-k}V - Q_V^\star. \tag{2.44}$$

For $f \in M_k\big(\Gamma(1)\big)$, consider the Eichler integral $\widehat{F}$ as defined in (2.22). Since $\widehat{F}$ is an Eichler integral, it satisfies its defining relation (2.9). The associated family of polynomials $P_M \in \mathbb{C}[X]_{k-2}$, $M \in \Gamma(1)$ can be expressed as

$$P_M(z) = \int_i^{M^{-1}i} f(w)\,(w-z)^{k-2}\,dw,$$

see Exercise 2.27. The family $\big(P_M\big)_{M\in\Gamma(1)}$ is a cocycle in the sense of Definition 2.36 since Lemma 2.15 shows that the P_M's satisfy (2.38). This argument establishes the following map:

$$M_k(\Gamma(1)) \to Z^1(\Gamma(1), \mathbb{C}[X]_{k-2}); \qquad f \overset{(2.22)}{\mapsto} \widehat{F} \overset{(2.9)}{\mapsto} \left(P_M\right)_{M\in\Gamma(1)}.$$

Using the fact that the cocycles are unique up to a coboundary, the above map induces a map α into the Eichler cohomology group.

Definition 2.43 Let $k \in 2\mathbb{N}$. We define the map α as follows:

$$\begin{aligned}\alpha : & M_k(\Gamma(1)) \longrightarrow H^1(\Gamma(1), \mathbb{C}[X]_{k-2}) \\ & f \overset{\substack{(2.22)\\(2.9)}}{\longmapsto} \left(P_M\right)_{M\in\Gamma(1)} \bmod B^1(\Gamma(1), \mathbb{C}[X]_{k-2}).\end{aligned} \tag{2.45}$$

Remark 2.44 If the realization of the Eichler integral in Definition 2.43 is replaced by another realization, for example by the $(k-1)$-fold antiderivative or the integral transformation for cusp forms, then the cocycle $\left(P_M\right)_{M\in\Gamma(1)}$ might change, but the associated cohomology class does not.

Using the same reasoning as above, we can define a map

$$M_k(\Gamma(1)) \to Z^1(\Gamma(1), \mathbb{C}[X]_{k-2}); \qquad g \overset{(2.28)}{\mapsto} \widehat{G}^- \overset{(2.29)}{\mapsto} \left(P_M\right)_{M\in\Gamma(1)}$$

using the Niebur integral instead of the Eichler integral. This leads again to the following.

Definition 2.45 Let $k \in 2\mathbb{N}$. We define the map β as

$$\begin{aligned}\beta : & M_k(\Gamma(1)) \longrightarrow H^1(\Gamma(1), \mathbb{C}[X]_{k-2}) \\ & f \overset{\substack{(2.28)\\(2.29)}}{\longmapsto} \left(\widehat{P}_M\right)_{M\in\Gamma(1)} \bmod B^1(\Gamma(1), \mathbb{C}[X]_{k-2}).\end{aligned} \tag{2.46}$$

Definition 2.46 Let $k \in 2\mathbb{N}$. The combination of the maps α and β leads to

$$\begin{aligned}\mu : & M_k(\Gamma(1)) \oplus S_k(\Gamma(1)) \longrightarrow H^1(\Gamma(1), \mathbb{C}[X]_{k-2}) \\ & (f,g) \overset{\substack{(2.44)\\(2.46)}}{\longmapsto} \alpha(f) + \beta(g).\end{aligned} \tag{2.47}$$

By construction, we have that the linear map μ maps the direct product space $M_k(\Gamma(1)) \oplus S_k(\Gamma(1))$ into the Eichler cohomology group. In what follows, we will discuss whether this map is an isomorphism or not.

Lemma 2.47 *Let* $k \in 2\mathbb{N}$.

1. *For* $f \in S_k(\Gamma(1))$, *recall that the image of* $\alpha(f) = \left(P_M\right)_{M\in\Gamma(1)}$ *is given by the transformation property* $P_M = \widehat{F}\big|_{2-k}M - \widehat{F}$ *of the Eichler integral* $\widehat{F}$ *defined in (2.22). We can express the values of the cocycle* $\left(P_M\right)_{M\in\Gamma(1)}$ *as*

$$P_M = \left(\tilde{F}\big|_{2-k}M - \tilde{F}\right) + \left((-P)\big|_{2-k}M - (-P)\right), \tag{2.48}$$

where $P \in \mathbb{C}[z]_{k-2}$ is the polynomial given in (2.27). We have in particular

$$P_S = \left(P\big|_{2-k}S - P\right),$$

which implies that the cocycle $\left(P_M\right)_{M\in\Gamma(1)}$ is parabolic and

$$\alpha : S_k\big(\Gamma(1)\big) \to H^1_{par}\big(\Gamma(1), \mathbb{C}[X]_{k-2}\big)$$

maps into the parabolic cohomology.

2. *For $g \in S_k\big(\Gamma(1)\big)$, recall that the image of $\beta(g) = \left(Q_M\right)_{M\in\Gamma(1)}$ is given by the transformation property $Q_M = \widehat{G}^-\big|_{2-k}M - \widehat{G}^-$ of the Niebur integral $\widehat{G}^-$ defined in (2.28). We can express the values of the cocycle $\left(Q_M\right)_{M\in\Gamma(1)}$ as*

$$Q_M = \left(\tilde{G}^-\big|_{2-k}M - \tilde{F}^-\right) + \left(P\big|_{2-k}M - P\right), \tag{2.49}$$

where $P \in \mathbb{C}[z]_{k-2}$ is the polynomial given in (2.36). We have in particular

$$P_S = \left(P\big|_{2-k}S - P\right),$$

which implies that the cocycle $\left(Q_M\right)_{M\in\Gamma(1)}$ is parabolic and

$$\beta : S_k\big(\Gamma(1)\big) \to H^1_{par}\big(\Gamma(1), \mathbb{C}[X]_{k-2}\big)$$

maps into the parabolic cohomology group.

Proof For $f \in S_k\big(\Gamma(1)\big)$, write the image of $\alpha(f) = \left(P_M\right)_{M\in\Gamma(1)}$ as given by transformation property $P_M = \widehat{F}\big|_{2-k}M - \widehat{F}$ of the Eichler integral $\widehat{F}$ defined in (2.22). Using Exercise 2.29, we see

$$\begin{aligned} P_M &\overset{(2.9)}{=} \widehat{F}\big|_{2-k}M - \widehat{F} \\ &\overset{(2.26)}{=} (\tilde{F} - P)\big|_{2-k}M - (\tilde{F} - P) \\ &= \left(\tilde{F}\big|_{2-k}M - \tilde{F}\right) + \left((-P)\big|_{2-k}M - (-P)\right) \end{aligned}$$

with $P \in \mathbb{C}[z]_{k-2}$ given by (2.27).

The fixed point condition $S\, i\infty = i\infty$ implies

$$\begin{aligned}\tilde{F}\big|_{2-k}S(z) &= \tilde{F}(z) + \int_{i\infty}^{S^{-1}i\infty} f(w)\,(w-z)^{k-2}\,dw \\ &= \tilde{F}(z) + \int_{i\infty}^{i\infty} f(w)\,(w-z)^{k-2}\,dw \\ &= \tilde{F}(z),\end{aligned}$$

following the proof of the first part of Lemma 2.23 for the parabolic element $S = \begin{pmatrix} 1 & 1 \\ 0 & 1 \end{pmatrix}$. This implies that the cocycle $\big(P_M\big)_{M\in\Gamma(1)}$ is in fact a parabolic cocycle. We have

$$\big(P_M\big)_{M\in\Gamma(1)} \in Z^1_{\text{par}}\big(\Gamma(1), \mathbb{C}[X]_{k-2}\big).$$

The map α maps $S_k\big(\Gamma(1)\big)$ into the parabolic cohomology group $H^1_{\text{par}}\big(\Gamma(1), \mathbb{C}[X]_{k-2}\big)$.

The second part is left to the reader as an exercise. □

Example 2.48 Let $k \in 2\mathbb{N}$ and let $f \in S_k\big(\Gamma(1)\big)$ be a cusp form. The period polynomial $P(X)$ associated with f by (2.1) and the parabolic cocycle $\big(P_M\big)_{M\in\Gamma(1)}$ attached to f by the Eichler integral $\tilde{F}$ given in (2.17) are related as follows: consider the cocycle given by its values P_T and P_S of the generators T and S of $\Gamma(1)$. Using the transformation property (2.9) of Eichler integrals, we find

$$\begin{aligned}P_S(X) &\overset{(2.9)}{=} \tilde{F}\big|_{2-k}S(X) - \tilde{F}(X) \\ &= \int_{i\infty}^{S^{-1}i\infty} f(w)\,(w-z)^{k-2}\,dw \qquad (2.50)\\ &\overset{Si\infty=i\infty}{=} \int_{i\infty}^{i\infty} f(w)\,(w-z)^{k-2}\,dw \\ &= 0,\end{aligned}$$

using the proof of the first part of Lemma 2.23 in (2.50), and

$$\begin{aligned}P_T(X) &\overset{(2.9)}{=} \tilde{F}\big|_{2-k}T(X) - \tilde{F}(X) \\ &= \int_{i\infty}^{T^{-1}i\infty} f(w)\,(w-z)^{k-2}\,dw \qquad (2.51)\\ &\overset{T^{-1}i\infty=0}{=} \int_{i\infty}^{0} f(w)\,(w-z)^{k-2}\,dw \\ &\overset{(2.1)}{=} -P(z),\end{aligned}$$

where we use the proof of the first part of Lemma 2.23 in (2.51) and where $P(z)$ is to f associated period function.

Hence the cocycle $\big(P_M\big)_{M\in\Gamma(1)}$ satisfies

$$P_S = 0 \qquad \text{and} \qquad P_T = -P. \tag{2.52}$$

We see in particular that period functions give a representative of the parabolic cocycle associated with cusp forms by the Eichler integral.

Lemma 2.49 *Let $k \in 2\mathbb{N}$.*

1. *Each element of parabolic cohomology group $H^1_{par}\big(\Gamma(1), k\big)$ can be uniquely represented (up to scalar multiplication) by a parabolic cocycle $\big(P_M\big)_{M\in\Gamma(1)} \in Z^1_{par}\big(\Gamma(1), k\big)$ chosen such that $P_S = 0$.*
2. *The map*

$$H^1_{par}\big(\Gamma(1), k\big) \to \mathbb{P}_{k-2}; \quad \Big\{\big(P_M\big)_{M\in\Gamma(1)} \text{ with } P_S = 0\Big\} \mapsto P_T$$

has codimension 1. The one-dimensional subspace in $\mathbb{P}_{k-2}$ is spanned by $X^{k-2} - 1 \in \mathbb{C}[X]_{k-2}$.

Proof

1. Recall Definition 2.41: each element of $H^1_{\text{par}}\big(\Gamma(1), k\big)$ is represented by a parabolic cocycle $\big(P_M\big)_{M\in\Gamma(1)} \in Z^1_{\text{par}}\big(\Gamma(1), k\big)$ that is unique up to a coboundary in $B^1\big(\Gamma(1), k\big)$. If $\big(Q_M\big)_{M\in\Gamma(1)} \in B^1\big(\Gamma(1), k\big)$ is a coboundary, then the cocycles $\big(P_M\big)_{M\in\Gamma(1)}$ and $\big(P_M + Q_M\big)_{M\in\Gamma(1)}$ represent the same cohomology class.

   Let $\big(P_M\big)_{M\in\Gamma(1)} \in Z^1_{\text{par}}\big(\Gamma(1), k\big)$ be a parabolic cocycle. Definition 2.41 implies that there exists $Q \in \mathbb{C}[X]_{k-2}$ such that

$$P_S = Q\big|_{2-k}(S-1).$$

   Define the coboundary $\big(Q_M\big)_{M\in\Gamma(1)}$ and a new cocycle $\big(P'_M\big)_{M\in\Gamma(1)}$ by

$$Q_M := -Q\big|_{2-k}(M-1) \quad \text{and} \quad P'_M := P_M + Q_M = P_M - Q\big|_{2-k}(M-1)$$

   for each $M \in \Gamma(1)$. Hence $\big(P_M\big)_{M\in\Gamma(1)}$ and $\big(P'_M\big)_{M\in\Gamma(1)}$ represent the same cohomology element in $H^1_{\text{par}}\big(\Gamma(1), k\big)$ since they differ by a coboundary. By construction, the cocycle $\big(P'_M\big)_{M\in\Gamma(1)}$ is parabolic and thus vanishes. We have

$$P'_S = P_S + Q_S = \Big(Q\big|_{2-k}(S-\mathbf{1})\Big) - \Big(Q\big|_{2-k}(S-\mathbf{1})\Big) = 0 = 0\big|_{2-k}(S-\mathbf{1}).$$

2. Using the first part of the lemma, we can represent each element of $H^1_{\text{par}}(\Gamma(1), k)$ with a parabolic cocycle $(P_M)_{M\in\Gamma(1)} \in Z^1_{\text{par}}(\Gamma(1), k)$ satisfying $P_S = 0$. We see, as in the proof of Proposition 2.16, that the polynomial P_T is a period function in the sense of Definition 2.7: $P_T \in \mathbb{P}_{k-2}$.

   Now, let $P \in \mathbb{P}_{k-2}$ be a period polynomial. We define

$$P_S := 0 \quad \text{and} \quad P_T := P.$$

   Since the matrices S and T generate $\Gamma(1)$, we see that the values of P_S and P_T determine the whole cocycle $(P_M)_{M\in\Gamma(1)}$, utilizing (2.44). Moreover, the cocycle is parabolic (since $P_S = 0 = 0|_{2-k}(S - \mathbf{1})$). Hence, we just constructed a map

$$\mathbb{P}_{k-2} \to Z^1_{\text{par}}(\Gamma(1), k); \quad P \mapsto (P_M)_{M\in\Gamma(1)} \quad \text{given by } P_S = 0 \text{ and } P_T = P,$$

   which induces a map

$$\mathbb{P}_{k-2} \to H^1_{\text{par}}(\Gamma(1), k).$$

   Next, we calculate the kernel of the above map. A $P \in \mathbb{P}_{k-2}$ lies in the kernel of above map if P is derived from a coboundary. There exists a $Q \in \mathbb{C}[X]_{k-2}$ such that $P = Q|_{2-k}(T - \mathbf{1})$ and $0 = Q|_{2-k}(S - \mathbf{1})$. Since the second condition on Q implies that Q is invariant under translation ($X \mapsto S\,X = X + 1$); hence, Q must be a constant function. (The constant functions are the only polynomials invariant under translation.) Put $Q := 1$ since $\mathbb{C}[X]_{k-2}$ is a $\mathbb{C}$-vector space. Hence,

$$\begin{aligned} P_T(X) &:= 1|_{2-k}(T - \mathbf{1})(X) = X^{k-2} - 1 \qquad \text{and} \\ P_S(X) &:= 1|_{2-k}(S - \mathbf{1})(X) = 1 - 1 = 0. \end{aligned}$$

   Since $X^{k-2} - 1 \in \mathbb{P}_{k-2}$, see Exercise 2.8, we conclude that the kernel of the map above is one dimensional and is spanned by $X^{k-2} - 1$. This implies

$$\dim \mathbb{P}_{k-2} = 1 + \dim H^1_{\text{par}}(\Gamma(1), k).$$

□

2.3 The Eichler Cohomology Theorem

Theorem 2.50 (Eichler cohomology theorem) *Let $k \in 2\mathbb{N}$. The spaces*

$$M_k(\Gamma(1)) \oplus S_k(\Gamma(1)) \quad \textit{and} \quad H^1(\Gamma(1), \mathbb{C}[X]_{k-2})$$

are isomorphic under the map μ defined in (2.47). *Here,* $\oplus$ *denotes the direct sum of two spaces.*

Remark 2.51 The isomorphism μ between the spaces $M_k\big(\Gamma(1)\big) \times S_k\big(\Gamma(1)\big)$ and $H^1\big(\Gamma(1), \mathbb{C}[X]_{k-2}\big)$, which was defined in (2.47), is also known as the *Eichler-Shimura isomorphism.*

Remark 2.52 The Eichler cohomology theorem goes back to Martin Eichler [73] and Goro Shimura [183]. Some details on the work of Martin Eichler can be found in [110] (article in German) and on the work of Shimura can be found in [154].

Corollary 2.53 *Let $k \in 2\mathbb{N}$. The spaces*

$$S_k\big(\Gamma(1)\big) \oplus S_k\big(\Gamma(1)\big) \quad \textit{and} \quad H^1_{par}\big(\Gamma(1), \mathbb{C}[X]_{k-2}\big)$$

are isomorphic under the map μ defined in (2.47).

Proof Consider $(f, g) \in S_k\big(\Gamma(1)\big) \oplus S_k\big(\Gamma(1)\big)$. The image

$$\mu(f, g) \in H^1\big(\Gamma(1), \mathbb{C}[X]_{k-2}\big)$$

is represented by a cocycle, see Theorem 2.50. Using Lemma 2.47, we see that the cocycle is in fact parabolic. Hence μ maps into the parabolic cocycle space

$$\mu : S_k\big(\Gamma(1)\big) \oplus S_k\big(\Gamma(1)\big) \to H^1_{\text{par}}\big(\Gamma(1), \mathbb{C}[X]_{k-2}\big)$$

The injectivity of the restricted map follows directly from the injectivity of the original map. Since $S_k\big(\Gamma(1)\big) \oplus S_k\big(\Gamma(1)\big)$ is a subspace of $M_k\big(\Gamma(1)\big) \oplus S_k\big(\Gamma(1)\big)$ of codimension 1 and also $H^1_{\text{par}}\big(\Gamma(1), \mathbb{C}[X]_{k-2}\big)$ is a subspace of $H^1\big(\Gamma(1), \mathbb{C}[X]_{k-2}\big)$ of codimension 1. This follows since $\Gamma(1) = \langle S, T\rangle$ and only S is parabolic. We see that the restricted map remains surjective. Hence the restricted map is also an isomorphism. □

Corollary 2.54 *Let $k \in 2\mathbb{N}$. The map*

$$\begin{aligned} &S_k\big(\Gamma(1)\big) \oplus S_k\big(\Gamma(1)\big) \to \mathbb{P}_{k_2} \\ &(f, g) \mapsto \int_0^{i\infty} f(w)\,(w - X)^{k-2}\,\mathrm{d}w + \overline{\int_0^{i\infty} g(w)\,(w - \overline{X})^{k-2}\,\mathrm{d}w} \end{aligned} \tag{2.53}$$

has codimension 1. The subspace in $\mathbb{P}_{k-2}$ is spanned by $X^{k-2} - 1$.

Proof Corollary 2.53 implies that the map

$$S_k\big(\Gamma(1)\big) \oplus S_k\big(\Gamma(1)\big) \to H^1_{\text{par}}\big(\Gamma(1), \mathbb{C}[X]_{k-2}\big)$$

induced by μ is an isomorphism. We may replace the Eichler and Niebur integrals $\widehat{F}$ and $\widehat{G}^-$ in the construction of μ in (2.47) by the respective integrals $\tilde{F}$ and $\tilde{G}^-$, see Lemma 2.47, since we assumed the spaces of cusp forms as the initial spaces. With this replacement, define the generating values P_S and P_T of the parabolic cocycle $\mu(f, g) = \left(P_M\right)_{M\in\Gamma(1)}$ by

$$\begin{aligned} P_S &= -\left(\int_{i\infty}^{S^{-1}i\infty} f(w)\,(w-X)^{k-2}\,\mathrm{d}w + \overline{\int_{i\infty}^{S^{-1}i\infty} g(w)\,(w-\overline{X})^{k-2}\,\mathrm{d}w}\right) \\ &= 0 \end{aligned}$$

and

$$\begin{aligned} P_T &= -\left(\int_{i\infty}^{T^{-1}i\infty} f(w)\,(w-X)^{k-2}\,\mathrm{d}w + \overline{\int_{i\infty}^{T^{-1}i\infty} g(w)\,(w-\overline{X})^{k-2}\,\mathrm{d}w}\right) \\ &= \int_0^{i\infty} f(w)\,(w-X)^{k-2}\,\mathrm{d}w + \overline{\int_0^{i\infty} g(w)\,(w-\overline{X})^{k-2}\,\mathrm{d}w}. \end{aligned}$$

Applying Lemma 2.49 concludes the proof. □

In what remains in this section, we present the proof of the Eichler cohomology theorem (Theorem 2.50).

2.3.1 Injectivity of the Map μ

We have the following variant of Green's Theorem:

Lemma 2.55 *Let $H : \mathbb{H} \to \mathbb{C}$ be a non-holomorphic function satisfying the grow estimate $H(z) = O\left(e^{-2\pi\delta\Im(z)}\right)$ as $\Im(z) \to \infty$ for some $\delta > 0$. Then H satisfies the identity*

$$\iint_{\mathcal{F}_{\Gamma(1)}} \frac{\partial}{\partial \overline{z}} H(z)\,\mathrm{d}x\,\mathrm{d}y = -\frac{i}{2}\int_{\partial\mathcal{F}_{\Gamma(1)}} H(z)\,\mathrm{d}z, \tag{2.54}$$

where the path of integration on the right-hand side is along the positively (counter-clockwise) oriented set-boundary $\partial\mathcal{F}_{\Gamma(1)}$ of the standard fundamental domain $\mathcal{F}_{\Gamma(1)}$.

Proof We give an outline of the proof here. Observe that the usual Green's Theorem does not apply directly to the fundamental domain $\mathcal{F}_{\Gamma(1)}$ since it contains a vertical half-strip. Hence, $\mathcal{F}_{\Gamma(1)}$ is not a bounded subset of $\mathbb{H}$. However, we can extend the theorem to cover this case by applying a simple limit argument and using the fact that $g \in S_k(\Gamma(1))$, being a cusp form, vanishes at the cusp $i\infty$. □

Lemma 2.56 *In the situation of Lemma 2.55, assume additionally that H satisfies the transformation property* (1.28) *with weight 2 (and trivial multiplier 1): $j(M,z)^{-k}\, H(M\,z) = H(z)$ for all $M \in \Gamma(1)$ and $z \in \mathbb{H}$.*

The integral $\int_{\partial\mathcal{F}_{\Gamma(1)}} H(z)\,\mathrm{d}z$ on the right-hand side of (2.54) *vanishes:*

$$\int_{\partial\mathcal{F}_{\Gamma(1)}} H(z)\,\mathrm{d}z = 0. \tag{2.55}$$

Proof Using Exercise 1.19, we can decompose $\partial\mathcal{F}_{\Gamma(1)}$ into the sets $C_1, \ldots, C_4$, as defined in the proof of Exercise 1.19. Hence the left-hand side of (2.55) can be written as

$$\begin{aligned}\int_{\partial\mathcal{F}_{\Gamma(1)}} H(z)\,\mathrm{d}z &= \int_{C_1} H(z)\,\mathrm{d}z + \int_{C_2} H(z)\,\mathrm{d}z \\ &\quad + \int_{C_3} H(z)\,\mathrm{d}z + \int_{C_4} H(z)\,\mathrm{d}z.\end{aligned} \tag{2.56}$$

Using the set-identities $S\,C_1 = C_2$ and $T\,C_3 = C_4$ from Exercise 1.19 and noticing that the orientation changes under the Möbius transformation, we conclude

$$\begin{aligned}\int_{C_2} H(z)\,\mathrm{d}z &= -\int_{SC_1} H(z)\,\mathrm{d}z \\ &= -\int_{C_1} H\big(S^{-1}z\big)\,\mathrm{d}(S^{-1}z) \\ &= -\int_{C_1} H\big|_2 S^{-1}(z)\,\mathrm{d}z \\ &= -\int_{C_1} H(z)\,\mathrm{d}z\end{aligned}$$

using the transformation property of the integrand. Similarly, we can show

$$\int_{C_4} H(z)\,\mathrm{d}z = -\int_{C_3} H\big|_2 S^{-1}(z)\,\mathrm{d}z.$$

This shows that the right-hand side of (2.56) vanishes and this concludes the proof. □

Proposition 2.57 *The map*

$$\mu : M_k\big(\Gamma(1)\big) \oplus S_k\big(\Gamma(1)\big) \longrightarrow H^1\big(\Gamma(1), \mathbb{C}[X]_{k-2}\big)$$

defined in (2.47) *is injective.*

We follow [108, §4] for the proof of this proposition.

Proof Since μ is a linear map, it suffices to show that the kernel of μ is trivial. Assume that $(f, g) \in M_k(\Gamma(1)) \oplus S_k(\Gamma(1))$ satisfies

$$\mu(f, g) = \{0\}$$

in $H^1(\Gamma(1), \mathbb{C}[X]_{k-2})$. This implies in particular that the cocycle image $P_M + \widehat{P}_M$ induced by μ must be a coboundary. There exists a $P \in \mathbb{C}[z]_{k-2}$ such that

$$P_M + \widehat{P}_M = P\big|_{2-k}M - P$$

for all $M \in \Gamma(1)$. Using Definitions 2.14 and 2.31 of Eichler and Niebur integrals respectively we see that the above condition implies that the function $\widehat{F} + \widehat{G}^- - P$ satisfies the transformation law (1.28) with weight $2 - k$:

$$\begin{aligned}
\left(\widehat{F} + \widehat{G}^- - P\right)\big|_{2-k}M &= \widehat{F}\big|_{2-k}M + \widehat{G}^-\big|_{2-k}M - P\big|_{2-k}M \\
&\overset{\substack{(2.9)\\(2.29)}}{=} \widehat{F} + P_M + \widehat{G}^- + \widehat{P}_M - P\big|_{2-k}M \\
&= \widehat{F} + \widehat{G}^- + \underbrace{P_M + \widehat{P}_M}_{=P|_{2-k}M-P} - P\big|_{-k}M \\
&= \widehat{F} + \widehat{G}^- - P
\end{aligned}$$

for all $M \in \Gamma(1)$. (Note that $\widehat{F} + \widehat{G}^- - P$ is not a modular form since $\widehat{G}^-$ and hence the whole function is not a holomorphic (or meromorphic) function. We see in particular that $\frac{\partial}{\partial \bar{z}}\widehat{G}^-(z)$ does not vanish, see Exercise 2.34.)

Recall that $g \in M_k(\Gamma(1))$ is holomorphic. With the help of Exercise 2.34 and the product rule, we find

$$\begin{aligned}
\frac{\partial}{\partial \bar{z}}\left(g(z)\,\widehat{G}^-(z)\right) &= g(z)\frac{\partial}{\partial \bar{z}}\,\widehat{G}^-(z) \\
&\overset{(2.32)}{=} g(z)\overline{g(z)}\left(z - \bar{z}\right)^{k-2} \\
&= |g(z)|\left(z - \bar{z}\right)^{k-2}.
\end{aligned} \tag{2.57}$$

Also, the function $g(z)\left(\widehat{F}(z) - P(z)\right)$ is holomorphic. Hence,

$$\frac{\partial}{\partial \bar{z}}\left(g(z)\left(\widehat{F}(z) - P(z)\right)\right) = 0. \tag{2.58}$$

Recall the fundamental domain $\mathcal{F}_{\Gamma(1)}$ as defined in (1.22). Its boundary is denoted by $\partial\mathcal{F}_{\Gamma(1)}$, oriented in the positive (counterclockwise) sense. Using (2.57) and (2.58) we find

$$\begin{aligned}
&\iint_{\mathcal{F}_{\Gamma(1)}} |g(z)| \left(z-\overline{z}\right)^{k-2} \mathrm{d}x\,\mathrm{d}y \\
&\overset{(2.57)}{=} \iint_{\mathcal{F}_{\Gamma(1)}} \frac{\partial}{\partial \overline{z}}\left(g(z)\,\widehat{G}^{-}(z)\right) \mathrm{d}x\,\mathrm{d}y \\
&\overset{(2.58)}{=} \iint_{\mathcal{F}_{\Gamma(1)}} \frac{\partial}{\partial \overline{z}}\left(g(z)\left(\widehat{F}(z)-P(z)\right)\right) + \frac{\partial}{\partial \overline{z}}\left(g(z)\,\widehat{G}^{-}(z)\right) \mathrm{d}x\,\mathrm{d}y \\
&= \iint_{\mathcal{F}_{\Gamma(1)}} \frac{\partial}{\partial \overline{z}}\left(g(z)\left(\widehat{F}(z)+\widehat{G}^{-}(z)-P(z)\right)\right) \mathrm{d}x\,\mathrm{d}y.
\end{aligned}$$

Recall that the g appearing as a multiplicative factor in the integrand is a cusp form. Exercise 1.67 shows that g itself satisfies the growth condition of the integrand as required by Lemma 2.55. The second multiplicative factor, composed of Eichler, respectively, Niebur integrals of entire modular forms and a polynomial, grows at most polynomially at the cusp ∞: there exists an $m \in \mathbb{Z}_{\geq 0}$ such that

$$F(z)+\widehat{G}^{-}(z)-P(z) = O\left(z^{k}\right) \qquad \text{as } \Im(z) \to \infty.$$

Hence, Lemma 2.55 applies to the integrand $g(z)\left(F(z)+\widehat{G}^{-}(z)-P(z)\right)$. We find

$$\begin{aligned}
&\iint_{\mathcal{F}_{\Gamma(1)}} |g(z)| \left(z-\overline{z}\right)^{k-2} \mathrm{d}x\,\mathrm{d}y \\
&\qquad = \iint_{\mathcal{F}_{\Gamma(1)}} \frac{\partial}{\partial \overline{z}}\left(g(z)\left(\widehat{F}(z)+\widehat{G}^{-}(z)-P(z)\right)\right) \mathrm{d}x\,\mathrm{d}y \\
&\qquad \overset{(2.54)}{=} -\frac{i}{2}\int_{\partial\mathcal{F}_{\Gamma(1)}} g(z)\left(F(z)+\widehat{G}^{-}(z)-P(z)\right) \mathrm{d}z.
\end{aligned} \tag{2.59}$$

Next, we use the modular transformation property (1.28) for g with weight k and $\widehat{F}+\widehat{G}^{-}-P$ with weight $2-k$. Hence

$$\begin{aligned}
g\left(F+\widehat{G}^{-}-P\right)\Big|_{2} M &= \left(g|_{k}M\right)\left(\left(F+\widehat{G}^{-}-P\right)\Big|_{2} M\right) \\
&\overset{(1.28)}{=} g\left(F+\widehat{G}^{-}-P\right)
\end{aligned}$$

for all $M \in \Gamma(1)$. This shows that the integrand in (2.59) satisfies the transformation law (1.28) with weight 2 (and trivial multiplier). Applying Lemma 2.56 to (2.59) we find

$$\iint_{\mathcal{F}_{\Gamma(1)}} |g(z)| \left(z - \overline{z}\right)^{k-2} \mathrm{d}x\, \mathrm{d}y \overset{(2.59)}{=} -\frac{i}{2} \int_{\partial\mathcal{F}_{\Gamma(1)}} g(z)\big(F(z) + \widehat{G}^{-}(z) - P(z)\big)\, \mathrm{d}z$$
$$\overset{(2.55)}{=} 0.$$

This can only happen if g vanishes.

To complete the proof, it suffices to show that $f = 0$ if $g = 0$. Since $g = 0$, we conclude $\widehat{G}^{-} = 0$, see (2.28). Thus the weight $2 - k$ transformation property of $\widehat{F} + \widehat{G}^{-} - P$ can be written as

$$\big(\widehat{F} - P\big)\big|_{2-k} = \widehat{F} - P.$$

Since f is an entire form we know that F is an entire function satisfying the modular transformation property for weight $2 - k \leq 0$. Hence $\widehat{F} - P \in M_{2-k}\big(\Gamma(1)\big)$ and, using Corollary 1.70 for negative $2 - k < 0$ and the fact that the space of holomorphic and bounded functions that are invariant under $\Gamma(1)$ consists of constant functions as stated in the introduction of §1.3.1 for $2 - k = 0$, $\widehat{F} - P$ is a constant function. Using (2.14) and recalling that $P(z) \in \mathbb{C}[z]_{k-2}$, we conclude that f indeed vanishes.

Summarizing, we just showed that $\mu(f, g) = \{0\}$ implies $f = g = 0$. Hence the map μ is injective. □

2.3.2 *Surjectivity of the Map μ*

In order to prove surjectivity, we need to show that the dimension of cohomology group is equal to the sum of the dimensions of the spaces of the entire forms and the cusp forms. Once we show surjectivity, we have the full proof of Eichler-Shimura isomorphism. As a result, we now calculate the dimension of the cohomology space $H^1\big(\Gamma(1), \mathbb{C}[X]_{k-2}\big)$.

Lemma 2.58 *Consider the operator*

$$\mathbb{C}[X]_1 \to \mathbb{C}[X]_1; \quad Q \mapsto Q\big|_{-1} TS.$$

acting on polynomials of degree at most 1 *and put* $\rho = \frac{1+\sqrt{3}i}{2}$. *This operator has two eigenspaces spanned by the polynomials*

$$\rho X + 1 \qquad \textit{and} \qquad \overline{\rho} X + 1$$

with eigenvalues $\overline{\rho}$ and ρ respectively.

Proof Since $TS \overset{(1.20)}{=} \begin{pmatrix} 0 & -1 \\ 1 & 0 \end{pmatrix}\begin{pmatrix} 1 & 1 \\ 0 & 1 \end{pmatrix} = \begin{pmatrix} 0 & -1 \\ 1 & 1 \end{pmatrix}$, we have

$$Q\big|_1(TS)(X) = (X+1)\Big(\frac{-\rho}{X+1}+1\Big) = X+1-\rho = \frac{1}{\rho}\Big(\rho X + (1-\rho)\rho\Big)$$

for $Q(X) = \rho X + 1 \in \mathbb{C}[X]_1$ with $\rho \neq 0$. Putting $\rho := \frac{1+\sqrt{3}i}{2}$, we easily calculate $|\rho| = 1, \overline{r} = \frac{1}{r}$, and $(1-r)r = 1$. Hence the above calculation shows that $Q(x) = \rho X + 1$ is an eigenpolynomial of degree 1 of the operator with eigenvalue $\overline{r} = \frac{1}{r}$:

$$Q\big|_1(TS)(X) = \overline{r}\, Q(X).$$

The analogous calculation shows that the polynomial $\overline{\rho}X + 1$ is also an eigenpolynomial with eigenvalue ρ of the operator. We just have seen that $\mathbb{C}[X]_1$ is spanned by two eigenspaces generated by $\rho X + 1$ and $\overline{\rho}X + 1$ with eigenvalues $\overline{\rho}$ and ρ respectively. □

We need to introduce the Gauss brackets before we give our next lemma.

Definition 2.59 We define the *Gauss brackets* $[\cdot] : \mathbb{R} \to \mathbb{Z}$ as the function assigning to each real number x, the largest integer less than or equal to x:

$$[x] := \max\big\{n \in \mathbb{Z}\colon\ n \le x\big\} \qquad (x \in \mathbb{R}). \tag{2.60}$$

Remark 2.60 The Gauss brackets $[\cdot]$ can be characterized by

$$[x] = \big\{n \in \mathbb{Z}\colon\ x-1 < n \le x\big\}.$$

It is also known as the *floor* function.

Lemma 2.61 ([116, p. 2, Lemma 2]) *Let $k \in 2\mathbb{N}$. We define subspaces $E, F \subset \mathbb{C}[X]_{k-2}$ by*

$$\begin{aligned} E &= \big\{P \in \mathbb{C}[X]_{k-2}\colon\ P + P\big|_{2-k}T = 0\big\} \quad \text{and} \\ F &= \big\{P \in \mathbb{C}[X]_{k-2}\colon\ P + P\big|_{2-k}TS + P\big|_{2-k}(TS)^2 = 0\big\}. \end{aligned} \tag{2.61}$$

The following statements hold.

1. *For $k \ge 4$, we have $E + F = \mathbb{C}[X]_{k-2}$ such that* $\dim(E+F) = k-1$, *and for $k = 2$, we have $E + F = \mathbb{C}[X]_0 = \mathbb{C}$ such that* $\dim(E+F) = 0$.
2. $\dim E = \begin{cases} \frac{k}{2} & \text{if } \frac{k-2}{2} \text{ is odd, and} \\ \frac{k}{2} - 1 & \text{if } \frac{k-2}{2} \text{ is even.} \end{cases}$

3. $\dim F = k - 1 - \begin{cases} 1 & \text{if } k = 2, \\ \left[\frac{k-2}{3}\right] + 1 & \text{if } 4 \leq k \equiv 1, 2 \pmod 3, \text{ and} \\ \left[\frac{k-2}{3}\right] & \text{if } 4 \leq k \equiv 0 \pmod 3). \end{cases}$

Here $[\cdot]$ *denotes the Gauss brackets given in Definition 2.59.*

Proof Recall the matrix identities

$$(\mathbf{1} - T)(\mathbf{1} + T) = \mathbf{1} - (-\mathbf{1}) \quad \text{and} \quad (\mathbf{1} - TS)\big(\mathbf{1} + TS + (TS)^2\big) = \mathbf{1} - (-\mathbf{1}).$$

Applying $\mathbf{1} - (\mathbf{1})$ with the slash operator to every $P \in \mathbb{C}[X]_{k-2}$ gives

$$\begin{aligned} P\big|_{2-k}\big(\mathbf{1} - (-\mathbf{1})\big)(X) &\overset{(1.132)}{=} \left(P\big|_{2-k}\mathbf{1}\right)(X) - \left(P\big|_{2-k}(-\mathbf{1})\right)(X) \\ &\overset{(1.131)}{=} P(X) - (-1)^{k-2}P(X) \\ &= 0. \end{aligned}$$

since k is an even integer. Hence, the above matrix identities imply

$$P\big|_{2-k}(\mathbf{1} - T)(\mathbf{1} + T) = 0 \quad \text{and} \quad P\big|_{2-k}(\mathbf{1} - TS)\big(\mathbf{1} + TS + (TS)^2\big) = 0$$

for every $P \in \mathbb{C}[X]_{k-2}$. Using (1.133), we rewrite this as

$$\big(P\big|_{2-k}(\mathbf{1}-T)\big)\big|_{2-k}(\mathbf{1}+T) = 0 \quad \text{and} \quad \big(P\big|_{2-k}(\mathbf{1}-TS)\big)\big|_{2-k}\big(\mathbf{1}+TS+(TS)^2\big) = 0$$

for every $P \in \mathbb{C}[X]_{k-2}$. Hence the polynomials $P\big|_{2-k}(\mathbf{1} - T)$ and $P\big|_{2-k}(\mathbf{1} - TS)$ satisfy

$$P\big|_{2-k}(\mathbf{1} - T) \in E \quad \text{and} \quad P\big|_{2-k}(\mathbf{1} - TS) \in F.$$

Even more, we get all elements of the spaces E and F in the same way. Summarizing, we have shown alternative representations of the spaces E and F:

$$E = \left\{P\big|_{2-k}(\mathbf{1}-T) \colon\ P \in \mathbb{C}[X]_{k-2}\right\} \quad \text{and} \quad F = \left\{P\big|_{2-k}(\mathbf{1}-TS) \colon\ P \in \mathbb{C}[X]_{k-2}\right\}. \tag{2.62}$$

1. Using (2.62), we can express each element of E as $P\big|_{2-k}(\mathbf{1} - T)$ and each element of F as $Q\big|_{2-k}(\mathbf{1} - TS)$ for $P, Q \in \mathbb{C}[X]_{k-2}$.

Then

$$\left(P\big|_{2-k}(\mathbf{1}-T)+Q\big|_{2-k}(\mathbf{1}-TS)\right)\Big|_{2-k}T$$
$$=P\big|_{2-k}\big(T-(-\mathbf{1})\big)+Q\big|_{2-k}(T-TST)$$
$$=P\big|_{2-k}\big(T-(-\mathbf{1})\big)+\big(Q\big|_{2-k}TS\big)\big|_{2-k}\big((-\mathbf{1})S-T\big).$$

Since $P\big|_{2-k}(-\mathbf{1})=P$, we have

$$\left(P\big|_{2-k}(\mathbf{1}-T)+Q\big|_{2-k}(\mathbf{1}-TS)\right)\Big|_{2-k}T=P\big|_{2-k}(T-\mathbf{1})+\big(Q\big|_{2-k}TS\big)\big|_{2-k}(S-T).$$

This implies the inclusion

$$\mathbb{C}[X]_{k-2}\supset E+F\supset\left\{P\big|_{2-k}(S-\mathbf{1}):\ P\in\mathbb{C}[X]_{k-2}\right\}.$$

Since the only polynomials invariant under the translation S are the constant functions, we see that $\left\{P\big|_{2-k}(S-\mathbf{1}):\ P\in\mathbb{C}[X]_{k-2}\right\}$ has codimension 1 in $\mathbb{C}[X]_{k-2}$. This shows that either $E+F=\mathbb{C}[X]_{k-2}$ or

$$E+F=\left\{P\big|_{2-k}(S-\mathbf{1}):\ P\in\mathbb{C}[X]_{k-2}\right\}.$$

Let us calculate the space

$$\left\{P\big|_{2-k}(S-\mathbf{1}):\ P\in\mathbb{C}[X]_{k-2}\right\}.$$

Let $k\geq 4$ and

$$P(X)=a\,X^{k-2}+\sum_{l=0}^{k-3}b_l\,X^l\in\mathbb{C}[X]_{k-2}.$$

We have

$$\begin{aligned}P\big|_{2-k}(S-\mathbf{1})&=P(X+1)-P(X)\\&=a\,(X+1)^{k-2}-a\,X^{k-2}+\sum_{l=0}^{k-3}b_l\left((X+1)^l-X^l\right)\\&=a\left(\sum_{m=0}^{k-3}\binom{k-2}{m}X^m\right)+\sum_{l=0}^{k-3}b_l\left((X+1)^l-X^l\right)\\&\in\mathbb{C}[X]_{k-3}.\end{aligned}$$

This shows that $P\big|_{2-k}(S-\mathbf{1}) \in \mathbb{C}[X]_{k-3}$ does not contain any term of order X^{k-2}. Since we have already shown that the space above has codimension 1 in $\mathbb{C}[X]_{k-2}$, we conclude that

$$\left\{P\big|_{2-k}(S-\mathbf{1}) \colon\; P \in \mathbb{C}[X]_{k-2}\right\} = \mathbb{C}[X]_{k-3}.$$

The case $k=2$ is easy. Since $\mathbb{C}[X]_{k-2} = \mathbb{C}$ is the space of constant polynomials which is translation invariant by construction, we see $P\big|_{2-k}(S-\mathbf{1}) = 0$ for every $P(X) \in \mathbb{C}[X]_0 = \mathbb{C}$. Hence

$$\left\{P\big|_{2-k}(S-\mathbf{1}) \colon\; P \in \mathbb{C}[X]_{k-2}\right\} = \{0\}$$

is the set containing only the zero function. Summarizing, we have

$$\left\{P\big|_{2-k}(S-\mathbf{1}) \colon\; P \in \mathbb{C}[X]_{k-2}\right\} = \begin{cases} \{0\} & \text{if } k=2 \text{ and} \\ \mathbb{C}[X]_{k-3} & \text{if } 4 \le k \in 2\mathbb{N}. \end{cases}$$

Now consider the space E again. Since $Q(X) := X^{k-2} - 1$ satisfies

$$Q\big|_{2-k}(\mathbf{1}+T)(X) = \left(X^{k-2} - 1\right) + X^{k-2}\left(\frac{1}{X^{k-2}} - 1\right) = 0,$$

we see that

$$X^{k-2} - 1 \in E.$$

Hence the space E and, therefore, $E+F$ contain polynomials of degree $k-2$. Therefore,

$$E + F = \mathbb{C}[X]_{k-2}$$

since we just showed that $E+F$ is strictly larger than $\left\{P\big|_{2-k}(S-\mathbf{1}) \colon\; P \in \mathbb{C}[X]_{k-2}\right\}$. This concludes the first part of the lemma since $\dim \mathbb{C}[X]_{k-2} = k-1$.

2. Assume that a polynomial

$$Q(X) := \sum_{l=0}^{k-2} a_l\, X^l \in \mathbb{C}[X]_{k-2}$$

satisfies

$$Q\big|_{k-2}(\mathbf{1}+T) = 0,$$

that is the defining property of the space E in (2.61). We have in particular

$$0 = \sum_{l=0}^{k-2} a_l\, X^l + \sum_{l=0}^{k-2} (-1)^l a_l\, X^{k-2-l} = \sum_{l=0}^{k-2} \left(a_l + (-1)^{k-2-l} a_{k-2-l} \right) X^l.$$

Comparing the coefficients of the monomial factors X^l shows that they satisfy the condition

$$a_l + (-1)^{k-2-l} a_{k-2-l} = 0.$$

Hence E has the dimension stated in the lemma.

3. First notice the set identity

$$\begin{aligned} F &\overset{(2.62)}{=} \left\{ P\big|_{k-2}(\mathbf{1} - TS) \colon\ P \in \mathbb{C}[X]_{k-2} \right\} \\ &= \mathbb{C}[X]_{k-2} \setminus \left\{ P \in \mathbb{C}[X]_{k-2} \colon\ P\big|_{k-2}(\mathbf{1} - TS) = 0 \right\}. \end{aligned}$$

Hence,

$$\dim F = k - 1 - \dim \left\{ P \in \mathbb{C}[X]_{k-2} \colon\ P\big|_{k-2}(\mathbf{1} - TS) = 0 \right\}$$

and we only need to calculate the dimension of the right-hand space. For $k = 2$, recall that constant functions are invariant under $\Gamma(1)$, see Proposition 1.74. Hence,

$$\dim \left\{ P \in \mathbb{C}[X]_0 \colon\ P\big|_{k-2}(\mathbf{1} - TS) = 0 \right\} = \dim \mathbb{C}[X]_0 = 1$$

and

$$\dim F = 2 - 1 - 1 = 0.$$

To calculate the dimension of the space $\left\{ P \in \mathbb{C}[X]_0 \colon\ P\big|_{k-2}(\mathbf{1} - TS) = 0 \right\}$ for $4 \le k \in 2\mathbb{N}$, we first consider the operator

$$\mathbb{C}[X]_1 \to \mathbb{C}[X]_1; \quad Q \mapsto Q\big|_{-1} TS.$$

Lemma 2.58 shows that $\mathbb{C}[X]_1$ can be decomposed into two linearly independent eigenspaces of the operator spanned by $\rho X + 1$ and $\overline{\rho} X + 1$ with eigenvalues $\overline{\rho}$ and ρ, respectively. Hence the operator

$$\mathbb{C}[X]_{k-2} \to \mathbb{C}[X]_{k-2}; \quad Q \mapsto Q\big|_{2-k} TS,$$

viewed as product operator, admits an eigenspace decomposition

$$\mathbb{C}[X]_{k-2} = \bigoplus_{l=0}^{k-2} \mathbb{C} \left(\overline{\rho} X + 1 \right)^l \left(\rho X + 1 \right)^{k-2-l}$$

with eigenvalues $\rho^l\,\overline{\rho}^{k-2-l}$, respectively. This allows us to consider the solution space of

$$P\big|_{2-k}(\mathbf{1}-TS).$$

It consists of the eigenspaces of the operator $Q \mapsto Q\big|_{2-k}TS$ discussed above provided that the eigenvalues satisfy

$$\rho^l\,\overline{\rho}^{k-2-l} = 1 \qquad \text{for suitable } 0 \le l \le k-2.$$

Since $\rho = \frac{1+\sqrt{3}i}{2}$, we have $\overline{\rho} = \rho^5$. This implies

$$\rho^l\,\rho^{5(k-2-l)} = 1$$

for suitable $0 \le l \le k-2$. On the level of the index l, this means that $0 \le l \le k-2$ has to satisfy

$$\begin{aligned} l + 5(k-2) - 5l \equiv 0 \bmod 6 \iff &\ 4l \equiv 5(k-2) \bmod 6 \\ \iff &\ l \equiv k-2 \bmod 3. \end{aligned}$$

Counting solutions of the above equality, we find

$$\begin{aligned} &\sharp\big\{0 \le l \le k-2\colon\ l \equiv k-2 \bmod 3\big\} \\ &= \begin{cases} \left\lceil \frac{k-2}{3} \right\rceil + 1 & \text{if } k-2 = 0, 2 \pmod 3 \text{ and} \\ \left\lceil \frac{k-2}{3} \right\rceil & \text{if } k-2 = 1 \pmod 3). \end{cases} \end{aligned}$$

Summarizing, we have shown

$$\begin{aligned} &\dim\big\{P \in \mathbb{C}[X]_{k-2}\colon\ P\big|_{k-2}(\mathbf{1}-TS) = 0\big\} \\ &= \begin{cases} \left\lceil \frac{k-2}{3} \right\rceil + 1 & \text{if } k-2 = 0, 2 \pmod 3 \text{ and} \\ \left\lceil \frac{k-2}{3} \right\rceil & \text{if } k-2 = 1 \pmod 3). \end{cases} \end{aligned}$$

and hence

$$\begin{aligned} \dim F &= k-1-\dim\big\{P \in \mathbb{C}[X]_{k-2}\colon\ P\big|_{k-2}(\mathbf{1}-TS) = 0\big\} \\ &= k-1-\begin{cases} \left\lceil \frac{k-2}{3} \right\rceil + 1 & \text{if } k-2 = 0, 2 \pmod 3 \text{ and} \\ \left\lceil \frac{k-2}{3} \right\rceil & \text{if } k-2 = 1 \pmod 3). \end{cases} \end{aligned}$$

This proves the third statement of the lemma.

□

Proposition 2.62 *The map* $\mu : M_k(\Gamma(1)) \oplus S_k(\Gamma(1)) \to H^1(\Gamma(1), \mathbb{C}[X]_{k-2})$ *defined in* (2.47) *is surjective.*

Proof Following the arguments in [98, §4] to prove the surjectivity of the linear map μ, we compare the dimensions of the spaces $M_k(\Gamma(1)) \oplus S_k(\Gamma(1))$ and $H^1(\Gamma(1), \mathbb{C}[X]_{k-2})$. In particular, we will show

$$\dim M_k(\Gamma(1)) + \dim S_k(\Gamma(1)) = \dim H^1(\Gamma(1), \mathbb{C}[X]_{k-2}).$$

The left-hand side is stated in (1.71) and (1.72). We have

$$\dim M_k(\Gamma(1)) \overset{(1.71)}{=} \begin{cases} \left[\frac{k}{12}\right] & \text{for } k \equiv 2 \bmod 12 \text{ and} \\ \left[\frac{k}{12}\right]+1 & \text{for } k \not\equiv 2 \bmod 12 \end{cases}$$

and

$$\dim S_k(\Gamma(1)) \overset{(1.72)}{\underset{}{\overset{(1.71)}{=}}} \begin{cases} 0 & \text{for } 2 \leq k \leq 10, \\ \left[\frac{k}{12}\right]-1 & \text{for } 12 \leq k \equiv 2 \bmod 12 \text{ and} \\ \left[\frac{k}{12}\right] & \text{for } 12 \leq k \not\equiv 2 \bmod 12. \end{cases}$$

Combining both results we get

$$\begin{aligned} \dim\left(M_k(\Gamma(1)) \oplus S_k(\Gamma(1))\right) &= \dim M_k(\Gamma(1)) + \dim S_k(\Gamma(1)) \\ &= \begin{cases} 0 & \text{for } k = 2, \\ 2\left[\frac{k}{12}\right]-1 & \text{for } 4 \leq k \equiv 2 \bmod 12 \text{ and} \\ 2\left[\frac{k}{12}\right]+1 & \text{for } 4 \leq k \not\equiv 2 \bmod 12. \end{cases} \end{aligned} \tag{2.63}$$

To calculate the dimension of $H^1(\Gamma(1), \mathbb{C}[X]_{k-2})$, we put D_3 equal to the dimension of the space of cocycles and D_4 equal to the space of coboundaries

$$D_3 := \dim Z^1(\Gamma(1), \mathbb{C}[X]_{k-2}) \quad \text{and} \quad D_4 := \dim B^1(\Gamma(1), \mathbb{C}[X]_{k-2}).$$

Definition 2.36 implies that the space of coboundaries is given by taking a polynomial $P \in \mathbb{C}[X]_{k-2}$ and forming the coboundary $\left(P\big|_{2-k}V - P\right)_{V\in\Gamma(1)}$. For such a coboundary to vanish means that P satisfies (1.28) for all $V \in \Gamma(1)$. In other words, $P \in M^!_{2-k}(\Gamma(1))$. The example stated in §1.3.1 and Proposition 1.74 implies

$$\dim M^!_{2-k}(\Gamma(1)) = \begin{cases} 1 & \text{for } k = 2 \text{ and} \\ 0 & \text{for } k > 2. \end{cases}$$

Hence,

$$\begin{aligned}
D_4 &= \dim B^1\big(\Gamma(1), \mathbb{C}[X]_{k-2}\big) \\
&= \dim \mathbb{C}[X]_{k-2} - \dim M^!_{2-k}\big(\Gamma(1)\big) \\
&= k-1-\begin{cases} 1 & \text{for } k=2 \text{ and} \\ 0 & \text{for } k>2 \end{cases} \\
&= \begin{cases} 0 & \text{for } k=2 \text{ and} \\ k-1 & \text{for } k>2. \end{cases}
\end{aligned} \tag{2.64}$$

To calculate D_3, we first observe the following: (2.38) implies that each cocycle $\big(P_M\big)_{M\in\Gamma(1)}$ is uniquely determined by its values on the generators S and T of the full modular group $\Gamma(1)$. Hence, we only need to assign polynomials $P_S, P_T \in \mathbb{C}[X]_{k-2}$ in such a way that the associated cocycle satisfies the relations related to $T^2 = -\mathbf{1} = (TS)^3$ (see Exercise 1.12):

$$0 \overset{(2.39)}{=} P_{-\mathbf{1}} = P_{T^2} \overset{(2.38)}{=} P_T\big|_{2-k}T + P_T = P_T\big|_{2-k}\big(T+\mathbf{1}\big)$$

and

$$\begin{aligned}
0 &\overset{(2.39)}{=} P_{-\mathbf{1}} = P_{(TS)^3} \overset{(2.38)}{=} P_T\big|_{2-k}S(TS)^2 + P_{S(TS)^2} \\
&\overset{(2.38)}{=} P_T\big|_{2-k}S(TS)^2 + P_S\big|_{2-k}(TS)^2 + P_T\big|_{2-k}S(TS) + \\
&\quad + P_S\big|_{2-k}(TS) + P_T\big|_{2-k}S + P_S \\
&= P_T\big|_{2-k}S\big((TS)^2+TS+\mathbf{1}\big) + P_S\big|_{2-k}\big((TS)^2+TS+\mathbf{1}\big) \\
&= \big(P_T\big|_{2-k}S + P_S\big)\big|_{2-k}\big((TS)^2+TS+\mathbf{1}\big).
\end{aligned}$$

By (2.61), this implies $P_T \in E$ and $P_T\big|_{2-k}S + P_S \in F$. In particular, we see that the space D_3 is isomorphic to the direct sum of the spaces $E \oplus F$ and the dimension of D_3 can be calculated as

$$\dim D_3 = \dim E + \dim F.$$

The terms appearing on the right-hand side are already calculated in Lemma 2.61. We have

$$\dim D_3 = \dim E + \dim F$$
$$= \begin{cases} \frac{k}{2} & \text{if } \frac{k-2}{2} \text{ is odd and} \\ \frac{k}{2} - 1 & \text{if } \frac{k-2}{2} \text{ is even,} \end{cases}$$
$$+ \ k - 1 - \begin{cases} 1 & \text{if } k = 2, \\ \left\lceil \frac{k-2}{3} \right\rceil + 1 & \text{if } 4 \leq k \equiv 1, 2 \pmod 3\text{, and} \\ \left\lceil \frac{k-2}{3} \right\rceil & \text{if } 4 \leq k \equiv 0 \pmod 3. \end{cases}$$
$$= k - 2 + \begin{cases} 0 & \text{if } k = 2, \\ 2\left[\frac{k}{12}\right] & \text{if } 4 \leq k \equiv 2 \pmod{12}\text{, and} \\ 2\left[\frac{k}{12}\right] + 2 & \text{if } 4 \leq k \not\equiv 2 \pmod{12}. \end{cases} \tag{2.65}$$

Summarizing, we see that the dimension of $H^1\big(\Gamma(1), \mathbb{C}[X]_{k-2}\big)$ is given by

$$\dim H^1\big(\Gamma(1), \mathbb{C}[X]_{k-2}\big) = \dim Z^1\big(\Gamma(1), \mathbb{C}[X]_{k-2}\big) - \dim B^1\big(\Gamma(1), \mathbb{C}[X]_{k-2}\big)$$
$$= \dim D_3 - \dim D_4$$
$$\overset{(2.64)}{\underset{(2.65)}{=}} \begin{cases} 0 & \text{if } k = 2, \\ 2\left[\frac{k}{12}\right] - 1 & \text{if } 4 \leq k \equiv 2 \pmod{12}\text{, and} \\ 2\left[\frac{k}{12}\right] + 1 & \text{if } 4 \leq k \not\equiv 2 \pmod{12}. \end{cases}$$

The dimension agrees with $\dim\Big(M_k\big(\Gamma(1)\big) \oplus S_k\big(\Gamma(1)\big)\Big)$ calculated in (2.63). We have

$$\dim H^1\big(\Gamma(1), \mathbb{C}[X]_{k-2}\big) = \dim\Big(M_k\big(\Gamma(1)\big) \oplus S_k\big(\Gamma(1)\big)\Big).$$

Hence the map μ defined in (2.47) is surjective. □

2.4 Hecke Operators and Period Polynomials

For the next subsection we follow the approach presented in [57] to describe Hecke operators acting on period functions. The approach is based on matrix identities within the formal polynomial ring of matrices $\mathbb{Z}\big[\Gamma(1)\big]$ modulo the relations imposed by the period polynomials.

There is an alternative approach using the integral representation (2.1) of period polynomials. This approach was originally used by Manin [131]. Although we present here the approach presented in [57], we also discuss the alternative approach in the context of period functions, using an integral representations of period functions in Section 5.3 of Chapter 5.

Recall the sets $\mathcal{R}_\star$ and $\mathcal{R}_n$, $n \in \mathbb{N}$, as introduced in Definition 1.170. Note that $\mathcal{R}_\star$ is a noncommutative ring with unit and is "multiplicatively graded" in the sense that $\mathcal{R}_m\mathcal{R}_n \subset \mathcal{R}_{mn}$. In particular, $\mathcal{R}_m$ is a left and right module over the group ring $\mathcal{R}_1 = \mathbb{Z}\big[\Gamma(1)\big]$. Also, we recall the slash operator introduced in (1.131): for each $k \in 2\mathbb{Z}$, $\sum_l \alpha_l M_l \in \mathcal{R}_n$, and $f : \mathbb{H} \to \mathbb{C}$, we have

$$f\big|_k\left(\sum_l \alpha_l M_l\right)(z) = \sum_l \alpha_l\, j(M_l, z)^k\, f(M_l\, z)$$

for $z \in \mathbb{H}$.

Recall the Hecke operator T_n, $n \in \mathbb{N}$, as given in Definition 1.172. The action of T_n on a function $f : \mathbb{H} \to \mathbb{C}$ can be described by the action of the element $\sum_{d|n}\sum_{b=0}^{d-1}\begin{pmatrix}\frac{n}{d} & b\\ 0 & d\end{pmatrix} \in \mathcal{R}_n$:

$$T_n\, f \overset{(1.137)}{=} n^{k-1}\sum_{d|n}\sum_{b=0}^{d-1} f\Big|_k\begin{pmatrix}\frac{n}{d} & b\\ 0 & d\end{pmatrix}.$$

We identify the Hecke operator T_n with its matrix representation. Hence, we denote the matrix representation also with T_n:

$$T_n := \sum_{d|n}\sum_{b=0}^{d-1}\begin{pmatrix}\frac{n}{d} & b\\ 0 & d\end{pmatrix} = \sum_{\substack{ad=n\\ 0\le b<d}}\begin{pmatrix}a & b\\ 0 & d\end{pmatrix} = \sum_{A\in\mathcal{T}(n)} A \in \mathcal{R}_n \tag{2.66}$$

where the set $\mathcal{T}(n)$ is defined in (1.139). The operator T_n is obtained by

$$T_n f = n^{k-1}\, f\big|_k T_n.$$

However, some elements of $\mathcal{R}_n$ act trivially on functions using the slash operator. For example, let $M = \begin{pmatrix}a & b\\ c & d\end{pmatrix}$ be a matrix in $\mathrm{GL}_2(\mathbb{Z})$ with determinant n. The formal difference $M - (-M) \in \mathcal{R}_n$ and the action of this element on functions gives

$$\begin{aligned} f\big|_k\big(M-(-M)\big)(z) &= \big(f\big|_k M\big)(z) - \big(f\big|_k(-M)\big)(z)\\ &= (cz+d)^{-k}\, f\left(\frac{az+b}{cz+d}\right) - (-cz-d)^{-k}\, f\left(\frac{-az-b}{-cz-d}\right)\\ &= 0 \end{aligned}$$

since $k \in 2\mathbb{Z}$. This shows that some elements of $\mathcal{R}_n$ lie in the kernel. To work around this issue, we make the following assumption for this section: *in §2.4, we write identities of elements in every $\mathcal{R}_n$ up to trivial actions by the slash operator. To mark this identification, we write $\doteq$ instead of $=$ for equality.* This means, for example, that we identify $(-\mathbf{1})$ with $\mathbf{1}$, and $M - (-M)$ with 0:

$$(-\mathbf{1}) \doteq \mathbf{1} \qquad \text{and} \qquad M - (-M) \doteq 0.$$

In [131], Manin gave the representation

$$T_n^* = \sum_{\substack{ad-bc=n \\ a>c>0 \\ d>-b>0}} \left[\begin{pmatrix} a & b \\ c & d \end{pmatrix} + \begin{pmatrix} a & -b \\ -c & d \end{pmatrix} \right] + \sum_{\substack{ad=n \\ -\frac{1}{2}d<b\leq\frac{1}{2}d}} \begin{pmatrix} a & b \\ 0 & d \end{pmatrix} + \sum_{\substack{ad=n \\ -\frac{1}{2}a<c\leq\frac{1}{2}a \\ c\neq 0}} \begin{pmatrix} a & 0 \\ c & d \end{pmatrix} \tag{2.67}$$

for the Hecke operators in $\mathcal{R}_m$ acting on period polynomials in the obvious sense. For P, a period polynomial in the sense of Definition 2.7, the element $\tilde{T}_n \in \mathcal{R}_n$ acts on P by

$$P \mapsto n^{k-1} \, P\big|_{2-k} T_n^*.$$

Choie and Zagier presented in [57] a simple proof that T_n^* indeed represents the Hecke operator of level n on period polynomials. To formulate this result, consider the matrices

$$S = \begin{pmatrix} 1 & 1 \\ 0 & 1 \end{pmatrix} \quad \text{and} \quad T = \begin{pmatrix} 0 & -1 \\ 1 & 0 \end{pmatrix} \tag{1.20}$$

and the right ideal $\mathcal{J}$ of $\mathcal{R}_1$ defined as

$$\mathcal{J} = (\mathbf{1} + T)\mathcal{R}_1 + \big(\mathbf{1} + TS + (TS)^2\big)\mathcal{R}_1. \tag{2.68}$$

Choie and Zagier proved:

Theorem 2.63 ([57, Theorem 2]) *For each integer $n \geq 1$, the matrix representation T_n of the nth Hecke operator satisfies the relations*

$$T_n(S - \mathbf{1}) \doteq 0, \qquad T_n(T - \mathbf{1}) \doteq (T - 1)\tilde{T}_n \quad (\mathrm{mod}\ (S - \mathbf{1})\mathcal{R}_n) \tag{2.69}$$

for a certain element $\tilde{T}_m \in \mathcal{R}_m$ that is unique modulo $J\mathcal{R}_n$ and satisfies $\tilde{T}_n\mathcal{J} \subset \mathcal{J}\mathcal{R}_n$. The elements $\tilde{T}_m, \tilde{T}_n \in \mathcal{R}_*$ satisfy the product formula

$$\tilde{T}_m \, \tilde{T}_n = \sum_{d|\gcd(n,m)} d^{-1} \begin{pmatrix} d & 0 \\ 0 & d \end{pmatrix} \tilde{T}_{\frac{nm}{d^2}} \quad (\mathrm{mod}\ \mathcal{J}\mathcal{R}_{nm}). \tag{2.70}$$

Theorem 2.64 ([57, Theorem 3]) *The set T_n^* satisfies the conditions put on $\tilde{T}_n$ in Theorem 2.63.*

Corollary 2.65 *Let $k \in 2\mathbb{N}$. If $P \in \mathbb{P}_{k-2}$ is a period polynomial and $n \in \mathbb{N}$, then the element $T_n^* \in \mathcal{R}_n$ given in (2.67) induces an operator on period polynomials, denoted by $\tilde{T}_n$ given by*

$$\tilde{T}_n P := P\big|_{2-k} T_n^*. \tag{2.71}$$

In particular $\tilde{T}_n P$ is a period polynomial. Moreover, the operators satisfy

$$\tilde{T}_m \tilde{T}_n P = \tilde{T}_n \tilde{T}_m P = \sum_{d \mid \gcd(m,n)} d^{2k-1}\, \tilde{T}^*_{\frac{mn}{d^2}} P \tag{2.72}$$

for all $n, m \in \mathbb{N}$.

Proof Let P be a period polynomial and $n \in \mathbb{N}$. Using (2.69) of Theorem 2.63, we see that there exists a $\tilde{T}_n \in \mathcal{R}_n$ such that $P\big|_{2-k}\tilde{T}_n$ corresponds to taking the Hecke operator T_n on the corresponding cusp form. Theorem 2.64 shows that such an $\tilde{T}_n \in \mathcal{R}_n$ is realized by T_n^*. □

The remainder of this section is used to prove Theorems 2.63 and 2.64 where we follow the arguments and reasonings presented in [57, §3] closely.

Before we explain Choie and Zagier's argument, we need some preliminary results. First, we need is a generalization of the identities (1.19) and (2.2).

Exercise 2.66 For $A \in \mathrm{Mat}_2(\mathbb{Z})$ with $\det A \neq 0$ and $k \in 2\mathbb{Z}$, prove

$$\mathrm{d}(Az) = \det(A)\, \frac{\mathrm{d}z}{j(A,z)^2} \tag{2.73}$$

and

$$\big(Aw - Az\big)^k = \det(A)^k\, \frac{(w-z)^k}{j(A,w)^k\, j(A,z)^k} \tag{2.74}$$

for all $z, w \in \mathbb{H}$.

Let $f \in S_k\big(\Gamma(1)\big)$. We have seen (e.g., in Example 2.48 and Lemma 2.49) that a period polynomial P can be written in terms of the Eichler integral $\tilde{F}$ (see (2.17)) of the cusp form f:

$$P = \tilde{F}\big|_{2-k}(T - \mathbf{1}) \quad \text{and} \quad 0 = \tilde{F}\big|_{2-k}(S - \mathbf{1}).$$

Denoting the matrix representation in $\mathcal{R}_n$ of the Hecke operator T_n of level n defined in (1.137) also by T_n, we have

$$T_n = \sum_{\substack{ad=n \\ 0 \le b < d}} \begin{pmatrix} a & b \\ 0 & d \end{pmatrix}. \tag{2.75}$$

Using this realization of the Hecke operator on cusp forms, we have for $A = \begin{pmatrix} a & b \\ 0 & d \end{pmatrix}$ with $ad = n$ and $0 \le b < d$:

$$\begin{aligned}
\tilde{F}\big|_{2-k} A(z) &= j(A, z)^{k-2}\, \tilde{F}(Az) \\
&\overset{(2.17)}{=} j(A, z)^{k-2} \int_{Az}^{i\infty} f(w)\,\big(w - Az\big)^{k-2}\, dw \\
&\overset{w \mapsto Aw}{=} j(A, z)^{k-2} \int_{z}^{A^{-1} i\infty} f(Aw)\,(Aw - Az)^{k-2}\, d(Aw) \\
&\overset{(2.73),(2.74)}{=} \det(A)^{k-1} \int_{z}^{A^{-1} i\infty} j(A, w)^{-k} f(Aw)\,(w - z)^{k-2}\, dw \\
&= \det(A)^{k-1} \int_{z}^{A^{-1} i\infty} \big(f\big|_k A\big)(w)\,(w - z)^{k-2}\, dw \\
&\overset{A^{-1} i\infty = i\infty}{=} \int_{z}^{i\infty} \det(A)^{k-1} \big(f\big|_k A\big)(w)\,(w - z)^{k-2}\, dw.
\end{aligned}$$

The last integral on the right-hand side is the Eichler integral of the cusp form $f\big|_k A$. This shows that the action of the Hecke operator T_n on the cusp form f can be lifted to an action on the associated Eichler integral $\tilde{F}$. The same arguments hold for the variant $\tilde{G}^-$ of the Niebur integral, see (2.33). We have just proven the following.

Lemma 2.67 *For $k \in 2\mathbb{N}$ and $f, g \in S_k\big(\Gamma(1)\big)$, denote by $\tilde{F}$ the associated Eichler integral given in* (2.17). *For $n \in \mathbb{N}$ taken as the level, we have*

$$f \mapsto T_n f = n^{k-1}\, f\big|_k T_n \iff \tilde{F} \mapsto \tilde{F}\big|_{2-k} T_n$$

and

$$g \mapsto T_n f = n^{k-1}\, g\big|_k T_n \iff \tilde{G}^- \mapsto \tilde{G}^-\big|_{2-k} T_n.$$

In other words, Lemma 2.67 shows that applying the Hecke operator of level n on a cusp form is the same as applying the map $\tilde{F} \mapsto \tilde{F}\big|_{2-k} T_n$ on the Eichler integral $\tilde{F}$ and the corresponding expression for the Niebur integral. Since the Eichler and Niebur integrals also satisfy $\tilde{F}\big|_{2-k}(S - \mathbf{1}) = 0$, we get the following algebraic description: suppose that there are elements X_n, $\tilde{T}_n$, and Y_n in $\mathcal{R}_n$ satisfying

$$T_n\,(S-\mathbf{1}) \doteq (S-\mathbf{1})\,X_n \quad \text{and} \quad T_n\,(T-\mathbf{1}) \doteq (T-\mathbf{1})\,\tilde{T}_n + (S-\mathbf{1})\,Y_n.$$

Then, these elements satisfy

$$\begin{aligned}\big(\tilde{F}\big|_{2-k}T_n\big)\big|_{2-k}(S-\mathbf{1}) &= \big(\tilde{F}\big|_{2-k}(S-\mathbf{1})\big)\big|_{2-k}X_n \\ &= 0\big|_{2-k}X_n = 0\end{aligned}$$

and

$$\begin{aligned}\big(\tilde{F}\big|_{2-k}T_n\big)\big|_{2-k}(T-\mathbf{1}) &= \big(\tilde{F}\big|_{2-k}(T-\mathbf{1})\big)\big|_{2-k}\tilde{T}_n \;+\; \big(\tilde{F}\big|_{2-k}(S-\mathbf{1})\big)\big|_{2-k}Y_n \\ &= \big(\tilde{F}\big|_{2-k}(T-\mathbf{1})\big)\big|_{2-k}\tilde{T}_n \;+\; 0 \\ &= P\big|_{2-k}\tilde{T}_n.\end{aligned}$$

This shows that $\tilde{F}\big|_{2-k}T_n$ and its corresponding Niebur integral are again modular integrals with associated period polynomial $\big|_{2-k}\tilde{T}_n$.

Lemma 2.68 *We have*

$$M-\mathbf{1} \in \big(T-\mathbf{1}\big)\mathcal{R}_1 + \big(S-\mathbf{1}\big)\mathcal{R}_1 \tag{2.76}$$

for every $M \in \Gamma(1)$.

Proof Assume that there exists $M \in \Gamma(1)$ satisfying (2.76). We could take $M = \mathbf{1}$ for example, since $\mathbf{1} - \mathbf{1} = 0$ is the neutral element of $\mathcal{R}_1$. Then, consider the elements

$$\begin{aligned}TM-\mathbf{1} &= \big(T-\mathbf{1}\big)M + \big(M-\mathbf{1}\big), \\ SM-\mathbf{1} &= \big(S-\mathbf{1}\big)M + \big(M-\mathbf{1}\big), \\ T^{-1}M-\mathbf{1} &= \big(T-\mathbf{1}\big)\big(-T^{-1}M\big) + \big(M-\mathbf{1}\big) \quad \text{and} \\ S^{-1}M-\mathbf{1} &= \big(S-\mathbf{1}\big)\big(-S^{-1}M\big) + \big(M-\mathbf{1}\big).\end{aligned}$$

This shows that the elements $TM-\mathbf{1}$, $TM-\mathbf{1}$, $TM-\mathbf{1}$ and $TM-\mathbf{1}$ are elements in $\big(T-\mathbf{1}\big)\mathcal{R}_1 + \big(S-\mathbf{1}\big)\mathcal{R}_1$. Using the fact that S and T generate $\Gamma(1)$, we conclude that all elements $M \in \Gamma(1)$ satisfy (2.76). □

Lemma 2.69 *The* T_n *defined in* (2.66) *satisfies* $T_n\big(T-\mathbf{1}\big) \in \big(T-\mathbf{1}\big)\mathcal{R}_n + \big(S-\mathbf{1}\big)\mathcal{R}_n$.

Proof Consider a term in the formal sum of matrices T_n. Each element has the form $A := \begin{pmatrix} a & b \\ 0 & d \end{pmatrix}$ with $ad = n$ and $0 \le b < d$. Consider

$$A\,T = \begin{pmatrix} a & b \\ 0 & d \end{pmatrix} T = \begin{pmatrix} b & -a \\ d & 0 \end{pmatrix} \in H(n).$$

Then $AT = M_A A'$ for some unique $A' \in \mathcal{T}(n)$ and $M_A \in \Gamma(1)$, both depending on A. Hence $T_n\,T$ can be written as

$$T_n\,T = \sum_{A \in cT(n)} A\,T = \sum_{A \in cT(n)} M_A\,A'.$$

This shows that we can write $T_n\,\big(T - \mathbf{1}\big)$ as

$$T_n\,\big(T - \mathbf{1}\big) = \sum_{A \in cT(n)} M_A\,A' - \sum_{A \in cT(n)} A = \sum_{A \in cT(n)} \big(M_A - \mathbf{1}\big)\,A'$$

since A' runs through all representatives of $\mathcal{T}(n)$ if A does. Using Lemma 2.68, we find

$$T_n\,\big(T - \mathbf{1}\big) = \sum_{A \in cT(n)} \big(M_A - \mathbf{1}\big)\,A' \in \big(T - \mathbf{1}\big)\mathcal{R}_n + \big(S - \mathbf{1}\big)\mathcal{R}_n.$$

□

Suppose that V is an abelian group on which $\Gamma(1)$ acts. Then V is a left $\mathcal{R}_1$-module. For $X \in \mathcal{R}_1$ write

$$\mathrm{Ker}(X) := \{v \in V\colon\ Xv = 0\}$$

and

$$\mathrm{Im}(X) := \{Xv\colon\ v \in V\}$$

for kernel and image of the map

$$V \mapsto V; \quad v \mapsto Xv.$$

We have the following.

Definition 2.70 We call V *acyclic* if

1. $\mathrm{Ker}(\mathbf{1} + T) \cap \mathrm{Ker}\big(\mathbf{1} + TS + (TS)^2\big) \doteq \{0\}$ (i.e., V acts trivially by the slash operator), and
2. $\mathrm{Ker}(\mathbf{1} - T) = \mathrm{Image}(\mathbf{1} + T)$ and $\mathrm{Ker}(\mathbf{1} - TS) = \mathrm{Image}\big(\mathbf{1} + TS + (TS)^2\big)$.

Remark 2.71

1. The first condition that $\mathrm{Ker}(\mathbf{1}+T)\cap\mathrm{Ker}\big(\mathbf{1}+TS+(TS)^2\big)$ acts trivially by the slash operator is required since we discuss $\Gamma(1) = \mathrm{SL}_2(\mathbb{Z})$. Choie and Zagier discuss the quotient group $\mathrm{PSL}_2(\mathbb{Z})$ that divides out the trivial action of $\Gamma(1) = \mathrm{SL}_2(\mathbb{Z})$ by Möbius transformations. In other words, the elements
$$\begin{pmatrix} a & b \\ c & d \end{pmatrix} \quad \text{and} \quad \begin{pmatrix} -a & -b \\ -c & -d \end{pmatrix}$$
in $\mathrm{PSL}_2(\mathbb{Z})$ are identified and treated as a single element.
2. The second defining property of acyclicity is listed for the sake of completeness since it is mentioned in [57, p. 13]. We do not necessarily need this extra assumption if we work over $\mathbb{Q}$ since $(\mathbf{1}-T)v = 0$ implies $v = (\mathbf{1}+T)\frac{v}{2}$ and $(\mathbf{1}-TS)v = 0$ implies $v = \big(\mathbf{1}+TS+(TS)^2\big)\frac{v}{3}$. This was already used in §2.3.2.

Example 2.72 The space of period functions $\mathbb{P}_{k-2}$ is given by the intersection of
$$\mathbb{P}_{k-2} = \mathrm{Ker}(\mathbf{1}+T) \cap \mathrm{Ker}\big(\mathbf{1}+TS+(TS)^2\big)$$
with $V = \mathbb{C}[z]_{k-2}$.

Choie and Zagier prove the following lemma using rational period functions. Since we did not introduce these functions, we just state this lemma as fact without proof.

Lemma 2.73 ([57, Lemma 2]) *$\mathcal{R}_n$ is an acyclic $\mathcal{R}_1$-module for all $n \in \mathbb{N}$.*

Using the acyclicity of $\mathcal{R}_n$, we are able to show:

Lemma 2.74 ([57, Lemma 3]) *If V is an acyclic $\mathcal{R}_1$-module and $v, v' \in V$ with $v \doteq v'$, then*
$$\big(\mathbf{1}-T\big)v \in \big(1-S\big)V \iff v' \in \big(\mathbf{1}+T\big)V+\big(\mathbf{1}+TS+(TS)^2\big)V = \mathcal{J}V \tag{2.77}$$
where the right ideal $\mathcal{J}$ of $\mathcal{R}_1$ is defined in (2.68).

Proof We consider the two directions separately.

$\Longrightarrow$: Consider $v \in \mathcal{J}V$. Then the form of $\mathcal{J}$ given in (2.68) implies that there exist $u, w \in V$ such that v can be written as
$$v \doteq v' = \big(\mathbf{1}+T\big)u + \big(\mathbf{1}+TS+(TS)^2\big)w.$$

Applying $\big(\mathbf{1}-T\big)$ from the left gives
$$\begin{aligned}\big(\mathbf{1}-T\big)v &\doteq \big(\mathbf{1}-T\big)\big(\mathbf{1}+T\big)u + \big(\mathbf{1}-T\big)\big(\mathbf{1}+TS+(TS)^2\big)w \\ &= \big(\mathbf{1}-T+T-(-\mathbf{1})\big)u+\big(\mathbf{1}+TS+(TS)^2-T-T^2S-T^2STS\big)w\end{aligned}$$

$$
\begin{aligned}
&\doteq \big(\mathbf{1} + TS + (TS)^2 - T - T^2S - T^2STS\big)w \\
&= \big(\mathbf{1} - S\big)\big(\mathbf{1} + TS + (TS)^2\big)w + \\
&\qquad + \big(S + STS + S(TS)^2 - T - T^2S - T^2STS\big)w
\end{aligned}
$$

using the identities

$$T^2 = (-\mathbf{1}) \iff (-T) = T^{-1} \quad \text{and} \quad (TS)^3 = (-\mathbf{1}).$$

(Recall that the $-$ sign in front of the matrix elements is not a "minus" in the sense of $\mathcal{R}_n$; it denotes the scalar multiplication of the matrix by -1. For example, $(-\mathbf{1}) = \begin{pmatrix} -1 & 0 \\ 0 & -0 \end{pmatrix}$ and $(-T) = \begin{pmatrix} 0 & 1 \\ -1 & 1 \end{pmatrix}$0). Calculating the right-hand term, we find

$$
\begin{aligned}
&S + STS + S(TS)^2 - T - T^2S - T^2STS \\
&= \begin{pmatrix} 1 & 1 \\ 0 & 1 \end{pmatrix} + \begin{pmatrix} 1 & 0 \\ 1 & 1 \end{pmatrix} + \begin{pmatrix} 0 & -1 \\ 1 & 0 \end{pmatrix} - \begin{pmatrix} 0 & -1 \\ 1 & 0 \end{pmatrix} - \begin{pmatrix} -1 & -1 \\ 0 & -1 \end{pmatrix} - \begin{pmatrix} -1 & 0 \\ -1 & -1 \end{pmatrix} \\
&= S + STS + T - T - (-S) - (-STS) \\
&\doteq 0.
\end{aligned}
\tag{2.78}
$$

Hence, $\big(\mathbf{1} - T\big)v$ can be written as

$$
\begin{aligned}
\big(\mathbf{1} - T\big)v &\doteq \big(\mathbf{1} - S\big)\big(\mathbf{1} + TS + (TS)^2\big)w + \\
&\qquad + \big(S + STS + S(TS)^2 - T - T^2S - T^2STS\big)w \\
&\doteq \big(\mathbf{1} - S\big)\big(\mathbf{1} + TS + (TS)^2\big)w.
\end{aligned}
$$

This shows that v satisfies

$$\big(\mathbf{1} - T\big)v \in \big(\mathbf{1} - S\big)V.$$

$\Longleftarrow$: Assume that V is acyclic and v satisfies

$$\big(\mathbf{1} - T\big)v = \big(\mathbf{1} - S\big)w$$

for some $w \in V$. Then, we have the identity

$$
\begin{aligned}
\big(\mathbf{1} - T\big)(v - w) &= \big(\mathbf{1} - T\big)v - \big(\mathbf{1} - T\big)w \\
&= \big(\mathbf{1} - S\big)w - \big(\mathbf{1} - T\big)w \\
&= \big(T - S\big)w.
\end{aligned}
$$

Multiplying both sides by T from the left gives

$$\begin{aligned}(T-\mathbf{1})(v-w) &\doteq (T-T^2)(v-w)\\ &= (T^2-TS)w\\ &\doteq (\mathbf{1}-TS)w.\end{aligned}$$

Hence, we have

$$X := (\mathbf{1}-T)(v-w) = -(\mathbf{1}-TS)w.$$

Since $(\mathbf{1}-T)(v-w) \in \operatorname{Im}(\mathbf{1}-T)$ and $-(\mathbf{1}-TS)w \in \operatorname{Im}(\mathbf{1}-TS)$, we see X lies in

$$\begin{aligned}X &\in \operatorname{Im}(\mathbf{1}-T) \cap \operatorname{Im}(\mathbf{1}-TS)\\ &= \operatorname{Ker}(\mathbf{1}-T) \cap \operatorname{Ker}(\mathbf{1}-TS)\\ &\doteq \{0\},\end{aligned}$$

using the second and successively the first property of Definition 2.70. Hence $X = (\mathbf{1}-T)(v-w) \doteq 0$ which implies that the element $v-w$ is in the kernel of $\mathbf{1}-T$ and, similarly, $-w$ is in the kernel of $\mathbf{1}-TS$. This gives the relation

$$v-w \in \operatorname{Im}(1+T) \quad \text{and} \quad -w \in \operatorname{Im}(1+TS+(TS)^2).$$

Hence v satisfies

$$\begin{aligned}v &= v - w \;+\; w\\ &\in \operatorname{Im}(1+T) \;\cap\; \operatorname{Im}(1+TS+(TS)^2)\\ &= (1+T)V \;+\; (1+TS+(TS)^2)V = \mathcal{J}V.\end{aligned}$$

□

Now, we proceed with the proof of Theorem 2.63.

Proof (of Theorem 2.63) Lemmas 2.73 and 2.74 show the following characterization of $\mathcal{J}\mathcal{R}_n$: We have

$$\mathcal{J}\mathcal{R}_n \doteq \{X \in \mathcal{R}_n \colon \; (\mathbf{1}-T)X \in (\mathbf{1}-S)\mathcal{R}_n\}.$$

The conditions in (2.69) on $\tilde{T}_n \in \mathcal{R}_n$ imply

$$
\begin{aligned}
(\mathbf{1}-T)\,\tilde{T}_n(\mathbf{1}+T) &\doteq \Big(T_n\,(\mathbf{1}-T)\;+\;(\mathbf{1}-S)w\Big)(\mathbf{1}+T)\\
&= T_n\,(\mathbf{1}-T)(\mathbf{1}+T)\;+\;(\mathbf{1}-S)w(\mathbf{1}+T)\\
&= T_n\,\big(\underbrace{\mathbf{1}-(-\mathbf{1})}_{\doteq 0}\big)\;+\;(\mathbf{1}-S)w(\mathbf{1}+T)\\
&= (\mathbf{1}-S)w(\mathbf{1}+T)\\
&\doteq 0 \qquad (\mathrm{mod}\ (S-\mathbf{1})\mathcal{R}_n)
\end{aligned}
$$

for some $w \in \mathcal{R}_n$. Similarly, we conclude

$$
\begin{aligned}
&(\mathbf{1}-T)\,\tilde{T}_n\big(\mathbf{1}+TS+(TS)^2\big)\\
&\quad\doteq \Big(T_n\,(\mathbf{1}-T)\;+\;(\mathbf{1}-S)w\Big)\big(\mathbf{1}+TS+(TS)^2\big)\\
&\quad= T_n\,(\mathbf{1}-T)\big(\mathbf{1}+TS+(TS)^2\big)\;+\;(\mathbf{1}-S)w\big(\mathbf{1}+TS+(TS)^2\big)\\
&\quad= T_n\,\big(\mathbf{1}+TS+(TS)^2-T-T^2S-T^2STS\big)\;+\;(\mathbf{1}-S)w\big(\mathbf{1}+TS+(TS)^2\big)\\
&\quad\doteq T_n\,\big(\mathbf{1}-S\big)\big(\mathbf{1}+TS+(TS)^2\big)\;+\;(\mathbf{1}-S)w\big(\mathbf{1}+TS+(TS)^2\big)\\
&\quad\doteq 0\ (\mathrm{mod}\ (S-\mathbf{1})\mathcal{R}_n)
\end{aligned}
$$

for some $w \in \mathcal{R}_n$ where we used the identity

$$
\begin{aligned}
&\big(\mathbf{1}+TS+(TS)^2-T-T^2S-T^2STS\big)\\
&\quad= \quad \big(\mathbf{1}-S\big)\big(\mathbf{1}+TS+(TS)^2\big)+\big(S+STS+S(TS)^2-T-T^2S-T^2STS\big)\\
&\quad\overset{(2.71)}{\doteq} \quad \big(\mathbf{1}-S\big)\big(\mathbf{1}+TS+(TS)^2\big).
\end{aligned}
$$

Property (2.70) follows from the following argument. We have

$$
\begin{aligned}
&(T-\mathbf{1})\left(\tilde{T}_m\,\tilde{T}_n-\sum_{d\mid\gcd(n,m)} d^{-1}\begin{pmatrix} d & 0\\ 0 & d\end{pmatrix}\tilde{T}_{\frac{nm}{d^2}}\right)\\
&\quad\doteq T_m(T-\mathbf{1})\tilde{T}_n-\sum_{d\mid\gcd(n,m)} d^{-1}\begin{pmatrix} d & 0\\ 0 & d\end{pmatrix}(T-\mathbf{1})\tilde{T}_{\frac{nm}{d^2}}\\
&\quad\doteq T_m\big[T_n(T-\mathbf{1})-(S-\mathbf{1})Y_n\big]-\sum_{d\mid\gcd(n,m)} d^{-1}\begin{pmatrix} d & 0\\ 0 & d\end{pmatrix}T_{\frac{nm}{d^2}}(T-\mathbf{1})\\
&\quad\doteq (T_m\,T_n-)\,(T-\mathbf{1})\sum_{d\mid\gcd(n,m)} d^{-1}\begin{pmatrix} d & 0\\ 0 & d\end{pmatrix}T_{\frac{nm}{d^2}}\ (\mathrm{mod}\ \mathcal{J}\mathcal{R}_{nm}).
\end{aligned}
$$

Since we have

$$T_m\,T_n - \sum_{d\,|\,\gcd(n,m)} d^{-1}\begin{pmatrix} d & 0 \\ 0 & d \end{pmatrix} T_{\frac{nm}{d^2}} \doteq 0 \pmod{(S-\mathbf{1})\mathcal{R}_n}$$

by (1.148), we conclude the proof. □

Continuing towards the proof of Theorem 2.64, consider the sets

$$\mathcal{A}_n := \left\{ \begin{pmatrix} a & b \\ c & d \end{pmatrix} \in H(n)\colon\ a > c > 0,\ d > -b \geq 0,\ \text{and if } b = 0 \text{ then } a \geq 2c \right\} \tag{2.79}$$

and

$$\mathcal{B}_n := \left\{ \begin{pmatrix} a & b \\ c & d \end{pmatrix} \in H(n)\colon\ a > -c \geq 0,\ d > b > 0,\ \text{and if } c = 0 \text{then } d \geq 2b \right\}. \tag{2.80}$$

Exercise 2.75 Show that the maps

$$\alpha:\quad A = \begin{pmatrix} a & b \\ c & d \end{pmatrix} \longmapsto TS^{-[\frac{a}{c}]}A = \begin{pmatrix} c & d \\ -(a - c[\frac{a}{c}]) & -b + d[\frac{a}{c}] \end{pmatrix}$$

and

$$\beta:\quad B = \begin{pmatrix} a & b \\ c & d \end{pmatrix} \longmapsto S^{\left[\frac{d}{b}\right]}T\,B = \begin{pmatrix} -c + a[\frac{d}{b}] & -(d - b[\frac{d}{b}]) \\ a & b \end{pmatrix}$$

give inverse bijections between the sets $\mathcal{A}_n$ and $\mathcal{B}_n$. Here, $[\cdot]$ denote the Gauss brackets defined in (2.60).

Following the arguments in [57, §3.2] closely, we now proceed with the proof.

Proof (of Theorem 2.64) Recall T_n^* as defined in (2.67). We have to show that T_n^* satisfies the right identity in (2.69):

$$T_n(T - \mathbf{1}) \doteq (T - \mathbf{1})T_n^* \qquad \pmod{(S-\mathbf{1})\mathcal{R}_n}.$$

To shorten the notation, we do not write $\pmod{(S-\mathbf{1})\mathcal{R}_n}$ into each congruence relation. We only indicate these by using the "$\doteq$" symbol.

Since the congruence relation $\pmod{(S-\mathbf{1})\mathcal{R}_n}$ implies $S^rA \doteq A$ for every $r \in \mathbb{Z}$ and $A \in H(n)$, we have, by Exercise 2.75,

$$\sum_{A\in\mathcal{A}_n} A \doteq T \sum_{B\in\mathcal{B}_n} B$$

$$\iff \sum_{\substack{a>c>0\\ d>-b>0\\ ad-bc=n}} \left[\begin{pmatrix} a & b\\ c & d\end{pmatrix} - T\begin{pmatrix} a & -b\\ -c & d\end{pmatrix}\right] \doteq \sum_{\substack{\frac{1}{2}d\geq b>0\\ ad=n}} \begin{pmatrix} 0 & -d\\ a & b\end{pmatrix} - \sum_{\substack{\frac{1}{2}a\geq c>0\\ ad=0}} \begin{pmatrix} a & 0\\ c & d\end{pmatrix}.$$

Conjugating this identity with $\begin{pmatrix} -1 & 0\\ 0 & 1\end{pmatrix}$ changes the sign of all the off-diagonal coefficients and preserves "$\doteq$." Adding this result to the original identity, we get

$$(\mathbf{1}-T) \sum_{\substack{a>c>0\\ d>-b>0\\ ad-bc=n}} \left[\begin{pmatrix} a & b\\ c & d\end{pmatrix} + \begin{pmatrix} a & -b\\ -c & d\end{pmatrix}\right] \doteq \sum_{\substack{0<|b|\leq\frac{1}{2}d\\ ad=n}} \left[\begin{pmatrix} 0 & -d\\ a & b\end{pmatrix} - \begin{pmatrix} d & 0\\ b & a\end{pmatrix}\right]. \tag{2.81}$$

(Recall that §2.4, we write identities of elements in every $\mathcal{R}_n$ up to trivial actions by the slash operator.) Identifying the sum on the left-hand side as part of T_n^*, see (2.67), we have

$$\begin{aligned}
&(\mathbf{1}-T)\, T_n^*\\
&\doteq (\mathbf{1}-T)\Bigg(\sum_{\substack{ad-bc=n\\ a>c>0\\ d>-b>0}} \left[\begin{pmatrix} a & b\\ c & d\end{pmatrix} + \begin{pmatrix} a & -b\\ -c & d\end{pmatrix}\right] +\\
&\qquad + \sum_{\substack{ad=n\\ -\frac{1}{2}d<b\leq\frac{1}{2}d}} \begin{pmatrix} a & b\\ 0 & d\end{pmatrix} + \sum_{\substack{ad=n\\ -\frac{1}{2}a<c\leq\frac{1}{2}a\\ c\neq 0}} \begin{pmatrix} a & 0\\ c & d\end{pmatrix}\Bigg)\\
&\doteq (\mathbf{1}-T)\left(\sum_{\substack{ad-bc=n\\ a>c>0\\ d>-b>0}} \left[\begin{pmatrix} a & b\\ c & d\end{pmatrix} + \begin{pmatrix} a & -b\\ -c & d\end{pmatrix}\right] \right)\\
&\quad + (\mathbf{1}-T)\left(\sum_{\substack{ad=n\\ -\frac{1}{2}d<b\leq\frac{1}{2}d}} \begin{pmatrix} a & b\\ 0 & d\end{pmatrix} + \sum_{\substack{ad=n\\ -\frac{1}{2}a<c\leq\frac{1}{2}a\\ c\neq 0}} \begin{pmatrix} a & 0\\ c & d\end{pmatrix} \right)
\end{aligned}$$

$$\overset{(2.81)}{\doteq} \sum_{\substack{0<|b|\leq \frac{1}{2}d\\ ad=n}} \left[\begin{pmatrix} 0 & -d \\ a & b \end{pmatrix} - \begin{pmatrix} d & 0 \\ b & a \end{pmatrix}\right] +$$

$$+ (\mathbf{1}-T)\left(\sum_{\substack{ad=n\\ -\frac{1}{2}d<b\leq \frac{1}{2}d}} \begin{pmatrix} a & b \\ 0 & d \end{pmatrix} + \sum_{\substack{ad=n\\ -\frac{1}{2}a<c\leq \frac{1}{2}a\\ c\neq 0}} \begin{pmatrix} a & 0 \\ c & d \end{pmatrix} \right)$$

$$\doteq \sum_{\substack{-\frac{1}{2}d<b\leq \frac{1}{2}d\\ ad=n}} \begin{pmatrix} a & b \\ 0 & d \end{pmatrix}(\mathbf{1}-T) + \sum_{\substack{a,d>0\\ ad=n\\ d \text{ even}}} \left[\begin{pmatrix} 0 & -d \\ a & -\frac{1}{2}d \end{pmatrix} - \begin{pmatrix} d & 0 \\ -\frac{1}{2}d & a \end{pmatrix}\right].$$

The first sum on the left-hand side satisfies

$$\sum_{\substack{-\frac{1}{2}d<b\leq \frac{1}{2}d\\ ad=n}} \begin{pmatrix} a & b \\ 0 & d \end{pmatrix}(\mathbf{1}-T) \doteq T_n(\mathbf{1}-T)$$

and the second evaluates as

$$\sum_{\substack{a,d>0\\ ad=n\\ d \text{ even}}} \left[\begin{pmatrix} 0 & -d \\ a & -\frac{1}{2}d \end{pmatrix} - \begin{pmatrix} d & 0 \\ -\frac{1}{2}d & a \end{pmatrix}\right] \doteq \sum_{\substack{a,d>0\\ ad=n\\ d \text{ even}}} (S^2-\mathbf{1})\begin{pmatrix} d & 0 \\ -\frac{1}{2}d & a \end{pmatrix}$$

$$\doteq (S-\mathbf{1}) \sum_{\substack{a,d>0\\ ad=n\\ d \text{ even}}} (S+\mathbf{1})\begin{pmatrix} d & 0 \\ -\frac{1}{2}d & a \end{pmatrix}$$

$$\doteq 0 \qquad (\mathrm{mod}\ (S-\mathbf{1})\mathcal{R}_n),$$

recalling again that §2.4 we write identities of elements in any $\mathcal{R}_n$ up to trivial actions by the slash operator. This means, in particular, that the matrices $\begin{pmatrix} a & b \\ c & d \end{pmatrix}$ and $\begin{pmatrix} -a & -b \\ -c & -d \end{pmatrix}$ are identified. □

Remark 2.76 There exists an alternative approach using the integral representation (2.1) of period polynomials. This approach was originally used by Manin [131]. Later it was adapted by [149] to work for period functions of Maass cusp forms too. The same author also extended Choie and Zagier's algebraic approach in [148] to the case of period functions.

2.5 Concluding Remarks

As mentioned earlier, the two initial chapters serve as a brief introduction to the theory of modular forms that is needed as a background to our discussion of Maass wave forms and related topics in the upcoming chapters. We restricted the presented topics and the completeness of their presentation to fit our needs.

There are several good textbooks available that cover modular forms in a more detailed manner. Some interesting textbooks focusing on the arithmetic properties of modular forms are Apostol [12], Serre [182], Shimura [184], Lang [116], Schoenberg [176], Koblitz [109], and Diamond and Shurman [66], as mentioned in the preface of this book.

Also, one can find good lecture notes on the internet. For example, see Ribet and Stein's lecture notes [168]. Their notes explain the connection between modular forms and algebraic curves, preparing the reader for some understanding of Wiles's proof of Fermat's last theorem with a focus on the computational aspects.

Another example of textbooks are by Stein [188, 189]. His books introduce the theory of modular forms, associated Dirichlet series, Hecke operators, and modular symbols in some detail and provides several computational examples.

Other generalizations of modular forms are also known. We mention only a few.

Hilbert modular forms: these are functions in n variables, each a complex number in the upper half-plane, satisfying a transformation property for 2×2 matrices with entries in a totally real number field. We refer to the textbooks [83, 84] for more details.

Siegel modular forms: these are modular forms associated with symplectic groups. Some reference textbooks are [82] (in German) and [104].

Jacobi forms: these forms are a mixture of modular forms and elliptic functions. Examples of such functions are the Jacobi theta functions (§1.5) and the Fourier coefficients of Siegel modular forms of genus two. They also play a role in the recent discussions of mock Jacobi modular forms (see §7.2.2). Further material on the topic can be found in the textbook [74] and the paper [185].

Automorphic forms: this notion extends the theory of modular forms to general Lie groups. Some references are the textbooks [17, 52, 85, 101]. The book [53] pushes the horizon even further and discusses the Langlands program.

Cohomology groups associated with modular forms: we briefly showed one classical result in this area, the Eichler cohomology group and the Eichler cohomology theorem, in Sections 2.2 and 2.3 of this chapter. However, the theory of cohomology groups associated with modular forms is still far from complete. For example, a cohomological description of the Petersson scalar product of modular cusp forms of positive even weight was announced by Bruggeman and Choie [32]. Classical books on this topic are for example Lang [116], introducing period functions, and Stein [188], working on modular symbols respectively Manin symbols. A short survey article [65] by Diamantis and Rolen reports on the theory of period polynomials attached to derivatives of L-functions from a cohomological perspective, including conjectures and results about zeros of period polynomials and of their counterparts for derivatives of L-functions.

2.5 Concluding Remarks

Chapter 3
Maass Wave Forms of Real Weight

We introduce Maass wave/cusp forms in this chapter. The idea of Maass wave forms goes back to the article [128] (in German) by Hans Maass where he introduced a "new kind of nonanalytic automorphic forms," see, e.g., [128, Satz 1]. There is now a vast literature on Maass forms, thanks to the works of many people, such as, e.g., Hejhal [94, 95], Iwaniec [100, 101], Maass [131], Roelcke [169], Selberg [179], Terras [194, 195], and Venkov [199].

We focus on an analytic introduction to Maass wave forms of real weight. Similar approaches are for example presented by Hejhal in [95, Chapter IX] or by Roelcke [169].

In this chapter, we introduce Maass wave/cusp forms of real weight on subgroups of finite index in the full modular group with compatible multiplier systems. After introducing multiplier systems and the unitary automorphic factor in §3.1 and discussing the hyperbolic Laplace operator and Maass operators in §3.2, we define Maass wave forms in §3.3. A few examples are given in §3.4. Then, we present a scalar product for Maass cusp forms in §3.5. We then use Hecke operators in the special case where the Maass wave form is of weight 0 and trivial multiplier system on the full group in §3.6. We conclude the chapter by mentioning some spectral theory for the Laplace operator and its Friedrichs extension in §3.7 and discuss briefly Selberg's Conjecture in 3.8. A few final remarks are given in 3.9.

We assume throughout this chapter that the subgroup $\Gamma \subset \Gamma(1)$ is of finite index.

3.1 Multiplier Systems and Unitary Automorphic Factors

One key requirement for a modular form is that it satisfies the transformation law (1.28). It made sense to define the automorphic factor $j(\cdot, \cdot)$ in (1.29) in a holomorphic way since we worked then with holomorphic functions. However, Maass wave forms are not holomorphic, and as a result we will replace this condition

T. Mühlenbruch, W. Raji, *On the Theory of Maass Wave Forms*, Universitext,
https://doi.org/10.1007/978-3-030-40475-8_3

by another. It therefore makes sense to adapt the notion of automorphic factors as well.

First, recall our argument convention in (1.30), since we will work again with real powers of complex numbers. We have

$$-\pi \leq \arg(\zeta) < \pi \qquad \text{for all } \zeta \in \mathbb{C}_{\neq 0} \tag{1.30}$$

which implies in particular the convention $\arg(-1) = -\pi$.

Informally speaking, we would like to consider real-analytic functions u on $\mathbb{H}$ which satisfy the transformation law

$$u(Mz) = v(M)\, j_u(M,z)^k\, u(z) \qquad \text{for all } z \in \mathbb{H} \text{ and for all } M \in \Gamma. \tag{3.1}$$

The terms $v(M)$ and $j_u(M,z)$ denote the multiplier system in the sense of Definition 1.44 and the unitary automorphic factor. The *unitary automorphic factor* is defined by

$$j_u : \mathrm{Mat}_2(\mathbb{R}) \times \mathbb{C} \to \mathbb{C}; \quad (M,z) \mapsto j_u(M,z) := e^{i \arg(cz+d)}, \tag{3.2}$$

where we write $M = \begin{pmatrix} a & b \\ c & d \end{pmatrix}$. The real parameter $k \in \mathbb{R}$ in (3.1) is the *weight* of u.

Exercise 3.1 Show that the unitary automorphic factor $j_u(\cdot,\cdot)$ in (3.1) and the automorphic factor $j(\cdot,\cdot)$ in (1.29) satisfies

$$j_u(M,z) = \frac{j(M,z)}{|j(M,z)|} \tag{3.3}$$

for all $M = \begin{pmatrix} \star & \star \\ c & d \end{pmatrix} \in \mathrm{Mat}_2(\mathbb{R})$ and $z \in \mathbb{C}_{\neq -\frac{d}{c}}$.

Lemma 3.2 *For $\Gamma \subset \Gamma(1)$ and $k \in \mathbb{R}$, let v denote a multiplier system with respect to (Γ, k). Let $f : \mathbb{H} \to \mathbb{C}$ be a function on the upper half-plane. Then the following two conditions are equivalent:*

1. *f satisfies the transformation property* (1.28)*:*

$$f(Mz) \overset{(1.28)}{=} v(M)\, j(M,z)^k\, f(z)$$

for all $M \in \Gamma$ and $z \in \mathbb{H}$.

2. *$u(z) := \Im(z)^{\frac{k}{2}} f(z)$ satisfies the transformation property*

$$u(Mz) = v(M)\, j_u(M,z)^k\, u(z) \tag{3.4}$$

for all $M \in \Gamma$ and $z \in \mathbb{H}$.

Proof Assume that f satisfies the transformation property (1.28). Using (1.18) and Exercise 3.1, we find

$$\begin{aligned}
u(Mz) &= \Im(Mz)^{\frac{k}{2}} f(Mz) \\
&\overset{(1.18)}{=} \frac{\Im(z)^{\frac{k}{2}}}{|j(M,z)|^k} f(Mz) \\
&\overset{(1.28)}{=} v(M) \left(\frac{j(M,z)}{|j(M,z)|}\right)^k \Im(z)^{\frac{k}{2}} f(z) \\
&\overset{(3.3)}{=} v(M)\, j_u(M,z)^k\, u(z).
\end{aligned}$$

This shows that u satisfies (3.4).

Now, assume that u satisfies the transformation property (3.4). Essentially, the same calculation as above shows that

$$\begin{aligned}
f(Mz) &= \frac{\Im(z)^{-\frac{k}{2}}}{|j(M,z)|^{-k}} u(Mz) \\
&\overset{(3.4)}{=} \frac{\Im(z)^{-\frac{k}{2}}}{|j(M,z)|^{-k}} v(M)\, j_u(M,z)^k\, u(z) \\
&= v(M)\, j(M,z)^k\, f(z).
\end{aligned}$$

This shows that f satisfies (1.28). □

In view of the slash operator $\big|_k$ defined in (1.131), we define a variant below which is better suited for the unitary automorphic factor appearing in the transformation law (3.1).

Definition 3.3 Let $k \in \mathbb{R}$ and let g be a function on $\mathbb{H}$ or on $\mathbb{R} \cup \{\infty\}$. We define the variant $\big\|_k$ of the *slash operator* by

$$g\big\|_k M(z) := j_u(M,z)^{-k}\, g(M\,z) \tag{3.5}$$

for all z and $M \in \mathrm{Mat}_2(\mathbb{R})$.

Definition 3.3 allows us to write the transformation property (3.4) as

$$u\big\|_k M = v(M)\, u \tag{3.6}$$

for all $M \in \Gamma$.

Exercise 3.4 For $\Gamma \subset \Gamma(1)$ and $k \in \mathbb{R}$, let v denote a multiplier system with respect to (Γ, k). Show that the following two statements are equivalent:

1. v satisfies (1.31).
2. v satisfies the condition

$$v(M_1M_2)\, j_u(M_1M_2, z)^k = v(M_1)v(M_2)\, j_u(M_1, M_2z)^k\, j_u(M_2, z)^k \tag{3.7}$$

for all $M \in \Gamma$ and $z \in \mathbb{H}$.

Exercise 3.5 For $\Gamma \subset \Gamma(1)$ and $k \in \mathbb{R}$, let v denote a multiplier system with respect to (Γ, k). Show that $\overline{v}$ is a multiplier system with respect to $(\Gamma, -k)$. (Here $\overline{v}$ denotes the complex conjugate multiplier system: $\overline{v}(M) := \overline{v(M)}$ for all $M \in \Gamma$.)

3.2 The Differential Operators Δ_k and E_k

First, we define the hyperbolic Laplace operator Δ_0 and its weighted variant Δ_k.

Definition 3.6 Let $k \in \mathbb{R}$. For a complex number $z = x + iy$, the *hyperbolic Laplace operator* Δ_0 *of weight* 0 is given by the second-order differential operator

$$\Delta_0 := -y^2 \left(\frac{\partial^2}{\partial x^2} + \frac{\partial^2}{\partial y^2} \right). \tag{3.8}$$

Similarly, the *hyperbolic Laplace operator* Δ_k *of weight* k given by the second-order differential operator

$$\Delta_k := -y^2 \left(\frac{\partial^2}{\partial x^2} + \frac{\partial^2}{\partial y^2} \right) + iky\frac{\partial}{\partial x}. \tag{3.9}$$

Remark 3.7 The above hyperbolic Laplace operator Δ_k of weight k was already considered by Hans Maass. He considered the operator

$$\Omega_{\alpha\beta} = y^2 \left(\frac{\partial^2}{\partial x^2} + \frac{\partial^2}{\partial y^2} \right) + i(\alpha - \beta)y\frac{\partial}{\partial x} - (\alpha + \beta)y\frac{\partial}{\partial y},$$

see [129, p. 175, Section "Non-Analytic Modular Forms", Equation (8)] with suitable conditions on the complex parameters α and β. For example, defining α and β by

$$\alpha - \beta = k \quad \text{and} \quad \alpha + \beta = 0$$

$$\alpha = \frac{k}{2} \quad \text{and} \quad \beta = -\frac{k}{2}$$

transforms the operator $\Omega_{\alpha\beta}$ into Δ_k.

Additionally, we define the *Maass operators* $E_k^{\pm}$ by the first-order differential operators

$$E_k^{\pm} := \pm 2iy\frac{\partial}{\partial x} + 2y\frac{\partial}{\partial y} \pm k. \tag{3.10}$$

Remark 3.8 Of course, to properly define an operator, we have to specify the space it acts on. Right now, we require that the functions, on which they act, are sufficiently smooth. We will discuss a bit more about function spaces in §3.7 where we construct self-adjoint extensions of Δ_k.

Lemma 3.9 *Let $k \in \mathbb{R}$.*

1. *We have the relations*

$$E_k^{+} = 4iy\frac{\partial}{\partial z} + k \quad \text{and} \quad E_k^{-} = -4iy\frac{\partial}{\partial \overline{z}} - k, \tag{3.11}$$

   where $\overline{z}$ denotes the complex conjugate of the complex variable z.

2. *The hyperbolic Laplace operator and the Maass operators satisfy*

$$\Delta_k = -\frac{1}{4}E_{k-2}^{+}E_k^{-} - \frac{k(k-2)}{4} = -\frac{1}{4}E_{k+2}^{-}E_k^{+} - \frac{k(k+2)}{4}. \tag{3.12}$$

Proof Recall the relations

$$\frac{\partial}{\partial z} = \frac{1}{2}\frac{\partial}{\partial x} - \frac{i}{2}\frac{\partial}{\partial y} \quad \text{and} \quad \frac{\partial}{\partial \overline{z}} = \frac{1}{2}\frac{\partial}{\partial x} + \frac{i}{2}\frac{\partial}{\partial y} \tag{3.13}$$

between the first-order differentials, see [207].

The first part is left to the reader as an exercise.

As for the second part, consider $E_{k-2}^{+}E_k^{-} f(x+iy)$. Using (3.10) and applying the product rule for differentiation, we get

$$\begin{aligned}
&E_{k-2}^{+}E_k^{-} f(x, iy) \\
&\overset{(3.10)}{=} \left(2iy\frac{\partial}{\partial x} + 2y\frac{\partial}{\partial y} + k - 2\right)\left(-2iy\frac{\partial}{\partial x} + 2y\frac{\partial}{\partial y} - k\right) f(x+iy) \\
&= 4y^2\left(\frac{\partial^2}{\partial x^2} f(x+iy)\right) - 2kiy\left(\frac{\partial}{\partial x} f(x+iy)\right) + 4y^2\left(\frac{\partial^2}{\partial y^2} f(x+iy)\right) \\
&\quad - 2iky\left(\frac{\partial}{\partial x} f(x+iy)\right) - (k-2)k\, f(x+iy) \\
&= 4y^2\left(\frac{\partial^2}{\partial x^2} f(x+iy)\right) + 4y^2\left(\frac{\partial^2}{\partial y^2} f(x+iy)\right)
\end{aligned}$$

$$- \; 2iky\left(\frac{\partial}{\partial x}f(x+iy)\right) \; - \; (k-2)k\, f(x+iy)$$

$$\overset{(3.9)}{=} \; 4\big(\Delta_k f\big)(x+iy) \; - \; (k-2)k\, f(x+iy).$$

Summarizing the above calculation, we have the relation

$$\Delta_k = -\frac{1}{4}\mathrm{E}_{k-2}^{+}\mathrm{E}_k^{-} - \frac{k(k-2)}{4}$$

between the hyperbolic Laplace operator and the Maass operators.

The remaining relation in (3.12) follows by similar calculation. This concludes the proof. □

The differential operators Δ_k and E_k are compatible with the slash operator $\|_k$ on the full modular group $\Gamma(1)$ as shown below.

Lemma 3.10 *Let $k \in \mathbb{R}$ and let $g : \mathbb{H} \to \mathbb{C}$ be a function that is twice continuously complex differentiable, i.e., $g \in C^2(\mathbb{H})$.*

We have

$$\big(\Delta_k g\big)\big\|_k M = \Delta_k\big(g\big\|_k M\big) \quad \text{and} \quad \big(\mathrm{E}_k^{\pm} g\big)\big\|_{k\pm 2} M = \mathrm{E}_k^{\pm}\big(g\big\|_k M\big) \tag{3.14}$$

for all $M \in \mathrm{SL}_2(\mathbb{R})$.

Proof First, we consider the Maass operator E_k^{+} and take a function $h : \mathbb{H} \to \mathbb{C}$ that is at least once differentiable. Recalling

$$\mathrm{E}_k^{+} = 4iy\frac{\partial}{\partial z} + k = 2\big(z-\overline{z}\big)\frac{\partial}{\partial z} + k$$

in (3.11) (with $z = x + iy$), we consider for a general element $M \in \begin{pmatrix} a & b \\ c & d \end{pmatrix} \in \mathrm{SL}_2(\mathbb{R})$ the term with the differential $2\big(z-\overline{z}\big)\frac{\partial}{\partial z}$. Using (3.3) and

$$e^{i\arg(cz+d)} \overset{(3.3)}{=} \frac{cz+d}{|cz+d|} = \frac{cz+d}{\sqrt{\big(cz+d\big)\big(c\overline{z}+d\big)}} = \left(\frac{cz+d}{c\overline{z}+d}\right)^{\frac{1}{2}},$$

we have

$$2\big(z-\overline{z}\big)\frac{\partial}{\partial z}\left[e^{-ik\arg(cz+d)}\, h\left(\frac{az+b}{cz+d}\right)\right]$$

$$\overset{(3.3)}{=} 2\big(z-\overline{z}\big)\frac{\partial}{\partial z}\left[\left(\frac{cz+d}{c\overline{z}+d}\right)^{-\frac{k}{2}} h\left(\frac{az+b}{cz+d}\right)\right]$$

$$= 2(z-\bar{z})\left[\frac{\partial}{\partial z}\left(\frac{cz+d}{c\bar{z}+d}\right)^{-\frac{k}{2}}\right]h\left(\frac{az+b}{cz+d}\right)$$

$$+ 2(z-\bar{z})\left(\frac{cz+d}{c\bar{z}+d}\right)^{-\frac{k}{2}}\left[\frac{\partial}{\partial z}h\left(\frac{az+b}{cz+d}\right)\right]$$

and using the product rule for differentiation, we have

$$= 2(z-\bar{z})\left(-\frac{k}{2}\frac{c}{c\bar{z}+d}\left(\frac{cz+d}{c\bar{z}+d}\right)^{-\frac{k}{2}-1}\right)h\left(\frac{az+b}{cz+d}\right)$$

$$+ 2(z-\bar{z})\left(\frac{cz+d}{c\bar{z}+d}\right)^{-\frac{k}{2}}\left[\frac{\partial}{\partial z}\left(\frac{az+b}{cz+d}\right)\right]\left[\frac{\partial}{\partial z}h\right]\left(\frac{az+b}{cz+d}\right)$$

$$= -k\frac{c}{c\bar{z}+d}\left(z-\bar{z}\right)e^{-i(k+2)\arg(cz+d)}\,h\left(\frac{az+b}{cz+d}\right)$$

$$+ 2\left(z-\bar{z}\right)e^{-ik\arg(cz+d)}\left(\frac{1}{(cz+d)^2}\right)\left[\frac{\partial}{\partial z}h\right]\left(\frac{az+b}{cz+d}\right).$$

Using

$$\left(z-\bar{z}\right) = \left(cz+d\right)\left(c\bar{z}+d\right)\left(\frac{az+b}{cz+d}-\frac{a\bar{z}+b}{c\bar{z}+d}\right), \tag{3.15}$$

we find

$$-k\frac{c}{c\bar{z}+d}\left(z-\bar{z}\right)e^{-i(k+2)\arg(cz+d)}\,h\left(\frac{az+b}{cz+d}\right)$$

$$+ 2\left(z-\bar{z}\right)e^{-ik\arg(cz+d)}\left(\frac{1}{(cz+d)^2}\right)\left[\frac{\partial}{\partial z}h\right]\left(\frac{az+b}{cz+d}\right)$$

$$\overset{(3.15)}{=} -k\frac{c}{c\bar{z}+d}\left(cz+d\right)\left(c\bar{z}+d\right)\left(\frac{az+b}{cz+d}-\frac{a\bar{z}+b}{c\bar{z}+d}\right)$$

$$\cdot\, e^{-i(k+2)\arg(cz+d)}\,h\left(\frac{az+b}{cz+d}\right)$$

$$+ 2\left(cz+d\right)\left(c\bar{z}+d\right)\left(\frac{az+b}{cz+d}-\frac{a\bar{z}+b}{c\bar{z}+d}\right)\cdot$$

$$\cdot\, e^{-ik\arg(cz+d)}\left(\frac{1}{(cz+d)^2}\right)\left[\frac{\partial}{\partial z}h\right]\left(\frac{az+b}{cz+d}\right)$$

$$= -kc(cz+d)\left(\frac{az+b}{cz+d}-\frac{a\bar{z}+b}{c\bar{z}+d}\right)e^{-i(k+2)\arg(cz+d)}\,h\left(\frac{az+b}{cz+d}\right)$$
$$+\ 2\left(\frac{az+b}{cz+d}-\frac{a\bar{z}+b}{c\bar{z}+d}\right)e^{-i(k+2)\arg(cz+d)}\left[\frac{\partial}{\partial z}h\right]\left(\frac{az+b}{cz+d}\right)$$

using (3.3) and $\frac{c\bar{z}+d}{cz+d}=\left(\frac{cz+d}{|cz+d|}\right)^{-2}\overset{(3.3)}{=}e^{-2i\arg cz+d}$.

This gives the following:

$$\mathrm{E}_k^+\left(h\|_k M\right)(z)$$
$$=2(z-\bar{z})\left(\frac{\partial}{\partial z}\left[e^{-ik\arg(cz+d)}\,h\left(\frac{az+b}{cz+d}\right)\right]\right)+k\left[e^{-ik\arg(cz+d)}\,h\left(\frac{az+b}{cz+d}\right)\right]$$
$$=-kc(cz+d)\left(\frac{az+b}{cz+d}-\frac{a\bar{z}+b}{c\bar{z}+d}\right)e^{-i(k+2)\arg(cz+d)}\,h\left(\frac{az+b}{cz+d}\right)$$
$$+\ 2\left(\frac{az+b}{cz+d}-\frac{a\bar{z}+b}{c\bar{z}+d}\right)e^{-i(k+2)\arg(cz+d)}\left[\frac{\partial}{\partial z}h\right]\left(\frac{az+b}{cz+d}\right)$$
$$+\ k\left[e^{-ik\arg(cz+d)}\,h\left(\frac{az+b}{cz+d}\right)\right].$$

On the other hand, we have

$$\left(\mathrm{E}_k^+h\right)\|_{k+2}M(z)$$
$$=2e^{-i(k+2)\arg(cz+d)}\left(\frac{az+b}{cz+d}-\frac{a\bar{z}+b}{c\bar{z}+d}\right)\left[\frac{\partial}{\partial z}h\right]\left(\frac{az+b}{cz+d}\right)$$
$$+\ ke^{-i(k+2)\arg(cz+d)}\,h\left(\frac{az+b}{cz+d}\right).$$

Subtracting both terms, we get

$$\mathrm{E}_k^+\left(g\|_k M\right)(z)-\left(\mathrm{E}_k^+g\right)\|_{k+2}M(z)$$
$$=-kc(cz+d)\left(\frac{az+b}{cz+d}-\frac{a\bar{z}+b}{c\bar{z}+d}\right)e^{-i(k+2)\arg(cz+d)}\,h\left(\frac{az+b}{cz+d}\right)$$
$$+\ k\left[e^{-ik\arg(cz+d)}\,h\left(\frac{az+b}{cz+d}\right)\right]$$
$$-\ k\left[e^{-i(k+2)\arg(cz+d)}\,h\left(\frac{az+b}{cz+d}\right)\right]$$

$$
\begin{aligned}
&= k\,h\left(\frac{az+b}{cz+d}\right)\Big[-c\big(cz+d\big)\left(\frac{az+b}{cz+d}-\frac{a\bar{z}+b}{c\bar{z}+d}\right)e^{-i(k+2)\arg(cz+d)} \\
&\qquad + e^{-ik\arg(cz+d)} - e^{-i(k+2)\arg(cz+d)}\Big] \\
&= k\,h\left(\frac{az+b}{cz+d}\right)\Big[-c\big(cz+d\big)\left(\frac{az+b}{cz+d}-\frac{a\bar{z}+b}{c\bar{z}+d}\right)e^{-i(k+2)\arg(cz+d)} \\
&\qquad + \underbrace{\left(\frac{cz+d}{c\bar{z}+d}\right)e^{-2i\arg(cz+d)}}_{=1}e^{-ik\arg(cz+d)} - e^{-i(k+2)\arg(cz+d)}\Big] \\
&= ke^{-i(k+2)\arg(cz+d)}\,h\left(\frac{az+b}{cz+d}\right) \\
&\qquad \cdot\Big[-c\big(cz+d\big)\left(\frac{az+b}{cz+d}-\frac{a\bar{z}+b}{c\bar{z}+d}\right)+\left(\frac{cz+d}{c\bar{z}+d}\right)-1\Big] \\
&= ke^{-i(k+2)\arg(cz+d)}\,h\left(\frac{az+b}{cz+d}\right) \\
&\qquad \cdot\Big[-c\big(az+b\big)+c\big(a\bar{z}+b\big)\frac{cz+d}{c\bar{z}+d}+\left(\frac{cz+d}{c\bar{z}+d}\right)-1\Big] \\
&= ke^{-i(k+2)\arg(cz+d)}\,h\left(\frac{az+b}{cz+d}\right)\Big[-acz-bc-1+\left(\frac{cz+d}{c\bar{z}+d}\right)(ac\bar{z}+bc+1)\Big]
\end{aligned}
$$

Using $\det M = ad - nc = 1$,

$$
\begin{aligned}
&= ke^{-i(k+2)\arg(cz+d)}\,h\left(\frac{az+b}{cz+d}\right)\Big[-a(cz+d)+\left(\frac{cz+d}{c\bar{z}+d}\right)a\big(c\bar{z}+d\big)\Big] \\
&= 0.
\end{aligned}
$$

This proves

$$
\big(\mathrm{E}_k^{+}h\big)\big\|_{k+2}M = \mathrm{E}_k^{+}\big(h\big\|_k M\big) \tag{3.16}
$$

that is one of the identities in (3.14).

The other identity

$$
\big(\mathrm{E}_k^{-}h\big)\big\|_{k-2}M = \mathrm{E}_k^{-}\big(h\big\|_k M\big) \tag{3.17}
$$

follows by similar direct calculations, which we leave to the reader.

The remaining identity in (3.14) follows from (3.12). We have

$$\begin{aligned}
(\Delta_k h)\|_k M &\overset{(3.12)}{=} \left(-\frac{1}{4}\mathrm{E}^+_{k-2}\mathrm{E}^-_k h - \frac{k(k-2)}{4}h\right)\|_k M \\
&= -\frac{1}{4}\left(\mathrm{E}^+_{k-2}(\mathrm{E}^-_k h)\right)\|_k M - \frac{k(k-2)}{4}h\|_k M \\
&\overset{(3.16)}{=} -\frac{1}{4}\mathrm{E}^+_{k-2}\left((\mathrm{E}^-_k h)\|_{k-2} M\right) - \frac{k(k-2)}{4}h\|_k M \\
&\overset{(3.17)}{=} -\frac{1}{4}\mathrm{E}^+_{k-2}\mathrm{E}^-_k (h\|_k M) - \frac{k(k-2)}{4}(h\|_k M) \\
&\overset{(3.12)}{=} (\Delta_k h)\|_k M.
\end{aligned}$$

This concludes the proof of Lemma 3.10. □

Remark 3.11 In addition to the "brute force" approach used in our proof of Lemma 3.10, there exists another approach using representation theory and pulling back everything to the (universal cover of the) Lie algebra $\mathfrak{sl}_2(\mathbb{R})$. The Casimir element in this Lie algebra corresponds to the hyperbolic Laplace operator Δ_k and the Maass operators $\mathrm{E}^\pm_k$ also have analogs there. A typical example of this approach for the analog of Lemma 3.10 can be found in [31, §2.2.3].

Exercise 3.12 Let $k \in \mathbb{R}$ and let $g : \mathbb{H} \to \mathbb{C}$ a twice continuously complex differentiable function, i.e., $g \in C^2(\mathbb{H})$.

1. Show that the Laplace operator Δ_k and the Maass operators $\mathrm{E}^\pm_k$ commute with diagonal matrices of the form $A = \begin{pmatrix} \alpha & 0 \\ 0 & \alpha \end{pmatrix}$ with $\alpha > 0$: $(\Delta_k g)\|_k A = \Delta_k (g\|_k A)$ and $(\mathrm{E}^\pm_k g)\|_{k\pm 2} A = \mathrm{E}^\pm_k (g\|_k A)$.
2. Show the relation

$$(\Delta_k g)\|_k M = \Delta_k (g\|_k M) \quad \text{and} \quad (\mathrm{E}^\pm_k g)\|_{k\pm 2} M = \mathrm{E}^\pm_k (g\|_k M) \tag{3.18}$$

for all $M \in \mathrm{GL}_2(\mathbb{R})$ with $\det M > 0$.

Exercise 3.13 Let $G : \mathbb{H} \to \mathbb{C}$ be a holomorphic function and $k \in \mathbb{R}$. Show that the function

$$g(z) := \Im(z)^{\frac{k}{2}}\, G(z)$$

satisfies

$$\mathrm{E}^-_k\, g = 0.$$

We fix $k \in \mathbb{R}$ and a parameter $\lambda \in \mathbb{C}$, where λ will take the role of an "eigenvalue" of Δ_k. Let us consider the differential equation

$$\Delta_k g(z) = \lambda\, g(z) \tag{3.19}$$

for a smooth function $g : \mathbb{H} \to \mathbb{C}$. Plugging in the definition of Δ_k in (3.9), we get

$$\left(-y^2 \frac{\partial^2}{\partial x^2} - y^2 \frac{\partial^2}{\partial y^2} + iky\frac{\partial}{\partial x} - \lambda\right) g(x+iy) = 0 \tag{3.20}$$

for $x = \Re(z) \in \mathbb{R}$ and $y = \Im(z) \in \mathbb{R}_{>0}$. This illustrates a change in our view of (3.19): We view the above differential equation as partial differential equation in the real variables x and y. Using standard results on "elliptic partial differential equations," we conclude that the function g itself is in fact a real-analytic function in the variables x and y, see for example [115, Appendix 4 ,§4 and §5] and [14, pp. 207–210].

Summarizing, we have the following.

Definition 3.14 A function $g : \mathbb{H} \to \mathbb{C}$ is called *real-analytic* in the real variables $x \in \mathbb{R}$ and $y \in \mathbb{R}_{>0}$ with $x + iy \in \mathbb{H}$ if it admits an absolutely convergent Taylor series expansion in the two variables x and y on $\mathbb{H}$.

Lemma 3.15 *If $g : \mathbb{H} \to \mathbb{C}$ is a smooth solution of the elliptic partial differential equation* (3.20)*, then g is a real-analytic function and admits a convergent Taylor series in the variables x and y.*

Proof This lemma follows directly from [14, pp. 207–210] or [115, Appendix 4 ,§4 and §5] which we already mentioned above. □

3.2.1 Whittaker Functions

Before we discuss eigenfunctions of the hyperbolic Laplace operator Δ_k, we give an introduction about Whittaker functions. These functions are special functions that represent solutions to a particular second-order ordinary differential equation and will be used later to give a natural expansion to eigenfunctions of the Laplace operator. The expansion will serve a similar purpose as the Fourier expansion serves for holomorphic modular forms discussed in §1.3.3.

Following the respective parts in [150] closely, we consider Whittaker's normalized differential equation and its solutions. *Whittaker's normalized differential equation* is the second-order ordinary differential equation

$$\frac{\partial^2}{\partial y^2} G(y) + \left(-\frac{1}{4} + \frac{k}{y} + \frac{\frac{1}{4} - \nu^2}{y^2}\right) G(y) = 0 \tag{3.21}$$

for smooth functions $G : (0, \infty) \to \mathbb{C}$ and $\nu \notin -\frac{1}{2}\mathbb{N}$. According to [153, (13.14.2), (13.14.3)], see also [130, Chapter VII], we have two solutions $M_{k,\nu}(y)$ and $W_{k,\nu}(y)$ with different behavior as $y \to \infty$:

$$M_{k,\nu}(y) \sim \frac{\Gamma(1+2\nu)}{\Gamma\left(\frac{1}{2}+\nu+k\right)} e^{\frac{1}{2}y} y^{-k} \qquad \text{and} \tag{3.22}$$

$$W_{k,\nu}(y) \sim e^{-\frac{1}{2}y} y^{k}. \tag{3.23}$$

The asymptotic behavior is valid for $k - \nu \notin \left\{\frac{1}{2}, \frac{3}{2}, \frac{5}{2}, \ldots\right\}$, see [153, (13.14.20), (13.14.21)]. These functions satisfy also the recurrence relations [153, (13.15.1), (13.15.11)] and differentiation relations [153, (13.15.17), (13.15.20) and (13.15.23), (13.15.26)] and [130, p. 302, §7.2.1].

Remark 3.16 Standard results on Whittaker functions can be found in [130, §7.1.1]. Some other books discussing the Whittaker functions are [186] and [153, §13].

For example [186, (1.9.10)] gives a relation to the generalized hypergeometric function ${}_1F_1$. We have [186, (1.9.7), (1.9.10)]

$$M_{k,\nu}(y) = {}_1F_1\left[\begin{matrix}\frac{1}{2}-k+\nu \\ 1+2\nu\end{matrix}\middle| y\right] e^{-\frac{y}{2}} y^{\frac{1}{2}+\nu}$$

and

$$W_{k,\nu}(y) = \Gamma(-2\nu)\frac{{}_1F_1\left[\begin{matrix}\frac{1}{2}-k+\nu \\ 1+2\nu\end{matrix}\middle| y\right]}{\Gamma(\frac{1}{2}-k-\nu)} e^{-\frac{y}{2}} y^{\frac{1}{2}+\nu} + \tag{3.24}$$

$$+ \Gamma(2\nu)\frac{{}_1F_1\left[\begin{matrix}\frac{1}{2}-k-\nu \\ 1-2\nu\end{matrix}\middle| y\right]}{\Gamma(\frac{1}{2}-k+\nu)} e^{-\frac{y}{2}} y^{\frac{1}{2}-\nu} \tag{3.25}$$

for $k \in \mathbb{R}$, $\nu \in \mathbb{C} \setminus \frac{1}{2}\mathbb{Z}$ and $y > 0$. We refer to Appendix A.10 for more details on the generalized hypergeometric functions.

Remark 3.17 The W-Whittaker function is not unconditionally defined for all parameters $\nu \in \mathbb{C}$. For example, the Γ-functions at the right side in (3.24) may have poles for half-integral values of ν. It is therefore crucial to use the analytic continuation of the right-hand side and this is a key issue in several constructions.

We consider a modified pair of solutions.

Definition 3.18 For $k + \nu \notin \frac{1}{2} - \mathbb{N}$, we define

$$\mathbb{W}_{k,\nu}(y) := W_{k,\nu}(y) \tag{3.26}$$

and

$$\mathbb{M}_{k,\nu}(y) := \frac{\Gamma\left(\frac{1}{2} + \nu + k\right)}{\Gamma(1 + 2\nu)} M_{k,\nu}(y) \tag{3.27}$$

for all $y \in (0, \infty)$.

Remark 3.19 The definition of $\mathbb{M}_{k,\nu}(\cdot)$ in (3.27) makes sense for $\nu \in -\frac{1}{2}\mathbb{N}$ since *Buchholts function* $\mathcal{M}_{k,\nu} : y \mapsto \frac{1}{\Gamma(1+2\nu)} M_{k,\nu}(y)$, $y > 0$, remains well defined at these values of ν, see [130, p. 297, §7.1.1].

Consider the action of the Maass operators $\mathrm{E}_k^{\pm}$ on $\mathbb{W}_{\frac{k}{2},\nu}$.

Exercise 3.20 Calculate $\mathrm{E}_k^{+}\, \mathbb{W}_{\frac{k}{2},\nu}(4\pi n y)\, e^{2\pi i n x}$ for $n > 0$.

All combinations of the action of the Maass operators $\mathrm{E}_k^{\pm}$ on $\mathbb{W}_{\frac{k}{2},\nu}$ are given in the following lemma.

Lemma 3.21 *Let $k, \nu \in \mathbb{C}$ such that $k \pm \nu \notin \frac{1}{2} + \mathbb{Z}$ and $\lambda = \frac{1}{4} - \nu^2$. We have for $n > 0$*

$$\mathrm{E}_k^{+}\, \mathbb{W}_{\frac{k}{2},\nu}(4\pi n y)\, e^{2\pi i n x} = -2\mathbb{W}_{\frac{k+2}{2},\nu}(4\pi n y)\, e^{2\pi i n x}$$

and

$$\mathrm{E}_k^{-}\, \mathbb{W}_{\frac{k}{2},\nu}(4\pi n y)\, e^{2\pi i n x} = \left(\frac{k(k-2)}{2} + 2\lambda\right) \mathbb{W}_{\frac{k-2}{2},\nu}(4\pi n y)\, e^{2\pi i n x}.$$

For $n < 0$, we have

$$\mathrm{E}_k^{+}\, \mathbb{W}_{-\frac{k}{2},\nu}(4\pi\, |n|\, y)\, e^{2\pi i n x} = \left(\frac{k(k+2)}{2} + 2\lambda\right) \mathbb{W}_{-\frac{k+2}{2},\nu}(4\pi\, |n|\, y)\, e^{2\pi i n x}$$

and

$$\mathrm{E}_k^{-}\, \mathbb{W}_{-\frac{k}{2},\nu}(4\pi\, |y|)\, e^{2\pi i n x} = -2\mathbb{W}_{-\frac{k-2}{2},\nu}(4\pi\, |n|\, y)\, e^{2\pi i n x}.$$

Proof See [147, proof of Lemma 4] or [31, p. 63, Table 4.1]. Alternatively, we can calculate the action of the Maass operators directly using Exercise 3.20 as an example of the arguments and using [186, (2.4.21) and (2.4.24]. □

Similar relations hold for the $M_{k,\nu}$-Whittaker function, using formulas on [130, p. 302].

Exercise 3.22 Calculate $\mathrm{E}_k^+ \, \mathbb{M}_{\frac{k}{2},\nu}(4\pi ny)\, e^{2\pi inx}$ for $n > 0$.

This exercise motivates the following.

Lemma 3.23 *Let $k, \nu \in \mathbb{C}$ such that $k \pm \nu \notin \frac{1}{2} + \mathbb{Z}$ and $\lambda = \frac{1}{4} - \nu^2$. We have for $n > 0$*

$$\mathrm{E}_k^+ \, \mathbb{M}_{\frac{k}{2},\nu}(4\pi ny)\, e^{2\pi inx} = -2\mathbb{M}_{\frac{k+2}{2},\nu}(4\pi ny)\, e^{2\pi inx}$$

and

$$\mathrm{E}_k^- \, \mathbb{M}_{\frac{k}{2},\nu}(4\pi ny)\, e^{2\pi inx} = \left(\frac{k(k-2)}{2} + 2\lambda\right) \mathbb{M}_{\frac{k-2}{2},\nu}(4\pi ny)\, e^{2\pi inx}.$$

For $n < 0$, we have

$$\mathrm{E}_k^+ \, \mathbb{M}_{-\frac{k}{2},\nu}(4\pi\, |n|\, y)\, e^{2\pi inx} = \left(\frac{k(k+2)}{2} + 2\lambda\right) \mathbb{M}_{-\frac{k+2}{2},\nu}(4\pi\, |n|\, y)\, e^{2\pi inx}$$

and

$$\mathrm{E}_k^- \, \mathbb{M}_{-\frac{k}{2},\nu}(4\pi\, |n|\, y)\, e^{2\pi inx} = -2\mathbb{M}_{-\frac{k-2}{2},\nu}(4\pi\, |n|\, y)\, e^{2\pi inx}.$$

Proof Using identities on [130, §7.2.1, p. 302] and the functional equation $x\Gamma(x) = \Gamma(x+1)$ of the Γ-function, we rewrite $M'_{\frac{k}{2},\nu}$. □

Exercise 3.24 Let $\nu \in \mathbb{C}$ be a complex number satisfying $\nu \neq \pm\frac{1}{2}$. Show the identities

$$\mathrm{E}_k^\pm \, y^{\frac{1}{2}+\nu} = \big(1 + 2\nu \pm k\big)\, y^{\frac{1}{2}+\nu} \quad \text{and} \quad \mathrm{E}_k^\pm \, y^{\frac{1}{2}-\nu} = \big(1 - 2\nu \pm k\big)\, y^{\frac{1}{2}-\nu}.$$

3.2.2 *Whittaker-Fourier Expansions on* $\Gamma(1)$

Let $k \in \mathbb{R}$ and let v denote a multiplier system with respect to $\big(\Gamma(1), k\big)$. Consider a smooth function g which satisfies the transformation property (3.6) and solves the partial differential equation (3.19) where we assume a parameterization of the eigenvalue λ by $\lambda = \frac{1}{4} - \nu^2$ for some $\nu \in \mathbb{C} \setminus -\frac{1}{2}\mathbb{N}$. In other words, g satisfies

$$g\,\big\|_k M = v(M)\, g \tag{3.6}$$

for all $M \in \Gamma(1)$ and

$$\Delta_k\, g = \left(\frac{1}{4} - \nu^2\right) g. \tag{3.19}$$

The action of the stabilizer S of the cusp ∞ implies that g is nearly periodic in the sense that

$$g(z+1) = v(S)\, g(z)$$

for all $z \in \mathbb{H}$. Similar to the proof of Theorem 1.56, we conclude that g has a Fourier expansion (in the real part $x = \Re(z)$ of the variable z) at the cusp ∞ of the form

$$g(x+iy) = \sum_{n\in\mathbb{Z}} a_n(y)\, e^{2\pi i (n+\kappa)x}, \tag{3.28}$$

where $0 \le \kappa < 1$ is given by $v(S) = e^{2\pi i\kappa}$, see (1.41) in Definition 1.52. The coefficients $a_n(y)$ still depend on y.

Since g solves the partial differential equation (3.19), we have

$$\left(-y^2\partial_x^2 - y^2\partial_y^2 + iky\partial_x - \left(\frac{1}{4} - \nu^2\right)\right) g(x+iy) = 0. \tag{3.20}$$

We use the method of separation of variables to solve this partial differential equation. We consider two cases;

Case $n+\kappa \neq 0$: Let sign $(\cdot)$ denote the *sign function*

$$\operatorname{sign}(x) := \begin{cases} -1 & \text{if } x < 0, \\ 0 & \text{if } x = 0 \text{ and} \\ 1 & \text{if } x > 0 \end{cases}$$

for real arguments $x \in \mathbb{R}$. Substituting

$$a_n(y) = h\big(4\pi\varepsilon(n+\kappa)y\big), \tag{3.29}$$

with $\varepsilon = \operatorname{sign}(n+\kappa)$ under the assumption $n+\kappa \neq 0$ in (3.20) and applying the separation of variables shows that h solves the ordinary differential equation

$$h''(t) + \left(-\frac{1}{4} + \frac{1}{2}\varepsilon k\frac{1}{t} + \frac{\frac{1}{4}-\nu^2}{t^2}\right) h(t) = 0 \tag{3.30}$$

which is the *Whittaker differential equation*. Solutions are the $\mathbb{W}_{\cdot,\cdot}$- and $\mathbb{M}_{\cdot,\cdot}$-Whittaker functions

$$t \mapsto \mathbb{W}_{\varepsilon\frac{k}{2},\nu}(t) \quad \text{and} \quad t \mapsto \mathbb{M}_{\varepsilon\frac{k}{2},\nu}(t). \tag{3.31}$$

Case $n = 0 = \kappa$ The separation of variables approach under the assumption $n + \kappa = 0$ shows that $a_0(y)$ solves the ordinary differential equation

$$t^2 h''(t) + \left(\frac{1}{4} - \nu^2\right) h(t) = 0. \tag{3.32}$$

Its independent solutions are

$$t \mapsto t^{\frac{1}{2}+\nu} \quad \text{and} \quad t \mapsto t^{\frac{1}{2}-\nu}. \tag{3.33}$$

This shows that g admits a *Fourier-Whittaker expansion* of the form

$$\begin{aligned} g(x+iy) = & \sum_{\substack{n\in\mathbb{Z}\\ n+\kappa\neq 0}} A_n \, \mathbb{W}_{\operatorname{sign}(n)\frac{k}{2},\nu}\big(4\pi\,|n+\kappa|\,y\big)\, e^{2\pi(n+\kappa)x} \\ & + \sum_{\substack{n\in\mathbb{Z}\\ n+\kappa\neq 0}} B_n \, \mathbb{M}_{\operatorname{sign}(n)\frac{k}{2},\nu}\big(4\pi\,|n+\kappa|\,y\big)\, e^{2\pi(n+\kappa)x} \\ & + C_+\, y^{\frac{1}{2}+\nu} + C_-\, y^{\frac{1}{2}-\nu}. \end{aligned} \tag{3.34}$$

The 0th-term coefficients C_+ and C_- vanish if $\kappa > 0$ is strictly positive.

The calculation above shows the following analogue of Theorem 1.56.

Proposition 3.25 *Let $k \in \mathbb{R}$ and let v denote a multiplier system with respect to $\big(\Gamma(1), k\big)$. Consider a real-analytic function $g : \mathbb{H} \to \mathbb{C}$ that satisfies the transformation property* (3.6) *and solves the partial differential equation* (3.19) *where we assume the parametrization of the eigenvalue λ by $\lambda = \frac{1}{4} - \nu^2$ for some $\nu \in \mathbb{C} \setminus -\frac{1}{2}\mathbb{N}$. Then $g(x+iy)$ admits a Whittaker-Fourier expansion of the form* (3.34)*. The 0th-term $C_+\, y^{\frac{1}{2}+\nu} + C_-\, y^{\frac{1}{2}-\nu}$ in* (3.34) *vanishes if $v(S) \neq 1$.*

3.2.3 Whittaker-Fourier Expansions for $\Gamma \subset \Gamma(1)$

We now present a theorem that gives the Fourier expansion of functions satisfying (3.6) and (3.19) for a given subgroup $\Gamma \subset \Gamma(1)$.

To do so, we need the following Lemma.

Lemma 3.26 *Let $q = \frac{a}{b} \in \mathbb{Q}$ with coprime integers $a \in \mathbb{Z}$ and $b \in \mathbb{Z}_{\neq 0}$. Recall the matrices $A = \begin{pmatrix} a & \star \\ b & \star \end{pmatrix}$ and M_q as given in Lemma 1.31. We assume that $M_q^l \in \Gamma(1)$ for some $l \in \mathbb{Z}$.*

1. *For every $k \in \mathbb{R}$, we have*

$$\frac{\left(M_q^l\, z-q\right)^k}{\left|M_q^l\, z-q\right|^k} = j_u\left(M_q^l, z\right)^{-k} \frac{(z-q)^k}{|z-q|^k}. \tag{3.35}$$

2. *We have the identity*

$$j\left(A^{-1}, z\right) = (-b)\left(z-q\right) \tag{3.36}$$

Proof The proof of this lemma will be left as an exercise for the reader. □

Theorem 3.27 *Let $\Gamma \subset \Gamma(1)$, $k \in \mathbb{R}$ and let v denote a multiplier system with respect to (Γ, k). Consider a smooth function g which satisfies the transformation property* (3.6) *and solves the partial differential equation* (3.19) *where we again assume a parameterization of the eigenvalue λ by $\lambda = \frac{1}{4} - \nu^2$ for some $\nu \in \mathbb{C} \setminus -\frac{1}{2}\mathbb{N}$. In other words, g satisfies*

$$g\,\big\|_k M \overset{(3.6)}{=} v(M)\, g \quad \text{for all } M \in \Gamma \quad \text{and} \quad \Delta_k\, g \overset{(3.19)}{=} \left(\frac{1}{4} - \nu^2\right) g.$$

Since Γ is a finite subset of $\Gamma(1)$, its standard fundamental domain $\mathcal{F}_\Gamma$ admits cusps $q_1 = \infty, q_2, \ldots, q_\mu \in \mathbb{Q} \cup \{\infty\}$. The matrices $A_1, A_2, \ldots, A_\mu \in \Gamma(1)$ denote representatives of the associated right cosets of Γ in $\Gamma(1)$ and that they satisfy $q_j = A_j\, \infty$. (See Lemma 1.28.) Let $l_1, l_2, \ldots, l_\mu$ be the width of the cusp q_i and $\kappa_1, \kappa_2, \ldots, \kappa_\mu$ be the corresponding real numbers in $[0, 1)$ *associated with q_j by* (1.41) *of Definition 1.52.*

Then for each j, $j \in \{1, \ldots, \mu\}$, there exists a nonnegative real y_j such that F has an expansion at the cusp q_j given by

$$\begin{aligned} g(z) & \\ = \tilde{\sigma}_{k,j}(z) & \sum_{\substack{n\in\mathbb{Z}\\ n+\kappa_j\neq 0}} A_j(n)\, \mathbb{W}_{\mathrm{sign}(n)\frac{k}{2},\nu}\left(4\pi\left|n+\kappa_j\right| \frac{\Im\left(A_j^{-1}\, z\right)}{l_j}\right) e^{2\pi(n+\kappa_j)\frac{\Re\left(A_j^{-1}\, z\right)}{l_j}} \\ + \tilde{\sigma}_{k,j}(z) & \sum_{\substack{n\in\mathbb{Z}\\ n+\kappa_j\neq 0}} B_j(n)\, \mathbb{M}_{\mathrm{sign}(n)\frac{k}{2},\nu}\left(4\pi\left|n+\kappa_j\right| \frac{\Im\left(A_j^{-1}\, z\right)}{l_j}\right) e^{2\pi(n+\kappa_j)\frac{\Re\left(A_j^{-1}\, z\right)}{l_j}} \\ + \tilde{\sigma}_{k,j}(z) & \left(C_{j,+}\left(\frac{\Im\left(A_j^{-1}\, z\right)}{l_j}\right)^{\frac{1}{2}+\nu} + C_{j,-}\left(\frac{\Im\left(A_j^{-1}\, z\right)}{l_j}\right)^{\frac{1}{2}-\nu}\right) \end{aligned} \tag{3.37}$$

valid for all $z \in \mathbb{H}$. The 0th-term coefficients $C_{j,+}$ and $C_{j,-}$ vanish if $\kappa_j > 0$ is strictly positive. The prefix factor $\tilde{\sigma}_{k,j}(z)$ is given by

$$\tilde{\sigma}_{k,j}(z) := \begin{cases} 1 & \text{if } q_j = \infty \text{ and} \\ \left(\frac{z-q_j}{|z-q_j|}\right)^{-k} & \text{if } q_j \neq \infty. \end{cases} \tag{3.38}$$

Remark 3.28 The above theorem is the analogue to Theorem 1.56 for modular forms. In particular, the factor $\tilde{\sigma}_(k, j)(z)$ defined in (3.28) is the unitary version of the factor $\sigma_{k,j}(z)$ defined in (1.45) of Theorem 1.56.

Proof (of Theorem 3.27) In a first step, we determine the expansion of g at the cusp $q_1 = \infty$. The cusp width l_1 and the value $\kappa_1 \in [0, 1)$ satisfy $v\big(S^{l_1}\big) = e^{2\pi i\kappa_1}$, see Definitions 1.36 and 1.52.

We apply (3.6) to get

$$g\big(S^{l_1}z\big) = g(z + \lambda_1) \overset{(3.6)}{=} e^{2\pi i\kappa_1}\, g(z)$$

for all $z \in \mathbb{H}$. Define the function $g_1 : \mathbb{H} \to \mathbb{C}$ by

$$g_1(z) := e^{-2\pi i\kappa_1\frac{z}{l_1}}\, g(z). \tag{3.39}$$

The function g_1 l_1-periodic on $\mathbb{H}$:

$$g_1(x + l_1 + iy) = g_1(x + iy) \qquad \text{for all } x \in \mathbb{R} \text{ and } y > 0.$$

Hence, g admits a Fourier expansion (in x) of the form

$$g_1(z) = \sum_{n=-\infty}^{\infty} a_1(n; y)\left(e^{2\pi ix}\right)^n = \sum_{n=-\infty}^{\infty} a_1(n; y)\, e^{2\pi inx},$$

where the coefficients $a_1(n; y)$ may depend on the imaginary part y of $z = x + iy$. Using (3.39), we see that g admits

$$g(z) = \sum_{n=-\infty}^{\infty} \tilde{a}_1\left(n; \frac{y}{l_1}\right) e^{2\pi i(n+\kappa_1)\frac{y}{l_1}}$$

for all $z = x + iy \in \mathbb{H}$.

We are now in a similar situation as in the proof of Proposition 3.25, g admits the above Fourier expansion in x and satisfies the partial differential equation (3.19). We use the method of separation of variables to solve this partial differential equation. This leads to two cases.

Case $n + \kappa \neq 0$ Put $\varepsilon := \mathrm{sign}\,(n+\kappa) \in \{-1, 1\}$. Substituting

$$a_1(n; y) = h\left(4\pi\varepsilon(n+\kappa)\frac{y}{l_1}\right),$$

under the assumption $n + \kappa \neq 0$ and using the separation of variables approach, shows that h solves the ordinary differential equation

$$h''(t) + \left(-\frac{1}{4} + \frac{1}{2}\varepsilon k\frac{1}{t} + \frac{\frac{1}{4} - \nu^2}{t^2}\right) h(t) = 0.$$

This second-order differential equation is again the *Whittaker differential equation*. Solutions are the $\mathbb{W}_{\cdot,\cdot}$- and $\mathbb{M}_{\cdot,\cdot}$-Whittaker functions $\mathbb{W}_{\varepsilon\frac{k}{2},\nu}(t)$ and $\mathbb{M}_{\varepsilon\frac{k}{2},\nu}(t)$, see (3.31).

Case $n + \kappa = 0$ Since $n \in \mathbb{Z}$ and $\kappa \in [0, 1)$, see Definition 1.52, we conclude directly that $\not\Leftarrow +\kappa = 0$ implies $n = 0$ and $\kappa = 0$. The separation of variables approach with $n = \kappa = 0$ shows that $a_1(0; y)$ solves the ordinary differential equation

$$t^2 h''(t) + \left(\frac{1}{4} - \nu^2\right) h(t) = 0. \tag{3.32}$$

Its independent solutions are $t^{\frac{1}{2}+\nu}$ and $t^{\frac{1}{2}-\nu}$, see (3.33).

The above arguments show that g admits a *Fourier-Whittaker expansion* of the form

$$\begin{aligned} g(x+iy) = &\sum_{\substack{n\in\mathbb{Z}\\ n+\kappa\neq 0}} A_1(n)\, \mathbb{W}_{\mathrm{sign}(n)\frac{k}{2},\nu}\left(\frac{4\pi|n+\kappa|y}{l_1}\right) e^{\frac{2\pi(n+\kappa)x}{l_1}} \\ &+ \sum_{\substack{n\in\mathbb{Z}\\ n+\kappa\neq 0}} B_1(n)\, \mathbb{M}_{\mathrm{sign}(n)\frac{k}{2},\nu}\left(\frac{4\pi|n+\kappa|y}{l_1}\right) e^{\frac{2\pi(n+\kappa)x}{l_1}} \\ &+ C_{1,+}\left(\frac{y}{l_1}\right)^{\frac{1}{2}+\nu} + C_{1,-}\left(\frac{y}{l_1}\right)^{\frac{1}{2}-\nu}. \end{aligned} \tag{3.40}$$

The 0th-term coefficients $C_1(+)$ and $C_1(-)$ vanish if $\kappa > 0$ is strictly positive.

In the next step, we determine the expansions at the other cusps $q_j \neq \infty$. We map the cusp $q_j = A_j\infty$ back to the cusp ∞ (by applying $z \mapsto A_j^{-1}z$) and follow the calculation above for the cusp $q_1 = \infty$. The action of the stabilizer $M_{q_j}^{l_j} = A_j S^{l_j} A_j^{-1}$ implies that g is nearly periodic in the following sense. Define

$$\tilde{g}_j(z) := \frac{(z-q_j)^k}{|z-q_j|^k}\, g(z) = \tilde{\sigma}_j(z)^{-1}\, g(z).$$

Since g satisfies (3.6) and in particular $g\|_k M_{q_j}^{l_j} = v\big(M_{q_j}^{l_j}\big)\, g(z)$, we see that $\tilde{g}_j$ satisfies

$$\begin{aligned}
\tilde{g}_j\big(M_{q_j}^{l_j} z\big) &= \frac{\big(M_{q_j}^{l_j} z - q_j\big)^k}{\big|M_{q_j}^{l_j} z - q_j\big|^k}\, g\big(M_{q_j}^{l_j} z\big) \\
&\overset{(3.35)}{=} j_u\big(M_q^l, z\big)^{-k}\, \frac{(z-q)^k}{|z-q|^k}\, g\big(M_{q_j}^{l_j} z\big) \\
&\overset{(3.6)}{=} j_u\big(M_q^l, z\big)^{-k}\, \frac{(z-q)^k}{|z-q|^k}\, j_u\big(M_q^l, z\big)^k\, v\big(j_u\big(M_q^l\big)\, g(z)\big) \\
&= v\big(M_q^l\big)\, \frac{(z-q)^k}{|z-q|^k}\, g(z) \\
&= v\big(M_q^l\big)\, \tilde{g}_j(z).
\end{aligned}$$

Using Definition 1.52, we have $v\big(M_q^l\big) = e^{2\pi i \kappa_{q_j}}$ with $0 \le \kappa_{q_j} < 1$. This shows that g_j satisfies

$$\tilde{g}_j\big(M_{q_j}^{l_j} z\big) = e^{2\pi i \kappa_{q_j}}\, \tilde{g}_j(z).$$

Defining

$$g_j(z) := \tilde{g}_j\big(A_j\, z\big)$$

and recalling $M_{q_j}^{l_j} = A_j\, S^{l_j}\, A_j^{-1}$, see Lemma 1.31, we find that g_j satisfies

$$\begin{aligned}
g_j\big(S^{l_j} z\big) &= \tilde{g}_j\big(A_j\, S^{l_j}\, z\big) \\
&= \tilde{g}_j\big(\underbrace{A_j\, S^{l_j}\, A_j^{-1}}_{=M_{q_j}^{l_j}}\, A_j\, z\big) \\
&= e^{2\pi i \kappa_{q_j}}\, \tilde{g}_j\big(A_j\, z\big) \\
&= e^{2\pi i \kappa_{q_j}}\, g_j(z).
\end{aligned}$$

Similar to the previous case $q_1 = \infty$, we conclude that $g_j(x+iy)$ admits a Fourier expansion

$$g_j(x+iy) = \sum_{n=-\infty}^{\infty} \tilde{a}_j\left(n; \frac{y}{l_j}\right) e^{2\pi i(n+\kappa_j)\frac{x}{l_j}}.$$

Following the arguments in the previous case shows also that g_j admits a *Fourier-Whittaker expansion* of the form

$$\begin{aligned} g_j(x+iy) = & \sum_{\substack{n\in\mathbb{Z}\\ n+\kappa_j\neq 0}} A_j(n)\, \mathbb{W}_{\mathrm{sign}(n)\frac{k}{2},\nu}\left(\frac{4\pi\left|n+\kappa_j\right|y}{l_j}\right) e^{\frac{2\pi i(n+\kappa_j)x}{l_j}} \\ & + \sum_{\substack{n\in\mathbb{Z}\\ n+\kappa_j\neq 0}} B_j(n)\, \mathbb{M}_{\mathrm{sign}(n)\frac{k}{2},\nu}\left(\frac{4\pi\left|n+\kappa_j\right|y}{l_1}\right) e^{\frac{2\pi i(n+\kappa_j)x}{l_j}} \\ & + C_{j,+}\left(\frac{y}{l_j}\right)^{\frac{1}{2}+\nu} + C_{j,-}\left(\frac{y}{l_j}\right)^{\frac{1}{2}-\nu}. \end{aligned} \tag{3.41}$$

Since

$$g_j(x+iy) = \tilde{g}_j\big(A_j(x+iy)\big) = \tilde{\sigma}_j\big(A_j(x+iy)\big)^{-1} g\big(A_j(x+iy)\big) \tag{3.42}$$

by construction of g_j, we see that g admits the expansion (3.37). This proves the theorem. □

3.2.4 *Maass Operators* $\mathrm{E}_k^{\pm}$ *and the Fourier-Whittaker Expansion*

Let $k \in \mathbb{R}$ and v a multiplier system compatible with $\big(\Gamma(1), k\big)$. In this subsection, we restrict ourselves to the full modular group $\Gamma(1)$.

Assume that g satisfies the transformation property (3.6) for the full modular group and satisfies (3.19) for some eigenvalue $\lambda = \frac{1}{4} - \nu^2$ for some $\nu \in \mathbb{C} \smallsetminus -\frac{1}{2}\mathbb{N}$. We are in the setting of §3.2.2.

Let us apply the Maass operators $\mathrm{E}_k^{\pm}$ to g. We know from Lemma 3.10 that $\mathrm{E}_k^{\pm} g$ satisfies (3.6) and (3.19) for weight $k \pm 2$. This means by Proposition 3.25 that g and $\mathrm{E}_k^{\pm} g$ admit Fourier-Whittaker expansions of the form given in (3.34), respectively. How do the coefficients change?

Using the results in §3.2.1, we get the following result.

Proposition 3.29 *Let $k \in \mathbb{R}$ and v a multiplier compatible with $\big(\Gamma(1), k\big)$. Assume that g satisfies the transformation property* (3.6) *for the full modular group and satisfies* (3.19) *for some eigenvalue $\lambda = \frac{1}{4} - \nu^2$ for some $\nu \in \mathbb{C} \smallsetminus -\frac{1}{2}\mathbb{N}$.*

Then $\mathrm{E}_k^{\pm} g$ admits the expansions

$$
\begin{aligned}
&\left(\mathrm{E}_k^+ g\right)(x+iy)\\
&= -2 \sum_{\substack{n\in\mathbb{Z}\\ n+\kappa>0}} \Big(A_n\, \mathbb{W}_{\frac{k+2}{2},\nu}\big(4\pi(n+\kappa)y\big)\, e^{2\pi(n+\kappa)x}\\
&\qquad\qquad + B_n\, \mathbb{M}_{\frac{k+2}{2},\nu}\big(4\pi(n+\kappa)y\big)\, e^{2\pi(n+\kappa)x}\Big)\\
&\quad + \left(\frac{k(k+2)}{2}+2\lambda\right) \sum_{\substack{n\in\mathbb{Z}\\ n+\kappa<0}} \Big(A_n\, \mathbb{W}_{-\frac{k+2}{2},\nu}\big(4\pi|n+\kappa|y\big)\, e^{2\pi(n+\kappa)x}\\
&\qquad\qquad + B_n\, \mathbb{M}_{-\frac{k+2}{2},\nu}\big(4\pi|n+\kappa|y\big)\, e^{2\pi(n+\kappa)x}\Big)\\
&\quad + \big(1+2\nu+k\big)\, C_+\, y^{\frac{1}{2}+\nu} + \big(1-2\nu+k\big)\, C_-\, y^{\frac{1}{2}-\nu}
\end{aligned}
\tag{3.43}
$$

and

$$
\begin{aligned}
&\left(\mathrm{E}_k^- g\right)(x+iy)\\
&= \left(\frac{k(k-2)}{2}+2\lambda\right) \sum_{\substack{n\in\mathbb{Z}\\ n+\kappa>0}} \Big(A(n)\, \mathbb{W}_{\frac{k-2}{2},\nu}\big(4\pi(n+\kappa)y\big)\, e^{2\pi(n+\kappa)x}\\
&\qquad\qquad + B(n)\, \mathbb{M}_{\frac{k-2}{2},\nu}\big(4\pi(n+\kappa)y\big)\, e^{2\pi(n+\kappa)x}\Big)\\
&\quad + (-2) \sum_{\substack{n\in\mathbb{Z}\\ n+\kappa<0}} \Big(A(n)\, \mathbb{W}_{-\frac{k-2}{2},\nu}\big(4\pi|n+\kappa|y\big)\, e^{2\pi(n+\kappa)x}\\
&\qquad\qquad + B(n)\, \mathbb{M}_{-\frac{k-2}{2},\nu}\big(4\pi|n+\kappa|y\big)\, e^{2\pi(n+\kappa)x}\Big)\\
&\quad + \big(1+2\nu-k\big)\, C_+\, y^{\frac{1}{2}+\nu} + \big(1-2\nu-k\big)\, C_-\, y^{\frac{1}{2}-\nu}
\end{aligned}
\tag{3.44}
$$

where the coefficients A_n, B_n, and $C_\pm$ are given by the Fourier-Whittaker expansion (3.34) *of g.*

Proof The proof is left to the reader as an exercise. □

Theorem 3.30 *Let $\Gamma \subset \Gamma(1)$, $k \in \mathbb{R}$, v a multiplier with respect to (Γ, k) and $\nu \in \mathbb{C} \setminus -\frac{1}{2}\mathbb{N}$ a spectral parameter. Moreover, denote by $q_1 = \infty, q_2, \ldots, q_\mu$ the cusps of the standard fundamental domain $\mathcal{F}_\Gamma$ and by $\mathrm{A}_1, \mathrm{A}_2, \ldots, \mathrm{A}_\mu \in \Gamma(1)$ the representatives of the associated right cosets of Γ in $\Gamma(1)$ and that they satisfy $q_j = \mathrm{A}_j\,\infty$, see Lemma 1.28.*

Consider a real-analytic function $g : \mathbb{H} \to \mathbb{C}$ that satisfies the transformation property (3.6) *for the group Γ and satisfies* (3.19) *with $\lambda = \frac{1}{4} - \nu^2$.*

Then $\mathrm{E}_k^{\pm} g$ *admits the expansions*

$$\begin{aligned}
&\left(\mathrm{E}_k^+ g\right)(x+iy) \\
&= -2\,\tilde{\sigma}_{k,j}(z) \sum_{\substack{n\in\mathbb{Z}\\ n+\kappa_j>0}} \Big(A_j(n)\, \mathbb{W}_{\frac{k+2}{2},\nu}\big(4\pi(n+\kappa_j)y\big)\, e^{2\pi(n+\kappa_j)x} \\
&\qquad + B_j(n)\, \mathbb{M}_{\frac{k+2}{2},\nu}\big(4\pi(n+\kappa_j)y\big)\, e^{2\pi(n+\kappa_j)x} \Big) \\
&\quad + \left(\frac{k(k+2)}{2} + 2\lambda\right) \tilde{\sigma}_{k,j}(z) \cdot \\
&\qquad \cdot \sum_{\substack{n\in\mathbb{Z}\\ n+\kappa_j<0}} \Big(A_j(n)\, \mathbb{W}_{-\frac{k+2}{2},\nu}\big(4\pi\,|n+\kappa_j|\,y\big)\, e^{2\pi(n+\kappa_j)x} \\
&\qquad\qquad + B_j(n)\, \mathbb{M}_{-\frac{k}{2},\nu}\big(4\pi\,|n+\kappa_j|\,y\big)\, e^{2\pi(n+\kappa_j)x} \Big) \\
&\quad + \big(1+2\nu+k\big)\, \tilde{\sigma}_{k,j}(z)\, C_{j,+}\, y^{\frac{1}{2}+\nu} + \big(1-2\nu+k\big)\, \tilde{\sigma}_{k,j}(z)\, C_{j,-}\, y^{\frac{1}{2}-\nu}
\end{aligned} \tag{3.45}$$

and

$$\begin{aligned}
&\left(\mathrm{E}_k^- g\right)(x+iy) \\
&= \left(\frac{k(k-2)}{2} + 2\lambda\right) \tilde{\sigma}_{k,j}(z) \sum_{\substack{n\in\mathbb{Z}\\ n+\kappa>0}} \Big(A_n\, \mathbb{W}_{\frac{k+2}{2},\nu}\big(4\pi(n+\kappa)y\big)\, e^{2\pi(n+\kappa)x} \\
&\qquad + B_n\, \mathbb{M}_{\frac{k+2}{2},\nu}\big(4\pi(n+\kappa)y\big)\, e^{2\pi(n+\kappa)x} \Big) \\
&\quad + (-2)\, \tilde{\sigma}_{k,j}(z) \sum_{\substack{n\in\mathbb{Z}\\ n+\kappa<0}} \Big(A_n\, \mathbb{W}_{-\frac{k+2}{2},\nu}\big(4\pi\,|n+\kappa|\,y\big)\, e^{2\pi(n+\kappa)x} \\
&\qquad + B_n\, \mathbb{M}_{-\frac{k}{2},\nu}\big(4\pi\,|n+\kappa|\,y\big)\, e^{2\pi(n+\kappa)x} \Big) \\
&\quad + \big(1+2\nu-k\big)\, \tilde{\sigma}_{k,j}(z)\, C_+\, y^{\frac{1}{2}+\nu} + \big(1-2\nu-k\big)\, \tilde{\sigma}_{k,j}(z)\, C_-\, y^{\frac{1}{2}-\nu},
\end{aligned} \tag{3.46}$$

where the coefficients A_n*,* B_n*, and* $C_\pm$ *are given by the Fourier-Whittaker expansion* (3.34) *of* g*. The prefix factor* $\tilde{\sigma}_{k,j}(z)$ *is defined in* (3.38).

Proof Following the arguments of the proof of Theorem 3.27, we construct the functions g_j. g and g_j are related by (3.42):

$$g_j(z) = \tilde{\sigma}_{k,j}(z)^{-1}\, g(A_j\, z).$$

Moreover, g_j admits a Fourier-Whittaker expansion of the form (3.41). On the other hand, consider $\mathrm{E}_k^{\pm} g$ and its associated function

$$h_j(z) = \tilde{\sigma}_{k\pm 2,j}(z)^{-1} \left(\mathrm{E}_k^{\pm} g(A_j\, z)\right)$$

by the same procedure. Since h_j admits a Fourier-Whittaker function of the form (3.41) for weight $k \pm 2$, we may compare the coefficients as done in the proof of Proposition 3.29. Performing this comparison implies the expansions (3.45) and (3.46) respectively. □

3.3 Maass Wave Forms of Real Weight

We follow a definition similar to Definition 1.59 for Maass wave forms.

Definition 3.31 Let $\Gamma \subset \Gamma(1)$, $k \in \mathbb{R}$, v a multiplier with respect to $\big(\Gamma, k\big)$ and $\nu \in \mathbb{C} \setminus -\frac{1}{2}\mathbb{N}$ a *spectral parameter*.

Since Γ is of finite index in $\Gamma(1)$, its standard fundamental domain $\mathcal{F}_\Gamma$ admits cusps $q_1 = \infty, q_2, \ldots, q_\mu$. The matrices $A_1, A_2, \ldots, A_\mu \in \Gamma(1)$ denote representatives of the associated right cosets of Γ in $\Gamma(1)$: they satisfy $q_j = A_j\, \infty$, see Lemma 1.28.

A function $u : \mathbb{H} \to \mathbb{C}$ is called a *Maass wave form of weight k and multiplier v for the group* Γ if u satisfies the following conditions:

1. u is a real-analytic function on $\mathbb{H}$ in the sense of Definition 3.14,
2. $\Delta_k\, u = \left(\frac{1}{4} - \nu^2\right) u$,
3. $u\big\|_k M = v(M)\, u$ for all $M \in \Gamma$ (i.e., u satisfies (3.6)) and
4. u admits an expansion of the form

$$\begin{aligned} u(z) &= \tilde{\sigma}_{k,j}(z) \sum_{\substack{n\in\mathbb{Z}\\ n+\kappa_j\neq 0}} A_j(n)\, \mathrm{W}_{\operatorname{sign}(n)\frac{k}{2},\nu}\left(4\pi\left|n+\kappa_j\right| \frac{\Im\left(A_j^{-1} z\right)}{l_j}\right) e^{2\pi i (n+\kappa_j)\frac{\Re\left(A_j^{-1} z\right)}{l_j}} \\ &\quad + \tilde{\sigma}_{k,j}(z) \left(C_{j,+} \left(\frac{\Im\left(A_j^{-1} z\right)}{l_j}\right)^{\frac{1}{2}+\nu} + C_{j,-} \left(\frac{\Im\left(A_j^{-1} z\right)}{l_j}\right)^{\frac{1}{2}-\nu}\right) \end{aligned} \tag{3.47}$$

for each cusp $q_j = A_j\, \infty$ where the factor $\tilde{\sigma}_{k,j}(z)$ is given in (3.38).

We denote the space of such Maass wave forms with $M_{k,v}^{\mathrm{Maass}}\big(\Gamma, \nu\big)$.

Definition 3.32 A function $u : \mathbb{H} \to \mathbb{C}$ is called a *Maass cusp form of weight k and multiplier v for the group* Γ if $u \in M_{k,v}^{\text{Maass}}\big(\Gamma(1), \nu\big)$ is a Maass wave form and it satisfies the stronger growth condition

(4′). Moreover, u admits an expansion of the form

$$u(z) = \tilde{\sigma}_{k,j}(z) \sum_{\substack{n\in\mathbb{Z}\\ n+\kappa_j\neq 0}} A_j(n) \cdot \cdot \mathbb{W}_{\text{sign}(n)\frac{k}{2},\nu}\left(4\pi\left|n+\kappa_j\right|\frac{\Im\left(A_j^{-1}z\right)}{l_j}\right) e^{2\pi i(n+\kappa_j)\frac{\Re\left(A_j^{-1}z\right)}{l_j}} \tag{3.48}$$

for each cusp q_j where $\tilde{\sigma}_{k,j}(z)$ and $0 \le \kappa_j < 1$ is given in Theorem 3.27.

We denote the space of such Maass cusp forms with $S_{k,v}^{\text{Maass}}\big(\Gamma, \nu\big)$.

Remark 3.33 Similar to Remark 1.62, we shorten the notation of the spaces of Maass forms in the case of the trivial multiplier $v \equiv 1$. In this case, we denote the spaces of Maass wave forms and Maass cusp forms respectively by $M_k^{\text{Maass}}(\Gamma, \nu)$ and $S_k^{\text{Maass}}(\Gamma, \nu)$ instead of $M_{k,v}^{\text{Maass}}(\Gamma, \nu)$ and $S_{k,v}^{\text{Maass}}(\Gamma, \nu)$.

Exercise 3.34 Show that a Maass cusp form $g \in S_{k,v}^{\text{Maass}}\big(\Gamma, \nu\big)$ admits the Fourier-Whittaker expansion

$$\begin{aligned}&\left(\tilde{\sigma}_{k,j}\big(A_j(l_j\, z)\big)\right)^{-1} u\big(A_j(l_j\, z)\big)\\ &\quad = \sum_{\substack{n\in\mathbb{Z}\\ n+\kappa\neq 0}} A_j(n)\, \mathbb{W}_{\text{sign}(n)\frac{k}{2},\nu}\big(4\pi\left|n+\kappa_j\right|\Im\left(z\right)\big)\, e^{2\pi i(n+\kappa_j)\Re(z)},\end{aligned} \tag{3.49}$$

where we use the notation introduced in Definition 3.31. We would like to point out that $A_j(n)$ denote the coefficients of the expansion and $A_j\,(x+iy)$ denotes the Möbius transformation of $x+iy \in \mathbb{H}$ under the matrix A_j corresponding to the cusp q_j.

Remark 3.35 Recall the identity $W_{0,\nu}(y) = \sqrt{\frac{y}{\pi}}\, K_\nu\left(\frac{1}{2}y\right)$, see, e.g., Appendix A.9 or [153, (13.18.9)]. Using this identity, the W-Fourier-Whittaker expansion in (3.37) leads to the usual Fourier-Bessel expansion of a classical Maass-cusp form $u \in M_0^{\text{Maass}}\big(\Gamma(1), \nu\big)$ in weight 0 and trivial multiplier system, which is well known in the literature. We have

$$u(x+iy) = C_+ y^{\frac{1}{2}+\nu} + C_- y^{\frac{1}{2}-\nu} + 2\sqrt{y} \sum_{n\in\mathbb{Z}_{\neq 0}} \sqrt{|n|} A_n\, K_\nu\big(2\pi\,|n|\,y\big)\, e^{2\pi i n x}. \tag{3.50}$$

We discussed extensively Fourier-Whittaker expansions of eigenfunctions of the Laplace Δ_k in the previous section §3.2. Similar to our approach in Theorem 1.68 for modular forms, we see that the growth estimate relates directly to the growth of the coefficients for Maass cusp forms.

Theorem 3.36 *Let $u \in S^{Maass}_{k,v}(\Gamma, v)$ with $\Gamma \subset \Gamma(1)$. Assume that u has an expansion at ∞ given by*

$$u(x+iy) = \sum_{\substack{n\in\mathbb{Z} \\ n+\kappa\neq 0}} A(n)\, \mathbb{W}_{\operatorname{sign}(n)\frac{k}{2},\nu}\left(4\pi\,|n+\kappa|\frac{y}{l}\right)\, e^{2\pi(n+\kappa)\frac{x}{l}}$$

with l denoting the cusp width at ∞ and κ satisfying $v\big(S^l\big) = e^{2\pi i\kappa}$.

Then the Fourier coefficients $A(n)$ satisfy

$$A(n) = O(1) \qquad \text{as } |n| \to \infty.$$

Proof Consider the expansion of the function $u(z)\, e^{-2\pi i |m+\kappa| \Re(z)}$. Since u as a Maass cusp form is real-analytic on $\mathbb{H}$, it is bounded on every compact subset of $\mathbb{H}$. Using this bound as constant integrable function on the compact set, we may use Lebesgue's dominated convergence theorem to interchange integration and summation. We obtain

$$\begin{aligned}
&\int_z^{z+l} u(\xi)\, e^{-2\pi i (m+\kappa)\Re(z)}\, d\xi \\
&= \int_z^{z+l} \sum_{\substack{n\in\mathbb{Z} \\ n+\kappa\neq 0}} A(n)\, \mathbb{W}_{\operatorname{sign}(n)\frac{k}{2},\nu}\big(4\pi\,|n+\kappa|\Im(\xi)\big)\, e^{2\pi i (n-m)\Re(\xi)}\, d\xi \\
&= \sum_{\substack{n\in\mathbb{Z} \\ n+\kappa\neq 0}} A(n)\, \mathbb{W}_{\operatorname{sign}(n)\frac{k}{2},\nu}\big(4\pi\,|n+\kappa|\Im(z)\big) \int_z^{z+l} e^{2\pi i (n-m)\Re(\xi)}\, d\xi \\
&= l\, A(m)\, \mathbb{W}_{\operatorname{sign}(m)\frac{k}{2},\nu}\big(4\pi\,|m+\kappa|\Im(z)\big)
\end{aligned}$$

for all z in the compact subset of $\mathbb{H}$ chosen above. We used the fact that

$$\int_z^{z+l} e^{2\pi i (m-n)\frac{\Re(\xi)}{l}}\, d\xi = \begin{cases} 0 & \text{if } m-n \neq 0 \text{ and} \\ l & \text{if } m-n = 0. \end{cases}$$

Thus, using the boundedness of $u(z)$ on every compact subset of $\mathbb{H}$ again, we get

$$l\, A(n)\, \mathbb{W}_{\operatorname{sign}(n)\frac{k}{2},\nu}\big(4\pi\,|n+\kappa|\Im(z)\big) = O(1)$$

is bounded for all $n \in \mathbb{Z}$ and z in a compact subset of $\mathbb{H}$. Since the function $\mathbb{W}_{\cdot,\cdot}(y)$ is also bounded (and exponentially small for $y \to \infty$), we find the estimate

$$A(n) = O(1).$$

□

Lemma 3.37 *Let $\kappa \in [0, 1)$ and $Y > 0$. Suppose that the function U given by the absolutely convergent series*

$$U(z) = \sum_{\substack{n\in\mathbb{Z}\\ n+\kappa\neq 0}} A(n)\, \mathbb{W}_{\mathrm{sign}(n)\frac{k}{2},\nu}\left(4\pi\,|n+\kappa|\Im(z)\right) e^{2\pi i(n+\kappa)\Re(z)} \tag{3.51}$$

for all $z \in \mathbb{H}$ with $\Im(z) \geq Y$.

Then there exists an $\varepsilon > 0$ such that U satisfies the growth condition

$$U(z) = O\left(\Im(z)^{\left|\frac{k}{2}\right|} e^{-\varepsilon\,\Im(z)}\right) \qquad \text{as } Y \leq \Im(z) \to \infty. \tag{3.52}$$

Proof Recall the asymptotic growth condition

$$W_{k,\nu}(y) \sim e^{-\frac{y}{2}}\, y^k \qquad \text{as } y \to \infty$$

given in (3.23). Using $\mathbb{W}_{k,\nu}(y) = W_{k,\nu}(y)$ in (3.26) implies immediately

$$\mathbb{W}_{\mathrm{sign}(n)\frac{k}{2},\nu}\left(4\pi\,|n+\kappa|\Im(z)\right) \sim \left(4\pi\,|n+\kappa|\Im(z)\right)^{\mathrm{sign}(n)\frac{k}{2}} e^{-2\pi\,|n+\kappa|\Im(z)}$$

as $y \to \infty$.

Now, we exploit the assumption that the series expansion of U in (3.51) converges absolutely for each $z \in \mathbb{H}$ with $\Im(z) > Y$.

This implies that for given fixed $z = it$ with $t > Y$, the series

$$\sum_{\substack{n\in\mathbb{Z}\\ n+\kappa\neq 0}} |A(n)| \left|\mathbb{W}_{\mathrm{sign}(n)\frac{k}{2},\nu}\left(4\pi\,|n+\kappa|t\right)\right|$$

converges absolutely. In particular, this shows that the coefficients $A(n)$ have to compensate the growth of the $\mathbb{W}_{\cdot,\cdot}(\cdot)$ term as $|n+\kappa| \to \infty$. The coefficients have to satisfy

$$A(n) \leq O\left(\left|\mathbb{W}_{\mathrm{sign}(n)\frac{k}{2},\nu}\left(4\pi\,|n+\kappa|t\right)\right|^{-1}\right)$$

as $|n+\kappa| \to \infty$. Combining this with the asymptotic estimate derived above for $\mathbb{W}_{\cdot,\cdot}(\cdot)$, we get

$$A(n) = O\left(\left(|n+\kappa|\right)^{-\operatorname{sign}(n)\frac{k}{2}} e^{2\pi|n+\kappa|t}\right)$$

as $|n+\kappa| \to \infty$ for given (fixed) $z = it$.

Substituting this particular growth estimate for $A(n)$ into the series definition of $U(z)$ for $z \in \mathbb{H}$ with $\Im(z) > t+1$ gives

$$\begin{aligned}
&|U(z)| \\
&= \sum_{\substack{n\in\mathbb{Z}\\ n+\kappa\neq 0}} O\left(|A(n)| \left|\mathbb{W}_{\operatorname{sign}(n)\frac{k}{2},\nu}\left(4\pi|n+\kappa|t\right)\right|\right) \\
&= \sum_{\substack{n\in\mathbb{Z}\\ n+\kappa\neq 0}} O\Bigg(\left(|n+\kappa|\right)^{-\operatorname{sign}(n)\frac{k}{2}} \cdot \\
&\qquad \cdot \left|e^{2\pi|n+\kappa|t}\left(4\pi|n+\kappa|\Im(z)\right)^{\operatorname{sign}(n)\frac{k}{2}} e^{-2\pi|n+\kappa|\Im(z)}\right|\Bigg) \\
&= \sum_{\substack{n\in\mathbb{Z}\\ n+\kappa\neq 0}} O\left(\left(|n+\kappa|\right)^{-\operatorname{sign}(n)\frac{k}{2}}\left(4\pi|n+\kappa|\Im(z)\right)^{\operatorname{sign}(n)\frac{k}{2}} e^{2\pi|n+\kappa|t} e^{-2\pi|n+\kappa|\Im(z)}\right) \\
&= \sum_{\substack{n\in\mathbb{Z}\\ n+\kappa\neq 0}} O\left(\Im(z)^{\operatorname{sign}(n)\frac{k}{2}} e^{-2\pi|n+\kappa|\left(\Im(z)-t\right)}\right).
\end{aligned}$$

Since $\Im(z) - t > 1$ by assumption we find that ε with $0 < \varepsilon < \min_{n+\kappa\neq 0}\left\{\pi|n+\kappa|\right\}$ implies

$$\begin{aligned}
|U(z)| &= \sum_{\substack{n\in\mathbb{Z}\\ n+\kappa\neq 0}} O\left(\Im(z)^{\operatorname{sign}(n)\frac{k}{2}} e^{-2\pi|n+\kappa|\left(\Im(z)-t\right)}\right) \\
&= O\left(\Im(z)^{\operatorname{sign}(n)\frac{k}{2}} e^{-\varepsilon\left(\Im(z)-t\right)}\right) \sum_{\substack{n\in\mathbb{Z}\\ n+\kappa\neq 0}} O\left(e^{\varepsilon\left(\Im(z)-t\right)-2\pi|n+\kappa|\left(\Im(z)-t\right)}\right) \\
&= O\left(\Im(z)^{\operatorname{sign}(n)\frac{k}{2}} e^{-\varepsilon\left(\Im(z)-t\right)}\right) \underbrace{\sum_{\substack{n\in\mathbb{Z}\\ n+\kappa\neq 0}} O\left(e^{-\pi|n+\kappa|\left(\Im(z)-t\right)}\right)}_{=O(1)} \\
&= O\left(\Im(z)^{\operatorname{sign}(n)\frac{k}{2}} e^{-\varepsilon\left(\Im(z)-t\right)}\right)
\end{aligned}$$

since the series

$$\sum_{\substack{n\in\mathbb{Z}\\ n+\kappa\neq 0}} e^{-\pi|n+\kappa|\left(\Im(z)-t\right)}$$

converges absolutely for $\Im(z) > t$. Hence, $U(z)$ admits the growth estimation

$$\begin{aligned}|U(z)| &= O\left(\Im(z)^{\mathrm{sign}(n)\frac{k}{2}}\, e^{-\varepsilon\left(\Im(z)-t\right)}\right)\\ &= O\left(\Im(z)^{\mathrm{sign}(n)\frac{k}{2}}\, e^{-\varepsilon\,\Im(z)}\right)\end{aligned}$$

since the factor $e^{\varepsilon t}$ does not depend on z and can be absorbed in the $O(\cdot)$-notation. This proves the estimate in (3.52). □

Lemma 3.37 gives us the technical tool to prove the following result.

Theorem 3.38 *Let $\Gamma \subset \Gamma(1)$, $k \in \mathbb{R}$, and v a multiplier system compatible with (Γ, k). Suppose that the standard fundamental domain $\mathcal{F}_\Gamma$ admits cusps $q_1 = \infty$, $q_2, \ldots, q_\mu$; the matrices $A_1, A_2, \ldots, A_\mu \in \Gamma(1)$ denote representatives of the associated right cosets of Γ in $\Gamma(1)$ and they satisfy $q_j = A_j\,\infty$, see Lemma 1.28.*

We have the following growth estimates:

1. *For $u \in S_{k,v}^{Maass}(\Gamma, \nu)$ a Maass cusp form and given cusp q_j, there exists an $\varepsilon > 0$ such that u satisfies*

$$\tilde{\sigma}_{k,j}\left(A_j\, z\right)^{-1} u\left(A_j\, z\right) = O\left(\Im(z)^{\left|\frac{k}{2}\right|} e^{-\varepsilon\Im(z)}\right) \qquad \text{as } y \to \infty \text{ and } x \text{ bounded.} \tag{3.53}$$

2. *For $u \in M_{k,v}^{Maass}(\Gamma, \nu)$ a Maass wave form and given cusp q_j, there exists a $C > 0$ such that u satisfies*

$$\tilde{\sigma}_{k,j}\left(A_j\, z\right)^{-1} u\left(A_j\, z\right) = O\left(y^C\right) \qquad \text{for } y \to \infty \text{ and } x \text{ bounded.} \tag{3.54}$$

Proof Let $u \in S_{k,v}^{\mathrm{Maass}}(\Gamma, \nu)$. Then $\tilde{\sigma}_{k,j}\left(A_j\, z\right)^{-1} u\left(A_j\, z\right)$ admits a Fourier-Whittaker expansion of the form

$$\begin{aligned}&\left(\tilde{\sigma}_{k,j}\left(A_j(l_j\, z)\right)\right)^{-1} u\left(A_j(l_j\, z)\right)\\ &\quad= \sum_{\substack{n\in\mathbb{Z}\\ n+\kappa\neq 0}} A_j(n)\,\mathbb{W}_{\mathrm{sign}(n)\frac{k}{2},\nu}\left(4\pi\left|n+\kappa_j\right|\Im(z)\right) e^{2\pi i(n+\kappa_j)\Re(z)}\end{aligned}$$

that converges absolutely on $\mathbb{H}$, see Exercise 3.34. Using Lemma 3.37 show that $\left(\tilde{\sigma}_{k,j}\left(A_j(l_j\, z)\right)\right)^{-1} u\left(A_j(l_j\, z)\right)$ satisfies the growth estimate stated in (3.53).

Now, consider $u \in M_{k,v}^{\mathrm{Maass}}(\Gamma, v)$. By Definition 3.31, u admits the Fourier-Whittaker expansion (3.47). Similar to Exercise 3.34, we write the expansion as

$$\begin{aligned}
&\left(\tilde{\sigma}_{k,j}\big(A_j(l_j\, z)\big)\right)^{-1} u\big(A_j(l_j\, z)\big) \\
&\quad = \sum_{\substack{n\in\mathbb{Z} \\ n+\kappa\neq 0}} A_j(n)\, \mathbb{W}_{\mathrm{sign}(n)\frac{k}{2},\nu}\big(4\pi \left|n+\kappa_j\right| \Im(z)\big)\, e^{2\pi i (n+\kappa_j)\Re(z)} \\
&\qquad + C_{j,+}\, \Im(z)^{\frac{1}{2}+\nu} + C_{j,-}\, \Im(z)^{\frac{1}{2}-\nu}
\end{aligned}$$

by multiplying (3.47) with $\left(\tilde{\sigma}_{k,j}\big(A_j(l_j\, z)\big)\right)^{-1}$ and substituting $z \mapsto A_j\big(l_j\, z\big)$. Now, applying Lemma 3.37 to the absolutely convergent series part shows that u satisfies the growth estimate

$$\begin{aligned}
&\left(\tilde{\sigma}_{k,j}\big(A_j(l_j\, z)\big)\right)^{-1} u\big(A_j(l_j\, z)\big) \\
&\quad = O\left(\Im(z)^{\left|\frac{k}{2}\right|} e^{-\varepsilon\Im(z)}\right) + C_{j,+}\, \Im(z)^{\frac{1}{2}+\nu} + C_{j,-}\, \Im(z)^{\frac{1}{2}-\nu} \\
&\quad = O\left(\Im(z)^{C}\right) \qquad \text{as } \Im(z) \to \infty
\end{aligned}$$

for some $C > \max\left\{\Re\left(\frac{1}{2}+\nu\right), \Re\left(\frac{1}{2}+\nu\right)\right\}$. This proves Estimate 3.54. □

Exercise 3.39 Let $k \in \mathbb{R}$. For u a function on $\mathbb{H}$ sufficiently smooth, define g by

$$g(z) = \Im(z)^{-\frac{k}{2}}\, u(z).$$

Show the statement

$$\mathrm{E}_k^-\, u = 0 \iff u(z) = \Im(z)^{\frac{k}{2}}\, g(z) \quad \text{with } \frac{\partial g(z)}{\partial \overline{z}} = 0.$$

Exercise 3.40 Let $\Gamma \subset \Gamma(1)$, $k \in \mathbb{R}$ and v a multiplier system compatible with (Γ, k). For u a function on $\mathbb{H}$, define g by

$$g(z) = \Im(z)^{-\frac{k}{2}}\, u(z).$$

The following two statements are equivalent

1. u satisfies the transformation property (3.6).
2. g satisfies the transformation property (1.28).

Corollary 3.41 *Let $\Gamma \subset \Gamma(1)$ be a subgroup of finite index, $k \in \mathbb{R}$, v a multiplier system compatible with (Γ, k), and $\nu \in \mathbb{C} \setminus -\frac{1}{2}\mathbb{N}$ a spectral value. The Maass operators $\mathrm{E}_k^{\pm}$ map the space of Maass cusp forms of weight k into the one of weight $k \pm 2$. We have*

$$\mathrm{E}_k^{\pm} u \in M_{k\pm 2,v}^{Maass}(\Gamma, \nu) \qquad \textit{for all } u \in M_{k,v}^{Maass}(\Gamma, \nu)$$

and

$$\mathrm{E}_k^{\pm} u \in S_{k\pm 2,v}^{Maass}(\Gamma, \nu) \qquad \textit{for all } u \in S_{k,v}^{Maass}(\Gamma, \nu).$$

Proof Take a Maass cusp form $u \in M_{k,v}^{\mathrm{Maass}}(\Gamma, \nu)$. To show $\mathrm{E}_k^{\pm} u \in M_{k\pm 2,v}^{\mathrm{Maass}}(\Gamma, \nu)$, we have to check the conditions in Definition 3.31 of Maass forms.

1. Recall the definition of $\mathrm{E}_k^{\pm}$ as differential operator in (3.10). u is real-analytic in the sense of Definition 3.14 by assumption. Hence, $\mathrm{E}_k^{\pm} u$ is too, since we only take partial derivatives in the real part and imaginary part.
2. This follows directly from Lemma 3.9. In case of E_k^{+}, we have

$$\begin{aligned}
\Delta_{k+2}\left(\mathrm{E}_k^{+} u\right) &\overset{(3.12)}{=} \left(-\frac{1}{4}\mathrm{E}_k^{+}\mathrm{E}_{k+2}^{-} - \frac{k(k+2)}{4}\right)\mathrm{E}_k^{+} u \\
&= -\frac{1}{4}\mathrm{E}_k^{+}\mathrm{E}_{k+2}^{-}\mathrm{E}_k^{+} u - \frac{k(k+2)}{4}\mathrm{E}_k^{+} u \\
&= \mathrm{E}_k^{+}\left(-\frac{1}{4}\mathrm{E}_{k+2}^{-}\mathrm{E}_k^{+} u - \frac{k(k+2)}{4}\right) u \\
&\overset{(3.12)}{=} \mathrm{E}_k^{+}\left(\Delta_k u\right).
\end{aligned}$$

   Since $\Delta_k u = \left(\frac{1}{4} - \nu^2\right) u$ by assumption, we find

$$\Delta_{k+2}\left(\mathrm{E}_k^{+} u\right) = \mathrm{E}_k^{+}\left(\Delta_k u\right) = \left(\frac{1}{4} - \nu^2\right)\mathrm{E}_k^{+} u.$$

   The other identity follows similarly, using the other option in (3.12).
3. This follows from Lemma 3.10:

$$\left(\mathrm{E}_k^{\pm} u\right)\big\|_{k\pm 2} M \overset{(3.14)}{=} \mathrm{E}_k^{\pm}\left(u\big\|_k M\right)$$

   for all $M \in \Gamma$.
4. The required Fourier-Whittaker expansion was shown in Proposition 3.29 in the case of the full modular group $\Gamma = \Gamma(1)$. Theorem 3.30 shows this for arbitrary subgroups Γ of $\Gamma(1)$.

Hence, $\mathrm{E}_k^{\pm} u \in M_{k\pm 2,\upsilon}^{\mathrm{Maass}}(\Gamma, \nu)$ is a Maass form of weight $k \pm 2$.

The same arguments show that $u \in M_{k,\upsilon}^{\mathrm{Maass}}(\Gamma, \nu)$ implies $\mathrm{E}_k^{\pm} u \in S_{k\pm 2,\upsilon}^{\mathrm{Maass}}(\Gamma, \nu)$ since vanishing "0th Fourier term" in u implies vanishing "0th Fourier term" in $\mathrm{E}_k^{\pm} u$, see Theorem 3.30. □

As an exercise, we would like to show that we can project Maass cusp forms onto the lower half-plane $\mathbb{H}^-$ in the following sense:

Exercise 3.42 Let u be a Maass cusp form of weight k, multiplier system υ, and eigenvalue λ for the full modular group $\Gamma(1)$.

Defining

$$\tilde{u} : \mathbb{H}^- \to \mathbb{C}; \quad z \mapsto \tilde{u}(z) := u(\bar{z}) \tag{3.55}$$

for a Maass cusp form u defines a real-analytic function on the lower half-plane which satisfies

1. $\tilde{u}\big\|_{-k}^{\upsilon}\gamma = \tilde{u}$ for every $\gamma \in \Gamma(1)$,
2. $\tilde{u}$ is an eigenfunction of the Laplace operator Δ_{-k} with eigenvalue λ, and
3. $\tilde{u}$ satisfies the growth condition $\tilde{u}(z) = O\left(|\Im(z)|^{\left|\frac{k}{2}\right|} e^{-\varepsilon|\Im(z)|}\right)$ as $\Im(z) \to -\infty$ for some $\varepsilon > 0$.

3.4 Examples of Maass Wave Forms

3.4.1 The Embedding of Classical Modular Forms

Let $\Gamma \subset \Gamma(1)$, $k \in \mathbb{R}$, and υ a multiplier system compatible with (Γ, k). For $F \in M_{k,\upsilon}(\Gamma)$, put

$$u_F(z) := \Im(z)^{\frac{k}{2}} F(z) \qquad z \in \mathbb{H}. \tag{3.56}$$

Then, u_F is indeed a Maass wave form for Γ with weight k and multiplier υ; its spectral parameter is $\pm\frac{1-k}{2}$.

Indeed, checking the properties in Definition 3.31, we find

1. F is holomorphic since it is a modular form. Hence u_F in (3.56) is real-analytic in the sense of Definition 3.14 since F admits a Taylor expansion in z (and hence in its real and imaginary parts).
2. Using the representation of Δ_k in (3.12) and Exercise 3.13, we find

$$\begin{aligned}\Delta_k\, u_F &\overset{(3.12)}{=} \frac{1}{4}\mathrm{E}_{k-2}^{+}\mathrm{E}_k^{-}\, u_F - \frac{k(k-2)}{4}\, u_F \\ &\overset{3.13}{=} 0 - \frac{k(k-2)}{4}\, u_F \qquad \text{using Exercise 3.13} \\ &= \frac{1-\left(1-2k+k^2\right)}{4}\, u_F = \left(\frac{1}{4} - \left(\pm\frac{1-k}{2}\right)^2\right) u_F.\end{aligned}$$

Hence, u_F satisfies $\Delta_k\, u_F = \left(\frac{1}{4} - \nu^2\right) u_F$ with spectral parameter $\nu = \pm\frac{k-1}{2}$.

3. Let $M \in \Gamma$ be an arbitrary element. We have

$$\begin{aligned}u_F(M\,z) &\overset{(3.56)}{=} \Im\,(M\,z)^{\frac{k}{2}}\, F(M\,z) \\ &\overset{(1.28)}{=} \Im\,(M\,z)^{\frac{k}{2}}\, \upsilon(M)\, j(M,z)^k\, F(z) \\ &\overset{(1.18)}{=} \frac{\Im\,(z)^{\frac{k}{2}}}{\left|j(M,z)^k\right|}\, \upsilon(M)\, j(M,z)^k\, F(z) \\ &\overset{(3.3)}{=} \upsilon(M)\, j_u(M,z)^k\, \Im\,(z)^{\frac{k}{2}}\, F(z) \\ &\overset{(3.56)}{=} \upsilon(M)\, j_u(M,z)^k\, u_F(z)\end{aligned}$$

for all $z \in \mathbb{H}$. Using Definition 3.3 of the double-slash operator, we find

$$u_F \big\|_k M = \upsilon(M)\, u_F$$

for all $M \in \Gamma$. (Alternatively, we could apply Exercise 3.40.)

4. Recall that F as an entire modular form admits a Fourier expansion of the form

$$F(z) = \sigma_j(z) \sum_{n=0}^{\infty} a_j(n)\, e^{2\pi i (n+\kappa_j) A_j^{-1} \frac{z}{\lambda_j}} \qquad z \in \mathbb{H}$$

see Definition 1.59 and Theorem 1.56 with Equation (1.44), where we use the notation from Theorem 1.56.

a. First, we consider the expansion at the cusp $q_1 = \infty$, assuming that F is a cusp form. The above expansion of $F(z)$ simplifies slightly to

$$F(z) = \sum_{\substack{n=0 \\ n+\kappa_1>0}}^{\infty} a_1(n)\, e^{2\pi i (n+\kappa_1) \frac{z}{\lambda_j}} \qquad (z \in \mathbb{H})$$

since $\sigma_1(z) = 1$. Using [153, (13.8.2)], see also Appendix A.9, we can rewrite the Fourier expansion as

$$\begin{aligned}
F(z) &= \sum_{\substack{n=0\\ n+\kappa_1>0}}^{\infty} a_1(n)\, e^{2\pi i (n+\kappa_1)\frac{z}{\lambda_1}} \\
&= \sum_{\substack{n=0\\ n+\kappa_1>0}}^{\infty} a_1(n)\, e^{-2\pi (n+\kappa_1)\frac{y}{\lambda_1}}\, e^{2\pi i (n+\kappa_1)\frac{x}{\lambda_1}} \\
&= \sum_{\substack{n=0\\ n+\kappa_1>0}}^{\infty} a_1(n) \left(4\pi (n+\kappa_1)\frac{y}{\lambda_1}\right)^{-\frac{k}{2}} \\
&\qquad \cdot\, W_{\frac{k}{2},\pm\frac{k-1}{2}}\left(4\pi (n+\kappa_1)\frac{y}{\lambda_1}\right)\, e^{2\pi i (n+\kappa_1)\frac{x}{\lambda_1}} \\
&= y^{-\frac{k}{2}} \sum_{\substack{n=0\\ n+\kappa_1>0}}^{\infty} a_1(n) \left(\frac{4\pi (n+\kappa_1)}{\lambda_1}\right)^{-\frac{k}{2}} \\
&\qquad \cdot\, W_{\frac{k}{2},\pm\frac{k-1}{2}}\left(4\pi (n+\kappa_1)\frac{y}{\lambda_1}\right)\, e^{2\pi i (n+\kappa_1)\frac{x}{\lambda_1}}
\end{aligned}$$

using $x = \Re(z)$ and $y = \Im(z)$ as usual. This implies that u_F has the expansion

$$\begin{aligned}
u_F(x+iy) &= y^{\frac{k}{2}}\, F(x+iy) \\
&= \sum_{\substack{n=0\\ n+\kappa_1>0}}^{\infty} a_1(n) \left(\frac{4\pi (n+\kappa_1)}{\lambda_1}\right)^{-\frac{k}{2}} \cdot \\
&\qquad \cdot\, W_{\frac{k}{2},\pm\frac{k-1}{2}}\left(4\pi (n+\kappa_1)\frac{y}{\lambda_1}\right)\, e^{2\pi i (n+\kappa_1)\frac{x}{\lambda_1}} \\
&\overset{(3.26)}{=} \sum_{\substack{n=0\\ n+\kappa_1>0}}^{\infty} a_1(n) \left(\frac{4\pi (n+\kappa_1)}{\lambda_1}\right)^{-\frac{k}{2}} \cdot \qquad (3.57) \\
&\qquad \cdot\, \mathbb{W}_{\mathrm{sign}(n)\frac{k}{2},\nu}\left(4\pi (n+\kappa_1)\frac{y}{\lambda_1}\right)\, e^{2\pi i (n+\kappa_1)\frac{x}{\lambda_1}}
\end{aligned}$$

at the cusp $q_1 = \infty$ with spectral parameter $\nu = \pm\frac{k-1}{2}$.

b. Next, we consider the expansion at the cusp $q_j \in \mathbb{R}$ (i.e., $\infty \neq q_j = \frac{p_j}{r_j}$ with coprime $p_j \in \mathbb{Z}$ and $r_j \in \mathbb{N}$). According to Theorem 1.56, we have

$$\sigma_j(z) = \left(z - q_j\right)^{-k}$$

with $q_j = A_j \infty$. We know in particular that the matrix A_j has the form $A_j = \begin{pmatrix} p_j & \star \\ r_j & \star \end{pmatrix}$. Hence, the automorphic factor $j\left(A_j^{-1}, z\right)$ is given by

$$j\left(A_j^{-1}, z\right) \overset{(3.36)}{=} \left(-r_j\right)\left(z - q_j\right)$$

using that $A_j^{-1} = \begin{pmatrix} \star & \star \\ -r_j & p_j \end{pmatrix}$ and that $r_j \in \mathbb{N}$ does not vanish.

Starting from the expansion of F at q_j, we use again [153, (13.8.2)], see also Appendix A.9, to rewrite the Fourier expansion. We get

$$\begin{aligned}
F(z) &= \sigma_j(z) \sum_{\substack{n=0 \\ n+\kappa_j>0}}^{\infty} a_j(n)\, e^{2\pi i (n+\kappa_j) A_j^{-1} \frac{z}{\lambda_j}} \\
&= \left(z - q_j\right)^{-k} \sum_{\substack{n=0 \\ n+\kappa_j>0}}^{\infty} a_j(n)\, e^{-2\pi (n+\kappa_j) A_j^{-1} \frac{y}{\lambda_j}}\, e^{2\pi i (n+\kappa_j) A_j^{-1} \frac{x}{\lambda_j}} \\
&= \left(z - q_j\right)^{-k} \sum_{\substack{n=0 \\ n+\kappa_j>0}}^{\infty} a_j(n) \left(4\pi (n+\kappa_j) A_j^{-1} \frac{y}{\lambda_j}\right)^{-\frac{k}{2}} \\
&\qquad \cdot W_{\frac{k}{2}, \pm\left(k-\frac{1}{2}\right)}\left(4\pi (n+\kappa_j) A_j^{-1} \frac{y}{\lambda_j}\right) e^{2\pi i (n+\kappa_j) A_j^{-1} \frac{x}{\lambda_1}} \\
&= \left(z - q_j\right)^{-k} \left(A_j^{-1} y\right)^{-\frac{k}{2}} \sum_{\substack{n=0 \\ n+\kappa_j>0}}^{\infty} a_j(n) \left(\frac{4\pi (n+\kappa_j)}{\lambda_j}\right)^{-\frac{k}{2}} \\
&\qquad \cdot W_{\frac{k}{2}, \pm\left(k-\frac{1}{2}\right)}\left(4\pi (n+\kappa_j) A_j^{-1} \frac{y}{\lambda_j}\right) e^{2\pi i (n+\kappa_j) A_j^{-1} \frac{x}{\lambda_j}} \\
&\overset{(1.18)}{=} \left(z - q_j\right)^{-k} \frac{y^{-\frac{k}{2}}}{\left|j\left(A_j^{-1}, z\right)\right|^{-k}} \sum_{\substack{n=0 \\ n+\kappa_j>0}}^{\infty} a_j(n) \left(\frac{4\pi (n+\kappa_j)}{\lambda_j}\right)^{-\frac{k}{2}} \\
&\qquad \cdot W_{\frac{k}{2}, \pm\left(k-\frac{1}{2}\right)}\left(4\pi (n+\kappa_j) A_j^{-1} \frac{y}{\lambda_j}\right) e^{2\pi i (n+\kappa_j) A_j^{-1} \frac{x}{\lambda_j}}.
\end{aligned}$$

Using the identity $j(A_j^{-1}, z) = (-r_j)(z - q_j)$ shown in Lemma 3.26, we get

$$F(z) = (z-q_j)^{-k} \frac{y^{-\frac{k}{2}}}{|(-r_j)(z-q_j)|^{-k}} \sum_{\substack{n=0\\ n+\kappa_j>0}}^{\infty} a_j(n) \left(\frac{4\pi(n+\kappa_j)}{\lambda_j}\right)^{-\frac{k}{2}}$$

$$\cdot W_{\frac{k}{2},\pm\left(k-\frac{1}{2}\right)}\left(4\pi(n+\kappa_j)A_j^{-1}\frac{y}{\lambda_j}\right) e^{2\pi i(n+\kappa_j)A_j^{-1}\frac{x}{\lambda_j}}$$

$$= y^{-\frac{k}{2}} \left(\frac{z-q_j}{|z-q_j|}\right)^{-k} \sum_{\substack{n=0\\ n+\kappa_j>0}}^{\infty} a_j(n) \left(\frac{4\pi(n+\kappa_j)}{\lambda_j r_j^2}\right)^{-\frac{k}{2}}$$

$$\cdot W_{\frac{k}{2},\pm\left(k-\frac{1}{2}\right)}\left(4\pi(n+\kappa_j)A_j^{-1}\frac{y}{\lambda_j}\right) e^{2\pi i(n+\kappa_j)A_j^{-1}\frac{x}{\lambda_j}}.$$

This implies that u_F has the expansion

$$u_F(x+iy) = y^{\frac{k}{2}} F(x+iy)$$

$$= \left(\frac{z-q_j}{|z-q_j|}\right)^{-k} \sum_{\substack{n=0\\ n+\kappa_j>0}}^{\infty} a_j(n) \left(\frac{4\pi(n+\kappa_j)}{\lambda_j r_j^2}\right)^{-\frac{k}{2}}$$

$$\cdot W_{\frac{k}{2},\pm\left(k-\frac{1}{2}\right)}\left(4\pi(n+\kappa_j)A_j^{-1}\frac{y}{\lambda_j}\right) e^{2\pi i(n+\kappa_j)A_j^{-1}\frac{x}{\lambda_j}}.$$

c. Now, consider the case when F is an entire modular form. Compared to the previous case, we may have an additional constant term or 0th Fourier term in the Fourier expansion of F. We only have to learn how to deal with this additional constant term. We may assume $\kappa_j = 0$ because there is no constant term in the Fourier expansion otherwise. u_F was already calculated above in this case.

For the cusp $q_1 = \infty$, we have (using the above calculation for the cusp form and assuming $\kappa_1 = 0$)

$$\begin{aligned} u_F(x+iy) &= y^{\frac{k}{2}} F(x+iy) \\ &= y^{\frac{k}{2}} \sum_{n=0}^{\infty} a_j(n)\, e^{2\pi i(n+\kappa_j)A_j^{-1}\frac{z}{\lambda_j}} \\ &= y^{\frac{k}{2}} a_j(0) \end{aligned}$$

$$+ \sum_{n=1}^{\infty} a_1(n) \left(\frac{4\pi n}{\lambda_1}\right)^{-k} W_{\frac{k}{2}, \pm\frac{k-1}{2}} \left(4\pi n \frac{y}{\lambda_1}\right) e^{2\pi i n \frac{x}{\lambda_1}}$$

$$\stackrel{(3.26)}{=} y^{\frac{k}{2}} a_j(0)$$

$$+ \sum_{n=1}^{\infty} a_1(n) \left(\frac{4\pi n}{\lambda_1}\right)^{-k} \mathbb{W}_{\operatorname{sign}(n)\frac{k}{2},\nu} \left(4\pi n \frac{y}{\lambda_1}\right) e^{2\pi i n \frac{x}{\lambda_1}}.$$

For the cusp $q_j \neq \infty$, we have (using the above calculation for the cusp form and assuming $\kappa_j = 0$) for the constant term

$$y^{\frac{k}{2}} \left(z - q_j\right)^{-k} a_j(0) = \frac{y^{\frac{k}{2}}}{\left|-r_j z + p_j\right|^k} \frac{\left|-r_j\right|^k \left|z - \frac{p_j}{r_j}\right|^k}{\left(z - q_j\right)^k} a_j(0)$$

$$\stackrel{(1.18)}{=} \left(A_j^{-1} y\right)^{\frac{k}{2}} \left(\frac{z - q_j}{\left|z - q_j\right|}\right)^{-k} \left|-r_j\right|^k a_j(0)$$

using again that $A_j^{-1} = \begin{pmatrix} \star & \star \\ -r_j & p_j \end{pmatrix}$ with $q_j = \frac{p_j}{r_j} = A_j \infty$. We get

$$u_F(x + iy) = y^{\frac{k}{2}} F(x + iy)$$

$$= y^{\frac{k}{2}} \left(z - q_j\right)^{-k} \sum_{n=0}^{\infty} a_j(n)\, e^{2\pi i (n+\kappa_j) A_j^{-1} \frac{z}{\lambda_j}}$$

$$= \left(\frac{z - q_j}{\left|z - q_j\right|}\right)^{-k} \left(A_j^{-1} y\right)^{\frac{k}{2}} \left|-r_j\right|^k a_j(0)$$

$$+ \left(\frac{z - q_j}{\left|z - q_j\right|}\right)^{-k} \sum_{\substack{n=0 \\ n+\kappa_j > 0}}^{\infty} a_j(n) \left(\frac{4\pi (n + \kappa_j)}{\lambda_j r_j^2}\right)^{-\frac{k}{2}}$$

$$\cdot\, \mathbb{W}_{\frac{k}{2}, \pm\left(k - \frac{1}{2}\right)} \left(4\pi (n + \kappa_j) A_j^{-1} \frac{y}{\lambda_j}\right) e^{2\pi i (n+\kappa_j) A_j^{-1} \frac{x}{\lambda_j}}.$$

We see that u_F admits a Fourier expansion of the form which is required in Definition 3.31 if we put

$$A_j(n) = \begin{cases} 0 & \text{if } n + \kappa y < 0 \text{ and} \\ a_j(n) \left(\frac{4\pi}{\lambda_j r_j^2}\right)^{-\frac{k}{2}} (n + \kappa_j)^{\frac{1-k}{2}} & \text{if } n + \kappa y > 0 \end{cases}$$

and

$$C_{j,\pm} = \begin{cases} 0 & \text{if } \nu = \mp\frac{k-1}{2} \text{ and} \\ \left|-r_j\right|^k a_j(0) & \text{if } \nu = \pm\frac{k-1}{2}. \end{cases}$$

3.4.2 Non-holomorphic Eisenstein Series for the Full Modular Group

We restrict ourselves to the full modular group $\Gamma(1)$ and we consider the vanishing weight $k = 0$ and trivial multiplier system $\upsilon = 1$ which is compatible with $(\Gamma(1), 0)$.

Exercise 3.43 Show that the function

$$h_\nu : \mathbb{H} \to \mathbb{C}; \quad z \mapsto \Im(z)^{\frac{1}{2}+\nu} \tag{3.58}$$

is an eigenfunction of Δ with eigenvalue $\frac{1}{4} - \nu^2$ for every $\nu \in \mathbb{C}$.

Inspired by Definition 1.84 of normalized holomorphic Eisenstein series and Definition 1.102 of holomorphic Poincaré series, we define Maass Eisenstein series as follows.

Definition 3.44 Let $\nu \in \mathbb{C}$ with $\Re(\nu) > \frac{1}{2}$. We define the normalized *non-holomorphic Eisenstein series* E_ν^{Maass} as

$$E^{\text{Maass}}(z, \nu) := \sum_{M \in \Gamma_\infty \backslash \Gamma(1)} \Im(M\,z)^{\frac{1}{2}+\nu} \tag{3.59}$$

for all $z \in \mathbb{H}$.

Remark 3.45 For $\Gamma \subset \Gamma(1)$ of finite index μ, let $q_1 = \infty, q_2, \ldots, q_\mu \in \mathbb{Q} \cup \{\infty\}$ be the cusps of Γ in the standard fundamental domain $\mathcal{F}_\Gamma$ and $A_1, A_2, \ldots, A_\mu$ be representatives of the associated right cosets of Γ in $\Gamma(1)$ satisfying $q_j = A_j\infty$, as introduced in Lemma 1.28. Recall the stabilizer groups Γ_{q_j} of the cusps q_j, see Definition 1.32 and its calculation in Proposition 1.34.

We can define non-holomorphic Eisenstein series $E_j(\cdot, \nu)$ for each cusp q_j as

$$E_j(z, \nu) := \sum_{M \in \Gamma_{q_j} \backslash \Gamma} \Im\left(A_j^{-1} M\,z\right)^{\frac{1}{2}+\nu} \tag{3.60}$$

for $\Re(\nu) > \frac{1}{2}$ and $z \in \mathbb{H}$. This generalizes naturally Definition 3.44 of non-holomorphic Eisenstein series on subgroups of $\Gamma(1)$.

Lemma 3.46 *Let $\nu \in \mathbb{C}$ with $\Re(\nu) > \frac{1}{2}$. The normalized non-holomorphic Eisenstein series $E^{Maass}(\cdot, \nu)$ is well defined on $\mathbb{H}$ and it is an eigenfunction of Δ_0 with eigenvalue $\frac{1}{4} - \nu^2$.*

Proof We show the two statements of the lemma separately.

To show that $E^{\text{Maass}}(\cdot, \nu)$ is well defined on $\mathbb{H}$, we write $E^{\text{Maass}}(\cdot, \nu)$ in (3.59) as

$$
\begin{aligned}
E^{\text{Maass}}(z, \nu) &= \sum_{M \in \Gamma_\infty \backslash \Gamma(1)} \Im(M z)^{\frac{1}{2}+\nu} \\
&\overset{(1.18)}{=} \sum_{M \in \Gamma_\infty \backslash \Gamma(1)} |j(M, z)|^{-1-2\nu} \Im(z)^{\frac{1}{2}+\nu} \\
&= \sum_{\substack{c \in \mathbb{Z}_{\geq 0}, d \in \mathbb{Z} \\ \gcd c, d = 1 \\ c = 0 \Rightarrow d = 1}} |cz + d|^{-1-2\nu} \Im(z)^{\frac{1}{2}+\nu} \\
&= \Im(z)^{\frac{1}{2}+\nu} \sum_{\substack{c \in \mathbb{Z}_{\geq 0}, d \in \mathbb{Z} \\ \gcd c, d = 1 \\ c = 0 \Rightarrow d = 1}} |cz + d|^{-1-2\nu}.
\end{aligned}
$$

for all $z \in \mathbb{H}$. Since $\Re(\nu) > \frac{1}{2}$, we have

$$
\begin{aligned}
\left|E^{\text{Maass}}(z, \nu)\right| &= \left|\Im(z)^{\frac{1}{2}+\nu}\right| \left| \sum_{\substack{c \in \mathbb{Z}_{\geq 0}, d \in \mathbb{Z} \\ \gcd c, d = 1 \\ c = 0 \Rightarrow d = 1}} |cz + d|^{-1-2\nu} \right| \\
&\leq \left|\Im(z)^{\frac{1}{2}+\nu}\right| \sum_{\substack{c \in \mathbb{Z}_{\geq 0}, d \in \mathbb{Z} \\ \gcd c, d = 1 \\ c = 0 \Rightarrow d = 1}} |cz + d|^{-1-2\Re(\nu)}.
\end{aligned}
$$

The series on the right-hand side converges absolutely for all $z \in \mathbb{H}$. It even converges uniformly on subsets $E_\alpha \subset \mathbb{H}$ for every $\alpha > 0$ where E_α is given by

$$
E_\alpha := \left\{ z \in \mathbb{H}\colon\ |\Re(z)| \leq \frac{1}{\alpha} \text{ and } \Im(z) \geq \alpha \right\}. \tag{3.61}
$$

Hence, $\left|E^{\text{Maass}}(z, \nu)\right|$ is well defined for all $z \in \mathbb{H}$ provided that $\Re(\mu) > \frac{1}{2}$.

To see that $E^{\text{Maass}}(\cdot, \nu)$ is an eigenfunction of Δ_0, we recall that $E^{\text{Maass}}(\cdot, \nu)$ can be written as

$$\begin{aligned}E^{\mathrm{Maass}}(z,\nu) &\overset{(3.59)}{=} \sum_{M\in\Gamma_\infty\backslash\Gamma(1)} \Im(Mz)^{\frac{1}{2}+\nu}\\ &\overset{(3.58)}{=} \sum_{M\in\Gamma_\infty\backslash\Gamma(1)} h_\nu\big(Mz\big)\\ &\overset{(3.5)}{=} \sum_{M\in\Gamma_\infty\backslash\Gamma(1)} \big(h_\nu\big\|_0 M\big)(z).\end{aligned}$$

Since h_ν is an eigenfunction of Δ_0 with eigenvalue $\frac{1}{4}-\nu^2$, see Exercise 3.43, and Δ_0 commutes with the double slash operator $\|_0$, see Lemma 3.10, we have

$$\begin{aligned}\Delta_0\, E^{\mathrm{Maass}}(z,\nu) &= \Delta_0\left(\sum_{M\in\Gamma_\infty\backslash\Gamma(1)} \big(h_\nu\big\|_0 M\big)(z)\right)\\ &= \sum_{M\in\Gamma_\infty\backslash\Gamma(1)} \big([\Delta_0 h_\nu]\big\|_0 M\big)(z)\\ &= \left(\frac{1}{4}-\nu^2\right)\sum_{M\in\Gamma_\infty\backslash\Gamma(1)} \big(h_\nu\big\|_0 M\big)(z)\\ &= \left(\frac{1}{4}-\nu^2\right) E^{\mathrm{Maass}}(z,\nu)\end{aligned}$$

for all $M\in\Gamma(1)$ and $z\in\mathbb{H}$. □

Using similar arguments as in the proof of Lemma 1.103, one can show the following transformation behavior of the normalized non-holomorphic Eisenstein series.

Exercise 3.47 Let $\nu\in\mathbb{C}$ with $\Re(\nu)>\frac{1}{2}$. The normalized non-holomorphic Eisenstein series $E^{\mathrm{Maass}}(\cdot,\nu)$ satisfies $E^{\mathrm{Maass}}(\cdot,\nu)\big\|_0 M = E^{\mathrm{Maass}}(\cdot,\nu)$ for all $M\in\Gamma(1)$.

It is known that the non-holomorphic Eisenstein series $E^{\mathrm{Maass}}(\cdot,\nu)$ extends meromorphically in ν to the full complex plane $\mathbb{C}$, admits a Fourier-Bessel expansion, and admits a functional equation in ν. Summarizing some known results from [101], we have the following.

Theorem 3.48 *Let $\nu\in\mathbb{C}$ be a complex number. We have the following:*

1. *Assume $\Re(\nu)>\frac{1}{2}$, the non-holomorphic Eisenstein series $E^{Maass}(\cdot,\nu)$ admits a Fourier-Bessel expansion of the form*

$$E(z,\nu) = y^{\frac{1}{2}+\nu} + \phi(\nu)\, y^{\frac{1}{2-\nu}} + \sqrt{y}\sum_{n\neq 0}\phi(n,\nu)\, K_\nu\big(2\pi|n|y\big)e^{2\pi nx} \tag{3.62}$$

with

$$\phi(\nu) = \sqrt{\pi}\, \frac{\Gamma(\nu)}{\Gamma\left(\nu + \frac{1}{2}\right)} \frac{\zeta(2\nu)}{\zeta(2\nu+1)} \quad \textit{and}$$
$$\phi(n,\nu) = (n\pi)^{\frac{1}{2}+\nu} \frac{1}{n\Gamma\left(\frac{1}{2}+\nu\right)} \sum_{c\in\mathbb{N}} c^{-2s-1}\, S(0,n;c). \tag{3.63}$$

2. *The non-holomorphic Eisenstein series $E^{Maass}(\cdot,\nu)$ extends to the right half-plane $\Re(\nu) \geq 0$ meromorphically with the only simple pole at $\nu = \frac{1}{2}$ with residue $res_{\nu=\frac{1}{2}} E^{Maass}(\cdot,\nu) = \frac{3}{\pi}$.*
3. *Put $\theta(\nu) = \pi^{-\nu-\frac{1}{2}}\Gamma\left(\nu+\frac{1}{2}\right)\zeta(2\nu+1)$. Then, $\theta(\nu)\,E^{Maass}(\cdot,\nu)$ admits a meromorphic extension to $\nu \in \mathbb{C}$ and satisfies the functional equation*

$$\theta(\nu)\,E(\cdot,\nu) = \theta(-\nu)\,E(\cdot,-\nu). \tag{3.64}$$

Proof The first statement follows directly from [101, Theorem 3.4] together with the Identity [101, (3.24)]. The second and third statements follow from the discussions on [101, p. 61]. □

Remark 3.49 An alternative expression for $\phi(n,s)$ is also given in [101, (3.25)]. Note that $\phi(\nu) = \frac{\theta\left(-\frac{1}{2}-\nu\right)}{\theta(\nu)}$, see [101, (3.28)].

Remark 3.50 A discussion of the non-holomorphic Eisenstein series for arbitrary horocyclic groups Γ is given in [101, Chapter VI]. Here, the factor $\phi(\nu)$ is replaced by a matrix valued function which is called the *scattering matrix* of the non-holomorphic Eisenstein series. Huxley calculated the entries of the scattering matrix explicitly for principal congruence subgroups $\Gamma(n)$, see [99].

3.4.3 Maass Cusp Forms Derived from the Dedekind η-Function

We consider the full modular group $\Gamma(1)$ and a positive weight $k > 0$.

The *Dedekind η-function* η was defined in Example 1.50 as

$$\eta : \mathbb{H} \to \mathbb{C}; \quad z \mapsto \eta(z) := e^{\frac{\pi i z}{12}} \prod_{n\in\mathbb{N}} \left(1 - e^{2\pi i n z}\right). \tag{3.65}$$

It is a holomorphic function on $\mathbb{H}$.

Just expanding the product in (3.65), we easily see that η admits a Fourier expansion of the form

$$\eta(z) = e^{\frac{\pi i z}{12}} \prod_{n\in\mathbb{N}} \left(1 - e^{2\pi i n z}\right) = e^{\frac{\pi i z}{12}} \left(1 + \sum_{n\in\mathbb{N}} A^\eta(n)\, e^{2\pi i n z}\right)$$
$$= \sum_{n=0}^{\infty} A^\eta(n)\, e^{2\pi i \left(n+\frac{1}{12}\right) z}, \tag{3.66}$$

where

$$A^\eta(n) = \begin{cases} 1 & \text{if } n = 0 \text{ and} \\ \sharp\left\{\{i_1, \ldots i_l\} \subset \mathbb{N}:\ i_1 < \ldots < i_l \text{ and } i_1 + \ldots + i_l = n\right\} & \text{if } n > 0 \end{cases} \tag{3.67}$$

denotes the number of ways to write the natural number $n \in \mathbb{N}$ as an ordered sum of positive integers. A few values of $A^\eta(n)$ for small n are given in Table 3.1.

As shown in Example 1.50, the $2k^{\text{th}}$ power of η satisfies the transformation property

$$\eta^{2k}\big|_k M = v_k(M)\, \eta^{2k} \tag{3.68}$$

with the η-multiplier v_k given in (1.40).

In view of the Fourier expansion of η^{2k}, we use (3.65) and we get

Table 3.1 Values of $A^\eta(n)$ for small n.

n	$A^\eta(n)$	Ordered sums
1	1	1
2	1	2
3	2	$1+2, 3$
4	2	$1+3, 4$
5	3	$1+4, 2+3, 5$
6	4	$1+2+3, 1+5, 2+4, 6$
7	5	$1+2+4, 1+6, 2+5, 3+4, 7$
8	6	$1+2+5, 1+3+4, 1+7, 2+6, 3+5, 8$
9	8	$1+2+6, 1+3+5, 2+3+4, 1+8, 2+7, 3+6, 4+5, 9$

$$\eta^{2k}(z) = \left(e^{\frac{\pi i z}{12}} \prod_{n\in\mathbb{N}} \left(1 - e^{2\pi i n z}\right)\right)^{2k}$$
$$= e^{\frac{\pi i k z}{6}} \prod_{n\in\mathbb{N}} \left(1 - e^{2\pi i n z}\right)^{2k} \tag{3.69}$$
$$= \sum_{n=0}^{\infty} A_k(n)\, e^{2\pi i \left(n+\frac{k}{12}\right) z}$$

for suitable coefficients $A_k(n)$ that can be calculated by expanding the product expression in (3.69) above. We have in particular

$$A_{\frac{1}{2}}(n) = A^{\eta}(n) \qquad \text{for all } n \in \mathbb{N}.$$

Using (3.56) of §3.4.1, we immediately see that

$$u_{\eta^{2k}}(z) :\overset{(3.56)}{=} \Im(z)^{\frac{k}{2}}\, \eta^{2k}(z) \tag{3.70}$$

is Maass cusp form of weight k and multiplier v_k for the full modular group $\Gamma(1)$. It admits the expansion

$$u_{\eta^{2k}}(z) \overset{\S 3.4.1}{=}$$
$$\sum_{n=0}^{\infty} A_k(n) \left(4\pi\left(n+\frac{k}{12}\right)\right)^{-\frac{k}{2}} \mathbb{W}_{\frac{k}{2},\pm\frac{k-1}{2}}\left(4\pi\left(n+\frac{k}{12}\right)y\right) e^{2\pi i\left(n+\frac{k}{12}\right)x} \tag{3.71}$$

with $z = x + iy \in \mathbb{H}$.

3.5 Scalar Product and the Hilbert Space of Cusp Forms

3.5.1 Definition of the Petersson Scalar Product

Recall the Petersson scalar product $\langle\cdot,\cdot\rangle_{\Gamma,k}$ for classical cusp forms discussed in §1.3.6, in particular in Theorem 1.98. We can introduce analogously a scalar product for Maass cusp forms.

Definition 3.51 Let Γ be a subgroup of the full modular group $\Gamma(1)$, $k \in \mathbb{R}$ a real weight, and v a multiplier system compatible with (Γ, k), and let $\nu \in \mathbb{C} \setminus -\frac{1}{2}\mathbb{N}$ be a spectral value.

We define the map $\langle\cdot,\cdot\rangle_{\Gamma}$ by

$$\begin{aligned}\langle\cdot,\cdot\rangle_\Gamma:\quad & M^{\text{Maass}}_{k,v}(\Gamma,\nu)\times S^{\text{Maass}}_{k,v}(\Gamma,\nu)\to\mathbb{C}\\ &(f,g)\mapsto\langle f,g\rangle_\Gamma:=\frac{1}{\mu}\int_{\mathcal{F}} f(z)\,\overline{g(z)}\,\Im(z)^{-2}\mathrm{d}\lambda_2(z).\end{aligned}\tag{3.72}$$

The hyperbolic measure $\Im(z)^{-2}\mathrm{d}\lambda_2(z)$ used in (3.72) was introduced in (1.59). It is invariant under coordinate changes by Möbius transformations, as stated in Lemma 1.93. Also $\langle\cdot,\cdot\rangle_\Gamma$ satisfies

$$\begin{aligned}\langle f,g\rangle_\Gamma &= \frac{1}{\mu}\int_{\mathcal{F}} f(z)\,\overline{g(z)}\,\Im(z)^{-2}\mathrm{d}\lambda_2(z)\\ &=\overline{\frac{1}{\mu}\int_{\mathcal{F}} g(z)\,\overline{f(z)}\,\Im(z)^{-2}\mathrm{d}\lambda_2(z)}\\ &=\overline{\langle g,f\rangle_\Gamma}\end{aligned}\tag{3.73}$$

by construction.

Similar to §1.3.6 where we discussed the Petersson scalar product for classical modular forms, we have to answer few naturally arising questions:

1. Is $\langle\cdot,\cdot\rangle_\Gamma$ well defined? In other words, is $\langle f,g\rangle_\Gamma$ finite for every pair $(f,g)\in M^{\text{Maass}}_{k,v}(\Gamma,\nu)\times S^{\text{Maass}}_{k,v}(\Gamma,\nu)$?
2. Does $\langle\cdot,\cdot\rangle_\Gamma$ depend on the choice of the fundamental domain $\mathcal{F}$?
3. How does $\langle\cdot,\cdot\rangle_\Gamma$ relate to transformations under Γ or even under $\mathrm{SL}_2(\mathbb{R})$?
4. For f,g are modular forms of a group Γ_1 with $\Gamma\subset\Gamma_1\subset\Gamma(1)$, how do $\langle f,g\rangle_\Gamma$ and $\langle f,g\rangle_{\Gamma_1}$ relate?

3.5.2 Properties of the Petersson Scalar Product

Lemma 3.52 *For $(f,g)\in M^{Maass}_{k,v}(\Gamma,\nu)\times S^{Maass}_{k,v}(\Gamma,\nu)$ the evaluation of $\langle\cdot,\cdot\rangle_\Gamma$ in (f,g) is finite: $\langle f,g\rangle_\Gamma\in\mathbb{C}$.*

Proof We consider $\langle f,g\rangle_\Gamma$. Using (3.72), we have

$$\langle f,g\rangle_\Gamma=\frac{1}{\mu}\int_{\mathcal{F}} f(z)\,\overline{g(z)}\,\Im(z)^{-2}\mathrm{d}\lambda_2(z).$$

According to Definition 3.31 f and g are continuous (since they are real-analytic on $\mathbb{H}$), and they satisfy the growth estimates in Theorem 3.38. In particular Theorem 3.38 shows that g in $S^{\text{Maass}}_{k,v}(\Gamma,\nu)$ implies $\left(\tilde{\sigma}_{k,j}(z)\right)^{-1}g(z)$ vanishes exponentially fast as z tends to the cusp q_j of the fundamental domain $\mathcal{F}_\Gamma$ whereas f is at most of polynomial growth. This shows that the integrand

$$f(z)\,\overline{g(z)}$$

is at least bounded on $\mathcal{F}_\Gamma$ and vanishes exponentially fast as we approach ∞ from within the fundamental domain.

For the full modular group $\Gamma(1)$, we have the estimate

$$\langle f, g\rangle_{\Gamma(1)} = O\left(\int_{\mathcal{F}_{\Gamma(1)}} \Im(z)^{-2} \mathrm{d}\lambda_2(z)\right) \overset{1.94}{=} O\left(\frac{\pi}{3}\right) = O(1),$$

see Exercise 1.94. Using the linearity of the integral and $\mathcal{F}_\Gamma$ given as in (1.56) as disjoint union of transformations of $\mathcal{F}_{\Gamma(1)}$, we calculate the hyperbolic area of the fundamental domain $\mathcal{F}_\Gamma$ as

$$\begin{aligned}
\frac{1}{\mu}\int_{\mathcal{F}_\Gamma} \Im(z)^{-2} \mathrm{d}\lambda_2(z) &\overset{(1.56)}{=} \sum_{i=1}^{\mu} \int_{\mathcal{F}_{A_i\Gamma(1)}} \Im(z)^{-2} \mathrm{d}\lambda_2(z) \\
&\overset{(1.62)}{=} \sum_{i=1}^{\mu} \int_{\mathcal{F}_{\Gamma(1)}} \Im(z)^{-2} \mathrm{d}\lambda_2(z) \\
&= \mu \int_{\mathcal{F}_{\Gamma(1)}} \Im(z)^{-2} \mathrm{d}\lambda_2(z) \\
&\overset{1.94}{=} \mu \frac{\pi}{3}.
\end{aligned}$$

Hence we have the estimate

$$\langle f, g\rangle_\Gamma = O(1).$$

This proves the lemma. □

We answer the question whether $\langle \cdot, \cdot\rangle_\Gamma$ depends on the choice of fundamental domain $\mathcal{F}_\Gamma$ and how it behaves under group actions, following the arguments presented in Lemma 1.96 and Remark 1.97.

Lemma 3.53 *For* $(f, g) \in M_{k,v}^{Maass}(\Gamma, v) \times S_{k,v}^{Maass}(\Gamma, v)$ *the map* $(f, g) \mapsto \langle f, g\rangle_\Gamma$ *defined in* (3.72) *is independent of the choice of the fundamental domain.*

Moreover, the map satisfies

$$\langle f, g\rangle_\Gamma = \frac{1}{\mu}\int_{M\mathcal{F}} f(z)\,\overline{g(z)}\,\Im(z)^{-2} \mathrm{d}\lambda_2(z) \tag{3.74}$$

for every $M \in \Gamma$.

Proof Consider

$$\begin{aligned}\int_{MF} f(z)\,\overline{g(z)}\,\Im(z)^{-2}\mathrm{d}\lambda_2(z) &= \int_F f(Mz)\,\overline{g(Mz)}\,\Im(Mz)^{-2}\mathrm{d}\lambda_2(Mz)\\ &\overset{(1.62)}{=} \int_F f(Mz)\,\overline{g(Mz)}\,\Im(z)^{-2}\mathrm{d}\lambda_2(z)\end{aligned} \tag{3.75}$$

for every $M \in \mathrm{SL}_2(\mathbb{R})$ and every domain $F \subset \mathbb{H}$.

Now, consider $M \in \Gamma$. We aim to prove the second part of the lemma. Applying this relation to the left-hand side of (3.74) and using the transformation property (3.6) of Maass wave cusp forms, we get

$$\begin{aligned}&\frac{1}{\mu}\int_{MF} f(z)\,\overline{g(z)}\,\Im(z)^{-2}\mathrm{d}\lambda_2(z)\\ &\overset{(3.75)}{=} \int_F f(Mz)\,\overline{g(Mz)}\,\Im(z)^{-2}\mathrm{d}\lambda_2(z)\\ &\overset{(3.6)}{=} \frac{1}{\mu}\int_{\mathcal{F}} v(M)\,j_u(M,z)^k\,f(z)\,\overline{v(M)\,j_u(M,z)^k\,g(z)}\,\Im(z)^{-2}\mathrm{d}\lambda_2(z)\\ &= \frac{1}{\mu}\int_{\mathcal{F}} f(z)\,\overline{g(z)}\,\Im(z)^{-2}\mathrm{d}\lambda_2(z)\end{aligned}$$

using the properties

$$j_u(M,z)^k\,\overline{j_u(M,z)^k} \overset{(3.2)}{=} e^{ik\arg(cz+d)}\,\overline{e^{ik\arg(cz+d)}} = 1 \qquad \text{with } M = \begin{pmatrix}\star & \star\\ c & d\end{pmatrix}$$

and

$$v(M)\,\overline{v(M)} = |v(M)|^2 = 1.$$

(The first identity is due to the definition of j_u in (3.2) and the second identity is due to Definition 1.44 of multiplier systems.) This proves (3.74).

Now, we continue with the proof of the first part of Lemma 3.53. Let $\mathcal{F}_\Gamma$ be a fundamental domain of Γ of the form given in (1.23):

$$\mathcal{F}_\Gamma = \bigcup_{i=1}^{\mu} A_i\,\mathcal{F}_{\Gamma(1)}$$

with $A_1, \ldots, A_\mu$ a right coset representation of Γ in $\Gamma(1)$. For $M_1, \ldots, M_\mu \in \Gamma$ consider the matrices $M_1A_1, \ldots, M_\mu A_\mu$; these also form a right coset representation of γ in $\Gamma(1)$. The fundamental domain $\mathcal{F}'_\Gamma$ associated with the new right coset representation is

$$\mathcal{F}'_\Gamma = \bigcup_{i=1}^{\mu} M_i A_i \mathcal{F}_{\Gamma(1)}.$$

Using (3.75) and the linearity of the integral we get for $\langle f, g\rangle_\Gamma$:

$$\begin{aligned}
&\frac{1}{\mu}\int_{\mathcal{F}'_\Gamma} f(z)\,\overline{g(z)}\,\Im(z)^{-2}\mathrm{d}\lambda_2(z)\\
&= \frac{1}{\mu}\sum_{i=1}^{\mu}\int_{M_i A_i \mathcal{F}_{\Gamma(1)}} f(z)\,\overline{g(z)}\,\Im(z)^{-2}\mathrm{d}\lambda_2(z)\\
&\overset{(3.75)}{=} \frac{1}{\mu}\sum_{i=1}^{\mu}\int_{A_i \mathcal{F}_{\Gamma(1)}} f(z)\,\overline{g(z)}\,\Im(z)^{-2}\mathrm{d}\lambda_2(z)\\
&= \frac{1}{\mu}\int_{\mathcal{F}_\Gamma} f(z)\,\overline{g(z)}\,\Im(z)^{-2}\mathrm{d}\lambda_2(z)\\
&\overset{(3.72)}{=} \langle f, g\rangle_\Gamma.
\end{aligned}$$

This shows that $\langle\cdot,\cdot\rangle_\Gamma$ is independent of the choice of the fundamental domain $\mathcal{F}_\Gamma$. □

Similar to Theorem 1.98, we have

Theorem 3.54 *Let $\Gamma \subset \Gamma(1)$ be a subgroup of finite index μ and let $k \in \mathbb{R}$ a weight and v a multiplier with respect to (Γ, k), and let $\nu \in \mathbb{C} \setminus -\frac{1}{2}\mathbb{N}$ be a spectral value. The map $\langle\cdot,\cdot\rangle_\Gamma$ defined in (3.72) satisfies*

1. $\langle\cdot,\cdot\rangle_\Gamma$ is a scalar product. It is well defined and satisfies

$$\begin{aligned}
\langle \alpha f, g\rangle_\Gamma &= \alpha\langle f, g\rangle_\Gamma = \langle f, \overline{\alpha} g\rangle_\Gamma,\\
\langle f_1 + f_2, g\rangle_\Gamma &= \langle f_1, g\rangle_\Gamma + \langle f_2, g\rangle_\Gamma \qquad \textit{and}\\
\langle f, g_1 + g_2\rangle_\Gamma &= \langle f, g_1\rangle_\Gamma + \langle f, g_2\rangle_\Gamma
\end{aligned}$$

for $f, f_1, f_2 \in M_{k,v}^{Maass}(\Gamma, \nu)$, $g, g_1, g_2 \in S_{k,v}^{Maass}(\Gamma, \nu)$ and $\alpha \in \mathbb{C}$. Moreover, we have

$$\langle f, g\rangle_\Gamma = \overline{\langle g, f\rangle_\Gamma}$$

for $f, g \in S_{k,v}^{Maass}(\Gamma, \nu)$ both cusp forms.

Proof The well-defined notion of the $\langle\cdot,\cdot\rangle_\Gamma$ follows directly from Lemma 3.52. It is independent of the choice of the fundamental domain, as can be seen from Remark 3.53. The linearity in the first component and anti-linearity in the second follows directly from the linearity of integrals. This shows that $\langle\cdot,\cdot\rangle_\Gamma$ indeed a scalar product. The Hermitian property also follows directly from the definition in (3.72), assuming that both forms f and g are cusp forms. □

Corollary 3.55 *Let $\Gamma \subset \Gamma(1)$ be a subgroup of finite index μ and let $k \in \mathbb{R}$ and v a multiplier with respect to (Γ, k), and let $\nu \in \mathbb{C} \setminus -\frac{1}{2}\mathbb{N}$ be a spectral value. The space $S_{k,v}^{Maass}(\Gamma, \nu)$ is a Hilbert space with the scalar product $\langle \cdot, \cdot \rangle_\Gamma$ as scalar product.*

Proof The proof of Corollary 3.55 is left to the reader as an exercise. □

3.5.3 The Petersson Scalar Product and Maass Cusp Forms

We would like to use the Petersson scalar product to show that the spectral value of Maass cusp forms has to be real or purely imaginary. But first, we discuss the growth behavior of Maass cusp forms under partial derivatives.

Lemma 3.56 *Let $u \in S_{k,v}^{Maass}(\Gamma, \nu)$ be a Maass cusp form. u decays exponentially fast as we approach the cusps of the fundamental domain $\mathcal{F}_\Gamma$ and vanishes at the cusps.*

Proof Recall the Maass operators defined in (3.10). We can use these operators to express the partial derivatives $\frac{\partial}{\partial x}$ and $\frac{\partial}{\partial y}$ as

$$\begin{aligned} \frac{\partial}{\partial x} &= \frac{1}{4iy}\left(\mathrm{E}_k^+ - \mathrm{E}_k^-\right) + i\frac{k}{2y} \quad \text{and} \\ \frac{\partial}{\partial y} &= \frac{1}{4y}\left(\mathrm{E}_k^+ + \mathrm{E}_k^-\right). \end{aligned} \tag{3.76}$$

Using this expression for the partial derivatives, we can easily conclude with Corollary 3.41: For $u \in S_{k,v}^{\text{Maass}}(\Gamma, \nu)$ a Maass cusp form of weight k, the partial derivatives

$$\begin{aligned} \frac{\partial}{\partial x}u(z) &= \frac{1}{4iy}\left(\mathrm{E}_k^+ u(z) - \mathrm{E}_k^- u(z)\right) + i\frac{k}{2y}u(z) \quad \text{and} \\ \frac{\partial}{\partial y}u(z) &= \frac{1}{4y}\left(\mathrm{E}_k^+ u(z) + \mathrm{E}_k^- u(z)\right) \end{aligned}$$

can be expressed in linear combinations and prefix factor $\frac{1}{y}$ of Maass cusp forms of weights $k-2$, k, and $k+2$. Using the estimate (3.53) of Theorem 3.38 for the Maass cusp forms of weights $k-2$, k, and $k+2$, we find that $\frac{\partial}{\partial x}u(z)$ and $\frac{\partial}{\partial y}u(z)$ are still exponentially decaying near the cusps of the fundamental domain and vanishes at the cusps. □

Lemma 3.57 *Let $u, h \in S_{k,\upsilon}^{Maass}(\Gamma, \nu)$ be two Maass cusp forms. We have*

$$\int_{\mathcal{F}_\Gamma} \frac{\partial u(z)}{\partial x} \overline{h(z)} \, d\lambda_2(z) = - \int_{\mathcal{F}_\Gamma} u(z) \overline{\frac{\partial h(z)}{\partial x}} \, d\lambda_2(z) \tag{3.77}$$

and

$$\int_{\mathcal{F}_\Gamma} \frac{\partial u(z)}{\partial y} \overline{h(z)} \, d\lambda_2(z) = - \int_{\mathcal{F}_\Gamma} u(z) \overline{\frac{\partial h(z)}{\partial y}} \, d\lambda_2(z), \tag{3.78}$$

$$\tag{3.79}$$

where $x = \Re(z)$ and $y = \Im(z)$ denote real and imaginary parts of the complex variable $z \in \mathbb{H}$.

Proof Since u, h, and its partial derivatives vanish at the cusps of the fundamental domain exponentially fast, see (3.53) of Theorem 3.38 and Lemma 3.56, this allows us to use the partial integration formula where the boundary term vanishes. Using the decomposition $z = x + iy$ of the complex variable into real and imaginary parts, we get

$$\begin{aligned}\int_{\mathcal{F}_\Gamma} \frac{\partial u}{\partial x}(z) \overline{h(z)} \, d\lambda_2(z) &= 0 - \int_{\mathcal{F}_\Gamma} u(z) \frac{\partial}{\partial x} \overline{h(z)} \, d\lambda_2(z) \\ &= - \int_{\mathcal{F}_\Gamma} u(z) \overline{\frac{\partial h}{\partial x}(z)} \, d\lambda_2(z).\end{aligned}$$

Similarly we calculate

$$\begin{aligned}\int_{\mathcal{F}_\Gamma} \frac{\partial u}{\partial y}(z) \overline{h(z)} \, d\lambda_2(z) &= 0 - \int_{\mathcal{F}_\Gamma} u(z) \frac{\partial}{\partial y} \overline{h(z)} \, d\lambda_2(z) \\ &= - \int_{\mathcal{F}_\Gamma} u(z) \overline{\frac{\partial h}{\partial y}(z)} \, d\lambda_2(z).\end{aligned}$$

This proves the identities (3.77) and (3.78). □

Proposition 3.58 *Let $u, h \in S_{k,\upsilon}^{Maass}(\Gamma, \nu)$ be Maass cusp forms of real weight k, multiplier υ, and spectral parameter ν for the group Γ. We have*

$$\langle \Delta_k u, h \rangle_\Gamma = \langle u, \Delta_k h \rangle_\Gamma. \tag{3.80}$$

Proof Definition 3.31 implies that u and h are real-analytic and vanish at the cusps. Also their partial derivatives vanish at the cusps as can be seen from Lemma 3.56. Hence, we can apply partial integration in x and y. The boundary terms at the cusps, that should appear using partial integration, vanish since Maass cusp forms and its partial derivatives vanish exponentially fast, see Theorem 3.38 and Lemma 3.56.

Using partial integration in x and y respectively, we get

$$\begin{aligned}
\langle u, \Delta_k h\rangle_\Gamma &\overset{(3.72)}{=} \int_{\mathcal{F}_\Gamma} u(z)\,\overline{\big(\Delta_k h\big)(z)}\,\Im(z)^{-2}\mathrm{d}\lambda_2(z) \\
&\overset{(3.9)}{=} -\int_{\mathcal{F}_\Gamma} u(z)\overline{\frac{\partial^2 h}{\partial^2 x}(z)} + u(z)\overline{\frac{\partial^2 h}{\partial y^2}(z)} - u(z)\overline{\frac{ik}{y}\frac{\partial h}{\partial x}(z)}\,\mathrm{d}\lambda_2(z) \\
&\overset{\text{L }3.57}{=} \int_{\mathcal{F}_\Gamma} \frac{\partial u}{\partial x}(z)\overline{\frac{\partial h}{\partial x}(z)}\,\mathrm{d}\lambda_2(z) \quad + \quad \int_{\mathcal{F}_\Gamma} \frac{\partial u}{\partial y}(z)\overline{\frac{\partial h}{\partial y}(z)}\,\mathrm{d}\lambda_2(z) \\
&\quad + \int_{\mathcal{F}_\Gamma} \frac{ik}{y}\frac{\partial u}{\partial x}(z)\,\overline{h(z)}\,\mathrm{d}\lambda_2(z)
\end{aligned}$$

(partial integration in x and y)

$$= \int_{\mathcal{F}_\Gamma} \frac{\partial u}{\partial x}(z)\overline{\frac{\partial h}{\partial x}(z)} + \frac{\partial u}{\partial y}(z)\overline{\frac{\partial h}{\partial y}(z)} + \frac{ik}{y}\frac{\partial u}{\partial x}(z)\,\overline{h(z)}\,\mathrm{d}\lambda_2(z).$$

An analogous calculation shows

$$\begin{aligned}
\langle \Delta_k u, h\rangle_\Gamma &\overset{(3.72)}{=} \int_{\mathcal{F}_\Gamma} \big(\Delta_k h\big)(z)\,\overline{u(z)}\,\Im(z)^{-2}\mathrm{d}\lambda_2(z) \\
&\overset{(3.9)}{=} \int_{\mathcal{F}_\Gamma} -\frac{\partial^2 u}{\partial^2 x}(z)\overline{h(z)} - \frac{\partial^2 u}{\partial y^2}(z)\overline{h(z)} + \frac{ik}{y}\frac{\partial u}{\partial x}(z)\overline{h(z)}\,\mathrm{d}\lambda_2(z) \\
&\overset{\text{L }3.57}{=} \int_{\mathcal{F}_\Gamma} \frac{\partial u}{\partial x}(z)\overline{\frac{\partial h}{\partial x}(z)} + \frac{\partial u}{\partial y}(z)\overline{\frac{\partial h}{\partial y}(z)} + \frac{ik}{y}\frac{\partial u}{\partial x}(z)\,\overline{h(z)}\,\mathrm{d}\lambda_2(z).
\end{aligned}$$

Hence, we find the identity

$$\begin{aligned}
\langle u, \Delta_k h\rangle_\Gamma &= \int_{\mathcal{F}_\Gamma} \frac{\partial u}{\partial x}(z)\overline{\frac{\partial h}{\partial x}(z)} + \frac{\partial u}{\partial y}(z)\overline{\frac{\partial h}{\partial y}(z)} + \frac{ik}{y}\frac{\partial u}{\partial x}(z)\,\overline{h(z)}\,\mathrm{d}\lambda_2(z) \\
&= \langle \Delta_k u, h\rangle_\Gamma.
\end{aligned}$$

□

Corollary 3.59 *Let $u \in S_{k,v}^{Maass}(\Gamma, \nu)$ be a Maass cusp form of weight k, multiplier v, and spectral parameter ν for the group Γ. Then, the spectral value satisfies $\nu \in \mathbb{R} \cup i\mathbb{R}$.*

Proof We have $\Delta_k u = \left(\frac{1}{4} - \nu^2\right) u$ by Definition 3.31. Using Theorem 3.54 and Proposition 3.58, we conclude

$$\begin{aligned}\left(\frac{1}{4} - \nu^2\right) \langle u, u\rangle_\Gamma &\overset{\text{T 3.54}}{=} \left\langle \left(\frac{1}{4} - \nu^2\right) u, u \right\rangle_\Gamma \\ &\overset{\text{D 3.31}}{=} \langle \Delta_k u, u\rangle_\Gamma \overset{(3.80)}{=} \langle u, \Delta_k u\rangle_\Gamma \\ &\overset{\text{D 3.31}}{=} \left\langle u, \left(\frac{1}{4} - \nu^2\right) u \right\rangle_\Gamma \overset{\text{T 3.54}}{=} \overline{\left(\frac{1}{4} - \nu^2\right)} \langle u, u\rangle_\Gamma .\end{aligned}$$

Since $\langle u, u\rangle_\Gamma$ does not vanish, we conclude that the value $\frac{1}{4} - \nu^2$ has to be real. This is equivalent to

$$\nu^2 \in \mathbb{R} \iff \nu \in \mathbb{R} \cup i\mathbb{R}.$$

□

3.6 Hecke Operators on Maass Cusp Forms

Recall §1.6.2 where we introduce the Hecke operators of level n for modular forms. Similar to the discussion in §1.6.2, we introduce Hecke operators for Maass cusp forms of weight $k = 0$ and trivial multiplier $v \equiv 1$ for the full modular group $\Gamma(1)$ only.

Let $n \in \mathbb{N}$ be the level. We recall Definition 1.172 of the Hecke operator T_n applied to a function $u : \mathbb{H} \to \mathbb{C}$ on $\mathbb{H}$. If the weight $k = 0$ vanishes, we get

$$\begin{aligned}\big(T_n u\big)(z) := n^{-1} \sum_{d|n} \sum_{b=0}^{d-1} u\left(\frac{nz + bd}{d^2}\right) &= n^{-1}\, u\big\|_0 \left(\sum_{d|n} \sum_{b=0}^{d-1} \begin{pmatrix} \frac{n}{d} & b \\ 0 & d \end{pmatrix}\right)(z) \\ &= n^{-1}\, u\big\|_0 \left(\sum_{A \in \mathcal{T}(n)} A\right)(z)\end{aligned} \tag{3.81}$$

for all $z \in \mathbb{H}$, where $\mathcal{T}(n)$ denotes the set of representatives given in (1.139).

Proposition 1.176 can be applied to the above operator. This leads to the following.

Corollary 3.60 *Let $n \in \mathbb{N}$. If a function $u : \mathbb{H} \to \mathbb{C}$ is periodic, then $T_n u$ is also periodic. In other words, we have*

$$u(z+1) = u(z) \quad \Longrightarrow \quad \big(T_n u\big)(z+1) = \big(T_n u)\big)(z) \tag{3.82}$$

for all $z \in \mathbb{H}$.

Proof The statement follows directly from Proposition 1.176 with $k = 0$. □

Similar to Corollary 1.181, we would like to show that the Hecke operators keep the space $S_{0,\nu}^{\text{Maass}}\big(\Gamma(1)\big)$ invariant. To do so, we assume $u \in S_{0,\nu}^{\text{Maass}}\big(\Gamma(1)\big)$ is a Maass cusp form and check whether $T_n u$ satisfies the defining properties of a Maass cusp form in Definition 3.31, too.

Lemma 3.61 *Let $n \in \mathbb{N}$ and $u : \mathbb{H} \to \mathbb{C}$ be a real-analytic function. Then the function*

$$\mathbb{H} \to \mathbb{C}; \quad z \mapsto \big(T_n u\big)(z)$$

is also real-analytic in the sense of Definition 3.14.

Proof We leave this proof to the reader as an exercise. □

Using the same arguments as in the Theorem 1.179, we obtain

Lemma 3.62 *If $u : \mathbb{H} \to \mathbb{C}$ satisfies*

$$u(M\,z) = u(z)$$

for all $M \in \Gamma(1)$ and $z \in \mathbb{H}$, then $T_n\, f$ defined in (3.31) satisfies

$$\big(T_n\, u\big)(M\,z) = \big(T_n\, u\big)(z)$$

for all $M \in \Gamma(1)$ and $z \in \mathbb{H}$, too.

Proof We leave the proof to the reader as an exercise. □

Proposition 3.63 *Let $\nu \in \mathbb{C} \smallsetminus -\frac{1}{2}\mathbb{N}$ be a spectral value and $n \in \mathbb{N}$. Suppose that the function $u : \mathbb{H} \to \mathbb{C}$ has an expansion of the from*

$$u(x+iy) = \sum_{m \in \mathbb{Z}_{\neq 0}} A(m)\, \mathbb{W}_{0,\nu}\big(4\pi\,|m|\,y\big)\, e^{2\pi m\,x}$$

for $x + iy \in \mathbb{H}$ with real part $x \in \mathbb{R}$ and imaginary part $y > 0$. (u admits an expansion of a Maass cusp form as given in (3.47) for the cusp ∞, cusp width $\lambda = 1$, and $\kappa = 0$.)

Then the function $T_n f$ admits the Fourier expansion

$$\big(T_n u\big)(z) = \sum_{m\in\mathbb{Z}_{\neq 0}} \gamma_n(m)\, \mathbb{W}_{0,\nu}\big(4\pi\,|m|\,y\big)\, e^{2\pi m\,x} \tag{3.83}$$

with

$$\gamma_n(m) := \sum_{0<d|\gcd(n,m)} d^{-1}\, A\left(\frac{mn}{d^2}\right) \tag{3.84}$$

for all $m \in \mathbb{Z}_{\neq 0}$.

Proof Consider $\frac{nz+bd}{d^2}$ which is a typical argument appearing the Hecke operator T_n in (3.81). Splitting it into real and imaginary parts, we get

$$\Re\left(\frac{nz+bd}{d^2}\right) = \frac{n\Re\,(z)+bd}{d^2} = \frac{n\,\Re\,(z)}{d^2} + \frac{b}{d} \quad \text{and}$$

$$\Im\left(\frac{nz+bd}{d^2}\right) = \frac{n\,\Im\,(z)}{d^2}$$

since the coefficients n, b, d are integers.

Plugging the Fourier expansion into (3.81), we get

$$\begin{aligned}
&\big(T_n u\big)(z)\\
&= n^{-1}\sum_{d|n}\sum_{b=0}^{d-1}\Big[\sum_{m\in\mathbb{Z}_{\neq 0}} A(m)\,\mathbb{W}_{0,\nu}\left(4\pi\,|m|\,\Im\left(\frac{nz+bd}{d^2}\right)\right)\Big]\, e^{2\pi i m\,\Re\left(\frac{nz+bd}{d^2}\right)}\\
&= \sum_{m\in\mathbb{Z}_{\neq 0}}\sum_{d|n}\left(\frac{n}{d}\right)^{-1}\sum_{b=0}^{d-1} A(m)\,\mathbb{W}_{0,\nu}\left(4\pi\,|m|\,\frac{ny}{d^2}\right)\, e^{2\pi i m\frac{nx}{d^2}}\,\frac{1}{d} e^{2\pi i m\frac{b}{d}}\\
&= \sum_{m\in\mathbb{Z}_{\neq 0}}\sum_{d|n}\left(\frac{n}{d}\right)^{-1} A(m)\,\mathbb{W}_{0,\nu}\left(4\pi\,|m|\,\frac{ny}{d^2}\right)\, e^{2\pi i m\frac{nx}{d^2}}\,\frac{1}{d}\sum_{b=0}^{d-1} e^{2\pi i m\frac{b}{d}}
\end{aligned}$$

with $z = x + iy \in \mathbb{H}$. Now, for $d \in \mathbb{N}$ and $m \in \mathbb{Z}$, we have

$$\sum_{b=0}^{d-1} e^{2\pi i m\frac{b}{d}} = \begin{cases} d & \text{if } d \mid m \text{ and} \\ 0 & \text{if } d \nmid m. \end{cases} \tag{3.85}$$

Using (3.85), we find that

$$\big(T_n u\big)(z) \;=\; \sum_{m\in\mathbb{Z}_{\neq 0}}\sum_{d|n}\left(\frac{n}{d}\right)^{-1} A(m)\,\mathbb{W}_{0,\nu}\left(4\pi\,|m|\,\frac{ny}{d^2}\right)\, e^{2\pi i m\frac{nx}{d^2}}\,\frac{1}{d}\sum_{b=0}^{d-1} e^{2\pi i m\frac{b}{d}}$$

$$
\begin{aligned}
&= \sum_{m\in\mathbb{Z}_{\neq 0}} \sum_{\substack{d|n \\ d|m}} \left(\frac{n}{d}\right)^{-1} A(m)\, \mathbb{W}_{0,\nu}\left(4\pi |m| \frac{ny}{d^2}\right) e^{2\pi i m \frac{nx}{d^2}} \\
&\overset{m=qd}{=} \sum_{q\in\mathbb{Z}_{\neq 0}} \sum_{d|n} \left(\frac{n}{d}\right)^{-1} A(qd)\, \mathbb{W}_{0,\nu}\left(4\pi |qd| \frac{ny}{d^2}\right) e^{2\pi i q d \frac{nx}{d^2}} \\
&= \sum_{q\in\mathbb{Z}_{\neq 0}} \sum_{d|n} \left(\frac{n}{d}\right)^{-1} A(qd)\, \mathbb{W}_{0,\nu}\left(4\pi |q| \frac{ny}{d}\right) e^{2\pi i d \frac{nx}{d}}.
\end{aligned}
$$

Observe that if d runs through all divisors of n, then so does $\frac{n}{d}$. Hence, we may substitute d by $\frac{n}{d}$ in the inner sum. This gives

$$
\begin{aligned}
\left(T_n u\right)(z) &= \sum_{q\in\mathbb{Z}_{\neq 0}} \sum_{d|n} \left(\frac{n}{d}\right)^{-1} A(qd)\, \mathbb{W}_{0,\nu}\left(4\pi |q| \frac{ny}{d}\right) e^{2\pi i d \frac{nx}{d}} \\
&\overset{d\mapsto \frac{n}{d}}{=} \sum_{q\in\mathbb{Z}_{\neq 0}} \sum_{d|n} \left(\frac{n}{\frac{n}{d}}\right)^{-1} A\left(q\frac{n}{d}\right) \mathbb{W}_{0,\nu}\left(4\pi |q| \frac{ny}{\frac{n}{d}}\right) e^{2\pi i d \frac{nx}{\frac{n}{d}}} \\
&= \sum_{q\in\mathbb{Z}_{\neq 0}} \sum_{d|n} d^{-1} A\left(\frac{qn}{d}\right) \mathbb{W}_{0,\nu}(4\pi |qd|\, y)\, e^{2\pi i q d\, x} \\
&\overset{m=qd}{=} \sum_{m\in\mathbb{Z}} \sum_{\substack{d|n \\ d|m}} d^{-1} A\left(\frac{mn}{d^2}\right) \mathbb{W}_{0,\nu}(4\pi |m|\, y)\, e^{2\pi i m\, x}.
\end{aligned}
$$

Summarizing, we just showed that $T_n f$ has the Fourier expansion

$$
\left(T_n u\right)(z) = \sum_{m\in\mathbb{Z}} \sum_{\substack{d|n \\ d|m}} d^{-1} A\left(\frac{mn}{d^2}\right) \mathbb{W}_{0,\nu}(4\pi |m|\, y)\, e^{2\pi i m\, x}.
$$

This proves Proposition 3.63. □

In complete analogy to Corollary 1.178, we deduce the following.

Corollary 3.64 *In the setting of Theorem 1.177, consider the coefficients $\gamma_n(m)$ as defined in* (1.144). *These satisfy*

$$
\gamma_n(m) = \gamma_m(n) \tag{3.86}
$$

for all $m, n \in \mathbb{N}$.

Proof Theorem 3.63 implies

$$\gamma_n(m) \overset{(3.84)}{=} \sum_{0<d|\gcd(n,m)} d^{k-1}\, A\left(\frac{mn}{d^2}\right) \overset{(3.84)}{=} \gamma_m(n)$$

for all $m, n \in \mathbb{N}$. □

We finally have our first main result about Hecke operators.

Corollary 3.65 *Let $\nu \in \mathbb{C} \smallsetminus -\frac{1}{2}\mathbb{N}$ be a spectral value and $n \in \mathbb{N}$. The Hecke operator T_n preserves the space $S_{0,\nu}^{Maass}(\Gamma(1))$.*

Proof Assume $u \in S_{0,\nu}^{\mathrm{Maass}}(\Gamma(1))$. We have to check whether $T_n u$ satisfies the defining properties of a Maass cusp form in Definition 3.6.

1. $T_n u$ is real-analytic and this was shown in Lemma 3.61.
2. That $T_n u$ is an eigenfunction of Δ with eigenvalue $\frac{1}{4}-\nu^2$ and this follows directly from (3.18) of Exercise 3.12 since the action of the matrices $\begin{pmatrix} \frac{n}{d} & b \\ 0 & d \end{pmatrix}$ commute with Δ.
3. The invariance property $(T_n u)\|_0 M = T_n u$ for all $m \in \Gamma(1)$ was shown in Corollary 3.62.
4. According to Definition 3.32 u admits an expansion of the form (3.48) for the cusp ∞ with $\kappa = 0$ and cusp length $l = 1$. Hence, u admits the expansion assumed in Proposition 3.63. Applying Proposition 3.63 shows that $T_n f$ admits the Fourier of the same form where the coefficients are replaced by $\gamma_n(m)$.

Hence $T_n u$ is a Maass cusp form of weight 0, spectral parameter ν, and trivial multiplier for the full modular group $\Gamma(1)$, as specified in Definition 3.6. □

Remark 3.66 Having established Proposition 3.63 and Corollary 3.62, we can immediately derive the multiplicative properties of Hecke operators as stated in Theorem 1.184 and Corollary 1.185 since the arguments of their proofs depend only on the representation of T_n as set $\mathcal{T}(n)$ of matrices defined in (1.139).

3.7 The Laplace Operator and Its Self-Adjoint Friedrichs Extension

We present the Friedrichs extension of Δ_0 and Δ_k in §3.7.1 and §3.7.2 respectively where we follow ideas presented in [118] and [28]. We refer to [103, VI.2.3] for a general treatment.

3.7.1 The Friedrichs Extension of Δ_0 for the Full Modular Group

Following the arguments presented in [118, §4], we establish the self-adjoint Friedrichs extension of the differential operator Δ_0.

Recall that we introduced the differential operator Δ_0 acting on 2 times differentiable functions $f : \mathbb{H} \to \mathbb{C}$ in Definition 3.6. Later, we consider Maass cusp forms in Definition 3.32, where we require additionally that the functions are real-analytic, are eigenfunctions of Δ_0, satisfy the transformation property (3.6) for $\Gamma(1)$, and admit a Fourier-Whittaker expansion (3.48).

We relax these assumptions a bit.

Definition 3.67 Let $L^2\big(\Gamma(1), 0, 1\big)$ denote the Hilbert space of square integrable functions on $\mathbb{H}$ which are invariant under the group $\Gamma(1)$:

$$L^2\big(\Gamma(1), 0, 1\big) := \Big\{ f : \mathbb{H} \to \mathbb{C}\colon\ f\big\|_0 M = f \quad \text{for all } M \in \Gamma(1), \\ \int_{\mathcal{F}_{\Gamma(1)}} |f(z)|^2\, \Im(z)^{-2} d\lambda_2(z) < \infty \Big\}. \tag{3.87}$$

Its scalar product and associated norm are given by

$$\langle f, g\rangle_{\Gamma(1)} \qquad \text{and} \qquad \|f\|_2 := \sqrt{\langle f, f\rangle_{\Gamma(1)}}. \tag{3.88}$$

The subset $C_0^2\big(\Gamma(1), 0, \mathbf{1}\big) \subset L^2\big(\Gamma(1), 0, 1\big)$ denotes the twice continuously differentiable functions f such that its restriction to the fundamental domain $\mathcal{F}_{\Gamma(1)}$ has compact support (in $\mathcal{F}_{\Gamma(1)}$)

$$C_0^2\big(\Gamma(1), 0, 1\big) := \Big\{ f \in L^2\big(\Gamma(1), 0, 1\big)\colon\ f \in C^2(\mathbb{H}) \quad \text{and} \\ f\big|_{\mathcal{F}_{\Gamma(1)}} \in C_0^2\big(\mathcal{F}_{\Gamma(1)}\big) \Big\}, \tag{3.89}$$

where $f\big|_{\mathcal{F}_{\Gamma(1)}}$ denotes the restriction of f to $\mathcal{F}_{\Gamma(1)}$.

Consider the scalar product $\langle \cdot, \cdot\rangle_{\Gamma(1)}$ defined in (3.72). We obviously notice that its defining relation

$$\langle f, g\rangle_{\Gamma(1)} = \int_{\mathcal{F}_{\Gamma(1)}} f(z)\, \overline{g(z)}\, \Im(z)^{-2} d\lambda_2(z)$$

is also well defined for $f, g \in C_0^2(\Gamma(1), 0, \mathbf{1})$. Also, $\Delta_0 f$ is well defined for all $f \in C_0^2(\Gamma(1), 0, 1)$, since it is a second-order elliptic differential operator. This leads to the following.

Lemma 3.68 *For $f, g \in C_0^2(\Gamma(1), 0, 1)$, we have*

$$\langle f, \Delta_0 g\rangle_{\Gamma(1)} = \int_{\mathcal{F}_{\Gamma(1)}} \frac{\partial f}{\partial x}(z)\overline{\frac{\partial g}{\partial x}(z)} + \frac{\partial f}{\partial y}(z)\overline{\frac{\partial g}{\partial y}(z)}\, d\lambda_2(z). \tag{3.90}$$

Here $x = \Re(z)$ denotes the real part and $y = \Im(z)$ denotes the imaginary part of a complex number z and $d\lambda_2(z)$ denotes the usual Euclidean surface Lebesgue measure introduced in (1.58).

Proof Recall that $f \in C_0^2(\Gamma(1), 0, 1)$ implies that f restricted to the fundamental domain $\mathcal{F}_{\Gamma(1)}$ has compact support. This implies in particular that $f(z)$ and its partial derivatives $\frac{\partial f}{\partial x}(z)$, $\frac{\partial f}{\partial y}(z)$, $\frac{\partial^2 f}{\partial x^2}(z)$, and $\frac{\partial^2 f}{\partial y^2}(z)$ vanish as z tends to the cusp $i\infty$.

Using partial integration in x and y respectively, we get

$$\begin{aligned}
\langle f, \Delta_0 g\rangle_{\Gamma(1)} &\overset{(3.72)}{=} \int_{\mathcal{F}_{\Gamma(1)}} f(z)\,\overline{(\Delta_0 g)(z)}\,\Im(z)^{-2} d\lambda_2(z) \\
&\overset{(3.8)}{=} -\int_{\mathcal{F}_{\Gamma(1)}} f(z)\overline{\frac{\partial^2 g}{\partial^2 x}(z)} + f(z)\overline{\frac{\partial^2 g}{\partial y^2}(z)}\, d\lambda_2(z) \\
&= \int_{\mathcal{F}_{\Gamma(1)}} \frac{\partial f}{\partial x}(z)\overline{\frac{\partial g}{\partial x}(z)} + \frac{\partial f}{\partial y}(z)\overline{\frac{\partial g}{\partial y}(z)}\, d\lambda_2(z).
\end{aligned}$$

This proves (3.90). □

This leads to the following properties of Δ_0 on $C_0^2(\Gamma(1), 0, 1)$.

Lemma 3.69 *The differential operator Δ_0 on the space $C_0^2(\Gamma(1), 0, 1)$ is symmetric and positive. This means*

1. $\langle f, \Delta_0 g\rangle_{\Gamma(1)} = \langle \Delta_0 f, g\rangle_{\Gamma(1)}$ (symmetric) *and*
2. $\langle f, \Delta_0 f\rangle_{\Gamma(1)} \geq 0$ (positive)

for all $f, g \in C_0^2(\Gamma(1), 0, \mathbf{1})$.

Proof We conclude directly from Lemma 3.68 and its proof that

$$\begin{aligned}
\langle f, \Delta_0 g\rangle_{\Gamma(1)} &= \int_{\mathcal{F}_{\Gamma(1)}} \frac{\partial f}{\partial x}(z)\overline{\frac{\partial g}{\partial x}(z)} + \frac{\partial f}{\partial y}(z)\overline{\frac{\partial g}{\partial y}(z)}\, d\lambda_2(z) \\
&= \langle \Delta_0 f, g\rangle_{\Gamma(1)}
\end{aligned}$$

since applying partial integration to the right-hand side above gives the second line, following the same arguments as in the proof of Lemma 3.68.

Similarly, we conclude the positivity of Δ_0. Again, using Lemma 3.68, we find

$$\begin{aligned}\langle f, \Delta_0 f\rangle_{\Gamma(1)} &= \int_{\mathcal{F}_{\Gamma(1)}} \frac{\partial f}{\partial x}(z)\overline{\frac{\partial f}{\partial x}(z)} + \frac{\partial f}{\partial y}(z)\overline{\frac{\partial f}{\partial y}(z)}\, \mathrm{d}\lambda_2(z) \\ &= \int_{\mathcal{F}_{\Gamma(1)}} \left|\frac{\partial f}{\partial x}(z)\right|^2 + \left|\frac{\partial f}{\partial y}(z)\right|^2 \mathrm{d}\lambda_2(z) \\ &\geq 0\end{aligned}$$

since the integrand $\left|\frac{\partial f}{\partial x}(z)\right|^2 + \left|\frac{\partial f}{\partial y}(z)\right|^2$ at the right-hand side is nonnegative. □

To construct the *Friedrichs extension* of the differential operator Δ_0 on the space $C_0^2\big(\Gamma(1), 0, 1\big)$, we follow the arguments outlined in [118, §4]. We define the sesquilinear form

$$s_0(f, g) := \langle f, g\rangle_{\Gamma(1)} + \langle f, \Delta_0\, g\rangle_{\Gamma(1)} \tag{3.91}$$

for all $f, g \in C_0^2\big(\Gamma(1), 0, 1\big)$. Adapting Lemma 3.69 to the sesquilinear form $C_0(\cdot, \cdot)$, we find the following properties.

Lemma 3.70 *Let* $f, g \in C_0^2\big(\Gamma(1), 0, 1\big)$.

The sesquilinear form $s_0(\cdot, \cdot)$ *satisfies:*

1. *The form is* symmetric, *i.e.*, $s_0(f, g) = \overline{s_0(g, f)}$ *and*
2. *is* bounded from below, *i.e.*, $s_0(f, f) \geq \|f\|_2$.
3. *Moreover, we have the identity*

$$s_0(f, g) = \int_{\mathcal{F}_{\Gamma(1)}} \frac{\partial f}{\partial x}(z)\overline{\frac{\partial g}{\partial x}(z)} + \frac{\partial f}{\partial y}(z)\overline{\frac{\partial g}{\partial y}(z)} + f(z)\overline{g(z)}\Im(z)^{-2}\, \mathrm{d}\lambda_2(z) \tag{3.92}$$

Proof We leave the proof to the reader. □

Using (3.92) as the defining equation of $s_0(f, g)$, we see that we can define $s_0(\cdot, \cdot)$ on a larger domain than $C_0^2\big(\Gamma(1), 0, \mathbf{1}\big)$. We can extend it to square integrable functions in $L^2\big(\Gamma(1), 0, \mathbf{1}\big)$ which have a square integrable partial derivatives. This is usually called the first Sobolev space $H^1\big(\Gamma(1), 0, \mathbf{1}\big) \subset L^2\big(\Gamma(1), 0, \mathbf{1}\big)$. In particular, $s_0(\cdot, \cdot)$ is closed in the sense of [103, VI, in particular Theorem 1.17]. Now applying [103, VI, Theorem 2.1], we see that there exists an operator A defined on a dense domain $D(A) \subset L^2\big(\Gamma(1), 0, \mathbf{1}\big)$ such that

$$s_0(f, g) = \langle f, Ag\rangle_{\Gamma(1)}$$

for all $f \in H^1(\Gamma(1), 0, \mathbf{1})$ and $g \in D(A)$. The operator A is self-adjoint. We usually denote this self-adjoint (Friedrichs)-extension of Δ_0 also by Δ_0.

Remark 3.71 The authors of [118] show a bit more in [118, §4]. Next to constructing the Friedrichs-extension of Δ_0, they also prove the following spectral result: The spectrum of Δ_0 contains $\left(\frac{1}{4}, \infty\right)$ and admits at most a finite number of eigenvalues with finite multiplicity outside of that interval. Moreover, they point out the following:

1. The continuous spectrum of Δ_0 contains $\left(\frac{1}{4}, \infty\right)$.
2. Δ_0 has no singular spectrum.
3. Δ_0 admits infinitely many eigenvalues embedded in $\left(\frac{1}{4}, \infty\right)$ with accumulation point ∞.
4. Δ_0 is positive (see Lemma 3.69), thus Δ_0 admits no negative spectrum.

3.7.2 The Friedrichs Extension of Δ_k

We slightly generalize the arguments of [118, §4] that we sketched above in §3.7.1. This follows loosely the idea presented in [28, §5].

As in §3.7.1, we first define the spaces on which the differential operator Δ_k will act.

Definition 3.72 Let $\Gamma \subset \Gamma(1)$ be of finite index μ, $k \in \mathbb{R}$, and v a multiplier with respect to (Γ, k).

We denote the Hilbert space of square integrable functions on $\mathbb{H}$ which satisfy the transformation property (3.6) by

$$L^2(\Gamma, k, v) := \Big\{ f : \mathbb{H} \to \mathbb{C} :\ f\|_k M = v(M) f \quad \text{for all } M \in \Gamma,$$

$$\int_{\mathcal{F}_{\Gamma(1)}} |f(z)|^2\, \Im(z)^{-2} \mathrm{d}\lambda_2(z) < \infty \Big\}. \tag{3.93}$$

Its scalar product and associated norm are given by

$$\langle f, g \rangle_\Gamma \qquad \text{and} \qquad \|f\|_2 := \sqrt{\langle f, f \rangle_\Gamma}. \tag{3.94}$$

The subset $C_0^2(\Gamma, k, v) \subset L^2(\Gamma, k, v)$ denotes the twice continuously differentiable functions f such that its restriction to the fundamental domain $\mathcal{F}_\Gamma$ has compact support (in $\mathcal{F}_{\Gamma(1)}$)

$$C_0^2(\Gamma, k, v) := \left\{ f \in L^2(\Gamma, k, v) : \ f \in C^2(\mathbb{H}) \ \text{ and } \ f\big|_{\mathcal{F}_\Gamma} \in C_0^2(\mathcal{F}_\Gamma) \right\}, \tag{3.95}$$

where $f\big|_{\mathcal{F}_\Gamma}$ denotes the restriction of f to the fundamental domain $\mathcal{F}_\Gamma$.

We extend $C_0^2(\Gamma, k, v)$ to square integrable functions in $L^2(\Gamma, k, v)$ that have a square integrable partial derivatives. This is usually called the first Sobolev space

$$H^1(\Gamma, k, v) := \left\{ f \in L^2(\Gamma, k, v) : \ \frac{\partial f}{\partial x} \text{ and } \frac{\partial f}{\partial x} \text{ are square-integrable} \right\}. \tag{3.96}$$

The three spaces obviously satisfy the inclusion

$$C_0^2(\Gamma, k, v) \subset H^1(\Gamma, k, v) \subset L^2(\Gamma, k, v).$$

Consider the scalar product $\langle \cdot, \cdot \rangle_\Gamma$ defined in (3.72). We obviously notice that its defining relation

$$\langle f, g \rangle_\Gamma = \frac{1}{\mu} \int_{\mathcal{F}_\Gamma} f(z)\, \overline{g(z)}\, \Im(z)^{-2} \mathrm{d}\lambda_2(z)$$

is also well defined for $f, g \in C_0^2(\Gamma, k, v)$. Here, μ denotes the index of Γ in $\Gamma(1)$. Also, $\Delta_k f$, $\mathrm{E}_k^+ f$ and $\mathrm{E}_k^- f$ are well defined for all $f \in C_0^2(\Gamma, k, v)$ since taking the derivatives is well defined. We have an associated norm given by

$$\|f\|_2 := \sqrt{\langle f, f \rangle_\Gamma} = \sqrt{\frac{1}{\mu} \int_{\mathcal{F}_\Gamma} |f(z)|\, \Im(z)^{-2} \mathrm{d}\lambda_2(z)}. \tag{3.97}$$

Before we introduce the suitable sesquilinear form $s_k(\cdot, \cdot)$, we need an auxiliary lemma which tells us how to integrate by parts.

Lemma 3.73 *Let $f, g : \mathbb{H} \to \mathbb{C}$ be two continuously differentiable functions such that their restrictions $f\big|_{\mathcal{F}}$ and $g\big|_{\mathcal{F}}$ to the fundamental domain $\mathcal{F}$ have compact support. We have*

$$\int_{\mathcal{F}_\Gamma} \frac{\partial f(z)}{\partial x}\, \overline{g(z)}\, \mathrm{d}\lambda_2(z) = -\int_{\mathcal{F}_\Gamma} f(z)\, \overline{\frac{\partial g(z)}{\partial x}}\, \mathrm{d}\lambda_2(z) \tag{3.98}$$

$$\tag{3.99}$$

and

$$\int_{\mathcal{F}_\Gamma} \frac{\partial f(z)}{\partial y}\, \overline{g(z)}\, \mathrm{d}\lambda_2(z) = -\int_{\mathcal{F}_\Gamma} f(z)\, \overline{\frac{\partial g(z)}{\partial y}}\, \mathrm{d}\lambda_2(z), \tag{3.100}$$

where $x = \Re(z)$ and $y = \Im(z)$ denote real and imaginary parts of the complex variable $z \in \mathbb{H}$.

Proof Since f and g restricted to $\mathcal{F}$ have compact support, we have that $f(z)$ and $g(z)$ and their partial derivatives vanish if z tends to each cusp of the fundamental domain $\mathcal{F}$. This allows us to use the partial integration formula where one term vanishes. Using the decomposition $z = x + iy$ of the complex variable into real and imaginary parts, we get

$$\begin{aligned}\int_{\mathcal{F}_\Gamma} \frac{\partial f}{\partial x}(z)\,\overline{g(z)}\,\mathrm{d}\lambda_2(z) &= 0 - \int_{\mathcal{F}_\Gamma} f(z)\,\frac{\partial}{\partial x}\overline{g(z)}\,\mathrm{d}\lambda_2(z)\\ &= -\int_{\mathcal{F}_\Gamma} f(z)\,\overline{\frac{\partial g}{\partial x}(z)}\,\mathrm{d}\lambda_2(z).\end{aligned}$$

This shows identity (3.98).

The proof of the remaining identity is left to the reader. □

Lemma 3.74 *Let $f, g : \mathbb{H} \to \mathbb{C}$ be two continuously differentiable functions such that their restrictions $f\big|_{\mathcal{F}}$ and $g\big|_{\mathcal{F}}$ to the fundamental domain $\mathcal{F}_\Gamma$ have compact support. We have*

$$\begin{aligned}&\langle \mathrm{E}_k^{\pm} f, \mathrm{E}_k^{\pm} g\rangle_\Gamma\\ &= \int_{\mathcal{F}_\Gamma} \left(4y^2 \frac{\partial f}{\partial x}(z)\overline{\frac{\partial g}{\partial x}(z)} \pm 4iy^2\frac{\partial f}{\partial x}(z)\overline{\frac{\partial g}{\partial y}(z)} + 2kiy\frac{\partial f}{\partial x}(z)\overline{g(z)}\right.\\ &\quad \mp 4iy^2\frac{\partial f}{\partial y}(z)\overline{\frac{\partial g}{\partial x}(z)} + 4y^2\frac{\partial f}{\partial y}(z)\overline{\frac{\partial g}{\partial y}(z)} \pm 2yk\frac{\partial f}{\partial y}(z)\overline{g(z)}\\ &\quad \left. - 2ikyf(z)\overline{\frac{\partial g}{\partial x}(z)} \pm 2ykf(z)\overline{\frac{\partial g}{\partial y}(z)} + k^2 f(z)\overline{g(z)}\right)\Im(z)^{-2}\mathrm{d}\lambda_2(z),\end{aligned} \tag{3.101}$$

where $x = \Re(z)$ and $y = \Im(z)$ denote real and imaginary parts of the complex variable $z \in \mathbb{H}$.

Proof We have

$$\begin{aligned}&\langle \mathrm{E}_k^{\pm} f, \mathrm{E}_k^{\pm} g\rangle_\Gamma\\ &\overset{(3.72)}{=} \int_{\mathcal{F}_\Gamma} (\mathrm{E}_k^{\pm} f)(z)\,\overline{(\mathrm{E}_k^{\pm} g)(z)}\,\Im(z)^{-2}\mathrm{d}\lambda_2(z)\\ &\overset{(3.10)}{=} \int_{\mathcal{F}_\Gamma} \left(\pm 2iy\frac{\partial f(z)}{\partial x} + 2y\frac{\partial f(z)}{\partial y} \pm kf(z)\right)\\ &\qquad \cdot \overline{\left(\pm 2iy\frac{\partial g(z)}{\partial x} + 2y\frac{\partial g(z)}{\partial y} \pm kg(z)\right)}\,\Im(z)^{-2}\mathrm{d}\lambda_2(z)\end{aligned}$$

$$= \int_{\mathcal{F}_\Gamma} \left(4y^2 \frac{\partial f}{\partial x}(z) \overline{\frac{\partial g}{\partial x}(z)} \pm 4iy^2 \frac{\partial f}{\partial x}(z) \overline{\frac{\partial g}{\partial y}(z)} + 2kiy \frac{\partial f}{\partial x}(z) \overline{g(z)} \right.$$

$$\mp 4iy^2 \frac{\partial f}{\partial y}(z) \overline{\frac{\partial g}{\partial x}(z)} + 4y^2 \frac{\partial f}{\partial y}(z) \overline{\frac{\partial g}{\partial y}(z)} \pm 2yk \frac{\partial f}{\partial y}(z) \overline{g(z)}$$

$$\left. - 2ikyf(z) \overline{\frac{\partial g}{\partial x}(z)} \pm 2ykf(z) \overline{\frac{\partial g}{\partial y}(z)} + k^2 f(z) \overline{g(z)} \right) \Im(z)^{-2} \mathrm{d}\lambda_2(z).$$

□

We define a sesquilinear form $s_k(\cdot, \cdot)$ on $H^1(\Gamma, k, v) \times H^1(\Gamma, k, v)$ by

$$s_k(f, g) := \frac{1}{8} \langle \mathrm{E}_k^+ f, \mathrm{E}_k^+ g \rangle_\Gamma + \frac{1}{8} \langle \mathrm{E}_k^- f, \mathrm{E}_k^- g \rangle_\Gamma - \frac{k^2}{4} \langle f, g \rangle_\Gamma. \tag{3.102}$$

This leads to the following properties of $s_k(\cdot, \cdot)$ on $H^1(\Gamma, k, v) \times H^1(\Gamma, k, v)$.

Lemma 3.75 *Let $f, g \in H^1(\Gamma, k, v) \times H^1(\Gamma, k, v)$.*
The sesquilinear form $s_k(\cdot, \cdot)$ satisfies:

1. *The form is* symmetric, *i.e.,* $s_k(f, g) = \overline{s_k(g, f)}$ *and*
2. *is* bounded from below, *i.e.,* $s_k(f, f) \geq -\frac{k^2}{4} \|f\|_2{}^2$.

Proof We leave the proof to the reader. □

This leads to the following.

Lemma 3.76 *For $f, g \in C_0^2(\Gamma, k, v)$, we have*

$$s_k(f, g) = \langle \Delta_k f, g \rangle_\Gamma. \tag{3.103}$$

Proof Recall that $f \in C_0^2(\Gamma, k, v)$ implies that f restricted to the fundamental domain $\mathcal{F}_\Gamma$ has compact support. This implies in particular that f and its partial derivatives $\frac{\partial f}{\partial x}(z)$, $\frac{\partial f}{\partial y}(z)$, $\frac{\partial^2 f}{\partial x^2}(z)$, and $\frac{\partial^2 f}{\partial y^2}(z)$ vanish as z tends to the cusps of the fundamental domain $\mathcal{F}_\Gamma$.

Applying Lemma 3.74, we get

$$8\mu\, s_k(f, g)$$

$$\overset{(3.102)}{=} \mu \langle \mathrm{E}_k^+ f, \mathrm{E}_k^+ g \rangle_\Gamma + \mu \langle \mathrm{E}_k^- f, \mathrm{E}_k^- g \rangle_\Gamma - 2k^2 \mu \langle f, g \rangle_\Gamma$$

$$\overset{(3.101)}{=} \int_{\mathcal{F}_\Gamma} \left(4y^2 \frac{\partial f}{\partial x}(z) \overline{\frac{\partial g}{\partial x}(z)} + 4iy^2 \frac{\partial f}{\partial x}(z) \overline{\frac{\partial g}{\partial y}(z)} + 2kiy \frac{\partial f}{\partial x}(z) \overline{g(z)} \right.$$

$$- 4iy^2 \frac{\partial f}{\partial y}(z) \overline{\frac{\partial g}{\partial x}(z)} + 4y^2 \frac{\partial f}{\partial y}(z) \overline{\frac{\partial g}{\partial y}(z)} + 2yk \frac{\partial f}{\partial y}(z) \overline{g(z)}$$

$$- 2iky f(z)\overline{\frac{\partial g}{\partial x}(z)} + 2ykf(z)\overline{\frac{\partial g}{\partial y}(z)} + k^2 f(z)\overline{g(z)}\Big) \, \Im(z)^{-2} \mathrm{d}\lambda_2(z)$$

$$+ \int_{\mathcal{F}_\Gamma} \Big(4y^2 \frac{\partial f}{\partial x}(z)\overline{\frac{\partial g}{\partial x}(z)} - 4iy^2 \frac{\partial f}{\partial x}(z)\overline{\frac{\partial g}{\partial y}(z)} + 2kiy\frac{\partial f}{\partial x}(z)\overline{g(z)}$$

$$+ 4iy^2 \frac{\partial f}{\partial y}(z)\overline{\frac{\partial g}{\partial x}(z)} + 4y^2 \frac{\partial f}{\partial y}(z)\overline{\frac{\partial g}{\partial y}(z)} - 2yk\frac{\partial f}{\partial y}(z)\overline{g(z)}$$

$$- 2iky f(z)\overline{\frac{\partial g}{\partial x}(z)} - 2ykf(z)\overline{\frac{\partial g}{\partial y}(z)} + k^2 f(z)\overline{g(z)}\Big) \, \Im(z)^{-2} \mathrm{d}\lambda_2(z)$$

$$- 2k^2 \int_{\mathcal{F}_\Gamma} f(z)\,\overline{g(z)}\,\Im(z)^{-2}\mathrm{d}\lambda_2(z)$$

$$= \int_{\mathcal{F}_\Gamma} \Big(8y^2 \frac{\partial f}{\partial x}(z)\overline{\frac{\partial g}{\partial x}(z)} + 8y^2 \frac{\partial f}{\partial y}(z)\overline{\frac{\partial g}{\partial y}(z)}$$

$$+ 4iky \left[\frac{\partial f}{\partial x}(z)\overline{g(z)} - f(z)\overline{\frac{\partial g}{\partial x}(z)} \right] \Big) \, \Im(z)^{-2}\mathrm{d}\lambda_2(z).$$

Now using integration by parts presented in Lemma 3.73, we find

$$8\mu\, s_k(f,g)$$

$$\overset{(3.98)}{=} \int_{\mathcal{F}_\Gamma} \Big(8y^2 \frac{\partial f}{\partial x}(z)\overline{\frac{\partial g}{\partial x}(z)} + 8y^2 \frac{\partial f}{\partial y}(z)\overline{\frac{\partial g}{\partial y}(z)}$$

$$+ 4iky\frac{\partial f}{\partial x}(z)\overline{g(z)}\Big) \, \Im(z)^{-2}\mathrm{d}\lambda_2(z) + \int_{\mathcal{F}_\Gamma} \frac{4ik}{y}\left[\frac{\partial f(z)}{\partial x}\overline{g(z)}\right] \mathrm{d}\lambda_2(z)$$

$$= 8 \int_{\mathcal{F}_\Gamma} \Big(y^2 \frac{\partial f}{\partial x}(z)\overline{\frac{\partial g}{\partial x}(z)} + y^2 \frac{\partial f}{\partial y}(z)\overline{\frac{\partial g}{\partial y}(z)}$$

$$+ iky\frac{\partial f}{\partial x}(z)\overline{g(z)}\Big) \, \Im(z)^{-2}\mathrm{d}\lambda_2(z)$$

$$\overset{(3.9)}{=} 8 \int_{\mathcal{F}_\Gamma} (\Delta_k f)(z)\,\overline{g(z)}\,\Im(z)^{-2}\mathrm{d}\lambda_2(z)$$

$$\overset{(3.72)}{=} 8\mu \langle \Delta_k f, g\rangle_\Gamma.$$

This proves (3.103). □

Notice that $s_k(\cdot,\cdot)$ is closed in the sense of [103, VI, in particular Theorem 1.17]. Now applying [103, VI, Theorem 2.1], we see that there exists an operator A_k defined on a dense domain $D(A_k) \subset L^2(\Gamma, k, v)$ such that

$$s_k(f, g) = \langle A_k f, g \rangle_\Gamma$$

for all $g \in H^1(\Gamma, k, v)$ and $f \in D(A_k)$. In particular, the operator A_k is self-adjoint. We usually denote this self-adjoint Friedrichs-extension of Δ_k also by Δ_k.

Remark 3.77 The focus of [28] and its subsequent papers along with [31] is to study perturbation theory in the weight parameter $k \in (-12, 12)$. Next to constructing the Friedrichs-extension of Δ_k, they also prove the following spectral result: The spectrum of Δ_k contains $\left(\frac{1}{4}, \infty\right)$ and admits at most a finite number of eigenvalues with finite multiplicity outside of that interval.

3.8 Selberg's Conjecture

We discussed in §3.7.1, and in particular in Remark 3.71, that the spectrum of Δ_0 and Δ_k respectively is real and positive. Moreover, its eigenvalues are contained in $\left(\frac{1}{4}, \infty\right)$ up to at most finitely many eigenvalues in the interval $\left(0, \frac{1}{4}\right)$. Selberg conjectured in [179, p. 13] that there are no such small eigenvalues. We translate the conjecture as follows:

Conjecture 3.78 ([179]) For $N \in \mathbb{N}$, consider the principal congruence subgroup $\Gamma(N)$ of level N, see (1.6). Let $u \in S_{0,1}^{\mathrm{Maass}}(\Gamma(1), \nu)$ be a nonvanishing Maass cusp form of weight zero and trivial multiplier. Then

$$\nu \in i\mathbb{R}. \tag{3.104}$$

Remark 3.79

1. The conjecture has to be seen in comparison to Corollary 3.59 where we could show that $\nu \in \mathbb{R} \cup i\mathbb{R}$ and also Remark 3.71 where the positivity of Δ_0 implies $\frac{1}{4} - \nu^2 \geq 0 \iff \nu \in \left[-\frac{1}{2}, \frac{1}{2}\right] \cup i\mathbb{R}$.
2. The conjectured relation (3.104) for the spectral parameter implies a more common formulation

$$\frac{1}{4} - \nu^2 \geq \frac{1}{4} \tag{3.105}$$

in terms of the eigenvalue $\frac{1}{4} - \nu^2$ of Δ_0.

Selberg himself already proved the following estimate.

Theorem 3.80 ([179]) *For $N \in \mathbb{N}$, consider the principal congruence subgroup $\Gamma(N)$ of level N and let $u \in S_{0,1}^{Maass}(\Gamma(1), \nu)$ be a nonvanishing Maass cusp form of weight zero and trivial multiplier. Then, the spectral value satisfies*

$$\frac{1}{4} - \nu^2 \geq \frac{3}{16}. \tag{3.106}$$

The authors of [127] show the following improvement of the above estimate, since $\frac{21}{100} > \frac{3}{16}$.

Theorem 3.81 ([127, Theorem 1.1]) *Let $\Gamma \subset \mathrm{SL}_2(\mathbb{Z})$ denote a congruence subgroup and let $u \in S_{0,1}^{Maass}(\Gamma, \nu)$ be a nonvanishing Maass cusp form of weight zero and trivial multiplier. Then, the spectral value ν of u satisfies*

$$\frac{1}{4} - \nu^2 \geq \frac{21}{100}. \tag{3.107}$$

Remark 3.82

1. Some progress towards Selberg's conjecture has been made. A short review on the progress (up to 1995) can be found in the review article [175].
2. According to [76, Selberg conjecture], the conjecture is known for the full modular group $\Gamma(1)$ and for the congruence subgroup $\Gamma_0(N)$ with small level N, see (1.8).

3.9 Concluding Remarks

This chapter introduced the Maass wave forms and some related topics. However, we focused the discussion on the forms themselves.

The starting point of Maass wave forms can be traced back to the work done by Hans Maass in his paper [128] published in 1949. Hans Maass studied some "non-holomorphic automorphic functions" and showed a relation to certain L-series satisfying a functional equation. We will discuss a relation between Maass cusp forms and the associated L-series in the following Chapter 4.

Nowadays, Maass wave forms appear in various places in the literature. For example, one important area is the representation theory of $\mathrm{GL}_2(\mathbb{R})$ or $\mathrm{SL}_2(\mathbb{R})$ and their associated Li-algebras. Here, Maass wave forms appear naturally as representations. A good starting point is for example the older research paper [163] by Pukánszky. Some more modern textbooks are Borel [17], Bump [52], Iwaniec [101], and Kubota [111].

The book [53] pushes the horizon even further and discusses the Langlands program. The Langlands program is a whole collection of far-reaching and influential conjectures about connections between number theory and geometry. It was proposed by Robert Langlands [117] and tries to relate Galois groups in algebraic number theory to automorphic forms and representation theory of algebraic groups over local fields and adeles. The lecture notes [86] from Gelbart might also be a good starting point.

Another interesting connection between the Fourier coefficients of Maass cusp forms (of weight 0, trivial character, and for the full modular group $SL_2(\mathbb{Z})$) is the Voronoi formula. This is an equality involving Fourier coefficients of Maass cusp forms, with the coefficients twisted by additive characters on either side. Sometimes the formula is also seen as a Poisson summation formula for non-abelian groups. We refer to Good's paper [88] as a starting point.

Other interesting modern extensions of Maass wave forms are mock modular forms and weak harmonic Maass wave forms. A short introduction to this topic is given in Chapter 7 of this book.

Of interest might also be the numerical calculation of Maass wave forms. Some starting points are here the PhD thesis [191] from Strömberg and subsequent publications like [192]. Other nice starting points are the papers [16] and [197]. Numerical data on Maass cusp forms can also be found in the *L-functions and Modular Forms Database* [124].

Chapter 4
Families of Maass Cusp Forms, L-Series, and Eichler Integrals

In the previous chapter, we introduced Maass wave forms of real weights k where we assumed that k is a given parameter. Moreover, we saw in Corollary 3.41 that Maass operators $\mathrm{E}_k^{\pm}$ map Maass cusp forms of weight k to Maass cusp forms of weight $k \pm 2$. Can we exploit this result and use Maass operators to get a whole family of Maass cusp forms by starting with a "reference" Maass wave form of weight k and generate all possible images by repeatedly applying one of the Maass operators?

Let's start with a Maass waveform u of weight k. We could simply "generate" a family of Maass wave forms $\left(u_l\right)_{l \equiv k \bmod 2}$ by "assigning" iteratively

$$
\begin{aligned}
u_k &: = u, \\
u_{k+2} &: = \mathrm{E}_k^{+}\, u, \quad u_{k+4} := \mathrm{E}_{k+2}^{+}\mathrm{E}_k^{+}\, u, \quad \ldots
\end{aligned}
$$

and

$$
u_{k-2} : = \mathrm{E}_k^{-}\, u, \quad u_{k-4} := \mathrm{E}_{k-2}^{-}\mathrm{E}_k^{-}\, u, \quad \ldots.
$$

However this approach is too simple. For example, generating a second family $\left(\tilde{u}_l\right)_{l \equiv k+2 \bmod 2}$ by starting with $\tilde{u} = \mathrm{E}_k^{+} u$ and "assigning" iteratively

$$
\begin{aligned}
\tilde{u}_{k+2} &: = \tilde{u} = \mathrm{E}_k^{+} u, \\
\tilde{u}_{k+4} &: = \mathrm{E}_{k+2}^{+}\tilde{u}, \quad u_{k+6} := \mathrm{E}_{k+4}^{+}\mathrm{E}_k^{+}\, \tilde{u}, \quad \ldots
\end{aligned}
$$

and

$$
\tilde{u}_k : = \mathrm{E}_{k+2}^{-}\, \tilde{u}, \quad \tilde{u}_{k-2} := \mathrm{E}_k^{-}\mathrm{E}_{k+2}^{-}\, \tilde{u}, \quad \ldots.
$$

T. Mühlenbruch, W. Raji, *On the Theory of Maass Wave Forms*, Universitext,
https://doi.org/10.1007/978-3-030-40475-8_4

produces a different family: We have

$$\left(u_l\right)_{l\equiv k \bmod 2} \neq \left(\tilde{u}_l\right)_{l\equiv k+2 \bmod 2}.$$

The good news is that we can modify the above method slightly to make it work. We do this in §4.1, where we introduce and discuss the concept of families of Maass wave forms. For simplicity, we restrict ourselves to Maass cusp forms on the full modular group $\Gamma(1)$.

Following similar concepts from Chapter 1 on modular cusp forms, we show in §4.2 that we can associate a pair of L-series with Maass cusp forms. We do this first on an individual weight basis, i.e., we associate an individual Maass cusp form of weight k with a pair of L-series. We then expand the concept to families of Maass cusp forms: we show that the pair of L-series is essentially the same for all modular cusp forms within a family. We also present an alternative representation of the pair of L-series, which we call a pair of D-functions.

The last object that we discuss in this chapter is a function called the "nearly periodic function." This is a function which is almost periodic on the upper half-plane, as the name suggests, and satisfies certain continuation and growth properties. In §4.3, we show that this function corresponds to a family of Maass cusp forms. Several connections are presented: The relation to D-functions, how nearly periodic functions can be obtained by an integral transform of an Maass cusp form, and how to relate the Eichler integrals of modular cusp forms, see §2.1.4.

4.1 Families of Maass Cusp Forms on the Full Modular Group

Let $k \in \mathbb{R}$ be a real weight, v a multiplier system that is compatible with $\left(\Gamma(1), k\right)$, and $\nu \in \mathbb{C} \smallsetminus -\frac{1}{2}\mathbb{N}$ a spectral parameter.

We know already from Chapter 3 that the Maass operators $\mathrm{E}_k^{\pm}$ map the space $S_{k,v}^{\mathrm{Maass}}\left(\Gamma(1), \nu\right)$ into $S_{k\pm 2,v}^{\mathrm{Maass}}\left(\Gamma(1), \nu\right)$, see Corollary 3.41. Hence, starting with a cusp form $u \in S_{k,v}^{\mathrm{Maass}}\left(\Gamma(1), \nu\right)$, we could construct a family of cusp forms $\left(u_l\right)_{l\in k+2\mathbb{Z}}$ given by

$$u_k := u \qquad \text{and} \quad u_{l\pm 2} = \mathrm{const}_l^{\pm}\, \mathrm{E}_l^{\pm}\, u_l \tag{4.1}$$

for all $l \in k + 2\mathbb{Z}$. We would like to have the unspecified constant depending only on l and on the $\pm$-choice of the Maass operator $\mathrm{E}_l^{\pm}$.

What is a good choice for the constant $\mathrm{const}_l^{\pm}$? We recall that each cusp form u in $S_{k,v}^{\mathrm{Maass}}\left(\Gamma(1), \nu\right)$ satisfies the eigenvalue equation

$$\Delta_k\, u = \lambda\, u \qquad \text{with } \lambda = \frac{1}{4} - \nu^2.$$

This allows us to rewrite Identity (3.12) as

$$\begin{aligned} \mathrm{E}^{\pm}_{k\mp 2}\,\mathrm{E}^{\mp}_{k}\, u &= -2\left(\frac{k(k\mp 2)}{2} + 2\lambda\right) u \\ &= \big(1+2\nu \pm (k\mp 2)\big)\big(1+2\nu \mp k\big) u \end{aligned} \tag{4.2}$$

for all $u \in S^{\mathrm{Maass}}_{k,v}\big(\Gamma(1), \nu\big)$.

On the other hand, assume that we have a family $\big(u_l\big)_{l\in k+2\mathbb{Z}}$ of Maass cusp forms as motivated in (4.1) above. If we start with a u_l of such a family, then $\mathrm{E}^{\pm}_{l\mp 2}\,\mathrm{E}^{\mp}_{l}\, u_l$ satisfies (4.2). Hence u_l should satisfy

$$\begin{aligned} \big(1+2\nu \pm (k\mp 2)\big)\big(1+2\nu \mp k\big) u_l &\overset{(4.2)}{=} \mathrm{E}^{\pm}_{l\mp 2}\,\mathrm{E}^{\mp}_{l}\, u_l \\ &\overset{(4.1)}{=} \mathrm{E}^{\pm}_{l\mp 2}\,\frac{1}{\mathrm{const}^{\mp}_{l}}\, u_{l\mp 2} \\ &\overset{(4.1)}{=} \frac{1}{\mathrm{const}^{\pm}_{l\mp 2}}\,\frac{1}{\mathrm{const}^{\mp}_{l}}\, u_l, \end{aligned}$$

assuming that the involved constants $\mathrm{const}^{\pm}_{l\mp 2}$ and $\mathrm{const}^{\mp}_{l}$ do not vanish. This calculation suggests the identity

$$\frac{1}{\mathrm{const}^{\pm}_{l\mp 2}}\,\frac{1}{\mathrm{const}^{\mp}_{l}} = \big(1+2\nu \pm (k\mp 2)\big)\big(1+2\nu \mp k\big). \tag{4.3}$$

Assume $1+2\nu \pm l \neq 0$. Then the choice

$$\mathrm{const}^{+}_{l} = \frac{-1}{1+2\nu+l} \quad \text{and} \quad \mathrm{const}^{-}_{l} = \frac{-1}{1+2\nu-l} \tag{4.4}$$

satisfies (4.3) and leads to the definition of families of Maass cusp forms.

Definition 4.1 Let $k \in \mathbb{R}$ be a real weight, v a multiplier system that is compatible with $\big(\Gamma(1), k\big)$, and $\nu \in \mathbb{C}$ a spectral parameter. A *family of Maass cusp forms* is a sequence $\big(u_l\big)_{l\in k+2\mathbb{Z}}$ of Maass cusp forms $u_l \in S^{\mathrm{Maass}}_{l,v}\big(\Gamma(1), \nu\big)$ such that

$$\mathrm{E}^{+}_{l}\, u_l = -\big(1+2\nu+l\big)\, u_{l+2} \quad \text{and} \quad \mathrm{E}^{-}_{l}\, u_l = -\big(1+2\nu-l\big)\, u_{l-2} \tag{4.5}$$

for all $l \in k+2\mathbb{Z}$.

We call a family of Maass cusp forms $\big(u_l\big)_{l\in k+2\mathbb{Z}}$ *trivial* if $u_l = 0$ for all $l \in k+2\mathbb{Z}$.

We call a family of Maass cusp forms $\left(u_l\right)_{l\in k+2\mathbb{Z}}$ *admissible*, if there exists an $l \in k+2\mathbb{Z}$ such that the three Maass cusp forms

$$u_{l-2} \neq 0, \quad u_l \neq 0 \quad \text{and} \quad u_{l+2} \neq 0.$$

For every admissible family of Maass cusp forms $\left(u_l\right)_{l\in k+2\mathbb{Z}}$, we define $l_{\min}, l_{\max} \in \left(k+2\mathbb{Z}\right) \cup \{\pm\infty\}$ such that

$$\begin{aligned} &\mathrm{E}_l^+\, u_l \neq 0 \qquad \text{for all } l \in k+2\mathbb{Z},\, l_{\min} \leq l < l_{\max} \text{ and} \\ &\mathrm{E}_l^-\, u_l \neq 0 \qquad \text{for all } l \in k+2\mathbb{Z},\, l_{\min} < l \leq l_{\max}. \end{aligned}$$

If $l_{\min} \neq -\infty$ or $l_{\max} \neq +\infty$, we require respectively

$$\mathrm{E}_{l_{\min}}^-\, u_{l_{\min}} = 0 \quad \text{or} \quad \mathrm{E}_{l_{\max}}^+\, u_{l_{\max}} = 0.$$

We call the set $\left[l_{\min}, l_{\max}\right] \cap \left(k+2\mathbb{Z}\right)$ the *range of admissible weights.*

Remark 4.2 If $l_{\min}$ and $l_{\max}$ are indeed finite, then the set of admissible weights can be written as

$$\left\{l_{\min}, l_{\min}+2, \ldots, l_{\max}-2, l_{\max}\right\}.$$

To simplify notation, we usually use the above notation, even if $l_{\min}$ or $l_{\max}$ are infinite. In this case, we understand the above set is not finite: If $l_{\min} = -\infty$, we understand that all $l \in k+2\mathbb{Z}$ with $l \leq l_{\max}$ are admissible weights, if $l_{\max} = \infty$, we understand that all $l \in k+2\mathbb{Z}$ with $l \geq l_{\min}$ are admissible weights, and if $l_{\min} = -\infty$ and $l_{\max} = \infty$, we understand that all $l \in k+2\mathbb{Z}$ are admissible weights.

Remark 4.3 Loosely speaking, an admissible family of Maass cusp forms admits a range of weights $l \in \left(k+2\mathbb{Z}\right)$ with $l_{\min} \leq l \leq l_{\max}$ such that the relation (4.5) is nontrivially satisfied in the sense that neither the Maass cusp form u_l vanishes nor the constant $-\left(1+2\nu \pm l\right)$ vanishes if the Maass operator $\mathrm{E}_l^\pm$ maps to a form of weight $l_{\min} \leq l \pm 2 \leq l_{\max}$. In particular the set of l in the range has a cardinality of at least 3, i.e., $\frac{|l_{\max}-l_{\min}|}{2} \geq 3$.

Remark 4.4 The terminology "families of Maass wave/cusp forms" appears also in [31] where its usage is different from our setup. In [31] Roelof Bruggeman investigates what happens to a Maass wave/cusp form if one perturbs the weight k. In particular, he shows that there are continuous functions $U : \mathbb{H} \times (0, 12) \to \mathbb{C}$ such that its restriction on the first coordinate $z \mapsto U(z, k)$ (for fixed $k \in (0, 12)$ is a Maass cusp form of weight k. The projection on the second coordinate, $k \mapsto U(\cdot, k)$, is a family of Maass cusp forms in the terminology of [31]. (A good introduction to this type of families of Maass wave/cusp forms may also be the papers [28–30] which precede [31].)

Lemma 4.5 *Let $k \in \mathbb{R}$ be a real weight, v a multiplier system which is compatible with $\big(\Gamma(1), k\big)$, and $\nu \in \mathbb{C}$ a spectral parameter.*

For $u \in S^{\mathrm{Maass}}_{k,v}\big(\Gamma(1), \nu\big)$, we recursively construct an associated family of Maass cusp forms $\big(u_l\big)_{l \in k+2\mathbb{Z}}$ by the following algorithm:

1. *Put $u_k := u$.*
2. *If u_l with $l \in k + 2\mathbb{Z}_{\geq 0}$ is already defined, put*

$$u_{l+2} := \begin{cases} -\big(1+2\nu+l\big)^{-1}\, \mathrm{E}^{+}_{l}\, u_l & \text{if } 1+2\nu+l \neq 0 \quad \text{and} \\ 0 & \text{if } 1+2\nu+l = 0. \end{cases} \tag{4.6}$$

3. *If u_l with $l \in k + 2\mathbb{Z}_{\leq 0}$ is already defined, put*

$$u_{l-2} := \begin{cases} -\big(1+2\nu-l\big)^{-1}\, \mathrm{E}^{-}_{l}\, u_l & \text{if } 1+2\nu-l \neq 0 \quad \text{and} \\ 0 & \text{if } 1+2\nu-l = 0. \end{cases} \tag{4.7}$$

The so defined family $\big(u_l\big)_{l \in k+2\mathbb{Z}}$ is indeed a family of Maass cusp forms with reference weight k.

If $u \neq 0$, then $\big(u_l\big)_{l \in k+2\mathbb{Z}}$ is nontrivial, in the sense of Definition 4.1: there exists an l with $u_l \neq 0$.

Proof We have to show that the family $\big(u_l\big)_{l \in k+2\mathbb{Z}}$ constructed recursively by the algorithm presented in the lemma is indeed a family of Maass cusp forms in the sense of Definition 4.1. Corollary 3.41 shows that all u_l are Maass cusp forms of weight l. To check condition (4.5), consider $\mathrm{E}^{\pm}_{l\mp 2}\, u_{l\mp 2}$. If $\big(1+2\nu \mp l\big) \neq 0$, we have

$$\begin{aligned} \mathrm{E}^{\pm}_{l\mp 2}\, u_{l\mp 2} &\overset{\substack{(4.6)\\(4.7)}}{=} -\big(1+2\nu \mp l\big)^{-1}\, \mathrm{E}^{\pm}_{l\mp 2}\, \mathrm{E}^{\mp}_{l}\, u_l \qquad \text{by recursive construction} \\ &\overset{(4.2)}{=} -\big(1+2\nu \mp l\big)^{-1}\, \big(1+2\nu \pm (l \mp 2)\big)\big(1+2\nu \mp l\big)\, u_l \\ &= -\big(1+2\nu \pm (l \mp 2)\big)\, u_l. \end{aligned}$$

If $\big(1+2\nu \mp l\big) = 0$, we have

$$\begin{aligned} \mathrm{E}^{\pm}_{l\mp 2}\, u_{l\mp 2} &= \mathrm{E}^{\pm}_{l\mp 2}\, 0 \qquad \text{since } u_{l\mp 2} := 0 \text{ in the algorithm} \\ &= 0 \\ &= \underbrace{\big(1+2\nu \mp l\big)}_{=0}\, u_l. \end{aligned}$$

Hence $\big(u_l\big)_{l \in k+2\mathbb{Z}}$ is indeed a family of Maass cusp forms.

That $(u_l)_{l\in k+2\mathbb{Z}}$ is nontrivial if $u \neq 0$ follows directly from the first assignment $u_k := u$ in the algorithm. □

Example 4.6 For $k \in \mathbb{R}$ a real weight and v a multiplier system compatible with $(\Gamma(1), k)$, consider a classical cusp form $0 \neq F \in S_{k,v}(\Gamma)$. Recall that we discussed its embedding into the space $S^{\text{Maass}}_{k,v}\left(\Gamma(1), \pm\frac{1-k}{2}\right)$ via (3.56) in §3.4.1. The algorithm in Lemma 4.5 constructs a nontrivial family of Maass cusp forms $(u_l)_{l\in k+2\mathbb{Z}}$. Since $\mathrm{E}_k^- F = 0$, see Exercise 3.13, we have $u_l = 0$ for all $l < k$.

Consider the constants $1 + 2\nu + l$ with $l \geq k$. If we plug in $\nu = -\frac{1-k}{2}$, we have on one hand

$$1 + 2\nu + l = 1 - 2\frac{1-k}{2} + l = k + l$$

that vanishes for $-k = l$. On the other hand, with $\nu = +\frac{1-k}{2}$, we have

$$1 + 2\nu + l = 1 + 2\frac{1-k}{2} + l = 2 - k + l$$

that vanishes only if $l = k - 2$. This is never the case since we assumed $l \geq k$.

Using Lemma 4.5 now shows that in the case of $\nu = +\frac{1-k}{2}$, we have $u_l \neq 0$ for all $k \leq l$. The family $(u_l)_{l\in k+2\mathbb{Z}}$ is admissible.

The case of $\nu = -\frac{1-k}{2}$ is a bit more involved. If $k \notin \mathbb{Z}_{\leq 0}$, we have $1 + 2\nu + l = k + l \neq 0$ for all $l \in k + 2\mathbb{Z}_{\geq 0}$ and hence $u_l \neq 0$ for all $l \in k + 2\mathbb{Z}_{\geq 0}$. The family $(u_l)_{l\in k+2\mathbb{Z}}$ is admissible. If $k \in \mathbb{Z}_{\leq 0}$ we have $1 + 2\nu + l = k + l = 0$ for some $l \in k + 2\mathbb{Z}_{\geq 0}$ and hence $u_l \neq 0$ for all $l \in k + 2\mathbb{Z}_{\geq 0}$ with $l \leq -k$ and $u_l = 0$ for all $l \in k+2\mathbb{Z}_{\geq 0}$ with $l > -k$. The family $(u_l)_{l\in k+2\mathbb{Z}}$ is only admissible if $k \leq -2$ since the inequality $k \leq l \leq -k$ does not allow three successive l values with $u_l \neq 0$.

The results of this example are summarized in Table 4.1.

Recall that Maass cusp forms $u \in S^{\text{Maass}}_{k,v}(\Gamma(1), \nu)$ have a Fourier-Whittaker expansion of the form

$$u(z) = \sum_{\substack{n\in\mathbb{Z}\\ n+\kappa\neq 0}} A(n)\, \mathbb{W}_{\operatorname{sign}(n)\frac{k}{2},\nu}\big(4\pi\, |n+\kappa| \Im(z)\big)\, e^{2\pi i (n+\kappa)\Re(z)}$$

Table 4.1 $l_{\min}$ and $l_{\max}$ of the family of Maass cusp forms induced by the embedding of classical cusp forms in $S_{k,v}(\Gamma(1))$.

Spectral value ν	Weight k	$l_{\min}$	$l_{\max}$	Admissible?
$+\frac{1-k}{2}$	$k \in \mathbb{R}$	k	∞	Yes
$-\frac{1-k}{2}$	$k \notin \mathbb{Z}_{\leq 0}$	k	∞	Yes
$-\frac{1-k}{2}$	$k \in \mathbb{Z}_{\leq 0}$	k	$\max\{l \in k + 2\mathbb{Z}\colon\ l \leq -k\}$	for $k \leq -2$

with $0 \le \kappa < 1$, see, e.g., Equation (3.48). This leads to the question on how the Fourier-Whittaker expansions of the associated family $(u_l)_{l\in k+2\mathbb{Z}}$ behave. In particular, can we specify the common coefficients and address the action of the Maass operators $\mathrm{E}_l^{\pm}$ on u_l in (4.5)?

We modify the Whittaker function $\mathbb{W}_{\cdot,\cdot}(\cdot)$ defined in (3.26).

Definition 4.7 For $k \in \mathbb{R}$ and $\nu \in \mathbb{C}$, we define

$$\begin{aligned}\psi_{k,\nu}(y) :&= \frac{\mathbb{W}_{\frac{k}{2},\nu}(4\pi y)}{\Gamma\left(\frac{1}{2}+\nu+\frac{k}{2}\right)} \\ &\overset{(3.26)}{=} \frac{W_{\frac{k}{2},\nu}(4\pi y)}{\Gamma\left(\frac{1}{2}+\nu+\frac{k}{2}\right)}\end{aligned} \tag{4.8}$$

for all $y > 0$.

Lemma 4.8 *Let $k \in \mathbb{R}$ and $\nu \in \mathbb{C}$. The function $\psi_{k,\nu}(y)$ is analytic for all $y > 0$. We have that*

$$\psi_{k,\nu}(y) = 0 \qquad \text{for all } y > 0 \ \text{ if } \frac{1}{2} - \nu - \frac{k}{2} \in \mathbb{N}. \tag{4.9}$$

The function $\psi_{k,\nu}$ satisfies the differential equation

$$\psi''_{k,\nu}(y) + \left(-4\pi^2 + \frac{2\pi k}{y} + \frac{\frac{1}{4}-\nu^2}{y^2}\right)\psi_{k,\nu}(y) = 0 \tag{4.10}$$

and has the asymptotic growth

$$\psi_{k,\nu}(y) \sim \frac{e^{-2\pi y}(4\pi y)^{\frac{k}{2}}}{\Gamma\left(\frac{1}{2}+\nu+\frac{k}{2}\right)} \qquad \text{as } y \to \infty. \tag{4.11}$$

We have

$$\psi_{k,\nu}(y) = O\left(y^{\frac{1}{2}-|\Re(\nu)|}\log y\right) \qquad \text{as } y \searrow 0. \tag{4.12}$$

The function $\psi_{k,\nu}$ satisfies the recurrence relations

$$\begin{aligned} y\psi'_{k,\nu}(y) &= \left(\frac{k}{2} - 2\pi y\right)\psi_{k,\nu}(y) - \left(\frac{1}{2}+\nu-\frac{k}{2}\right)\psi_{k-2,\nu}(y) & (4.13)\\ &= \left(2\pi y - \frac{k}{2}\right)\psi_{k,\nu}(y) - \left(\frac{1}{2}+\nu+\frac{k}{2}\right)\psi_{k+2,\nu}(y). & (4.14)\end{aligned}$$

For $\Re(\nu) > 0$, we have the following integral representation:

$$\psi_{k,\nu}(y) = \pi^{-\frac{1}{2}-\nu} y^{\frac{1}{2}-\nu} e^{-\frac{\pi i k}{2}} \int_{\mathbb{R}} (\xi - i)^{-\frac{1}{2}-\nu-\frac{k}{2}} (\xi + i)^{-\frac{1}{2}-\nu+\frac{k}{2}} e^{2\pi i y \xi}\, \mathrm{d}\xi \tag{4.15}$$

for $y > 0$.

Proof All properties mentioned in this lemma are direct consequences of the properties of the Whittaker $W_{k,\nu}(y)$ function that decreases exponentially for $y \to \infty$. The function $W_{k,\nu}$ satisfies the differential equation

$$W''_{k,\nu}(y) + \left(-\frac{1}{4} + \frac{k}{y} + \frac{\frac{1}{4} - \nu^2}{y^2} \right) W_{k,\nu}(y) = 0 \tag{4.16}$$

which implies that $\psi_{k,\nu}(y)$ satisfies (4.10).

The Whittaker function has the asymptotic growth estimate

$$W_{k,\nu}(y) \sim e^{-\frac{y}{2}} y^k \qquad \text{for } y \to \infty \tag{4.17}$$

and satisfies the growth condition

$$W_{k,\nu}(y) = O\left(y^{\frac{1}{2}-|\Re(\nu)|} \log y \right) \qquad \text{for } y \searrow 0. \tag{4.18}$$

See [186, (4.1.22)] for the first estimate. The second estimate follows from the differential equation (4.16), or, alternatively, it can be found in [130, 6.8.1 and 7.2.3]. Hence, the function $\psi_{k,\nu}(y)$ satisfies the growth conditions given in the lemma.

The Whittaker function $W_{k,\nu}(s)$ also satisfies several recurrence relations, in particular the one stated in [186, (2.4.19-24)] or [130, 7.2.2]. We have for example [186, (2.4.21) and (2.4.24)]:

$$\begin{aligned} y W'_{k,\nu}(y) &= \left(k - \frac{1}{2} y \right) W_{k,\nu}(y) + \left(\frac{1}{2} + \nu - k \right) \left(\frac{1}{2} - \nu - k \right) W_{k-1,\nu}(y) \\ &= \left(\frac{1}{2} y - k \right) W_{k,\nu}(y) - W_{k+1,\nu}(y), \end{aligned}$$

which imply the recurrence formulas (4.13) and (4.14).

The Whittaker function has the following representations as a Pochammer integral and its variants

$$\frac{W_{k,\nu}(y)}{\Gamma(\frac{1}{2}+\nu+k)}$$

$$= \frac{-1}{2\pi i} e^{-\frac{1}{2}y} y^k \int (-t)^{-\frac{1}{2}-\nu-k} \left(1+\frac{t}{y}\right)^{-\frac{1}{2}-\nu+k} e^{-t}\, \mathrm{d}t \tag{4.19}$$

$$= \frac{1}{\pi}\left(\frac{y}{4}\right)^{\frac{1}{2}-\nu} e^{-\pi i k} \int (\tau - i)^{-\frac{1}{2}-\nu-k} (\tau+i)^{-\frac{1}{2}-\nu+k} e^{\frac{1}{2}iy\tau}\, \mathrm{d}\tau \tag{4.20}$$

$$= \frac{1}{\pi}\left(\frac{y}{4}\right)^{\frac{1}{2}-\nu} e^{-\pi i k} \int_{-\infty}^{\infty} (\xi - i)^{-\frac{1}{2}-\nu-k} (\xi+i)^{-\frac{1}{2}-\nu+k} e^{\frac{1}{2}iy\xi}\, \mathrm{d}\xi. \tag{4.21}$$

Slater states the integral (4.19) (with $-\nu$ instead of ν) in [186, (3.5.18)] only for $\frac{1}{2}-\nu-k \notin \mathbb{N}$. The equation (4.19) holds also for the excluded parameters since the integral in (4.19) is 0 in this case. The path of integration in (4.19) is over the left path of Figure 4.1 and in (4.20) is over the right path. We used the substitution $e^{-\frac{\pi i}{2}}(\tau) = \frac{2t}{y}+1$ from (4.19) to (4.20). The integral (4.21) is defined for $\Re(\nu) > 0$ and $\Im(y) = 0$ and the path of integration is the real axis. The integral representation for $\psi_{k,\nu}(y)$ follows directly from the integral expression in (4.21) and the definition of $\psi_{k,\nu}(y)$ in (4.8).

The Γ-function $\Gamma\left(\frac{1}{2}+\nu+\frac{k}{2}\right)$ has a simple pole exactly for $-\frac{1}{2}-\nu-\frac{k}{2} \in \mathbb{Z}_{\geq 0}$, see Appendix A.7. This implies that the function $\psi_{k,\nu}$ vanishes for $\frac{1}{2}-\nu-\frac{k}{2} \in \mathbb{N}$ since the just described Γ-function appears in the denominator of the defining equation (4.8) of ψ. □

Exercise 4.9 Show that the integral in (4.15) vanishes for $\frac{1}{2}-\nu-\frac{k}{2} \in \mathbb{N}$.

Remark 4.10 We would like to point out that the Pochammer integral in (4.19) is not invariant under $\nu \mapsto -\nu$. This implies directly that $\psi_{k,\nu}$ is also not invariant under this substitution. However, the W-Whitaker function $W_{\frac{k}{2},\nu}(y)$ is invariant under $\nu \mapsto -\nu$, see the discussion in §3.2.1. Hence, (4.8) implies

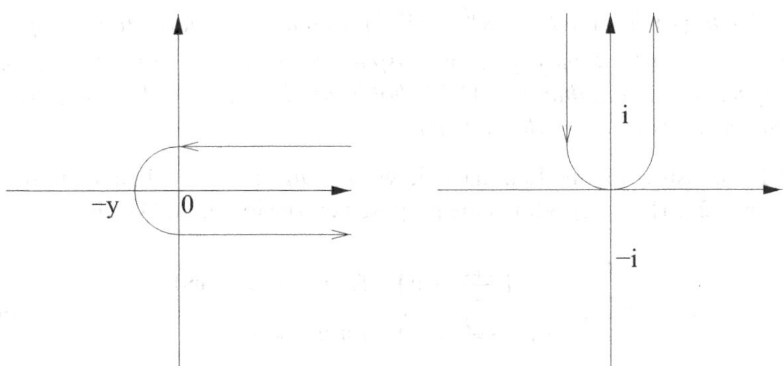

Fig. 4.1 Paths of integration in (4.19) (left) and (4.20) (right).

$$\Gamma\left(\frac{1}{2}+\nu+\frac{k}{2}\right)\,\psi_{k,\nu}(y)=\Gamma\left(\frac{1}{2}-\nu+\frac{k}{2}\right)\,\psi_{k,-\nu}(y). \tag{4.22}$$

Using the special function $\psi_{\cdot,\cdot}$ in the expansion (3.48), we obtain

$$u(z)=\sum_{\substack{n\in\mathbb{Z}\\ n+\kappa>0}}\tilde{A}(n)\,\mathbb{W}_{\operatorname{sign}(n)\frac{k}{2},\nu}\big(4\pi\,|n+\kappa|\,\Im(z)\big)\,e^{2\pi i(n+\kappa)\Re(z)} \tag{4.23}$$

of the Maass cusp form $u\in S^{\mathrm{Maass}}_{k,v}\big(\Gamma(1),\nu\big)$ (with coefficients $\tilde{A}(n)$). This gives us a proposed expansion of u_l of the form

$$\begin{aligned}u_l(z)&=\frac{1}{\Gamma\left(\frac{1+l}{2}+\nu\right)}\sum_{\substack{n\in\mathbb{Z}\\ n+\kappa>0}}A(n)\,\mathbb{W}_{\frac{l}{2},\nu}\big(4\pi(n+\kappa)\Im(z)\big)\,e^{2\pi i(n+\kappa)\Re(z)}+\\&\quad+\frac{1}{\Gamma\left(\frac{1-l}{2}+\nu\right)}\sum_{\substack{n\in\mathbb{Z}\\ n+\kappa<0}}A(n)\,\mathbb{W}_{-\frac{l}{2},\nu}\big(4\pi\,|n+\kappa|\,\Im(z)\big)\,e^{2\pi i(n+\kappa)\Re(z)}\\&=\sum_{\substack{n\in\mathbb{Z}\\ n+\kappa\neq 0}}A(n)\,\psi_{\operatorname{sign}(n)l,\nu}\big(\,|n+\kappa|\,\Im(z)\,\big)e^{2\pi i(n+\kappa)\Re(z)}\end{aligned} \tag{4.24}$$

for $z\in\mathbb{H}$ and $l\in k+2\mathbb{Z}$ with $0\le\kappa<1$ associated with the cusp $i\infty$ by (1.41). The coefficients $A(n)$ are given by

$$A(n):=\begin{cases}\tilde{A}(n)\,\Gamma\left(\frac{1+k}{2}+\nu\right) & \text{for } n+\kappa>0 \text{ and}\\ \tilde{A}(n)\,\Gamma\left(\frac{1-k}{2}+\nu\right) & \text{for } n+\kappa<0\end{cases} \tag{4.25}$$

with $\tilde{A}(n)$ are the coefficients of the expansion of u in (4.23).

Lemma 4.11 *For $k\in\mathbb{R}$ and $\nu\in\mathbb{C}$ with $\nu\pm\frac{k}{2}\notin-\frac{1}{2}\mathbb{N}$ and v a multiplier system compatible to $\big(\Gamma(1),k\big)$, let $u\in S^{Maass}_{k,v}\big(\Gamma(1),\nu\big)$ a nonvanishing Maass cusp form.*

If $\big(u_l\big)_{l\in k+2\mathbb{Z}}$ is the family of Maass cusp forms associated with u by Lemma 4.5 and the family is admissible, then (4.24) *holds for all $l_{\min}\le l\le l_{\max}$. In particular, the coefficients $A(n)$ do not depend on l.*

Proof By construction in Lemma 4.5, we put $u_k:=u$ with u admitting the expansion 4.23. Hence u_k admits the proposed expansion in (4.24) with

$$A(n):=\begin{cases}\tilde{A}(n)\,\Gamma\left(\frac{1+k}{2}+\nu\right) & \text{for } n+\kappa>0 \text{ and}\\ \tilde{A}(n)\,\Gamma\left(\frac{1-k}{2}+\nu\right) & \text{for } n+\kappa<0.\end{cases} \tag{4.25}$$

Now consider $l \geq k$. Lemma 4.5 shows that we define u_{l+2} by (4.6). In particular we have

$$u_{l+2} := -\left(1+2\nu+l\right)^{-1} \mathrm{E}_l^+ \, u_l$$

for $k \leq l < l_{\max}$. The assumption on k and ν imply that $1+2\nu+l \neq 0$ for all $l \in k+2\mathbb{Z}$ that lie in the admissible range: $l_{\min} \leq l \leq l_{\max}$. If u_l admits an expansion of the form (4.24), we apply Proposition 3.29 and find that $\mathrm{E}_l^+ \, u_l$ admits

$$\begin{aligned}
&\mathrm{E}_l^+ \, u_l \\
&\overset{(3.43)}{=} \frac{-2}{\Gamma\left(\frac{1+l}{2}+\nu\right)} \sum_{\substack{n\in\mathbb{Z}\\ n+\kappa>0}} A(n)\, \mathbb{W}_{\frac{l+2}{2},\nu}\big(4\pi(n+\kappa)\Im(z)\big)\, e^{2\pi i(n+\kappa)\Re(z)} + \\
&\quad + \frac{\frac{l(l+2)}{2}+\frac{1}{2}-2\nu^2}{\Gamma\left(\frac{1-l}{2}+\nu\right)} \sum_{\substack{n\in\mathbb{Z}\\ n+\kappa<0}} A(n)\, \mathbb{W}_{-\frac{l-2}{2},\nu}\big(4\pi\,|n+\kappa|\,\Im(z)\big)\, e^{2\pi i(n+\kappa)\Re(z)} \\
&= \frac{-2\left(\frac{1+l}{2}+\nu\right)}{\Gamma\left(\frac{1+(l+2)}{2}+\nu\right)} \sum_{\substack{n\in\mathbb{Z}\\ n+\kappa>0}} A(n)\, \mathbb{W}_{\frac{l+2}{2},\nu}\big(4\pi(n+\kappa)\Im(z)\big)\, e^{2\pi i(n+\kappa)\Re(z)} + \\
&\quad + \frac{\frac{l(l+2)}{2}+\frac{1}{2}-2\nu^2}{\left(\frac{1-(l+2)}{2}+\nu\right)\Gamma\left(\frac{1-(l+2)}{2}+\nu\right)} \cdot \\
&\qquad \cdot \sum_{\substack{n\in\mathbb{Z}\\ n+\kappa<0}} A(n)\, \mathbb{W}_{-\frac{l-2}{2},\nu}\big(4\pi\,|n+\kappa|\,\Im(z)\big)\, e^{2\pi i(n+\kappa)\Re(z)} \\
&= -\left(1+2\nu+l\right) \frac{1}{\Gamma\left(\frac{1+(l+2)}{2}+\nu\right)} \cdot \\
&\qquad \cdot \sum_{\substack{n\in\mathbb{Z}\\ n+\kappa>0}} A(n)\, \mathbb{W}_{\frac{l+2}{2},\nu}\big(4\pi(n+\kappa)\Im(z)\big)\, e^{2\pi i(n+\kappa)\Re(z)} - \\
&\quad - \left(1+2\nu+l\right) \frac{1}{\Gamma\left(\frac{1-(l+2)}{2}+\nu\right)} \cdot \\
&\qquad \cdot \sum_{\substack{n\in\mathbb{Z}\\ n+\kappa<0}} A(n)\, \mathbb{W}_{-\frac{l-2}{2},\nu}\big(4\pi\,|n+\kappa|\,\Im(z)\big)\, e^{2\pi i(n+\kappa)\Re(z)} \\
&\overset{(4.8)}{=} -\left(1+2\nu+l\right) \sum_{\substack{n\in\mathbb{Z}\\ n+\kappa\neq 0}} A(n)\, \psi_{\mathrm{sign}(n)(l+2),\nu}\big(\,|n+\kappa|\,\Im(z)\,\big) e^{2\pi i n\Re(z)},
\end{aligned}$$

where we used the identity $x\Gamma(x) = \Gamma(x+1)$ for Γ-functions and

$$-2\left(\frac{l(l+2)}{2} + 2\left(\frac{1}{4} - \nu^2\right)\right) = \big(1 + 2\nu - (l+2)\big)\big(1 + 2\nu + l\big)$$

that was already calculated in (4.2). The left defining relation in (4.5) for families of Maass cusp forms shows that u_{l+2} indeed admits the expansion

$$u_{l+2} = \sum_{\substack{n\in\mathbb{Z}\\ n+\kappa\neq 0}} A(n)\, \psi_{\mathrm{sign}(n)(l+2),\nu}\big(|n+\kappa|\Im(z)\big)e^{2\pi i n \Re(z)}.$$

Similarly, we calculate the expansion of u_l for $l_{\min} \leq l \leq k$. We leave this to the reader, see Exercise 4.12 below. □

Exercise 4.12 Calculate the missing case in the proof of Lemma 4.11.

Remark 4.13 Lemma 4.11 above shows in particular that every admissible family of Maass cusp forms $(u_l)_{l\in k+2\mathbb{Z}}$ admits an expansion of the form (4.24) in its admissible range: (4.24) is valid for all $l_{\min} \leq l \leq l_{\max}$. The coefficients $A(n)$ in the expansion depend only on the family, the reference weight k in the admissible range, the parameter $0 \leq \kappa < 1$, and the spectral parameter ν. They do not however depend on the particular weight l. Moreover the reference weight k and the parameter $0 \leq \kappa < 1$ satisfy

$$\frac{k}{12} \equiv \kappa \bmod 1 \iff k \in 12\kappa + 12\mathbb{Z}. \tag{4.26}$$

Exercise 4.14 Let $(u_l)_{l\in k+2\mathbb{Z}}$ be an admissible family of Maass cusp forms with parameter $0 \leq \kappa < 1$, reference weight k, and spectral parameter ν. Show that k and κ satisfy (4.26) and prove that the multiplier v is already given by (1.38).

Example 4.15 Let us compute an example. Assume that $k > 0$ is given. We assign $\nu = \frac{1-k}{2}$ and $0 < \kappa < 1$ with $\frac{k}{12} - \kappa \in \mathbb{Z}$ (i.e., $\frac{k}{12} \equiv \kappa \pmod 1$). (The assignment $\nu = \frac{1-k}{2}$ was derived in Example 4.6 for modular cusp forms that are lifted to families of Maass cusp forms.)

Let η^{2k} be the cusp form and $u_{\eta^{2k}}$ the Maass cusp form discussed in §3.4.3. The Whittaker-Fourier expansion of $u_{\eta^{2k}}$ is given by (3.71)

$$u_{\eta^{2k}}(z) \overset{(3.71)}{=} \sum_{n=0}^{\infty} A_k(n) \left(4\pi\left(n + \frac{k}{12}\right)\right)^{-\frac{k}{2}} \cdot \mathrm{W}_{\frac{k}{2},\pm\frac{k-1}{2}}\left(4\pi\left(n+\frac{k}{12}\right)y\right) e^{2\pi i\left(n+\frac{k}{12}\right)x}.$$

Let denote the associated family to $u_{\eta^{2k}}$ by $(u_l)_{l\in k+e\mathbb{Z}}$ where the u_l are given by Lemma 4.5. We have in particular

$$u_l = 0 \quad \text{for all } l \in k+2\mathbb{Z}, l < k$$

since $1+2\nu+(k-2) = 1+2\frac{1-k}{2}+k-2 = 0$. The admissible range is $l_{\min} = k$ and $l_{\max} = \infty$.

Equations (4.23)–(4.25) and Lemma 4.11 show that u_l has the expansion

$$\begin{aligned} u_l(z) &= \sum_{\substack{n\in\mathbb{Z}\\ n+\frac{k}{12}>0}} A_k(n)\, \underbrace{\Gamma\left(\frac{1+k}{2}+\frac{1-k}{2}\right)}_{=\Gamma(1)=1} \left(4\pi\left(n+\frac{k}{12}\right)\right)^{-\frac{k}{2}} \cdot \\ &\quad \cdot \psi_{l,\frac{1-k}{2}}\left(\left(n+\frac{k}{12}\right)\Im(z)\right) e^{2\pi i\left(n+\frac{k}{12}\right)\Re(z)} \qquad (4.27) \\ &= \sum_{n=0}^{\infty} A_k(n) \left(4\pi\left(n+\frac{k}{12}\right)\right)^{-\frac{k}{2}} \psi_{l,\frac{1-k}{2}}\left(\left(n+\frac{k}{12}\right)\Im(z)\right) e^{2\pi i\left(n+\frac{k}{12}\right)\Re(z)}. \end{aligned}$$

4.2 L-Functions Associated with Maass Cusp Forms

This subsection describes the connection between Maass cusp forms and L-series. Hans Maass connected Maass cusp forms of real weight k, multiplier system v, and spectral parameter ν on the full modular group $\Gamma(1)$ to a pair of L-series. We refer to [128] for the case of vanishing weight $k = 0$ and trivial multiplier $v = 1$ and to [129, Chapter V] for real weight k and multiplier v compatible to $(\Gamma(1), k)$. He generalized the classical bijection between holomorphic modular forms and one L-series, see Corollary 1.217 in §1.7.2. The Mellin transforms of the W-Whittaker functions lead to complicated factors expressible in generalized hypergeometric ${}_2F_1$ functions replacing the simple Γ-factors appearing in Corollary 1.217. (Some details on generalized hypergeometric functions can be found in Appendix A.10.)

In §4.2.1, we present Maass' result [129, p. 225, Theorem 35] in a slightly modified version to reflect the underlying structure of our Maass cusp forms. Recently, V. A. Bykovskii published in [54] a result that simplifies the hypergeometric ${}_2F_1$ functions found by H. Maass to factors containing only Γ-functions.

Our own result, which is presented in §4.2.1 and §4.2.2, is based on the families of automorphic forms. The attached L-series satisfy a vector-valued functional equation. The involved factors are just simple trigonometric functions. It can also be derived from V. A. Bykovskii's [54, Theorem 2], but we think that our proof better reflects the extra structure encoded in the families of automorphic forms.

The next paragraph, §4.2.3, connects families of cusp forms with two-sided series of the form $\sum_{\substack{n\in\mathbb{Z}\\ n+\kappa\neq 0}} A_n\big(\pm i(n+\kappa)\big)^{-s}$. It will turn out in § 4.3 and §5.1.1 that this representation is useful in the context of connecting families of Maass cusp forms and L-series with period functions.

4.2.1 L-Functions for $S_{k,v}\big(\Gamma(1), v\big)$

As already mentioned, Maass derived a correspondence between individual Maass cusp forms and pairs of L-series in [129, p. 225, Theorem 35]. However, before we state and prove Maass' theorem on L-series we need to discuss the Mellin transform of the special function ψ.

Let $k \in \mathbb{R}$ and $\nu \in \mathbb{C}$ through this section. We denote the Mellin transform of ψ by

$$\begin{aligned} \tilde{\psi}_{k,\nu}(s) &:= \mathcal{M}\big(\psi_{k,\nu}\big)(s) \\ &\overset{(1.154)}{\underset{(4.8)}{=}} \int_0^\infty \psi_{k,\nu}(y)\, y^{s-1}\, dy \end{aligned} \tag{4.28}$$

for $\Re(s) > |\Re(\nu)| - \frac{1}{2}$. The growth estimate (4.12) for $\psi_{k,\nu}$ in Lemma 4.8 imply that the Mellin integral transform is well defined and holomorphic in s in that region.

Lemma 4.16 *Let $k \in \mathbb{R}$ and $\nu \in \mathbb{C}$. The function $\tilde{\psi}_{k,\nu}(s)$ defined in* (4.28) *extends meromorphically to the complex plane with poles at most in $s \in \frac{1}{2} \pm \nu + \mathbb{Z}_{\leq 0}$. The orders of those poles are at most 2. If $\nu \notin \frac{1}{2}\mathbb{Z}$, then the poles are simple.*

The function $\tilde{\psi}_{k,\nu}(s)$ satisfies the recurrence formula

$$\left(\frac{1}{2}+s-\nu\right)\left(\frac{1}{2}+s+\nu\right)\tilde{\psi}_{k,\nu}(s) = 4\pi^2\tilde{\psi}_{k,\nu}(s+2) - 2\pi k\tilde{\psi}_{k,\nu}(s+1). \tag{4.29}$$

Furthermore, $\psi_{k,\nu}$ has the representation

$$\tilde{\psi}_{k,\nu}(s) = 2^{\frac{k}{2}-s-1}\pi^{-s}\frac{\Gamma\big(\frac{1}{2}+s+\nu\big)\Gamma\big(\frac{1}{2}+s-\nu\big)}{\Gamma\big(\frac{1}{2}+\nu+\frac{k}{2}\big)\Gamma\big(1+s-\frac{k}{2}\big)}\,{}_2F_1\left[\begin{matrix}\frac{1}{2}+\nu-\frac{k}{2},\ \frac{1}{2}-\nu-\frac{k}{2}\\ 1+s-\frac{k}{2}\end{matrix}\middle|\frac{1}{2}\right]. \tag{4.30}$$

The function $\tilde{\psi}$ vanishes for special choices of k. We have

$$\tilde{\psi}_{-1-2\nu-2N,\nu}(s) = 0 \tag{4.31}$$

for all $s \in \mathbb{C}$ and $N \in \mathbb{Z}_{\geq 0}$.

We also have the identity

$$\mathcal{M}\left(y\psi'_{k,\nu}(y)\right)(s) = -s\,\tilde{\psi}_{k,\nu}(s) \tag{4.32}$$

for $\Re(s) > |\Re(\nu)| - \frac{1}{2}$.

Proof We show all the statements separately.

1. The growth estimates for $\psi_{k,\nu}(y)$ in (4.11) and (4.12) of Lemma 4.8 show that $\psi_{k,\nu}(y)$ satisfies the growth estimates

$$\begin{aligned} \psi_{k,\nu}(y) &= O\left(y^C e^{-2\pi y}\right) && \text{as } y \to \infty \text{ for some } C \in \mathbb{R} \text{ and} \\ \psi_{k,\nu}(y) &= O\left(y^{\frac{1}{2}-|\Re(\nu)|}\log y\right) && \text{as } y \searrow 0. \end{aligned} \tag{4.33}$$

   Applying Lemma 1.201, we see that the Mellin transform

$$\tilde{\psi}_{k,\nu}(s) = \mathcal{M}(\psi_{k,\nu})(s)$$

   in (4.28) is well defined for $\Re(s) > |\Re(\nu)| - \frac{1}{2}$.

2. Let $\Re(s) > |\Re(\nu)| - \frac{1}{2}$. The differential equation (4.10) together with (4.13), (4.14), and the growth estimates for $\psi_{k,\nu}(y)$ in (4.11) and (4.12) of Lemma 4.8 show that $y\psi'_{k,\nu}(y)$ and $y^2\psi''_{k,\nu}(y)$ are of order $O\left(y^C e^{-2\pi y}\right)$ as $y \to \infty$ for some $C \in \mathbb{R}$, and of order $O\left(y^{\frac{1}{2}-|\Re(\nu)|}\log y\right)$ as $y \searrow 0$. Using the growth estimates and applying Lemma 1.201 as above shows that the Mellin transforms

$$\mathcal{M}\left(y\psi'_{k,\nu}(y)\right)(s) \quad \text{and} \quad \mathcal{M}\left(y^2\psi''_{k,\nu}(y)\right)(s)$$

   of $y\psi'_{k,\nu}(y)$ and $y^2\psi''_{k,\nu}(y)$ are well defined for $\Re(s) > |\Re(\nu)| - \frac{1}{2}$.

   Also the growth estimate implies that the following use of integration by parts is valid: Using partial integration twice and the differential equation (4.10) once, we see that

$$\begin{aligned} \tilde{\psi}_{k,\nu}(s) &\overset{(4.28)}{=} \int_0^\infty \psi_{k,\nu}(y)\, y^{s-1}\, dy \\ &= \left[\psi_{k,\nu}(y)\frac{y^s}{s}\right]_{y=0}^{y=\infty} - \frac{1}{s}\int_0^\infty \psi'_{k,\nu}(y)\, y^s\, dy \\ &= \frac{1}{s(s+1)}\int_0^\infty \psi''_{k,\nu}(y)\, y^{s+1}\, dy \\ &\overset{(4.10)}{=} \frac{1}{s(s+1)}\int_0^\infty \left(4\pi^2 - \frac{2\pi k}{y} - \frac{\frac{1}{4}-\nu^2}{y^2}\right)\psi_{k,\nu}(y)\, y^{s+1}\, dy \\ &\overset{(4.28)}{=} \frac{4\pi^2}{s(s+1)}\tilde{\psi}_{k,\nu}(s+2) - \frac{2\pi k}{s(s+1)}\tilde{\psi}_{k,\nu}(s+1) - \frac{\frac{1}{4}-\nu^2}{s(s+1)}\tilde{\psi}_{k,\nu}(s). \end{aligned} \tag{4.34}$$

Hence $\tilde{\psi}_{k,\nu}(s)$ satisfies the recurrence formula (4.29) for $\Re(s) > |\Re(\nu)| - \frac{1}{2}$. In particular (4.34) shows (4.32).

The meromorphic continuation of $\tilde{\psi}_{k,\nu}(s)$ to $s \in \mathbb{C}$ follows immediately from the recurrence formula (4.29). The iterative use of the recurrence formula can lead to singularities due to the factor

$$\left(\frac{1}{2} + s - \nu\right)\left(\frac{1}{2} + s + \nu\right);$$

the order of the singularities is at most two for $\nu \in \frac{1}{2}\mathbb{Z}$ and at most one otherwise.

3. Let $\Re(s) > |\Re(\nu)| - \frac{1}{2}$ and $\Re\left(\frac{1}{2} - \nu - \frac{k}{2}\right) > 0$. We start from the integral representation

$$\Gamma\left(\frac{1}{2} - \nu - \frac{k}{2}\right) W_{\frac{k}{2},-\nu}(y) = y^{\frac{1}{2}-\nu} 2^{2\nu} \cdot \\ \cdot \int_1^\infty e^{-\frac{1}{2}yt} (t-1)^{-\frac{1}{2}-\nu-\frac{k}{2}} (t+1)^{-\frac{1}{2}-\nu+\frac{k}{2}}\, dt$$

for $y > 0$, see [186, (3.5.12)] or Appendix A.9. (Slater states the formula only under the condition $\Re\left(\frac{1}{2} - \nu \pm k\right) > 0$; but it is obvious that the integral converges also under the slightly relaxed condition given here.) Taking the Mellin transform of $\psi_{k,\nu}(y) \overset{(4.8)}{=} \frac{W_{\frac{k}{2},-\nu}(4\pi y)}{\Gamma\left(\frac{1}{2}+\nu+\frac{k}{2}\right)}$ and interchanging the integrals – this is allowed under the condition $\Re(s) > |\Re(\nu)|$ – we get

$$\tilde{\psi}_{k,\nu}(s) = \frac{2^{\frac{1}{2}+\nu-s}\pi^{-s}\Gamma\left(\frac{1}{2}+s-\nu\right)}{\Gamma\left(\frac{1}{2}+\nu+\frac{k}{2}\right)\Gamma\left(\frac{1}{2}-\nu-\frac{k}{2}\right)} \int_1^\infty (t-1)^{-\frac{1}{2}-\nu-\frac{k}{2}} (t+1)^{-\frac{1}{2}-\nu+\frac{k}{2}} t^{\nu-s-\frac{1}{2}}\, dt.$$

Substituting $t = \frac{1}{1-v}$, we obtain

$$\tilde{\psi}_{k,\nu}(s) = \frac{2^{-1-s+\frac{k}{2}}\pi^{-s}\Gamma\left(\frac{1}{2}+s-\nu\right)}{\Gamma\left(\frac{1}{2}+\nu+\frac{k}{2}\right)\Gamma\left(\frac{1}{2}-\nu-\frac{k}{2}\right)} \int_0^1 v^{-\frac{1}{2}-\nu-\frac{k}{2}} (1-v)^{s+\nu-\frac{1}{2}} \left(1-\frac{v}{2}\right)^{-\frac{1}{2}-\nu+\frac{k}{2}} dv.$$

Using [89, 90, (9.111), (8.384.1)], we replace the integral by a hypergeometric function. Hence,

$$\tilde{\psi}_{k,\nu}(s) = 2^{\frac{k}{2}-s-1}\pi^{-s} \frac{\Gamma\left(\frac{1}{2}+s+\nu\right)\Gamma\left(\frac{1}{2}+s-\nu\right)}{\Gamma\left(\frac{1}{2}+\nu+\frac{k}{2}\right)\Gamma\left(1+s-\frac{k}{2}\right)} {}_2F_1\left[\begin{matrix}\frac{1}{2}+\nu-\frac{k}{2}, \frac{1}{2}-\nu-\frac{k}{2} \\ 1+s-\frac{k}{2}\end{matrix} \middle| \frac{1}{2}\right].$$

Since the left and the right sides of the last equation are meromorphic in s, we can extend the formula. The singularities of the hypergeometric function in $1+$

$s - \frac{k}{2} \in \mathbb{Z}_{\leq 0}$ are cancelled by the poles of the gamma function $\Gamma\left(1+s-\frac{k}{2}\right)$ in $1+s-\frac{k}{2} \in \mathbb{Z}_{\leq 0}$.

4. The function $\psi_{-1-2\nu-2N,\nu}(y) = 0$ for all $y > 0$ and $N \in \mathbb{Z}_{\geq 0}$, see (4.9). Hence its Mellin transform $\tilde{\psi}_{-1-2\nu-2N,\nu}$ vanishes, too.

□

Remark 4.17 Using the standard identities for hypergeometric functions, we can rewrite the representation of $\tilde{\psi}_{k,\nu}(s)$. We have

$$\begin{aligned}\tilde{\psi}_{k,\nu}(s) &= 2^{\frac{1}{2}-\nu-s}\pi^{-s}\frac{\Gamma\left(\frac{1}{2}+s+\nu\right)\Gamma\left(\frac{1}{2}+s-\nu\right)}{\Gamma\left(\frac{1}{2}+\nu+\frac{k}{2}\right)\Gamma\left(1+s-\frac{k}{2}\right)}\,{}_2F_1\left[\begin{matrix}\frac{1}{2}-\nu-\frac{k}{2},\frac{1}{2}+s-\nu\\ 1+s-\frac{k}{2}\end{matrix}\middle|-1\right]\\ &= (4\pi)^{-s}\frac{\Gamma\left(\frac{1}{2}+s+\nu\right)\Gamma\left(\frac{1}{2}+s-\nu\right)}{\Gamma\left(\frac{1}{2}+\nu+\frac{k}{2}\right)\Gamma\left(1+s-\frac{k}{2}\right)}\,{}_2F_1\left[\begin{matrix}\frac{1}{2}+s-\nu,\frac{1}{2}+s+\nu\\ 1+s-\frac{k}{2}\end{matrix}\middle|\frac{1}{2}\right]\end{aligned} \tag{4.35}$$

for $s \in \mathbb{C}$, see [186, (3.6.11,12)] or use [89, 90, (9.131.1)] on the representation given in Lemma 4.16.

The singularities of the hypergeometric function in the numerator and those of the Γ-factors in the denominator cancel each other.

Replacing the hypergeometric function by its power series in the last representation gives

$$\tilde{\psi}_{k,\nu}(s) = \frac{(4\pi)^{-s}}{\Gamma\left(\frac{1}{2}+\nu+\frac{1}{2}k\right)}\sum_{n=0}^{\infty}\frac{\Gamma\left(\frac{1}{2}+s+n+\nu\right)\Gamma\left(\frac{1}{2}+s+n-\nu\right)}{n!\,2^n\,\Gamma\left(1+s+n-\frac{k}{2}\right)}. \tag{4.36}$$

Replacing the hypergeometric function ${}_2F_1$ in (4.30) by its power series

$${}_2F_1\left[\begin{matrix}a,b\\ c\end{matrix}\middle|z\right] = \sum_{n=0}^{\infty}\frac{(a)_n\,(b)_n\,z^n}{(c)_n\,n!},$$

see Appendix A.10, we obtain

$$\tilde{\psi}_{k,\nu}(s) = 2^{\frac{k}{2}-s}\pi^{-s}\frac{\Gamma\left(\frac{1}{2}+s+\nu\right)\Gamma\left(\frac{1}{2}+s-\nu\right)}{\Gamma\left(\frac{1}{2}+\nu+\frac{k}{2}\right)}\sum_{l=0}^{\infty}\frac{\left(\frac{1}{2}+\nu-\frac{k}{2}\right)_l\left(\frac{1}{2}-\nu-\frac{k}{2}\right)_l}{2^l\,l!\,\Gamma\left(1+s+l-\frac{k}{2}\right)},$$

where the *Pochammer symbol* is defined as

$$(\alpha)_l = \frac{\Gamma(\alpha+l)}{\Gamma(\alpha)}.$$

We have the special cases

$$\tilde{\psi}_{0,\nu}(s) = \frac{1}{2\pi^{s+\frac{1}{2}}\Gamma\left(\frac{1}{2}+\nu\right)}\Gamma\left(\frac{\frac{1}{2}+s-\nu}{2}\right)\Gamma\left(\frac{\frac{1}{2}+s+\nu}{2}\right) \quad \text{and} \tag{4.37}$$

$$\tilde{\psi}_{\frac{1}{2},\nu}(s) = \frac{1}{2\pi^{s+\frac{1}{2}}\Gamma(1+\nu)}\left[\Gamma\left(\frac{s+\frac{3}{2}-\nu}{2}\right)\Gamma\left(\frac{\frac{s+1}{2}+s+\nu}{2}\right)+\Gamma\left(\frac{\frac{1}{2}+s-\nu}{2}\right)\Gamma\left(\frac{s+\frac{3}{2}+\nu}{2}\right)\right].$$

We used (4.35), [89, 90, (9.122.1), (9.133)] and identities of the Γ function for the first identity. The second identity follows from (4.35), [89, 90, (9.122.1), (9.133), (9.137.4)], and also identities of the Γ function.

Remark 4.18 In view of the period functions, which we will introduce in §5.1, we mention that the factor $\Gamma\left(\frac{\frac{1}{2}+s-\nu}{2}\right)\Gamma\left(\frac{\frac{1}{2}+s+\nu}{2}\right)$ in (4.37) appeared already in [122, Theorem I.1].

Next, we give yet another integral representation of $\tilde{\psi}_{k,\nu}$ which we will use later in Lemma 4.37. The idea for this argument is due to Motohashi who used the statements of Lemma 4.20 and 4.37 together in the [146, proof of Theorem 2].

Lemma 4.19 *Let $\zeta \in \mathbb{C}$ with $\Re(\zeta) > 0$. The Γ-function $\Gamma(s)$ defined in* (1.155) *admits the integral representation*

$$\Gamma(s) = \int_{\zeta\mathbb{R}_{>0}} e^{-z}\, z^{s-1}\, dz \qquad (\Re(s) > 0). \tag{4.38}$$

The path of integration $\zeta\mathbb{R}_{>0} = \{r\zeta\colon\ r > 0\}$ is the rotated half-line starting at the origin and running through ζ towards infinity.

Proof Let $\zeta \in \mathbb{C}$ be a complex number with positive real part $\Re(\zeta) > 0$ and assume $\Re(s) \geq 1$. For given $R > 0$, consider the triangle D_R given by the origin 0, the point R on the positive real axis, and the point $\zeta_R \in \zeta\mathbb{R}_{\geq 0}$ that is uniquely defined by $\Re(\zeta_R) = R$. The triangle is contained in the closed right half-plane: $D_R \subset \{z \in \mathbb{C}\colon\ \Re(z) > 0\}$. Its boundary δD_R can be decomposed into the three line segment $[0, R] \subset \mathbb{R}_{\geq 0}$ on the real line, $L_R := \{z \in \zeta\mathbb{R}_{\geq 0};\ \Re(z) \leq R\}$ and the vertical line segment V_R. We have

$$\delta D_R = [0, R] \cup L_R \cup V_R.$$

The integrand $e^{-z}\, z^{s-1}$ in (4.38) is a holomorphic function on the right half-plane $\{z \in \mathbb{C}\colon\ \Re(z) \geq 0\}$ (since the integrand has no singularity at $z = 0$ for $\Re(s) \geq 1$). Hence Cauchy's integral theorem, see Appendix A.5, implies that the integral

$$\int_{D_R} e^{-z}\, z^{s-1}\, dz = 0.$$

Decomposing D_R into its three line segments and applying the linearity of the integral show

$$\begin{aligned} 0 &= \int_{D_R} e^{-z} z^{s-1}\, \mathrm{d}z \\ &= \int_{[0,R]} e^{-z} z^{s-1}\, \mathrm{d}z + \int_{L_R} e^{-z} z^{s-1}\, \mathrm{d}z + \int_{V_R} e^{-z} z^{s-1}\, \mathrm{d}z. \end{aligned}$$

Here the direction of the integration over L_R is from the common point with V_R towards the origin 0. Taking the limit $R \to \infty$, we see that

$$\lim_{R\to\infty} \int_{V_R} e^{-z} z^{s-1}\, \mathrm{d}z = 0.$$

(The integrand is absolutely bounded by

$$e^{-z} z^{s-1} = O\left(e^{-R} |z|^{\Re(s)-1}\right) = O\left(e^{\delta-R}\right) \qquad \text{as } R \to \infty$$

for some $\delta > 0$.) This bound decays faster than the length of the path or line segment V_R, which is of $O\left(R^2\right)$ by Pythagoras theorem. Hence, we have

$$\begin{aligned} 0 &= \int_{[0,R]} e^{-z} z^{s-1}\, \mathrm{d}z + \int_{L_R} e^{-z} z^{s-1}\, \mathrm{d}z + \underbrace{\int_{V_R} e^{-z} z^{s-1}\, \mathrm{d}z}_{\to 0} \\ &\longrightarrow \int_{[0,\infty)} e^{-z} z^{s-1}\, \mathrm{d}z + \int_{L_\infty} e^{-z} z^{s-1}\, \mathrm{d}z \qquad \text{as } R \to \infty, \end{aligned}$$

where $L_\infty = \zeta\mathbb{R}_{\geq 0}$ as set but the path of integration is the line coming from infinity towards the origin 0 via the point ζ. Recall that we have

$$\int_{[0,\infty)} e^{-z} z^{s-1}\, \mathrm{d}z \overset{(1.155)}{=} \Gamma(s)$$

and

$$\int_{L_\infty} e^{-z} z^{s-1}\, \mathrm{d}z = -\int_{\zeta\mathbb{R}_{\geq 0}} e^{-z} z^{s-1}\, \mathrm{d}z,$$

where the paths of integration L_∞ and over $\zeta\mathbb{R}_{\geq 0}$ change the orientation: L_∞ is from infinity towards the origin 0 via the point ζ and $\zeta\mathbb{R}_{>0} = \left\{r\xi:\ r > 0\right\}$ is the rotated half-line starting at the origin 0 and running through ζ towards infinity.

Summarizing, we have shown

$$\Gamma(s) = \int_{\zeta \mathbb{R}_{\geq 0}} e^{-z}\, z^{s-1}\, \mathrm{d}z$$

for all $\Re(s) \geq 1$. Of course, we can remove the origin from the path of integration. This leads to

$$\Gamma(s) = \int_{\zeta \mathbb{R}_{> 0}} e^{-z}\, z^{s-1}\, \mathrm{d}z$$

which is valid (by analytic continuation) for all $\Re(s) > 0$. □

Lemma 4.20 *Let $k \in \mathbb{R}$ and $\Re(\nu) > -\frac{1}{2}$. The function $\tilde{\psi}_{k,\nu}$ has the integral representation*

$$\begin{aligned} \tilde{\psi}_{k,\nu}(s) = {} & 2^{\nu-s-\frac{1}{2}}\pi^{-s-1}e^{-\frac{\pi i k}{2}}\,\Gamma\left(\frac{1}{2}+s-\nu\right)\cdot \\ & \cdot \int_{\mathbb{R}} (-i\xi)^{\nu-s-\frac{1}{2}}(\xi-i)^{-\frac{1}{2}-\nu-\frac{k}{2}}(\xi+i)^{-\frac{1}{2}-\nu+\frac{k}{2}}\,\mathrm{d}\xi \end{aligned} \tag{4.39}$$

for s in the vertical strip $-\Re(\nu) - \frac{1}{2} < \Re(s) < \Re(\nu) + \frac{1}{2}$.

Proof Assume the stronger assumptions $\Re(\nu) > 0$ and $\Re(\nu) - \frac{1}{2} < \Re(s) < \Re(\nu) + \frac{1}{2}$.

Consider the integral representation (4.15) of $\psi_{k,\nu}$. Then the Mellin transform $\tilde{\psi}_{k,\nu}$ of $\psi_{k,\nu}$ can be written as

$$\begin{aligned} \tilde{\psi}_{k,\nu}(s) &\overset{(4.28)}{=} \int_0^\infty \psi_{k,\nu}(y)\, y^{s-1}\, \mathrm{d}y \\ &\overset{(4.15)}{=} \pi^{-\frac{1}{2}-\nu}e^{-\frac{\pi i k}{2}} \int_0^\infty y^{s-\nu-\frac{1}{2}}\cdot \\ &\qquad \cdot \int_{\mathbb{R}} (\xi-i)^{-\frac{1}{2}-\nu-\frac{k}{2}}(\xi+i)^{-\frac{1}{2}-\nu+\frac{k}{2}}e^{2\pi i y\xi}\,\mathrm{d}\xi\,\mathrm{d}y. \end{aligned}$$

We move the path of the inner integral from $\mathbb{R}$ to $\mathbb{R} + \frac{1}{2}i$. Then the double integral converges absolutely due to the exponential decaying factor $\left|e^{2\pi i y\xi}\right| = e^{-\pi y}$ with $\Im(\xi) = \frac{1}{2}$. Interchanging the order of integration gives

$$\begin{aligned} &\tilde{\psi}_{k,\nu}(s) \\ &= \pi^{-\frac{1}{2}-\nu}e^{-\frac{\pi i k}{2}} \int_0^\infty y^{s-\nu-\frac{1}{2}} \int_{\mathbb{R}+\frac{i}{2}} (\xi-i)^{-\frac{1}{2}-\nu-\frac{k}{2}}(\xi+i)^{-\frac{1}{2}-\nu+\frac{k}{2}}e^{2\pi i y\xi}\,\mathrm{d}\xi\,\mathrm{d}y \\ &= \pi^{-\frac{1}{2}-\nu}e^{-\frac{\pi i k}{2}} \int_{\mathbb{R}+\frac{i}{2}} (\xi-i)^{-\frac{1}{2}-\nu-\frac{k}{2}}(\xi+i)^{-\frac{1}{2}-\nu+\frac{k}{2}} \int_0^\infty e^{2\pi i y\xi}\, y^{s-\nu-\frac{1}{2}}\,\mathrm{d}y\,\mathrm{d}\xi. \end{aligned}$$

Consider now the inner integral $\int_0^\infty e^{2\pi iy\xi}\, y^{s-\nu-\frac{1}{2}}\, \mathrm{d}y$ with $\xi \in \mathbb{R} + \frac{1}{2}i$. We would like to apply Lemma 4.19. Using the substitution $-t = 2\pi i\xi\, y$, we find

$$\begin{aligned}\int_0^\infty e^{2\pi iy\xi}\, y^{s-\nu-\frac{1}{2}}\, \mathrm{d}y &\overset{-t=2\pi i\xi\, y}{=} \left(-2\pi i\xi\right)^{\nu-s-\frac{1}{2}} \int_{(-2\pi i\xi)\mathbb{R}_{>0}} e^{-t}\, t^{s-\nu-\frac{1}{2}}\, \mathrm{d}t \\ &\overset{(4.38)}{=} \left(-2\pi i\xi\right)^{\nu-s-\frac{1}{2}}\, \Gamma\left(\frac{1}{2}+s-\nu\right),\end{aligned}$$

where we used $(-2\pi i\xi)\mathbb{R}_{>0}$ for the integration path from the origin 0 via $-2\pi i\xi$ towards infinity as introduced in Lemma 4.19. We still have to check whether $\Re\,(-2\pi i\xi) > 0$ which is an assumption used in Lemma 4.19: Using $\xi \in \mathbb{R} + \frac{i}{2}$, we find

$$\Re\,(-2\pi i\xi) = \Re\left(-2\pi i\,\frac{i}{2}\right) = \pi > 0.$$

Putting into the double integral, we get

$$\begin{aligned}&\tilde{\psi}_{k,\nu}(s) \\ &\quad= \pi^{-\frac{1}{2}-\nu} e^{-\frac{\pi ik}{2}} \int_0^\infty y^{s-\nu-\frac{1}{2}} \int_{\mathbb{R}} (\xi-i)^{-\frac{1}{2}-\nu-\frac{k}{2}} (\xi+i)^{-\frac{1}{2}-\nu+\frac{k}{2}} e^{2\pi iy\xi}\, \mathrm{d}\xi\, \mathrm{d}y \\ &\quad= \pi^{-\frac{1}{2}-\nu} e^{-\frac{\pi ik}{2}} \int_{\mathbb{R}+\frac{i}{2}} (\xi-i)^{-\frac{1}{2}-\nu-\frac{k}{2}} (\xi+i)^{-\frac{1}{2}-\nu+\frac{k}{2}} \\ &\qquad\cdot \int_0^\infty y^{s-\nu-\frac{1}{2}}\, e^{2\pi iy\xi}\, \mathrm{d}y\, \mathrm{d}\xi \\ &\quad= \pi^{-s-1} 2^{\nu-s-\frac{1}{2}}\, e^{-\frac{\pi ik}{2}}\, \Gamma\left(\frac{1}{2}+s-\nu\right) \\ &\qquad\cdot \int_{\mathbb{R}+\frac{1}{2}i} (\xi-i)^{-\frac{1}{2}-\nu-\frac{k}{2}} (\xi+i)^{-\frac{1}{2}-\nu+\frac{k}{2}}\, (-i\xi)^{\nu-s-\frac{1}{2}}\, \mathrm{d}\xi \\ &\quad= \pi^{-s-1} 2^{\nu-s-\frac{1}{2}}\, e^{-\frac{\pi ik}{2}}\, \Gamma\left(\frac{1}{2}+s-\nu\right) \\ &\qquad\cdot \int_{\mathbb{R}} (\xi-i)^{-\frac{1}{2}-\nu-\frac{k}{2}} (\xi+i)^{-\frac{1}{2}-\nu+\frac{k}{2}}\, (-i\xi)^{\nu-s-\frac{1}{2}}\, \mathrm{d}\xi,\end{aligned}$$

where we moved the path of integration over $\mathbb{R} + \frac{1}{2i}i$ back to the real line $\mathbb{R}$. The integral is valid since the singularity $\xi^{\nu-s-\frac{1}{2}}$ in the integrand is integrable for $\Re\left(\nu - s - \frac{1}{2}\right) > -1$ (which is satisfied by our assumption $\Re\,(s) < \Re\,(\nu) + \frac{1}{2}$).

Summarizing, we have

$$\tilde{\psi}_{k,\nu}(s) = 2^{\nu-s-\frac{1}{2}}\pi^{-s-1}e^{-\frac{\pi ik}{2}}\,\Gamma\left(\frac{1}{2}+s-\nu\right) \cdot \int_{\mathbb{R}} (-i\xi)^{\nu-s-\frac{1}{2}}(\xi-i)^{-\frac{1}{2}-\nu-\frac{k}{2}}(\xi+i)^{-\frac{1}{2}-\nu+\frac{k}{2}}\,\mathrm{d}\xi \tag{4.40}$$

under the assumptions $\Re(\nu) > 0$ and $\Re(\nu) - \frac{1}{2} < \Re(s) < \Re(\nu) + \frac{1}{2}$.

The integral in (4.40) converges for all s in the slightly larger vertical strip

$$-\Re(\nu) - \frac{1}{2} < \Re(s) < \Re(\nu) + \frac{1}{2},$$

as the following arguments show: We require

$$\Re\left(\nu - s - \frac{1}{2}\right) > -1 \iff \Re(s) < \Re(\nu) + \frac{1}{2}$$

to integrate the possible singularity of the integrand around 0 and

$$\begin{aligned}
&\Re\left(\left(\nu - s - \frac{1}{2}\right) + \left(-\frac{1}{2} - \nu - \frac{k}{2}\right) + \left(-\frac{1}{2} - \nu + \frac{k}{2}\right)\right) < -1\\
&\iff \Re\left(-\nu - s - \frac{3}{2}\right) < -1\\
&\iff -\Re(\nu) - \frac{1}{2} < \Re(s)
\end{aligned}$$

to guarantee that the limits of the integral

$$\lim_{\substack{b\to-\infty\\ c\to\infty}} \int_b^c \xi^{\nu-s-\frac{1}{2}}(\xi-i)^{-\frac{1}{2}-\nu-\frac{k}{2}}(\xi+i)^{-\frac{1}{2}-\nu+\frac{k}{2}}\,\mathrm{d}\xi$$

exist.

Next, we can drop the condition $\Re(\nu) > 0$ since the integral converges as long as $\Re\left(\nu - s - \frac{1}{2}\right) > -1$ and $\Re\left(-\nu - s - \frac{3}{2}\right) < -1$. The integral representation is valid for $\Re(\nu) > -\frac{1}{2}$ and s in the vertical strip

$$-\Re(\nu) - \frac{1}{2} < \Re(s) < \Re(\nu) + \frac{1}{2}.$$

□

The recurrence relations for the Whittaker $W_{\frac{k}{2},\nu}$ function, for example found in [186, 2.4.19-24], imply relations for $\tilde{\psi}_{k,\nu}$. The following lemma gives two relations related to the action of the Maass operators $\mathrm{E}_k^{\pm}$ given in Lemmas 3.21 and 3.23.

Lemma 4.21 *Let $k \in \mathbb{R}$ and $\nu \in \mathbb{C}$. The function $\tilde{\psi}_{k,\nu}$ satisfies*

$$2\pi\,\tilde{\psi}_{k,\nu}(s+1) = \left(\frac{k}{2}+s\right)\tilde{\psi}_{k,\nu}(s) - \left(\frac{1}{2}+\nu-\frac{k}{2}\right)\tilde{\psi}_{k-2,\nu}(s), \tag{4.41}$$

$$= \left(\frac{k}{2}-s\right)\tilde{\psi}_{k,\nu}(s) + \left(\frac{1}{2}+\nu+\frac{k}{2}\right)\tilde{\psi}_{k+2,\nu}(s) \tag{4.42}$$

for $s \in \mathbb{C}$.

Proof We can easily verify the identities (4.41) and (4.42) from the identities (4.13) and (4.14) for the $\psi_{k,\nu}$ function together with (4.32) in Lemma 4.16.

Consider the identity (4.13) for $\psi_{k,\nu}$. Using (4.32) on the left-hand side of (4.13), we get

$$\mathcal{M}\left(y\psi_{k,\nu}'(y)\right)(s) \overset{(4.32)}{=} -s\tilde{\psi}_{k,\nu}(s).$$

On the right-hand side of (4.13), we calculate the Mellin transform and we get

$$\begin{aligned}
&\mathcal{M}\left(y\psi_{k,\nu}'(y)\right)(s)\\
&\overset{(4.13)}{=} \mathcal{M}\left(\left(\frac{k}{2}-2\pi y\right)\psi_{k,\nu}(y) - \left(\frac{1}{2}+\nu-\frac{k}{2}\right)\psi_{k-2,\nu}(y)\right)(s)\\
&= \frac{k}{2}\tilde{\psi}_{k,\nu}(s) - 2\pi\,\tilde{\psi}_{k,\nu}(s+1) - \left(\frac{1}{2}+\nu-\frac{k}{2}\right)\tilde{\psi}_{k-2,\nu}(s).
\end{aligned}$$

Hence, we have the identity

$$-s\tilde{\psi}_{k,\nu}(s) = \frac{k}{2}\tilde{\psi}_{k,\nu}(s) - 2\pi\,\tilde{\psi}_{k,\nu}(s+1) - \left(\frac{1}{2}+\nu-\frac{k}{2}\right)\tilde{\psi}_{k-2,\nu}(s+1).$$

Resorting the terms gives

$$2\pi\,\tilde{\psi}_{k,\nu}(s+1) = \left(\frac{k}{2}+s\right)\tilde{\psi}_{k,\nu}(s) - \left(\frac{1}{2}+\nu-\frac{k}{2}\right)\tilde{\psi}_{k-2,\nu}(s1).$$

A similar calculation with the use of (4.14) in place of (4.13) gives (4.42). □

Exercise 4.22 Show the following identity of meromorphic functions

$$\left(\frac{1}{2}+\nu+\frac{k}{2}\right)\tilde{\psi}_{k+2,\nu}(s) + \left(\frac{1}{2}+\nu-\frac{k}{2}\right)\tilde{\psi}_{k-2,\nu}(s) = 2s\tilde{\psi}_{k,\nu}(s) \tag{4.43}$$

for all $s \in \mathbb{C}$.

Lemma 4.23 *Let $k \in \mathbb{R}$ and $\nu \in \mathbb{C}$. There exists a $C \in \mathbb{R}$ such that*

$$\tilde{\psi}_{k,\nu}(s) = O\left(|\Im(s)|^C e^{-\frac{\pi}{2}|\Im(s)|}\right), \tag{4.44}$$

uniformly for $|\Im(s)| \to \infty$ and $\Re(s)$ bounded.

There exists a $C_1 \in \mathbb{R}$ such that

$$\tilde{\psi}_{k,\nu}(s) = O\left(e^{|s|^{C_1}}\right) \tag{4.45}$$

for $|s| \to \infty$, $\Re(s) \geq |\Re(\nu)|$.

Proof Recall Stirling's formula in Appendix A.3. In particular, we have the estimates

$$|\Gamma(s)| \sim \sqrt{2\pi}\, e^{-\frac{\pi}{2}|\Im(s)|}\, |\Im(s)|^{\Re(s)-\frac{1}{2}}$$

uniformly for $|\Im(s)| \to \infty$ and $\Re(s)$ bounded and

$$\Gamma(s) \sim \sqrt{2\pi}\, e^{(s-\frac{1}{2})\log s - s} \tag{4.46}$$

as $|s| \to \infty$ and $\arg(s) < \frac{\pi}{2}$.

We need also

$${}_2F_1\left[\begin{matrix}\alpha, \beta \\ \gamma\end{matrix}\middle| z\right] = 1 + O\left(\gamma^{-1}\right) \tag{4.47}$$

for $\alpha, \beta \in \mathbb{C}$, $\Re(z) \leq \frac{1}{2}$, and $|\arg(\gamma)| \leq \pi - \delta$, where the implied constant depends on α, β, z, and δ, see [95, p. 614, Appendix A, Theorem 1].

Combining these estimates, there exists a constant $C \in \mathbb{R}$ such that

$$\left|\frac{\Gamma\left(\frac{1}{2}+s+\nu\right)\Gamma\left(\frac{1}{2}+s-\nu\right)}{\Gamma\left(1+s-\frac{k}{2}\right)}\right| = O\left(|\Im(s)|^C e^{-\frac{\pi}{2}|\Im(s)|}\right),$$

uniformly for $|\Im(s)| \to \infty$ and $\Re(s)$ bounded. Since

$$\lim_{|\Im(s)|\to\infty} {}_2F_1\left[\begin{matrix}\frac{1}{2}+\nu-\frac{k}{2}, \frac{1}{2}-\nu-\frac{k}{2} \\ 1+s-\frac{k}{2}\end{matrix}\middle| \frac{1}{2}\right] = O(1) \qquad \text{for } |\Im(s)| \to \infty$$

the first part of the lemma follows from the representation (4.30) of $\tilde{\psi}_{k,\nu}$ stated in Lemma 4.16.

For the second part, we apply again Stirling's formula (4.46). It implies that there exists a constant $C_1 \in \mathbb{R}$ such that

$$\left| \frac{\Gamma\left(\frac{1}{2}+s+\nu\right)\Gamma\left(\frac{1}{2}+s-\nu\right)}{\Gamma\left(1+s-\frac{k}{2}\right)} \right| = O\left(e^{|s|^{C_1}}\right)$$

for $|s| \to \infty$, $\Re\left(1+s-\frac{k}{2}\right) > 0$ and $\Re\left(\frac{1}{2}+s\pm\nu\right) > 0$. Applying $s\Gamma(s) = \Gamma(s+1)$, we see that the estimate is also valid for $|s| \to \infty$, $\Re(s) \geq 0$. Plugging all estimates into the representation (4.30) of $\tilde{\psi}_{k,\nu}$ shows the second result. □

Maass proved the following lemma in [129, Chapter V, §1].

Definition 4.24 Let $k \in \mathbb{R}$, $\nu \in \mathbb{C}$. We define the matrix $\tilde{\Psi}_{k,\nu}(s)$ by

$$\tilde{\Psi}_{k,\nu}(s) := \begin{pmatrix} \tilde{\psi}_{k,\nu}(s) & \tilde{\psi}_{-k,\nu}(s) \\ \tilde{\psi}_{k,\nu}(s+1) & -\tilde{\psi}_{-k,\nu}(s+1) \end{pmatrix}. \tag{4.48}$$

for $s \in \mathbb{C}$.

Lemma 4.25 *Let $k \in \mathbb{R}$, $\nu \in \mathbb{C}$. The matrix $\tilde{\Psi}_{k,\nu}(s)$ defined in* (4.48) *satisfies*

$$\det \tilde{\Psi}_{k,\nu}(s) = \frac{-2}{(2\pi)^{2s+1}} \frac{\Gamma\left(\frac{1}{2}+s+\nu\right)\Gamma\left(\frac{1}{2}+s-\nu\right)}{\Gamma\left(\frac{1}{2}+\nu+\frac{k}{2}\right)\Gamma\left(\frac{1}{2}+\nu-\frac{k}{2}\right)}. \tag{4.49}$$

Moreover, the matrix $\tilde{\Psi}_{k,\nu}$ satisfies the recursion

$$\tilde{\Psi}_{k,\nu}(s+1) = \begin{pmatrix} 0 & 1 \\ \frac{\left(\left(\frac{1}{2}+s-\nu\right)\left(\frac{1}{2}+s+\nu\right)\right.}{4\pi^2} & \frac{k}{2\pi} \end{pmatrix} \tilde{\Psi}_{k,\nu}(s) \begin{pmatrix} 1 & 0 \\ 0 & -1 \end{pmatrix}. \tag{4.50}$$

Proof Let us first compute $\tilde{\Psi}_{k,\nu}(s+1)$. We have

$$\begin{aligned} \tilde{\Psi}_{k,\nu}(s+1) &\overset{(4.48)}{=} \begin{pmatrix} \tilde{\psi}_{k,\nu}(s+1) & \tilde{\psi}_{-k,\nu}(s+1) \\ \tilde{\psi}_{k,\nu}(s+2) & -\tilde{\psi}_{-k,\nu}(s+2) \end{pmatrix} \\ &\overset{(4.29)}{=} \begin{pmatrix} 0 & 1 \\ \frac{\left(\frac{1}{2}+s-\nu\right)\left(\frac{1}{2}+s+\nu\right)}{4\pi^2} & \frac{k}{2\pi} \end{pmatrix} \tilde{\Psi}_{k,\nu}(s) \begin{pmatrix} 1 & 0 \\ 0 & -1 \end{pmatrix}, \end{aligned}$$

where we used the recurrence relation (4.50) in Lemma 4.16.

Let $\frac{1}{2}+\nu\pm\frac{k}{2} \notin -\mathbb{Z}_{\geq 0}$ and put

$$D_{k,\nu}(s) := (2\pi)^{2s+1} \frac{\Gamma\left(\frac{1}{2}+\nu+\frac{k}{2}\right)\Gamma\left(\frac{1}{2}+\nu-\frac{k}{2}\right)}{\Gamma\left(\frac{1}{2}+s+\nu\right)\Gamma\left(\frac{1}{2}+s-\nu\right)} \det \tilde{\Psi}_{k,\nu}(s).$$

With the recurrence formula (4.50) for $\tilde{\Psi}_{k,\nu}(s)$ above and the well-known identity $s\Gamma(s) = \Gamma(s+1)$, see Appendix A.7, we easily conclude that $D_{k,\nu}$ has period 1 in s. We have

$$\begin{aligned}
D_{k,\nu}(s+1) &= (2\pi)^{2s+3} \frac{\Gamma\left(\frac{1}{2}+\nu+\frac{k}{2}\right)\Gamma\left(\frac{1}{2}+\nu-\frac{k}{2}\right)}{\Gamma\left(\frac{3}{2}+s+\nu\right)\Gamma\left(\frac{3}{2}+s-\nu\right)} \det \tilde{\Psi}_{k,\nu}(s+1) \\
&\overset{(4.50)}{=} (2\pi)^{2s+1} \frac{\Gamma\left(\frac{1}{2}+\nu+\frac{k}{2}\right)\Gamma\left(\frac{1}{2}+\nu-\frac{k}{2}\right)}{\Gamma\left(\frac{1}{2}+s+\nu\right)\Gamma\left(\frac{1}{2}+s-\nu\right)} \frac{(2\pi)^2}{\left(\frac{1}{2}+s-\nu\right)\left(\frac{1}{2}+s+\nu\right)} \\
&\qquad \cdot \det\left[\begin{pmatrix} 0 & 1 \\ \frac{\left(\frac{1}{2}+s-\nu\right)\left(\frac{1}{2}+s+\nu\right)}{4\pi^2} & \frac{k}{2\pi} \end{pmatrix} \tilde{\Psi}_{k,\nu}(s) \begin{pmatrix} 1 & 0 \\ 0 & -1 \end{pmatrix}\right] \\
&= (2\pi)^{2s} \frac{\Gamma\left(\frac{1}{2}+\nu+\frac{k}{2}\right)\Gamma\left(\frac{1}{2}+\nu-\frac{k}{2}\right)}{\Gamma\left(\frac{1}{2}+s+\nu\right)\Gamma\left(\frac{1}{2}+s-\nu\right)} \det \tilde{\Psi}_{k,\nu}(s) \\
&= D_{k,\nu}(s).
\end{aligned}$$

We rewrite $D_{k,\nu}(s)$ in terms of Γ and hypergeometric functions where we use the representation (4.30) of $\tilde{\psi}_{k,\nu}(s)$:

$$\begin{aligned}
D_{k,\nu}(s) = &-\frac{\Gamma\left(\frac{1}{2}+s+\nu\right)\Gamma\left(\frac{3}{2}+s-\nu\right)}{\Gamma\left(1+s-\frac{k}{2}\right)\Gamma\left(1+s+\frac{k}{2}\right)} \\
&\cdot \left\{ \frac{\frac{1}{2}+s+\nu}{1+s+\frac{k}{2}}\, {}_2F_1\left[\begin{matrix} \frac{1}{2}-\nu-\frac{k}{2},\ \frac{1}{2}+\nu-\frac{k}{2} \\ 1+s-\frac{k}{2} \end{matrix}\middle| \frac{1}{2}\right] \right. \\
&\qquad \cdot {}_2F_1\left[\begin{matrix} \frac{1}{2}-\nu+\frac{k}{2},\ \frac{1}{2}+\nu+\frac{k}{2} \\ 2+s+\frac{k}{2} \end{matrix}\middle| \frac{1}{2}\right] \\
&+ \frac{\frac{1}{2}+s+\nu}{1+s-\frac{k}{2}}\, {}_2F_1\left[\begin{matrix} \frac{1}{2}-\nu+\frac{k}{2},\ \frac{1}{2}+\nu+\frac{k}{2} \\ 1+s+\frac{k}{2} \end{matrix}\middle| \frac{1}{2}\right] \cdot \\
&\qquad \left. \cdot {}_2F_1\left[\begin{matrix} \frac{1}{2}-\nu-\frac{k}{2},\ \frac{1}{2}+\nu-\frac{k}{2} \\ 2+s-\frac{k}{2} \end{matrix}\middle| \frac{1}{2}\right] \right\}.
\end{aligned}$$

It remains to compute the limit $\Re(s) \to \infty$, $\Im(s)$ constant, of this expression. Equation (4.47) shows that the limit of the part between the big brackets is 2. The limit for the remaining gamma factors can be calculated with Stirling's formula in (4.46). We find

$$\lim_{\substack{\Re(s)\to\infty \\ \Im(s)=\text{const}}} \frac{\Gamma\left(\frac{1}{2}+s+\nu\right)\Gamma\left(\frac{3}{2}+s-\nu\right)}{\Gamma\left(1+s-\frac{k}{2}\right)\Gamma\left(1+s+\frac{k}{2}\right)} = 1. \tag{4.51}$$

Summarizing, we have

$$D_{k,\nu}(s) \to -2$$

for $\Re(s) \to \infty$, $\Im(s)$ constant.

For the remaining case $\frac{1}{2}+\nu\pm\frac{k}{2} \in -\mathbb{Z}_{\geq 0}$, Lemma 4.16 implies that $\tilde{\psi}_{k,\nu}$ or $\tilde{\psi}_{-k,\nu}$ vanish. Hence $\det \tilde{\Psi}_{\pm k,\nu}(s) = 0$. This implies that (4.49) is valid, since we have $\frac{1}{\Gamma\left(\frac{1}{2}+\nu\pm\frac{k}{2}\right)} = 0$. □

Exercise 4.26 Give an alternative proof of (4.51) based on Equation (4.47).

A direct consequence of Lemma 4.25 is the following.

Lemma 4.27 *Let $k \in \mathbb{R}$ and $\nu \in \mathbb{C}$ such that if $\nu = 0$, then $k \neq -1$.*

We have the matrix relation

$$\begin{pmatrix} \tilde{\psi}_{k,\nu}(s) & \tilde{\psi}_{-k,\nu}(s) \\ \tilde{\psi}_{k+2,\nu}(s) & \tilde{\psi}_{-k-2,\nu}(s) \end{pmatrix} = \begin{pmatrix} 1 & 0 \\ -\frac{\frac{k}{2}-s}{\frac{1}{2}+\nu+\frac{k}{2}} & \frac{2\pi}{\frac{1}{2}+\nu+\frac{k}{2}} \end{pmatrix} \cdot \tilde{\Psi}_{k,\nu}(s), \tag{4.52}$$

which implies in particular that

$$\det\begin{pmatrix} \tilde{\psi}_{k,\nu}(s) & \tilde{\psi}_{-k,\nu}(s) \\ \tilde{\psi}_{k+2,\nu}(s) & \tilde{\psi}_{-k-2,\nu}(s) \end{pmatrix} = -2(2\pi)^{-2s}\frac{\Gamma\left(\frac{1}{2}+s+\nu\right)\Gamma\left(\frac{1}{2}+s-\nu\right)}{\Gamma\left(\frac{3}{2}+\nu+\frac{k}{2}\right)\Gamma\left(\frac{1}{2}+\nu-\frac{k}{2}\right)} \tag{4.53}$$

as an identity of meromorphic functions in $s \in \mathbb{C}$.

Proof Assume first that $\frac{1}{2}+\nu+\frac{k}{2} \neq 0$. Equations (4.41) and (4.42) imply

$$\tilde{\Psi}_{k,\nu}(s) = \begin{pmatrix} 1 & 0 \\ \frac{\frac{k}{2}-s}{2\pi} & \frac{\frac{1}{2}+\nu+\frac{k}{2}}{2\pi} \end{pmatrix} \cdot \begin{pmatrix} \tilde{\psi}_{k,\nu}(s) & \tilde{\psi}_{-k,\nu}(s) \\ \tilde{\psi}_{k+2,\nu}(s) & \tilde{\psi}_{-k-2,\nu}(s) \end{pmatrix},$$

where $\tilde{\Psi}_{k,\nu}$ is given in Definition 4.24. Inverting the equation we have

$$\begin{aligned} \begin{pmatrix} \tilde{\psi}_{k,\nu}(s) & \tilde{\psi}_{-k,\nu}(s) \\ \tilde{\psi}_{k+2,\nu}(s) & \tilde{\psi}_{-k-2,\nu}(s) \end{pmatrix} &= \frac{2\pi}{\frac{1}{2}+\nu+\frac{k}{2}} \begin{pmatrix} \frac{\frac{1}{2}+\nu+\frac{k}{2}}{2\pi} & 0 \\ \frac{s-\frac{k}{2}}{2\pi} & 1 \end{pmatrix} \cdot \tilde{\Psi}_{k,\nu}(s) \\ &= \begin{pmatrix} 1 & 0 \\ \frac{s-\frac{k}{2}}{\frac{1}{2}+\nu+\frac{k}{2}} & \frac{2\pi}{\frac{1}{2}+\nu+\frac{k}{2}} \end{pmatrix} \cdot \tilde{\Psi}_{k,\nu}(s). \end{aligned}$$

This shows the identity (4.52) of the lemma.

Taking the determinant, we have

$$\begin{aligned}
\det\begin{pmatrix} \tilde{\psi}_{k,\nu}(s) & \tilde{\psi}_{-k,\nu}(s) \\ \tilde{\psi}_{k+2,\nu}(s) & \tilde{\psi}_{-k-2,\nu}(s) \end{pmatrix}
&\overset{(4.52)}{=} \frac{2\pi}{\frac{1}{2}+\nu+\frac{k}{2}} \det \tilde{\Psi}_{k,\nu}(s) \\
&\overset{(4.49)}{=} \frac{2\pi}{\frac{1}{2}+\nu+\frac{k}{2}} \frac{-2}{(2\pi)^{2s+1}} \frac{\Gamma\left(\frac{1}{2}+s+\nu\right)\Gamma\left(\frac{1}{2}+s-\nu\right)}{\Gamma\left(\frac{1}{2}+\nu+\frac{k}{2}\right)\Gamma\left(\frac{1}{2}+\nu-\frac{k}{2}\right)} \\
&= -2(2\pi)^{-2s} \frac{\Gamma\left(\frac{1}{2}+s+\nu\right)\Gamma\left(\frac{1}{2}+s-\nu\right)}{\Gamma\left(\frac{3}{2}+\nu+\frac{k}{2}\right)\Gamma\left(\frac{1}{2}+\nu-\frac{k}{2}\right)},
\end{aligned}$$

proving (4.53).

Now, assume $\frac{1}{2}+\nu+\frac{k}{2}=0$. We write

$$\begin{pmatrix} \tilde{\psi}_{k,\nu}(s) & \tilde{\psi}_{-k,\nu}(s) \\ \tilde{\psi}_{k+2,\nu}(s) & \tilde{\psi}_{-k-2,\nu}(s) \end{pmatrix} = \begin{pmatrix} -\frac{\frac{k+2}{2}-s}{\frac{1}{2}-\nu+\frac{k}{2}} & \frac{2\pi}{\frac{1}{2}-\nu+\frac{k}{2}} \\ 1 & 0 \end{pmatrix} \cdot \tilde{\Psi}_{k+2,\nu}(s),$$

where we used equations (4.41) and (4.42). The determinant can be calculated using Lemma 4.25. □

We are finally able to state a modified version of Maass' theorem on Maass L-series attached to families of Maass cusp forms after this long technical excursion on special functions.

Theorem 4.28 (H. Maass) *Let $k \in \mathbb{R}$ and $\nu \in \mathbb{C}$ such that if $2\nu+k$ or $2\nu-k \in 1+2\mathbb{Z}$, then $k \notin \mathbb{Z}$. Then there is a canonical correspondence between the following:*

1. *An admissible family of Maass cusp forms $(u_l)_{l\in k+2\mathbb{Z}}$ of cusp forms as discussed in §4.1.*
2. *A pair of L-series L_+ and L_- such that*

    a. *$L_\pm$ have a series representation*

$$L_\pm(s) = \sum_{\substack{n\in\mathbb{Z} \\ n+\kappa \gtrless 0}} A(n)\, |n+\kappa|^{-s} \tag{4.54}$$

that converges on a right half-plane.

    b. *The functions $L_l^\star$, $l \in k+2\mathbb{Z}$ defined by*

$$L_l^\star(s) = \tilde{\psi}_{l,\nu}(s)L_+(s) + \tilde{\psi}_{-l,\nu}(s)L_-(s) \tag{4.55}$$

are entire functions of s of finite order and satisfy the functional equations

$$L_l^\star(-s) = (-1)^{\frac{k-l}{2}} L_l^\star(s). \tag{4.56}$$

(This implies that the L-series $L_\pm$ extend to entire functions on $\mathbb{C}$ as well.)

The correspondence $u_l \mapsto L_\pm$ is given by the Mellin transform

$$L_l^\star(s) = \mathcal{M}\Big(u_l(i\cdot)\Big)(s) \overset{(1.154)}{=} \int_0^\infty u_l(iy)\, y^{s-1}\, dy. \tag{4.57}$$

Remark 4.29 The coefficients $A(n)$ in (4.54) are in fact the coefficients $A(n)$ in (4.25) of the associated family of Maass cusp forms Lemma 4.11 as the proof of Theorem 4.28 will show.

The proof of Theorem 4.28 contains several steps of interchanging summation and integration. The following lemma ensures that the interchange is valid.

Lemma 4.30 *Let $k \in \mathbb{R}$, $0 \le \kappa < 1$, and $\nu \in \mathbb{C}$. Let A_n, $n \in \mathbb{Z}$ such that $n+\kappa \neq 0$, be a sequence of complex numbers satisfying $A_n = O\left(|n+\kappa|^C\right)$ for some $C \in \mathbb{R}$.*

The Mellin transform

$$\tilde{F}(s) := \mathcal{M}\big(F\big)(s) \overset{(1.154)}{=} \int_0^\infty F(y)\, y^{s-1}\, dy$$

of the function

$$F(y) = \sum_{\substack{n\in\mathbb{Z}\\ n+\kappa\neq 0}} A_n\, \psi_{\mathrm{sign}(n)k,\nu}\big(|n+\kappa|y\big) \qquad (y>0)$$

is well defined for $\Re(s)$ large enough.

We have that

$$\tilde{F}(s) = \tilde{\psi}_{k,\nu}(s) \sum_{\substack{n\in\mathbb{Z}\\ n+\kappa>0}} A_n\,(n+\kappa)^{-s} + \tilde{\psi}_{-k,\nu}(s) \sum_{\substack{n\in\mathbb{Z}\\ n+\kappa<0}} A_n\, |n+\kappa|^{-s}$$

for $\Re(s)$ large enough. The inverse Mellin transform

$$F(y) = \frac{1}{2\pi i}\int_{\Re(s)} = \sigma \tilde{F}(s)\, y^{-s}\, ds \qquad (y>0,\ \sigma \text{ large enough})$$

can also be computed term by term.

Proof The estimates for $\psi_{k,\nu}(y)$ in Lemma 4.8 imply the growth estimates (4.33) stated in the proof of Lemma 4.16.

Recall that A_n satisfies $A_n = O\left(|n+\kappa|^C\right)$ for some $C \in \mathbb{R}$ by assumption. We define

$$G(y) := \sum_{\substack{n\in\mathbb{Z}\\ n+\kappa\neq 0}} |n+\kappa|^{C+\frac{1}{2}-|\Re(\nu)|} y^{\frac{1}{2}-|\Re(\nu)|} e^{-2\pi|n+\kappa|y} \qquad (y>0),$$

where C is the same constant as in the growth estimate of A_n. Hence, the function F, given in the lemma above, satisfies

$$F(y) = O(G(y)) \qquad \text{for all } y>0,$$

using the growth estimates on the special function $\psi_{\pm k,\nu}\big(|n+\kappa|\,y\big)$ in (4.33). The dominated convergence theorem (Theorem of Beppo-Levi), see [202, p. 331, IX.13], allows us to compute the Mellin transform

$$\int_0^\infty G(y)\, y^{s-1}\, \mathrm{d}y$$

term by term for $\Re(s) > K > |\Re(\nu)| - \frac{1}{2}$, see Lemma 1.201. Since we have

$$\begin{aligned}
\tilde{G}(s) &= \mathcal{M}\big(G\big)(s) \\
&\overset{(1.154)}{=} \int_0^\infty G(y)\, y^{s-1}\, \mathrm{d}y \\
&= \int_0^\infty \left(\sum_{\substack{n\in\mathbb{Z}\\ n+\kappa\neq 0}} |n+\kappa|^{C+\frac{1}{2}-|\Re(\nu)|}\, y^{\frac{1}{2}-|\Re(\nu)|} e^{-2\pi|n+\kappa|y} \right) y^{s-1}\, \mathrm{d}y \\
&= \sum_{\substack{n\in\mathbb{Z}\\ n+\kappa\neq 0}} |n+\kappa|^{C+\frac{1}{2}-|\Re(\nu)|} \int_0^\infty e^{-2\pi|n+\kappa|y}\, y^{s-|\Re(\nu)|-\frac{1}{2}}\, \mathrm{d}y \\
&\overset{(1.155)}{=} (2\pi)^{\frac{1}{2}+|\Re(\nu)|-s}\, \Gamma\left(\frac{1}{2}+|\Re(\nu)|-s\right) \sum_{\substack{n\in\mathbb{Z}\\ n+\kappa\neq 0}} |n+\kappa|^{C+\frac{1}{2}+|\Re(\nu)|-s}
\end{aligned}$$

for $\Re(s) > \max\left\{|\Re(\nu)| + C + \frac{1}{2},\, K\right\}$, the dominated convergence theorem, [202, p. 332, IX.14], allows us again to compute the Mellin transform of $F(y)$ term by term. Hence, we have

$$
\begin{aligned}
\tilde{F}(s) &= \mathcal{M}(F)(s) \\
&\overset{(1.154)}{=} \int_0^\infty F(y)\, y^{s-1}\, dy \\
&= \sum_{\substack{n\in\mathbb{Z}\\ n+\kappa\neq 0}} A_n \int_0^\infty \psi_{\operatorname{sign}(n)k,\nu}\big(|n+\kappa|\, y\big)\, y^{s-1}\, dy \\
&= \sum_{\substack{n\in\mathbb{Z}\\ n+\kappa\neq 0}} A_n\, |n+\kappa|^{-s} \int_0^\infty \psi_{\operatorname{sign}(n)k,\nu}(y)\, y^{s-1}\, dy \qquad (4.58)\\
&\overset{(4.28)}{=} \tilde{\psi}_{k,\nu}(s) \sum_{\substack{n\in\mathbb{Z}\\ n+\kappa>0}} A_n\, (n+\kappa)^{-s} + \tilde{\psi}_{-k,\nu}(s) \sum_{\substack{n\in\mathbb{Z}\\ n+\kappa<0}} A_n\, |n+\kappa|^{-s} \qquad (4.59)
\end{aligned}
$$

for $\Re(s) > \max\left\{|\Re(\nu)| + C + \frac{1}{2},\, K\right\}$.

To show that (4.59) can be inverted, we note that $\tilde{F}$ satisfies an estimate

$$
\tilde{F}(s) = O\left(e^{\varepsilon|\Im(s)|}\right) \qquad \text{as } |\Im(s)| \to \infty,
$$

for some $\varepsilon > 0$ uniform on vertical strips satisfying at least $\Re(s) > \max\left\{|\Re(\nu)| + C + 1, K\right\}$. This follows from the estimate for $\tilde{\psi}_{k,\nu}$ given in Lemma 4.23 and the fact that the L-series $\sum_{\substack{n\in\mathbb{Z}\\ n+\kappa>0}} A_n\,(n+\kappa)^{-s}$ and $\sum_{\substack{n\in\mathbb{Z}\\ n+\kappa<0}} A_n\,|n+\kappa|^{-s}$ can be dominated by multiples of Hurwitz zeta functions. This allows us to apply Mellin inversion formula in Proposition 1.204. We have

$$
F(y) = \frac{1}{2\pi i} \int_{\Re(s)=\sigma} \tilde{F}(y)\, y^{-s}\, ds \qquad (4.60)
$$

for all $\sigma := \Re(s) > \max\left\{|\Re(\nu)| + C + \frac{1}{2},\, K\right\}$ and $y > 0$. Now, Fubini's theorem allows us to calculate the right-hand side of (4.60). It can be calculated by term-wise integration of the series representation of $\tilde{F}(s)$. □

Proof *(of Theorem 4.28)* First, we show that the conditions for the two L-series L_+ and L_- in Item 2 of the Theorem imply that L_+ and L_- extend to entire functions on $\mathbb{C}$.

Let $k \in \mathbb{R}$ and $\nu \in \mathbb{C}$ such that if $2\nu + k$ or $2\nu - k \in 1 + 2\mathbb{Z}$, then $k \notin \mathbb{Z}$. This implies that have at least one $l \in k + 2\mathbb{Z}$ in the admissible range that satisfies $l \neq -1$, $\frac{3}{2} + \nu + \frac{l}{2} \notin -\mathbb{Z}_{\geq 0}$, and $\frac{1}{2} + \nu - \frac{l}{2} \notin -\mathbb{Z}_{\geq 0}$. Writing (4.55) for l and $l+2$ in vector notation, we have

$$\begin{pmatrix} L_l^\star(s) \\ L_{l+2}^\star(s) \end{pmatrix} \overset{(4.55)}{=} \begin{pmatrix} \tilde{\psi}_{l,\nu}(s) & \tilde{\psi}_{-l,\nu}(s) \\ \tilde{\psi}_{l+2,\nu}(s) & \tilde{\psi}_{-l-2,\nu}(s) \end{pmatrix} \begin{pmatrix} L_+(s) \\ L_-(s) \end{pmatrix}. \tag{4.61}$$

Inverting the vector equation, we get

$$\begin{aligned} \begin{pmatrix} L_+(s) \\ L_-(s) \end{pmatrix} &= \begin{pmatrix} \tilde{\psi}_{l,\nu}(s) & \tilde{\psi}_{-l,\nu}(s) \\ \tilde{\psi}_{l+2,\nu}(s) & \tilde{\psi}_{-l-2,\nu}(s) \end{pmatrix}^{-1} \begin{pmatrix} L_l^\star(s) \\ L_{l+2}^\star(s) \end{pmatrix} \\ &\overset{(4.53)}{=} -\frac{1}{2}(2\pi)^{2s} \frac{\Gamma\left(\frac{3}{2}+\nu+\frac{l}{2}\right)\Gamma\left(\frac{1}{2}+\nu-\frac{l}{2}\right)}{\Gamma\left(\frac{1}{2}+s+\nu\right)\Gamma\left(\frac{1}{2}+s-\nu\right)} \\ &\quad \cdot \begin{pmatrix} \tilde{\psi}_{-l-2,\nu}(s) & -\tilde{\psi}_{-l,\nu}(s) \\ -\tilde{\psi}_{l+2,\nu}(s) & \tilde{\psi}_{l,\nu}(s) \end{pmatrix} \begin{pmatrix} L_l^\star(s) \\ L_{l+2}^\star(s) \end{pmatrix}, \end{aligned} \tag{4.62}$$

where we used Lemma 4.27 to compute the determinate of the matrix. (The determinant is nonzero for our choice of l.)

Using representation (4.30) in Lemma 4.16 for the functions $\tilde{\psi}_{k,\nu}(s)$, we see that the L-series $L_\pm$ extend to $\mathbb{C}$ as holomorphic functions, since the Γ-factors $\Gamma\left(\frac{1}{2}+s+\nu\right)\Gamma\left(\frac{1}{2}+s-\nu\right)$ in the denominator and in the coefficients of the matrix cancel and the functions $L_l^\star(s)$ and $L_{l+2}^\star(s)$ are entire by assumption.

Next, we show the equivalence of the two statements in Theorem 4.28 by showing each direction separately.

$1 \Longrightarrow 2$ Let $k \in \mathbb{R}$, $\nu \in \mathbb{C}$ and take a $l \in k + 2\mathbb{Z}$ in the admissible range. We assume that u_l has the expansion (4.24) with coefficients $A(n)$. Theorem 3.36 implies $A(n) = O(1)$.

We define L-series $L_\pm$ by

$$\begin{aligned} L_+(s) &= \sum_{\substack{n\in\mathbb{Z} \\ n+\kappa>0}} A(n)\,(n+\kappa)^{-s} \quad \text{and} \\ L_-(s) &= \sum_{\substack{n\in\mathbb{Z} \\ n+\kappa<0}} A(n)\,|n+\kappa|^{-s}, \end{aligned} \tag{4.63}$$

that converge for $\Re(s) > 1$ absolutely. Define $L_l^\star$ by (4.57). Lemma 4.30 shows that the Mellin transform $\mathcal{M}\big(u_l(i\cdot)\big)(s)$ of $u_l(iy)$ can be calculated term by term integration of the expansion (4.24) of u_l for all s in a certain right half-plane. We find

$$\begin{aligned}
L_l^\star(s) &= \sum_{\substack{n\in\mathbb{Z}\\ n+\kappa\neq 0}} A(n) \int_0^\infty \psi_{\mathrm{sign}(n)k,\nu}\big(|n+\kappa|\, y\big)\, y^{s-1}\, \mathrm{d}y \\
&= \tilde{\psi}_{k,\nu}(s) \sum_{\substack{n\in\mathbb{Z}\\ n+\kappa>0}} A(n)\,(n+\kappa)^{-s} \;+\; \tilde{\psi}_{-k,\nu}(s) \sum_{\substack{n\in\mathbb{Z}\\ n+\kappa<0}} A(n)\,|n+\kappa|^{-s} \\
&= \tilde{\psi}_{k,\nu}(s)\, L_+(s) \;+\; \tilde{\psi}_{-k,\nu}(s)\, L_-(s)
\end{aligned}$$

for $\Re(s) > 0$ large enough and $L_\pm(s)$ as in (4.63).

Consider now the transformation formula (3.6) for u_l. For $T \overset{(1.20)}{=} \begin{pmatrix} 0 & -1 \\ 1 & 0 \end{pmatrix} \in \Gamma(1)$ and $z = iy$, we find the relation

$$\begin{aligned}
u_l\left(\frac{i}{y}\right) &\overset{(1.20)}{=} u_l\big(T\, iy\big) \\
&\overset{(3.6)}{=} v(T)\, e^{il \arg\big(j(T,iy)\big)}\, u_l(iy) \\
&\overset{(3.2)}{=} v(T)\, e^{\frac{\pi i}{2} l}\, u_l(iy).
\end{aligned}$$

Recalling that the multiplier v is compatible with $\big(\Gamma(1), k\big)$ and using $v(T) = e^{-\frac{i\pi}{2}k}$, see Exercise 1.48, gives

$$\begin{aligned}
u_l\left(\frac{i}{y}\right) &= e^{\frac{\pi i}{2}(l-k)}\, u_l(iy) \\
&= (-1)^{\frac{l-k}{2}}\, u_l(iy),
\end{aligned} \tag{4.64}$$

where we use the argument convention (1.30) to write $e^{-\pi i} = -1$. The factor $(-1)^{\frac{l-k}{2}}$ is well defined since $l \in k + 2\mathbb{Z}$ and hence the exponent $\frac{l-k}{2} \in \mathbb{Z}$ is an integer.

Splitting the integral in (4.57) at 1 and using the substitution $y \mapsto \frac{1}{y}$ in one part of the integral transformation, it follows that

$$\begin{aligned}
L_l^\star(s) &\overset{(4.57)}{=} \int_0^\infty u_l(iy)\, y^{s-1}\, \mathrm{d}y \\
&= \int_0^1 u_l(iy)\, y^{s-1}\, \mathrm{d}y + \int_1^\infty u_l(iy)\, y^{s-1}\, \mathrm{d}y \\
&\overset{y\mapsto\frac{1}{y}}{=} \int_\infty^1 u_l\left(\frac{i}{y}\right) \left(\frac{1}{y}\right)^{s-1} \frac{-1}{y^2}\mathrm{d}y + \int_1^\infty u_l(iy)\, y^{s-1}\, \mathrm{d}y
\end{aligned}$$

$$
\begin{aligned}
&= \int_1^\infty u_l\left(\frac{i}{y}\right) y^{-s-1}\, dy + \int_1^\infty u_l(iy)\, y^{s-1}\, dy \\
&\overset{(4.64)}{=} \int_1^\infty (-1)^{\frac{l-k}{2}}\, u_l(iy)\, y^{-s-1}\, dy + \int_1^\infty u_l(iy)\, y^{s-1}\, dy \\
&= \int_1^\infty u_l(iy) \left((-1)^{\frac{l-k}{2}}\, y^{-s-1} + y^{s-1} \right) dy \qquad (4.65)
\end{aligned}
$$

for $\Re(s) \gg 0$ large enough.

The integral in (4.65) converges for all $s \in \mathbb{C}$ since u_l as Maass cusp form decays exponentially fast at the cusp $i\infty$, see (3.53) in Theorem 3.38. Hence, the function $L_k^\star(s)$ extends as an entire function on $\mathbb{C}$. Moreover, the integral representation (4.65) gives us the desired functional equation for $s \mapsto -s$:

$$
\begin{aligned}
(-1)^{\frac{l-k}{2}}\, L_l^\star(-s) &\overset{(4.65)}{=} (-1)^{\frac{l-k}{2}} \int_1^\infty u_l(iy) \left((-1)^{\frac{l-k}{2}}\, y^{-(-s)-1} + y^{(-s)-1} \right) dy \\
&= \int_1^\infty u_l(iy) \left((-1)^{\frac{l-k}{2}}\, y^{-s-1} + y^{s-1} \right) dy \\
&\overset{(4.65)}{=} L_l^\star(s)
\end{aligned}
$$

for all $s \in \mathbb{C}$.

The integral representation (4.65) implies

$$
\begin{aligned}
M_l(r) :&= \max_{|s|=r} \left| L_l^\star(s) \right| \\
&\leq \max_{|s|=r} \int_1^\infty |u_l(iy)| \left(y^{\Re(s)-1} + (-1)^{l-k}\, y^{-\Re(s)-1} \right) dy \\
&\leq 2 \int_1^\infty |u_l(iy)|\, y^{r-1}\, dy
\end{aligned}
$$

for $r \geq 1$. It follows that

$$
M_l(r) = O\left(C^{-r}\, \Gamma(r) \right)
$$

for all $r \geq 1$ since the cusp form u_l satisfies the growth condition $u_l(iy) = O\left(e^{-Cy}\right)$, $y \to \infty$ for some $C > 0$, see (3.53) of Theorem 3.38. This implies together with Stirling's formula, see (4.46) and Appendix A.3, that the order of the function $L_l^\star(s)$ is finite. In other words, we have

$$
\limsup_{r \to \infty} \frac{\log\log M_l(r)}{\log r} < \infty.
$$

The finiteness of the lim sup expression is equivalent with $L_l^\star(s) = O\left(e^{|s|^c}\right)$ for some $c > 0$.

$2 \Longrightarrow 1$ The L-series $L_\pm(s)$ converge on a certain right half-plane. This implies that the coefficients $A(n)$ in (4.54) satisfy

$$A(n) = O\left(|n+\kappa|^C\right)$$

for all $n \in \mathbb{Z}$ with $n + \kappa \neq 0$ and some $C > 0$. Lemma 4.30 implies that the inverse Mellin transformation

$$\frac{1}{2\pi i} \int_{\Re(s)=\sigma} L_l^\star(s)\, y^{-s}\, ds \qquad (y > 0)$$

exists for suitable σ large enough and for all $l \in k + 2\mathbb{Z}$ in the admissible range. The inverse Mellin transformation can be calculated term by term. We get

$$\begin{aligned}
&\frac{1}{2\pi i} \int_{\Re(s)=\sigma} L_l^\star(s)\, y^{-s}\, ds \\
&\overset{(4.55)}{=} \frac{1}{2\pi i} \int_{\Re(s)=\sigma} \tilde{\psi}_{l,\nu}(s)\, L_+(s)\, y^{-s}\, ds \, + \qquad (4.66)\\
&\quad + \frac{1}{2\pi i} \int_{\Re(s)=\sigma} \tilde{\psi}_{-l,\nu}(s)\, L_-(s)\, y^{-s}\, ds \\
&\overset{(4.54)}{=} \sum_{\substack{n\in\mathbb{Z}\\ n+\kappa>0}} A(n) \frac{1}{2\pi i} \int_{\Re(s)=\sigma} \tilde{\psi}_{l,\nu}(s)\, (n+\kappa)^{-s}\, y^{-s}\, ds \, + \\
&\quad + \sum_{\substack{n\in\mathbb{Z}\\ n+\kappa<0}} A(n) \frac{1}{2\pi i} \int_{\Re(s)=\sigma} \tilde{\psi}_{-l,\nu}(s)\, |n+\kappa|^{-s}\, y^{-s}\, ds \\
&\overset{L\,4.30}{=} \sum_{\substack{n\in\mathbb{Z}\\ n+\kappa\neq 0}} A(n)\, \psi_{\mathrm{sign}(n)l,\nu}\big(\,|n+\kappa|\, y\big) \qquad (4.67)
\end{aligned}$$

for all $y > 0$.
We define $u_l : \mathbb{H} \to \mathbb{C}$ by the Fourier expansion (4.24), i.e.,

$$u_l(z) = \sum_{\substack{n\in\mathbb{Z}\\ n+\kappa\neq 0}} A(n)\, \psi_{\mathrm{sign}(n)k,\nu}\left(|n+\kappa|\Im(z)\right) e^{2\pi i n \Re(z)}$$

for all $z \in \mathbb{H}$. The polynomial growth of $A(n)$ for $|n + \kappa| \to \infty$ implies that the series converges, and that u_l satisfies the growth condition

$$u_l(z) = O\left(\Im(z)^A\right)$$

for all $A \in \mathbb{R}$ and for $y \to \infty$, uniform in $\Re(z)$. We have

$$u_l(iy) = \frac{1}{2\pi i} \int_{\Re(s)=\sigma} L_l^\star(s)\, y^{-s}\, ds \qquad (y > 0) \tag{4.68}$$

as calculated in (4.67).

Now, we calculate $\mathrm{E}_l^\pm u_l(z)$, with l in the admissible range, term by term in the Fourier expansion of u_l, since the differential operator $\mathrm{E}_k^\pm$, defined in (3.10), commutes with the operator "taking the n-th Fourier term." Applying (4.13) and (4.14) of Lemma 4.8, we find

$$\begin{aligned}
&\mathrm{E}_l^\pm u_l(x+iy) \\
&\overset{(3.10)}{=} \sum_{\substack{n\in\mathbb{Z}\\ n+\kappa\neq 0}} A(n)\, e^{2\pi i(n+\kappa)x} \Big(\mp 4\pi n y\, \psi_{\mathrm{sign}(n)l,\nu}\big(|n+\kappa|\, y\big) + \\
&\qquad + 2\,|n+\kappa|\, y\, \psi'_{\mathrm{sign}(n)l,\nu}\big(|n+\kappa|\, y\big) \pm l\, \psi_{\mathrm{sign}(n)l,\nu}\big(|n+\kappa|\, y\big)\Big) \\
&\overset{(4.13)\,(4.14)}{=} -(1+2\nu\pm l) \sum_{\substack{n\in\mathbb{Z}\\ n+\kappa\neq 0}} A(n)\, \psi_{\mathrm{sign}(n)(l\pm 2),\nu}\big(|n+\kappa|\, y\big)\, e^{2\pi i(n+\kappa)x} \\
&= -(1+2\nu\pm k)\, u_{l\pm 2}(z).
\end{aligned}$$

Hence the system $\left(u_l\right)_{l\in k+2\mathbb{Z}}$ satisfies the two relations in (4.5) in the definition of families of Maass cusp forms. It also shows that u_l is an eigenfunction of Δ_l with eigenvalue $\frac{1}{4} - \nu^2$ since Δ_l and $\mathrm{E}_l^\pm$ are connected by (3.12).

The functions u_l are smooth functions in x and y since we can express every partial derivative in the functions u_{l+m}, $m \in 2\mathbb{Z}$ via the Maass operators. We have for instance

$$4iy\, \partial_x u_l = \left(\mathrm{E}_l^+ - \mathrm{E}_l^- - l\right) u_l. \tag{4.69}$$

We find in particular that u_l are in fact real-analytic functions since they are eigenfunctions of Δ_l.

We still have to show that u_l with $l \in k + 2\mathbb{Z}$ in the admissible range satisfy (3.6) with weight k and multiplier system v in the sense of Definition 3.31. It is sufficient to check this for generators S and T of $\Gamma(1)$, since every element of $\Gamma(1)$ can be expressed in terms of S and T, and using the consistency condition (3.7) of the multiplier v.

It follows directly from the Fourier expansion (4.24) of u_l that

$$\begin{aligned} u_l(Sz) &\overset{(1.20)}{=} u_l(z+1) \\ &\overset{(4.24)}{=} \sum_{\substack{n\in\mathbb{Z}\\ n+\kappa\neq 0}} A(n)\, \psi_{\mathrm{sign}(n)l,\nu}\big(\,|n+\kappa|\, \Im\,(z+1)\,\big) e^{2\pi i (n+\kappa)\Re(z+1)} \\ &\overset{(4.24)}{=} e^{2\pi i\kappa}\, u_l(z) \\ &\overset{(1.41)}{=} v(S)\, u_l(z) \qquad (z\in\mathbb{H}). \end{aligned}$$

with $0\le\kappa<1$ given by $v(S)=e^{2\pi i\kappa}$, see (1.41) in Definition 1.52.
It is more complicated to calculate $u_l(Tz)$. First, we show that (4.68) is satisfied for all $\sigma\in\mathbb{R}$. Indeed, (4.55) together with the growth estimate for $\tilde{\psi}_{k,\nu}$ in Lemma 4.23 implies

$$L_l^\star(s) = O\Big(\big(1+\Im\,(s)\big)^{-c}\Big) \qquad \text{for all } c>1 \tag{4.70}$$

for all s with $\Re\,(s)=\sigma$ and σ large enough. The functional equation (4.56) shows that (4.70) also holds for $-\sigma$ large.
The assumption that $L_l^\star$ has finite order together with the Phragmén-Lindelöf theorem, see, e.g., [164, Chapter IV, §29, Theorem] or Appendix A.2, implies that $L_k^\star(s)$ satisfies (4.70) for all s in the vertical strip $|\Re\,(s)|\le\sigma$. Hence, we have

$$\begin{aligned} u_l\left(\frac{i}{y}\right) &\overset{(4.68)}{=} \frac{1}{2\pi i}\int_{\Re(s)=\sigma} L_l^\star(s)\left(\frac{1}{y}\right)^{-s} \mathrm{d}s \\ &= \frac{1}{2\pi i}\int_{\sigma-i\infty}^{\sigma+i\infty} L_l^\star(s)\, y^s\, \mathrm{d}s \\ &\overset{s\mapsto -s}{=} -\frac{1}{2\pi i}\int_{-\sigma+i\infty}^{-\sigma-i\infty} L_l^\star(-s)\, y^{-s}\, \mathrm{d}s \\ &\overset{(4.56)}{=} \frac{(-1)^{\frac{k-l}{2}}}{2\pi i}\int_{-\sigma-i\infty}^{-\sigma+i\infty} L_l^\star(-s)\, y^{-s}\, \mathrm{d}s \\ &= \frac{(-1)^{\frac{k-l}{2}}}{2\pi i}\int_{\Re(s)=-\sigma} L_l^\star(s)\, y^{-s}\, \mathrm{d}s \\ &\overset{(4.68)}{=} (-1)^{\frac{k-l}{2}}\, u_l(iy) = v(T)\, e^{il\arg(iy)} u_l(iy). \end{aligned}$$

for $y>0$ and $l\in k+2\mathbb{Z}$. Now, we use the unitary automorphic factor in (3.2) and $T=\begin{pmatrix}0&-1\\1&0\end{pmatrix}$ in (1.20) to write

$$u_l\left(\frac{i}{y}\right) = (-1)^{\frac{k-l}{2}}\, u_l(iy) \overset{(3.2)}{\underset{(1.20)}{=}} v(T)\, j_u(T, iy)^l\, u_l(iy)$$

that implies

$$u_l\big(T\,iy\big) - v(T)\, j_u(T, iy)^l\, u_l(iy) = 0.$$

Define

$$f_l(z) := u_l(Tz) - v(T)\, e^{il\arg(z)}\, u_k(z).$$

We showed above that f vanishes on the upper imaginary line: $f(iy) = 0$ for all $y > 0$. On the other hand, we have

$$\mathrm{E}_l^{\pm} f_l = -(1 + 2\nu \pm l)\, f_{l\pm 2}.$$

Then (4.69) implies that

$$\begin{aligned} y\partial_x f(z) &\overset{(4.69)}{=} \frac{1}{4i}\Big((1 + 2\nu + l)\, f_{l+2}(z) - (1 + 2\nu - l)\, f_{l-2}(z) - l\, f_l(z)\Big) \\ &= 0 \end{aligned}$$

for $z = iy$, $y > 0$. Since we can express every partial derivative of f_l via the Maass operators, we see that $(\partial_x)^\alpha (\partial_y)^\beta f_l(z)$ vanishes for $z = iy$, $y > 0$, and all $\alpha, \beta \in \mathbb{Z}_{\geq 0}$. Hence, the Taylor expansion of f_l, centered in $z = i$, vanishes. Hence f_l vanishes everywhere, since f_l is real-analytic. Summarizing, we just showed that u_l satisfies

$$u_l(Tz) = v(T)\, e^{il\arg(z)}\, u_l(z)$$

for all $l \in k + 2\mathbb{Z}$ and $z \in \mathbb{H}$. Thus u_l satisfies (3.6) for the generators S and T of $\Gamma(1)$.

□

Remark 4.31 Maass's original correspondence, see [129, p. 225, Theorem 35], is between one automorphic form u_l in a certain weight l on one side and the two L-series $L_\pm$ together with two entire functions satisfying a functional equation on the other side. One of the entire functions is $L_l^\star$. The other is a linear combination of $L_{l-2}^\star$ and $L_{l+2}^\star$ and is related to the partial derivative $\partial_x u_l$.

We can derive Maass's original correspondence if we replace the matrix

$$\begin{pmatrix} \tilde{\psi}_{l,\nu}(s) & \tilde{\psi}_{-l,\nu}(s) \\ \tilde{\psi}_{l+2,\nu}(s) & \tilde{\psi}_{-l-2,\nu}(s) \end{pmatrix}^{-1}$$

in (4.62) by

$$\left(\begin{pmatrix} 1 & 0 \\ -\frac{\frac{1}{2}-s+\frac{k}{2}}{\frac{1}{2}+\nu+\frac{k}{2}} & \frac{2\pi}{\frac{1}{2}+\nu+\frac{k}{2}} \end{pmatrix} \cdot \tilde{\Psi}_{k,\nu}(s) \right)^{-1},$$

see (4.52).

Maass includes also the special case $\frac{1\pm l}{2} + \nu \in \mathbb{Z}$ and $l \in \mathbb{Z}$.

Example 4.32 Assume that $k > 0$ is given. We assign $\nu = \frac{1-k}{2}$ and $0 < \kappa < 1$ with $\frac{k}{12} - \kappa \in \mathbb{Z}$ (i.e., $\frac{k}{12} \equiv \kappa \pmod 1$). (The assignment $\nu = \frac{1-k}{2}$ was derived in Example 4.6 for modular cusp forms that are lifted to families of Maass cusp forms.)

Let η^{2k} be the cusp form and $u_{\eta^{2k}}$ the Maass cusp form discussed in §3.4.3. In Example 4.15, we show that u_l has the expansion

$$u_l(z) = \sum_{n=0}^{\infty} A_k(n) \left(4\pi \left(n + \frac{k}{12}\right)\right)^{-\frac{k}{2}} \psi_{l,\frac{1-k}{2}} \left(\left(n + \frac{k}{12}\right) \Im(z)\right) e^{2\pi i \left(n+\frac{k}{12}\right)\Re(z)}.$$

Then Theorem 4.28 associates L series with the system of cusp forms. We have that $L_-(s) = 0$ and

$$\begin{aligned} L_+(s) &= \sum_{\substack{n\in\mathbb{Z} \\ n+\frac{k}{12}>0}} A_{\eta^{2k}}(n) \left(4\pi \left(n + \frac{k}{12}\right)\right)^{-\frac{k}{2}} \left(n + \frac{k}{12}\right)^{-s} \\ &= (4\pi)^{-\frac{k}{2}} \sum_{n=0}^{\infty} A_{\eta^{2k}}(n) \left(n + \frac{k}{12}\right)^{-s-\frac{k}{2}}. \end{aligned} \tag{4.71}$$

The function $L_k^\star$ in (4.55) simplifies to

$$\begin{aligned} L_k^\star(s) &= \tilde{\psi}_{k,\frac{1-k}{2}}(s) \sum_{\substack{n\in\mathbb{Z} \\ n+\frac{k}{12}>0}} A_k(n) \left(4\pi \left(n + \frac{k}{12}\right)\right)^{-\frac{k}{2}} \left(n + \frac{k}{12}\right)^{\frac{1}{2}-s} \\ &= 2^{\frac{k}{2}-s-1}\pi^{-s}\Gamma\left(s + \frac{k}{2}\right) (4\pi)^{-\frac{k}{2}} \sum_{n=0}^{\infty} A_{\eta^{2k}}(n) \left(n + \frac{k}{12}\right)^{-s-\frac{k}{2}} \\ &= \frac{1}{2}(2\pi)^{-s-\frac{k}{2}}\Gamma\left(s + \frac{k}{2}\right) \sum_{n=0}^{\infty} A_{\eta^{2k}}(n) \left(n + \frac{k}{12}\right)^{-s-\frac{k}{2}}, \end{aligned} \tag{4.72}$$

where we used (4.30) in Lemma 4.16 and ${}_2F_1\left[\begin{matrix}0,\, b\\ c\end{matrix}\middle| w\right] = 1$, see Appendix A.10, to calculate

$$\begin{aligned}
\tilde{\psi}_{k,\frac{1-k}{2}}(s) &\overset{(4.30)}{=} 2^{\frac{k}{2}-s-1}\pi^{-s}\frac{\Gamma\left(\frac12+s+\frac{1-k}{2}\right)\Gamma\left(\frac12+s-\frac{1-k}{2}\right)}{\Gamma\left(\frac12+\frac{1-k}{2}+\frac{k}{2}\right)\Gamma\left(1+s-\frac{k}{2}\right)} \\
&\qquad \cdot {}_2F_1\left[\begin{matrix}\frac12+\frac{1-1}{2}-\frac{k}{2},\, \frac12-\frac{1-k}{2}-\frac{k}{2}\\ 1+s-\frac{k}{2}\end{matrix}\middle| \frac12\right] \\
&= 2^{\frac{k}{2}-s-1}\pi^{-s}\frac{\Gamma\left(1+s-\frac{k}{2}\right)\Gamma\left(s+\frac{k}{2}\right)}{\Gamma(1)\,\Gamma\left(1+s-\frac{k}{2}\right)}\underbrace{{}_2F_1\left[\begin{matrix}1-k,0\\ 1+s-\frac{k}{2}\end{matrix}\middle| \frac12\right]}_{=1} \\
&= 2^{\frac{k}{2}-s-1}\pi^{-s}\Gamma\left(s+\frac{k}{2}\right).
\end{aligned}$$

Since $L_k^\star$ is invariant under $s\mapsto -s$, see (4.56) of Theorem 2, we find that

$$\begin{aligned}
&\frac12 (2\pi)^{-s-\frac{k}{2}}\Gamma\left(s+\frac{k}{2}\right)\sum_{n=0}^{\infty} A_{\eta^{2k}}(n)\left(n+\frac{k}{12}\right)^{-s-\frac{k}{2}} \\
&\overset{(4.72)}{=} L_k^\star(s) \overset{(4.56)}{=} L_k^\star(-s) \\
&\overset{(4.72)}{=} \frac12 (2\pi)^{s-\frac{k}{2}}\Gamma\left(-s+\frac{k}{2}\right)\sum_{n=0}^{\infty} A_{\eta^{2k}}(n)\left(n+\frac{k}{12}\right)^{s-\frac{k}{2}}
\end{aligned}$$

holds. In other words, the function

$$2\,L_k^\star\left(s-\frac{k}{2}\right) = (2\pi)^{-s}\,\Gamma(s)\sum_{n=0}^{\infty} A_k(n)\left(n+\frac{k}{12}\right)^{-s} \tag{4.73}$$

is invariant under $s\mapsto k-s$.

For weight $k=12$, this agrees with Hecke's classical result stated in Corollary 1.217. For $k=12$, we have $v=1$ the trivial multiplier and $\eta^{2k}=\Delta$ the discriminant function given Definition (1.86) of §1.3.5, and that is a modular cusp form of weight 12 and trivial multiplier. The L-function associated by Corollary 1.217 is given by

$$\begin{aligned} L(s) &\overset{(1.153)}{=} \sum_{n=1} \tau(n)\, n^s \\ &= \sum_{n=0} A_{12}(n)\,(n+1)^s \\ &\overset{(4.71)}{=} L_+\left(s - \frac{k}{2}\right) \end{aligned}$$

converging for $\Re(s) > \frac{k}{2} + 1$ with $\tau(n) = A_{12}(n-1)$ by construction of A_k. The associated $L^\star$ function

$$L^\star(s) \overset{(1.160)}{=} (2\pi)^{-s}\,\Gamma(s)\,L(s) \overset{(4.73)}{=} L^\star_{12}(s-6)$$

extends to an entire function and satisfies the functional equation

$$L^\star(s) = (-1)^{\frac{12}{2}}\, L^\star(12-s) \quad \Longleftrightarrow \quad L^\star_{12}(s) = L^\star_{12}(-s).$$

4.2.2 A Simpler Representation of Families of Cusp Forms

In the previous subsection, we studied the function $\tilde{\psi}_{l,\nu}$. We have seen that $\tilde{\psi}_{l,\nu}$, which appears in the expansion of the families of cusp forms $\left(u_l\right)_{l\in k+2\mathbb{Z}}$, depends in particular on the weight $l \in k + 2\mathbb{Z}$. Based on an idea of Motohashi in [146, p. 39], we write $\tilde{\psi}_{l,\nu}$ as a combination of generalized beta functions $B_{\pm l,\nu}$, which we introduce below. This will enable us to isolate the essential factor of the L-series in Theorem 4.28 from the influence of the weight l.

Our main result on generalized beta functions is given in Lemma 4.36. Then, we state an alternative version of Maass theorem on L-series in Theorem 4.39 where we factor out the dependency on weight $l \in k + 2\mathbb{Z}$.

Let $l \in \mathbb{R}$ and $\Re(\nu) > -\frac{1}{2}$ in this subsection. We define the function

$$B_{l,\nu} : \left\{ s:\ -\frac{1}{2} - \Re(\nu) < \Re(s) < \frac{1}{2} + \Re(\nu) \right\} \longrightarrow \mathbb{C}$$

by

$$\begin{aligned} B_{l,\nu}(s) &= \int_0^\infty \left(t + t^{-1}\right)^{-\frac{1}{2}-\nu} \left(\frac{t+i}{t-i}\right)^{\frac{l}{2}} t^{s-1}\, \mathrm{d}t \\ &= \int_0^\infty (t+i)^{-\frac{1}{2}-\nu+\frac{l}{2}} (t-i)^{-\frac{1}{2}-\nu-\frac{l}{2}} t^{s+\nu-\frac{1}{2}}\, \mathrm{d}t, \end{aligned} \tag{4.74}$$

where we choose the branch

$$\lim_{t\to\infty}\left(\frac{t+i}{t-i}\right)^l = 1.$$

The integral converges absolutely for $-\Re(\nu)-\frac{1}{2} < \Re(s) < \Re(\nu)+\frac{1}{2}$, and $B_{k,\nu}(s)$ is holomorphic on this vertical strip.

Remark 4.33 The function $B_{l,\nu}$ can be expressed in terms of the hypergeometric functions. We have

$$B_{l,\nu}(s) = i\frac{2^{-\frac{1}{2}-\nu}\pi\Gamma\left(\frac{1}{2}+s+\nu\right)}{\cos\left(\pi(s-\nu)\right)}\cdot \tag{4.75}$$

$$\cdot\left(\frac{2^{\frac{l}{2}}e^{-\frac{\pi i}{2}\left(\frac{1}{2}+s-\nu\right)}}{\Gamma\left(\frac{1+l}{2}+\nu\right)\Gamma\left(1+s-\frac{l}{2}\right)}\,{}_2F_1\left[\begin{matrix}\frac{1}{2}+\nu-\frac{l}{2},\ \frac{1}{2}-\nu-\frac{l}{2}\\ 1+s-\frac{l}{2}\end{matrix}\middle|\frac{1}{2}\right]-\right.$$

$$\left.-\frac{2^{-\frac{l}{2}}e^{\frac{\pi i}{2}\left(\frac{1}{2}+s-\nu\right)}}{\Gamma\left(\frac{1}{2}+\nu-\frac{l}{2}\right)\Gamma\left(1+s+\frac{l}{2}\right)}\,{}_2F_1\left[\begin{matrix}\frac{1}{2}+\nu+\frac{l}{2},\ \frac{1}{2}-\nu+\frac{l}{2}\\ 1+s+\frac{l}{2}\end{matrix}\middle|\frac{1}{2}\right]\right).$$

It is not hard to see that (4.75) implies that $B_{l,\nu}$ continues meromorphically to all $s\in\mathbb{C}$.

Remark 4.34 A special case of the $B_{l,\nu}$-function is related directly to the Euler beta function. We have

$$\begin{aligned}2\,B_{0,\nu}(s) &\overset{(4.74)}{=} 2\int_0^\infty\left(t+t^{-1}\right)^{-\frac{1}{2}-\nu}t^{s-1}\,\mathrm{d}t\\ &= 2\int_0^\infty\left(t^2+1\right)^{-\frac{1}{2}-\nu}t^{s-\frac{1}{2}+\nu}\,\mathrm{d}t\\ &= B\left(\frac{1+2\nu+2s}{4},\ \frac{1+2\nu-2s}{4}\right),\end{aligned}$$

using [89, 90, 8.380.3] in the last step. The function $B(x,y)=\frac{\Gamma(x)\Gamma(y)}{\Gamma(x+y)}$ is Euler's Beta function, see Appendix A.7.

Exercise 4.35 Show that $B_{l,\nu}$ satisfies the recurrence relations

$$B_{l,\nu}(s) = e^{\pi i\frac{l}{2}}\,B_{-l,\nu}(-s), \tag{4.76}$$

$$B_{l,\nu-1}(s) = B_{l,\nu}(s-1)+B_{l,\nu}(s+1)\qquad\text{and} \tag{4.77}$$

$$B_{l+2,\nu+1}(s) = B_{l,\nu}(s+1)+2i\,B_{l,\nu}(s)-B_{l,\nu}(s-1) \tag{4.78}$$

for $-\Re(\nu)-\frac{1}{2}<\Re(s)<\Re(\nu)+\frac{1}{2}$.

Lemma 4.36 *Let $l \in \mathbb{R}$, $\Re(\nu) > -\frac{1}{2}$ and $-\Re(\nu) - \frac{1}{2} < \Re(s) < \Re(\nu) + \frac{1}{2}$. The two functions*

$$B_{l,\nu}(s) \quad \text{and} \quad B_{-l,\nu}(s)$$

are linearly independent for all $l \in k + 2\mathbb{Z}$. In other words, if there are $\alpha, \beta \in \mathbb{C}$ satisfying

$$\alpha\, B_{l,\nu}(s) + \beta\, B_{-l,\nu}(s) = 0 \qquad \text{for all } s \text{ and } l \in k + 2\mathbb{Z},$$

then

$$\alpha = \beta = 0.$$

Proof Let $-\Re(\nu) - \frac{1}{2} < \Re(s) < \Re(\nu) + \frac{1}{2}$. We start with the integral representation (4.74) of $B_{l+4m,\nu}$ for $m \in \mathbb{Z}$. Substituting $t = \tan\theta = \frac{\sin\theta}{\cos\theta}$, we get

$$\begin{aligned}
&B_{l+4m,\nu}(s)\\
&\overset{(4.74)}{=} \int_0^\infty \left(t + t^{-1}\right)^{-\frac{1}{2}-\nu} \left(\frac{t+i}{t-i}\right)^{\frac{l}{2}+2m} t^{s-1}\, dt\\
&\overset{t=\tan\theta}{=} \int_0^{\frac{\pi}{2}} \left(\frac{\sin\theta}{\cos\theta} + \frac{\cos\theta}{\sin\theta}\right)^{-\frac{1}{2}-\nu} \left(\frac{\frac{\sin\theta}{\cos\theta} + i}{\frac{\sin\theta}{\cos\theta} - i}\right)^{\frac{l}{2}+2m} \left(\frac{\sin\theta}{\cos\theta}\right)^{s-1} \frac{1}{\cos^2\theta}\, d\theta\\
&= \int_0^{\frac{\pi}{2}} \left(\cos\theta\right)^{-\frac{1}{2}-s+\nu} \left(\sin\theta\right)^{-\frac{1}{2}+s+\nu} e^{\pi i \frac{l}{2}}\, e^{-il\theta}\, e^{-4im\theta}\, d\theta.
\end{aligned}$$

For $-\frac{1}{2} - \Re(\nu) < \Re(s) < \frac{1}{2} + \Re(\nu)$, define the function

$$g_{l,\nu,s}(\theta) := \left(\cos\theta\right)^{-\frac{1}{2}-s+\nu} \left(\sin\theta\right)^{-\frac{1}{2}+s+\nu} e^{\pi i \frac{l}{2}}\, e^{-il\theta} \qquad (0 \le \theta \le \frac{\pi}{2}),$$

that is a factor of the integrand above. $g_{l,\nu,s}$ is an L^1-function on the interval $\left[0, \frac{\pi}{2}\right]$ and continuous on $\left(0, \frac{\pi}{2}\right)$. Hence, we can view $B_{l+4m,\nu}$ as the m-th Fourier coefficient of $g_{k,\nu,s}$ with $m \in \mathbb{Z}$. Moreover the $g_{l,\nu,s}$ satisfies the relation

$$g_{l+2,\nu,s}(\theta) = e^{\pi i}\, e^{-2i\theta}\, g_{l,\nu,s}(\theta), \tag{4.79}$$

since we have

$$\begin{aligned}
g_{l+2,\nu,s}(\theta) &= \left(\cos\theta\right)^{-\frac{1}{2}-s+\nu} \left(\sin\theta\right)^{-\frac{1}{2}+s+\nu} e^{\pi i \frac{l+2}{2}}\, e^{-i(l+2)\theta}\\
&= e^{\pi i}\, e^{-2i\theta} \left(\cos\theta\right)^{-\frac{1}{2}-s+\nu} \left(\sin\theta\right)^{-\frac{1}{2}+s+\nu} e^{\pi i \frac{l}{2}}\, e^{-il\theta}\\
&= e^{\pi i}\, e^{-2i\theta}\, g_{l,\nu,s}(\theta).
\end{aligned}$$

We want to show that the sequences $\left(B_{l+2m,\nu}(s)\right)_{m\in\mathbb{Z}}$ and $\left(B_{-(l+2m),\nu}(s)\right)_{m\in\mathbb{Z}}$ are linearly independent. Assume that we have $\alpha, \beta \in \mathbb{C}$ such that

$$\alpha B_{l+2m,\nu}(s) + \beta B_{-(l+2m),\nu}(s) = 0 \qquad \text{for all } m \in \mathbb{Z}. \tag{4.80}$$

If we show that this implies $\alpha = \beta = 0$, then we can conclude that $B_{l+2m,\nu}(s)$ and $B_{-(l+2m),\nu}(s)$ are linearly independent.

Using (4.76) and our assumption (4.80), we have

$$\begin{aligned}
&\alpha \begin{pmatrix} B_{l+4m,\nu}(s) \\ B_{k+4m+2,\nu}(s) \end{pmatrix} + \beta \begin{pmatrix} e^{-\frac{\pi i l}{2}} B_{l+4m,\nu}(-s) \\ e^{-\frac{\pi i (l+2)}{2}} B_{l+4m+2,\nu}(-s) \end{pmatrix} \\
&\overset{(4.76)}{=} \alpha \begin{pmatrix} B_{l+4m,\nu}(s) \\ B_{k+4m+2,\nu}(s) \end{pmatrix} + \beta \begin{pmatrix} e^{-\frac{\pi i l}{2}} e^{\pi i \frac{l+4m}{2}} B_{-(l+4m),\nu}(s) \\ e^{-\frac{\pi i (l+2)}{2}} e^{\pi i \frac{l+4m+2}{2}} B_{-(l+4m+2),\nu}(s) \end{pmatrix} \\
&= \alpha \begin{pmatrix} B_{l+4m,\nu}(s) \\ B_{k+4m+2,\nu}(s) \end{pmatrix} + \beta \begin{pmatrix} B_{-(l+4m),\nu}(s) \\ B_{-(l+4m+2),\nu}(s) \end{pmatrix} \\
&\overset{(4.80)}{=} 0
\end{aligned}$$

for all $m \in \mathbb{Z}$. This implies

$$\alpha \begin{pmatrix} g_{l,\nu,s}(\theta) \\ g_{l+2,\nu,s}(\theta) \end{pmatrix} + \beta \begin{pmatrix} e^{-\frac{\pi i l}{2}} g_{l,\nu,-s}(\theta) \\ e^{-\frac{\pi i l+2}{2}} g_{l+2,\nu,-s}(\theta) \end{pmatrix} = 0 \tag{4.81}$$

as L^1-functions since all Fourier coefficients of the function above vanish. All functions in (4.81) are continuous on the open interval $\left(0, \frac{\pi}{4}\right)$. Hence equation (4.81) holds for all $0 < \theta < \frac{\pi}{2}$. Using (4.79), we simplify (4.81) to

$$\begin{aligned}
&\alpha \begin{pmatrix} 1 \\ e^{\pi i}\, e^{-2i\theta} \end{pmatrix} g_{l,\nu,s}(\theta) + \beta \begin{pmatrix} e^{-\frac{\pi i l}{2}} \\ e^{-\frac{\pi i l+2}{2}} e^{\pi i}\, e^{-2i\theta} \end{pmatrix} g_{l,\nu,-s}(\theta) \\
&= \alpha \begin{pmatrix} 1 \\ e^{\pi i}\, e^{-2i\theta} \end{pmatrix} g_{l,\nu,s}(\theta) + \beta \begin{pmatrix} 1 \\ e^{-2i\theta} \end{pmatrix} e^{-\frac{\pi i l}{2}} g_{l,\nu,-s}(\theta) \\
&= 0.
\end{aligned}$$

Choosing $\theta = \frac{\pi}{4}$, we have $\cos\frac{\pi}{4} = \sin\frac{\pi}{4} = \frac{1}{\sqrt{2}}$ which implies

$$g_{l,\nu,s}\left(\frac{\pi}{4}\right) = 2^{\frac{1}{2}-\nu}\, e^{\pi i \frac{l}{2}}\, e^{-\frac{\pi i}{4} l}.$$

Hence, the above vector identity gives

$$0 \overset{\theta=\frac{\pi}{4}}{=} \alpha \begin{pmatrix} 1 \\ e^{\pi i}\, e^{-\frac{\pi i}{2}} \end{pmatrix} g_{l,\nu,s}\left(\frac{\pi}{4}\right) + \beta \begin{pmatrix} 1 \\ e^{-\frac{\pi i}{2}} \end{pmatrix} e^{-\frac{\pi i l}{2}}\, g_{l,\nu,-s}\left(\frac{\pi}{4}\right)$$

$$= \alpha \begin{pmatrix} 1 \\ i \end{pmatrix} 2^{\frac{1}{2}-\nu} e^{\pi i \frac{l}{4}} + \beta \begin{pmatrix} 1 \\ -i \end{pmatrix} 2^{\frac{1}{2}-\nu} e^{-\frac{\pi i}{4} l}.$$

This can only happen for $\alpha = \beta = 0$. □

Lemma 4.37 *Let $l \in \mathbb{R}$ and $\Re(\nu) > -\frac{1}{2}$. The function $\tilde{\psi}_{l,\nu}$ can be written in terms of $B_{l,\nu}$: We have*

$$\begin{aligned} \tilde{\psi}_{l,\nu}(s) &= 2^{\nu-s-\frac{1}{2}} \pi^{-s-1}\, \Gamma\left(\frac{1}{2}+s-\nu\right) \\ &\quad \cdot \left[e^{-\frac{\pi i}{2}\left(\frac{1}{2}+s-\nu\right)} B_{l,\nu}(s) + e^{\frac{\pi i}{2}\left(\frac{1}{2}+s-\nu\right)} B_{-l,\nu}(s) \right] \end{aligned} \tag{4.82}$$

for $-\frac{1}{2} - \Re(\nu) < \Re(s) < \frac{1}{2} + \Re(\nu)$.

Proof Let $\Re(\nu) > -\frac{1}{2}$ and $-\frac{1}{2} - \Re(\nu) < \Re(s) < \frac{1}{2} + \Re(\nu)$. We consider the integral representation (4.39) of $\tilde{\psi}_{l,\nu}$ in Lemma 4.20:

$$\begin{aligned} \tilde{\psi}_{l,\nu}(s) &\overset{(4.39)}{=} 2^{\nu-s-\frac{1}{2}} \pi^{-s-1} e^{-\frac{\pi i}{2} l}\, \Gamma\left(\frac{1}{2}+s-\nu\right) \\ &\quad \cdot \int_{\mathbb{R}} (-i\xi)^{\nu-s-\frac{1}{2}} (\xi - i)^{-\frac{1}{2}-\nu-\frac{l}{2}} (\xi + i)^{-\frac{1}{2}-\nu+\frac{l}{2}}\, d\xi. \end{aligned}$$

Splitting the integral at $\xi = 0$, and applying the substitution $\xi \to -\xi$ to the integral over $(-\infty, 0]$, we obtain

$$\begin{aligned} \tilde{\psi}_{l,\nu}(s) &= 2^{\nu-s-\frac{1}{2}} \pi^{-s-1} e^{-\frac{\pi i}{2} l}\, \Gamma\left(\frac{1}{2}+s-\nu\right) \\ &\quad \cdot \left[\int_{-\infty}^{0} (-i\xi)^{\nu-s-\frac{1}{2}} (\xi - i)^{-\frac{1}{2}-\nu-\frac{l}{2}} (\xi + i)^{-\frac{1}{2}-\nu+\frac{l}{2}}\, d\xi \right. \\ &\qquad \left. + \int_{0}^{\infty} (-i\xi)^{\nu-s-\frac{1}{2}} (\xi - i)^{-\frac{1}{2}-\nu-\frac{l}{2}} (\xi + i)^{-\frac{1}{2}-\nu+\frac{l}{2}}\, d\xi \right] \\ &\overset{\xi \mapsto -\xi}{=} 2^{\nu-s-\frac{1}{2}} \pi^{-s-1} e^{-\frac{\pi i}{2} l}\, \Gamma\left(\frac{1}{2}+s-\nu\right) \\ &\quad \cdot \left[\int_{0}^{\infty} (i\xi)^{\nu-s-\frac{1}{2}} (-\xi - i)^{-\frac{1}{2}-\nu-\frac{l}{2}} (-\xi + i)^{-\frac{1}{2}-\nu+\frac{l}{2}}\, d\xi \right. \\ &\qquad \left. + \int_{0}^{\infty} (-i\xi)^{\nu-s-\frac{1}{2}} (\xi - i)^{-\frac{1}{2}-\nu-\frac{l}{2}} (\xi + i)^{-\frac{1}{2}-\nu+\frac{l}{2}}\, d\xi \right]. \end{aligned}$$

Looking at the first integrand and recalling the argument convention (1.30), we have written

$$-i\xi = e^{-\frac{\pi i}{2}}\xi \quad \text{and} \quad i\xi = e^{\frac{\pi i}{2}}\xi$$

for all $\xi > 0$. We would like to factor out the "$-$"-sign in the factors $-\xi - i$ and $-\xi + i$ in such a way that we do not cross the jump in the argument on the negative real axis. We write

$$-\xi - i = e^{-\pi i}(\xi + i) \quad \text{and} \quad -\xi + i = e^{\pi i}(\xi - i)$$

for all $\xi > 0$. Hence, we can rewrite the first integrand as

$$\begin{aligned}
&(i\xi)^{\nu-s-\frac{1}{2}}(-\xi - i)^{-\frac{1}{2}-\nu-\frac{l}{2}}(-\xi + i)^{-\frac{1}{2}-\nu+\frac{l}{2}} \\
&= e^{-\frac{\pi i}{2}\left(\frac{1}{2}+s-\nu-2l\right)}\,\xi^{\nu-s-\frac{1}{2}}\,(\xi + i)^{-\frac{1}{2}-\nu-\frac{l}{2}}\,(\xi - i)^{-\frac{1}{2}-\nu+\frac{l}{2}}.
\end{aligned}$$

Plugging in, we get

$$\begin{aligned}
\tilde{\psi}_{l,\nu}(s) &= 2^{\nu-s-\frac{1}{2}}\pi^{-s-1}\,\Gamma\left(\frac{1}{2}+s-\nu\right) \\
&\quad\cdot\left[e^{-\frac{\pi i}{2}\left(\frac{1}{2}+s-\nu-l\right)}\int_0^\infty \xi^{\nu-s-\frac{1}{2}}\,(\xi + i)^{-\frac{1}{2}-\nu-\frac{l}{2}}\,(\xi - i)^{-\frac{1}{2}-\nu+\frac{l}{2}}\,d\xi\right. \\
&\qquad\left. +\, e^{\frac{\pi i}{2}\left(\frac{1}{2}+s-\nu-l\right)}\int_0^\infty \xi^{\nu-s-\frac{1}{2}}(\xi - i)^{-\frac{1}{2}-\nu-\frac{l}{2}}(\xi + i)^{-\frac{1}{2}-\nu+\frac{l}{2}}\,d\xi\right] \\
&\overset{(4.74)}{=} 2^{\nu-s-\frac{1}{2}}\pi^{-s-1}\,\Gamma\left(\frac{1}{2}+s-\nu\right) \\
&\quad\cdot\left[e^{-\frac{\pi i}{2}\left(\frac{1}{2}+s-\nu-l\right)}B_{-l,\nu}(-s) + e^{\frac{\pi i}{2}\left(\frac{1}{2}+s-\nu-l\right)}B_{l,\nu}(-s)\right] \\
&\overset{(4.76)}{=} 2^{\nu-s-\frac{1}{2}}\pi^{-s-1}\,\Gamma\left(\frac{1}{2}+s-\nu\right) \\
&\quad\cdot\left[e^{-\frac{\pi i}{2}\left(\frac{1}{2}+s-\nu-l\right)}e^{-\frac{\pi i}{2}l}B_{l,\nu}(s) + e^{\frac{\pi i}{2}\left(\frac{1}{2}+s-\nu-l\right)}e^{\frac{\pi i}{2}l}B_{-l,\nu}(s)\right] \\
&= 2^{\nu-s-\frac{1}{2}}\pi^{-s-1}\,\Gamma\left(\frac{1}{2}+s-\nu\right) \\
&\quad\cdot\left[e^{-\frac{\pi i}{2}\left(\frac{1}{2}+s-\nu\right)}B_{l,\nu}(s) + e^{\frac{\pi i}{2}\left(\frac{1}{2}+s-\nu\right)}B_{-l,\nu}(s)\right].
\end{aligned}$$

This proves the lemma. □

Remark 4.38 Equation (4.75) is essentially the inverse of (4.82): The Equation (4.82) for l and $-l$ gives an invertible system of equations which allows us to express $B_{l,\nu}(s)$ in terms of $\tilde{\psi}_{l,\nu}(s)$ and $\tilde{\psi}_{-l,\nu}(s)$. Applying (4.30) gives (4.75).

Let us focus again on the L-series representation of an admissible family of cusp forms $\big(u_l\big)_{l\in k+2\mathbb{Z}}$. The result given in Theorem 4.28 is a bit complicated in the sense that we have infinitely many entire functions $L_l^\star$ to consider. On the other hand, we only have one pair (L_+, L_-) of L-series.

We now derive another representation in terms of the L-series (L_+, L_-) and a vector-valued functional equation relating this pair to itself.

Theorem 4.39 *Let $k \in \mathbb{R}$, $0 \leq \kappa < 1$ and $\nu \in \mathbb{C}$ with $\Re(\nu) > -\frac{1}{2}$ such that $2\nu \notin \mathbb{Z}_{\neq 0}$ and such that if $2\nu + k$ or $2\nu - k \in 1 + 2\mathbb{Z}$, then $k \notin \mathbb{Z}$.*

There is a canonical correspondence between the following:

1. *An admissible family of Maass cusp forms $(u_l)_{l\in k+2\mathbb{Z}}$ as discussed in §4.1. Here, we denote the "index-shift" in the Whittaker-Fourier expansion (4.24) of u_l by $0 \leq \kappa < 1$.*
2. *A pair of functions L_+ and L_- satisfying*
   a. *$L_\pm$ have a series representation*

$$\begin{aligned} L_+(s) &= \sum_{\substack{n\in\mathbb{Z}\\ n+\kappa>0}} A(n)\,(n+\kappa)^{-s} \\ L_-(s) &= \sum_{\substack{n\in\mathbb{Z}\\ n+\kappa<0}} A(n)\,|n+\kappa|^{-s} \end{aligned} \tag{4.83}$$

   that converges on a right half-plane.

   b. *The two functions L_+ and L_- extend to entire functions of s of finite order. They satisfy the functional equation*

$$\begin{pmatrix} L_+(-s) \\ L_-(-s) \end{pmatrix} = -\,(2\pi)^{-2s}\,\frac{\Gamma\left(\frac{1}{2}+s+\nu\right)\Gamma\left(\frac{1}{2}+s-\nu\right)}{\pi}\,\cdot \\ \cdot \begin{pmatrix} \sin\left(\pi\left(s-\frac{k}{2}\right)\right) & \sin\left(\pi\left(\nu-\frac{k+1}{2}\right)\right) \\ \sin\left(\pi\left(\nu+\frac{k-1}{2}\right)\right) & \sin\left(\pi\left(s+\frac{k}{2}\right)\right) \end{pmatrix} \begin{pmatrix} L_+(s) \\ L_-(s) \end{pmatrix} \tag{4.84}$$

   for all $s \in \mathbb{C}$.

Remark 4.40 The two restrictions on ν in the beginning of the theorem are technical restrictions. The first restriction is necessary for Lemma 4.37. We need the second restriction in the proof of the direction from 2 to 1, where we want to show that (4.91) is an identity of holomorphic functions. We could remove the second condition, since the condition "if $2\nu + k$ or $2\nu - k \in 1 + 2\mathbb{Z}$ then $k \notin \mathbb{Z}$" implies $2\nu \notin \mathbb{Z} \setminus \{0\}$.

Before we prove Theorem 4.39, we show the following trigonometric identities.

Lemma 4.41 *For $\frac{1}{2} - s - \nu \notin \mathbb{Z}$, we have the matrix identity*

$$\frac{1}{\sin\left(\pi\left(s+\nu-\frac{1}{2}\right)\right)}\begin{pmatrix}\sin\left(\pi\left(s-\frac{l}{2}\right)\right) & \sin\left(\pi\left(\nu-\frac{l+1}{2}\right)\right)\\ \sin\left(\pi\left(\nu+\frac{l-1}{2}\right)\right) & \sin\left(\pi\left(s+\frac{l}{2}\right)\right)\end{pmatrix}$$
$$= \begin{pmatrix} e^{-\frac{\pi i}{2}\left(\frac{1}{2}-s-\nu\right)} & e^{\frac{\pi i}{2}\left(\frac{1}{2}-s-\nu\right)} \\ e^{\frac{\pi i}{2}\left(\frac{1}{2}-s-\nu\right)} & e^{-\frac{\pi i}{2}\left(\frac{1}{2}-s-\nu\right)}\end{pmatrix}^{-1} \begin{pmatrix} 0 & e^{-\frac{\pi i l}{2}} \\ e^{\frac{\pi i l}{2}} & 0\end{pmatrix}^{-1} \cdot \tag{4.85}$$
$$\cdot \begin{pmatrix} e^{-\frac{\pi i}{2}\left(\frac{1}{2}+s-\nu\right)} & e^{\frac{\pi i}{2}\left(\frac{1}{2}+s-\nu\right)} \\ e^{\frac{\pi i}{2}\left(\frac{1}{2}+s-\nu\right)} & e^{-\frac{\pi i}{2}\left(\frac{1}{2}+s-\nu\right)}\end{pmatrix}.$$

Moreover, we have

$$\begin{pmatrix}\sin\left(\pi\left(s+\frac{l}{2}\right)\right) & \sin\left(\pi\left(\nu+\frac{l-1}{2}\right)\right)\\ \sin\left(\pi\left(\nu-\frac{l+1}{2}\right)\right) & \sin\left(\pi\left(s-\frac{l}{2}\right)\right)\end{pmatrix}$$
$$= (-1)^{k-l}\begin{pmatrix}\sin\left(\pi\left(s+\frac{k}{2}\right)\right) & \sin\left(\pi\left(\nu+\frac{k-1}{2}\right)\right)\\ \sin\left(\pi\left(\nu-\frac{k+1}{2}\right)\right) & \sin\left(\pi\left(s-\frac{k}{2}\right)\right)\end{pmatrix}. \tag{4.86}$$

Proof Identity (4.85) can be shown directly and will be left as an exercise. To show identity (4.86), consider the trigonometric identity

$$\sin\left(\pi\left(a\pm\frac{l}{2}\right)\right) = \sin\left(-\pi\left(a\pm\frac{k}{2}\mp\underbrace{\frac{l-k}{2}}_{\in\mathbb{Z}}\right)\right)$$
$$= (-1)^{\frac{k-l}{2}}\sin\left(-\pi\left(a\pm\frac{k}{2}\right)\right)$$

for all $l \in k + 2\mathbb{Z}$ and $a \in \mathbb{C}$ and apply it to each entry of the right-hand side matrix in (4.86). □

Proof *(of Theorem 4.39)* We apply Theorem 4.28. All we have to show is the correspondence between the L-series in Theorem 4.28, part 2, and part 2 of Theorem 4.39.

Theorem 4.28, part 2 $\Longrightarrow$ Theorem 4.39, Item 2 We have a pair of L-series $L_\pm$ given by 4.54 in Theorem 4.28. Then, $L_\pm$ has the form given in (4.83) of Theorem 4.39, part 2. The L-series $L_\pm$ extend to entire functions, as stated in Theorem 4.28.

We want to show that $L_\pm$ are entire functions of finite order. At the beginning of the proof of Theorem 4.28, we have shown that $L_\pm$ extend to entire functions. In particular, we have shown equation (4.62). The second part of Lemma 4.23 implies that $L_\pm(s)$ satisfies the growth condition

$$L_\pm(s) = O\left(e^{|s|^C}\right) \qquad \text{for some } C > 0 \tag{4.87}$$

for $\Re(s) \ge |\Re(\nu)|$ since the functions $L_l^\star$ are of finite order for all $l \in k + 2\mathbb{Z}$. Recall that the poles of $\Gamma\left(\frac{1}{2}+s+\nu\right)\Gamma\left(\frac{1}{2}+s-\nu\right)$ cancel the poles of $\tilde{\psi}_{\pm l,\nu}(s)$ and $\tilde{\psi}_{\pm l\pm 2,\nu}(s)$. Then the first part of Lemma 4.23 implies that (4.87) is satisfied for s in the larger right half-plane $\Re(s) > -|\Re(\nu)|$. If $L_\pm$ satisfies the functional equation (4.84), we see that $L_\pm$ are indeed entire functions of finite order.

Let us restrict s to the vertical strip $-\frac{1}{2}-\Re(\nu) < \Re(s) < \frac{1}{2}+\Re(\nu)$. Applying Equation (4.82) of Lemma 4.37 to the functions $\tilde{\psi}_{\pm l,\nu}$ in (4.55), we find

$$\begin{aligned} L_l^\star(s) &\overset{(4.55)}{=} \left(\tilde{\psi}_{l,\nu}(s),\ \tilde{\psi}_{-l,\nu}(s)\right)\begin{pmatrix} L_+(s) \\ L_-(s)\end{pmatrix} \\ &\overset{(4.82)}{=} 2^{\nu-s-\frac{1}{2}}\pi^{-s-1}\,\Gamma\left(\frac{1}{2}+s-\nu\right) \\ &\quad \cdot \left(B_{l,\nu}(s),\ B_{-l,\nu}(s)\right)\begin{pmatrix} e^{-\frac{\pi i}{2}\left(\frac{1}{2}+s-\nu\right)} & e^{\frac{\pi i}{2}\left(\frac{1}{2}+s-\nu\right)} \\ e^{\frac{\pi i}{2}\left(\frac{1}{2}+s-\nu\right)} & e^{-\frac{\pi i}{2}\left(\frac{1}{2}+s-\nu\right)}\end{pmatrix}\begin{pmatrix} L_+(s) \\ L_-(s)\end{pmatrix} \end{aligned} \tag{4.88}$$

for all admissible $l \in k + 2\mathbb{Z}$.

The functional equation (4.56) implies that this is equal to

$$\begin{aligned} &(-1)^{\frac{l-k}{2}}\,L_l^\star(-s) \\ &\overset{(4.55)}{=} (-1)^{\frac{l-k}{2}}\left(\tilde{\psi}_{l,\nu}(-s),\ \tilde{\psi}_{-l,\nu}(-s)\right)\begin{pmatrix} L_+(-s) \\ L_-(-s)\end{pmatrix} \\ &\overset{(4.82)}{=} (-1)^{\frac{l-k}{2}}\,2^{\nu+s-\frac{1}{2}}\pi^{s-1}\,\Gamma\left(\frac{1}{2}-s-\nu\right) \\ &\quad \cdot \left(B_{l,\nu}(-s),\ B_{-l,\nu}(-s)\right)\begin{pmatrix} e^{-\frac{\pi i}{2}\left(\frac{1}{2}-s-\nu\right)} & e^{\frac{\pi i}{2}\left(\frac{1}{2}-s-\nu\right)} \\ e^{\frac{\pi i}{2}\left(\frac{1}{2}-s-\nu\right)} & e^{-\frac{\pi i}{2}\left(\frac{1}{2}-s-\nu\right)}\end{pmatrix}\begin{pmatrix} L_+(-s) \\ L_-(-s)\end{pmatrix} \end{aligned}$$

$$\overset{(4.76)}{=} (-1)^{\frac{l-k}{2}}\, 2^{\nu+s-\frac{1}{2}}\pi^{s-1}\,\Gamma\left(\frac{1}{2}-s-\nu\right)\,\big(B_{l,\nu}(s),\,B_{-l,\nu}(s)\big)\begin{pmatrix}0 & e^{-\frac{\pi il}{2}}\\ e^{\frac{\pi il}{2}} & 0\end{pmatrix}$$
$$\cdot \begin{pmatrix} e^{-\frac{\pi i}{2}\left(\frac{1}{2}-s-\nu\right)} & e^{\frac{\pi i}{2}\left(\frac{1}{2}-s-\nu\right)}\\ e^{\frac{\pi i}{2}\left(\frac{1}{2}-s-\nu\right)} & e^{-\frac{\pi i}{2}\left(\frac{1}{2}-s-\nu\right)}\end{pmatrix}\begin{pmatrix}L_+(-s)\\ L_-(-s)\end{pmatrix}$$

for all $l \in k+2\mathbb{Z}$, where we used identity (4.82).
Since this relation between $L_\pm(s)$ and $L_\pm(-s)$ holds for all admissible $l \in k+2\mathbb{Z}$, and since the entries of

$$\big(B_{l,\nu}(s),\,B_{-l,\nu}(s)\big)_{l\in k+2\mathbb{Z}}$$

are linearly independent, see Lemma 4.36, we have the following vector identity:

$$\begin{pmatrix}0 & e^{-\frac{\pi il}{2}}\\ e^{\frac{\pi il}{2}} & 0\end{pmatrix}\begin{pmatrix} e^{-\frac{\pi i}{2}\left(\frac{1}{2}-s-\nu\right)} & e^{\frac{\pi i}{2}\left(\frac{1}{2}-s-\nu\right)}\\ e^{\frac{\pi i}{2}\left(\frac{1}{2}-s-\nu\right)} & e^{-\frac{\pi i}{2}\left(\frac{1}{2}-s-\nu\right)}\end{pmatrix}\begin{pmatrix}L_+(-s)\\ L_-(-s)\end{pmatrix}$$
$$= (-1)^{\frac{k-l}{2}}\,(2\pi)^{-2s}\,\frac{\Gamma\left(\frac{1}{2}+s-\nu\right)}{\Gamma\left(\frac{1}{2}-s-\nu\right)}\begin{pmatrix} e^{-\frac{\pi i}{2}\left(\frac{1}{2}+s-\nu\right)} & e^{\frac{\pi i}{2}\left(\frac{1}{2}+s-\nu\right)}\\ e^{\frac{\pi i}{2}\left(\frac{1}{2}+s-\nu\right)} & e^{-\frac{\pi i}{2}\left(\frac{1}{2}+s-\nu\right)}\end{pmatrix}\begin{pmatrix}L_+(s)\\ L_-(s)\end{pmatrix}.$$

Assume that $\frac{1}{2}-s-\nu \notin \mathbb{Z}$. Then, the matrix

$$\begin{pmatrix}0 & e^{-\frac{\pi il}{2}}\\ e^{\frac{\pi il}{2}} & 0\end{pmatrix}\begin{pmatrix} e^{-\frac{\pi i}{2}\left(\frac{1}{2}-s-\nu\right)} & e^{\frac{\pi i}{2}\left(\frac{1}{2}-s-\nu\right)}\\ e^{\frac{\pi i}{2}\left(\frac{1}{2}-s-\nu\right)} & e^{-\frac{\pi i}{2}\left(\frac{1}{2}-s-\nu\right)}\end{pmatrix}$$

on the left-hand side is invertible and we get

$$\begin{pmatrix}L_+(-s)\\ L_-(-s)\end{pmatrix}$$
$$= (-1)^{\frac{k-l}{2}}\,(2\pi)^{-2s}\,\frac{\Gamma\left(\frac{1}{2}+s-\nu\right)}{\Gamma\left(\frac{1}{2}-s-\nu\right)}$$
$$\cdot \begin{pmatrix} e^{-\frac{\pi i}{2}\left(\frac{1}{2}-s-\nu\right)} & e^{\frac{\pi i}{2}\left(\frac{1}{2}-s-\nu\right)}\\ e^{\frac{\pi i}{2}\left(\frac{1}{2}-s-\nu\right)} & e^{-\frac{\pi i}{2}\left(\frac{1}{2}-s-\nu\right)}\end{pmatrix}^{-1}\begin{pmatrix}0 & e^{-\frac{\pi il}{2}}\\ e^{\frac{\pi il}{2}} & 0\end{pmatrix}^{-1}$$
$$\cdot \begin{pmatrix} e^{-\frac{\pi i}{2}\left(\frac{1}{2}+s-\nu\right)} & e^{\frac{\pi i}{2}\left(\frac{1}{2}+s-\nu\right)}\\ e^{\frac{\pi i}{2}\left(\frac{1}{2}+s-\nu\right)} & e^{-\frac{\pi i}{2}\left(\frac{1}{2}+s-\nu\right)}\end{pmatrix}\begin{pmatrix}L_+(s)\\ L_-(s)\end{pmatrix}$$

$$\overset{(4.85)}{=} (-1)^{\frac{k-l}{2}} (2\pi)^{-2s} \frac{\Gamma\left(\frac{1}{2}+s-\nu\right)}{\Gamma\left(\frac{1}{2}-s-\nu\right)} \cdot \frac{1}{\sin\left(\pi\left(s+\nu-\frac{1}{2}\right)\right)} \begin{pmatrix} \sin\left(\pi\left(s-\frac{l}{2}\right)\right) & \sin\left(\pi\left(\nu-\frac{l+1}{2}\right)\right) \\ \sin\left(\pi\left(\nu+\frac{l-1}{2}\right)\right) & \sin\left(\pi\left(s+\frac{l}{2}\right)\right) \end{pmatrix} \begin{pmatrix} L_+(s) \\ L_-(s) \end{pmatrix},$$

using Lemma 4.41. Recalling the identities $\Gamma(s)\,\Gamma(1-s) = \frac{\pi}{\sin(\pi s)}$, we rewrite

$$\begin{aligned}
&\frac{\Gamma\left(\frac{1}{2}+s-\nu\right)}{\Gamma\left(\frac{1}{2}-s-\nu\right)} \frac{1}{\sin\left(\pi\left(s+\nu-\frac{1}{2}\right)\right)} \\
&= \frac{\Gamma\left(\frac{1}{2}+s-\nu\right)}{\Gamma\left(\frac{1}{2}-s-\nu\right)} \frac{\Gamma\left(s+\nu-\frac{1}{2}\right)\Gamma\left(\frac{3}{2}-s-\nu\right)}{\pi} \\
&= \frac{\Gamma\left(\frac{1}{2}+s-\nu\right)}{\Gamma\left(\frac{1}{2}-s-\nu\right)} \frac{\Gamma\left(s+\nu-\frac{1}{2}\right)\left(\frac{1}{2}-s-\nu\right)\Gamma\left(\frac{1}{2}-s-\nu\right)}{\pi} \qquad (4.89) \\
&= \left(\frac{1}{2}-s-\nu\right) \frac{\Gamma\left(s+\nu-\frac{1}{2}\right)\Gamma\left(\frac{1}{2}+s-\nu\right)}{\pi} \\
&= -\frac{\left(s+\nu-\frac{1}{2}\right)\Gamma\left(s+\nu-\frac{1}{2}\right)\Gamma\left(\frac{1}{2}+s-\nu\right)}{\pi} \\
&= -\frac{\Gamma\left(\frac{1}{2}+s+\nu\right)\Gamma\left(\frac{1}{2}+s-\nu\right)}{\pi}.
\end{aligned}$$

Hence, the above vector-valued identity reads

$$\begin{pmatrix} L_+(-s) \\ L_-(-s) \end{pmatrix} = -(-1)^{\frac{k-l}{2}} (2\pi)^{-2s} \frac{\Gamma\left(\frac{1}{2}+s+\nu\right)\Gamma\left(\frac{1}{2}+s-\nu\right)}{\pi} \cdot \begin{pmatrix} \sin\left(\pi\left(s-\frac{l}{2}\right)\right) & \sin\left(\pi\left(\nu-\frac{l+1}{2}\right)\right) \\ \sin\left(\pi\left(\nu+\frac{l-1}{2}\right)\right) & \sin\left(\pi\left(s+\frac{l}{2}\right)\right) \end{pmatrix} \begin{pmatrix} L_+(s) \\ L_-(s) \end{pmatrix}.$$

For $k = l$, the matrix equation above simplifies to (4.84) for $-\frac{1}{2} - \Re(\nu) < \Re(s) < \frac{1}{2} + \Re(\nu)$ and $\frac{1}{2} + s \pm \nu \notin \mathbb{Z}_{\leq 0}$. The vector $\begin{pmatrix} L_+(-s) \\ L_-(-s) \end{pmatrix}$ on the left-hand side of (4.84) contains only entire functions is s. This implies that (4.84) is true for all $s \in \mathbb{C}$ as identity on vectors of entire functions.

Theorem 4.39, part 2 $\Longrightarrow$ Theorem 4.28, Item 2 Let $l \in k + 2\mathbb{Z}$. We define the meromorphic function

$$L_l^\star : \mathbb{C} \to \mathbb{C}$$

by

$$\begin{aligned} L_l^\star(s) :&= \tilde{\psi}_{l,\nu}(s)\, L_+(s) + \tilde{\psi}_{-l,\nu}(s)\, L_-(s) \\ &= \left(\tilde{\psi}_{l,\nu}(s),\, \tilde{\psi}_{-l,\nu}(s)\right) \begin{pmatrix} L_+(s) \\ L_-(s) \end{pmatrix}. \end{aligned}$$

The function $L_l^\star$ may have poles only in $s \in \frac{1}{2} \pm \nu - \mathbb{Z}_{\geq 0}$ since $L_+(s)$ and $L_+(s)$ are entire and the functions $\tilde{\psi}_{\pm l,\nu}$ have poles in $s \in \frac{1}{2} \pm \nu - \mathbb{Z}_{\geq 0}$, see Lemma 4.16. We restrict s temporarily to the strip $-\frac{1}{2} - \Re(\nu) < \Re(s) < \frac{1}{2} + \Re(\nu)$. Then (4.82) in Lemma 4.37 implies

$$\begin{aligned} L_l^\star(s) &= \left(\tilde{\psi}_{l,\nu}(s),\, \tilde{\psi}_{-l,\nu}(s)\right) \begin{pmatrix} L_+(s) \\ L_-(s) \end{pmatrix} \\ &\overset{(4.82)}{=} 2^{\nu-s-\frac{1}{2}}\, \pi^{-s-1}\, \Gamma\left(\frac{1}{2} + s - \nu\right) \left(B_{l,\nu}(s),\, B_{-l,\nu}(s)\right) \\ &\quad \cdot \begin{pmatrix} e^{-\frac{\pi i}{2}(1+s-\nu)} & e^{\frac{\pi i}{2}(1+s-\nu)} \\ e^{\frac{\pi i}{2}(1+s-\nu)} & e^{-\frac{\pi i}{2}(1+s-\nu)} \end{pmatrix} \begin{pmatrix} L_+(s) \\ L_-(s) \end{pmatrix} \end{aligned} \tag{4.90}$$

and

$$\begin{aligned} L_l^\star(-s) &= \left(\tilde{\psi}_{l,\nu}(-s),\, \tilde{\psi}_{-l,\nu}(-s)\right) \begin{pmatrix} L_+(-s) \\ L_-(-s) \end{pmatrix} \\ &\overset{(4.82)}{=} 2^{\nu+s-\frac{1}{2}}\, \pi^{s-1}\, \Gamma\left(\frac{1}{2} - s - \nu\right) \left(B_{l,\nu}(-s),\, B_{-l,\nu}(-s)\right) \cdot \\ &\quad \cdot \begin{pmatrix} e^{-\frac{\pi i}{2}(1-s-\nu)} & e^{\frac{\pi i}{2}(1-s-\nu)} \\ e^{\frac{\pi i}{2}(1-s-\nu)} & e^{-\frac{\pi i}{2}(1-s-\nu)} \end{pmatrix} \begin{pmatrix} L_+(-s) \\ L_-(-s) \end{pmatrix}. \end{aligned}$$

Applying the functional equation (4.84), we get

$$
\begin{aligned}
&(-1)^{\frac{l-k}{2}} L_l^\star(-s)\\
&= (-1)^{\frac{l-k}{2}} 2^{\nu+s-\frac{1}{2}} \pi^{s-1} \Gamma\left(\frac{1}{2}-s-\nu\right) \left(B_{l,\nu}(-s),\ B_{-l,\nu}(-s)\right)\\
&\quad \cdot \begin{pmatrix} e^{-\frac{\pi i}{2}(1-s-\nu)} & e^{\frac{\pi i}{2}(1-s-\nu)} \\ e^{\frac{\pi i}{2}(1-s-\nu)} & e^{-\frac{\pi i}{2}(1-s-\nu)} \end{pmatrix} \begin{pmatrix} L_+(-s) \\ L_-(-s) \end{pmatrix}\\
&\overset{(4.84)}{=} 2^{\nu+s-\frac{1}{2}} \pi^{s-1} \Gamma\left(\frac{1}{2}-s-\nu\right) \left(B_{l,\nu}(-s),\ B_{-l,\nu}(-s)\right)\\
&\quad \cdot \begin{pmatrix} e^{-\frac{\pi i}{2}(1-s-\nu)} & e^{\frac{\pi i}{2}(1-s-\nu)} \\ e^{\frac{\pi i}{2}(1-s-\nu)} & e^{-\frac{\pi i}{2}(1-s-\nu)} \end{pmatrix}\\
&\quad \cdot (-1)\,(2\pi)^{-2s} \frac{\Gamma\left(\frac{1}{2}+s+\nu\right)\Gamma\left(\frac{1}{2}+s-\nu\right)}{\pi}\\
&\quad \cdot \underbrace{(-1)^{\frac{l-k}{2}} \begin{pmatrix} \sin\left(\pi\left(s-\frac{k}{2}\right)\right) & \sin\left(\pi\left(\nu-\frac{k+1}{2}\right)\right) \\ \sin\left(\pi\left(\nu+\frac{k-1}{2}\right)\right) & \sin\left(\pi\left(s+\frac{k}{2}\right)\right) \end{pmatrix}}_{(4.86)} \begin{pmatrix} L_+(s) \\ L_-(s) \end{pmatrix}\\
&\overset{(4.86)}{=} 2^{\nu+s-\frac{1}{2}} \pi^{s-1} \Gamma\left(\frac{1}{2}-s-\nu\right) \left(B_{l,\nu}(-s),\ B_{-l,\nu}(-s)\right)\\
&\quad \cdot \begin{pmatrix} e^{-\frac{\pi i}{2}(1-s-\nu)} & e^{\frac{\pi i}{2}(1-s-\nu)} \\ e^{\frac{\pi i}{2}(1-s-\nu)} & e^{-\frac{\pi i}{2}(1-s-\nu)} \end{pmatrix}\\
&\quad \cdot (-1)\,(2\pi)^{-2s} \frac{\Gamma\left(\frac{1}{2}+s+\nu\right)\Gamma\left(\frac{1}{2}+s-\nu\right)}{\pi}\\
&\quad \cdot \begin{pmatrix} \sin\left(\pi\left(s-\frac{l}{2}\right)\right) & \sin\left(\pi\left(\nu-\frac{l+1}{2}\right)\right) \\ \sin\left(\pi\left(\nu+\frac{l-1}{2}\right)\right) & \sin\left(\pi\left(s+\frac{l}{2}\right)\right) \end{pmatrix} \begin{pmatrix} L_+(s) \\ L_-(s) \end{pmatrix}\\
&= -2^{\nu-s-\frac{1}{2}} \pi^{-s-1} \frac{\Gamma\left(\frac{1}{2}-s-\nu\right)\Gamma\left(\frac{1}{2}+s+\nu\right)\Gamma\left(\frac{1}{2}+s-\nu\right)}{\pi}\\
&\quad \cdot \left(B_{l,\nu}(-s),\ B_{-l,\nu}(-s)\right) \begin{pmatrix} e^{-\frac{\pi i}{2}(1-s-\nu)} & e^{\frac{\pi i}{2}(1-s-\nu)} \\ e^{\frac{\pi i}{2}(1-s-\nu)} & e^{-\frac{\pi i}{2}(1-s-\nu)} \end{pmatrix}\\
&\quad \cdot \begin{pmatrix} \sin\left(\pi\left(s-\frac{l}{2}\right)\right) & \sin\left(\pi\left(\nu-\frac{l+1}{2}\right)\right) \\ \sin\left(\pi\left(\nu+\frac{l-1}{2}\right)\right) & \sin\left(\pi\left(s+\frac{l}{2}\right)\right) \end{pmatrix} \begin{pmatrix} L_+(s) \\ L_-(s) \end{pmatrix}
\end{aligned}
$$

$$\stackrel{(4.85)}{=} 2^{\nu-s-\frac{1}{2}}\pi^{-s-1}\Gamma\left(\frac{1}{2}-s-\nu\right)$$

$$\cdot \underbrace{(-1)\frac{\Gamma\left(\frac{1}{2}+s+\nu\right)\Gamma\left(\frac{1}{2}+s-\nu\right)}{\pi}\sin\left(\pi\left(s+\nu-\frac{1}{2}\right)\right)}_{\stackrel{(4.89)}{=}\frac{\Gamma\left(\frac{1}{2}+s-\nu\right)}{\Gamma\left(\frac{1}{2}-s-\nu\right)}}$$

$$\cdot \big(B_{l,\nu}(-s),\, B_{-l,\nu}(-s)\big)\begin{pmatrix} e^{\frac{\pi i}{2}(1-s-\nu)} & e^{-\frac{\pi i}{2}(1-s-\nu)} \\ e^{-\frac{\pi i}{2}(1-s-\nu)} & e^{\frac{\pi i}{2}(1-s-\nu)} \end{pmatrix}$$

$$\cdot \begin{pmatrix} e^{-\frac{\pi i}{2}\left(\frac{1}{2}-s-\nu\right)} & e^{\frac{\pi i}{2}\left(\frac{1}{2}-s-\nu\right)} \\ e^{\frac{\pi i}{2}\left(\frac{1}{2}-s-\nu\right)} & e^{-\frac{\pi i}{2}\left(\frac{1}{2}-s-\nu\right)} \end{pmatrix}^{-1}\begin{pmatrix} 0 & e^{-\frac{\pi i l}{2}} \\ e^{\frac{\pi i l}{2}} & 0 \end{pmatrix}^{-1}$$

$$\cdot \begin{pmatrix} e^{-\frac{\pi i}{2}\left(\frac{1}{2}+s-\nu\right)} & e^{\frac{\pi i}{2}\left(\frac{1}{2}+s-\nu\right)} \\ e^{\frac{\pi i}{2}\left(\frac{1}{2}+s-\nu\right)} & e^{-\frac{\pi i}{2}\left(\frac{1}{2}+s-\nu\right)} \end{pmatrix}\begin{pmatrix} L_+(s) \\ L_-(s) \end{pmatrix}$$

$$\stackrel{(4.89)}{=} 2^{\nu-s-\frac{1}{2}}\pi^{-s-1}\Gamma\left(\frac{1}{2}+s-\nu\right)$$

$$\cdot \underbrace{\big(B_{l,\nu}(-s),\, B_{-l,\nu}(-s)\big)\begin{pmatrix} 0 & e^{-\frac{\pi i l}{2}} \\ e^{\frac{\pi i l}{2}} & 0 \end{pmatrix}^{-1}}_{\stackrel{(4.76)}{=}\big(B_{l,\nu}(s),\, B_{-l,\nu}(s)\big)}$$

$$\cdot \begin{pmatrix} e^{-\frac{\pi i}{2}\left(\frac{1}{2}+s-\nu\right)} & e^{\frac{\pi i}{2}\left(\frac{1}{2}+s-\nu\right)} \\ e^{\frac{\pi i}{2}\left(\frac{1}{2}+s-\nu\right)} & e^{-\frac{\pi i}{2}\left(\frac{1}{2}+s-\nu\right)} \end{pmatrix}\begin{pmatrix} L_+(s) \\ L_-(s) \end{pmatrix}$$

$$\stackrel{(4.76)}{=} 2^{\nu-s-\frac{1}{2}}\pi^{-s-1}\Gamma\left(\frac{1}{2}+s-\nu\right)$$

$$\cdot \big(B_{l,\nu}(s),\, B_{-l,\nu}(s)\big)\begin{pmatrix} e^{-\frac{\pi i}{2}\left(\frac{1}{2}+s-\nu\right)} & e^{\frac{\pi i}{2}\left(\frac{1}{2}+s-\nu\right)} \\ e^{\frac{\pi i}{2}\left(\frac{1}{2}+s-\nu\right)} & e^{-\frac{\pi i}{2}\left(\frac{1}{2}+s-\nu\right)} \end{pmatrix}\begin{pmatrix} L_+(s) \\ L_-(s) \end{pmatrix}$$

$$\stackrel{(4.90)}{=} L^\star(s).$$

We obtain

$$(-1)^{\frac{l-k}{2}}\, L^\star(-s) = L^\star(s) \tag{4.91}$$

for all $-\frac{1}{2} - \Re(\nu) < \Re(s) < \frac{1}{2} + \Re(\nu)$. We can meromorphically extend (4.91) to all $s \in \mathbb{C}$. The functional equation (4.91) also implies that $L_k^\star$ is entire, since the location of possible poles in the left side of 4.91 is in $\pm\nu - \mathbb{Z}_{\geq 0}$ and the location of possible poles in the right side of (4.91) is in $\mp\nu + \mathbb{N}$. The two sets are disjoint. This shows that $L^\star(s)$ defined by (4.55) satisfies (4.56).

To complete the proof, we have to show that $L_l^\star$ is of finite order. Using the functional equation, we can restrict ourselves to the right half-plane $\Re(s) \geq 0$. Using the second part of Lemma 4.23, we see that $L_l^\star$ satisfies indeed the necessary growth condition

$$L_l^\star(s) = O\left(e^{|s|^C}\right) \qquad \text{for some } C > 0$$

on the right half-plane $\Re(s) > |\Re(\nu)|$, since $L_\pm$ has this property. The first part of Lemma 4.23 and the fact that $L_l^\star$ is entire imply that the growth condition also holds for the strip $0 \leq \Re(s) \leq |\Re(\nu)|$.

□

Example 4.42 Recall Example 4.32 in §4.2.1. As discussed there, we assume that $k > 0$ is given and assign $\nu = \frac{1-k}{2}$ and $0 \leq \kappa < 1$ with $\frac{k}{12} - \kappa \in \mathbb{Z}$ (i.e., $\frac{k}{12} \equiv \kappa \pmod 1$).

For η^{2k} being the cusp form and $u_{\eta^{2k}}$ the Maass cusp form discussed in §3.4.3, we consider its Whittaker-Fourier expansion

$$u_{\eta^{2k}}(z) \overset{(3.71)}{=} \sum_{n=0}^{\infty} A_k(n) \left(4\pi\left(n + \frac{k}{12}\right)\right)^{-\frac{k}{2}} \cdot \mathrm{W}_{\frac{k}{2}, \pm\frac{k-1}{2}}\left(4\pi\left(n + \frac{k}{12}\right)y\right) e^{2\pi i\left(n+\frac{k}{12}\right)x}.$$

The $u_{\eta^{2k}}$-associated family $(u_l)_{l\in k+2\mathbb{Z}}$ are given by Lemma 4.5. We have in particular

$$u_l = 0 \quad \text{for all } l \in k + 2\mathbb{Z},\, l < k$$

since $1+2\nu+(k-2) = 1+2\frac{1-k}{2}+k-2 = 0$. The admissible range is $l_{\min} = k$ and $l_{\max} = \infty$. Equations (4.23)–(4.25) and Lemma 4.11 show that u_l has the expansion

$$u_l(z) = \sum_{n=0}^{\infty} A_k(n) \left(4\pi\left(n + \frac{k}{12}\right)\right)^{-\frac{k}{2}} \psi_{l,\frac{1-k}{2}}\left(\left(n + \frac{k}{12}\right)\Im(z)\right) e^{2\pi i\left(n+\frac{k}{12}\right)\Re(z)}.$$

Then Theorem 4.39 associates L series with the system of cusp forms. We have

$$L_-(s) = 0$$

and

$$L_+(s) \stackrel{(4.71)}{=} (4\pi)^{-\frac{k}{2}} \sum_{n=0}^{\infty} A_{\eta^{2k}}(n) \left(n + \frac{k}{12}\right)^{-s-\frac{k}{2}}.$$

Up to here, everything is the same as in Example 4.32.

Theorem 4.39 shows that $L_\pm$ satisfies the functional equation (4.84):

$$\begin{aligned}
\begin{pmatrix} L_+(-s) \\ L_-(-s) \end{pmatrix} &\stackrel{(4.84)}{=} -(2\pi)^{-2s} \frac{\Gamma\left(\frac{1}{2}+s+\nu\right)\Gamma\left(\frac{1}{2}+s-\nu\right)}{\pi} \\
&\quad \cdot \begin{pmatrix} \sin\left(\pi\left(s-\frac{k}{2}\right)\right) & \sin\left(\pi\left(\nu-\frac{k+1}{2}\right)\right) \\ \sin\left(\pi\left(\nu+\frac{k-1}{2}\right)\right) & \sin\left(\pi\left(s+\frac{k}{2}\right)\right) \end{pmatrix} \begin{pmatrix} L_+(s) \\ L_-(s) \end{pmatrix} \\
&\stackrel{\nu=\frac{1-k}{2}}{=} -(2\pi)^{-2s} \frac{\Gamma\left(\frac{1}{2}+s+\frac{1-k}{2}\right)\Gamma\left(\frac{1}{2}+s-\frac{1-k}{2}\right)}{\pi} \\
&\quad \cdot \begin{pmatrix} \sin\left(\pi\left(s-\frac{k}{2}\right)\right) & \sin\left(\pi\left(\frac{1-k}{2}-\frac{k+1}{2}\right)\right) \\ \sin\left(\pi\left(\frac{1-k}{2}+\frac{k-1}{2}\right)\right) & \sin\left(\pi\left(s+\frac{k}{2}\right)\right) \end{pmatrix} \begin{pmatrix} L_+(s) \\ L_-(s) \end{pmatrix} \\
&= -(2\pi)^{-2s} \frac{\Gamma\left(1+s-\frac{k}{2}\right)\Gamma\left(s+\frac{k}{2}\right)}{\pi} \\
&\quad \cdot \begin{pmatrix} \sin\left(\pi\left(s-\frac{k}{2}\right)\right) & \sin\left(-\pi k\right) \\ 0 & \sin\left(\pi\left(s+\frac{k}{2}\right)\right) \end{pmatrix} \begin{pmatrix} L_+(s) \\ L_-(s) \end{pmatrix}
\end{aligned}$$

using $\nu = \frac{k-1}{2}$ as stated above. Recalling $L_-(s) = 0$ the above vector-valued identity simplifies to

$$L_+(-s) = (2\pi)^{-2s}\,\Gamma\left(s+\frac{k}{2}\right)\Gamma\left(1+s-\frac{k}{2}\right) \frac{\sin\left(\pi\left(\frac{k}{2}-s\right)\right)}{\pi}\, L_+(s).$$

Using the identity

$$\Gamma\left(\frac{k}{2}-s\right)\Gamma\left(1+s-\frac{k}{2}\right) = \frac{\pi}{\sin\left(\pi\left(\frac{k}{2}-s\right)\right)},$$

see Appendix A.7, we again simplify above expression to

$$\begin{aligned}L_+(-s) &= (2\pi)^{-2s}\,\Gamma\left(s+\frac{k}{2}\right)\Gamma\left(1+s-\frac{k}{2}\right)\frac{\sin\left(\pi\left(\frac{k}{2}-s\right)\right)}{\pi}\,L_+(s)\\ &= (2\pi)^{-2s}\,\frac{\Gamma\left(s+\frac{k}{2}\right)\Gamma\left(1+s-\frac{k}{2}\right)}{\Gamma\left(\frac{k}{2}-s\right)\Gamma\left(1+s-\frac{k}{2}\right)}\,L_+(s)\\ &= (2\pi)^{-2s}\,\frac{\Gamma\left(s-\frac{k}{2}\right)}{\Gamma\left(-s+\frac{k}{2}\right)}\,L_+(s).\end{aligned}$$

Symmetrizing gives

$$(2\pi)^s\,\Gamma\left(-s+\frac{k}{2}\right)\,L_+(-s) = (2\pi)^{-s}\,\Gamma\left(s+\frac{k}{2}\right)\,L_+(s).$$

Next, we use (4.71) to express $L_+(s)$ into its L-series expansion. We find that the above identity reads as

$$\begin{aligned}&(2\pi)^s\,\Gamma\left(-s+\frac{k}{2}\right)\sum_{n=0}^{\infty} A_{\eta^{2k}}(n)\left(n+\frac{k}{12}\right)^{s-\frac{k}{2}}\\ &\qquad\qquad = (2\pi)^{-s}\,\Gamma\left(s+\frac{k}{2}\right)\sum_{n=0}^{\infty} A_{\eta^{2k}}(n)\left(n+\frac{k}{12}\right)^{-s-\frac{k}{2}}.\end{aligned}$$

Substituting $s+\frac{k}{2}=t$ shows that the function

$$s \mapsto (2\pi)^{-s}\,\Gamma(s)\sum_{n=0}^{\infty} A_{\eta^{2k}}(n)\left(n+\frac{k}{12}\right)^{-s} \tag{4.92}$$

is invariant under $s \mapsto k-s \overset{\nu=\frac{k-1}{2}}{=} 1+2\nu-s$.

This confirms the result (4.73) of Example (4.32). For weight $k=12$, it also agrees with Hecke's classical result stated in Corollary 1.217.

In 2000, V.A. Bykovskii published an interesting result [54, Theorem 2], where he states that every cusp form $u \in S^{\mathrm{Maass}}_{k,v_k}\big(\Gamma(1),\,\nu\big)$ is equivalent to a pair of L-series satisfying a pair of functional equations. Here v_k denotes the η-multiplier defined in (1.40) of Example 1.50. The special functions involved in Bykovskii's result are Γ-functions that are much simpler than our $\tilde{\psi}_{\pm k,\nu}(s)$ appearing in Theorem 4.28. How does Bykovskii's result compare to our Theorem 4.39?

We first state Bykovskii's result in our notation.

Theorem 4.43 ([54, Theorem 2]) *Let $k \in \mathbb{R}$ and $\nu \in \mathbb{C}$ with real value $\lambda = \frac{1}{4}-\nu^2$. There exists a canonical correspondence between the following.*

1. *A cusp form* $u \in S^{Maass}_{k,v_k}\big(\Gamma(1),\nu\big)$.
2. *A pair of L-series* Λ_+ *and* Λ_- *such that*

   a. $\Lambda_\pm$ *have a series representation*

$$\Lambda_\pm(s) = \sum_{\substack{n\in\mathbb{Z}\\ n+\frac{k}{12}\gtrless 0}} \Phi_n \left| n + \frac{k}{12}\right|^{-s} \tag{4.93}$$

   that converges on a right half-plane.

   b. *The two functions* Λ_+ *and* Λ_- *extend to entire functions of* s *of finite order. They satisfy the functional equation*

$$\begin{pmatrix}\Lambda_+(-s)\\ \Lambda_-(-s)\end{pmatrix} = (2\pi)^{-2s}\,\frac{\Gamma\left(\frac12+s-\nu\right)\Gamma\left(\frac12+s+\nu\right)}{\pi} \tag{4.94}$$

$$\cdot\begin{pmatrix} -\sin\pi\left(s-\frac{k}{2}\right) & \frac{\pi}{\Gamma(\frac{1+k}{2}+\nu)\Gamma(\frac{1+k}{2}-\nu)}\\ \frac{\pi}{\Gamma(\frac{1-k}{2}+\nu)\Gamma(\frac{1-k}{2}-\nu)} & -\sin\pi\left(s+\frac{k}{2}\right)\end{pmatrix}\begin{pmatrix}\Lambda_+(s)\\ \Lambda_-(s)\end{pmatrix}$$

   for all $s \in \mathbb{C}$.

The coefficients Φ_n in (4.83) are in fact the coefficients $A(n)$ given in Theorem 3.36 of the Maass cusp form u.

Remark 4.44 Bykovskii stated his result using a different notation. We sketch briefly, how the notations can be translated: Let $u \in S^{\mathrm{Maass}}_{k,v_k}\big(\Gamma(1),\nu\big)$ a Maass cusp form of weight k, multiplier v_k for the full modular group $\Gamma(1)$. Then, u has the Whittaker-Fourier expansion of the form stated in Theorem 3.36. This is identical with [54, (0.2)]. Our L-series $\Lambda^\pm$ in (4.93) are shifted compared to Bykovskii's L-series $\xi_u^{(\pm)}(S:s)$. We have

$$\xi_u^{(\pm)}(S:s) = (2\pi)^{-s}\Lambda^\pm\left(s+\frac12\right).$$

Then, [54, Theorem 2] can be rewritten in Theorem 4.43.

Of course, our Theorem 4.39 follows from Bykovskii's Theorem 4.43, if the cusp form u is an element of an admissible system of cusp forms. (Write u in the form of expansion 4.23.) This gives the relation (4.25) between the Fourier coefficients $\tilde{A}(n)$ of u in the expansion 4.23 and $A(n)$ in expansion (4.24) of u_k.

Let us compare the methods used in the proofs of Theorem 4.39 and [54, Theorem 2]: Our key point in proving Theorem 4.39 is Lemma 4.37, where we split $\tilde{\psi}_{k,\nu}$ in a factor independent of weight k and a remaining term, depending on the weight k. The idea goes back to Motohashi where he uses it in the proof of [146, Theorem 2]. By studying families, i.e., all weights $l \in k + 2\mathbb{Z}$ together, we see that Maass's Theorem 4.28, simplifies to Theorem 4.39.

Bykovskii, on the other hand, generalized the Mellin transform of the W-Whittaker function. He studied the integral transform

$$\int_0^\infty u\big((\tau+i)y\big)\, y^{s-1}\, dy$$

with an extra parameter $\tau \in \mathbb{R}$, in place of

$$\int_0^\infty u(iy)\, y^{s-1}\, dy \quad \text{and} \quad \int_0^\infty \partial_x u(iy)\, y^{s-1}\, dy.$$

Compared to our results, this implies that our function $\tilde{\psi}_{k,\nu}(s)$ is replaced by

$$\int_0^\infty W_{\frac{k}{2},\nu}(2y)\, e^{wy}\, y^{s-1}\, dy = \frac{\Gamma\left(\frac{1}{2}+\nu+s\right)\Gamma\left(\frac{1}{2}-\nu+s\right)}{2^s\,\Gamma\left(1+s-\frac{k}{2}\right)} \cdot {}_2F_1\left[\begin{matrix}\frac{1}{2}+\nu+s,\ \frac{1}{2}-\nu+s\\ 1+s-\frac{k}{2}\end{matrix}\,\middle|\,\frac{1+w}{2}\right],$$

see [54, (1.10)]. The extra parameter enabled Bykovskii to use Kummer relations instead of Lemmas 4.36 and 4.37.

4.2.3 D-Functions for $S_{k,v}(\Gamma(1), v)$

Reviewing the proof of Theorem 4.39 carefully, we see that the vector

$$\begin{pmatrix} e^{-\frac{\pi i}{2}\left(\frac{1}{2}+s-\nu\right)} & e^{\frac{\pi i}{2}\left(\frac{1}{2}+s-\nu\right)} \\ e^{\frac{\pi i}{2}\left(\frac{1}{2}+s-\nu\right)} & e^{-\frac{\pi i}{2}\left(\frac{1}{2}+s-\nu\right)} \end{pmatrix} \begin{pmatrix} L_+(s) \\ L_-(s) \end{pmatrix}$$

is an essential object of the proof. Using this vector gives us another L-series representation of families of cusp forms.

Theorem 4.45 *Let $k \in \mathbb{R}$, $0 \le \kappa < 1$, and $|\Re(\nu)| < \frac{1}{2}$ such that if $2\nu + k$ or $2\nu - k \in 1 + 2\mathbb{Z}$, then $k \notin \mathbb{Z}$. κ and k satisfy* (4.26). *There is a canonical correspondence between the following objects.*

1. *An admissible family $(u_l)_{l\in k+2\mathbb{Z}}$ of Maass cusp forms as discussed in §4.1.*
2. *A pair of functions $D_0(s)$ and $D_1(s)$ satisfying*

    a. *D_δ, $\delta = 0, 1$, have a series representation of the form*

$$D_0(s) = \sum_{\substack{n\in\mathbb{Z}\\ n+\kappa\neq 0}} b_n\bigl(i(n+\kappa)\bigr)^{-s}, \quad and \quad D_1(s) = \sum_{\substack{n\in\mathbb{Z}\\ n+\kappa\neq 0}} b_n\bigl(-i(n+\kappa)\bigr)^{-s} \tag{4.95}$$

that converge absolutely on a right half-plane.

b. *For both choices of* $\pm$*, the function*

$$s \mapsto \frac{1}{\sin(\pi s)}\Bigl(D_0(s) - e^{\pm\pi i s}\, D_1(s)\Bigr) \tag{4.96}$$

extends to an entire function of finite order.

c. *The two functions*

$$D_\delta^\star(s) = (2\pi)^{-s}\,\Gamma(s)\,D_\delta(s) \qquad (\delta = 0, 1) \tag{4.97}$$

are entire. They satisfy the functional equation

$$D_0^\star(s) = e^{-\frac{\pi i k}{2}}\, D_1^\star(1-2\nu-s) \tag{4.98}$$

for all $s \in \mathbb{C}$.

Remark 4.46 In this theorem, we assume an extra condition $\Re(\nu) < \frac{1}{2}$ compared to Theorem 4.39. This condition ensures that $D_0^\star$ and $D_1^\star$ are entire functions. We show that $D_0^\star$ and $D_1^\star$ are holomorphic functions for $\Re(s) > -\frac{1}{2}$. Then, we use the functional equation 4.98 to show that both functions are also holomorphic for $\Re(s) < \frac{1}{2} - 2\Re(\nu)$. These two regions do not intersect if $\Re(\nu) \geq \frac{1}{2}$. The condition $\Re(\nu) > -\frac{1}{2}$ is necessary to apply Theorem 4.39.

Remark 4.47 The proof of Theorem 4.45 will show that the coefficients $A(n)$ in the expansion (4.24) of u_l, with $l \in k + 2\mathbb{Z}$ in the admissible range, and the coefficients b_n are related by

$$b_n = A(n)\ |n+\kappa|^{\frac{1}{2}-\nu} \tag{4.99}$$

for all $n \in \mathbb{Z}$ with $n+\kappa \neq 0$. The parameter $0 \leq \kappa < 1$ is related to k by (4.26), see Remark 4.13.

Remark 4.48 Note also that either (b) or (c) of part 2 of Theorem 4.45 imply that D_0 and D_1 extend to entire functions. (b) also shows that D_0 and D_1 are of finite order. Both parts together show

$$\begin{aligned} D_0(-n) = D_1(-n) = 0 \quad &\text{for } n \in \mathbb{Z}_{\leq 0} \quad \text{and} \\ D_0(n) - (-1)^n\, D_1(n) = 0 \quad &\text{for } n \in \mathbb{Z}. \end{aligned} \tag{4.100}$$

Exercise 4.49 Let (D_0, D_1) be a pair of D-functions given by series representations (4.95). We assume that the series representations converge absolutely on a right half-plane, as stated in part 2 of Theorem 4.45.

Show that the coefficients b_n satisfy a polynomial growth condition: There exists a $C > 0$ such that

$$b_n = O\left(|n+\kappa|^C\right) \tag{4.101}$$

for all $n \in \mathbb{Z}$ with $n + \kappa \neq 0$.

Proof *(of Theorem 4.45)* First, we show that the conditions in Item 2 of Theorem 4.45 imply that the functions D_0, D_1, $D_0^\star$, and $D_1^\star$ are entire and of finite order. Write

$$h_\pm(s) = \frac{\mp e^{\mp\frac{\pi i s}{2}}}{2i\sin(\pi s)}\left(D_0(s) - e^{\pm\pi i s} D_1(s)\right).$$

The condition (b) in part 2 says that the functions $h_\pm(s)$ are entire and are of finite order. Writing the equation in vector notation, we have

$$\begin{pmatrix} h_+(s) \\ h_-(s) \end{pmatrix} = \frac{1}{2i\sin(\pi s)}\begin{pmatrix} -e^{-\frac{\pi i s}{2}} & e^{\frac{\pi i s}{2}} \\ e^{\frac{\pi i s}{2}} & -e^{-\frac{\pi i s}{2}} \end{pmatrix}\begin{pmatrix} D_0(s) \\ D_1(s) \end{pmatrix}.$$

Multiplying both sides with $\begin{pmatrix} e^{-\frac{\pi i s}{2}} & e^{\frac{\pi i s}{2}} \\ e^{\frac{\pi i s}{2}} & e^{-\frac{\pi i s}{2}} \end{pmatrix}$ from the left, we get

$$\begin{pmatrix} e^{-\frac{\pi i s}{2}} & e^{\frac{\pi i s}{2}} \\ e^{\frac{\pi i s}{2}} & e^{-\frac{\pi i s}{2}} \end{pmatrix}\begin{pmatrix} h_+(s) \\ h_-(s) \end{pmatrix} = \begin{pmatrix} D_0(s) \\ D_1(s), \end{pmatrix}$$

which implies that D_δ, $\delta = 0, 1$ are entire and of finite order.

Then equation (4.97) and the functional equation (4.98) imply that $D_0^\star$ and $D_1^\star$ are entire functions of finite order.

Theorem 4.39 reduces the proof to showing that there is a canonical correspondence between Item 2 of Theorem 4.39 and Item 2 of this theorem.

Theorem 4.39 part 2 $\Longrightarrow$ Theorem 4.45 part 2 We define the functions D_0 and D_1 by

$$\begin{pmatrix} D_0(s) \\ D_1(s) \end{pmatrix} = \begin{pmatrix} e^{-\frac{\pi i}{2}s} & e^{\frac{\pi i}{2}s} \\ e^{\frac{\pi i}{2}s} & e^{-\frac{\pi i}{2}s} \end{pmatrix}\begin{pmatrix} L_+\left(s+\nu-\frac{1}{2}\right) \\ L_-\left(s+\nu-\frac{1}{2}\right) \end{pmatrix}. \tag{4.102}$$

Hence, D_0 and D_1 are entire functions of finite order. Moreover, the functions $D_\delta^\star$ defined by (4.97) are meromorphic functions in s with at most simple poles in $s \in \mathbb{Z}_{\leq 0}$.

Inserting the series representation (4.83) of $L_\pm$ in (4.102), we see that the function D_0 has a series representation

$$\begin{aligned} D_0(s) &\overset{(4.102)}{=} e^{-\frac{\pi i}{2}s} L_+\left(s+\nu-\frac{1}{2}\right) + e^{\frac{\pi i}{2}s} L_-\left(s+\nu-\frac{1}{2}\right) \\ &= \left(e^{\frac{\pi i}{2}}\right)^{-s} \sum_{\substack{n\in\mathbb{Z} \\ n+\kappa>0}} A(n)\,(n+\kappa)^{\frac{1}{2}-s-\nu} + \\ &\qquad + \left(e^{-\frac{\pi i}{2}}\right)^{-s} \sum_{\substack{n\in\mathbb{Z} \\ n+\kappa<0}} A(n)(-n-\kappa)^{\frac{1}{2}-s-\nu} \\ &= \sum_{\substack{n\in\mathbb{Z} \\ n+\kappa\neq 0}} A(n)\,|n+\kappa|^{\frac{1}{2}-\nu}\,\big(i(n+\kappa)\big)^{-s} \end{aligned}$$

that converges absolutely for s in a certain right half-plane. The same calculation for D_1 shows

$$D_1(s) = \sum_{\substack{n\in\mathbb{Z} \\ n+\kappa\neq 0}} A(n)\,|n+\kappa|^{\frac{1}{2}-\nu}\,\big(-i(n+\kappa)\big)^{-s}.$$

Using (4.85) of Lemma 4.41 and applying the functional equation (4.84) in part 2 of Theorem 4.39 to (4.102), we get

$$\begin{aligned} &\begin{pmatrix} D_0(1-2\nu-s) \\ D_1(1-2\nu-s) \end{pmatrix} \\ &\overset{(4.102)}{=} \begin{pmatrix} e^{-\frac{\pi i}{2}(1-2\nu-s)} & e^{\frac{\pi i}{2}(1-2\nu-s)} \\ e^{\frac{\pi i}{2}(1-2\nu-s)} & e^{-\frac{\pi i}{2}(1-2\nu-s)} \end{pmatrix} \begin{pmatrix} L_+\left(\frac{1}{2}-\nu-s\right) \\ L_-\left(\frac{1}{2}-\nu-s\right) \end{pmatrix} \\ &\overset{(4.84)}{=} -(2\pi)^{-2\left(s+\nu-\frac{1}{2}\right)} \frac{\Gamma\left(\frac{1}{2}+\left(s+\nu-\frac{1}{2}\right)+\nu\right)\Gamma\left(\frac{1}{2}+\left(s+\nu-\frac{1}{2}\right)-\nu\right)}{\pi} \\ &\quad\cdot \begin{pmatrix} e^{-\frac{\pi i}{2}(1-2\nu-s)} & e^{\frac{\pi i}{2}(1-2\nu-s)} \\ e^{\frac{\pi i}{2}(1-2\nu-s)} & e^{-\frac{\pi i}{2}(1-2\nu-s)} \end{pmatrix} \cdot \\ &\quad\cdot \begin{pmatrix} \sin\left(\pi\left(s+\nu-\frac{1}{2}-\frac{k}{2}\right)\right) & \sin\left(\pi\left(\nu-\frac{k+1}{2}\right)\right) \\ \sin\left(\pi\left(\nu+\frac{k-1}{2}\right)\right) & \sin\left(\pi\left(s+\nu-\frac{1}{2}+\frac{k}{2}\right)\right) \end{pmatrix} \begin{pmatrix} L_+\left(s+\nu-\frac{1}{2}\right) \\ L_-\left(s+\nu-\frac{1}{2}\right) \end{pmatrix} \end{aligned}$$

$$= -(2\pi)^{1-2s-2\nu}\frac{\Gamma(s+2\nu)\,\Gamma(s)}{\pi}$$
$$\cdot\begin{pmatrix} e^{-\frac{\pi i}{2}(1-2\nu-s)} & e^{\frac{\pi i}{2}(1-2\nu-s)} \\ e^{\frac{\pi i}{2}(1-2\nu-s)} & e^{-\frac{\pi i}{2}(1-2\nu-s)} \end{pmatrix}\cdot$$
$$\cdot\begin{pmatrix} \sin\left(\pi\left(s+\nu-\frac{k+1}{2}\right)\right) & \sin\left(\pi\left(\nu-\frac{k+1}{2}\right)\right) \\ \sin\left(\pi\left(\nu+\frac{k-1}{2}\right)\right) & \sin\left(\pi\left(s+\nu+\frac{k-1}{2}\right)\right) \end{pmatrix}\begin{pmatrix} L_+\left(s+\nu-\frac{1}{2}\right) \\ L_-\left(s+\nu-\frac{1}{2}\right)\end{pmatrix}$$
$$\overset{(4.85)}{=} (2\pi)^{1-2s-2\nu}\,\Gamma(s+2\nu)\,\Gamma(s)\,(-1)\,\frac{\sin\left(\pi\left(s+2\nu-1\right)\right)}{\pi}$$
$$\cdot\begin{pmatrix} 0 & e^{-\frac{\pi i k}{2}} \\ e^{\frac{\pi i k}{2}} & 0 \end{pmatrix}^{-1}\begin{pmatrix} e^{-\frac{\pi i}{2}s} & e^{\frac{\pi i}{2}s} \\ e^{\frac{\pi i}{2}s} & e^{-\frac{\pi i}{2}s} \end{pmatrix}\begin{pmatrix} L_+\left(s+\nu-\frac{1}{2}\right) \\ L_-\left(s+\nu-\frac{1}{2}\right)\end{pmatrix}$$
$$\overset{(4.102)}{=} (2\pi)^{1-2\nu-2s}\,\frac{\Gamma(s)}{\Gamma(1-2\nu-s)}\begin{pmatrix} 0 & e^{-\frac{\pi i k}{2}} \\ e^{\frac{\pi i k}{2}} & 0 \end{pmatrix}\begin{pmatrix} D_0(s) \\ D_1(s)\end{pmatrix}.$$

We used also the identity $\Gamma(s)\Gamma(1-s)=\frac{\pi}{\sin(\pi s)}$, see Appendix A.7.
The calculation shows that the functions $D^\star_\delta$ defined in (4.97) satisfy the functional equation (4.98). This implies in particular that the functions $D^\star_\delta$ are entire, since we already have established that $D^\star_\delta(s)$ is holomorphic in the right half-plane $\Re(s)>0$.
We still have to show that the functions in (4.96) are entire and of finite order. Inverting (4.102)) gives

$$\begin{pmatrix} L_+\left(s+\nu-\frac{1}{2}\right) \\ L_-\left(s+\nu-\frac{1}{2}\right)\end{pmatrix} = \frac{1}{2i\sin(\pi s)}\begin{pmatrix} -e^{-\frac{\pi i}{2}s} & e^{\frac{\pi i}{2}s} \\ e^{\frac{\pi i}{2}s} & -e^{-\frac{\pi i}{2}s} \end{pmatrix}\begin{pmatrix} D_0(s) \\ D_1(s)\end{pmatrix}. \tag{4.103}$$

Since $L_\pm$ are entire and of finite order, we see that

$$\frac{1}{2i\sin(\pi s)}\left(\mp e^{\mp\frac{\pi i}{2}s}D_0(s)\pm e^{\pm\frac{\pi i}{2}s}D_1(s)\right)$$

are entire and of finite order that is equivalent to Condition (b) of Part 2 of Theorem 4.45.
Theorem 4.45 part 2 $\Longrightarrow$ Theorem 4.39 part 2 Denote by $L_\pm(s)$ the series

$$L_\pm(s)=\sum_{\substack{n\in\mathbb{Z}\\ n+\kappa\gtrless 0}} b_n\,|n+\kappa|^{\nu-s-\frac{1}{2}}.$$

(Equation (4.95) implies that the series converge absolutely on a right half-plane.) Then (4.95) implies

$$\begin{pmatrix} D_0(s) \\ D_1(s) \end{pmatrix} = \begin{pmatrix} e^{-\frac{\pi i}{2}s} & e^{\frac{\pi i}{2}s} \\ e^{\frac{\pi i}{2}s} & e^{-\frac{\pi i}{2}s} \end{pmatrix} \begin{pmatrix} L_+\left(s+\nu-\frac{1}{2}\right) \\ L_-\left(s+\nu-\frac{1}{2}\right) \end{pmatrix}. \tag{4.104}$$

Inverting the equation and shifting the argument give

$$\begin{pmatrix} L_+(s) \\ L_-(s) \end{pmatrix} = \begin{pmatrix} e^{-\frac{\pi i}{2}\left(\frac{1}{2}+s-\nu\right)} & e^{\frac{\pi i}{2}\left(\frac{1}{2}+s-\nu\right)} \\ e^{\frac{\pi i}{2}\left(\frac{1}{2}+s-\nu\right)} & e^{-\frac{\pi i}{2}\left(\frac{1}{2}+s-\nu\right)} \end{pmatrix}^{-1} \begin{pmatrix} D_0\left(\frac{1}{2}+s-\nu\right) \\ D_1\left(\frac{1}{2}+s-\nu\right) \end{pmatrix}. \tag{4.105}$$

This shows that $L_\pm(s)$ are meromorphic in $s \in \mathbb{C}$. They are holomorphic on $\mathbb{C}$ except in $s \in \nu - \frac{1}{2} + \mathbb{Z}$ where $L_\pm(s)$ may have simple poles. However, (4.96) implies that $L_\pm(s)$ are entire functions of finite order.

It follows from the next argument that $L_\pm$ satisfy the functional equation given in Theorem 4.39 part 2: Writing $L_\pm(-s)$ in terms of $D_0^\star$ and $D_1^\star$, see (4.104) and (4.97), we have

$$\begin{aligned} \begin{pmatrix} L_+(-s) \\ L_-(-s) \end{pmatrix} &\overset{(4.105)}{=} \begin{pmatrix} e^{-\frac{\pi i}{2}\left(\frac{1}{2}-s-\nu\right)} & e^{\frac{\pi i}{2}\left(\frac{1}{2}-s-\nu\right)} \\ e^{\frac{\pi i}{2}\left(\frac{1}{2}-s-\nu\right)} & e^{-\frac{\pi i}{2}\left(\frac{1}{2}-s-\nu\right)} \end{pmatrix}^{-1} \begin{pmatrix} D_0\left(\frac{1}{2}-s-\nu\right) \\ D_1\left(\frac{1}{2}-s-\nu\right) \end{pmatrix} \\ &\overset{(4.97)}{=} \frac{(2\pi)^{\frac{1}{2}-s-\nu}}{\Gamma\left(\frac{1}{2}-s-\nu\right)} \cdot \\ &\quad \cdot \begin{pmatrix} e^{-\frac{\pi i}{2}\left(\frac{1}{2}-s-\nu\right)} & e^{\frac{\pi i}{2}\left(\frac{1}{2}-s-\nu\right)} \\ e^{\frac{\pi i}{2}\left(\frac{1}{2}-s-\nu\right)} & e^{-\frac{\pi i}{2}\left(\frac{1}{2}-s-\nu\right)} \end{pmatrix}^{-1} \begin{pmatrix} D_0^\star\left(\frac{1}{2}-s-\nu\right) \\ D_1^\star\left(\frac{1}{2}-s-\nu\right) \end{pmatrix}. \end{aligned}$$

Applying (4.98), we see that

$$\begin{aligned} \begin{pmatrix} L_+(-s) \\ L_-(-s) \end{pmatrix} &\overset{(4.98)}{=} \frac{(2\pi)^{\frac{1}{2}-s-\nu}}{\Gamma\left(\frac{1}{2}-s-\nu\right)} \begin{pmatrix} e^{-\frac{\pi i}{2}\left(\frac{1}{2}-s-\nu\right)} & e^{\frac{\pi i}{2}\left(\frac{1}{2}-s-\nu\right)} \\ e^{\frac{\pi i}{2}\left(\frac{1}{2}-s-\nu\right)} & e^{-\frac{\pi i}{2}\left(\frac{1}{2}-s-\nu\right)} \end{pmatrix}^{-1} \cdot \\ &\quad \cdot \begin{pmatrix} 0 & e^{-\frac{\pi i k}{2}} \\ e^{\frac{\pi i k}{2}} & 0 \end{pmatrix} \begin{pmatrix} D_0^\star\left(\frac{1}{2}+s-\nu\right) \\ D_1^\star\left(\frac{1}{2}+s-\nu\right) \end{pmatrix} \\ &\overset{(4.97)}{=} (2\pi)^{-2s} \frac{\Gamma\left(\frac{1}{2}+s-\nu\right)}{\Gamma\left(\frac{1}{2}-s-\nu\right)} \begin{pmatrix} e^{-\frac{\pi i}{2}\left(\frac{1}{2}-s-\nu\right)} & e^{\frac{\pi i}{2}\left(\frac{1}{2}-s-\nu\right)} \\ e^{\frac{\pi i}{2}\left(\frac{1}{2}-s-\nu\right)} & e^{-\frac{\pi i}{2}\left(\frac{1}{2}-s-\nu\right)} \end{pmatrix}^{-1} \cdot \end{aligned}$$

$$
\cdot \begin{pmatrix} 0 & e^{-\frac{\pi ik}{2}} \\ e^{\frac{\pi ik}{2}} & 0 \end{pmatrix} \begin{pmatrix} D_0\left(\frac{1}{2}+s-\nu\right) \\ D_1\left(\frac{1}{2}+s-\nu\right) \end{pmatrix}
$$

$$
\overset{(4.104)}{=} (2\pi)^{-2s} \frac{\Gamma\left(\frac{1}{2}+s-\nu\right)}{\Gamma\left(\frac{1}{2}-s-\nu\right)} \begin{pmatrix} e^{-\frac{\pi i}{2}\left(\frac{1}{2}-s-\nu\right)} & e^{\frac{\pi i}{2}\left(\frac{1}{2}-s-\nu\right)} \\ e^{\frac{\pi i}{2}\left(\frac{1}{2}-s-\nu\right)} & e^{-\frac{\pi i}{2}\left(\frac{1}{2}-s-\nu\right)} \end{pmatrix}^{-1} \cdot
$$

$$
\cdot \begin{pmatrix} 0 & e^{-\frac{\pi ik}{2}} \\ e^{\frac{\pi ik}{2}} & 0 \end{pmatrix} \begin{pmatrix} e^{-\frac{\pi i}{2}\left(\frac{1}{2}+s-\nu\right)} & e^{\frac{\pi i}{2}\left(\frac{1}{2}+s-\nu\right)} \\ e^{\frac{\pi i}{2}\left(\frac{1}{2}+s-\nu\right)} & e^{-\frac{\pi i}{2}\left(\frac{1}{2}+s-\nu\right)} \end{pmatrix} \begin{pmatrix} L_+(s) \\ L_-(s) \end{pmatrix}
$$

$$
\overset{(4.85)}{=} (2\pi)^{-2s} \frac{\Gamma\left(\frac{1}{2}+s-\nu\right)}{\Gamma\left(\frac{1}{2}-s-\nu\right)} \frac{1}{\sin\left(\pi\left(s+\nu-\frac{1}{2}\right)\right)}
$$

$$
\cdot \begin{pmatrix} \sin\left(\pi\left(s-\frac{l}{2}\right)\right) & \sin\left(\pi\left(\nu-\frac{l+1}{2}\right)\right) \\ \sin\left(\pi\left(\nu+\frac{l-1}{2}\right)\right) & \sin\left(\pi\left(s+\frac{l}{2}\right)\right) \end{pmatrix} \begin{pmatrix} L_+(s) \\ L_-(s) \end{pmatrix}
$$

$$
\overset{(4.89)}{=} -(2\pi)^{-2s} \frac{\Gamma\left(\frac{1}{2}+s+\nu\right)\Gamma\left(\frac{1}{2}+s-\nu\right)}{\pi} \frac{1}{\sin\left(\pi\left(s+\nu-\frac{1}{2}\right)\right)}
$$

$$
\cdot \begin{pmatrix} \sin\left(\pi\left(s-\frac{l}{2}\right)\right) & \sin\left(\pi\left(\nu-\frac{l+1}{2}\right)\right) \\ \sin\left(\pi\left(\nu+\frac{l-1}{2}\right)\right) & \sin\left(\pi\left(s+\frac{l}{2}\right)\right) \end{pmatrix} \begin{pmatrix} L_+(s) \\ L_-(s) \end{pmatrix}.
$$

This shows that $L_\pm$ satisfies the functional equation (4.84) in Theorem 4.39 part 2.

□

Remark 4.50 In the situation of Theorem 4.45, we find that the L-functions $L_\pm$ given in part 2 of Theorem 4.28 and the functions D_0 and D_1 in part 2 of Theorem 4.45 are entire functions and satisfy

$$
\begin{pmatrix} L_+(s) \\ L_-(s) \end{pmatrix} \overset{(4.103)}{=} \frac{1}{2i\sin\left(\pi\left(\frac{1}{2}+s-\nu\right)\right)} \cdot \begin{pmatrix} -e^{-\frac{\pi i}{2}\left(\frac{1}{2}+s-\nu\right)} & e^{\frac{\pi i}{2}\left(\frac{1}{2}+s-\nu\right)} \\ e^{\frac{\pi i}{2}\left(\frac{1}{2}+s-\nu\right)} & -e^{-\frac{\pi i}{2}\left(\frac{1}{2}+s-\nu\right)} \end{pmatrix} \begin{pmatrix} D_0\left(\frac{1}{2}+s-\nu\right) \\ D_1\left(\frac{1}{2}+s-\nu\right) \end{pmatrix} \tag{4.106}
$$

for all $s \in \mathbb{C}$.

Since the $L_\pm$ are entire, see Theorem 4.45, Item 2 Equation (4.106) shows that the functions

$$s \mapsto \frac{1}{\sin(\pi s)} \left(e^{\pm \frac{\pi i s}{2}} D_1(s) - e^{\mp \frac{\pi i s}{2}} D_0(s) \right) \tag{4.107}$$

on the right side of (4.107) are entire.

Example 4.51 Recall Examples 4.15 in §4.1, 4.32 in §4.2.1, and 4.42 in §4.2.2:

Assume that $k > 0$ is given. We assign $\nu = \frac{1-k}{2}$ and $0 < \kappa < 1$ with $\frac{k}{12} - \kappa \in \mathbb{Z}$ (i.e., $\frac{k}{12} \equiv \kappa \pmod 1$, see (4.26)).

Let η^{2k} be the cusp form and $u_{\eta^{2k}}$ the Maass cusp form discussed in §3.4.3. The Whittaker-Fourier expansion of $u_{\eta^{2k}}$ is given by (3.70).

As in Example 4.15, we denote the associated admissible family of $u_{\eta^{2k}}$ by $(u_l)_{l \in k+2\mathbb{Z}}$ where the u_l are given by Lemma 4.5. Equations (4.23)–(4.25) and Lemma 4.11 show that u_l has the expansion (4.27). Theorem 4.45 associates D-series to the system of cusp forms. Calculating D_0 and D_1 in (4.95) for $b_n = A_k(n)\, |n+\kappa|^{-\nu}$, $n \in \mathbb{Z}$ with $n + \kappa \neq 0$ and $A(n)$, $A_k(n)$ as in (4.27), we find

$$\begin{aligned} D_0(s) &= e^{-\frac{\pi i s}{2}} \sum_{\substack{n \in \mathbb{Z} \\ n+\kappa > 0}} A(n)\, (n+\kappa)^{-s-\frac{1-k}{2}} \\ &= (4\pi)^{-\frac{k}{2}}\, e^{-\frac{\pi i s}{2}} \sum_{n=0}^{\infty} A_k(n) \left(n + \frac{k}{12}\right)^{-s} \end{aligned} \tag{4.108}$$

and

$$\begin{aligned} D_1(s) &= e^{\frac{\pi i s}{2}} \sum_{\substack{n \in \mathbb{Z} \\ n+\kappa > 0}} A(n)\, (n+\kappa)^{-s-\frac{1-k}{2}} \\ &= (4\pi)^{-\frac{k}{2}}\, e^{\frac{\pi i s}{2}} \sum_{n=0}^{\infty} A_k(n) \left(n + \frac{k}{12}\right)^{-s}. \end{aligned} \tag{4.109}$$

Using (4.97), we have the expressions

$$\begin{aligned} D_0^\star(s) &= \frac{(4\pi)^{-s-\frac{k}{2}}}{\sqrt{2}} \Gamma(s)\, e^{-\frac{\pi i s}{2}} \sum_{n=0}^{\infty} A_k(n) \left(n + \frac{k}{12}\right)^{-s} \quad \text{and} \\ D_1^\star(s) &= \frac{(4\pi)^{-s-\frac{k}{2}}}{\sqrt{2}} \Gamma(s)\, e^{\frac{\pi i s}{2}} \sum_{n=0}^{\infty} A_k(n) \left(n + \frac{k}{12}\right)^{-s} \end{aligned} \tag{4.110}$$

for $D_0^\star$ and $D_1^\star$. The functional equation for $D_0^\star$ and $D_1^\star$, see (4.97) and (4.98), implies

$$s \mapsto (2\pi)^{-s}\,\Gamma(s)\,\sum_{n=0}^{\infty} A_{\eta^{2k}}(n)\,\left(n+\frac{k}{12}\right)^{-s} \tag{4.111}$$

is invariant under $s \mapsto k-s = 1-2\nu-s$.

Again, (4.73), (4.92), and (4.111) are identical.

Remark 4.52 We would like to point out that the Examples 4.32 in §4.2.1, 4.42 in §4.2.2, and 4.51 above are illustrating as general property. For every admissible family of Maass cusp forms $(u_l)_{l\in k+2\mathbb{Z}}$, we find two pairs of series $\big(L_+(s), L_-(s)\big)$ via Theorem 4.28 or Theorem 4.39 and $\big(D_0(s), D_1(s)\big)$ via Theorem 4.45 that are related in a simple way, see, e.g., (4.104), but satisfy different functional equations.

In the setting of our examples in 4.32, 4.42, and 4.51, we have that the series L_- vanishes. This implies in particular

$$e^{\frac{\pi i s}{2}}\,D_0(s) = L_+\left(\frac{1}{2}+s+\nu\right) = e^{-\frac{\pi i s}{2}}\,D_1(s)$$

and the different functional equations are essentially the same.

We close the discussion about L-series by giving a technical lemma that will be of use later on.

Lemma 4.53 *Let $k \in \mathbb{R}$ and $\nu \in \mathbb{C}$. For each $\varepsilon > 0$ and each bounded interval $I \subset \mathbb{R}$, the functions D_δ, $\delta = 0, 1$ in part 2 of Theorem 4.45 satisfy the estimate*

$$D_\delta^\star(s) = O\left(e^{\varepsilon|\Im(s)|}\right) \qquad (\delta = 0, 1) \tag{4.112}$$

uniformly on the vertical strip $\Re(s) \in I$. The implied constant depends only on I and ε.

Proof Let $\delta \in \{0, 1\}$ and $\sigma > 0$ large enough such that the series representation (4.95) of $D_\delta(s)$ converges absolutely for all s with $\Re(s) > \sigma$. This implies that

$$D_\delta(s) = O\left(e^{\frac{\pi}{2}|\Im(s)|}\right)$$

uniformly on the right half-plane $\Re(s) > \sigma$. (The factor $O\left(e^{\frac{\pi}{2}|\Im(s)|}\right)$ comes from estimating $(\pm i)^{-s}$.)

Let $M > \sigma$ large. Then Stirling's formula, see (4.46) and Appendix A.3, and Equation (4.97) implies

$$\begin{aligned} D_\delta^\star(s) &= O\left(\left((2\pi)^{-\Re(s)}\,|\Im(s)|^{\Re(s)-\frac{1}{2}}\right)e^{-\frac{\pi}{2}|\Im(s)|}\right)\,O\left(e^{\frac{\pi}{2}|\Im(s)|}\right) \\ &= O\left(e^{\varepsilon|\Im(s)|}\right) \end{aligned}$$

for every $\varepsilon > 0$ uniformly on the vertical strip $\sigma < \Re(s) < M$. Using the functional equation (4.98) for $D_\delta^\star$, we see that $D_\delta^\star(s) = O(1)$ uniformly on the two vertical strips $\sigma < |\Re(s)| < M$. The Phragmén-Lindelöf theorem, see, e.g., [164, Section 6, Theorem 12] or Appendix A.2, implies the statement of the lemma. □

4.3 Nearly Periodic Functions and Maass Cusp Forms

We discuss the relation between admissible family of Maass cusp forms and certain nearly periodic functions (where "nearly periodic" is given in Definition 4.54 and the precise meaning of the correspondence is stated in Theorem 4.55 below). This connection was first presented by Lewis and Zagier in [122, Theorem 1, Items (a) and (c)] for Maass cusp forms of weight 0. We extend their result and embed it into the concept of admissible families of Maass cusp forms.

Similar to the previous §4.2 we divide this subsection into several parts. The first two parts are devoted to the proof of the connection between "certain nearly periodic functions," as we called them above, and the pair of D-functions connected to the admissible family of Maass cusp forms by Theorem 4.45. The next part, §4.3.3, discusses an integral transform which connects admissible families of Maass cusp forms directly with "certain nearly periodic functions." The last part of this subsection discusses the relation to Eichler integrals in the special case that the admissible family of Maass cusp forms is in fact derived by embedding (3.56) of a classical cusp form and subsequent discussion in Example 4.6.

Definition 4.54 We call a function g *nearly periodic* if there exists $a \in \mathbb{C}$ with $|a| = 1$ such that $g(z+1) = a\, g(z)$ for all z.

Theorem 4.55 *Let ν be a complex number with $|\Re(\nu)| < \frac{1}{2}$ and $k \in \mathbb{R}$ such that $k \pm (1-2\nu) \notin 2\mathbb{Z}$ for both choices of $\pm$. Moreover, let $0 \leq \kappa < 1$.*

There are natural bijective correspondences between objects of the following four types:

1. *An admissible family of Maass cusp forms $(u_l)_{l \in k+2\mathbb{Z}}$ of cusp forms as discussed in §4.1, where k is the reference weight within the admissible range.*
2. *A pair of L-series L_+ and L_- as specified in the second part of Theorem 4.39;*
3. *A pair of functions D_0 and D_1 as specified in the second part of Theorem 4.45;*
4. *A holomorphic function $f : \mathbb{C} \smallsetminus \mathbb{R} \to \mathbb{C}$ that satisfies $f(z+1) = e^{2\pi i \kappa} f(z)$, $z \in \mathbb{C} \smallsetminus \mathbb{R}$, and is bounded by a multiple of $|\Im(z)|^{-C}$ for some $C > 0$, such that the function $f(z) - e^{\frac{\pi i k}{2}} z^{2\nu-1} f\left(-\frac{1}{z}\right)$ extends holomorphically across the right half-plane and is bounded by a multiple of $\min\left\{1, |z|^{2\Re(\nu)-1}\right\}$ in the right half-plane.*

Remark 4.56 The interesting part of above Theorem 4.55 is the statement in Item 4. It describes a class of certain nearly periodic functions that correspond to admissible families of Maass cusp forms. This particular class of nearly periodic functions will serve as a first way toward period functions, once we introduce these functions in the following Chapter 5.

We already have established the correspondence between the first three parts of Theorem 4.55. Theorem 4.28 establishes the correspondence between the admissible family of Maass cusp forms and the pair of L-series L_+ and L_- where Theorem 4.39 gives a simpler representation of the pair of L-series without the function $L^\star$; the correspondence between the admissible family of Maass cusp forms and the pair of D-functions is given in Theorem 4.45.

The remaining part of proving the connection between the pair of D-functions and the nearly periodic function f in the fourth part of Theorem 4.55 is shown below in §4.3.1 and §4.3.2.

Remark 4.57 We can interpret the expressions in the fourth part on the nearly periodic function f in Theorem 4.55 as follows: Recalling (1.38) of Exercise 1.49, we see that $f(z+1) = e^{2\pi i\kappa} f(z)$, $z \in \mathbb{C} \smallsetminus \mathbb{R}$ can be written as

$$\begin{aligned} & f(z+1) = e^{2\pi i\kappa} f(z) \qquad \text{for all } z \in \mathbb{C} \smallsetminus \mathbb{R} \\ \Longleftrightarrow\ & v(S)^{-1} f\big|_{2\nu-1}(S) = f \end{aligned}$$

using $v(S) = e^{\frac{\pi i k}{6}} = e^{2\pi i\kappa}$ since $k \in \mathbb{R}$ and $0 \le \kappa < 1$ satisfy (4.26).

Similarly, we can write $f(z) - e^{\frac{\pi i k}{2}} z^{2\nu-1} f\left(-\frac{1}{z}\right)$ as

$$\begin{aligned} & f(z) - e^{\frac{\pi i k}{2}} z^{2\nu-1} f\left(-\frac{1}{z}\right) \qquad \text{for all } z \in \mathbb{C} \smallsetminus \mathbb{R} \\ \Longleftrightarrow\ & f - v(T)^{-1} f\big|_{2\nu-1}(S) \end{aligned}$$

using $v(S) = e^{-\frac{\pi i k}{2}}$, see (1.38).

4.3.1 D-Functions Giving Rise to Nearly Periodic Functions

Consider a pair of D-functions (D_0, D_1) as described in §4.2.3, Theorem 4.45. The associated sequence $(b_n)_{\substack{n\in\mathbb{Z}\\ n+\kappa\neq 0}}$ of coefficients is related to the D-functions by the series representation (4.95) of D_0 and D_1. Using these b_n's, we define the function $f: \mathbb{C} \smallsetminus \mathbb{R} \to \mathbb{C}$ in the following way:

$$f(z) := \begin{cases} \sqrt{2\pi} \displaystyle\sum_{\substack{n\in\mathbb{Z}\\ n+\kappa>0}} b_n\, e^{2\pi i(n+\kappa)z} & \text{for } \Im(z) > 0 \text{ and,} \\ -\sqrt{2\pi} \displaystyle\sum_{\substack{n\in\mathbb{Z}\\ n+\kappa<0}} b_n\, e^{2\pi i(n+\kappa)z} & \text{for } \Im(z) < 0. \end{cases} \tag{4.113}$$

Exercise 4.58 Assume that the sequence $\left(b_n\right)_{\substack{n\in\mathbb{Z}\\ n+\kappa\neq 0}}$ satisfies a polynomial growth condition: $b_n = O\left(|n+\kappa|^C\right)$ for some $C > 0$. Then the function f defined in (4.113) is well defined and is bounded uniformly by a power of $|\Im(z)|$ for $|\Im(z)| \to 0$ and is exponentially small for $|\Im(z)| \to \infty$.

Exercise 4.59 Show that the function f in (4.113) is nearly periodic in the sense of Definition 4.54: it satisfies

$$f(z+1) = e^{2\pi i\kappa}\, f(z) \tag{4.114}$$

for all $z \in \mathbb{C} \smallsetminus \mathbb{R}$.

Remark 4.60 The Exercises 4.58 and 4.59 above (except the part on the growth estimates) are analog to the results in Lemma 1.215 for L-series associated with classical L-series in §1.7.2.

Lemma 4.61 *Let $k \in \mathbb{R}$ and $\nu \in \mathbb{C}$ with $|\Re(\nu)| < \frac{1}{2}$ such that if $2\nu + k$ or $2\nu - k \in 1 + 2\mathbb{Z}$, then $k \notin \mathbb{Z}$. We consider functions $D^\star_\delta$, $\delta = 0, 1$, in* (4.97) *of Theorem 4.45, functions $L_\pm$ in* (4.54) *of Theorem 4.28 where the coefficients b_n and $A(n)$ are related by* (4.99) *in Remark 4.47.*

We define the nearly periodic function $f : \mathbb{C} \smallsetminus \mathbb{R} \to \mathbb{C}$ associated with (D_0, D_1) in (4.113).

The function f satisfies the following.

1. *The function $f(z)$ is bounded by a power of $\Im(z)$ for all $z \in \mathbb{C} \smallsetminus \mathbb{R}$.*
2. *The Mellin transform*

$$\tilde{f}_\pm(s) := \int_0^\infty f(\pm iy)\, y^{s-1}\, dy \tag{4.115}$$

extends to meromorphic functions on $\mathbb{C}$ with simple poles in $\mathbb{Z}_{\leq 0}$. Moreover, the quotients

$$\frac{(2\pi)^{\frac{1}{2}-s}}{2i}\,\frac{\tilde{f}_\pm(s)}{\Gamma(s)}$$

are entire and have finite order.

3. *We have*

$$\tilde{f}_\pm(s) = \frac{\sqrt{2\pi}}{2i\sin(\pi s)}\left(e^{\pm\frac{\pi i s}{2}}\, D^\star_1(s) - e^{\mp\frac{\pi i s}{2}}\, D^\star_0(s)\right) \tag{4.116}$$

$$= \pm(2\pi)^{\frac{1}{2}-s}\Gamma(s)\, L_\pm\left(s+\nu-\frac{1}{2}\right) \tag{4.117}$$

for $s \in \mathbb{C}$, $s \notin \mathbb{Z}_{\leq 0}$.

4. *For each* $\varepsilon > 0$ *and each bounded interval* I

$$\tilde{f}_{\pm}(s) = O\left(e^{(\varepsilon - \frac{\pi}{2})|\Im(s)|}\right) \qquad \text{as } |\Im(s)| \to \infty \text{ with } \Re(s) \in I. \tag{4.118}$$

The implied constant depends only on ε *and* I.

Proof In the situation of the Lemma, let D_0 and D_1 as in Theorem 4.45, Item 2. The coefficients $A(n)$ of $L_{\pm}$ in (4.54) and b_n of D_0, D_1 in (4.95) are related by (4.99) in Remark 4.47. This allows us to use both types of series expansions and simultaneously use only the coefficients b_n.

The coefficients b_n with $n \in \mathbb{Z}, n+\kappa \neq 0$, satisfy a polynomial growth condition, see Exercise 4.49. This implies that the function f, given in (4.113), is well defined, see Exercise 4.58. We also conclude in Exercise 4.58 that f is bounded uniformly by a power of $|\Im(z)|$ for $|\Im(z)| \to 0$ and is exponentially small for $|\Im(z)| \to \infty$. Exercise 4.59 shows that f is nearly periodic as required. This proves the first part of the lemma.

The growth estimates imply also that the Mellin transforms

$$\tilde{f}_{\pm}(s) := \mathcal{M}\big(f(\pm i\cdot)\big)(s) \overset{(1.154)}{=} \int_0^\infty f(\pm iy)\, y^{s-1}\, \mathrm{d}y$$

are well defined, see Lemma 1.201, and can be computed term by term for sufficiently large $\Re(s)$, see the proof of Lemma 1.207. We find

$$\begin{aligned}
\tilde{f}_{\pm}(s) &= \int_0^\infty f(\pm iy)\, y^{s-1}\, \mathrm{d}y \\
&\overset{(4.113)}{=} \pm\sqrt{2\pi} \sum_{\substack{n\in\mathbb{Z}\\ n+\kappa \gtrless 0}} b_n \int_0^\infty e^{-2\pi|n+\kappa|y}\, y^{s-1}\, \mathrm{d}y \\
&= \pm\sqrt{2\pi} \sum_{\substack{n\in\mathbb{Z}\\ n+\kappa \gtrless 0}} b_n \int_0^\infty e^{-y} \left(\frac{y}{2\pi\,|n+\kappa|}\right)^{s-1} \frac{\mathrm{d}y}{2\pi\,|n+\kappa|} \\
&\overset{(1.155)}{=} \pm(2\pi)^{\frac{1}{2}-s}\,\Gamma(s) \sum_{\substack{n\in\mathbb{Z}\\ n+\kappa \gtrless 0}} b_n\, |n+\kappa|^{-s} \\
&\overset{(4.99)}{=} \pm(2\pi)^{\frac{1}{2}-s}\,\Gamma(s) \sum_{\substack{n\in\mathbb{Z}\\ n+\kappa \gtrless 0}} A(n)\, |n+\kappa|^{\frac{1}{2}-\nu-s} \\
&\overset{(4.54)}{=} \pm(2\pi)^{\frac{1}{2}-s}\Gamma(s)\, L_{\pm}\left(s+\nu-\frac{1}{2}\right),
\end{aligned}$$

using the integral representation of the Γ-function in (1.155) and Appendix A.7. This proves (4.117).

Using (4.106) and (4.97),

$$\begin{aligned}
\tilde{f}_\pm(s) &= \pm(2\pi)^{\frac{1}{2}-s}\Gamma(s)\,L_\pm\left(s+\nu-\frac{1}{2}\right) \\
&\overset{(4.106)}{=} \pm\frac{(2\pi)^{\frac{1}{2}-s}\Gamma(s)}{2i\,\sin(\pi s)}\left(\mp e^{\mp\frac{\pi i s}{2}}D_0(s) \pm e^{\pm\frac{\pi i s}{2}}D_1(s)\right) \\
&= \frac{(2\pi)^{\frac{1}{2}-s}\Gamma(s)}{2i\,\sin(\pi s)}\left(e^{\pm\frac{\pi i s}{2}}D_1(s) - e^{\mp\frac{\pi i s}{2}}D_0(s)\right) \qquad (4.119) \\
&\overset{(4.97)}{=} \frac{\sqrt{2\pi}}{2i\,\sin(\pi s)}\left(e^{\pm\frac{\pi i s}{2}}D_1^\star(s) - e^{\mp\frac{\pi i s}{2}}D_0^\star(s)\right)
\end{aligned}$$

for all s such that $\Re(s)$ is sufficiently large. Applying that "the function in (4.96) is entire of finite order" in Theorem 4.45 to (4.119) shows that the function $s \mapsto \frac{(2\pi)^{\frac{1}{2}-s}}{2i}\frac{\tilde{f}_\pm(s)}{\Gamma(s)}$ is entire and of finite order. Hence, $\tilde{f}_\pm$ extends to a meromorphic function on $\mathbb{C}$ with at most simple poles in $s \in \mathbb{Z}_{\leq 0}$. (The possible poles are derived from the poles of the Γ-function, see Appendix A.7.) This completes the proof of parts (2) and (3) of the lemma.

Lemma 4.53 states that for each $\varepsilon > 0$ and bounded interval I, we have $D_\delta^\star(s) = O\left(e^{\varepsilon|\Im(s)|}\right)$ uniformly for s in the vertical strip $\Re(s) \in I$. Hence, the growth of $\tilde{f}_\pm$ is determined by the other factors in (4.116). We have

$$\tilde{f}_\pm(s) = O\left(e^{(\varepsilon-\frac{\pi}{2})|\Im(s)|}\right) \qquad \text{as } |\Im(s)| \to \infty$$

for all $\varepsilon > 0$ uniformly for s in the vertical strip $\Re(s) \in I$, due to

$$\begin{aligned}
\frac{e^{\pm\frac{\pi i s}{2}}}{2i\,\sin(\pi s)} &= O\left(\frac{e^{\pm\frac{\pi i s}{2}}}{e^{\pi i s} - e^{-\pi i s}}\right) \\
&= \begin{cases} O\left(e^{\pm\frac{\pi i s}{2}+\pi i s}\,\frac{1}{e^{2\pi i s}-1}\right) & \text{as } \Im(s) \to +\infty \text{ and} \\ O\left(e^{\pm\frac{\pi i s}{2}-\pi i s}\,\frac{1}{1-e^{-2\pi i s}}\right) & \text{as } \Im(s) \to -\infty \end{cases} \\
&= O\left(e^{-\frac{\pi}{2}|\Im(s)|}\right) \qquad \text{as } |\Im(s)| \to \infty.
\end{aligned}$$

The implied constant depends on ε and I. This shows (4.118) of the lemma. □

Remark 4.62 The Mellin transforms $\tilde{f}_\pm(s)$ in Equation (4.115) satisfy the slightly sharper estimate

$$\tilde{f}_\pm(s) = O\left(e^{-\frac{\pi}{2}|\Im(s)|}\right) \qquad \text{as } |\Im(s)| \to \infty$$

for all $\Re(s)$ large enough using Stirling's estimate of the Γ-function, see Appendix A.3, and that the L-series $L_\pm$ in (4.54) converge absolutely on some right half-plane.

Lemma 4.63 *Let $k \in \mathbb{R}$ and $\nu \in \mathbb{C}$, such that $|\Re(\nu)| < \frac{1}{2}$ and such that if $2\nu + k$ or $2\nu - k \in 1 + 2\mathbb{Z}$, then $k \notin \mathbb{Z}$. Let $D_\delta^\star$, $\delta = 0, 1$ as in the second part of Theorem 4.45 and let $L_l^\star$, $l \in k + 2\mathbb{Z}$ in the admissible range, with k the reference weight as part 2 of Theorem 4.28. We denote by $f : \mathbb{C} \setminus \mathbb{R} \to \mathbb{C}$ the function associated with (D_0, D_1) in (4.113) and by $\tilde{f}_\pm$ its Mellin transforms in (4.115).*

Then the function

$$A_{k,\nu}^\pm(s) := \tilde{f}_\pm(s)e^{\pm\frac{\pi i s}{2}} - \tilde{f}_\pm(1 - 2\nu - s)e^{\frac{\pi i}{2}(k \mp (1-2\nu-s))}, \tag{4.120}$$

defined for $0 < \Re(s) < 1 - 2\Re(\nu)$, is independent of the choice of $\pm$ and satisfies the growth condition

$$A_{k,\nu}^\pm(s) = O\left(e^{(\varepsilon-\pi)|\Im(s)|}\right) \tag{4.121}$$

for all $\varepsilon > 0$ on vertical lines $\Re(s) = \sigma$ with $0 < \sigma < 1 - 2\Re(\nu)$. The implied constant depends only on ε and $\Re(\nu)$.

We have in particular that

$$\begin{aligned} \frac{A_{k,\nu}^\pm(s)}{\sqrt{2\pi}} &= D_1^\star(s)\left(\frac{\sin(2\pi\nu)}{2i\sin(\pi s)\sin(\pi(1-2\nu-s))}\right) - \\ &- D_0^\star(s)\left(\frac{1}{2i\sin(\pi s)} + \frac{e^{\pi i k}}{2i\sin(\pi(1-2\nu-s))}\right) \end{aligned} \tag{4.122}$$

Proof Recall that $\tilde{f}_\pm(s)$ is holomorphic for all $\Re(s) > 0$, see the second part of Lemma 4.61. This implies that the function $A_{k,\nu}^\pm$, as defined in (4.120), is holomorphic on the strip $0 < \Re(s) < 1 - 2\Re(\nu)$.

Applying (4.116), we get

$$\begin{aligned} A_{k,\nu}^\pm(s) &\overset{(4.120)}{=} \tilde{f}_\pm(s)e^{\pm\frac{\pi i s}{2}} - \tilde{f}_\pm(1-2\nu-s)e^{\frac{\pi i}{2}(k\mp(1-2\nu-s))} \\ &\overset{(4.116)}{=} \sqrt{2\pi}\,\frac{e^{\pm\pi i s}D_1^\star(s) - D_0^\star(s)}{2i\sin(\pi s)} \\ &\quad - \sqrt{2\pi}\,e^{\frac{\pi i k}{2}}\,\frac{D_1^\star(1-2\nu-s) - e^{\mp\pi i(1-2\nu-s)}D_0^\star(1-2\nu-s)}{2i\sin(\pi(1-2\nu-s))}. \end{aligned}$$

Using the functional equation (4.98) for $D_0^\star$ and $D_1^\star$, we obtain

$$\frac{A^{\pm}_{k,\nu}(s)}{\sqrt{2\pi}} = D^{\star}_{1}(s)\left(\frac{e^{\pm\pi i s}}{2i\sin(\pi s)} + \frac{e^{\mp\pi i(1-2\nu-s)}}{2i\sin\big(\pi(1-2\nu-s)\big)}\right) - D^{\star}_{0}(s)\left(\frac{1}{2i\sin(\pi s)} + \frac{e^{\pi i k}}{2i\sin\big(\pi(1-2\nu-s)\big)}\right). \tag{4.123}$$

This identity extends to all $s \in \mathbb{C}$ as an identity of meromorphic functions. The trigonometric identity

$$\frac{e^{\pm\pi i s}}{2i\sin(\pi s)} + \frac{e^{\mp\pi i(1-2\nu-s)}}{2i\sin(\pi(1-2\nu-s))} = \frac{\sin(2\pi\nu)}{2i\,\sin(\pi s)\,\sin(\pi(1-2\nu-s))} \tag{4.124}$$

implies that $A^{\pm}_{k,\nu}$ does not depend on the choice of $\pm$ and proves (4.122). We have

$$\frac{A^{\pm}_{k,\nu}(s)}{\sqrt{2\pi}} = D^{\star}_{1}(s)\left(\frac{\sin(2\pi\nu)}{2i\sin(\pi s)\sin\big(\pi(1-2\nu-s)\big)}\right) - - D^{\star}_{0}(s)\left(\frac{1}{2i\sin(\pi s)} + \frac{e^{\pi i k}}{2i\sin\big(\pi(1-2\nu-s)\big)}\right). \tag{4.122}$$

We still have to prove that $A^{\pm}_{\kappa,\nu}$ satisfies the growth estimate (4.121). Starting from (4.122), we find that the sin-factors determine the growth on vertical lines, since Lemma 4.53 implies that $D^{\star}_{\delta}(s) = O\left(e^{\varepsilon|\Im(s)|}\right)$ for all $\varepsilon > 0$ and $\Re(s) \in I$, I a bounded interval. The implied constant depends only on ε and I. Since

$$\sin^{-1}(\pi s) = O\left(e^{-\pi|\Im(s)|}\right) \qquad \text{as } |\Im(s)| \to \infty,$$

the growth estimate (4.121) follows. □

Exercise 4.64 Show the trigonometric identity (4.124) that is used in the proof of Lemma 4.63.

In §5.1.4, we need an expression of $A^{\pm}_{k,\nu}(s)$ in terms of the pair $\big(L^{\star}_{k-2}(s), L^{\star}_{k}(s)\big)$. We show this relation in the following lemma. However, the proof is quite long.

Lemma 4.65 *Let $k \in \mathbb{R}$ and $\nu \in \mathbb{C}$, such that $|\Re(\nu)| < \frac{1}{2}$ and the conditions $2\nu + k \notin 1 + 2\mathbb{Z}$ and $2\nu - k \notin 1 + 2\mathbb{Z}$ hold. Let $L^{\star}_{k}$ and $L^{\star}_{k-2}$ two L-functions as given in part 2 of Theorem 4.28 and $A^{\pm}_{k,\nu}(s)$ as defined in* (4.120) *of Lemma 4.63. Then the L-series and the term $A^{\pm}_{k,\nu}(s)$ satisfy*

$$A^{\pm}_{k,\nu}(s) = -2^{-\frac{3}{2}}\pi^{\nu}e^{\frac{\pi i}{2}k} \cdot \Big[\big(1+2\nu-k\big)\,B_{-k,-\nu}\left(s+\nu-\frac{1}{2}\right)\,L^{\star}_{k-2}\left(s+\nu-\frac{1}{2}\right) - \big(1-2\nu-k\big)\,B_{2-k,-\nu}\left(s+\nu-\frac{1}{2}\right)\,L^{\star}_{k}\left(s+\nu-\frac{1}{2}\right)\Big] \tag{4.125}$$

for all $0 < \Re(s) < 1 - 2\Re(\nu)$.

Remark 4.66 The identity (4.125) also shows that the term $A^{\pm}_{k,\nu}(s)$ does not depend on the choice of the sign $\pm$. This confirms the same statement in Lemma 4.63 by an independent calculation.

Proof *(of Lemma 4.65)* Let us start with (4.117). We have

$$\tilde{f}_{\pm}(s) = \pm(2\pi)^{\frac{1}{2}-s}\Gamma(s)\,L_{\pm}\left(s+\nu-\frac{1}{2}\right). \tag{4.117}$$

To express $L_{\pm}(\cdot)$ in terms of $\begin{pmatrix} L^{\star}_{k-2}(\cdot) \\ L^{\star}_{k}(\cdot) \end{pmatrix}$, we use (4.55) in the matrix-vector form given in (4.62). We have

$$\begin{pmatrix} L_{+}\left(s+\nu-\frac{1}{2}\right) \\ L_{-}\left(s+\nu-\frac{1}{2}\right) \end{pmatrix} \overset{(4.62)}{=} -\frac{1}{2}(2\pi)^{2s+2\nu-1}\frac{\Gamma\left(\frac{1}{2}+\nu+\frac{k}{2}\right)\Gamma\left(\frac{3}{2}+\nu-\frac{k}{2}\right)}{\Gamma(s+2\nu)\,\Gamma(s)} \cdot \begin{pmatrix} \tilde{\psi}_{-k,\nu}\left(s+\nu-\frac{1}{2}\right) & -\tilde{\psi}_{2-k,\nu}\left(s+\nu-\frac{1}{2}\right) \\ -\tilde{\psi}_{k,\nu}\left(s+\nu-\frac{1}{2}\right) & \tilde{\psi}_{k-2,\nu}\left(s+\nu-\frac{1}{2}\right) \end{pmatrix} \begin{pmatrix} L^{\star}_{k-2}\left(s+\nu-\frac{1}{2}\right) \\ L^{\star}_{k}\left(s+\nu-\frac{1}{2}\right) \end{pmatrix}. \tag{4.126}$$

Now consider the matrix $\begin{pmatrix} \tilde{\psi}_{-k,\nu}\left(s+\nu-\frac{1}{2}\right) & -\tilde{\psi}_{2-k,\nu}\left(s+\nu-\frac{1}{2}\right) \\ -\tilde{\psi}_{k,\nu}\left(s+\nu-\frac{1}{2}\right) & \tilde{\psi}_{k-2,\nu}\left(s+\nu-\frac{1}{2}\right) \end{pmatrix}$. Using (4.22) and using the functional equation $x\,\Gamma(x) = \Gamma(x+1)$, see Appendix A.7, we can rewrite the following matrix as

$$\begin{pmatrix} \tilde{\psi}_{-k,\nu}\left(s+\nu-\frac{1}{2}\right) & -\tilde{\psi}_{2-k,\nu}\left(s+\nu-\frac{1}{2}\right) \\ -\tilde{\psi}_{k,\nu}\left(s+\nu-\frac{1}{2}\right) & \tilde{\psi}_{k-2,\nu}\left(s+\nu-\frac{1}{2}\right) \end{pmatrix}$$

$$\overset{(4.22)}{=} \begin{pmatrix} \frac{\Gamma\left(\frac{1}{2}-\nu-\frac{k}{2}\right)}{\Gamma\left(\frac{3}{2}+\nu-\frac{k}{2}\right)} & 0 \\ 0 & \frac{\Gamma\left(-\frac{1}{2}-\nu+\frac{k}{2}\right)}{\Gamma\left(\frac{1}{2}+\nu+\frac{k}{2}\right)} \end{pmatrix} \begin{pmatrix} \tilde{\psi}_{-k,-\nu}\left(s+\nu-\frac{1}{2}\right) & -\tilde{\psi}_{2-k,-\nu}\left(s+\nu-\frac{1}{2}\right) \\ \tilde{\psi}_{k,-\nu}\left(s+\nu-\frac{1}{2}\right) & -\tilde{\psi}_{k-2,-\nu}\left(s+\nu-\frac{1}{2}\right) \end{pmatrix} \cdot \begin{pmatrix} \left(\frac{1}{2}+\nu-\frac{k}{2}\right) & 0 \\ 0 & \left(\frac{1}{2}-\nu-\frac{k}{2}\right) \end{pmatrix}.$$

Hence, we get

$$\begin{pmatrix} L_+\left(s+\nu-\frac{1}{2}\right) \\ L_-\left(s+\nu-\frac{1}{2}\right) \end{pmatrix} \overset{(4.126)}{=} -\frac{1}{2}(2\pi)^{2s+2\nu-1} \frac{1}{\Gamma(s+2\nu)\,\Gamma(s)} \cdot \begin{pmatrix} \frac{\pi}{\cos\left(\pi\left(\frac{k}{2}+\nu\right)\right)} & 0 \\ 0 & -\frac{\pi}{\cos\left(\pi\left(\frac{k}{2}-\nu\right)\right)} \end{pmatrix} \begin{pmatrix} \tilde{\psi}_{-k,-\nu}\left(s+\nu-\frac{1}{2}\right) & -\tilde{\psi}_{2-k,-\nu}\left(s+\nu-\frac{1}{2}\right) \\ \tilde{\psi}_{k,-\nu}\left(s+\nu-\frac{1}{2}\right) & -\tilde{\psi}_{k-2,-\nu}\left(s+\nu-\frac{1}{2}\right) \end{pmatrix} \cdot \begin{pmatrix} \left(\frac{1}{2}+\nu-\frac{k}{2}\right) L^\star_{k-2}\left(s+\nu-\frac{1}{2}\right) \\ \left(\frac{1}{2}-\nu-\frac{k}{2}\right) L^\star_{k}\left(s+\nu-\frac{1}{2}\right) \end{pmatrix}$$

using $\Gamma(s)\Gamma(1-s) = \frac{\pi}{\sin(\pi s)}$ in Appendix A.7.

As a result, we get

$$\begin{pmatrix} \tilde{f}_+(s) \\ \tilde{f}_-(s) \end{pmatrix} = -\frac{1}{2}(2\pi)^{s+2\nu-\frac{1}{2}} \frac{1}{\Gamma(s+2\nu)} \cdot \begin{pmatrix} \frac{\pi}{\cos\left(\pi\left(\frac{k}{2}+\nu\right)\right)} & 0 \\ 0 & \frac{\pi}{\cos\left(\pi\left(\frac{k}{2}-\nu\right)\right)} \end{pmatrix} \cdot \begin{pmatrix} \tilde{\psi}_{-k,-\nu}\left(s+\nu-\frac{1}{2}\right) & -\tilde{\psi}_{2-k,-\nu}\left(s+\nu-\frac{1}{2}\right) \\ \tilde{\psi}_{k,-\nu}\left(s+\nu-\frac{1}{2}\right) & -\tilde{\psi}_{k-2,-\nu}\left(s+\nu-\frac{1}{2}\right) \end{pmatrix} \cdot \begin{pmatrix} \left(\frac{1}{2}+\nu-\frac{k}{2}\right) L^\star_{k-2}\left(s+\nu-\frac{1}{2}\right) \\ \left(\frac{1}{2}-\nu-\frac{k}{2}\right) L^\star_{k}\left(s+\nu-\frac{1}{2}\right) \end{pmatrix}. \tag{4.127}$$

The next step is to substitute (4.82) given by

$$\tilde{\psi}_{k,-\nu}(s) \overset{(4.82)}{=} 2^{-\nu-s-\frac{1}{2}}\pi^{-s-1}\,\Gamma\left(\frac{1}{2}+s+\nu\right) \cdot \left[e^{-\frac{\pi i}{2}\left(\frac{1}{2}+s+\nu\right)} B_{k,-\nu}(s) + e^{\frac{\pi i}{2}\left(\frac{1}{2}+s+\nu\right)} B_{-k,-\nu}(s)\right].$$

As a result, we have

$$\begin{pmatrix} e^{\frac{\pi i}{2}s}\,\tilde{f}_+(s) \\ e^{-\frac{\pi i}{2}s}\,\tilde{f}_-(s) \end{pmatrix} = \begin{pmatrix} e^{\frac{\pi i}{2}s} & 0 \\ 0 & e^{-\frac{\pi i}{2}s} \end{pmatrix} \begin{pmatrix} \tilde{f}_+(s) \\ \tilde{f}_-(s) \end{pmatrix}$$
$$= -2^{-\frac{3}{2}}\pi^{\nu} \begin{pmatrix} \frac{1}{\cos\left(\pi\left(\frac{k}{2}+\nu\right)\right)} & 0 \\ 0 & \frac{1}{\cos\left(\pi\left(\frac{k}{2}-\nu\right)\right)} \end{pmatrix} \begin{pmatrix} e^{\pi i(s+\nu)} & e^{-\pi i\nu} \\ e^{-\pi i(s+\nu)} & e^{\pi i\nu} \end{pmatrix} \cdot \begin{pmatrix} B_{k,-\nu}\left(s+\nu-\frac{1}{2}\right) & -B_{k-2,-\nu}\left(s+\nu-\frac{1}{2}\right) \\ B_{-k,-\nu}\left(s+\nu-\frac{1}{2}\right) & -B_{2-k,-\nu}\left(s+\nu-\frac{1}{2}\right) \end{pmatrix} \cdot \begin{pmatrix} \left(\frac{1}{2}+\nu-\frac{k}{2}\right) L^{\star}_{k-2}\left(s+\nu-\frac{1}{2}\right) \\ \left(\frac{1}{2}-\nu-\frac{k}{2}\right) L^{\star}_{k}\left(s+\nu-\frac{1}{2}\right) \end{pmatrix}. \tag{4.128}$$

We also need similar expressions for $\tilde{f}_{\pm}(1-2\nu-s)e^{\frac{\pi i}{2}(\kappa\mp(1-2\nu-s))}$. Similarly, we find that

$$\begin{pmatrix} e^{\frac{\pi i}{2}(\kappa-(1-2\nu-s))}\,\tilde{f}_+(1-2\nu-s) \\ e^{\frac{\pi i}{2}(\kappa+(1-2\nu-s))}\,\tilde{f}_-(1-2\nu-s) \end{pmatrix}$$
$$= -2^{-\frac{3}{2}}\pi^{\nu} \begin{pmatrix} \frac{e^{\frac{\pi i}{2}(\kappa-(1-2\nu-s))}}{\cos\left(\pi\left(\frac{k}{2}+\nu\right)\right)} & 0 \\ 0 & \frac{e^{\frac{\pi i}{2}(\kappa+(1-2\nu-s))}}{\cos\left(\pi\left(\frac{k}{2}-\nu\right)\right)} \end{pmatrix} \begin{pmatrix} e^{\frac{\pi i}{2}(1-s)} & e^{-\frac{\pi i}{2}(1-s)} \\ e^{-\frac{\pi i}{2}(1-s)} & e^{\frac{\pi i}{2}(1-s)} \end{pmatrix} \cdot \begin{pmatrix} -e^{\frac{\pi ik}{2}} B_{-k,-\nu}\left(s+\nu-\frac{1}{2}\right) & e^{\frac{\pi ik}{2}} B_{2-k,-\nu}\left(s+\nu-\frac{1}{2}\right) \\ -e^{-\frac{\pi ik}{2}} B_{k,-\nu}\left(s+\nu-\frac{1}{2}\right) & e^{-\frac{\pi ik}{2}} B_{k-2,-\nu}\left(s+\nu-\frac{1}{2}\right) \end{pmatrix} \cdot \begin{pmatrix} \left(\frac{1}{2}+\nu-\frac{k}{2}\right) L^{\star}_{k-2}\left(s+\nu-\frac{1}{2}\right) \\ \left(\frac{1}{2}-\nu-\frac{k}{2}\right) L^{\star}_{k}\left(s+\nu-\frac{1}{2}\right) \end{pmatrix} \tag{4.129}$$

$$= -2^{-\frac{3}{2}}\pi^{\nu} \begin{pmatrix} \frac{e^{\frac{\pi i}{2}(\kappa-(1-2\nu-s))}}{\cos\left(\pi\left(\frac{k}{2}+\nu\right)\right)} & 0 \\ 0 & \frac{e^{\frac{\pi i}{2}(\kappa+(1-2\nu-s))}}{\cos\left(\pi\left(\frac{k}{2}-\nu\right)\right)} \end{pmatrix} \begin{pmatrix} -e^{-\frac{\pi i}{2}(1-s+k)} & -e^{\frac{\pi i}{2}(1-s+k)} \\ -e^{\frac{\pi i}{2}(1-s-k)} & -e^{-\frac{\pi i}{2}(1-s-k)} \end{pmatrix}$$
$$\cdot \begin{pmatrix} B_{k,-\nu}\left(s+\nu-\frac{1}{2}\right) & -B_{k-2,-\nu}\left(s+\nu-\frac{1}{2}\right) \\ B_{-k,-\nu}\left(s+\nu-\frac{1}{2}\right) & -B_{2-k,-\nu}\left(s+\nu-\frac{1}{2}\right) \end{pmatrix} \cdot$$
$$\cdot \begin{pmatrix} \left(\frac{1}{2}+\nu-\frac{k}{2}\right) L^{\star}_{k-2}\left(s+\nu-\frac{1}{2}\right) \\ \left(\frac{1}{2}-\nu-\frac{k}{2}\right) L^{\star}_{k}\left(s+\nu-\frac{1}{2}\right) \end{pmatrix}. \tag{4.130}$$

Note that we can write

$$- \begin{pmatrix} \frac{e^{\frac{\pi i}{2}(k-(1-2\nu-s))}}{\cos\left(\pi\left(\frac{k}{2}+\nu\right)\right)} & 0 \\ 0 & \frac{e^{\frac{\pi i}{2}(k+(1-2\nu-s))}}{\cos\left(\pi\left(\frac{k}{2}-\nu\right)\right)} \end{pmatrix} \begin{pmatrix} -e^{-\frac{\pi i}{2}(1-s+k)} & -e^{\frac{\pi i}{2}(1-s+k)} \\ -e^{\frac{\pi i}{2}(1-s-k)} & -e^{-\frac{\pi i}{2}(1-s-k)} \end{pmatrix}$$
$$= \begin{pmatrix} \frac{1}{\cos\left(\pi\left(\frac{k}{2}+\nu\right)\right)} & 0 \\ 0 & \frac{1}{\cos\left(\pi\left(\frac{k}{2}-\nu\right)\right)} \end{pmatrix} \begin{pmatrix} e^{-\pi i(1-\nu-s)} & e^{\pi i(k+\nu)} \\ e^{\pi i(1-\nu-s)} & e^{\pi i(k-\nu)} \end{pmatrix}.$$

Plugging this identity into (4.130), we get

$$\begin{pmatrix} e^{\frac{\pi i}{2}(\kappa-(1-2\nu-s))}\tilde{f}_{+}(1-2\nu-s) \\ e^{\frac{\pi i}{2}(\kappa+(1-2\nu-s))}\tilde{f}_{-}(1-2\nu-s) \end{pmatrix}$$
$$= 2^{-\frac{3}{2}}\pi^{\nu} \begin{pmatrix} \frac{1}{\cos\left(\pi\left(\frac{k}{2}+\nu\right)\right)} & 0 \\ 0 & \frac{1}{\cos\left(\pi\left(\frac{k}{2}-\nu\right)\right)} \end{pmatrix} \begin{pmatrix} e^{-\pi i(1-\nu-s)} & e^{\pi i(k+\nu)} \\ e^{\pi i(1-\nu-s)} & e^{\pi i(k-\nu)} \end{pmatrix} \cdot$$
$$\cdot \begin{pmatrix} B_{k,-\nu}\left(s+\nu-\frac{1}{2}\right) & -B_{k-2,-\nu}\left(s+\nu-\frac{1}{2}\right) \\ B_{-k,-\nu}\left(s+\nu-\frac{1}{2}\right) & -B_{2-k,-\nu}\left(s+\nu-\frac{1}{2}\right) \end{pmatrix} \cdot$$
$$\cdot \begin{pmatrix} \left(\frac{1}{2}+\nu-\frac{k}{2}\right) L^{\star}_{k-2}\left(s+\nu-\frac{1}{2}\right) \\ \left(\frac{1}{2}-\nu-\frac{k}{2}\right) L^{\star}_{k}\left(s+\nu-\frac{1}{2}\right) \end{pmatrix}. \tag{4.131}$$

Recall $A^{\pm}_{\kappa}(s)$ in (4.120), we find that

$$\begin{pmatrix} A^+_{k,\nu}(s) \\ A^-_{k,\nu}(s) \end{pmatrix} \overset{(4.120)}{=} \begin{pmatrix} e^{\frac{\pi i}{2}s}\,\tilde{f}_+(s) \\ e^{-\frac{\pi i}{2}s}\,\tilde{f}_-(s) \end{pmatrix} - \begin{pmatrix} e^{\frac{\pi i}{2}(k-(1-2\nu-s))}\,\tilde{f}_+(1-2\nu-s) \\ e^{\frac{\pi i}{2}(k+(1-2\nu-s))}\,\tilde{f}_-(1-2\nu-s) \end{pmatrix}$$

$$= \; -2^{-\frac{3}{2}}\pi^{\nu} \begin{pmatrix} \frac{1}{\cos\left(\pi\left(\frac{k}{2}+\nu\right)\right)} & 0 \\ 0 & \frac{1}{\cos\left(\pi\left(\frac{k}{2}-\nu\right)\right)} \end{pmatrix}$$

$$\cdot \left[\begin{pmatrix} e^{\pi i(s+\nu)} & e^{-\pi i\nu} \\ e^{-\pi i(s+\nu)} & e^{\pi i\nu} \end{pmatrix} + \begin{pmatrix} e^{-\pi i(1-\nu-s)} & e^{\pi i(k+\nu)} \\ e^{\pi i(1-\nu-s)} & e^{\pi i(k-\nu)} \end{pmatrix} \right]$$

$$\cdot \begin{pmatrix} B_{k,-\nu}\left(s+\nu-\frac{1}{2}\right) & -B_{k-2,-\nu}\left(s+\nu-\frac{1}{2}\right) \\ B_{-k,-\nu}\left(s+\nu-\frac{1}{2}\right) & -B_{2-k,-\nu}\left(s+\nu-\frac{1}{2}\right) \end{pmatrix} \cdot$$

$$\cdot \begin{pmatrix} \left(\frac{1}{2}+\nu-\frac{k}{2}\right) L^{\star}_{k-2}\left(s+\nu-\frac{1}{2}\right) \\ \left(\frac{1}{2}-\nu-\frac{k}{2}\right) L^{\star}_{k}\left(s+\nu-\frac{1}{2}\right) \end{pmatrix}.$$

We now simplify the following matrix and we get

$$\begin{pmatrix} e^{\pi i(s+\nu)} & e^{-\pi i\nu} \\ e^{-\pi i(s+\nu)} & e^{\pi i\nu} \end{pmatrix} + \begin{pmatrix} e^{-\pi i(1-\nu-s)} & e^{\pi i(k+\nu)} \\ e^{\pi i(1-\nu-s)} & e^{\pi i(k-\nu)} \end{pmatrix}$$

$$= 2e^{\frac{\pi i k}{2}} \begin{pmatrix} 0 & \cos\left(\pi\left(\frac{k}{2}+\nu\right)\right) \\ 0 & \cos\left(\pi\left(\frac{k}{2}-\nu\right)\right) \end{pmatrix},$$

using $\cos x = \frac{1}{2}\left(e^{ix}+e^{-ix}\right)$. Plugging in, we get

$$\begin{pmatrix} A^+_{k,\nu}(s) \\ A^-_{k,\nu}(s) \end{pmatrix} = -2^{-\frac{1}{2}}\pi^{\nu} e^{\frac{\pi i}{2}k} \begin{pmatrix} \frac{1}{\cos\left(\pi\left(\frac{k}{2}+\nu\right)\right)} & 0 \\ 0 & \frac{1}{\cos\left(\pi\left(\frac{k}{2}-\nu\right)\right)} \end{pmatrix} \begin{pmatrix} 0 & \cos\left(\pi\left(\frac{k}{2}+\nu\right)\right) \\ 0 & \cos\left(\pi\left(\frac{k}{2}-\nu\right)\right) \end{pmatrix}.$$

$$\cdot \begin{pmatrix} B_{k,-\nu}\left(s+\nu-\frac{1}{2}\right) & -B_{k-2,-\nu}\left(s+\nu-\frac{1}{2}\right) \\ B_{-k,-\nu}\left(s+\nu-\frac{1}{2}\right) & -B_{2-k,-\nu}\left(s+\nu-\frac{1}{2}\right) \end{pmatrix} \cdot$$

$$\cdot \begin{pmatrix} \left(\frac{1}{2}+\nu-\frac{k}{2}\right) L^{\star}_{k-2}\left(s+\nu-\frac{1}{2}\right) \\ \left(\frac{1}{2}-\nu-\frac{k}{2}\right) L^{\star}_{k}\left(s+\nu-\frac{1}{2}\right) \end{pmatrix}$$

$$=-2^{-\frac{1}{2}}\pi^{\nu} e^{\frac{\pi i}{2}k} \begin{pmatrix} 0 & 1 \\ 0 & 1 \end{pmatrix} \begin{pmatrix} B_{-k,-\nu}\left(s+\nu-\frac{1}{2}\right) & -B_{2-k,-\nu}\left(s+\nu-\frac{1}{2}\right) \\ B_{-k,-\nu}\left(s+\nu-\frac{1}{2}\right) & -B_{2-k,-\nu}\left(s+\nu-\frac{1}{2}\right) \end{pmatrix}$$

$$
\cdot \begin{pmatrix} \left(\frac{1}{2}+\nu-\frac{k}{2}\right) L^\star_{k-2}\left(s+\nu-\frac{1}{2}\right) \\ \left(\frac{1}{2}-\nu-\frac{k}{2}\right) L^\star_{k}\left(s+\nu-\frac{1}{2}\right) \end{pmatrix}
$$

$$
= -2^{-\frac{1}{2}} \pi^{\nu} e^{\frac{\pi i}{2} k}
$$

$$
\cdot \Bigg[\left(\frac{1}{2}+\nu-\frac{k}{2}\right) B_{-k,-\nu}\left(s+\nu-\frac{1}{2}\right) L^\star_{k-2}\left(s+\nu-\frac{1}{2}\right)
$$

$$
- \left(\frac{1}{2}-\nu-\frac{k}{2}\right) B_{2-k,-\nu}\left(s+\nu-\frac{1}{2}\right) L^\star_{k}\left(s+\nu-\frac{1}{2}\right)\Bigg].
$$

This shows (4.125) and concludes the proof of Lemma 4.65. □

Lemma 4.67 *In the situation of Lemma 4.61, the inverse Mellin transforms*

$$
\frac{1}{2\pi i} \int_{\sigma-i\infty}^{\sigma+i\infty} \tilde{f}_{\pm}(s)\, y^{-s}\, \mathrm{d}s \qquad (y > 0) \tag{4.132}
$$

are well defined for all $\sigma > 0$*. The integrals in* (4.132) *converge for all* $\sigma \in \mathbb{R} \setminus \mathbb{Z}_{\leq 0}$*. We have*

$$
\frac{1}{2\pi i} \int_{\Re(s)=-m-\frac{1}{2}} \tilde{f}_{\pm}(s)\, y^{-s}\, \mathrm{d}s = \frac{1}{2\pi i} \int_{\Re(s)=-m+\frac{1}{2}} \tilde{f}_{\pm}(s)\, y^{-s} \mathrm{d}s - \operatorname*{res}_{s=-m} \left(\tilde{f}_{\pm}(s)\, y^{-s}\right) \tag{4.133}
$$

for all $m \in \mathbb{Z}_{\geq 0}$ *and* $y > 0$*.*

Proof The functions $\tilde{f}_{\pm}(s)$ are meromorphic functions with simple poles in $s \in \mathbb{Z}_{\leq 0}$, see Lemma 4.61. The growth estimate in (4.118) of Lemma 4.61 for $\tilde{f}_{\pm}(s)$ imply the convergence of the integral in (4.132) for all σ except $\sigma \in \mathbb{Z}_{\leq 0}$.

Assuming $\sigma > 0$, we immediately see that the integral (4.132) is in fact the inverse Mellin transforms of $\tilde{f}_{\pm}$.

The remaining part of the lemma is a direct consequence of the residue theorem, e.g., [152, p. 75, §3.2, Theorem 1] or Appendix A.13. □

Lemma 4.68 *In the situation of Lemma 4.61, the function*

$$
F : \mathbb{C} \setminus \mathbb{R} \to \mathbb{C}; \quad z \mapsto F(z) := f(z) - e^{\frac{\pi i k}{2}} z^{2\nu-1} f\left(-\frac{1}{z}\right) \tag{4.134}
$$

extends to a holomorphic function on the cut-plane

$$
\mathbb{C}' := \mathbb{C} \setminus \mathbb{R}_{\leq 0} \tag{4.135}
$$

and is bounded by an multiple of $\min\left\{1, |z|^{2\Re(\nu)-1}\right\}$ *in the right half-plane* $\Re(z) > 0$*. The Mellin transform of F satisfies*

$$\int_0^{\infty} F(z)\, z^{s-1}\, dz = A^{\pm}_{k,\nu}(s) \tag{4.136}$$

for $0 < \Re(s) < 1 - 2\Re(\nu)$, *where* $A^{\pm}_{k,\nu}(s)$ *is given in Lemma 4.63.*

Proof Let $k \in \mathbb{R}$ and $\nu \in \mathbb{C}$ with $|\Re(\nu)| < \frac{1}{2}$.

Consider the inverse Mellin transformations $\mathcal{M}^{-1}(\tilde{f}_{\pm})$ of $\tilde{f}_{\pm}$ where the inverse Mellin transform is defined in (1.156) with some of its properties presented in Lemma 1.203 and Proposition 1.204. The growth estimate (4.118) in Lemma 4.61 together with Proposition 1.204 implies that the inverse Mellin transformations exists and satisfies

$$f(\pm iy) = \frac{1}{2\pi i} \int_{\sigma - i\infty}^{\sigma + i\infty} \tilde{f}_{\pm}(s) y^{-s}\, ds$$

for all $\sigma > 0$ and $y > 0$. Recall that the condition $\sigma > 0$ ensures that $\tilde{f}_{\pm}(s)$ are only evaluated in points away from the possible pole at $s = 0$, see the second part of Lemma 4.61.

The inverse Mellin transforms extend analytically to all y with $|\arg(y)| < \frac{\pi}{2} - \varepsilon$ for some $\varepsilon > 0$. This gives us

$$f(z) = \frac{1}{2\pi i} \int_{\sigma - i\infty}^{\sigma + i\infty} \tilde{f}_{\pm}(s)\, e^{\pm \frac{\pi i s}{2}}\, z^{-s}\, ds$$

for $\Im(z) \gtrless 0$ and $\sigma > 0$ (where the choice of $\gtrless$ and $\pm$ in the equation above is connected in the sense that either take simultaneously "$>$" and "$+$" or take "$<$" and "$-$"). Hence we have

$$\begin{aligned}
F(z) &= f(z) - e^{\frac{\pi i k}{2}} z^{2\nu - 1} f\left(-\frac{1}{z}\right) \\
&= \frac{1}{2\pi i} \int_{\sigma - i\infty}^{\sigma + i\infty} \tilde{f}_{\pm}(s)\, e^{\pm \frac{\pi i s}{2}}\, z^{-s}\, ds \\
&\qquad - \frac{1}{2\pi i} \int_{\sigma - i\infty}^{\sigma + i\infty} \tilde{f}_{\pm}(s)\, e^{\frac{\pi i}{2}(k \pm s)} z^{2\nu - 1} \left(\frac{-1}{z}\right)^{-s} ds \\
&= \frac{1}{2\pi i} \int_{\sigma - i\infty}^{\sigma + i\infty} \tilde{f}_{\pm}(s)\, e^{\pm \frac{\pi i s}{2}}\, z^{-s}\, ds \\
&\qquad - \frac{1}{2\pi i} \int_{\sigma - i\infty}^{\sigma + i\infty} \tilde{f}_{\pm}(s)\, e^{\frac{\pi i}{2}(k \pm s)} z^{s + 2\nu - 1} \underbrace{(-1)^{-s}}_{= e^{\mp \pi i s}} ds \\
&= \frac{1}{2\pi i} \int_{\sigma - i\infty}^{\sigma + i\infty} \tilde{f}_{\pm}(s) \left(e^{\pm \frac{\pi i s}{2}} z^{-s} - e^{\frac{\pi i}{2}(k \mp s)} z^{s + 2\nu - 1}\right) ds
\end{aligned}$$

for all $\Im(z) \gtrless 0$. Now, we substitute $s \mapsto 1 - 2\nu - s$ in the second part of the integrand and move the path of integration to $\sigma + i\mathbb{R}$. We find

$$\begin{aligned}
&- \int_{\sigma-i\infty}^{\sigma+i\infty} \tilde{f}_\pm(s)\, e^{\frac{\pi i}{2}(k\mp s)}\, z^{2\nu-1+s}\, ds \\
&\quad = \int_{1-2\nu-\sigma+i\infty}^{1-2\nu-\sigma-i\infty} \tilde{f}_\pm(1-2\nu-s)\, e^{\frac{\pi i}{2}\left(k\mp(1-2\nu-s)\right)}\, z^{-s}\, ds \\
&\quad = - \int_{\sigma-i\infty}^{\sigma+i\infty} \tilde{f}_\pm(1-2\nu-s)\, e^{\frac{\pi i}{2}\left(k\mp(1-2\nu-s)\right)}\, z^{-s}\, ds.
\end{aligned}$$

We can write $F(z)$ as

$$\begin{aligned}
F(z) &= \frac{1}{2\pi i} \int_{\sigma-i\infty}^{\sigma+i\infty} \tilde{f}_\pm(s) e^{\pm\frac{\pi i s}{2}}\, z^{-s}\, ds \\
&\quad - \frac{1}{2\pi i} \int_{\sigma-i\infty}^{\sigma+i\infty} \tilde{f}_\pm(1-2\nu-s) e^{\frac{\pi i}{2}(k\mp(1-2\nu-s))}\, z^{-s}\, ds \\
&\overset{(4.120)}{=} \frac{1}{2\pi i} \int_{\sigma-i\infty}^{\sigma+i\infty} A^\pm_{k,\nu}(s)\, z^{-s}\, ds \qquad (\Im(z) \gtrless 0)
\end{aligned} \tag{4.137}$$

for $0 < \sigma < 1 - 2\Re(\nu)$, where $A^\pm_{k,\nu}(s)$ is given in (4.120). Hence, we have expressed the function $F(z)$ in (4.134) as an integral that converges for all $z \in \mathbb{C}\setminus\mathbb{R}$. Moreover the integral in (4.137) is independent of the choice of $\pm$, see (4.122) in Lemma 4.63.

Recalling the growth estimate (4.121) of $A^\pm_{k,\nu}$ in Lemma 4.63, we conclude that the integral in (4.137) converges also for $z \in \mathbb{R}_{>0}$. This establishes that the integral representation in (4.137) of $F(z)$ exists for all $z \in \mathbb{C}'$. Hence $F(z)$ extends to the complex cut plane $\mathbb{C}'$ holomorphically.

The estimates for $F(z)$ in the right half-plane $\Re(z) > 0$ follow again from the integral representation in (4.137). Indeed, the integral immediately gives a (uniform) bound $O\left(|z|^{-\sigma}\right)$ in the half-plane $\Re(z) > 0$ for every σ between 0 and $1 - 2\Re(\nu)$. Since the integrand is meromorphic with simple poles at $s = 0$ and $s = 1 - 2\nu$, we can even move the path of integration to a vertical line $\Re(s) = \sigma$ with σ slightly to the left of 0 or slightly to the right of $1-2\nu$, picking up a residue term proportional to 1 or $z^{2\nu-1}$. (Moving the path of integration even further should give us the complete asymptotic expansion of $F(z)$.)

Finally, these growth estimates allow us to compute the Mellin transform

$$\int_0^\infty F(z)\, z^{s-1} dz \qquad (0 < \Re(s) < 1 - 2\Re(\nu))$$

over the positive real axis, and (4.136) follows. □

Writing (4.116) in vector form, we have

$$\begin{pmatrix} D_0^\star(s) \\ D_1^*(s) \end{pmatrix} = \frac{\sqrt{2\pi}}{2i\ \sin(\pi s)} \begin{pmatrix} e^{-\frac{\pi i s}{2}} & -e^{\frac{\pi i s}{2}} \\ e^{\frac{\pi i s}{2}} & -e^{-\frac{\pi i s}{2}} \end{pmatrix} \begin{pmatrix} \tilde{f}_+(s) \\ \tilde{f}_-(s) \end{pmatrix}, \tag{4.138}$$

which is valid for all $s \notin \mathbb{Z}_{\leq 0}$.

Since the second part of Lemma 4.61 – see below (4.115) – shows that

$$\frac{(2\pi)^{\frac{1}{2}-s}}{2i} \frac{\tilde{f}_\pm(s)}{\Gamma(s)}$$

is entire in s, we conclude that

$$e^{\frac{\pi i s}{2}} \tilde{f}_+(s) - e^{-\frac{\pi i s}{2}} \tilde{f}_-(s) \quad \text{and} \quad e^{-\frac{\pi i s}{2}} \tilde{f}_+(s) - e^{\frac{\pi i s}{2}} \tilde{f}_-(s)$$

are also entire functions (in s).

Lemma 4.69 *Consider the situation of Lemma 4.61. The function f in Equation* (4.113)*, restricted to $i\mathbb{R}_{\neq 0}$, has an asymptotic expansion in* 0 *of the form*

$$f(iy) = \frac{1}{2\pi i} \sum_{n=0}^{\infty} \left(i^{-n} D_0^\star(-n) - i^n D_1^\star(-n) \right) y^n \tag{4.139}$$

as $|y| \to 0$.

Proof We begin with a computation of the residues in the poles of the Mellin transforms $\tilde{f}_\pm(s)$ of f, given in Lemma 4.61. The second part of Lemma 4.61 states that $\tilde{f}_\pm(s)$ has only simple poles in $s \in \mathbb{Z}_{\leq 0}$. Moreover, the $D^\star_\cdot(s)$-functions are entire as stated around (4.97) of Theorem 4.45. Hence

$$\tilde{f}_\pm(s) \overset{(4.116)}{=} \frac{\sqrt{2\pi}}{2i \sin(\pi s)} \left(e^{\pm\frac{\pi i s}{2}} D_1^\star(s) - e^{\mp\frac{\pi i s}{2}} D_0^\star(s) \right)$$

has only simple poles in $s \in \mathbb{Z}_{\leq 0}$ with residues

$$\operatorname{res}_{s=-n} \tilde{f}_\pm(s) = \frac{\sqrt{2\pi}}{2i} \left(\operatorname{res}_{s=n} \frac{1}{\sin(\pi s)} \right) \left(e^{\mp\frac{\pi i n}{2}} D_1^\star(-n) - e^{\pm\frac{\pi i n}{2}} D_0^\star(-n) \right).$$

Using

$$\operatorname{res}_{s=n} \frac{1}{\sin(\pi s)} = (-1)^n \pi^{-1} \qquad \text{for all } n \in \mathbb{Z},$$

we find

$$\begin{aligned}\operatorname*{res}_{s=-n} \tilde{f}_{\pm}(s) &= \frac{(-1)^n}{\sqrt{2\pi i}} \left(e^{\mp\frac{\pi i n}{2}} D_1^\star(-n) - e^{\pm\frac{\pi i n}{2}} D_0^\star(-n) \right) \\ &= \frac{(\mp i)^n}{\sqrt{2\pi i}} \left((-1)^n D_1^\star(-n) - D_0^\star(-n) \right)\end{aligned}$$

for all $n \in \mathbb{Z}_{\geq 0}$.

To prove the lemma, we start with the inverse Mellin transforms of $\tilde{f}_{\pm}$. The inverse Mellin transforms are well defined, see Lemma 4.67. We have

$$f(\pm iy) = \frac{1}{2\pi i} \int_{\Re(s)=\frac{1}{2}} \tilde{f}_{\pm}(s)\, y^{-s}\, ds$$

for $y > 0$ since the functions $\tilde{f}_{\pm}$ are Mellin transforms of $f(\pm y)$. Moving the path of integration from $\Re(s) = \frac{1}{2}$ to $\Re(s) = -\frac{1}{2} - N$, $N \in \mathbb{Z}_{\geq 0}$, Lemma 4.67 gives us

$$\begin{aligned}f(\pm iy) &= \frac{1}{2\pi i} \int_{\Re(s)=-\frac{1}{2}-N} \tilde{f}_{\pm}(s)\, y^{-s}\, ds - \sum_{n=0}^{N} \operatorname*{res}_{s=-n} \left(\tilde{f}_{\pm}(s)\, y^{-s} \right) \\ &= \frac{1}{2\pi i} \int_{\Re(s)=-\frac{1}{2}-N} \tilde{f}_{\pm}(s)\, y^{-s} ds \\ &\quad - \frac{1}{\sqrt{2\pi i}} \sum_{n=0}^{N} \left((-1)^n D_1^\star(-n) - D_0^\star(-n) \right) (\mp iy)^n\end{aligned}$$

for $y > 0$. The remaining integrals satisfy

$$\int_{\Re(s)=-\frac{1}{2}-N} \tilde{f}_{\pm}(s)\, y^{-s}\, ds = O\left(y^{N+\frac{1}{2}} \right)$$

for $y \searrow 0$.

Now consider $y \in \mathbb{R}_{\neq 0}$ with $|y|$ small. Taking $N \to \infty$ in the above expression shows the asymptotic expansion

$$\begin{aligned}f(iy) &= f\big(\operatorname{sign}(y)\, i\, |y|\big) \\ &= \lim_{N\to\infty} \left(\frac{1}{2\pi i} \int_{\Re(s)=-\frac{1}{2}-N} \tilde{f}_{\pm}(s)\, y^{-s} ds \right. \\ &\qquad \left. - \frac{1}{\sqrt{2\pi i}} \sum_{n=0}^{N} \left((-1)^n D_1^\star(-n) - D_0^\star(-n) \right) \big(-\operatorname{sign}(y)\, iy \big)^n \right)\end{aligned}$$

$$= \frac{-1}{\sqrt{2\pi}\, i} \sum_{n=0}^{\infty} \left((-1)^n D_1^\star(-n) - D_0^\star(-n)\right)(-i)^n \left(\operatorname{sign}(y)\, y\right)^n$$

$$= \frac{1}{\sqrt{2\pi}\, i} \sum_{n=0}^{\infty} \left(i^{-n}\, D_0^\star(-n) - i^n D_1^\star(-n)\right) y^n$$

for $y \to 0$. This concludes the proof. □

Summarizing the lemmata above we have the following.

Proposition 4.70 *Let k, κ, and ν be as in Theorem 4.55 and let D_0 and D_1 be a pair of functions satisfying the second part of Theorem 4.45.*

Then we have:

1. *the function f, defined in* (4.113)*, satisfies the fourth part of Theorem 4.55;*
2. *the restriction of f to $i\mathbb{R}_{\neq 0}$ has an asymptotic expansion in* 0 *of the form*

$$f(iy) = \frac{1}{\sqrt{2\pi}\, i} \sum_{n=0}^{\infty} \left(i^{-n} D_0^\star(-n) - i^n D_1^\star(-n)\right) y^n \tag{4.139}$$

as $|y| \to 0$.

Proof Consider the function f, defined in (4.113). We have shown in Exercise 4.59 that f is a nearly periodic function with factor $e^{2\pi i\kappa}$. Lemma 4.61 implies that $f(z)$, $z \in \mathbb{C} \smallsetminus \mathbb{R}$, is bounded by a multiple of $|\Im(z)|^{-C}$ for some $C > 0$. We obtain from Lemma 4.68 that the function

$$\mathbb{C} \smallsetminus \mathbb{R}; \qquad z \mapsto f(z) - e^{\frac{\pi i k}{2}} z^{2\nu-1} f\left(-\frac{1}{z}\right)$$

extends holomorphically across the right half-plane and is bounded by a multiple of $\min\left\{1, |z|^{2\Re(\nu)-1}\right\}$ in the right half-plane.

The second part of the proposition was already shown in Lemma 4.69. □

Example 4.71 Continuing the discussions in Examples 4.15, 4.32, and 4.51, we assume $k > 0$ and assign $\nu = \frac{1-k}{2}$ and $0 \le \kappa < 1$ with $\frac{k}{12} - \kappa \in \mathbb{Z}$ (i.e., $\frac{k}{12} \equiv \kappa \pmod 1$).

Following Example 4.15, let η^{2k} be the cusp form and $u_{\eta^{2k}}$ the Maass cusp form discussed in §3.4.3. The Whittaker-Fourier expansion of $u_{\eta^{2k}}$ is given by (3.71). The associated admissible family of cusp forms $(u_l)_{l \in k+2\mathbb{Z}}$ with its admissible range $l_{\min} = k$ and $l_{\max} = \infty$ is discussed in Example 4.15.

The next step was done in §4.2.3 where we associated a pair of D-functions (D_0, D_1), see Theorem 4.45.

Now, we calculate f in (4.113) using (4.99), $b_n = A(n)\, |n+\kappa|^{\frac{1}{2}-\nu}$ to relate the coefficients $A(n)$ of the admissible family $(u_l)_l$ and b_n appearing in the defining expansion (4.95) of the D-functions. We find

$$f(z) := \begin{cases} \frac{(2\pi)^{\frac{1-k}{2}}}{\sqrt{2}} \sum_{n=0}^{\infty} A_k(n)\,(n+k)^{\frac{1}{2}-\nu-\frac{k}{2}}\, e^{2\pi i(n+k)z} & \text{for } \Im(z) > 0 \text{ and,} \\ 0 & \text{for } \Im(z) < 0 \end{cases}$$

$$= \begin{cases} \frac{(2\pi)^{\frac{1-k}{2}}}{\sqrt{2}} \sum_{n=0}^{\infty} A_k(n)\, e^{2\pi i(n+k)z} & \text{for } \Im(z) > 0 \text{ and,} \\ 0 & \text{for } \Im(z) < 0 \end{cases} \tag{4.140}$$

using $\nu = \frac{1-k}{2}$ which implies $(n+k)^{\frac{1}{2}-\nu-\frac{k}{2}} = (n+k)^0 = 1$. Comparing above expansion with (3.69), we find in particular that

$$f(z) = \begin{cases} \frac{(2\pi)^{\frac{1-k}{2}}}{\sqrt{2}}\, \eta^{2k} & \text{for } \Im(z) > 0 \text{ and,} \\ 0 & \text{for } \Im(z) < 0. \end{cases} \tag{4.141}$$

We may calculate the function $A^{\pm}_{k,\nu}$ introduced in (4.120) of Lemma 4.63 explicitly. Using (4.122), we express $A^{\pm}_{k,\nu}$ in terms of $D^{\star}_0$ and $D^{\star}_1$. We find

$$\begin{aligned} A^{\pm}_{k,\nu}(s) &\overset{(4.122)}{=} D^{\star}_1(s) \left(\frac{\sin(2\pi\nu)}{2i\sin(\pi s)\sin(\pi(1-2\nu-s))} \right) \\ &\quad - D^{\star}_0(s) \left(\frac{1}{2i\sin(\pi s)} + \frac{e^{\pi i k}}{2i\sin(\pi(1-2\nu-s))} \right) \\ &\overset{(4.110)}{=} \frac{(4\pi)^{-s-\frac{k}{2}}}{\sqrt{2}} \Gamma(s) \sum_{n=0}^{\infty} A_k(n)\,(n+k)^{-s} \\ &\quad \cdot \left[e^{\frac{\pi i s}{2}} \frac{\sin(\pi(1-k))}{2i\sin(\pi s)\sin(\pi(k-s))} - \right. \\ &\quad \left. - e^{-\frac{\pi i s}{2}} \left(\frac{1}{2i\sin(\pi s)} + \frac{e^{\pi i k}}{2i\sin(\pi(k-s))} \right) \right] \\ &= \frac{(4\pi)^{-s-\frac{k}{2}}}{\sqrt{2}} \Gamma(s) \sum_{n=0}^{\infty} A_k(n)\,(n+k)^{-s} \\ &\quad \cdot \frac{e^{\frac{\pi i s}{2}} \sin(\pi(1-k)) - e^{-\frac{\pi i s}{2}} \left[\sin(\pi(k-s)) + e^{\pi i k}\sin(\pi s) \right]}{2i\sin(\pi s)\sin(\pi(k-s))} \\ &= 0 \end{aligned}$$

This implies that the Mellin transform of the function

$$F(z) = f(z) - e^{\frac{\pi ik}{2}} z^{2\nu-1} f\left(-\frac{1}{z}\right)$$

defined in (4.134) vanishes for all $0 < \Re(s) < k$, see Lemma 4.68. Hence, F itself vanishes everywhere on $\mathbb{C} \setminus \mathbb{R}$.

It is not surprising that the function F vanishes, as the following argument shows: Recalling that f is closely related to the Dedekind η-function by (4.141), we find

$$\begin{aligned} F(z) &= f(z) - e^{\frac{\pi ik}{2}} z^{-k} f\left(\frac{-1}{z}\right) \\ &\overset{(4.141)}{=} \frac{(2\pi)^{\frac{1-k}{2}}}{\sqrt{2}}\, \eta^{2k}\left(\eta^{2k}(z) - e^{\frac{\pi ik}{2}} z^{-k} \eta^{2k}\left(\frac{-1}{z}\right)\right) \end{aligned}$$

for $\Im(z) > 0$. (Note that η^{2k} is nearly periodic due to the translation property of η under S, see (1.39).) Using the transformation rule (1.39) of the $(2r)^{\text{th}}$ power of the Dedekind η-function, see Example (1.50), we obtain directly

$$\eta^{2k}(z) - e^{\frac{\pi ik}{2}} z^{-k} \eta^{2k}\left(\frac{-1}{z}\right) = 0.$$

Hence, we have

$$F(z) = f(z) - e^{\frac{\pi ik}{2}} z^{-k} f\left(\frac{-1}{z}\right) = 0$$

for $\Im(z) > 0$. Since $f(z) = 0$ for $\Im(z) < 0$, we obtain $F(z) = 0$ for all $z \in \mathbb{C} \setminus \mathbb{R}$.

Remark 4.72 We explained a result by Bykovskii [54, Theorem 2] in Theorem 4.43. Our readers might wonder, why did we use families of Maass cusp forms $(u_l)_l$ and functions D_0 and D_1 in Theorem 4.45 instead of working directly with Bykovskii's Theorem 4.43?

We think that our arguments leading to the periodic expansion and the expression of $A_{k,\pm}(s)$ in (4.120) are a bit simpler. In particular, we separate the influence of the weight k of the individual form from the systematic influence visible in the associated admissible family. Also working with the individual Maass cusp form u does not give more insight into cusp forms as obtained by our notion of families.

Also, we can show that using Bykovskii result directly, we still have to use the Fourier-Whittaker expansion for elements of a family of Maass cusp forms in (4.24). A brief sketch of our arguments for this assertion is presented below.

Let $u \in S_{k,v}^{\text{Maass}}(\Gamma(1), \nu)$ be a nonvanishing cusp form of weight k and multiplier system v and spectral parameter ν. The cusp form u has the Fourier-Whittaker expansion (3.48) at the cusp ∞:

$$u(x+iy) = \sum_{\substack{n\in\mathbb{Z}\\ n+\kappa\neq 0}} A(n)\, \mathbb{W}_{\mathrm{sign}(n)\frac{k}{2},\nu}\big(4\pi\,|n+\kappa|\,y\big)\, e^{2\pi i(n+\kappa)x}.$$

As a first step, we derive the functions $D_{k,0}$ and $D_{k,1}$ that correspond to D_0 and D_1 in Theorem 4.45. We factorize the matrix $\begin{pmatrix} -\sin\pi\left(s-\frac{k}{2}\right) & \frac{\pi}{\Gamma(\frac{1+k}{2}+\nu)\Gamma(\frac{1+k}{2}-\nu)} \\ \frac{\pi}{\Gamma(\frac{1-k}{2}+\nu)\Gamma(\frac{1-k}{2}-\nu)} & -\sin\pi\left(s+\frac{k}{2}\right) \end{pmatrix}$ in (4.94) of Bykovskii's Theorem 4.43 as

$$\begin{pmatrix} \Gamma(\frac{1+k}{2}+\nu) & 0 \\ 0 & \Gamma(\frac{1-k}{2}+\nu) \end{pmatrix}^{-1} \begin{pmatrix} -\sin\pi\left(s-\frac{k}{2}\right) & \cos\pi\left(\nu-\frac{k}{2}\right) \\ \cos\pi\left(\nu+\frac{k}{2}\right) & -\sin\pi\left(s+\frac{k}{2}\right) \end{pmatrix} \cdot \begin{pmatrix} \Gamma(\frac{1+k}{2}+\nu) & 0 \\ 0 & \Gamma(\frac{1-k}{2}+\nu) \end{pmatrix},$$

using the well-known identity $\cos(\pi\tau) = \frac{\pi}{\Gamma(\frac{1}{2}+\tau)\Gamma(\frac{1}{2}-\tau)}$ (see Appendix A.7). Let

$$\begin{pmatrix} D_{k,0} \\ D_{k,1} \end{pmatrix} := \begin{pmatrix} e^{-\frac{\pi i}{2}(s-\nu)} & e^{\frac{\pi i}{2}(s-\nu)} \\ e^{\frac{\pi i}{2}(s-\nu)} & e^{-\frac{\pi i}{2}(s-\nu)} \end{pmatrix} \begin{pmatrix} \Gamma(\frac{1+k}{2}+\nu) & 0 \\ 0 & \Gamma(\frac{1-k}{2}+\nu) \end{pmatrix} \begin{pmatrix} \Lambda_+(s) \\ \Lambda_-(s) \end{pmatrix}$$

for all $s \in \mathbb{C}$. Using the matrix identity

$$\begin{aligned} &-\begin{pmatrix} -\sin\pi\left(s-\frac{l}{2}\right) & \cos\pi\left(\nu-\frac{l}{2}\right) \\ \cos\pi\left(\nu+\frac{l}{2}\right) & -\sin\pi\left(s+\frac{l}{2}\right) \end{pmatrix} \\ &= \begin{pmatrix} \sin\left(\pi\left(s-\frac{l}{2}\right)\right) & \sin\left(\pi\left(\nu-\frac{l+1}{2}\right)\right) \\ \sin\left(\pi\left(\nu+\frac{l-1}{2}\right)\right) & \sin\left(\pi\left(s+\frac{l}{2}\right)\right) \end{pmatrix} \\ &\overset{(4.85)}{=} \begin{pmatrix} e^{-\frac{\pi i}{2}\left(\frac{1}{2}-s-\nu\right)} & e^{\frac{\pi i}{2}\left(\frac{1}{2}-s-\nu\right)} \\ e^{\frac{\pi i}{2}\left(\frac{1}{2}-s-\nu\right)} & e^{-\frac{\pi i}{2}\left(\frac{1}{2}-s-\nu\right)} \end{pmatrix}^{-1} \begin{pmatrix} 0 & e^{-\frac{\pi i l}{2}} \\ e^{\frac{\pi i l}{2}} & 0 \end{pmatrix}^{-1} \cdot \\ &\qquad \cdot \begin{pmatrix} e^{-\frac{\pi i}{2}\left(\frac{1}{2}+s-\nu\right)} & e^{\frac{\pi i}{2}\left(\frac{1}{2}+s-\nu\right)} \\ e^{\frac{\pi i}{2}\left(\frac{1}{2}+s-\nu\right)} & e^{-\frac{\pi i}{2}\left(\frac{1}{2}+s-\nu\right)} \end{pmatrix} \end{aligned}$$

used in the proof of Theorem 4.39, the direction "Theorem 4.39, part 2 $\Longrightarrow$ Theorem 4.28, Item 2," we can rewrite (4.94) as

$$\begin{pmatrix} D_{k,0}(1-2\nu-s) \\ D_{k,1}(1-2\nu-s) \end{pmatrix} = \frac{(2\pi)^{1-2\nu-2s}\,\Gamma(s)}{\Gamma(1-2\nu-s)} \begin{pmatrix} 0 & e^{-\frac{\pi i k}{2}} \\ e^{\frac{\pi i k}{2}} & 0 \end{pmatrix} \begin{pmatrix} D_{k,0}(s) \\ D_{k,1}(s) \end{pmatrix}.$$

The functional equation simplifies to

$$\begin{pmatrix} D^\star_{k,0}(1-2\nu-s) \\ D^\star_{k,1}(1-2\nu-s) \end{pmatrix} = \begin{pmatrix} 0 & e^{-\frac{\pi i k}{2}} \\ e^{\frac{\pi i k}{2}} & 0 \end{pmatrix} \begin{pmatrix} D^\star_{k,0}(s) \\ D^\star_{k,1}(s) \end{pmatrix}$$

if we write $D^\star_{k,\delta}(s) = (2\pi)^{\frac{1}{2}-s}\,\Gamma(s)\,D_{k,\delta}(s)$, $\delta = 0, 1$, as in (4.97) of Theorem 4.45.

We introduce the nearly periodic function $f_k : \mathbb{C} \smallsetminus \mathbb{R} \to \mathbb{C}$ by

$$f_k(z) = \begin{cases} \sqrt{2\pi}\,\sum_{\substack{n\in\mathbb{Z} \\ n+\kappa>0}} A_n\, e^{2\pi i n z} & \text{for } \Im(z) > 0 \text{ and} \\ -\sqrt{2\pi}\,\sum_{\substack{n\in\mathbb{Z} \\ n+\kappa<0}} A_n\, e^{2\pi i n z} & \text{for } \Im(z) < 0, \end{cases} \tag{4.142}$$

where A_n are the coefficients of the Fourier expansion (3.48) of the cusp form u. The definition of f_k corresponds to the function f in (4.113).

Similar to Lemma 4.68, we want to show that the function

$$\mathbb{C} \smallsetminus \mathbb{R} \to \mathbb{C}; \quad z \mapsto f_k(z) - e^{\frac{\pi i k}{2}}\, z^{2\nu-1} f_k\Big(-\frac{1}{z}\Big)$$

extends to a holomorphic function on $\mathbb{C}' = \mathbb{C} \smallsetminus \mathbb{R}_{\leq 0}$ and is bounded by a multiple of $\min\left\{1, |z|^{2\Re(\nu)-1}\right\}$ in the right half-plane $\Re(z) > 0$. Following the argument in the proof of Lemma 4.68, we write

$$\tilde{f}_{k,\pm}(s) := \int_0^\infty f_k(\pm i y)\, y^{s-1}\, dy.$$

To show that f_k extends holomorphically to $\mathbb{C}'$, we have to consider the expressions

$$A^\pm_{f_k,\nu}(s) := \tilde{f}_{k,\pm}(s) e^{\pm\frac{\pi i s}{2}} - \tilde{f}_{k\pm}(1-2\nu-s) e^{\frac{\pi i}{2}(k\mp(1-2\nu-s))}.$$

The functions $A^\pm_{f_k,\nu}(s)$ correspond to $A^\pm_{k,\nu}(s)$ in (4.121). In particular, we have to show that $A^\pm_{f_k,\nu}(s)$ do not depend on the choice of $\pm$. Similar to (4.122), we show that the $A^\pm_{f_k,\nu}(s)$ are equal to

$$\begin{aligned} &D^\star_{k,1}(s) \left(\frac{e^{\pm\pi i s}}{2i\Gamma(\frac{1-k}{2}+\nu)\sin(\pi s)} + \frac{e^{\mp\pi i(1-2\nu-s)}}{2i\Gamma(\frac{1+k}{2}+\nu)\sin(\pi(1-2\nu-s))} \right) \\ &- D^\star_{k,0}(s) \left(\frac{1}{2i\Gamma(\frac{1+k}{2}+\nu)\sin(\pi s)} + \frac{e^{\pi i k}}{2i\Gamma(\frac{1-k}{2}+\nu)\sin(\pi(1-2\nu-s))} \right). \end{aligned} \tag{4.143}$$

Here, the expressions in the first big brackets do depend on the choice of $\pm$ for all $k \neq 0$.

To correct this approach, we can insert suitable Γ-factors $\Gamma\left(\frac{1\pm k}{2}+\nu\right)$ in (4.142) such that (4.143) is indeed independent of the choice of $\pm$. If we include the Γ-factors in the Fourier-Whittaker expansion (3.48) of the Maass form u instead in the definition of f_k in (4.142), we see that we work in fact with the Fourier-Whittaker expansion (4.24) of the associated admissible system $(u_l)_l$ in the sense of Lemma 4.5 and Lemma 4.11.

The arguments sketched above give evidence that using Bykovskii's Theorem 4.43 in this subsection is equivalent to working with admissible families of cusp forms.

4.3.2 Nearly Periodic Functions Giving Rise to D-Functions

Now, we start with a holomorphic function $f : \mathbb{C} \smallsetminus \mathbb{R} \to \mathbb{C}$, defined on the upper and lower complex half-plane, which satisfies the fourth part of Theorem 4.55: f is nearly periodic in the sense of $f(z+1) = e^{2\pi i\kappa} f(z)$ for all $z \in \mathbb{C} \smallsetminus \mathbb{R}$, f is bounded by a multiple of $|\Im(z)|^{-C}$ for some $C > 0$, and the function $f(z) - e^{\frac{\pi ik}{2}} z^{2\nu-1} f\left(-\frac{1}{z}\right)$ extends holomorphically across the right half-plane and is bounded by a multiple of $\min\left\{1, |z|^{2\Re(\nu)-1}\right\}$ in the right half-plane.

Our goal here is to derive a pair of D-functions satisfying the second part of Theorem 4.45. We approach our goal by splitting the arguments into several lemmata below.

Lemma 4.73 *Let $\nu \in \mathbb{C}$, $\Re(\nu) < \frac{1}{2}$, $k \in \mathbb{R}$, $0 \le \kappa < 1$, and let $f : \mathbb{C} \smallsetminus \mathbb{R} \to \mathbb{R}$ be a holomorphic nearly periodic function satisfying the conditions in part 4 of Theorem 4.55: f is nearly periodic in the sense of $f(z+1) = e^{2\pi i\kappa} f(z)$ for all $z \in \mathbb{C} \smallsetminus \mathbb{R}$, f is bounded by a multiple of $|\Im(z)|^{-C}$ for some $C > 0$, and the function $f(z) - e^{\frac{\pi ik}{2}} z^{2\nu-1} f\left(-\frac{1}{z}\right)$ extends holomorphically across the right half-plane and is bounded by a multiple of $\min\left\{1, |z|^{2\Re(\nu)-1}\right\}$ in the right half-plane.*

Then f admits a Fourier expansion of the form

$$f(z) = \begin{cases} \sqrt{2\pi} \sum_{\substack{n \in \mathbb{Z} \\ n+\kappa>0}} B_n\, e^{2\pi i(n+\kappa)z} & \text{for } \Im(z) > 0 \text{ and} \\ -\sqrt{2\pi} \sum_{\substack{n \in \mathbb{Z} \\ n+\kappa<0}} B_n\, e^{2\pi i(n+\kappa)z} & \text{for } \Im(z) < 0, \end{cases} \tag{4.144}$$

where coefficients B_n are of at most polynomial growth.

The function f satisfies the growth estimate

$$f(\pm iy) = \begin{cases} O\left(e^{-\varepsilon y}\right) & \text{as } y \to \infty \text{ and} \\ O(1) & \text{as } y \to 0 \end{cases} \tag{4.145}$$

for some $\varepsilon > 0$.

Proof Let f be a function satisfying the fourth part of Theorem 4.55. This means in particular that f is a holomorphic function on $\mathbb{C} \setminus \mathbb{R}$ which is nearly periodic in the sense of Definition 4.54 with factor $e^{2\pi i\kappa}$. Hence f can be expanded in a Fourier expansion of the form

$$f(z) = \sum_{\substack{n\in\mathbb{Z}\\ n+\kappa\neq 0}} B_n^{\pm}\, e^{2\pi i(n+\kappa)z} \qquad (\Im(z) \gtrless 0) \tag{4.146}$$

with Fourier coefficients B_n^+ for $\Im(z) > 0$ and B_n^- for $\Im(z) < 0$. The nth Fourier term can be calculated by

$$B_n^{\pm}\, e^{-2\pi(n+\kappa)y} = \int_0^1 f(x+iy)\, e^{-2\pi i(n+\kappa)x}\, \mathrm{d}x$$

for every $y \in \mathbb{R}_{\neq 0}$ and $n \in \mathbb{Z}$ with $n+\kappa \neq 0$. Since $f(z) = O\left(|\Im(z)|^{-C}\right)$ for some $C > 0$, we obtain the estimate

$$B_n^{\pm}\, e^{-2\pi(n+\kappa)y} = O\left(|y|^{-C}\right) \qquad (y \in \mathbb{R}_{\neq 0}) \tag{4.147}$$

for all $n \in \mathbb{Z}$ with $n+\kappa \neq 0$.

The above growth estimate implies in particular $B_n^{\pm} = 0$ for all $\pm n \leq 0$. (The growth estimate would not hold otherwise.) This allows us to define

$$B_n := \begin{cases} B_n^+ & \text{if } n+\kappa > 0 \text{ and} \\ B_n^- & \text{if } n+\kappa < 0, \end{cases}$$

where we can write (4.146) as

$$f(z) = \sum_{\substack{n\in\mathbb{Z}\\ n+\kappa\gtrless 0}} B_n\, e^{2\pi i(n+\kappa)z} \qquad (\Im(z) \gtrless 0).$$

Substituting $y = \pm\frac{1}{n}$ for $\operatorname{sign}(n) \gtrless 0$ in (4.147), we obtain

$$B_n = O\left(|n|^C\right) \qquad (n \in \mathbb{Z}, n+\kappa \neq 0)$$

for some $C > 0$, i.e., B_n has at most polynomial growth. In particular, the function f satisfies the growth estimate

$$f(\pm iy) = O\left(e^{-\varepsilon y}\right) \qquad (y \to \infty) \tag{4.148}$$

for some $\varepsilon > 0$.

To establish the remaining growth estimate for $|y| \searrow 0$, consider the function

$$F(z) := f(z) - e^{\frac{\pi ik}{2}} z^{2\nu-1} f\left(-\frac{1}{z}\right). \tag{4.149}$$

F is holomorphic on $\mathbb{C} \setminus \mathbb{R}$ since f is holomorphic there. By assumption, F is bounded by a multiple of $\min\left\{1, |z|^{2\Re(\nu)-1}\right\}$ on the right half-plane $\Re(z) > 0$.

Using the fact that f is continuous, since f is holomorphic, we can extend the bound on F on the right half-plane $\Re(z) > 0$ to the imaginary axis. Hence, F satisfies the growth condition

$$F(iy) = O\left(\min\left\{1, |y|^{2\Re(\nu)-1}\right\}\right)$$

for all $y \in \mathbb{R}_{\neq 0}$. Using the assumption $\Re(\nu) < \frac{1}{2}$, we conclude that $F(iy) = O(1)$ for $y \to 0$, $y \neq 0$. Hence, (4.149) implies

$$\begin{aligned} f(iy) &= F(iy) + e^{\frac{\pi i\kappa}{2}} (iy)^{2\nu-1} f\left(\frac{i}{y}\right) \\ &= O(1) + O\left(y^{2\Re(\nu)-1} e^{-\frac{\varepsilon}{|y|}}\right) \\ &= O(1) \end{aligned}$$

as $y \to 0$, $y \neq 0$. Combining this growth estimate with (4.148), we find the second part of (4.145). □

The following lemma interchanges the order of summation and integration in some cases.

Lemma 4.74 *Let $\kappa \in \mathbb{R}$ be a real number and $(B_n)_{\substack{n\in\mathbb{Z} \\ n+\kappa>0}}$ be a sequence of numbers satisfying the growth estimate $B_n = O\left(|n+\kappa|^C\right)$ for some $C > 0$.*

The Mellin transform

$$\begin{aligned} \tilde{g}(s) &:= \mathcal{M}\big(g\big)(s) \\ &\overset{(1.154)}{=} \int_0^\infty g(y)\, y^{s-1}\, dy \end{aligned}$$

of the function $g(y) = \sum_{\substack{n\in\mathbb{Z} \\ n+\kappa>0}} B_n\, e^{-2\pi(n+\kappa)y}$ for $y > 0$ is well defined on the right half-plane $\Re(s) > C + 1$ and can be calculated term by term: We have

$$\tilde{g}(s) = \sum_{\substack{n\in\mathbb{Z} \\ n+\kappa>0}} B_n \int_0^\infty e^{-2\pi(n+\kappa)y}\, y^{s-1}\, dy.$$

Proof Let

$$G(y) = \sum_{\substack{n\in\mathbb{Z}\\ n+\kappa>0}} |n+\kappa|^C \, e^{2\pi(n+\kappa)y}.$$

The theorem of Fubini allows us to calculate the Mellin transform

$$\tilde{G}(s) = \int_0^\infty G(y)\, y^{s-1}\, dy$$

term by term for $\Re(s)$ large enough. We have

$$\tilde{G}(s) = (2\pi)^{-s}\Gamma(s) \sum_{\substack{n\in\mathbb{Z}\\ n+\kappa>0}} (n+\kappa)^{C-s}$$

that converges for $\Re(s) > C+1$. We obtain from the dominated convergence theorem, see, e.g., [202, 9.14, p. 332], that $\tilde{g}(s)$ is well defined for $\Re(s) > C+1$ and that we can calculate the Mellin transform $\tilde{g}(s)$ term by term. □

Now, we can prove the following proposition which attaches the D-functions to the nearly periodic function f.

Proposition 4.75 *Let $\nu \in \mathbb{C}$ with $\Re(\nu) < \frac{1}{2}$, $0 \le \kappa < 1$, $k \in \mathbb{R}$, and let $f : \mathbb{C}\setminus\mathbb{R} \to \mathbb{R}$ be a holomorphic and nearly periodic function with factor $e^{2\pi i\kappa}$ satisfying the conditions in part 4 of Theorem 4.55.*

Then there exists a pair of functions D_0 and D_1 that satisfy the conditions of Part 3 in Theorem 4.55.

Proof Let f be a function satisfying part 4 of Theorem 4.55. Put

$$F(z) := f(z) - e^{\frac{\pi i k}{2}}\, z^{2\nu-1}\, f\left(-\frac{1}{z}\right) \tag{4.150}$$

for all $z \in \mathbb{C}'$. The conditions on f in part 4 of Theorem 4.55 state that the function F is holomorphic on $\mathbb{C}'$ and that F satisfies the growth condition

$$F(z) = \begin{cases} O(1) & \text{if } \Re(z) \ge 0,\ |z| \le 1 \text{ and} \\ O\left(|z|^{2\Re(\nu)-1}\right) & \text{if } \Re(z) \ge 0,\ |z| \ge 1. \end{cases} \tag{4.151}$$

We included the imaginary line $i\mathbb{R}_{\neq 0}$ in the estimate above since f is in particular continuous on $\mathbb{C}\setminus\mathbb{R}$.

Applying Lemma 4.73 to f, we find that f admits a Fourier expansion of the form (4.144). The Fourier coefficients B_n of the expansion (4.144) have at most polynomial growth. Hence the functions

$$D_0(s) := \sum_{\substack{n\in\mathbb{Z}\\ n+\kappa\neq 0}} B_n\,\big(i(n+\kappa)\big)^{-s} \qquad \text{and}$$

$$D_1(s) := \sum_{\substack{n\in\mathbb{Z}\\ n+\kappa\neq 0}} B_n\,\big(-i(n+\kappa)\big)^{-s}$$

converge on some right half-plane.

Lemma 4.73 also implies that the Mellin transformations

$$\begin{aligned} \tilde{f}_\pm(s) &:= \mathcal{M}(f)(s) \\ &\overset{(1.154)}{=} \int_0^\infty f(\pm iy)\,y^{s-1}\,\mathrm{d}y \end{aligned} \tag{4.152}$$

converge for all $\Re(s) > 0$. Moreover $\tilde{f}_\pm(s)$ is a holomorphic function on this right half-plane. This means in particular that the two functions

$$e^{\frac{\pi is}{2}}\,\tilde{f}_+(s) - e^{-\frac{\pi is}{2}}\,\tilde{f}_-(s) \quad \text{and} \quad e^{-\frac{\pi is}{2}}\,\tilde{f}_+(s) - e^{\frac{\pi is}{2}}\,\tilde{f}_-(s) \tag{4.153}$$

are holomorphic functions for $\Re(s) > 0$.

Lemma 4.74 allows us to calculate $\tilde{f}_\pm(s)$ term by term for $\Re(s)$ sufficiently large. We have

$$\begin{aligned} \tilde{f}_\pm(s) &= \int_0^\infty f(\pm iy)\,y^{s-1}\,\mathrm{d}y \\ &\overset{(4.144)}{=} \pm\int_0^\infty \sqrt{2\pi}\sum_{\substack{n\in\mathbb{Z}\\ n+\kappa\gtrless 0}} B_n\,e^{-2\pi|n+\kappa|y}\,y^{s-1}\,\mathrm{d}y \\ &= \pm\sqrt{2\pi}\sum_{\substack{n\in\mathbb{Z}\\ n+\kappa\gtrless 0}} B_n\,\big(2\pi\,|n+\kappa|\big)^{-s}\int_0^\infty e^{-y}\,y^{s-1}\,\mathrm{d}y \\ &\overset{(1.155)}{=} \pm(2\pi)^{\frac{1}{2}-s}\,\Gamma(s)\sum_{\substack{n\in\mathbb{Z}\\ n+\kappa\gtrless 0}} B_n\,|n+\kappa|^{-s} \end{aligned}$$

for $\Re(s)$ large. Calculating the functions in (4.153) term-wise for large $\Re(s)$, we find

$$\begin{aligned} &e^{\frac{\pi is}{2}}\,\tilde{f}_+(s) - e^{-\frac{\pi is}{2}}\,\tilde{f}_-(s) \\ &\quad = (2\pi)^{\frac{1}{2}-s}\,\Gamma(s)\,(-i)^{-s}\sum_{\substack{n\in\mathbb{Z}\\ n+\kappa>0}} B_n\,(n+\kappa)^{-s} + \end{aligned}$$

$$+ \ (2\pi)^{\frac{1}{2}-s}\,\Gamma(s)\,i^{-s} \sum_{\substack{n\in\mathbb{Z}\\ n+\kappa<0}} B_n\left((-1)(n+\kappa)\right)^{-s}$$

$$= (2\pi)^{\frac{1}{2}-s}\,\Gamma(s)\,D_1(s)$$

and similarly

$$e^{-\frac{\pi i s}{2}}\,\tilde{f}_+(s) - e^{\frac{\pi i s}{2}}\,\tilde{f}_-(s) = (2\pi)^{\frac{1}{2}-s}\,\Gamma(s)\,D_1(s)$$

for $\Re(s)$ large. In matrix-vector notation, we have

$$\begin{pmatrix} e^{\frac{\pi i s}{2}} & -e^{-\frac{\pi i s}{2}} \\ e^{-\frac{\pi i s}{2}} & -e^{\frac{\pi i s}{2}} \end{pmatrix} \begin{pmatrix} \tilde{f}_+(s) \\ \tilde{f}_-(s) \end{pmatrix} = (2\pi)^{\frac{1}{2}-s}\,\Gamma(s) \begin{pmatrix} D_1(s) \\ D_0(s) \end{pmatrix}. \tag{4.154}$$

Since $\tilde{f}_\pm(s)$ are holomorphic functions on the right half-plane $\Re(s) > 0$, we see that

$$D^\star_\delta(s) \overset{(4.97)}{:=} (2\pi)^{-s}\,\Gamma(s)\,D_\delta(s),$$

$\delta \in \{0,1\}$, extend to holomorphic functions on the right half-plane $\Re(s) > 0$. Equation (4.154) reads then

$$\sqrt{2\pi} \begin{pmatrix} D^\star_1(s) \\ D^\star_0(s) \end{pmatrix} = \begin{pmatrix} e^{\frac{\pi i s}{2}} & -e^{-\frac{\pi i s}{2}} \\ e^{-\frac{\pi i s}{2}} & -e^{\frac{\pi i s}{2}} \end{pmatrix} \begin{pmatrix} \tilde{f}_+(s) \\ \tilde{f}_-(s) \end{pmatrix}. \tag{4.155}$$

for $\Re(s) > 0$.

We now discuss some properties of the function F. The growth condition (4.151) implies that the Mellin transforms

$$\tilde{F}(s) = \int_0^\infty F(x)\,x^{s-1}\,\mathrm{d}x \quad \text{and} \quad \int_0^\infty F(\pm iy)\,y^{s-1}\mathrm{d}y \tag{4.156}$$

converge for every s in the vertical strip $0 < \Re(s) < 1 - 2\Re(\nu)$. Rotating the integration path of the left integral in (4.156) to the upper or lower imaginary axis gives

$$i^s \int_0^\infty F(+iy)y^{s-1}\mathrm{d}y = \tilde{F}(s) = i^{-s} \int_0^\infty F(-iy)y^{s-1}\mathrm{d}y. \tag{4.157}$$

(4.150) implies that

$$\int_0^\infty F(\pm iy)\,y^{s-1}\,\mathrm{d}y \overset{(4.150)}{=} \int_0^\infty \left[f(\pm iy) - e^{\frac{\pi i k}{2}}\,(\pm iy)^{2\nu-1}\,y^{2\nu-1}\,f\left(\pm\frac{i}{y}\right)\right] y^{s-1}\,\mathrm{d}y$$

$$
\begin{aligned}
&= \int_0^\infty f(\pm iy)\, y^{s-1}\, dy - \\
&\qquad - e^{\frac{\pi i}{2}\left(k\pm(2\nu-1)\right)} \int_0^\infty f\left(\pm\frac{i}{y}\right) y^{s+2\nu-2}\, dy \\
&\overset{\substack{(4.152)\\ y\mapsto\frac{1}{y}}}{=} \tilde{f}_\pm(s) - e^{\frac{\pi i}{2}\left(k\pm(2\nu-1)\right)} \int_0^\infty f(\pm iy)\, y^{2-2\nu-s}\, \frac{1}{y^2} dy \\
&= \tilde{f}_\pm(s) - e^{\frac{\pi i}{2}\left(k\pm(2\nu-1)\right)}\, \tilde{f}_\pm(1-2\nu-s)
\end{aligned}
$$

for s in the strip $0 < \Re(s) < 1 - 2\Re(\nu)$. Using (4.157), we conclude

$$
\begin{aligned}
&e^{\frac{\pi is}{2}} \left[\tilde{f}_+(s) - e^{\frac{\pi i}{2}(k+2\nu-1)}\, \tilde{f}_+(1-2\nu-s)\right] \\
&\qquad = e^{-\frac{\pi is}{2}} \left[\tilde{f}_-(s) - e^{\frac{\pi i}{2}(k+1-2\nu)}\, \tilde{f}_-(1-2\nu-s)\right].
\end{aligned}
$$

Using the row and column vector notation, we see that the above identity is equivalent to

$$
\begin{aligned}
&\begin{pmatrix} e^{\frac{\pi is}{2}} & -e^{-\frac{\pi is}{2}} \end{pmatrix} \begin{pmatrix} \tilde{f}_+(s) \\ \tilde{f}_-(s) \end{pmatrix} \\
&\qquad = e^{\frac{\pi ik}{2}} \begin{pmatrix} e^{-\frac{\pi i}{2}(1-2\nu-s)} & -e^{\frac{\pi i}{2}(1-2\nu-s)} \end{pmatrix} \begin{pmatrix} \tilde{f}_+(1-2\nu-s) \\ \tilde{f}_-(1-2\nu-s) \end{pmatrix}
\end{aligned}
$$

in the strip $0 < \Re(s) < 1 - 2\Re(\nu)$. Applying (4.155) to this identity gives

$$
D_1^\star(s) = e^{\frac{\pi ik}{2}}\, D_0^\star(1-2\nu-s) \tag{4.158}
$$

in the strip $0 < \Re(s) < 1 - 2\Re(\nu)$. This gives the extension of the two functions $D_0^\star(s)$ and $D_1^\star(s)$ to entire functions on $\mathbb{C}$ and shows also that they satisfy the desired functional equation.

Inverting (4.155) gives the functions $\tilde{f}_\pm(s)$ in terms of entire functions $D_\delta^\star(s)$, $\delta \in \{0, 1\}$. We have

$$
\begin{pmatrix} \tilde{f}_+(s) \\ \tilde{f}_-(s) \end{pmatrix} = \frac{\sqrt{2\pi}}{2i\sin(\pi s)} \begin{pmatrix} e^{\frac{\pi is}{2}} & -e^{-\frac{\pi is}{2}} \\ e^{-\frac{\pi is}{2}} & -e^{\frac{\pi is}{2}} \end{pmatrix} \begin{pmatrix} D_1^\star(s) \\ D_0^\star(s) \end{pmatrix}. \tag{4.159}
$$

We have shown that $\tilde{f}_\pm(s)$ are holomorphic functions on the right half-plane $\Re(s) > 0$. The equation above gives a meromorphic continuation to the complex plane such that the functions $\frac{\tilde{f}_\pm(s)}{\Gamma(s)}$ are entire. Here we used the identity $\Gamma(s)\Gamma(1-s) = \frac{\pi}{\sin(\pi s)}$, see Appendix A.7.

Consider the auxiliary function

$$h_\pm(s) := \frac{1}{\sin(\pi s)} \left(e^{\pm\frac{\pi i s}{2}} D_1(s) - e^{\mp\frac{\pi i s}{2}} D_0(s) \right).$$

We have

$$\begin{aligned}\begin{pmatrix} h_+(s) \\ h_-(s) \end{pmatrix} &= \frac{1}{\sin(\pi s)} \begin{pmatrix} e^{\frac{\pi i s}{2}} & -e^{-\frac{\pi i s}{2}} \\ e^{-\frac{\pi i s}{2}} & -e^{\frac{\pi i s}{2}} \end{pmatrix} \begin{pmatrix} D_1(s) \\ D_0(s) \end{pmatrix} \\ &= \frac{(2\pi)^s}{\Gamma(s)\,\sin(\pi s)} \begin{pmatrix} e^{\frac{\pi i s}{2}} & -e^{-\frac{\pi i s}{2}} \\ e^{-\frac{\pi i s}{2}} & -e^{\frac{\pi i s}{2}} \end{pmatrix} \begin{pmatrix} D_1^\star(s) \\ D_0^\star(s) \end{pmatrix} \\ &= \frac{(2\pi)^s}{\Gamma(s)} \begin{pmatrix} \tilde{f}_+(s) \\ \tilde{f}_-(s) \end{pmatrix} \end{aligned} \tag{4.160}$$

We used (4.159) in the last step. The calculation shows that the auxiliary functions $h_\pm(s)$ are entire since we have already established that the functions $\frac{\tilde{f}_\pm(s)}{\Gamma(s)}$ are entire.

We still have to check that the functions $h_\pm(s)$ are of finite order. First, we will study $\tilde{f}_\pm(s)$ on the right half-plane $\Re(s) > 0$. We will show that $D_0^\star(s)$ and $D_1^\star(s)$ are entire functions of finite order. We will also show that this implies that $\frac{\tilde{f}_\pm(s)}{\Gamma(s)}$ are of finite order. (4.160) implying that $h_\pm(s)$ are of finite order. Using the growth estimate for $f(\pm i y)$ in Lemma 4.73, we have

$$\begin{aligned} \left|\tilde{f}_\pm(s)\right| &\le \int_0^\infty |f(\pm i y)|\; y^{\Re(s)-1}\, dy \\ &\ll \int_0^\infty e^{-\varepsilon y}\, y^{\Re(s)-1}\, dy \\ &= \varepsilon^{-\Re(s)}\, \Gamma\big(\Re(s)\big) \end{aligned}$$

for $\Re(s) > 0$. Using Stirling's formula, see Appendix A.3, we see that

$$\tilde{f}_\pm(s) = O\left(e^{|s|^c}\right)$$

for some $c > 0$ on the right half-plane $\Re(s) > 0$.

(4.155) implies that $D_\delta^\star$, $\delta \in \{0, 1\}$, satisfies

$$D_\delta^\star(s) = O\left(e^{|s|^C}\right) \qquad (\Re(s) > 0)$$

for some $C > 0$. The functional equation (4.158) implies that the estimate also holds for all $s \in \mathbb{C}$. Hence the functions $D_\delta^\star$, $\delta \in \{0, 1\}$, are entire functions of finite order. (4.159) gives

$$\frac{1}{\Gamma(s)}\begin{pmatrix}\tilde{f}_+(s)\\ \tilde{f}_-(s)\end{pmatrix}=\frac{\Gamma(1-s)}{2\pi i}\begin{pmatrix}e^{\frac{\pi i s}{2}} & -e^{-\frac{\pi i s}{2}}\\ e^{-\frac{\pi i s}{2}} & -e^{\frac{\pi i s}{2}}\end{pmatrix}\begin{pmatrix}D_1^\star(s)\\ D_0^\star(s)\end{pmatrix}.$$

Since $D_\delta^\star(s)$, $\delta\in\{0,1\}$, are entire and of finite order, we see that $\frac{\tilde{f}_\pm(s)}{\Gamma(s)}$ satisfy the growth estimate

$$\frac{\tilde{f}_\pm(s)}{\Gamma(s)}=O\left(e^{|s|^c}\right)$$

for some $c>0$ on the left half-plane $\Re(s)<1$. Hence the two functions $\frac{\tilde{f}_\pm(s)}{\Gamma(s)}$ are of finite order.

We already mentioned that (4.160) implies now that the functions $h_\pm(s)$ are of finite order. Hence

$$\frac{1}{\sin(\pi s)}\,e^{\pm\frac{\pi i s}{2}}\,h_\mp(s)=\left(D_1(s)-e^{\pm\pi i s}\,D_0(s)\right)$$

are entire and of finite order. □

4.3.3 *Nearly Periodic Functions via Integral Transforms*

We want to construct nearly periodic functions as integral transformations of Maass cusp forms, based on ideas presented in [151]. To do so, we need to define the Maass-Selberg differential form that will be used later in order to define the kernel of the associated integrals. First, we define what is known as the R-function. It will play an important role in the construction of the kernel.

4.3.3.1 The *R*-Function

We define

$$h(z):=\Im(z) \tag{4.161}$$

for $z\in\mathbb{H}$.

Exercise 4.76 Show that $h(z)$ is a real-analytic function on the upper half-plane $\mathbb{H}$, is positive for all $z\in\mathbb{H}$, and that h satisfies the differential equations

$$\Delta_k\,h^{\frac{1}{2}-\nu}=\left(\frac{1}{4}-\nu^2\right)h^{\frac{1}{2}-\nu}\quad\text{and}\quad \mathrm{E}_k^\pm h^{\frac{1}{2}-\nu}=(1-2\nu\pm k)\,h^{\frac{1}{2}-\nu} \tag{4.162}$$

for $k \in \mathbb{R}$ and $\nu \in \mathbb{C}$. The hyperbolic Laplace operator Δ_k and Maass operators E_k are defined in (3.9) and (3.10), respectively.

We define

$$R_{k,\nu}(z,\zeta) := \left(\frac{\sqrt{\zeta - z}}{\sqrt{\zeta - \overline{z}}}\right)^{-k} \left(\frac{|\Im(z)|}{(\zeta - z)(\zeta - \overline{z})}\right)^{\frac{1}{2}-\nu}. \tag{4.163}$$

for all $\zeta, z \in \mathbb{C}$ satisfying

$$\zeta - z,\ \zeta - \overline{z} \notin \mathbb{R}_{\leq 0}. \tag{4.164}$$

The square roots $\sqrt{\zeta - z}$ and $\sqrt{\zeta - \overline{z}}$ in (4.163) are well defined since we require that $\zeta - z$ and $\zeta - \overline{z}$ are in $\mathbb{C} \setminus \mathbb{R}_{\leq 0}$. This implies in particular that these square roots, interpreted as principal square roots, are holomorphic.

The R-function has the following properties:

Proposition 4.77 *1. For given $\zeta \in \mathbb{C}$, consider the function*

$$\left\{z \in \mathbb{C}:\ z \textit{ satisfies } (4.164)\right\} \to \mathbb{C}; \qquad z \mapsto R_{k,\nu}(z,\zeta).$$

This function is smooth in the real and imaginary part of z.

2. For given $z \in \mathbb{C}$, consider the function

$$\mathbb{C} \setminus \{z - r, \overline{z} - r:\ r \geq 0\} \to \mathbb{C}; \qquad \zeta \mapsto R_{k,\nu}(z,\zeta).$$

This function is holomorphic on $\mathbb{C} \setminus \{z - r, \overline{z} - r:\ r \geq 0\}$.

3. $R_{k,\nu}$ has the form

$$\begin{aligned} R_{k,\nu}(z,\zeta) &= e^{-ik\arg(\zeta - z)} \left(\frac{\Im(z)}{(\zeta - z)(\zeta - \overline{z})}\right)^{\frac{1}{2}-\nu} \\ &= \left(h^{\frac{1}{2}-\nu}\Big\|_k \begin{pmatrix} 0 & 1 \\ -1 & \zeta \end{pmatrix}\right)(z) \end{aligned} \tag{4.165}$$

for real ζ and $z \in \mathbb{H}$. Here, the double-slash notation $\|_k$ is defined in (3.5).

4. Let $\zeta \in \mathbb{C}$. The function

$$\left\{z \in \mathbb{H}:\ z \textit{ satisfies } (4.164)\right\} \to \mathbb{C}; \qquad z \mapsto R_{k,\nu}(z,\zeta)$$

satisfies the differential equations

$$\begin{aligned} &\Delta_k R_{k,\nu}(\cdot,\zeta) = \left(\frac{1}{4} - \nu^2\right) R_{k,\nu}(\cdot,\zeta) \quad \textit{and} \\ &\mathrm{E}_k^{\pm} R_{k,\nu}(\cdot,\zeta) = (1 - 2\nu \pm k)\, R_{k\pm 2,\nu}(\cdot,\zeta). \end{aligned} \tag{4.166}$$

The function

$$\{z \in \mathbb{H}^- :\ z \text{ satisfies } (4.164)\} \to \mathbb{C}; \qquad z \mapsto R_{k,\nu}(z,\zeta)$$

satisfies

$$\Delta_{-k}\, R_{k,\nu}(\cdot,\zeta) = \left(\frac{1}{4} - \nu^2\right) R_{k,\nu}(\cdot,\zeta) \quad \text{and}$$

$$\mathrm{E}^{\pm}_{-k}\, R_{k,\nu}(\cdot,\zeta) = (1 - 2\nu \pm k)\, R_{k\pm 2,\nu}(\cdot,\zeta).$$

The set $\mathbb{H}^-$ *is the lower complex half-plane defined in* (1.1).

Proof

1. For fixed $\zeta \in \mathbb{C}$, assume that $z \in \mathbb{C}$ satisfies (4.164). It is then obvious from (4.163) that the function $z \mapsto R_{k,\nu}(z,\zeta)$ is smooth in the real and the imaginary parts of z. (Observe that the values under the square-roots are never negative by condition (4.164).)
2. Fix $z \in \mathbb{C}$. Again, it is obvious from (4.163) that the function $\zeta \mapsto R_{k,\nu}(z,\zeta)$ is holomorphic for all $\zeta \in \mathbb{C}$ satisfying condition (4.164). Noticing that ζ satisfies (4.164) is equivalent to the fact that ζ is an element of the "two-cut plane" $\mathbb{C} \smallsetminus \left(z + \mathbb{R}_{\le 0} \cup \overline{z} + \mathbb{R}_{\le 0}\right)$. This shows the second part of the proposition.
3. The first equality follows by rewriting the right-hand side of (4.163) using the identity

$$\frac{\sqrt{\zeta - z}}{\sqrt{\zeta - \overline{z}}} = \frac{\zeta - z}{\sqrt{\zeta - z}\sqrt{\zeta - \overline{z}}} = e^{i \arg(\zeta - z)}, \tag{4.167}$$

   that is correct under the given assumptions $\zeta \in \mathbb{R}$ and $z \in \mathbb{H}$. The second equality follows by the double-slash operator defined in (3.5).
4. We assume real ζ and $z \in \mathbb{H}$. Combining the last expression of $R_{k,\nu}(z,\zeta)$ in (4.165) with (3.14) and then with (4.162) gives

$$\begin{aligned}\Delta_k R_{k,\nu}(\cdot,\zeta) &= \Delta_k \left(h^{\frac{1}{2}-\nu} \Big\|_k \begin{pmatrix} 0 & 1 \\ -1 & \zeta \end{pmatrix} \right) \\ &= \left(\Delta_k h^{\frac{1}{2}-\nu}\right) \Big\|_k \begin{pmatrix} 0 & 1 \\ -1 & \zeta \end{pmatrix} \\ &= \left(\frac{1}{4} - \nu^2\right) h^{\frac{1}{2}-\nu} \Big\|_k \begin{pmatrix} 0 & 1 \\ -1 & \zeta \end{pmatrix} \\ &= \left(\frac{1}{4} - \nu^2\right) R_{k,\nu}(\cdot,\zeta).\end{aligned}$$

   Analogously, using again (3.14) shows

$$\mathrm{E}_k^{\pm} R_{k,\nu}(\cdot,\zeta) = (1-2\nu \pm k)\, R_{k\pm 2,\nu}(\cdot,\zeta).$$

Next, using the substitution $y \mapsto -y$ for the imaginary part $y = \Im(z)$ (i.e., $z \mapsto \overline{z}$) shows that $R_{k,\nu}(z,\zeta)$ satisfies the stated properties for $z \in \mathbb{H}^-$.

The last step is to extend ζ from real values to complex values. This can be done since ζ is just a constant for the differential operators. This shows that the stated differential equations hold also for complex ζ as long as z and ζ satisfy the condition (4.164).

□

Remark 4.78 The R-function appeared first in [122], where Lewis and Zagier introduced

$$\begin{aligned} R_{0,-\frac{1}{2}}(z,\zeta) &= \frac{y}{(x-\zeta)^2+y^2} \\ &= \frac{i}{2}\left(\frac{1}{z-\zeta} - \frac{1}{\overline{z}-\zeta}\right) \end{aligned}$$

in [122, p. 211, above (2.6)]. Their notation for the above expression was $R_\zeta(z)$.

Before we describe the transformation law of $R_{k,\nu}(z,\zeta)$, we need one trivial auxiliary result that will allow us to perform a certain factorization in the proof of the forthcoming Lemma 4.81.

Exercise 4.79 Let $z, w \in \mathbb{C}'$, where the cut-plane $\mathbb{C}'$ is defined in (4.135), and let $\alpha \in \mathbb{C}$ satisfying either

1. $z \in \mathbb{R}_{>0}$ or
2. the product $zw \in \mathbb{R}_{>0}$.

Then we have the identity

$$\left(z\,w\right)^\alpha = z^\alpha\, w^\alpha.$$

Remark 4.80 We would like to emphasize that we assumed $z, w \notin \mathbb{R}_{<0}$ in the second case of Exercise 4.79.

If z, w are both real and negative, then zw itself is positive. However, the stated result in the exercise would not hold: Due to the choice involved in the argument function, see (1.30), the arguments $\arg(z) = -\pi$ and $\arg(w) = -\pi$ are not anymore of opposite sign. Hence, the factorization in the solution of the Exercise 4.79 does not work anymore.

Next, we discuss the transformation law of $R_{k,\nu}(\zeta, z)$.

Lemma 4.81 *Let $M \in \mathrm{SL}_2(\mathbb{Z})$, $\zeta, z \in \mathbb{C}$ satisfying (4.164) and $j(M,\zeta)$, $j(M,z) \in \mathbb{C}'$ with $\Re(j(,m\zeta)) > 0$. (Here, $j\left(\begin{pmatrix}\star & \star\\ c & d\end{pmatrix}, z\right) = cz+d$ is the automorphic factor given in (1.29) and $\mathbb{C}'$ is the complex cut-plane defined* *in* (4.135).*) Moreover, assume that* ζ *and* z *satisfy one of the following three conditions:*

1. $j(M,\zeta) \in \mathbb{R}_{>0}$,
2. $\zeta \in \mathbb{H}$ *and* $Mz \in M\zeta + i\mathbb{R}_{>0}$ *or*
3. $\zeta \in \mathbb{H}^-$ *and* $M\overline{z} \in M\overline{\zeta} + i\mathbb{R}_{>0}$.

Then, the function $(\zeta, z) \mapsto R_{k,\nu}(z,\zeta)$ *satisfies the transformation formula*

$$R_{k,\nu}(Mz, M\zeta) = e^{ik \arg\left(j(M,z)\right)} \left(j(M,\zeta)\right)^{1-2\nu} R_{k,\nu}(z,\zeta). \tag{4.168}$$

Proof Take $M \in \mathrm{SL}_2(\mathbb{Z})$, take $\zeta, z \in \mathbb{C}$ satisfying (4.164) and $j(M,z) \in \mathbb{C}'$.

One key observation is the factorization

$$\left(\frac{|\Im(Mz)|}{(M\zeta - Mz)(M\zeta - M\overline{z})}\right)^{\frac{1}{2}-\nu} = \left(j(M,\zeta)\right)^{1-2\nu} \left(\frac{|\Im(z)|}{(\zeta - z)(\zeta - \overline{z})}\right)^{\frac{1}{2}-\nu} \tag{4.169}$$

if ζ and z satisfy one of the additional assumptions.

First, we assume $\mu(\gamma,\zeta) \in \mathbb{R}_{>0}$. Using identities in (1.18) and Exercise 4.79 gives

$$\begin{aligned}\left(\frac{|\Im(Mz)|}{(M\zeta - Mz)(M\zeta - M\overline{z})}\right)^{\frac{1}{2}-\nu} &= \left(\frac{|\Im(z)|}{\frac{\zeta - z}{j(M,\zeta)}\frac{\zeta-\overline{z}}{j(M,\zeta)}}\right)^{\frac{1}{2}-\nu} \\ &= \left(j(M,\zeta)\right)^{1-2\nu} \left(\frac{|\Im(z)|}{(\zeta - z)(\zeta - \overline{z})}\right)^{\frac{1}{2}-\nu}.\end{aligned}$$

Next, we assume the second case $\zeta \in \mathbb{H}$ and $Mz \in M\zeta + i\mathbb{R}_{>0}$. In particular, we have that $M\zeta - Mz$ and also $M\zeta - M\overline{z}$ are nonvanishing purely imaginary:

$$M\zeta - Mz = -it \qquad \text{and} \qquad M\zeta - Mz = it'$$

for some $t, t' \in \mathbb{R}_{\geq 0}$. We have in fact $t' = t + 2\Re(M\zeta)$. Hence the expression

$$\begin{aligned}\frac{|\Im(Mz)|}{(M\zeta - Mz)(M\zeta - M\overline{z})} &= \frac{|\Im(Mz)|}{(-it)it'} \\ &= \frac{|\Im(Mz)|}{tt'}\end{aligned}$$

is positive. On the other hand, we have

$$\frac{|\Im(Mz)|}{(M\zeta - Mz)(M\zeta - M\overline{z})} = \frac{|\Im(z)|}{(\zeta - z)(\zeta - \overline{z})}\left(j(M,\zeta)\right)^2.$$

We may apply Exercise 4.79 to $j(M,\zeta)^2$ and $\frac{|\Im(z)|}{(\zeta-z)(\zeta-\overline{z})}$ since the assumption $\Re\,(j(M,\zeta)) > 0$ implies $\left|\arg\left(j(M,\zeta)\right)\right| \leq \frac{\pi}{2}$ and hence $\left|\arg\left(j(M,\zeta)\right)^2\right| < \pi$:

$$\left(\frac{|\Im\,(Mz)|}{(M\zeta - Mz)(M\zeta - M\overline{z})}\right)^{\frac{1}{2}-\nu} = \left(j(M,\zeta)\right)^{1-2\nu} \left(\frac{|\Im\,(z)|}{(\zeta - z)(\zeta - \overline{z})}\right)^{\frac{1}{2}-\nu}.$$

Finally, we assume the third case $\zeta \in \mathbb{H}^-$ and $Mz \in M\zeta + i\mathbb{R}_{>0}$. We show that (4.169) holds by interchanging the role of z and $\overline{z}$ in the calculation above.

This proves the identity (4.169) for all three cases.

We need also the identity

$$\left(\frac{\sqrt{M\zeta - Mz}}{\sqrt{M\zeta - M\overline{z}}}\right)^{-k} = e^{ik\arg j(M,z)} \left(\frac{\sqrt{\zeta - z}}{\sqrt{\zeta - \overline{z}}}\right)^{-k}. \tag{4.170}$$

Indeed, it follows also by applying (1.18) and the assumption $j(M,z) \in \mathbb{C}'$ as the following calculation shows:

$$\begin{aligned}\left(\frac{\sqrt{M\zeta - Mz}}{\sqrt{M\zeta - M\overline{z}}}\right)^{-k} &= \left(\frac{\sqrt{j(M,\overline{z})}}{\sqrt{j(M,z)}}\right)^{-k} \left(\frac{\sqrt{\zeta - z}}{\sqrt{\zeta - \overline{z}}}\right)^{-k} \\ &= e^{ik\arg j(M,z)} \left(\frac{\sqrt{\zeta - z}}{\sqrt{\zeta - \overline{z}}}\right)^{-k}.\end{aligned}$$

We also used (4.167) and that $k \in \mathbb{R}$ is real.

To finally prove the lemma, we combine the identities (4.169) and (4.170). We have

$$\begin{aligned}&R_{k,\nu}(Mz, M\zeta) \\ &= \left(\frac{\sqrt{M\zeta - Mz}}{\sqrt{M\zeta - M\overline{z}}}\right)^{-k} \left(\frac{|\Im\,(Mz)|}{(M\zeta - Mz)(M\zeta - M\overline{z})}\right)^{\frac{1}{2}-\nu} \\ &= e^{ik\arg\mu(\gamma,z)} \left(\mu(\gamma,\zeta)\right)^{1-2\nu} \left(\frac{\sqrt{\zeta - z}}{\sqrt{\zeta - \overline{z}}}\right)^{-k} \left(\frac{|\Im\,(z)|}{(\zeta - z)(\zeta - \overline{z})}\right)^{\frac{1}{2}-\nu} \\ &= e^{ik\arg j(M,z)} \left(j(M,\zeta)\right)^{1-2\nu} R_{k,\nu}(z,\zeta).\end{aligned}$$

□

Remark 4.82 The second assumption on ζ and z in Lemma 4.81 is "$\zeta \in \mathbb{H}$ and $Mz \in M\zeta + i\mathbb{R}_{>0}$." This is equivalent to saying that "z lies in the open geodesic ray connecting $\zeta \in \mathbb{H}$ to the cusp $\gamma^{-1}(i\infty)$."

The third assumption can be rephrased analogously: The third condition is equivalent with saying that "$\overline{z}$ lies in the geodesic ray connecting $\overline{\zeta} \in \mathbb{H}$ to the cusp $M^{-1}(i\infty)$."

4.3.3.2 The Maass-Selberg Differential Form

We recall differential forms presented in [147] and [151] and observe the action of Maass raising and lowering operators applied to those differential forms.

Let f, g be real-analytic functions and write again $z = x + iy$. We define

$$\{f, g\}^+(z) := f(z)g(z)\frac{\mathrm{d}z}{y} \qquad \text{and} \qquad \{f, g\}^-(z) := f(z)g(z)\frac{\mathrm{d}\overline{z}}{y}. \tag{4.171}$$

We extend the slash operator notation in (1.131) to linear combinations of the differential forms $\{f, g\}^\pm$ in the following way: Let $k \in \mathbb{R}$ be a real weight, v a compatible multiplier system, and $M \in \mathrm{SL}_2(\mathbb{Z})$. We define

$$\begin{aligned} \{f, g\}^+ \|_k^v M(z) &:= e^{-ik \arg j(M,z)}\, v(M)^{-1}\, f(Mz)\, g(Mz)\, \frac{\mathrm{d}(Mz)}{\Im(Mz)} \qquad \text{and} \\ \{f, g\}^- \|_k^v M(z) &:= e^{-ik \arg j(M,z)}\, v(M)^{-1}\, f(Mz)\, g(Mz)\, \frac{\mathrm{d}\overline{(Mz)}}{\Im(Mz)}, \end{aligned} \tag{4.172}$$

and extend it in the obvious way to linear combinations. For example we have

$$\Big(\{f, g\}^+ - \{f, g\}^-\Big)\Big\|_k^v M(z) = \{f, g\}^+ \|_k^v M(z) - \{f, g\}^- \|_k^v M(z).$$

Lemma 4.83 *Let v be a multiplier and $k, q \in \mathbb{R}$.*

1. *We have for every matrix $M \in \mathrm{SL}_2(\mathbb{Z})$ that*

$$\{f, g\}^\pm \|_{k+q\mp 2}^v \gamma = \begin{cases} \{v^{-1}(M)\, f \|_k \gamma,\, g \|_q \gamma\}^\pm & \text{and} \\ \{f \|_k \gamma,\, v^{-1}(M)\, g \|_q \gamma\}^\pm. \end{cases} \tag{4.173}$$

2. $\{f, g\}^\pm = \{g, f\}^\pm$.

Proof The last property follows directly from the definition in (4.171).

We prove (4.173) by direct calculation, using the identities in (1.18) and (1.19). For example, we have

$$\begin{aligned} \{f, g\}^+ \|_{k+q-2}^v M(z) &= v(M)^{-1}\, e^{(-k-q+2)i \arg j(M,z)}\, f(Mz)\, g(Mz)\, \frac{\mathrm{d}(Mz)}{\Im(Mz)} \\ &= \{v^{-1}(M)\, f \|_k M,\, g \|_q M\}^+(z) \end{aligned}$$

for all $z \in \mathbb{H}$. The other identities follow analogously. □

Exercise 4.84 Prove the identities in (4.173) analogous to the identities in the proof of Lemma 4.83.

Combining the property (4.173) of the 1-forms in (4.171) with Maass operators, we see that

$$\begin{aligned} \left\{\mathrm{E}_k^{\pm}\left(v^{-1}(M)\, f\left\|_k M\right), g\right\|_q M\right\}^{\pm} &= \left\{\mathrm{E}_k^{\pm}\left(f\left\|_k M\right), v^{-1}(M)\, g\right\|_q M\right\}^{\pm} \\ &= \left\{\mathrm{E}_k^{\pm} f, g\right\}^{\pm} \|_{k+q}^{v} M \end{aligned} \tag{4.174}$$

for every $M \in \mathrm{SL}_2(\mathbb{R})$ and $k, q \in \mathbb{R}$. We also have the relations

$$\begin{aligned} \left\{\mathrm{E}_k^{+} f, g\right\}^{+} &= -\left\{f, \mathrm{E}_{-k}^{+} g\right\}^{+} + 4i\,\partial_z(fg)\mathrm{d}z \quad \text{and} \\ \left\{\mathrm{E}_k^{-} f, g\right\}^{-} &= -\left\{f, \mathrm{E}_{-k}^{-} g\right\}^{-} - 4i\,\partial_{\overline{z}}(fg)\mathrm{d}\overline{z}. \end{aligned} \tag{4.175}$$

and

$$\begin{aligned} \left\{\mathrm{E}_k^{-} f, g\right\}^{+} &= -\left\{f, \mathrm{E}_{-k}^{-} g\right\}^{+} - 4i\,\partial_{\overline{z}}(fg)\mathrm{d}z \quad \text{and} \\ \left\{\mathrm{E}_k^{+} f, g\right\}^{-} &= -\left\{f, \mathrm{E}_{-k}^{+} g\right\}^{-} + 4i\,\partial_z(fg)\mathrm{d}\overline{z}. \end{aligned}$$

We are now able to define the Maass-Selberg form.

Definition 4.85 Let f, g be real-analytic and $k \in \mathbb{R}$. We define the *Maass-Selberg form* η_k by

$$\eta_k(f, g) := \left\{\mathrm{E}_k^{+} f, g\right\}^{+} - \left\{f, \mathrm{E}_{-k}^{-} g\right\}^{-}. \tag{4.176}$$

Lemma 4.86 *Let f, g be real-analytic and $k \in \mathbb{R}$. The Maass-Selberg form has the following properties:*

1. *We have the equations*

$$\eta_k(f, g) + \eta_{-k}(g, f) = 4i \cdot \mathrm{d}(fg) \tag{4.177}$$

   and

$$\eta_k(f, g) - \eta_{-k}(g, f) = 4\left([g f_y - f g_y]\mathrm{d}x + \left[f g_x - g f_x + \frac{ik}{y} f g\right]\mathrm{d}y\right), \tag{4.178}$$

   where f_x denotes $\partial_x f$, $f_y = \partial_y f$, $g_x = \partial_x g$, and $g_y = \partial_y g$, respectively.
2. *Assume that there exists a $\lambda \in \mathbb{R}$ such that f and g satisfy $\Delta_k f = \lambda f$ and $\Delta_{-k} g = \lambda g$. Then the Maass-Selberg form $\eta_k(f, g)$ is closed.*
3. *We have that*

$$\begin{aligned}\eta_k(f,g)\big\|_0^v M &= \eta_k\big(v^{-1}(M)\, f\big\|_k M,\ g\big\|_{-k} M\big)\\ &= \eta_k\big(f\big\|_k M,\ v^{-1}(M)\, g\big\|_{-k}^{v} \gamma\big)\end{aligned} \tag{4.179}$$

for every multiplier system v.

4. *Let $\nu \in \mathbb{C}$ and assume that f and g are eigenfunctions of the operators Δ_k and Δ_{-k}, respectively, both with eigenvalue $\frac{1}{4}-\nu^2$. Then, we have*

$$\begin{aligned}\eta_{k+2}(\mathrm{E}_k^+ f,\, \mathrm{E}_{-k}^- g) &= (1+2\nu+k)(1-2\nu+k)\,\eta_k(f,\, g) + \\ &\quad + 4i\,\mathrm{d}\big((\mathrm{E}_k^+ f)(\mathrm{E}_{-k}^- g)\big)\,.\end{aligned} \tag{4.180}$$

Proof Recall the identities $\partial_z = \frac{1}{2}\partial_x - \frac{i}{2}\partial_y$, $\partial_{\overline{z}} = \frac{1}{2}\partial_x + \frac{i}{2}\partial_y$, $\mathrm{d}z = \mathrm{d}x + i\mathrm{d}dy$, $\mathrm{d}\overline{z} = \mathrm{d}x - i\mathrm{d}y$ and $\mathrm{d}f = \partial_z f\mathrm{d}z + \partial_{\overline{z}} f\mathrm{d}\overline{z}$ for every function f which is smooth in x and y.

1. We prove the first item via direct calculation:

$$\begin{aligned}\eta_k(f,g) + \eta_{-k}(g,f) &= -4i\big[(fg_z + gf_z)\mathrm{d}z + \big(fg_{\overline{z}} + gf_{\overline{z}}\big)\mathrm{d}\overline{z}\big]\\ &= 4i\mathrm{d}(fg)\end{aligned}$$

and

$$\begin{aligned}\eta_k(f,g) - \eta_{-k}(g,f) &= 4i\bigg[\Big(gf_z - fg_z - \tfrac{ik}{2y} fg\Big)\mathrm{d}z - \\ &\qquad - \Big(gf_{\overline{z}} - fg_{\overline{z}} - \frac{ik}{2y} fg\Big)\mathrm{d}\overline{z}\bigg]\\ &= 4\bigg[\big(gf_y - fg_y\big)\mathrm{d}x + \Big(fg_x - gf_x + i\frac{k}{y} fg\Big)\mathrm{d}y\bigg],\end{aligned}$$

where we used (3.11) for the Maass operators.

2. We want to show that $\mathrm{d}\,\eta_k(f,g) = 0$ under the given conditions. Since

$$2\eta_k(f,g) = \big(\eta_k(f,g) + \eta_{-k}(g,f)\big) + \big(\eta_k(f,g) - \eta_{-k}(g,f)\big),$$

and by the first part of the lemma, it is enough to show that $\eta_k(f,g) - \eta_{-k}(g,f)$ is closed. We find, after some computations, that

$$\mathrm{d}\big(\eta_k(f,g) - \eta_{-k}(g,f)\big) = \big[f\Delta_{-k} g - g\Delta_k f\big]\frac{\mathrm{d}x \wedge \mathrm{d}y}{y^2}.$$

3. This follows directly from (4.174).
4. It follows directly from the equations in (4.175) that

$$\eta_{k+2}(\mathrm{E}_k^+ f, \mathrm{E}_{-k}^- g) = 4i\mathrm{d}\left[(\mathrm{E}_k^+ f)(\mathrm{E}_{-k}^- g)\right] - \\ - \left\{\mathrm{E}_k^+ f, \mathrm{E}_{-k-2}^+ \mathrm{E}_{-k}^- g\right\}^+ + \left\{\mathrm{E}_{k+2}^- \mathrm{E}_k^+ f, \mathrm{E}_{-k}^- g\right\}^+.$$

We may apply (4.2), since we assume that f and g are eigenfunctions of the Laplace operators Δ_k and Δ_{-k},respectively, with the same eigenvalue $\frac{1}{4} - \nu^2$. The statement in the lemma follows.

□

Remark 4.87 The first three items of Lemma 4.86 are generalizations of the lemma given in [122, II.2]. We have

$$\eta_0(f, g) = [f, g],$$

where $[\cdot, \cdot]$ is defined in [122, Chapter II, §2]. Our form $\{\cdot, \cdot\}^\pm$ in (4.171) differs from the form $\{\cdot, \cdot\}$ in [122, II.§2, (2.5)], contrary to the notational resemblance. Lewis and Zagier call their form $\{\cdot, \cdot\}$ the Green form in the line above [122, II.§2, (2.5)].

Lemma 4.88 *Let f and g be smooth functions (in x and y) on $\mathbb{H} \cup \mathbb{H}^-$ satisfying $f(z) = f(\overline{z})$ and $g(z) = g(\overline{z})$.*

1. *We can extend the Maass-Selberg form to smooth (in x and y) functions f, g defined on the lower half-plane.*
2. *The Maass-Selberg form satisfies*

$$\eta_k(f, g)(z) = \eta_k(g, f)(\overline{z}). \tag{4.181}$$

Proof

1. All differentials and other components used in the definition of the Maass-Selberg form are well defined for smooth (in x and y) functions on the lower half-plane $\mathbb{H}^-$. Hence the extension makes sense.
2. This follows by direct calculations. First use the relations (4.176), (4.171), and (3.10) or (3.11) to rewrite everything depending on the pair $(z, \overline{z})$ respectively (x, y). Then, use the substitution $(z, \overline{z}) \mapsto (\overline{z}, z)$ respectively $(x, y) \mapsto (x, -y)$. As a final step, use the above relations in reverse order.

□

Exercise 4.89 Perform the explicit calculation indicated in the second part of the proof above.

4.3.3.3 Everything Combined

We will now insert the function $R_{k,\nu}$ into the Maass-Selberg form and use the form to define nearly periodic functions in §4.3.3.4.

Lemma 4.90 *Let v be a multiplier that is compatible with the real weight k, $\nu \in \mathbb{C}$, u a Maass cusp form with weight k multiplier v and eigenvalue $\frac{1}{4} - \nu^2$ for the full modular group. Moreover, let $M \in \mathrm{SL}_2(\mathbb{Z})$ and let $\zeta, z \in \mathbb{C}$ satisfying the assumptions of Lemma 4.81.*

It follows

$$v^{-1}(M)\, \eta_k(u, R_{-k,\nu}(\cdot, M\zeta))\big\|_0 M(z) = \big(j(M,\zeta)\big)^{1-2\nu} \eta_k\big(u, R_{-k,\nu}(\cdot, \zeta)\big)(z) \text{ and}$$
$$v^{-1}(M)\, \eta_{-k}(R_{-k,\nu}(\cdot, M\zeta), u)\big\|_0 M(z) = \big(j(M,\zeta)\big)^{1-2\nu} \eta_{-k}\big(R_{-k,\nu}(\cdot, \zeta), u\big)(z). \tag{4.182}$$

Proof We show the second identity:

$$\begin{aligned} &v^{-1}(M)\, \eta_{-k}\big(R_{-k,\nu}(\cdot, M\zeta), u\big)\big\|_0 M(z) \\ &= \eta_{-k}\big(R_{-k,\nu}(\cdot, M\zeta)\big\|^1_{-k} M, v^{-1}(M) u \big\|_k M\big)(z) \qquad \text{using (4.179)} \\ &= \eta_{-k}\big(R_{-k,\nu}(\cdot, M\zeta)\big\|^1_{-k} M, u\big)(z) \\ &= \big(j(M,\zeta)\big)^{1-2\nu}\, \eta_k\big(R_{-k,\nu}(\cdot, \zeta), u\big)(z) \qquad \text{using (4.168).} \end{aligned}$$

The use of the transformation formula (4.168) in the calculation above is allowed since z and ζ satisfy the assumptions of Lemma 4.81 and since $\mathrm{E}^+_{-k} R_{-k,\nu}\big(z, \zeta\big)$ appearing in the construction of $\eta_{\pm k}\big(R_{-k,\nu}(\cdot, \zeta), u\big)(z)$ satisfies (4.166).

The first identity follows by the same arguments. □

4.3.3.4 Nearly Periodic Functions via Integral Transformations

Let u be a Maass cusp form of weight k, multiplier v, and spectral value ν for the full modular group $\Gamma(1)$ as defined in Definition 3.31. Recalling its defining properties, u satisfies:

1. u is a real-analytic function on $\mathbb{H}$ in the sense of Definition 3.14,
2. $\Delta_k\, u = \left(\frac{1}{4} - \nu^2\right) u$,
3. $u\big\|_k M = v(M)\, u$ for all $M \in \Gamma$ (i.e., u satisfies (3.6)) and
4. u admits an expansion of the form (3.48) at the cusp ∞.

u also satisfies the growth condition (3.53) of Theorem 3.38 at the cusp ∞: There exists an $\varepsilon > 0$ such that

$$u(z) = O\left(\Im(z)^{\left|\frac{k}{2}\right|} e^{-\varepsilon\, \Im(z)}\right) \qquad \text{as } \Im(z) \to \infty.$$

Now, we can define the function $f : \mathbb{C} \smallsetminus \mathbb{R} \to \mathbb{C}$ as follows:

$$\zeta \mapsto f(\zeta) := \begin{cases} \int_\zeta^{i\infty} \eta_{-k}\big(R_{-k,\nu}(\cdot,\zeta), u\big)(z) & \text{if } \Im(\zeta) > 0 \text{ and} \\ -\int_\zeta^{-i\infty} \eta_k\big(R_{-k,\nu}(\cdot,\zeta), \tilde{u}\big)(z) & \text{if } \Im(\zeta) < 0, \end{cases} \tag{4.183}$$

where we define

$$\tilde{u}(z) = u(\overline{z}) \tag{3.55}$$

as in Exercise 3.42. The path of integration is the geodesic ray connecting ζ and the cusp $i\infty$ for ζ in the upper half-plane or the geodesic ray between ζ and $-i\infty$ for ζ in the lower half-plane, respectively.

Remark 4.91

1. We have made a choice in (4.183) between integrating over the forms

$$-\eta_k\big(u, R_{-k,\nu}(\cdot,\zeta)\big) \quad \text{or} \quad (1b) \quad \eta_{-k}\big(R_{-k,\nu}(\cdot,\zeta), u\big) \tag{4.184}$$

on $\mathbb{H}$ and

$$-\eta_k\big(R_{-k,\nu}(\cdot,\zeta), \tilde{u}\big) \quad \text{or} \quad (2b) \quad \eta_{-k}\big(\tilde{u}, R_{-k,\nu}(\cdot,\zeta)\big) \tag{4.185}$$

on $\mathbb{H}^-$.

We will see later in Remark 5.15 that each choice leads to the same period function on $\mathbb{H}$ and respectively on $\mathbb{H}^-$. Remark 4.94 compares our choice to the situation discussed in [122].

2. The reason why we extended Maass cusp forms and the R-function to the lower half-plane $\mathbb{H}^-$ and also extended the Maass-Selberg form η_k to functions on the lower half-plane, see, e.g., Exercise 3.42, Proposition 4.77, and Lemma 4.88 respectively, is the second case of (4.183). We want to be able to integrate along the geodesic ray $\zeta - i\mathbb{R}_{>0}$ connecting $\zeta \in \mathbb{H}^-$ and $-i\infty$ in the lower half-plane. In our opinion, this representation illustrates better how the function f is defined on $\mathbb{H}^-$ compared to the (on $\mathbb{H}^-$ equivalent) integral representation in Lemma 4.93 below.

Lemma 4.92 *For* $|\Re(\nu)| < \frac{1}{2}$, *the integration in* (4.183) *is well defined along the geodesic paths connecting* ζ *to* $i\infty$ *in* $\mathbb{H}$ *respectively* ζ *to* $-i\infty$ *in* $\mathbb{H}^-$. *Moreover, the defined function* $f(\zeta)$ *is holomorphic in* ζ.

Proof The singularity of $R_{k,\nu}(z,\zeta)$ for $z \to \zeta \in \mathbb{H}$ respectively $z \to \overline{\zeta} \in \mathbb{H}$ is of the form $(\zeta - z)^{\nu-\frac{1}{2}}$ respectively $\big(\zeta - \overline{z}\big)^{\nu-\frac{1}{2}}$. The whole integrand has at most the same singularity since $R_{k,\nu}$ is an eigenfunction of the Maass operators, see (4.166).

The weight argument $\arg \zeta - z$ has also a well-defined limit. Hence the integration is well defined for ν values satisfying $|\Re(\nu)| < \frac{1}{2}$.

The fact that $f(\zeta)$ is holomorphic follows directly from the fact that the R-function defined in (4.163) is in ζ holomorphic, see, e.g., Proposition 4.77. □

Lemma 4.93 *For $\zeta \in \mathbb{H}^-$, we have*

$$f(\zeta) = -\int_{\overline{\zeta}}^{i\infty} \eta_k\big(u, R_{-k,\nu}(\cdot, \zeta)\big)(z). \tag{4.186}$$

Proof For $\zeta \in \mathbb{H}^-$ we have

$$\begin{aligned} f(\zeta) &= -\int_{\zeta}^{-i\infty} \eta_k\big(R_{-k,\nu}(\cdot, \zeta), \tilde{u}\big)(z) \\ &= -\int_{\zeta}^{-i\infty} \eta_k\big(u, R_{-k,\nu}(\cdot, \zeta)\big)(\overline{z}) \qquad \text{using (4.181)} \\ &= -\int_{\overline{\zeta}}^{i\infty} \eta_k\big(u, R_{-k,\nu}(\cdot, \zeta)\big)(z), \end{aligned}$$

where we used $u(z) = \tilde{u}\big(\overline{z}\big)$ for $z \in \mathbb{H}$, see (3.55) in Exercise 3.42. □

Remark 4.94 For $k = 0$, we may compare our approach to the original one presented in [122]. Comparing the definition of f in (4.183) to the one in [122, p. 212], we find $\eta_0(f, g) = [f, g]$, see also Remark 4.87.

We see that [122] uses exactly the opposite choice: They use the integral transformation $\int_{\zeta}^{i\infty} \eta_0\big(u, R_{-0,\nu}(\cdot, \zeta)\big)$ for $\zeta \in \mathbb{H}$ (compared to $\int_{\zeta}^{i\infty} \eta_{-0}\big(R_{-0,\nu}(\cdot, \zeta), u\big)$ in (4.183)). On the lower half-plane they use $-\int_{\overline{\zeta}}^{i\infty} \eta_{-0}\big(R_{-0,\nu}(\cdot, \zeta), u\big)$ for $\zeta \in \mathbb{H}^-$ (compared to $-\int_{\overline{\zeta}}^{i\infty} \eta_0\big(u, R_{-0,\nu}(\cdot, \zeta)\big)$ in (4.186)).

Since we discuss nearly periodic functions associated with Maass cusp forms, we have to establish that f given in (4.183) is indeed nearly periodic in the sense of Definition 4.54. To do so, we start with a lemma describing the transformation property of function f.

Lemma 4.95 *Let $|\Re(\nu)| < \frac{1}{2}$. The function f defined in* (4.183) *satisfies*

$$v(M)^{-1} f\big|_{1-2\nu}M(z) \overset{(1.131)}{=} v(M)^{-1} \big(j(M,\zeta)\big)^{2\nu-1} f(M\zeta)$$
$$= \begin{cases} \displaystyle\int_{\zeta}^{M^{-1}i\infty} \eta_{-k}\big(R_{-k,\nu}(\cdot,\zeta),u\big)(z) & \textit{if } \zeta \in \mathbb{H} \textit{ and} \\ \displaystyle -\int_{\overline{\zeta}}^{M^{-1}i\infty} \eta_{k}\big(u,R_{-k,\nu}(\cdot,\zeta)\big)(z) & \textit{if } \zeta \in \mathbb{H}^- \end{cases} \tag{4.187}$$

for every $\zeta \in \mathbb{H} \cup \mathbb{H}^-$ *and* $M \in \Gamma(1)$ *satisfying* $\Re\,(j(M,\zeta)) > 0$. *The path of integration on the right-hand side is the geodesic ray connecting* ζ *respectively* $\overline{\zeta}$ *and* $\gamma^{-1}(i\infty)$.

Proof Let $\zeta \in \mathbb{H}$ and $M \in \Gamma(1)$ such that $\Re\,(j(M,\zeta)) > 0$. We have

$$\begin{aligned} f(M\zeta) &= \int_{M\zeta}^{i\infty} \eta_{-k}\big(R_{-k,\nu}(\cdot,M\zeta),u\big)(z) \\ &\overset{M\,z\mapsto z}{=} \int_{\zeta}^{M^{-1}i\infty} \eta_{-k}\big(R_{-k,\nu}(\cdot,M\zeta),u\big)(Mz) \\ &= v(M) \int_{\zeta}^{M^{-1}i\infty} v(M)^{-1}\, \eta_{-k}\big(R_{-k,\nu}(\cdot,M\zeta),u\big)\big\|_0 M(z). \end{aligned}$$

Now, we would like to apply Lemma 4.90. Therefore, we must check if all z of the integration path satisfy the conditions on ζ and z as given in Lemma 4.81. Remark 4.167 implies that we have to verify if the integration path is the geodesic ray connecting ζ and $M^{-1}(i\infty)$. This is indeed the case. Hence the second condition of Lemma 4.81 is satisfied since we assume $\Re\,(j(M,\zeta)) > 0$. Using the transformation formula (4.182) in Lemma 4.90 gives

$$\begin{aligned} f(\gamma\zeta) &= v(\gamma) \int_{\zeta}^{M^{-1}i\infty} v(M)^{-1}\, \eta_{-k}\big(R_{-k,\nu}(\cdot,M\zeta),u\big)\big\|_0 M(z) \\ &\overset{(4.182)}{=} v(M)\,\big(j(M,\zeta)\big)^{1-2\nu} \int_{\zeta}^{M^{-1}i\infty} \eta_{-k}\big(R_{-k,\nu}(\cdot,\zeta),u\big)(z). \end{aligned}$$

The same calculation for $\zeta \in \mathbb{H}^-$, using the integral for f in (4.186), gives

$$v(M)^{-1}\big(j(M,\zeta)\big)^{2\nu-1} f(M\zeta) = -\int_{\overline{\zeta}}^{M^{-1}i\infty} \eta_{k}\big(u,R_{-k,\nu}(\cdot,\zeta)\big)(z).$$

□

To check whether $f(\zeta)$ is nearly periodic is now an application of Lemma 4.95: For $\zeta \in \mathbb{H}$, we find

$$\begin{aligned} v(S)^{-1} f(S\zeta) &\overset{(4.187)}{=} \int_\zeta^{S^{-1}i\infty} \eta_{-k}\big(R_{-k,\nu}(\cdot,\zeta),u\big)(z) \\ &= \int_\zeta^{i\infty} \eta_{-k}\big(R_{-k,\nu}(\cdot,\zeta),u\big)(z) \\ &\overset{(4.183)}{=} f(\zeta), \end{aligned}$$

where we use the invariance of the cusp $i\infty$ under translation $\zeta \mapsto S\zeta = \zeta + 1$ and the trivial fact that $\Re(j(S,\zeta)) = 1$. The same arguments hold for $\zeta \in \mathbb{H}^-$. Hence, we just proved the following lemma.

Lemma 4.96 *The function f defined in* (4.183) *satisfies*

$$v(S)^{-1} f(\zeta+1) = f(\zeta) \qquad \textit{for every } \zeta \in \mathbb{C} \smallsetminus \mathbb{R}. \tag{4.188}$$

Written in the slash notation (1.131)*, we have*

$$v(S)^{-1} f\big|_{1-2\nu} S = f \qquad \textit{on } \mathbb{C} \smallsetminus \mathbb{R}. \tag{4.189}$$

Summarizing, we have established the following result.

Proposition 4.97 *Let u be a Maass cusp form of real weight $k \in \mathbb{R}$, multiplier v, and spectral value ν for the full modular group $\Gamma(1)$ as defined in Definition 3.31. We assume that the spectral value satisfies $|\Re(\nu)| < \frac{1}{2}$.*

The function $f : \mathbb{C} \smallsetminus \mathbb{R} \to \mathbb{C}$ defined in (4.183) *is well defined and holomorphic. Moreover, f is nearly periodic in the sense of Definition 4.54 with factor $v(T)$.*

Remark 4.98 Proposition 4.97 does not show that the nearly periodic function f defined in (4.183) satisfies the conditions in part 4 of Theorem 4.55. This will be done below in Theorem 5.21 (and Lemma 5.20).

4.3.4 Nearly Periodic Functions and Eichler Integrals

Consider a classical cusp form and its associated period polynomial via (2.1) in §2.1.1. On the other hand, we can embed the classical cusp form into the space of Maass cusp forms by (3.56) in §3.4.1 and then attach a periodic function via the integral transformation (4.183). How are these two functions, the nearly periodic function and the Eichler integral, related?

Let F be a modular cusp form of even weight $k \in 2\mathbb{N}$ as introduced in Definition 1.59. We attach a Maass cusp form $u_F : \mathbb{H} \to \mathbb{C}$ to F by (3.56):

$$u_F(z) := \Im(z)^{\frac{k}{2}} F(z). \tag{3.56}$$

As shown in §3.4.1, u is indeed a Maass cusp form of weight k, trivial multiplier $v \equiv 1$, and eigenvalue $\frac{k}{2}\left(1-\frac{k}{2}\right)$. Hence, u has spectral values $\nu \in \left\{\frac{k-1}{2}, \frac{1-k}{2}\right\}$.

The following proposition compares the nearly periodic function associated with u_F by (4.183) and the Eichler integral (2.17) associated with F.

Proposition 4.99 *Let F be a modular cusp form of even weight $k \in 2\mathbb{N}$, trivial multiplier for the full modular group $\Gamma(1)$ as introduced in Definition 1.59, and let u_F be the F associated with Maass cusp form by embedding* (3.56).

1. *The integral transformation* (4.183) *defining $f(\zeta)$ associated with u_F for $\zeta \in \mathbb{H}$ is well defined for both spectral values $\nu \in \left\{\frac{1-k}{2}, \frac{k-1}{2}\right\}$.*
2. *The function f associated with u_F by* (4.183) *with weight k and spectral value $\nu = \frac{1-k}{2}$ vanishes everywhere.*
3. *The function f associated with u_F by* (4.183) *with weight k and spectral value $\nu = \frac{k-1}{2}$ and restricted to the upper half-plane $\mathbb{H}$ is a multiple of $\tilde{F}$ associated with F by* (2.17)*:*

$$f = (2-2k)\,\tilde{F}. \tag{4.190}$$

Proof For $\zeta \in \mathbb{H}$, consider the nearly periodic function f associated with u_F given by (4.183). We have

$$f(\zeta) = \int_\zeta^{i\infty} \eta_{-k}\big(R_{-k,\nu}(\cdot,\zeta), u\big)(z) \qquad \text{for } \zeta \in \mathbb{H}. \tag{4.183}$$

Using (4.176) and (4.171), we then get

$$f(\zeta) = \int_\zeta^{i\infty} \left[\left(E^+_{-k} R_{-k,\nu}(\cdot,\zeta)\right)(z)\, u(z)\frac{\mathrm{d}z}{y} - R_{-k,\nu}(z,\zeta)\,\left(E^-_k u\right)(z)\frac{\mathrm{d}\bar{z}}{y}\right].$$

Recalling $\mathrm{E}^+_{-k} R_{-k,\nu} = (1-2\nu-k)R_{2-k,\nu}$ in (4.166) and $\mathrm{E}^-_k u = 0$ as shown in Exercise 3.13, we find

$$f(\zeta) = (1-2\nu-k)\int_\zeta^{i\infty} R_{2-k,\nu}(z,\zeta)\,u(z)\frac{\mathrm{d}z}{y} \tag{4.191}$$

for $\zeta \in \mathbb{H}$.

To prove the second part of the Proposition, we assume $\nu = \frac{1-k}{2}$. Then, the factor $1-2\nu-k = 1-2\frac{1-k}{2}-k = 0$ in (4.191) vanishes, implying $f(\zeta) = 0$.

To prove the third part of the Proposition, we assume $\nu = \frac{k-1}{2}$. Using (5.32) and (3.56) in (4.191) gives

$$f(\zeta) = (2-2k)\int_{\zeta}^{i\infty} (\zeta - z)^{k-2}\, u_{\mathrm{h}}(z)\mathrm{d}z \overset{(2.17)}{=} (2-2k)\,\tilde{F}(\zeta)$$

for every $\zeta \in \mathbb{H}$.

The first part follows also from the above calculations. Even if the calculations above are a priori formal, the well definiteness of the results shows that the original integral transforms are also well defined. □

4.4 Concluding Remarks

The starting point of Maass wave forms and its relation to L-series can be traced back to the work done by Hans Maass in his paper [128] published in 1949. A more refined version appeared in [129] in his lecture notes. For example, Hans Maass result between Maass cusp forms and L-series is given in [129, Theorem 35].

Our approach to this connection is based on the concept of "families of Maass cusp forms" as defined in Definition 4.1 in this chapter. These families focus on the fact that the raising and lowering Maass operators $\mathrm{E}_k^{\pm}$, see (3.10) in Section 3.2, which raise (respectively lower) the weight of a Maass wave form of weight k by 2. Using correctly scaled/weighted expansions of the Maass wave form, we see that the coefficients in the expansion do not change, see Lemma 4.11 in Section 4.1 of this chapter.

We used this property in Section 4.2: We can change the explicit weight of an element in a family of Maass cusp forms without changing the associated L-series since the L-series only depends on the coefficients in the expansion. This leads to a proof of Theorem 4.28 (a slightly modified version of Hans Maass [129, Theorem 35]) where we state the relation between a family of Maass cups forms and the two L-series $L_{\pm}(s)$ in (4.54). We also present simplified versions of this result.

L-series are interesting objects of its own. Often, one can show that L-series and certain values of it encode interesting arithmetic properties.

Of interest are also certain values of the L-series, for example the central L-value. If the L-function $L(s)$ satisfies a functional equation relating $L(s)$ and $L(w-s)$, then $L\left(\frac{w}{2}\right)$ is called the *central L-value*. For example, in the setting of Theorem 4.28, the central L-values are $L_l^{\star}(0)$. The nonvanishing of such central L-values and their derivatives of L-functions associated with Maass wave forms and more general automorphic forms is an important research topic, due to the connection between such values and various aspects of mathematics, such as arithmetic geometry, spectral deformation theory, and analytic number theory. The combination of the method of moments and the mollification method, initiated by Iwaniec and Sarnak [102], has been a very fruitful approach in yielding positive-proportional nonvanishing results on central L-values and their derivatives in some collection of automorphic forms. We refer to recent research paper [123] and the references therein for more information on this method.

Section 4.3 introduced the nearly periodic functions associated with families of Maass cusp forms. These functions were introduced by Lewis and Zagier in [122] in the context of Maass cusp forms of weight 0 and trivial multiplier for the full modular group $\Gamma(1)$. We can say in some sense that the nearly periodic functions fulfill a similar role to Eichler integrals for modular cusp forms: According to Definition 2.14 Eichler integrals satisfy the transformation property

$$j(M,z)^{k-2}\, F\big(M\,z\big) \overset{(2.9)}{=} F(z) + P_M(z)$$

for all $M \in \Gamma(1)$ and $z \in \mathbb{C}$ with P_M a polynomial. On the other hand, a nearly periodic function f associated with a family of Maass cusp forms satisfies

$$f(S\,s) = c\,f(z) \quad \text{and} \quad -\,e^{\frac{\pi i k}{2}}\,z^{2\nu-1}\,f\,(T\,z) = f(z) + \text{nice function}$$

for all $z \in \mathbb{C} \smallsetminus \mathbb{R}$ where the constant c is of absolute value 1 and depends on the family of Maass cusp forms, see Theorem 4.55. In some sense, the "nice function" has some nicer properties as the nearly periodic function: for example, the "nice function" extends holomorphically across the right half-plane. We see here a similar situation as in the modular form case where P_M has the nice property of being a polynomial compared to the holomorphic Eichler integral F. A short look ahead: we discuss this "nice function" in §5.1.1 of Chapter 5.

Chapter 5
Period Functions

Recall the period polynomials that we discussed in Chapter 2. According to Definition 2.7 period polynomials are polynomials $P \in \mathbb{C}[X]_{k-2}$ which satisfy the two functional equations

$$P + P\big|_{2-k}T = 0 \quad \text{and} \quad P + P\big|_{2-k}(TS) + P\big|_{2-k}(TS)^2 = 0.$$

The different explicit constructions in §2.1 together with the Eichler-Shimura Theorem 2.50 show that period polynomials contain all information of the associated modular forms. In principle, we could calculate everything about a modular form or cusp form by studying the associated period polynomial. This approach has the nice advantage that period polynomials as polynomials are much simpler objects compared to modular forms that are holomorphic functions on the upper half-plane $\mathbb{H}$.

Now, we consider families of Maass cusp forms, introduced in Chapter 4. Can we find objects associated with (families of) Maass cusp forms, similar to the period polynomials associated with modular forms? This question was positively answered by Lewis [121] in 1997 for even Maass cusp forms (of weight 0 and trivial multiplier for the full modular group). Lewis and Zagier then extended this result to all Maass cusp forms (of weight 0 and trivial multiplier for the full modular group) in [122]. Lewis and Zagier used also the name "period functions" for holomorphic functions P on $\mathbb{C} \smallsetminus \mathbb{R}_{\leq 0}$ satisfying the functional equation

$$P = P\big|_{1-2\nu}S + P\big|_{2\nu-1}STS$$

and certain growth conditions, where ν denotes the spectral parameter of the associated Maass cusp form.

In Section 5.1 we introduce period functions and give several relations to (families of) Maass cusp forms. Section 5.2 briefly presents some remarks about Zagier's cohomology theorem (presented in [36]) that generalizes the Eichler

T. Mühlenbruch, W. Raji, *On the Theory of Maass Wave Forms*, Universitext,
https://doi.org/10.1007/978-3-030-40475-8_5

cohomology theorem in §2.3. We conclude this chapter with a discussion of Hecke operators acting on period functions in Section 5.3 and some remarks in Section 5.4. As in Chapter 4 we restrict ourselves to the full modular group for simplicity reasons.

5.1 Period Functions Associated with Maass Cusp Forms

We discussed in Chapter 2 period polynomials associated with modular cusp forms. In particular, we showed that period polynomials associated with modular cusp forms on the full modular group satisfy the functional equations in (2.4).

This inspires us to give the following definition of period functions.

Definition 5.1 Let $\nu \in \mathbb{C}$. We call a holomorphic function $P : \mathbb{C}' \to \mathbb{C}$ on the complex *cut-plane* $\mathbb{C}' := \mathbb{C} \setminus (-\infty, 0]$ a *period function* if P satisfies the *three-term equation*

$$P(\zeta) = v\left(\begin{pmatrix} 1 & 1 \\ 0 & 1 \end{pmatrix}\right)^{-1} P(\zeta+1) + v\left(\begin{pmatrix} 1 & 0 \\ 1 & 1 \end{pmatrix}\right)^{-1} (\zeta+1)^{2\nu-1} P\left(\frac{\zeta}{\zeta+1}\right) \tag{5.1}$$

on $\mathbb{C}'$ and satisfies the growth condition

$$P(\zeta) = \begin{cases} O\left(z^{\max\{0, 2\Re(\nu)-1\}}\right) & \text{as } \Im(z) = 0,\ \zeta \searrow 0 \text{ and} \\ O\left(z^{\min\{0, 2\Re(\nu)-1\}}\right) & \text{as } \Im(z) = 0,\ \zeta \to \infty. \end{cases} \tag{5.2}$$

Remark 5.2

1. The complex cut-plane $\mathbb{C}'$ defined above was already introduced in (4.135).
2. The three-term equation (5.1) will appear again in (5.11), expressed in terms of the matrices $S \overset{(1.20)}{=} \begin{pmatrix} 1 & 1 \\ 0 & 1 \end{pmatrix}$ and $S' := \begin{pmatrix} 1 & 0 \\ 1 & 1 \end{pmatrix}$.

Remark 5.3 The three-term equation (5.1) above is related to the function equations (2.4) in the following sense: Assume that $k \in 2\mathbb{Z}$ and $v = 1$ is the trivial multiplier and assume that a function P satisfies (2.4). Then we have

$$\begin{aligned} P &\overset{(2.4)}{=} -P\big|_{2-k}(TS) - P\big|_{2-k}(TS)^2 \\ &= \left(-P\big|_{2-k}T\right)\big|_{2-k}S + \left(-P\big|_{2-k}T\right)\big|_{2-k}(STS) \\ &\overset{(2.4)}{=} P\big|_{2-k}S + P\big|_{2-k}(STS). \end{aligned}$$

Since $STS = \begin{pmatrix} 1 & 0 \\ 1 & 1 \end{pmatrix}$, we see that P satisfies the three-term equation (5.1) with $v = 1$ and $\nu = \frac{k-1}{2}$.

Exercise 5.4 Let $k \in 2\mathbb{Z}$, $v = 1$ be the trivial multiplier and put $\nu = \frac{k-1}{2}$. We assume that a function $P : \mathbb{C} \mapsto \mathbb{C}$ satisfies the three-term equation (5.1):

$$P = P\big|_{2-k} S + P\big|_{2-k} (STS). \tag{5.3}$$

Show that if P satisfies additionally

$$P + P\big|_{2-k} T = 0, \tag{5.4}$$

then P satisfies the equations given in (2.4).

Example 5.5 This example goes back to [122, III.1]. Let ν be a complex number. The function $p_1 : \mathbb{C}' \to \mathbb{C}$ given by

$$p_1(\zeta) := 1 - \zeta^{2\nu-1} \qquad (\zeta \in \mathbb{C}') \tag{5.5}$$

satisfies the three-term equation (5.1) with trivial multiplier $v = 1$.

Exercise 5.6 Show that p_1 given in (5.5) satisfies the three-term equation (5.1) with trivial multiplier $v = 1$. In other words, p_1 satisfies

$$p_1(\zeta) = p_1(\zeta + 1) + (\zeta + 1)^{2\nu-1}\, p_1\left(\frac{\zeta}{\zeta + 1}\right) \tag{5.6}$$

for all $\zeta \in \mathbb{C}'$.

Is p_1 a period function in the sense of Definition 5.1?

Remark 5.7 Lewis and Zagier call functions which satisfy the three-term equation (5.1) (for the trivial multiplier 1) and do not satisfy the growth condition (5.2) "periodlike functions" [122, III]. They give several examples in [122, III.1] and discuss several properties of those periodlike functions in the subsequent sections.

Exercise 5.8 We discuss another example that goes back to [122, III.1].

Let $\nu \in \mathbb{C}$ satisfy $\Re(\nu) < -\frac{1}{2}$. We define the function $p_2 : \mathbb{C}' \to \mathbb{C}$ by

$$\begin{aligned} p_2(\zeta) :&= \sum_{m,n \geq 0}{}^{\star} \frac{1}{(mz + n)^{1-2\nu}} \qquad (\zeta \in \mathbb{C}') \\ &= \sum_{m,n>0} \frac{1}{(mz + n)^{1-2\nu}} + \frac{1}{2} \sum_{m>0} \frac{1}{(mz)^{1-2\nu}} + \frac{1}{2} \sum_{n>0} \frac{1}{n^{1-2\nu}}, \end{aligned} \tag{5.7}$$

where the *star summation* $\sum^\star$ means that the term $m = n = 0$, i.e., the "corner term," is omitted and that the terms with either $m = 0$ or $n = 0$, i.e., the "edge terms," are only counted with multiplicity $\frac{1}{2}$. (The star summation was already used in [122, III.1].) Show that p_2 satisfies the three-term equation (5.1) with trivial multiplier $v = 1$.

5.1.1 Period Functions and Nearly Periodic Functions

Similar to [122, Proposition 2] and [151, Lemma 5.8], we continue to prove an algebraic correspondence between nearly periodic functions f in the sense of Definition 4.54 and a solution P of a suitable three-term equation (5.1) on $\mathbb{C} \setminus \mathbb{R}$. But first, we extend the slash operator introduced in Definition 1.168 slightly.

Definition 5.9 Let h be a meromorphic function on $\mathbb{C}$ (or a suitable subset) and let k be a real weight with a compatible multiplier system v for a group Γ. We define the *slash operator of weight k and multiplier v* by

$$\begin{aligned} h\big|_k^v M(z) : &= v(M)^{-1}\, h\big|_k M(z) \\ &\overset{(1.131)}{=} v(M)^{-1}\, j(M,z)^{-k}\, h(Mz) \end{aligned} \tag{5.8}$$

for all $M \in \Gamma$. We extend the slash operator to formal sums and linear combinations of elements in Γ as

$$h\big|_k^v\big(\alpha M + V\big) := \alpha\Big(h\big|_k^v M\Big) + \Big(h\big|_k^v V\Big) \tag{5.9}$$

for all $M, V \in \Gamma$ and $\alpha \in \mathbb{C}$, similar to (1.132).

Lemma 5.10 *Assume that k and v satisfy* $e^{\mp\pi i(2v-1)} \neq e^{\pi i k}$. *Put*

$$c_\pm^\star = 1 - e^{\pi i k} e^{\pm\pi i(2v-1)}. \tag{5.10}$$

Then, there exists a bijection between nearly periodic functions f satisfying (4.188) *and solutions P of the* three-term equation

$$P(\zeta) = v(S)^{-1} P(\zeta+1) + v(S')^{-1}(\zeta+1)^{2v-1} P\left(\frac{\zeta}{\zeta+1}\right) \tag{5.11}$$

i.e. $P\big|_{1-2v}^v(\mathbf{1} - S - S')(\zeta) = 0$ *for* $\zeta \in \mathbb{C} \setminus \mathbb{R}$.

Here, S and S′ are matrices given by

$$S \overset{(1.20)}{=} \begin{pmatrix} 1 & 1 \\ 0 & 1 \end{pmatrix} \quad \text{and} \quad S' := \begin{pmatrix} 1 & 0 \\ 1 & 1 \end{pmatrix} = STS. \tag{5.12}$$

The bijection is given by the formulas

$$\begin{aligned} c_{\pm}^{\star} f(\zeta) &= P(\zeta) + v(T)^{-1} \zeta^{2\nu-1} P(T\zeta) \qquad (\Im(\zeta) \gtrless 0) \\ &= P\big|_{1-2\nu}^{v} (\mathbf{1} + T)(\zeta) \end{aligned} \tag{5.13}$$

and

$$\begin{aligned} P(\zeta) &= f(\zeta) - v(T)^{-1} \zeta^{2\nu-1} f(T\zeta) \qquad (\zeta \in \mathbb{C} \smallsetminus \mathbb{R}) \\ &= f\big|_{1-2\nu}^{v} (\mathbf{1} - T)(\zeta). \end{aligned} \tag{5.14}$$

Remark 5.11 Observe that P, as defined in (5.14), satisfies the three-term functional equation (5.11), as the following formal calculation shows:

$$\begin{aligned} P\big|_{1-2\nu}^{v} &(\mathbf{1} - S - S') \\ &= f\big|_{1-2\nu}^{v} (\mathbf{1} - T)\big|_{1-2\nu}^{v} (\mathbf{1} - S - STS) \qquad (STS = S' \text{ by } (5.12)) \\ &= f\big|_{1-2\nu}^{v} (\mathbf{1} - T - S + TS - STS + S^{-1}T) \qquad (TSTS = S^{-1}T) \\ &= f\big|_{1-2\nu}^{v} \big((\mathbf{1} - S) + (S^{-1}T - T) + (TS - STS)\big) \\ &= f\big|_{1-2\nu}^{v} (\mathbf{1} - S)\big|_{1-2\nu}^{v} (\mathbf{1} + S^{-1}T + TS) \\ &= 0. \end{aligned}$$

However, the calculation is only formal, since the slash notation $h\big|_{1-2\nu}^{v} M$ defined in (5.8) just hides the weight factors and the multipliers. In general, we do not know whether they match since the slash notation is not a group action for arbitrary real weight k. We have to check them on each occasion. In particular, we have to check whether Mz is still in the set of definition of the function h.

Proof (of Lemma 5.10) We adapt the proof of [147, Lemma 34] to our situation.
Let $z \in \mathbb{C} \smallsetminus \mathbb{R}$. First, we compute

$$v(T)^{-1} v(T)^{-1} \zeta^{2\nu-1} \left(\frac{-1}{\zeta}\right)^{2\nu-1}.$$

We have

$$\begin{aligned} \zeta^{2\nu-1} \left(\frac{-1}{\zeta}\right)^{2\nu-1} &= e^{(2\nu-1)i\left(\arg\zeta + \arg\frac{-1}{\zeta}\right)} \\ &= e^{\pm\pi i(2\nu-1)} \qquad (\Im(\zeta) \gtrless 0) \end{aligned}$$

since $\arg\zeta + \arg -\frac{1}{\zeta} = \pm\pi$ for $\Im(\zeta) \gtrless 0$. The choices $+$ and $>$, respectively $-$ and $<$ correspond to each other. The consistency relation (1.31) for multipliers implies

$$v(T)\, v(T) = e^{-ik\pi},$$

see (1.37) of Exercise 1.48. Hence,

$$v(T)^{-1} v(T)^{-1}\, \zeta^{2\nu-1}\Big(\frac{-1}{\zeta}\Big)^{2\nu-1} = e^{\pi ik} e^{\pm\pi i(2\nu-1)}. \tag{5.15}$$

Next, we show that (5.13) and (5.14) are inverses of each other. On one hand, we have

$$\begin{aligned}
c_{\pm}^{\star} f(\zeta) &\overset{(5.13)}{=} P(\zeta) + v(T)^{-1}\zeta^{2\nu-1} P\Big(\frac{-1}{\zeta}\Big) \\
&\overset{(5.14)}{=} f(\zeta) - v(T)^{-1}\zeta^{2\nu-1} f\Big(\frac{-1}{\zeta}\Big) \\
&\qquad + v(T)^{-1}\zeta^{2\nu-1}\Big[f\Big(\frac{-1}{\zeta}\Big) - v(T)^{-1}\Big(\frac{-1}{\zeta}\Big)^{2\nu-1} f(\zeta)\Big] \\
&= f(\zeta)\Big[1 - v(T)^{-1} v(T)^{-1}\zeta^{2\nu-1}\Big(\frac{-1}{\zeta}\Big)^{2\nu-1}\Big] \\
&\overset{(5.15)}{=} f(\zeta)\big[1 - e^{\pi ik} e^{\pm\pi i(2\nu-1)}\big] \qquad (\text{for } \Im(\zeta) \gtrless 0).
\end{aligned}$$

On the other hand, we have

$$\begin{aligned}
c_{\pm}^{\star} P(\zeta) &\overset{(5.14)}{=} c_{\pm}^{\star}\Big[f(\zeta) - v(T)^{-1}\zeta^{2\nu-1} f\Big(\frac{-1}{\zeta}\Big)\Big] \\
&\overset{(5.13)}{=} P(\zeta) + v(T)^{-1}\zeta^{2\nu-1} P\Big(\frac{-1}{\zeta}\Big) - \\
&\qquad - v(T)^{-1}\zeta^{2\nu-1}\Big[P\Big(\frac{-1}{\zeta}\Big) + v(T)^{-1}\Big(\frac{-1}{\zeta}\Big)^{2\nu-1} P(\zeta)\Big] \\
&\overset{(5.15)}{=} P(\zeta)\big[1 - e^{\pi ik} e^{\pm\pi i(2\nu-1)}\big] \qquad (\text{for } \Im(\zeta) \gtrless 0),
\end{aligned}$$

using the same argument calculations as above.

We now show that f being nearly periodic corresponds to P satisfying the three-term equation. Let f be a nearly periodic function satisfying (4.188) and P function given by (5.14). Then, we find (for $\zeta \in \mathbb{H} \cup \mathbb{H}^-$)

$$
\begin{aligned}
& P\big|_{1-2\nu}^{v}(\mathbf{1}-S-S')(\zeta) \\
& \overset{(5.14)}{=} \Big(f(\zeta)-v(T)^{-1}\zeta^{2\nu-1}\,f(T\zeta)\Big)- \\
& \quad - v(S)^{-1}\Big(f(S\zeta)-v(T)^{-1}(S\zeta)^{2\nu-1}\,f(TS\zeta)\Big)- \\
& \quad - -v(S')^{-1}(\zeta+1)^{2\nu-1}\cdot \\
& \qquad \cdot\Big(f(S'\zeta)-v(T)^{-1}(S'\zeta)^{2\nu-1}\,f(TS'\zeta)\Big) \\
& = \Big(f(\zeta)-v(S)^{-1}\,f(S\zeta)\Big)+ \\
& \quad +\Big(v(S')^{-1}v(T)^{-1}\,(\zeta+1)^{2\nu-1}(S'\zeta)^{2\nu-1}\,f(TSTS\zeta)- \\
& \qquad -v(T)^{-1}\,\zeta^{2\nu-1}\,f(T\zeta)\Big)+ \\
& \quad +\Big(v(S)^{-1}v(T)^{-1}\,(S\zeta)^{2\nu-1}\,f(TS\zeta)- \\
& \qquad -v(S')^{-1}\,(\zeta+1)^{2\nu-1}\,f(STS\zeta)\Big)
\end{aligned}
$$

(f is nearly periodic and $TSTS = S^{-1}T$)

$$
\begin{aligned}
& = 0+v(T)^{-1}\,\zeta^{2\nu-1}\cdot \\
& \quad \cdot\Big(v(S')^{-1}\,\frac{1}{\zeta^{2\nu-1}}(\zeta+1)^{2\nu-1}(S'\zeta)^{2\nu-1}f(S^{-1}T\zeta)-f(T\zeta)\Big) \\
& \quad +v(S)v(S')^{-1}\,(\zeta+1)^{2\nu-1}\Big(v(S')v(S)^{-2}v(T)^{-1}f(TS\zeta) \\
& \qquad -v(S)^{-1}\,f(STS\zeta)\Big) \\
& = v(T)^{-1}\,\zeta^{2\nu-1}\Big(v(S)\,f(S^{-1}T\zeta)-f(T\zeta)\Big)+ \\
& \quad +v(S)v(S')^{-1}\,(\zeta+1)^{2\nu-1}\Big(f(TS\zeta)-v(S)^{-1}\,f(STS\zeta)\Big) \\
& = 0.
\end{aligned}
$$

We used the multiplier identities based on the consistency relation (1.31). Hence, P satisfies the three-term equation (5.11).

Conversely, let us assume that the function P satisfies the three-term equation (5.11) on $\mathbb{C} \smallsetminus \mathbb{R}$. We have to show that f attached by (5.13) is indeed nearly periodic. Applying the three-term equation to P in ζ and $TS\,\zeta = \frac{-1}{\zeta+1}$, we obtain:

$$
\begin{aligned}
0 &= \left(P|^{v}_{1-2\nu}\left[-\mathbf{1} + S + S'\right]\right)|^{v}_{1-2\nu}[\mathbf{1} - TS](\zeta)\\
&= \left[-P(z) + v(S)^{-1}\,P(\zeta+1) + v(S')^{-1}\,(\zeta+1)^{2\nu-1}\,P\big(\frac{\zeta}{\zeta+1}\big)\right]\\
&\quad - v(TS)^{-1}\,(\zeta+1)^{2\nu-1}\left[-P\big(\frac{-1}{\zeta+1}\big) + v(S)^{-1}P\big(\frac{\zeta}{\zeta+1}\big)\right.\\
&\qquad \left. + v(S')^{-1}\left(\frac{-1}{\zeta+1}+1\right)^{2\nu-1} P\left(\frac{\frac{-1}{\zeta+1}}{\frac{-1}{\zeta+1}+1}\right)\right]\\
&= -\left[P(\zeta) + v(TS)^{-1}v(S')^{-1}\,(\zeta+1)^{2\nu-1}\left(\frac{\zeta}{\zeta+1}\right)^{2\nu-1} P\big(\frac{-1}{\zeta}\big)\right]\\
&\quad + \left[v(S)^{-1}P(\zeta+1) + v(Ts)^{-1}\,(\zeta+1)^{2\nu-1}P\big(\frac{-1}{\zeta+1}\big)\right]\\
&\quad + \left[v(S')^{-1}\,(\zeta+1)^{2\nu-1}P(S'\zeta) - \right.\\
&\qquad \left. v(TS)^{-1}v(S)^{-1}\,(\zeta+1)^{2\nu-1}P\big(\frac{\zeta}{\zeta+1}\big)\right]\\
&= -\left[P(\zeta) + v(T)^{-1}\,z^{2\nu-1}\,P\big(\frac{-1}{\zeta}\big)\right]\\
&\quad + v(S)^{-1}\left[P(S\zeta) + v(T)^{-1}\,(S\zeta)^{2\nu-1}P\big(\frac{-1}{S\zeta}\big)\right] \quad +0\\
&= -c^\star_\pm f(\zeta) \;+\; v(S)^{-1}\,c^\star_\pm f(\zeta+1)\\
&= c^\star_\pm\left(v(S)^{-1}\,f(\zeta+1) - f(\zeta)\right) \qquad\qquad (\text{for } \Im(\zeta) \gtrless 0),
\end{aligned}
$$

using again the multiplier identities derived from the consistency relation (1.31). This shows that if P satisfies the three-term equation, then f is nearly periodic. □

What happens if we apply Lemma 5.10 to a nearly periodic function f satisfying the conditions of part 4 of Theorem 4.55?

Let $\nu \in \mathbb{C}$ and $\kappa \in \mathbb{R}$ satisfying $|\Re(\nu)| < \frac{1}{2}$ and $\kappa \pm (1-2\nu) \neq 2\mathbb{Z}$. Assume that $f : \mathbb{C}\smallsetminus\mathbb{R} \to \mathbb{C}$ is a nearly periodic function satisfying part (4) of Theorem 4.55. We have in particular that f is a holomorphic function f on $\mathbb{C} \smallsetminus \mathbb{R}$, satisfies $f(z+1) = e^{2\pi i\kappa} f(z)$ for all $z \in \mathbb{C} \smallsetminus \mathbb{R}$, and is bounded by a multiple of $|\Im(z)|^{-C}$ for some $C > 0$, such that the function $f(z) - e^{\frac{\pi i\kappa}{2}} z^{2\nu-1} f\left(-\frac{1}{z}\right)$ extends holomorphically across the right half-plane and is bounded by a multiple of $\min\left\{1, |z|^{2\Re(\nu)-1}\right\}$ in the right half-plane. This means in particular that the function P associated by (5.14),

$$P(\zeta) \overset{(5.14)}{=} f(\zeta) - v(T)^{-1}\zeta^{2\nu-1} f(T\zeta) \overset{(1.20)}{\underset{(1.37)}{=}} f(\zeta) - e^{\frac{\pi ik}{2}} \zeta^{2\nu-1} f\left(\frac{-1}{\zeta}\right)$$

extends holomorphically to the cut-plane $\mathbb{C}'$ (where $\mathbb{C}' = \mathbb{C} \smallsetminus \mathbb{R}_{\leq 0}$ is already defined in (4.135)). Moreover P satisfies the growth estimate

$$P(\zeta) = O\left(\min\left\{1, |\zeta|^{2\Re(\nu)-1}\right\}\right) = \begin{cases} \zeta^{2\nu-1} & \text{for } \Re(\zeta) \to \infty \text{ and} \\ 1 & \text{for } 0 < \Re(\zeta) \searrow 0. \end{cases}$$

Hence P is a period function in the sense of Definition 5.1.

This argument already shows one direction of the following result.

Theorem 5.12 *Let $\nu \in \mathbb{C}$ and $\kappa \in \mathbb{R}$ satisfying $|\Re(\nu)| < \frac{1}{2}$ and $\kappa \pm (1-2\nu) \neq 2\mathbb{Z}$. There are natural bijective correspondences between the following:*

1. *nearly periodic functions f on $\mathbb{C} \smallsetminus \mathbb{R}$ satisfying the conditions of part 4 of Theorem 4.55,*
2. *period functions P in the sense of Definition 5.1.*

The bijections are as described by the formulas (5.13) *and* (5.14) *in Lemma 5.10.*

Proof

(1) $\Longrightarrow$ (2) This direction was already shown in the arguments presented a priori. We have seen that the period function P is given by f with the help of (5.13).

(2) $\Longrightarrow$ (1) We use (5.13) to define the function f on $\mathbb{C} \smallsetminus \mathbb{R}$. Lemma 5.10 implies that f is nearly periodic with factor $e^{2\pi i\kappa}$ Moreover f is holomorphic on $\mathbb{C} \smallsetminus \mathbb{R}$ since P is holomorphic there, and $f(z) - e^{\frac{\pi i\kappa}{2}} z^{2\nu-1} f\left(-\frac{1}{z}\right)$ extends holomorphically across the right half-plane and is bounded by a multiple of $\min\left\{1, |z|^{2\Re(\nu)-1}\right\}$ in the right half-plane, since P does this.

□

5.1.2 Period Functions via Integral Transformations

We base this part on [151, §6.1], which is an extension of [122, Chapter II, §2]. Let us define the following integral transformation of a Maass cusp form.

Definition 5.13 Let $\zeta \in (0, \infty)$, $\nu \in \mathbb{C}$ a spectral parameter, and $k \in \mathbb{R}$ a real weight with a compatible multiplier v. Let u be a Maass cusp form of weight k, multiplier v, and eigenvalue $\frac{1}{4} - \nu^2$ for the full modular group $\Gamma(1)$ in the sense of Definition 3.31.

We associate a function $P_{k,\nu} : (0, \infty) \to \mathbb{C}; \zeta \mapsto P_{k,\nu}(\zeta)$ with the cusp form u by the integral transform

$$P_{k,\nu}(\zeta) = \int_0^{i\infty} \eta_{-k}\big(R_{-k,\nu}(\cdot, \zeta), u\big)(z), \tag{5.16}$$

where the path of integration is the upper imaginary axis, i.e., the geodesic connecting 0 and $i\infty$.

Lemma 5.14 *The function $P_{k,\nu}$ in* (5.16) *is well defined.*

Proof Let $\zeta \in (0, \infty)$ and consider the function $R_{-k,\nu}(z, \zeta)$ discussed in §4.3.3.1.

On one hand, we conclude from the definition of $R_{-k}(\cdot, \zeta)$ in (4.163) that $z \mapsto R_{-k}(z, \zeta)$ admits at most polynomial growth for $\Im(z) \to \infty$ and $\Im(z) \searrow 0$. On the other hand, the Maass cusp form u decays quicker than every polynomial at the cusps 0 and ∞, see (3.53) in Theorem 3.38. Since the Maass-Selberg form $\eta_k(\cdot, \cdot)$ defined in (4.176) does not change this growth behavior, we see that the integral $\int_0^{i\infty} \eta_{-k}\big(R_{-k,\nu}(\cdot, \zeta), u\big)(z)$ is well defined along the geodesic integration path connecting 0 and $i\infty$. □

Remark 5.15 The notion of $P_{k,\nu}$ appeared nearly verbatim in [147]. However, the definition of $P_{k,\nu}$ in [147, Definition 41] is

$$P_{k,\nu}(\zeta) = \int_0^{i\infty} \eta_k\big(u, R_{-k,\nu}(\cdot, \zeta)\big)(z)$$

which seems to differ from the one we use in (5.16).

However, u and $R_{-k,\nu}$ are eigenfunctions of Δ_k and Δ_{-k} respectively. This implies that the Maass-Selberg form is closed, see Lemma 4.86. We have

$$\begin{aligned}\int_0^{i\infty} &\eta_{-k}\big(R_{-k,\nu}(\cdot, \zeta), u\big)(z) \\ &= \int_0^{i\infty} \eta_k\big(u, R_{-k,\nu}(\cdot, \zeta)\big)(z) + \int_0^{i\infty} \mathrm{d}\big(u(\cdot)\, R_{-k,\nu}(\cdot, \zeta)\big).\end{aligned}$$

Due to u being cuspidal, and hence vanishing in 0 and $i\infty$, we have

$$\int_0^{i\infty} \mathrm{d}\big(u(\cdot)\, R_{-k,\nu}(\cdot,\zeta)\big) = 0.$$

Hence, the definitions of $P_{k,\nu}$ in (5.16) and in [147, Definition 41] agree. This also shows that the choice mentioned in Remark 4.91 does not matter for the period functions.

Let $X \subset \mathbb{C} \cup \{\infty\}$ be a set and $M \in \mathrm{SL}_2(\mathbb{R})$ be a matrix. We extend the Möbius transformation to act on sets in the following way: We define

$$M\,X := \big\{Mz\colon\ z \in X\big\}. \tag{5.17}$$

Lemma 5.16 *Let k, v, ν, and u be as in Definition 5.13, let $\zeta \in (0,\infty)$ and $M \in \Gamma(1)$ such that $j(M,\zeta) > 0$ and $M(0,\infty) \subset (0,\infty)$. The function $P_{k,\nu}$ defined in* (5.16) *satisfies*

$$\int_{M^{-1}0}^{M^{-1}\infty} \eta_{-k}\big(R_{-k,\nu}(\cdot,\zeta), u\big)(z) = \big(P_{k,\nu}\big|^{v}_{1-2\nu} M\big)(\zeta), \tag{5.18}$$

where the path of integration is the geodesic connecting $M^{-1}0$ and $M^{-1}\infty$.

Proof We have

$$\begin{aligned}
\big(P_{k,\nu}\big|^{v}_{1-2\nu} M\big)(\zeta) &\overset{(5.8)}{=} v(M)^{-1}\, j(M,\zeta)^{2\nu-1}\, P_{k,\nu}(M\zeta)\\
&\overset{(5.16)}{=} v(M)^{-1}\, j(M,\zeta)^{2\nu-1} \int_0^{\infty} \eta_{-k}\big(R_{-k,\nu}(\cdot, M\zeta), u\big)(z)\\
&\overset{(4.182)}{=} \int_0^{\infty} \eta_{-k}\big(R_{-k,\nu}(\cdot,\zeta), u\big)(M^{-1}z) \qquad \text{using Lemma 4.90}\\
&\overset{M^{-1}z\mapsto z}{=} \int_{M^{-1}0}^{M^{-1}\infty} \eta_{-k}\big(R_{-k,\nu}(\cdot,\zeta), u\big)(z),
\end{aligned}$$

where we use the substitution $M^{-1}z \mapsto z$ for the last line. The use of Lemma 4.90 is valid since ζ and $j(M,\zeta)$ are both positive. The path of integration of the last integral is the geodesic connecting the cusps $M^{-1}0$ and $M^{-1}\infty$ and lies in the upper left quadrant $\{z \in \mathbb{C}\colon\ \Re(z) \le 0, \Im(z) \ge 0\}$ of $\mathbb{C}$. □

We show next that $P_{k,\nu}$ satisfies the *three-term equation* on $\mathbb{R}_+$.

Lemma 5.17 *Let ν, k, and v as in Definition 5.13 and u a Maass cusp form with weight k compatible multiplier v and eigenvalue $\frac{1}{4} - \nu^2$ for the full modular group $\Gamma(1)$. The function $P_{k,\nu}$ defined in* (5.16) *satisfies the* three-term equation

$$0 = P_{k,\nu}\big|^{v}_{1-2\nu}\big(\mathbf{1} - S - S'\big) \qquad \textit{on } (0,\infty). \tag{5.19}$$

Proof Let $\zeta > 0$. Recall the matrices S and S' in (5.12). We have

$$S^{-1}0 = -1, \quad S^{-1}\infty = \infty, \quad \left(S'\right)^{-1}0 = 0, \quad \text{nd} \quad \left(S'\right)^{-1}\infty = -1.$$

Lemma 5.16 allows us to write

$$\begin{aligned} 0 &= \left(\int_0^\infty - \int_{-1}^\infty - \int_0^{-1}\right)\eta_{-k}\big(R_{-k,\nu}(\cdot,\zeta),u\big)(z) \\ &\overset{(5.18)}{=} P_{k,\nu}(\zeta) - v(S)^{-1}\,P_{k,\nu}(S\zeta) - v(S')^{-1}(\zeta+1)^{2\nu-1}\,P_{k,\nu}(S'\zeta) \\ &= P_{k,\nu}\big\|^{v}_{1-2\nu}(\mathbf{1}-S-S')(\zeta). \end{aligned}$$

This proves the lemma. □

Remark 5.18 We have seen three versions of the three-term equation in (5.1), (5.11), and (5.19). The difference between those three versions lies in the domains on which the functions are defined. For example, in (5.1) we require P to fulfill the $P\big|^{v}_{1-2\nu}(\mathbf{1}-S-S') = 0$ on $\mathbb{C}'$, whereas (5.11) requires it only on $\mathbb{C} \setminus \mathbb{R}$ and (5.19) on $(0,\infty)$.

The next step is to extend $P_{k,\nu}(\zeta)$ to the right half-plane $\{\zeta \in \mathbb{C};\, \Re(\zeta) > 0\}$. Let ζ be in the right half-plane and recall that $R_{-k,\nu}(z,\zeta)$ is holomorphic in ζ if $\Re(z) \le 0$. Hence, the function $P_{k,\nu}(\zeta)$ given by the integral transform (5.16) extends holomorphically to $\{\zeta \in \mathbb{C};\, \Re(\zeta) > 0\}$. It is easily checked that $P_{k,\nu}(\zeta+1)$ and $P_{k,\nu}\left(\frac{\zeta}{\zeta+1}\right)$ have also holomorphic extensions to this right half-plane.

The last step is to extend $P_{k,\nu}$ to the cut plane $\mathbb{C}' = \mathbb{C} \setminus (-\infty, 0]$. Assume $\Re(\zeta) > 0$. Since the differential form $\eta_k\big(u, R_{-k,\nu}(\cdot,\zeta)\big)$ is closed, see Lemma 4.86, we replace vertical path of integration in (5.16) by a path that connects 0 and $i\infty$ in the upper left quadrant and which passes to the left of either ζ or $\bar{\zeta}$. We then may move ζ to any point for which either ζ or $\bar{\zeta}$ is still right of the new integration path. This procedure extends $P_{k,\nu}$ to a holomorphic function on $\mathbb{C}'$.

Summarizing we have the following.

Theorem 5.19 *Under the assumptions of Definition 5.13, the function $P_{k,\nu}$ associated with the Maass cusp form u by* (5.16) *extends to a holomorphic function on the cut plane $\mathbb{C}'$ that satisfies the three-term equation* (5.1) *on $\mathbb{C}' = \mathbb{C} \setminus (-\infty, 0]$. It also satisfies the growth conditions*

$$P_{k,\nu}(\zeta) = \begin{cases} O\left(z^{\max\{0,2\Re(\nu)-1\}}\right) & \text{as } \Im(z) = 0,\ \zeta \searrow 0 \text{ and} \\ O\left(z^{\min\{0,2\Re(\nu)-1\}}\right) & \text{as } \Im(z) = 0,\ \zeta \to \infty. \end{cases} \tag{5.20}$$

In other words, P is a period function in the sense of Definition 5.1.

Proof The first part of the proposition follows from the discussion above.

The cusp form u is bounded on $\mathbb{H}$ since a cusp form vanishes at all cusps $\mathbb{Q} \cup i\infty$ and u is real-analytic on $\mathbb{H}$. Also, $\mathrm{E}_k^+ u$ is bounded since the Maass operator maps cusp forms of weight k to cusp forms of weight $k+2$. Applying successively (5.16), (4.176), (4.171), (4.166), and (4.165), we find

$$\begin{aligned}
&P_{k,\nu}(\zeta) \\
&\overset{(5.16)}{=} \int_0^{i\infty} \eta_{-k}\big(R_{-k,\nu}(\cdot,\zeta), u\big)(z) \\
&\overset{\substack{(4.176)\\(4.171)}}{=} \int_0^{i\infty} \left[\big(\mathrm{E}_{-k}^+ R_{-k,\nu}(\cdot,\zeta)\big)(z)\, u(z)\, \frac{\mathrm{d}z}{y} \;-\; R_{-k,\nu}(z,\zeta)\, \big(\mathrm{E}_k^- u\big)(z)\, \frac{\mathrm{d}\bar{z}}{y} \right] \\
&\overset{(4.166)}{=} \int_0^{i\infty} \left[(1-2\nu-k)\, R_{2-k,\nu}(z,\zeta)\, u(z)\, \frac{\mathrm{d}z}{y} - R_{-k,\nu}(z,\zeta)\, \big(\mathrm{E}_k^- u\big)(z)\, \frac{\mathrm{d}\bar{z}}{y} \right] \\
&\overset{(4.165)}{=} i \int_0^{\infty} e^{ik\arg\zeta - iy} \left(\frac{y}{(\zeta - iy)(\zeta+iy)} \right)^{\frac{1}{2}-\nu} \\
&\qquad \left[(1-2\nu-k)\, e^{2i\arg\zeta - iy} u(iy) - \big(\mathrm{E}_k^- u\big)(iy) \right] \frac{\mathrm{d}y}{y}
\end{aligned}$$

for $\zeta > 0$. Using the notation $f(z) \ll g(z)$ for $f(z) = O(g(z))$, we find the estimate

$$\big|P_{k,\nu}(\zeta)\big| \ll \int_0^{\infty} \left| \frac{y}{\zeta^2+y^2} \right|^{\frac{1}{2}-\Re(\nu)} \cdot \max\left\{ \big|\big(\mathrm{E}_k^- u\big)(iy)\big|, \big|(1-2\nu-k)\, u(iy)\big| \right\} \frac{\mathrm{d}y}{y} \tag{5.21}$$

for $\zeta > 0$. The integral converges since u and hence $u(iy)$ and $\big(E_k^- u\big)(iy)$ decay quickly as $y \to \infty$ and as $y \searrow 0$.

Using the estimate

$$\frac{y}{\zeta^2+y^2} \le \zeta^{-2}\, y$$

in (5.21) gives

$$P_{k,\nu}(\zeta) = O\left(\zeta^{2\Re(\nu)-1}\right) \qquad \text{for every } \zeta > 0.$$

We have

$$P_{k,\nu}(\zeta) = O(1) \qquad \text{for every } \zeta > 0$$

if we use

$$\frac{y}{\zeta^2+y^2}\le y^{-1}$$

in (5.18). This proves the stated growth condition. □

5.1.3 Period Functions via Nearly Periodic Functions Given by an Integral Transform

Let us start with a Maass cusp form u of weight k, multiplier v, and spectral value ν as in Definition 3.31. We associated in §4.3.3 a nearly periodic function f by the integral transform (4.183):

$$\mathbb{C}\smallsetminus\mathbb{R}\to\mathbb{C};$$

$$\zeta\mapsto f(\zeta):=\begin{cases}\displaystyle\int_\zeta^{i\infty}\eta_{-k}\big(R_{-k,\nu}(\cdot,\zeta),u\big)(z) & \text{if } \zeta\in\mathbb{H} \text{ and}\\ \displaystyle -\int_\zeta^{-i\infty}\eta_{k}\big(R_{-k,\nu}(\cdot,\zeta),\tilde{u}\big)(z) & \text{if } \zeta\in\mathbb{H}^-.\end{cases} \tag{4.183}$$

Then, we attached a period-function P by (5.14):

$$P=f\big|_{1-2\nu}^{v}(\mathbf{1}-T)\qquad (\text{on } \mathbb{H}\cup\mathbb{H}^-) \tag{5.14}$$

which satisfies the three-term equation

$$0=P\big|_{1-2\nu}^{v}(\mathbf{1}-S-S')\qquad (\text{on } \mathbb{H}\cup\mathbb{H}^-). \tag{5.11}$$

On the other hand, we have the integral transformation (5.16) from the Maass cusp form u to the period function $P_{k,\nu}$:

$$P_{k,\nu}(\zeta)=\int_0^{i\infty}\eta_{-k}\big(R_{-k,\nu}(\cdot,\zeta),u\big)(z)\qquad (\text{on } \mathbb{R}_{>0}) \tag{5.16}$$

which satisfies the three-term equation

$$0=P_{k,\nu}\big|_{1-2\nu}^{v}(\mathbf{1}-S-S')\qquad (\text{on } \mathbb{R}_{>0}) \tag{5.19}$$

and extends to $\mathbb{C}'$ (Theorem 5.19).

Are both directions compatible? In other words, do we get the same function P on $\mathbb{H}\cup\mathbb{H}^-$, regardless of using the intermediate periodic function via (4.183) and (5.14) or taking the formula (5.16)?

Lemma 5.20 *Let k, v, ν, and u be as in Definition 5.13 with $|\Re(\nu)| < \frac{1}{2}$. The maps*

$$u \overset{(4.183)}{\longmapsto} f \overset{(5.14)}{\longmapsto} P \quad \text{and} \quad u \overset{(5.16)}{\longmapsto} P_{k,\nu}$$

give rise to the same function $P = P_{k,\nu}$ on $\{\zeta \in \mathbb{C}\colon\ \Re(\zeta) > 0,\ \Im(\zeta) \neq 0\}$.

Proof For $\zeta \in \mathbb{H}$ with $\Re(\zeta) > 0$ in the upper half-plane, we find

$$\begin{aligned}
P_{k,\nu}(\zeta) &\overset{(5.16)}{=} \int_0^{i\infty} \eta_{-k}\big(R_{-k,\nu}(\cdot,\zeta), u\big)(z) \\
&= \int_\zeta^{i\infty} \eta_{-k}\big(R_{-k,\nu}(\cdot,\zeta), u\big)(z) + \int_0^{\zeta} \eta_{-k}\big(R_{-k,\nu}(\cdot,\zeta), u\big)(z) \\
&= \int_\zeta^{i\infty} \eta_{-k}\big(R_{-k,\nu}(\cdot,\zeta), u\big)(z) - \int_\zeta^{S^{-1}i\infty} \eta_{-k}\big(R_{-k,\nu}(\cdot,\zeta), u\big)(z) \\
&\overset{(4.187)}{=} f(\zeta) - v(S)^{-1}\,\zeta^{2\nu-1}\, f(S\,\zeta) \qquad \text{using Lemma 4.95} \\
&\overset{(5.14)}{=} P(\zeta).
\end{aligned}$$

A similar calculation holds for $\zeta \in \mathbb{H}^-$ with $\Re(\zeta) > 0$:

$$\begin{aligned}
P_{k,\nu}(\zeta) &\overset{(5.16)}{=} \int_0^{i\infty} \eta_{-k}\big(R_{-k,\nu}(\cdot,\zeta), u\big)(z) \\
&\overset{(4.177)}{=} -\int_0^{i\infty} \eta_{k}\big(u, R_{-k,\nu}(\cdot,\zeta)\big)(z) \qquad \text{using Lemma 4.86} \\
&= -\int_{\overline{\zeta}}^{i\infty} \eta_{k}\big(u, R_{-k,\nu}(\cdot,\zeta)\big)(z) - \int_0^{\overline{\zeta}} \eta_{k}\big(u, R_{-k,\nu}(\cdot,\zeta)\big)(z) \\
&= -\int_{\overline{\zeta}}^{i\infty} \eta_{k}\big(u, R_{-k,\nu}(\cdot,\zeta)\big)(z) + \int_{\overline{\zeta}}^{S^{-1}i\infty} \eta_{k}\big(u, R_{-k,\nu}(\cdot,\zeta)\big)(z) \\
&\overset{(4.187)}{=} f(\zeta) - v(S)^{-1}\,\zeta^{2\nu-1}\, f(S\,\zeta) \qquad \text{using Lemma 4.95} \\
&\overset{(5.14)}{=} P(\zeta).
\end{aligned}$$

□

Theorem 5.21 *The function P given by* (5.14) *on $\mathbb{H} \cup \mathbb{H}^-$ is holomorphic, extends holomorphically to the cut-plane $\mathbb{C}' = \mathbb{C} \setminus (-\infty, 0]$, satisfies the three-term equation* (5.1) *on $\mathbb{C}'$, and satisfies the growth condition* (5.20).

Proof Lemma 5.20 shows that P agrees on $\{\zeta \in \mathbb{C}: \Re(\zeta) > 0, \Im(\zeta) \neq 0\}$ with $P_{k,\nu}$ given by (5.16). The latter extends holomorphically to $\mathbb{C}'$ and satisfies the growth condition (5.20) by Theorem 5.19. □

5.1.4 Period Functions and Families of Maass Cusp Forms

On one hand, we have the period function P associated with families of Maass cusp forms via Theorem 4.55 and Theorem 5.12; on the other hand, we have period functions $P_{k,\nu}$ for individual cusp forms given by Definition 5.13. How are these two period functions related?

Lemma 5.22 *Let ν be a complex number with $|\Re(\nu)| < \frac{1}{2}$ and $k \in \mathbb{R}$ such that $k \pm (1-2\nu) \notin 2\mathbb{Z}$ for both choices of $\pm$. Take an admissible family of Maass cusp forms $(u_l)_{l\in k+2\mathbb{Z}}$ as discussed in §4.1 and denote by $P_{l,\nu}$ the period function associated with each individual Maass cusp forms u_l as defined in* (5.16).

If $l \in k+2\mathbb{Z}$ with l and $l-2$ are in the range of admissible weights, then

$$P_{l-2,\nu} = -P_{l,\nu}. \tag{5.22}$$

Proof For $f = R_{-l,\nu}(\cdot,\zeta)$ and $g = u_l$, the left and right sides of Equation (4.180) in Lemma 4.86 simplify to

$$\begin{aligned}
&-(1-2\nu-l)(1+2\nu-l)\,\eta_{2-l}\big(R_{2-l,\nu}(\cdot,\zeta),u_{l-2}\big)(z)\\
&\overset{(4.166)}{\underset{(4.5)}{=}}\ \eta_{2-l}\big(\mathrm{E}^+_{-l}R_{-l,\nu}(\cdot,\zeta),\mathrm{E}^-_l u_l\big)(z)\\
&\overset{(4.180)}{=}\ (1+2\nu-l)(1-2\nu-l)\,\eta_{-l}\big(R_{-l,\nu}(\cdot,\zeta),u_l\big)(z)\\
&\qquad + 4i\mathrm{d}\Big(\big(\mathrm{E}^+_{-l}R_{-l,\nu}(\cdot,\zeta)\big)\big(\mathrm{E}^-_l u_l\big)\Big)\\
&\overset{(4.166)}{\underset{(4.5)}{=}}\ (1+2\nu-l)(1-2\nu-l)\,\eta_{-l}\big(R_{-l,\nu}(\cdot,\zeta),u_l\big)(z)\\
&\qquad - 4i(1-2\nu-l)(1+2\nu-l)\,\mathrm{d}\big(R_{2-l,\nu}(\cdot,\zeta)\,u_{l-2}\big).
\end{aligned}$$

Since the family $(u_l)_l$ is admissible, we see that the factor

$$(1-2\nu-l)(1+2\nu-l) \neq 0$$

does not vanish in the range of admissible weights. We have

$$\eta_{2-l}\big(R_{2-l,\nu}(\cdot,\zeta),u_{l-2}\big)(z) = -\eta_{-l}\big(R_{-l,\nu}(\cdot,\zeta),u_l\big)(z) + 4i\,\mathrm{d}\big(R_{2-l,\nu}(\cdot,\zeta)\,u_{l-2}\big).$$

Applying this identity to (5.16) in the Definition 5.13 of the period function $P_{l,\nu}$, we find

$$\begin{aligned} P_{l-2,\nu}(\zeta) &\overset{(5.16)}{=} \int_0^\infty \big(R_{2-l,\nu}(\cdot,\zeta), u_{l-2}\big)(z) \\ &= 4i \int_0^\infty \mathrm{d}\big(R_{2-l,\nu}(\cdot,\zeta)\, u_{l-2}\big) - \int_0^\infty \eta_{-l}\big(R_{-l,\nu}(\cdot,\zeta), u_l\big)(z) \\ &\overset{(5.16)}{=} -P_{l,\nu}(\zeta). \end{aligned}$$

Here, we used that the exponential decay of the Maass cusp forms u_l and u_{l-2}, see estimate (3.53) in Theorem 3.38, and that at most polynomial growth of the R-function, see the argument in the proof of Lemma 5.14, to calculate

$$\int_0^\infty \mathrm{d}\big(R_{2-l,\nu}(\cdot,\zeta)\, u_{l-2}\big) = 0.$$

This proves the lemma. □

Lemma 5.23 *Let $k \in \mathbb{R}$, $l \in k + 2\mathbb{Z}$, $\nu \in \mathbb{C}$ with $\Re(\nu) < \frac{1}{2}$ such that $2\nu + k$ and $2\nu - k$ are not both in $1 + 2\mathbb{Z}$. Assume that $\big(u_l\big)_{l\in k+2\mathbb{Z}}$ is an admissible family of Maass cusp forms with k and $k-2$ in the admissible range of weights and denote by $P_{k,\nu}$ the period function associated with u_k by* (5.16)*. Then, the Mellin transform*

$$\begin{aligned} \tilde{P}_{k,\nu}(s) &:= \mathcal{M}\big(P_{k,\nu}\big)(s) \\ &\overset{(1.154)}{=} \int_0^\infty P_{k,\nu}(\zeta)\, \zeta^{s-1}\, \mathrm{d}\zeta \end{aligned} \tag{5.23}$$

exists for $0 < \Re(s) < 1 - 2\Re(\nu)$. It satisfies

$$\begin{aligned} \tilde{P}_{k,\nu}(s) = {} & i(1-2\nu-k)\, B_{2-k,-\nu}\left(s+\nu-\frac{1}{2}\right) L_k^\star\left(s+\nu-\frac{1}{2}\right) - \\ & - i(1+2\nu-k)\, B_{-k,-\nu}\left(s+\nu-\frac{1}{2}\right) L_{k-2}^\star\left(s+\nu-\frac{1}{2}\right). \end{aligned} \tag{5.24}$$

Remark 5.24

1. The functions $L_{k-2}^\star$ and $L_k^\star$ are given by (4.55) in Theorem 4.28.
2. The condition "$2\nu + k$ and $2\nu - k$ are not both in $1 + 2\mathbb{Z}$" is needed for the existence of $L_{k-2}^\star$ and $L_k^\star$ in Theorem 4.28.

Proof *(of Lemma 5.23)* The growth estimates in (5.20) of Theorem 5.19 implies that the Mellin transform $\tilde{P}(s)$ is well defined for $0 < \Re(s) < 1 - 2\Re(\nu)$. Calculating the Mellin transform, we have

$$
\begin{aligned}
&\tilde{P}_{k,\nu}(s)\\
&\overset{(5.23)}{=} \int_0^\infty P_{k,\nu}(\zeta)\,\zeta^{s-1}\,\mathrm{d}\zeta\\
&\overset{(5.16)}{=} \int_0^\infty \int_0^{i\infty} \eta_{-k}\big(R_{-k,\nu}(\cdot,\zeta),u_k\big)(z)\,\zeta^{s-1}\mathrm{d}\zeta\\
&\overset{(4.176)}{=} \int_0^\infty \int_0^{i\infty} \Big(\big\{\mathrm{E}^+_{-k}R_{-k,\nu}(\cdot,\zeta),u_k\big\}^+(z)\\
&\qquad\qquad - \big\{R_{-k,\nu}(\cdot,\zeta),\mathrm{E}^-ku_k\big\}^-(z)\Big)\,\zeta^{s-1}\mathrm{d}\zeta\\
&\overset{(4.171)}{=} \int_0^\infty \int_0^{i\infty} \big(\mathrm{E}^+_{-k}R_{-k,\nu}(z,\zeta)\big)\,u_k(z)\,\frac{\mathrm{d}z}{y}\\
&\qquad\qquad - \int_0^{i\infty} R_{-k,\nu}(z,\zeta)\,\big(\mathrm{E}^-_k u_k(z)\big)\,\frac{\mathrm{d}\bar{z}}{y}\,\zeta^{s-1}\mathrm{d}\zeta\\
&\overset{\substack{(4.5)\\(4.166)}}{=} \int_0^\infty (1-2\nu-k)\int_0^{i\infty} R_{2-k,\nu}(z,\zeta)\,u_k(z)\,\frac{\mathrm{d}z}{y}\\
&\qquad\qquad + (1+2\nu-k)\int_0^{i\infty} R_{-k,\nu}(z,\zeta)\,u_{k-2}(z)\,\frac{\mathrm{d}\bar{z}}{y}\,\zeta^{s-1}\,\mathrm{d}\zeta\\
&\overset{z\mapsto iy}{=} i(1-2\nu-k)\int_0^\infty\int_0^\infty R_{2-k,\nu}(iy,\zeta)\,u_k(iy)\,\frac{\mathrm{d}y}{y}\\
&\qquad\qquad - i(1+2\nu-k)\int_0^\infty R_{-k,\nu}(iy,\zeta)\,u_{k-2}(iy)\,\frac{\mathrm{d}y}{y}\,\zeta^{s-1}\,\mathrm{d}\zeta.
\end{aligned}
\tag{5.25}
$$

Applying (4.163) and interchanging the integrals in the first integral in (5.25) give

$$
\begin{aligned}
&\int_0^\infty\int_0^\infty R_{2-k,\nu}(iy,\zeta)\,u_k(iy)\,\frac{\mathrm{d}y}{y}\,\zeta^{s-1}\,\mathrm{d}\zeta\\
&= \int_0^\infty u_k(iy)\left(\int_0^\infty R_{2-k,\nu}(iy,\zeta)\,\zeta^{s-1}\,\mathrm{d}\zeta\right)\frac{\mathrm{d}y}{y}\\
&\overset{(4.163)}{=} \int_0^\infty u_k(iy)\,y^{-1}\left(\int_0^\infty\left(\frac{\sqrt{\zeta-iy}}{\sqrt{\zeta+iy}}\right)^{k-2}\left(\frac{y}{(\zeta-iy)(\zeta+iy)}\right)^{\frac{1}{2}-\nu}\zeta^{s-1}\,\mathrm{d}\zeta\right)\mathrm{d}y\\
&= \int_0^\infty u_k(iy)\,y^{-\frac{1}{2}-\nu}\left(\int_0^\infty(\zeta-iy)^{\nu-\frac{1}{2}+\frac{k-2}{2}}\,(\zeta+iy)^{\nu-\frac{1}{2}-\frac{k-2}{2}}\,\zeta^{s-1}\,\mathrm{d}\zeta\right)\mathrm{d}y\\
&\overset{\zeta=y\tau}{=} \int_0^\infty u_k(iy)\,y^{-\frac{1}{2}-\nu}\\
&\qquad\cdot\left(\int_0^\infty(y\tau-iy)^{\nu-\frac{1}{2}+\frac{k-2}{2}}\,(y\tau+iy)^{\nu-\frac{1}{2}-\frac{k-2}{2}}\,(y\tau)^{s-1}\,y\,\mathrm{d}\tau\right)\mathrm{d}y
\end{aligned}
$$

$$= \int_0^\infty (\tau - i)^{\nu-\frac{1}{2}+\frac{k-2}{2}} (\tau + i)^{\nu-\frac{1}{2}-\frac{k-2}{2}} \tau^{s-1} \,\mathrm{d}\tau \int_0^\infty u_k(iy)\, y^{s-\frac{3}{2}+\nu} \,\mathrm{d}y$$

$$\overset{(4.74)}{=} B_{2-k,-\nu}\left(s+\nu-\frac{1}{2}\right) \int_0^\infty u_k(iy)\, y^{s+\nu-\frac{3}{2}} \mathrm{d}y$$

$$\overset{(4.57)}{=} B_{2-k,-\nu}\left(s+\nu-\frac{1}{2}\right) L_k^\star\left(s+\nu-\frac{1}{2}\right).$$

Similarly, we calculate the second integral in (5.25) as

$$\begin{aligned}\int_0^\infty \int_0^\infty & R_{-k,\nu}(iy,\zeta)\, u_{k-2}(iy) \,\frac{\mathrm{d}y}{y}\, \zeta^{s-1} \,\mathrm{d}\zeta \\ &= B_{-k,-\nu}\left(s+\nu-\frac{1}{2}\right) L_{k-2}^\star\left(s+\nu-\frac{1}{2}\right).\end{aligned} \tag{5.26}$$

Summarizing, we have

$$\begin{aligned}&\tilde{P}_{k,\nu}(s) \\ &\overset{(5.25)}{=} i(1-2\nu-k) \int_0^\infty \int_0^\infty R_{2-k,\nu}(iy,\zeta)\, u_k(iy)\, \frac{\mathrm{d}y}{y}\, \zeta^{s-1}\,\mathrm{d}\zeta \\ &\quad - i(1+2\nu-k) \int_0^\infty R_{-k,\nu}(iy,\zeta)\, u_{k-2}(iy)\, \frac{\mathrm{d}y}{y}\, \zeta^{s-1}\,\mathrm{d}\zeta \\ &= i(1-2\nu-k)\, B_{2-k,-\nu}\left(s+\nu-\frac{1}{2}\right) L_k^\star\left(s+\nu-\frac{1}{2}\right) \\ &\quad - i(1+2\nu-k)\, B_{-k,-\nu}\left(s+\nu-\frac{1}{2}\right) L_{k-2}^\star\left(s+\nu-\frac{1}{2}\right).\end{aligned}$$

This proves the lemma. □

Exercise 5.25 Show the identity (5.26) in the proof of Lemma 5.23.

Theorem 5.26 *Let $k \in \mathbb{R}$, $\nu \in \mathbb{C}$ with $|\Re(\nu)| < \frac{1}{2}$ and such that $2\nu + k$ and $2\nu - k$ are both not in $1 + 2\mathbb{Z}$. Assume that $\left(u_l\right)_{l \in k+2\mathbb{Z}}$ is an admissible family of Maass cusp forms with k and $k-2$ in the admissible range of weights and denote by $P_{k,\nu}$ the period function associated with u_k by* (5.16).

The function $f : \mathbb{C} \smallsetminus \mathbb{R} \to \mathbb{C}$ is the nearly periodic function defined in (4.113) *with factor $e^{2\pi i\kappa}$ associated with $(u_k)_k$ by Theorem 4.55.*

We have

$$f(z) - e^{\frac{\pi i k}{2}} z^{2\nu-1} f\left(-\frac{1}{z}\right) = i(-1)^{\frac{k-\kappa}{2}} e^{\frac{\pi i \kappa}{2}} 2^{-\frac{3}{2}} \pi^\nu P_{k,\nu}(z) \tag{5.27}$$

for all $z \in \mathbb{C} \smallsetminus \mathbb{R}$.

Proof Let $f : \mathbb{C} \setminus \mathbb{R}$ be the nearly periodic function defined in (4.113) associated with a system of cusp forms $(u_l)_l$ via Theorem 4.55. In particular, f has the property that the function

$$F(z) \overset{(4.134)}{=} f(z) - e^{\frac{\pi i \kappa}{2}} z^{2\nu-1} f\left(-\frac{1}{z}\right) \qquad (z \in \mathbb{C} \setminus \mathbb{R}) \tag{5.28}$$

extends holomorphically across the right half-plane and is bounded by a multiple of $\min(1, |z|^{2\Re(\nu)-1})$ in the right half-plane, see Lemma 4.68. We have in particular

$$\int_0^\infty F(z)\, z^{s-1}\, dz \overset{(4.136)}{=} A^{\pm}_{\kappa,\nu}(s)$$

for all $0 < \Re(s) < 1 - 2\Re(\nu)$, where $A^{\pm}_{\kappa,\nu}$ is defined in (4.120) of Lemma 4.63.

To prove the theorem, we show that $P_{k,\nu}$ and F have essentially the same Mellin transform. They differ only by a multiplicative constant.

Lemma 5.23 gives the Mellin transform $\tilde{P}_{k,\nu}$ of $P_{k,\nu}$. We have

$$\begin{aligned}\int_0^\infty P(z)\, z^{s-1}\, dz &\overset{(5.23)}{=} \tilde{P}_{k,\nu}(s)\\ &\overset{(5.24)}{=} i(1-2\nu-k)\, B_{2-k,-\nu}\left(s+\nu-\frac{1}{2}\right) L_k^\star\left(s+\nu-\frac{1}{2}\right)\\ &\quad - i(1+2\nu-k)\, B_{-k,-\nu}\left(s+\nu-\frac{1}{2}\right) L_{k-2}^\star\left(s+\nu-\frac{1}{2}\right).\end{aligned}$$

Combining Lemma 4.65 with Lemma 4.68 in §4.3.1 gives the Mellin transform of F as

$$\begin{aligned}\int_0^\infty F(z)\, z^{s-1}\, dz &\overset{(4.136)}{=} A^{\pm}_{\kappa,\nu}(s)\\ &\overset{(4.125)}{=} -2^{-\frac{3}{2}}\pi^{\nu} e^{\frac{\pi i}{2}k}\\ &\quad \cdot \Big[\left(1+2\nu-k\right) B_{-k,-\nu}\left(s+\nu-\frac{1}{2}\right) L_{k-2}^\star\left(s+\nu-\frac{1}{2}\right)\\ &\quad - \left(1-2\nu-k\right) B_{2-k,-\nu}\left(s+\nu-\frac{1}{2}\right) L_k^\star\left(s+\nu-\frac{1}{2}\right)\Big].\end{aligned}$$

Both expressions are equal up to a nonzero multiplicative scalar. We find explicitly that

$$\int_0^\infty F(z)\, z^{s-1}\, dz = i^{-1}\, 2^{-\frac{3}{2}}\pi^{\nu} e^{\frac{\pi i}{2}k} \int_0^\infty P(z)\, z^{s-1}\, dz \tag{5.29}$$

for $0 < \Re(s) < 1 - 2\Re(\nu)$ and all $k-2$ and k in the admissible range.

The inverse Mellin transform is well defined on the terms above, since

$$\frac{1}{2\pi i}\int_{\sigma-i\infty}^{\sigma+i\infty} A^{\pm}_{\kappa,\nu}(s)\, z^{-s}\mathrm{d}s \qquad (0<\sigma<1-2\Re(\nu))$$

is well defined for all $z\in\mathbb{R}_{>0}$, as can be seen by Proposition 1.204 and the growth condition (4.121) of $A^{\pm}_{\kappa,\nu}(s)$. Calculating the inverse Mellin transform on both sides of (5.29), we find

$$F(z)=i^{-1}\,2^{-\frac{3}{2}}\pi^{\nu}e^{\frac{\pi i k}{2}}\;P_{k,\nu}(z)$$

for all $z\in\mathbb{R}_{>0}$. Replacing F by the expression in (5.28), we show (5.27) of Theorem 5.26 for $\Re(z)>0$, $\Im(z)\neq 0$. Both sides of equation in (5.27) extend holomorphically to $z\in\mathbb{C}\smallsetminus\mathbb{R}$. □

Corollary 5.27 *In the situation of Theorem 5.26, let $P:\mathbb{C}'\to\mathbb{C}$ be the period function associated with the system $(u_l)_l$ via Theorem 4.55.*

The functions P and $P_{l,\nu}$, $l\in k+2\mathbb{Z}$ in the admissible range are scalar multiples of each other.

Proof Let $l\in k+2\mathbb{Z}$ be in the admissible range. We have that the nearly periodic function f in Theorem 5.26 and the period function P are related to each other by (5.14) of Lemma 5.10 in §5.1.1. For $\Re(z)>0$, the right side in (5.14) is the function F in Theorem 5.26. The corollary follows. □

5.1.5 Period Functions and Period Polynomials

Consider a classical cusp form and its associated period polynomial via (2.1) in §2.1.1. We can also embed the classical cusp form into the space of Maass cusp forms by (3.56) in §3.4.1 and then attach a period function via the integral transformation (5.16). How are these two functions, the period polynomial and the period function, related?

Let F be a modular cusp form of even weight $k\in 2\mathbb{N}$ as introduced in Definition 1.59. Let $u_F:\mathbb{H}\to\mathbb{C}$ be the Maass cusp form associated with F by (3.56):

$$u_F(z):=\Im(z)^{\frac{k}{2}}\;F(z). \qquad (3.56)$$

As shown in §3.4.1, u is indeed a Maass cusp form of weight k, trivial multiplier $v\equiv 1$, and eigenvalue $\frac{k}{2}\left(1-\frac{k}{2}\right)$. Hence, u has spectral values $\nu\in\left\{\frac{k-1}{2},\frac{1-k}{2}\right\}$.

The following proposition compares the period functions attached to u_F and the period polynomial attached to F.

Proposition 5.28 *Let F be a modular cusp form of even weight $k \in 2\mathbb{N}$, trivial multiplier for the full modular group $\Gamma(1)$ as introduced in Definition 1.59 and let u_F be the F associated with Maass cusp form by embedding* (3.56).

1. *The function $P_{k,\frac{1-k}{2}}$ associated with u_F by* (5.16) *vanishes everywhere.*
2. *The function $P_{k,\frac{k-1}{2}}$ associated with u_F by* (5.16) *restricted to the right half-plane $\{\zeta \in \mathbb{C}\colon \Re(\zeta) > 0\}$ is a multiple of the period polynomial P associated with F by* (2.1)*:*

$$P_{k,\frac{k-1}{2}} = (2-2k)\,P. \tag{5.30}$$

Proof Since

$$\left(\frac{k-1}{2}\right)\left(\frac{1-k}{2}\right) = \frac{k}{2}\left(1-\frac{k}{2}\right),$$

we see that $\frac{k-1}{2}$ and $\frac{1-k}{2}$ are spectral values of u_F. Moreover, $\mathrm{E}_k^- u_F = 0$ as shown in Exercise 3.13.

Let $P_{k,\nu}$ be the period function associated with u via (5.16):

$$P_{k,\nu}(\zeta) \overset{(5.16)}{=} \int_0^{i\infty} \eta_{-k}(R_{-k,\nu}(\cdot,\zeta),\, u_F)(z).$$

Using (4.176) and (4.171), we find

$$P_{k,\nu}(\zeta) = \int_0^{i\infty} \left[\left(E_{-k}^{+} R_{-k,\nu}(\cdot,\zeta)\right)(z)\, u_F(z)\frac{\mathrm{d}z}{y} - R_{-k,\nu}(z,\zeta)\,\left(E_k^- u_F\right)(z)\frac{\mathrm{d}\bar z}{y}\right].$$

Recalling that $\mathrm{E}_{-k}^{+} R_{-k,\nu} = (1-2\nu-k)R_{2-k,\nu}$ in (4.166) and $\mathrm{E}_k^- u_F = 0$ shown above, we find

$$P_{k,\nu}(\zeta) = (1-2\nu-k)\int_0^{i\infty} R_{2-k,\nu}(z,\zeta)\, u_F(z)\frac{\mathrm{d}z}{y} \tag{5.31}$$

for $\Re(\zeta) > 0$.

To prove the first part of the Proposition, we assume $\nu = \frac{1-k}{2}$. Then, the factor $1-2\nu-k = 1-2\frac{1-k}{2}-k = 0$ in (5.31) vanishes, implying $P_{k,\frac{1-k}{2}} = 0$.

To prove the second part of the Proposition, we assume $\nu = \frac{k-1}{2}$. By (4.163) we have

$$\begin{aligned} R_{2-k,\frac{k-1}{2}}(z,\zeta) &= \left(\frac{\sqrt{\zeta-z}}{\sqrt{\zeta-\bar z}}\right)^{k-2}\left(\frac{|\Im(z)|}{(\zeta-z)(\zeta-\bar z)}\right)^{\frac{2-k}{2}} \\ &= (\zeta-z)^{k-2}\,|\Im(z)|^{\frac{2-k}{2}} \end{aligned} \tag{5.32}$$

for every ζ and z with $\zeta - z, \zeta - \bar{z} \neq \mathbb{R}_{\leq 0}$. Combining this with (3.56) in (5.31) gives

$$\begin{aligned}
P_{k,\frac{k-1}{2}}(\zeta) &\overset{(5.31)}{=} \left(1 - 2\frac{k-1}{2} - k\right) \int_0^{i\infty} R_{2-k,\frac{k-1}{2}}(z,\zeta)\, u_F(z) \frac{\mathrm{d}z}{y} \\
&\overset{(5.32)}{=} (2-2k) \int_0^{i\infty} (\zeta - z)^{k-2}\, |\Im(z)|^{\frac{2-k}{2}}\, u_F(z)\, \frac{\mathrm{d}z}{y} \\
&\overset{(3.56)}{=} (2-2k) \int_0^{i\infty} (\zeta - z)^{k-2}\, F(z)\, \mathrm{d}z \\
&\overset{(2.1)}{=} (2-2k)\, P(\zeta)
\end{aligned}$$

for at least all ζ in the right half-plane. □

5.1.6 Some Remarks on Period Functions

We would like to mention a few things at the end of §5.1.

The subsections §5.1.1–§5.1.4 above present several ways to attach period functions to Maass wave forms using different intermediate steps. In the end, we show that all different approaches towards period functions are compatible in the following sense: if we start with an admissible family of Maass cusp forms $(u_l)_l$ we get the same period function (up to scalar multiple) using the different calculation schemes above, see, e.g., Corollary 5.27.

We see the analog to the classical setting of modular cusp forms and period polynomials where we also have several different calculation schemes to associate a period polynomial with modular cusp forms, see, e.g., Section 2.1. It is even possible to calculate a period polynomial using the indirect way of embedding the modular cusp form to a Maass cusp form and then calculate the associate period function. This period function is in fact the original period polynomial (up to a scalar multiple) as shown in §5.1.5 above.

Since all presented calculation schemes are consistent with each other we may use the scheme which fits best the intended application. For example we will calculate representations of Hecke operators that act on period functions using the integral transformation approach in §5.1.2.

Recall for example our discussion of the Eichler cohomology group associated with period polynomials in §2.2 and the related Eichler cohomology theorem discussed in §2.3. One may wonder if some equivalent concept exists for period functions? This was positively answered in [36]. The authors of [36] show how a cohomology group somehow similar to $H^1\big(\Gamma(1), \mathbb{C}[X]_{k-2}\big)$ defined in (2.41) are associated with the space of Maass forms for arbitrary Fuchsian groups. The methods rely on methods developed in [35] by the same authors where they deal with the representation theory of the group $PSL_2(\mathbb{R})$ and describe several models of the principal series. These two publications continue the discussion of period functions as introduced by [122]. A brief introduction to Zagier's cohomology theorem in [36] is given in the following Section 5.2.

5.2 A Visit to the Semi-Analytic Cohomology Group

The Eichler cohomology group was introduced in Definition 2.40 in Section 2.2. Their associated space of cocycles and coboundaries are also introduced in Definition 2.36. One key ingredient in the classical modular forms case was the space of polynomials $\mathbb{C}[X]_{k-2}$ in which the values P_M of a cocycle $\{P_M\}_{M\in\Gamma(1)}$ lived. Following Bruggeman, Lewis, and Zagier in [36], we extend this space in the context of Maass cusp forms. This leads to the line model [36, §2.1].

5.2.1 *The Spaces* $\mathcal{V}_\nu^\infty$, $\mathcal{V}_\nu^\omega$, *and* $\mathcal{V}_\nu^{\omega^*,\infty}$

Let $\nu \in \mathbb{C}$ with $|\Re(\nu)| < \frac{1}{2}$ be a purely imaginary number.

Following [36, §2.1], we introduce the space $\mathcal{V}_\nu$ as the function space consisting of functions $\phi : \mathbb{R} \to \mathbb{C}$ such that $\lim_{|x|\to\infty}\phi(x)$ exists. In other words, we require

$$\lim_{x\to-\infty}\phi(x) = \lim_{x\to+\infty}\phi(x).$$

Loosely speaking, we can assign ϕ a function value at $x = \infty$.

The subspace $\mathcal{V}_\nu^\infty \subset \mathcal{V}_\nu$ contains all smooth functions $\phi \in \mathcal{V}_\nu$ that also behave well at ∞. Here "behaves well" means that $\phi \in \mathcal{V}_\nu^\infty$ admits an asymptotic expansion

$$\phi(x) = |t|^{1-2\nu} \sum_{n=0}^{\infty} c_n\, x^{-n} \tag{5.33}$$

as $|t| \to \infty$.

Finally, we define the subspace $\mathcal{V}_\nu^\omega \subset \mathcal{V}_\nu$ as the space of all analytic functions $\phi \in \mathcal{V}_\nu^\infty$ that satisfy additionally that the asymptotic expansion in (5.33) converges absolutely near ∞. In other words, ϕ is analytic and admits an absolute convergent expansion of the form (5.33) for all $|x| > r_0$ for some $r_0 > 0$.

On each of the above three function spaces $\mathcal{V}_\nu$, $\mathcal{V}_\nu^\infty$, and $\mathcal{V}_\nu^\omega$ operate the following modified *slash operator*

$$\phi\tilde{\Big|}_\nu \begin{pmatrix} a & b \\ c & d \end{pmatrix}(x) := |cx+d|^{2\nu-1}\,\phi\left(\frac{ax+b}{cx+d}\right) \tag{5.34}$$

for all $x \in \mathbb{R}$, $x \neq -\frac{d}{c}$ and $\begin{pmatrix} a & b \\ c & d \end{pmatrix} \in \Gamma(1)$. The action of the slash operator at $x = -\frac{d}{c}$ and $x = \infty$ is also well defined in the sense that the limit towards both values is well defined. This shows that the full modular group acts on the three function spaces mentioned above.

We introduce another function space $\mathcal{V}_\nu^{\omega^*,\infty} \subset \mathcal{V}_\nu^\infty$ as the space of smooth functions $\phi \in \mathcal{V}_\nu^\infty$ that are additionally analytic except at finitely many points. Again, the full modular group $\Gamma(1)$ acts on $\mathcal{V}_\nu^{\omega^*,\infty} \subset \mathcal{V}_\nu^\infty$ using the slash operator in (5.34).

5.2.2 *Mixed Parabolic Cohomology Group* $H^1_{par}\left(\Gamma(1), \mathcal{V}_\nu^\omega, \mathcal{V}_\nu^{\omega^*,\infty}\right)$

Recall the definition of the parabolic cohomology group in Definition 2.41 in Section 2.2. Summarizing, the parabolic cohomology group $H^1_{\text{par}}\left(\Gamma(1), \mathbb{C}[X]_{k-2}\right)$ is given by parabolic cocycles $\left(\phi_M\right)_{M\in\Gamma(1)}$ that satisfy

$$P_S = Q\big|_{2-k} S - Q$$

for some polynomial $Q \in \mathbb{C}[X]_{k-2}$. This identification is unique up to a coboundary.

We extend Definition 2.41 of the mixed parabolic cohomology group in the following way: Let $V \supset W$ be two spaces. The mixed parabolic cohomology group $H^1_{\text{par}}(\Gamma(1), V, W)$ consists of parabolic cocycle $\left(P_M\right)_{M\in\Gamma(1)}$ with entries $P_M \in V$ for all $M \in \Gamma(1)$. The "coboundary" Q, such that $P_S = Q\big|\tilde{}_\nu S - Q$ holds for the parabolic element S, is an element in the larger space $W \supset V$. The cocycles are unique up to coboundaries.

Following the discussion in §2.2, we define mixed parabolic cocycles with values in $\mathcal{V}_\nu^\omega$ as follows.

Definition 5.29 Let $\nu \in \mathbb{C}$ with $|\Re(\nu)| < \frac{1}{2}$ and $\left(P_M\right)_{M\in\Gamma(1)}$ a collection of functions $P_M \in \mathcal{V}_\nu^\omega$.

1. We call the collection $\left(P_M\right)_{M\in\Gamma(1)}$ a *cocycle* if its elements satisfy

$$P_{MV} = P_M\tilde{\big|}_\nu V + P_V \qquad \text{for all } M, V \in \Gamma(1). \tag{5.35}$$

2. We call the collection $\left(P_M\right)_{M\in\Gamma(1)}$ a *coboundary* if there exists a fixed function $P \in c\mathcal{V}_\nu^\omega$ such that $P_M = P\tilde{\big|}_\mu M - P$ holds for all $M \in \Gamma(1)$.

3. We call the collection $\left(P_M\right)_{M\in\Gamma(1)}$ a *mixed parabolic cocycle* if there exists a fixed function $P \in c\mathcal{V}_\nu^{\omega^*,\infty}$ such that $P_M = P\big|_\mu^{\sim} M - P$ holds for all parabolic $M \in \Gamma(1)$.

The sets of such cocycles, coboundaries, and mixed parabolic cocycles are denoted by $Z^1\left(\Gamma(1), \mathcal{V}_\nu^{\omega}\right)$, $B^1\left(\Gamma(1), \mathcal{V}_\nu^{\omega}\right)$, and $Z^1_{\text{par}}\left(\Gamma(1), \mathcal{V}_\nu^{\omega}, \mathcal{V}_\nu^{\omega^*,\infty}\right)$ respectively.

Definition 5.30 Let $V \subset W$ be two function spaces on which the slash operator $\tilde{\phi}$ operates in the sense of §5.2.1. We define the *mixed parabolic cohomology group* as the quotient group

$$H^1_{\text{par}}\left(\Gamma(1), \mathcal{V}_\nu^{\omega}, \mathcal{V}_\nu^{\omega^*,\infty}\right) = \frac{Z^1_{\text{par}}\left(\Gamma(1), \mathcal{V}_\nu^{\omega}, \mathcal{V}_\nu^{\omega^*,\infty}\right)}{B^1\left(\Gamma(1), \mathcal{V}_\nu^{\omega}\right)} \tag{5.36}$$

where $Z^1_{\text{par}}\left(\Gamma(1), \mathcal{V}_\nu^{\omega}, \mathcal{V}_\nu^{\omega^*,\infty}\right)$ and $B^1\left(\Gamma(1), \mathcal{V}_\nu^{\omega}\right)$ denote the set of mixed parabolic cocycles and the set of coboundaries in the sense of Definition 5.30.

Remark 5.31 Why do we need mixed parabolic cocycles? The Eichler cohomology in Section 2.2 only considers polynomials. Isn't that enough?

Recall the Eichler cohomology Theorem 2.50 in §2.3. We saw that the cocycles are essentially given by the period polynomials associated with modular cusp forms, see for example the maps in (2.1) in Section 2.1.1 and in (2.53) of Corollary 2.54.

Now, consider period functions that are associated with Maass cusp forms. Take a Maass cusp form u of weight 0, trivial multiplier ($v = 1$), and spectral value ν. The associated function $P_{0,\nu} : (0, \infty) \to \mathbb{C}$ given in (5.16) in Section 5.1.2 satisfies the three-term equation

$$P_{0,\nu}(\zeta) = P_{0,\nu}(\zeta+1) + (\zeta+1)^{2\nu-1}\, P_{0,\nu}\left(\frac{\zeta}{\zeta+1}\right) \tag{5.37}$$

for all $\zeta \in (0, \infty)$, see Lemma 5.17. The authors of [36] realized that this is in fact a cocycle condition, since $P_{0,\nu}$ can be written as

$$P_{0,\nu}(\zeta) = h(\zeta) - \zeta^{2\nu-1}\, h\left(\frac{-1}{\zeta}\right) \tag{5.38}$$

for some real-analytic function $h : \mathbb{R} \to \mathbb{C}$, i.e., $h \in \mathcal{V}_\nu^{\omega}$. We point out that the function h is not unique. The associated cocycle can be given by the mapping

$$P_T = 0 \quad \text{and} \quad P_S(\zeta) = h(\zeta+1) - h(\zeta).$$

We refer to [36, (3)–(5) and Chapter IV] for more details.

This discussion motivates the use of the space $\mathcal{V}_\nu^{\omega}$. Reformulating everything into a parabolic cocycle makes the use of the space $\mathcal{V}_\nu^{\omega^*,\infty}$ necessary.

5.2.3 Zagier's Cohomology Theorem

We now state some other interesting results from [36].

Theorem 5.32 ([36, Theorem B]) *Let $\nu \in \mathbb{C}$, $|\Re(\nu)| < \frac{1}{2}$ be a spectral value. The space $M_{0,1}^{Maass}\big(\Gamma(1), \nu\big)$ of Maass wave forms of weight 0 and trivial multiplier $v = 1$ for the full modular group $\Gamma(1)$ is canonically isomorphic to the mixed parabolic cohomology $H^1_{par}\big(\Gamma(1), \mathcal{V}^{\omega}_{\nu}, \mathcal{V}^{\omega^*,\infty}_{\nu}\big)$.*

Remark 5.33 We now state some of the interesting results in [36] besides the above-mentioned theorem.

1. In [36], the authors establish Theorem 5.32 for all cofinite discrete subgroups Γ of the full modular group $\mathrm{SL}_2(\mathbb{R})$. It was also shown that the parabolic cohomology groups $H^1_{\mathrm{par}}\big(\Gamma(1), \mathcal{V}^{\omega^*,\infty}_{\nu}\big)$ and $H^1_{\mathrm{par}}\big(\Gamma(1), \mathcal{V}^{\infty}_{\nu}\big)$ are canonically isomorphic to the space of Maass cusp forms $M_{0,1}^{\mathrm{Maass}}\big(\Gamma(1), \nu\big)$.
2. [36, Theorem A] establishes a canonical isomorphism between "generalized Maass wave forms" of real weight 0 and trivial multiplier $v = 1$ that are functions u satisfying conditions (1)–(3) of Definition 3.31, with cohomology groups $H^1\big(\Gamma(1), \mathcal{V}^{\infty}_{\nu}\big)$ and $H^1\big(\Gamma(1), \mathcal{V}^{\infty}_{\nu}\big)$ for cocompact subgroups Γ of $\mathrm{SL}_2(\mathbb{R})$.

Remark 5.34 The authors establish an isomorphism between the Maass wave forms and the cohomology groups by studying a map similar to the map (4.183). They consider

$$r_M(t) = \int_{M^{-1}z_0}^{z_0} \eta_0\left(u, \left(R_{0,-\frac{1}{2}}(\cdot, t)\right)^{\frac{1}{2}-\nu}\right),$$

where the function R is discussed in §4.3.3.1 and the differential form $\eta_0(\cdot, \cdot)$ is introduced in §4.3.3.2. The relation to the notation used in [122] and similarly in [36] is pointed out in Remarks 4.78 and 4.87.

Remark 5.35 There exists also the recent monograph [37] from Bruggeman, Choie, and Diamantis, where the authors discuss the Theorem 5.32 further and discuss also complex weights.

For example, take a cofinite discrete subgroups Γ of the full modular group $\mathrm{SL}_2(\mathbb{R})$ with cusps and let k be a complex weight with multiplier system v (in the sense of a suitable extended version of Definition 1.44). The authors consider the space $A_k(\Gamma, v)$ which denotes space of holomorphic functions satisfying (1.28) and the map

$$A_k(\Gamma, v) \longrightarrow H^1_{\mathrm{par}}\big(\Gamma, *, *\big)$$

to the associated mixed parabolic cohomology group.

5.3 Hecke Operators on Period Functions

Recall §2.4 where we introduced Hecke operators acting on period polynomials. Can we construct Hecke operators in the context of period functions? One approach would be the extension of Choie and Zagier's approach [57], see also Theorem 2.63, to period functions.

Recall Definition 5.1 of period functions; in particular recall that period functions are defined on the cup-plane $\mathbb{C}' = \mathbb{C} \smallsetminus (-\infty, 0]$. We cannot apply Theorem 2.63 directly to period functions since some matrices in the matrix identity (2.69) of Theorem 2.63 do not map the cut-plane into itself. The matrix identity (2.69) has to be suitably modified to work on period functions. This work was done in [148], where its author presents a modified version of Choie and Zagier's result in [57] (i.e., a modified version of Theorem 2.63) suitable in particular in the context of period functions.

There exists an alternative approach for Hecke operators on period polynomials using the integral representations, as we mentioned in Remark 2.76. This alternative approach seems possible since we just discussed period functions given by an integral transform of Maass cusp forms (of weight 0) in §5.1.2. We already have

1. Hecke operators acting on Maass cusp forms, which are described by the action of the set $\mathcal{T}(n)$ of upper triangle matrices with determinant n, see (1.139) and (3.81), and
2. the integral transform (5.16), which calculates the period function.

The main question is how to move the Hecke operator on the Maass cusp form u (of weight 0) into one involving the period function $P_{0,\nu}$. In other words and informally speaking, for each $n \in \mathbb{N}$ find an operator $H_n^\star$ acting on period functions such that for all Maass cusp forms $u \in S_{0,1}^{\mathrm{Maass}}\big(\Gamma(1), \nu\big)$ we have:

$$P_{0,\nu}(\zeta) \overset{(5.16)}{=} \int_0^{i\infty} \eta_0\big(R_{0,\nu}(\cdot, \zeta), u\big)(z)$$

implies

$$H_n^\star P_{0,\nu}(\zeta) \overset{!}{=} \int_0^{i\infty} \eta_0\big(R_{0,\nu}(\cdot, \zeta), T_n u\big)(z).$$

The diagram in Figure 5.1 commutes.

We approach the above question using the following structure: In §5.3.1, we introduce briefly Farey sequences with some properties. Based on these Farey sequences, we construct a left neighbor map and a left neighbor sequence in §5.3.2 which has the property that the fractions in the left neighbor sequence are related by $\mathrm{SL}_2(\mathbb{Z})$-matrices. In §5.3.4, we calculate a representation of the Hecke operators which act on period functions. By doing so we answer the above motivating question.

$$\begin{array}{ccc} u & \longrightarrow & P_{0,\nu} \\ \downarrow & & \downarrow \\ T_n\, u & \longrightarrow & H_n^\star\, P_{0,\nu} \end{array}$$

Fig. 5.1 The operator $H_n^\star$ acts on period functions $P_{0,\nu}$ as the Hecke operator T_n on Maass cusp forms u.

5.3.1 On Farey Sequences

Let us recall the theory of Farey sequences. If we try to describe Farey sequences in one sentence, we could say the following: The Farey sequence of order n is the ordered sequence of completely reduced fractions between 0 and 1 where the reduced fractions have denominators $\leq n$. The concept dates back to at least the end of the 18th century since most of the properties mentioned below can be found in [97].

We adhere to the convention of denoting infinity in rational form as $\infty = \frac{1}{0}$ and $-\infty = \frac{-1}{0}$ and denoting rational numbers $\frac{p}{q} \in \mathbb{Q}$ with coprime numerator $p \in \mathbb{Z}$ and denominator $q \in \mathbb{N}$. This is quite a common convention in the context of Farey sequences.

Definition 5.36 Let $n \in \mathbb{N}$. The *Farey sequence* F_n of *level* n is the finite sequence

$$F_n := \left(\frac{u}{v} \colon\; u, v \in \mathbb{Z},\, |u| \leq n,\, 0 \leq v \leq n\right). \tag{5.39}$$

ordered by the standard order $<$ of $\mathbb{R}$. We define F_0 as

$$F_0 := \left(\frac{-1}{0}, \frac{0}{1}, \frac{1}{0}\right). \tag{5.40}$$

The *level function* $\mathrm{lev} : \mathbb{Q} \to \mathbb{Z}$ is defined by

$$\mathrm{lev}\left(\frac{a}{b}\right) := \begin{cases} 0 & \text{if } \frac{a}{b} \in \left\{\frac{-1}{0}, \frac{0}{1}, \frac{1}{0}\right\} \text{ and} \\ \max\left\{\, |a|\,, |b|\,\right\} & \text{otherwise.} \end{cases} \tag{5.41}$$

Example 5.37 The Farey sequences of level ≤ 3 are

$$F_0 = \left(\frac{-1}{0}, \frac{0}{1}, \frac{1}{0}\right)$$

$$F_1 = \left(\frac{-1}{0}, \frac{-1}{1}, \frac{0}{1}, \frac{1}{1}, \frac{1}{0}\right)$$

$$F_2 = \left(\frac{-1}{0}, \frac{-2}{1}, \frac{-1}{1}, \frac{-1}{2}, \frac{0}{1}, \frac{1}{2}, \frac{1}{1}, \frac{2}{1}, \frac{1}{0}\right)$$

$$F_3 = \left(\frac{-1}{0}, \frac{-3}{1}, \frac{-2}{1}, \frac{-3}{2}, \frac{-1}{1}, \frac{-2}{3}, \frac{-1}{2}, \frac{-1}{3}, \frac{0}{1}, \frac{1}{3}, \frac{1}{2}, \frac{2}{3}, \frac{1}{1}, \frac{3}{2}, \frac{2}{1}, \frac{3}{0}, \frac{1}{0}\right).$$

Exercise 5.38 Show the following statement:

$$\frac{u}{v} \in F_n \quad \Longrightarrow \quad \frac{u}{v} \in F_{n+1}$$

for all $n \in \mathbb{Z}_{\geq 0}$.

Remark 5.39 Sometimes, the Farey sequences discussed in the literature are the F_n as defined above but with the extra restriction that the fractions are between 0 and 1.

Remark 5.40 Let $\frac{a}{c}$ and $\frac{b}{d}$ be two neighbors in the Farey sequence F_n. Hurwitz showed that the matrix $\begin{pmatrix} a & b \\ c & d \end{pmatrix}$ satisfies $\det \begin{pmatrix} a & b \\ c & d \end{pmatrix} = \pm 1$, see [97, Satz 1].

Lemma 5.41 *Let $\frac{a}{c}$ and $\frac{b}{d}$ be two neighbors of the Farey sequence F_n. We have*

$$\det \begin{pmatrix} a & b \\ c & d \end{pmatrix} = -1 \quad \Longleftrightarrow \quad \frac{a}{c} < \frac{b}{d}. \tag{5.42}$$

Proof For $c = 0$ (resp. $d = 0$), we have $\frac{-1}{0} < \frac{-n}{1}$ (resp. $\frac{n}{1} < \frac{1}{0}$) and $\det \begin{pmatrix} -1 & -n \\ 0 & 1 \end{pmatrix} = -1$ (resp. $\det \begin{pmatrix} n & 1 \\ 1 & 0 \end{pmatrix} = -1$).

Assume that $c, d > 0$. Since $\frac{a}{c} < \frac{b}{d}$ is equivalent to $\frac{ad}{cd} < \frac{bc}{cd}$ and hence to $ad - bc < 0$, the statement of the lemma follows from [97, Satz 1]. □

Exercise 5.42 Use Lemma 5.41 to show the following: If $\frac{p}{q}$ has the neighbors $\frac{a}{c}$ and $\frac{b}{d}$ in a Farey sequence of some level n with order $\frac{a}{c} < \frac{p}{q} < \frac{b}{d}$, then we have $\frac{p}{q} = \frac{a+b}{c+d}$.

Remark 5.43 Our applications of the Farey sequences deal mostly with the case $\det \begin{pmatrix} a & b \\ c & d \end{pmatrix} = -1$. However, we prefer matrices that are in $\Gamma(1)$. To achieve this, we replace the matrix $\begin{pmatrix} a & b \\ c & d \end{pmatrix}$ by $\begin{pmatrix} -a & b \\ -c & d \end{pmatrix}$. This obviously does not change the rational numbers $\frac{a}{c}$ and $\frac{b}{d}$, but it flips the sign of the determinant of the matrix.

We need also the following result in [97]:

Lemma 5.44 ([97]) *For $\frac{a}{c}$ and $\frac{b}{d}$ with $a, b, c, d \in \mathbb{Z}$, $c, d \geq 0$, and $ad - bc = \pm 1$ define*

$$n := \max\left\{\operatorname{lev}\left(\frac{a}{c}\right), \operatorname{lev}\left(\frac{b}{d}\right)\right\}. \tag{5.43}$$

Then, the fractions $\frac{a}{c}$ and $\frac{b}{d}$ are neighbors in the Farey sequence F_n of level n.

5.3.2 Left Neighbor Sequences

We define the *left neighbor map*

$$\mathrm{LN} : \mathbb{Q} \cup \{+\infty\} \to \mathbb{Q} \cup \{-\infty\}$$

such that $\mathrm{LN}(q)$ is the left neighbor of q in the Farey sequence $F_{\operatorname{lev}(q)}$, that is

$$\mathrm{LN}(q) = \max\left\{r \in F_{\operatorname{lev}(q)} : \; r < q\right\}. \tag{5.44}$$

For $q = \frac{b}{d} \in \mathbb{Q} \cup \{+\infty\}$, write $\frac{a}{c} = \mathrm{LN}\left(\frac{b}{d}\right) \in \mathbb{Q} \cup \{-\infty\}$ in its normalized rational form (p and q are coprime, $p \in \mathbb{Z}$ and $q \in \mathbb{Z}_{\geq 0}$). Then, by construction of the map LN, we have $\operatorname{lev}\left(\frac{a}{c}\right) \leq \operatorname{lev}\left(\frac{b}{d}\right)$ and by Lemma 5.41 $\det\begin{pmatrix} a & b \\ c & d \end{pmatrix} = -1$.

Lemma 5.45 *For $q \in \mathbb{Q} \cup \{+\infty\}$ and $\operatorname{lev}(q) > 0$, we have $\operatorname{lev}\big(\mathrm{LN}(q)\big) < \operatorname{lev}(q)$.*

Proof If $\operatorname{lev}(q) = 1$, then the set of rational numbers of level 1 is $\{\pm 1\}$. The statement of the lemma then follows since $\mathrm{LN}(1) = 0$ and $\mathrm{LN}(-1) = -\infty$ are of level 0, see (5.40).

If $N = \operatorname{lev}(q) > 1$ write $q = \frac{b}{d}$ with $\gcd(b, d) = 1$, $d \geq 0$ and $\mathrm{LN}(q) = \frac{a}{c}$ with $\gcd(a, c) = 1$, $c \geq 0$. Lemma 5.41 implies that $\begin{pmatrix} a & b \\ c & d \end{pmatrix}$ $ad - bc = -1$. Assume that $\operatorname{lev}\big(\mathrm{LN}(q)\big) = \operatorname{lev}(q)$. Then,

$$\max\left\{\, |a|\,, c\right\} = \max\left\{\, |b|\,, dbig\right\} = N \quad \text{and} \quad |a| \neq c, \; |b| \neq d.$$

There are four cases to consider:

$|a| = |b| = N$ we find $-1 = ad - bc = \operatorname{sign}(a)\, Nd - \operatorname{sign}(b)\, Nc$. Hence, $\operatorname{sign}(b)\, c - \operatorname{sign}(a)\, d = \frac{1}{N}$ which contradicts $c, d \in \mathbb{Z}$,

$c = d = N$ we find $-1 = ad - bc = aN - bN$. Hence, $b - a = \frac{1}{N}$ which contradicts $a, b \in \mathbb{Z}$.

$|a| = d = N$ we find $-1 = ad - bc = \operatorname{sign}(a) N^2 - bc$. Hence, $bc = 1 + \operatorname{sign}(a) N^2$ and $|b|, c \leq N - 1$. The second estimate implies $|bc| < N^2 - 1$ which is a contradiction.

$|b| = c = N$ we find $-1 = ad - bc = ad - \operatorname{sign}(b) N^2$. Hence, $ad = -1 + \operatorname{sign}(b) N^2$ and $|a|, d \leq N - 1$. The second estimate implies $|ad| < N^2 - 1$ which is a contradiction.

Hence the assumption $\operatorname{lev}\big(\operatorname{LN}(q)\big) = \operatorname{lev}(q)$ was wrong and

$$\operatorname{lev}\big(\operatorname{LN}(q)\big) < \operatorname{lev}(q)$$

must hold. □

Definition 5.46 Let $q \in \mathbb{Q} \cup \{+\infty\}$ and $L = L_q \in \mathbb{N}$ such that

$$\operatorname{LN}^L(q) = -\infty \qquad \text{and} \qquad \operatorname{LN}^l(q) > -\infty \quad \text{for all } l = 1, \ldots, L-1.$$

The *left neighbor sequence* $\operatorname{LNS}(q)$ of q is the finite sequence

$$\operatorname{LNS}(q) := \Big(\operatorname{LN}^L(q), \operatorname{LN}^{L-1}(q), \ldots, \operatorname{LN}^1(q), q\Big), \tag{5.45}$$

where we use the notation $\operatorname{LN}^l(q) := \underbrace{\operatorname{LN}\Big(\operatorname{LN}\big(\cdots \operatorname{LN}(q)\big)\cdots\Big)}_{l \text{ times}}$.

Remark 5.47 The number L in Definition 5.46 is well defined and unique (for each q).

Following some notations in [96], we introduce minimal partitions of $x \in \mathbb{Q}$.

Definition 5.48 A *partition* $P = P(x)$ of $x \in \mathbb{Q}$ is a sequence

$$P = \left(\frac{p_0}{q_0}, \ldots, \frac{p_L}{q_L}\right)$$

for some $L \in \mathbb{N}$ such that the fractions $\frac{p_i}{q_i}$ in normal form ($\gcd(p_i, q_i) = 1$, $p_i \in \mathbb{Z}$ and $q_i \in \mathbb{Z}_{\geq 0}$) satisfying

1. $-\infty = \frac{p_0}{q_0} < \frac{p_1}{q_1} < \cdots < \frac{p_L}{q_L} = x$ and
2. $\det \begin{pmatrix} p_{i-1} & p_i \\ q_{i-1} & q_i \end{pmatrix} = -1$ for all $i \in \{1, \ldots, L\}$.

The number $L + 1$ is called the *length* of the partition P.

A partition P of x is called a *minimal partition* if the denominators satisfy $0 = q_0 < q_1 < \ldots < q_{L-1} < q_L$.

Lemma 5.49 *For $q \in \mathbb{Q}$, consider the left neighbor sequence*

$$LNS(q) = (y_0, \ldots, y_L)$$

with $y_0 = -\infty$ and $y_L = q$, as defined in (5.45). *The sequence $(y_L, \ldots, y_0)$ is a* partition *of q. For $0 \leq q < 1$, the partition is also* minimal *in the sense of Definition 5.48.*

Proof We assume the elements of LNS(q) to be given as $y_l = \frac{a_l}{b_l}$ with $\gcd(a_l, b_l) = 1$ and $b_l \geq 0$. By construction, $y_{l-1} < y_l$ and both numbers are neighbors in the Farey sequence $F_{\text{lev}(y_l)}$. Lemma 5.41 implies $\det \begin{pmatrix} a_{l-1} & a_l \\ b_{l-1} & b_l \end{pmatrix} = -1$. Hence, $(-\infty = y_0, \ldots, y_L = q)$ is a partition of q of length $L + 1$ in the sense of Definition 5.48.

We consider the case $0 \leq q < 1$. We have to show that the partition is minimal in the sense of Definition 5.48: we have to show that the denominators b_l of y_l have to satisfy

$$0 = b_0 < b_1 < \ldots < b_{L-1} < b_L.$$

The construction of the left neighbor map and the fact $q > 0$ implies that $y_1 = 0$ and $b_1 = 1$. Since $0 = y_1 < y_l < 1$ for all $l = 2, \ldots, L$ forces the denominator b_l of y_l to be larger than 1 for all $l = 2, \ldots, L$ we have to check the inequalities above only for the indices $l \geq 2$. Obviously for $0 < \frac{a}{b} < 1$ with $\gcd(a, b) = 1$ and $a, b \geq 0$ one has $b > a$ and hence $\text{lev}\left(\frac{a}{b}\right) = b$.

Consider y_l, $l = 2, \ldots, L$, with denominator $b_l = \text{lev}\left(\frac{a_l}{b_l}\right) > 0$. Lemma 5.45 implies that $y_{l-1} = \text{LN}(y_l)$ satisfies $b_{l-1} = \text{lev}\big(y_{l-1}\big) < b_l$. The partition $(y_L, \ldots, y_0)$ is indeed minimal. □

Lemma 5.50 *For rational $0 < q < 1$, the two sequences LNS(q) $= \big(y_0, \ldots, y_L\big)$ with $y_0 = -\infty$ and $y_L = q$ and the sequence defined by the modified continued fraction expansion $(x_0, \ldots, x_{L'})$ with $x_0 = q$ and $x_{L'} = -\infty$ given in [96] coincide. Indeed $L = L'$ and $y_l = x_{L-l}$ for all $l = 0, \ldots, L$.*

Proof For $0 \leq q < 1$ rational, the left neighbor sequence LNS(q) is a minimal partition by Lemma 5.49. According to [96] this partition is unique and hence Lemma 5.50 holds. □

Definition 5.51 To $q \in [0, 1)$ rational and LNS$(q) = \left(\frac{a_0}{b_0}, \ldots, \frac{a_L}{b_L}\right)$ with $\gcd(a_l, b_l) = 1$ and $b_l \geq 0$, $l = 0, \ldots, L$, we attach the element $M(q) = \sum_{l=1}^{L} m_l \in \mathcal{R}_1$:

$$M(q) = \begin{pmatrix} -a_0 & a_1 \\ -b_0 & b_1 \end{pmatrix}^{-1} + \ldots + \begin{pmatrix} -a_{l-1} & a_l \\ -b_{l-1} & b_l \end{pmatrix}^{-1} + \ldots + \begin{pmatrix} -a_{L-1} & a_L \\ -b_{L-1} & b_L \end{pmatrix}^{-1}. \tag{5.46}$$

Remark 5.52

1. The number $L = L(q)$ in Definition 5.51 depends on q since it is the same L as in (5.45).
2. Recall the set $\mathcal{R}_n$ of finite linear combinations of elements $\mathrm{Mat}_2(\mathbb{Z})$ with fixed determinant n that was defined in (1.135) of §1.6.1. We have obviously $M(q) \in \mathcal{R}_1$.

In the following, we need some properties of the matrices in $M(q)A$ for $A \in \mathrm{Mat}_2(\mathbb{Z})$:

Lemma 5.53 *For* $0 \le q < 1$ *rational and* $M(q) = \sum_{l=1}^{L} \begin{pmatrix} * & * \\ c_l & d_l \end{pmatrix}$, *one has* $c_l\zeta + d_l > 0$ *for all* $\zeta \ge q$.

Proof By construction, the lth summand of $M(q)$ in (5.46) has the form

$$\begin{pmatrix} b_l & -a_l \\ b_{l-1} & -a_{l-1} \end{pmatrix} \qquad \text{with} \qquad \frac{a_{l-1}}{b_{l-1}} < \frac{a_l}{b_l}.$$

Since $\frac{a_0}{b_0} = -\infty$ and $\frac{a_L}{b_L} = q$ we find $\frac{a_{l-1}}{b_{l-1}} < q$ for $l = 1, \ldots, L$ and therefore $b_{l-1}q - a_{l-1} > 0$. Since $\zeta \ge q$ and $c_l = b_{l-1} \ge 0$ the lemma follows immediately. □

Lemma 5.54 *Let* $A = \begin{pmatrix} a & b \\ 0 & d \end{pmatrix} \in \mathrm{Mat}_2(\mathbb{Z})$ *be such that* $a, b \in \mathbb{N}$, $0 \le b < d$ *and* $M\left(\frac{b}{d}\right) = \sum_{l=1}^{L} m_l$. *The matrices* $m_l A$ *contain only nonnegative integer entries.*

Proof By construction, $m_l = \begin{pmatrix} b_l & -a_l \\ b_{l-1} & -a_{l-1} \end{pmatrix}$ with $\frac{a_{l-1}}{b_{l-1}} < \frac{a_l}{b_l}$ for all $l = 1, \ldots, L$ and $\frac{a_0}{b_0} = -\infty$ and $\frac{a_L}{b_L} = \frac{b}{d}$. Hence, we have

$$\frac{a_l}{b_l} \le \frac{b}{d} \qquad \Longleftrightarrow \qquad a_l d \le b b_l$$

for all $l = 0, \ldots, L$. We conclude that $m_l A = \begin{pmatrix} ab_l & bb_l - da_l \\ ab_{l-1} & bb_{l-1} - da_{l-1} \end{pmatrix}$ has only nonnegative entries. □

Lemma 5.55 *Let* A *and* m_l *be as in Lemma 5.54. Then the entries of the matrix* $m_l A = \begin{pmatrix} a' & b' \\ c' & d' \end{pmatrix}$ *satisfy*

$$a' > c' \ge 0 \qquad \text{and} \qquad d' > b' \ge 0.$$

Proof For $q = \frac{b}{d}$, the sequence LNS $\left(\frac{b}{d}\right)$ in reversed order is minimal according to Lemma 5.49. This allows us to use [96, statement (6.3)] which is formulated there only for certain upper triangular matrices $A_{[c:d]}$. The proof of this statement extends however also to the upper triangular matrix A. □

5.3.3 *Paths in the Upper Half-Plane*

Recall that $\Gamma(1)$ acts on $\mathbb{H}$ and on its boundary $\mathbb{R} \cup \{i\infty\}$ by Möbius transformations (1.16). Here we use the one point compactification of $\mathbb{R}$ by identifying $+\infty$ and $-\infty$ with the cusp $i\infty$. We denote by $\mathbb{H}^\star$ the union of the upper half-plane $\mathbb{H}$ with its boundary $\mathbb{R} \cup \{i\infty\}$:

$$\mathbb{H}^\star := \mathbb{H} \cup \mathbb{R} \cup \{i\infty\}. \tag{5.47}$$

By a *simple path* L connecting points $z_0, z_1 \in \mathbb{H}^\star$, we understand a piecewise smooth curve that lies inside $\mathbb{H}$ except possibly for the initial and end point z_0, z_1 and is analytic in all points $\mathbb{H}^\star \setminus \mathbb{H}$ in the sense of [116, p. 58]. Two simple paths L_{z_0,z_1} and L'_{z_0,z_1} are always homotopic, see for example [116]. A *path* L connecting points $z_0, z_1 \in \mathbb{H}^\star$ is given by the union of finitely many simple paths L_n, $n = 1, \ldots, N$ connecting the points $z_{0,n}, z_{1,n} \in \mathbb{H}^\star$ such that $z_{0,1} = z_0$, $z_{1,n} = z_{0,n+1}$ and $z_{1,N} = z_1$. We say that a path L *lies in the first quadrant* respectively *in the second quadrant* if $\Re(z) \geq 0$ respectively $\Re(z) \leq 0$ for almost all $z \in L$. For distinct $z_0, z_1 \in \mathbb{H}^\star \setminus \mathbb{H}$, the *standard path* L_{z_0,z_1} is the geodesic connecting z_0 and z_1. A standard path L is also a simple path. Figure 5.2b illustrates two paths connecting q and $i\infty$, one is a simple path and the other is the union of three simple paths.

Lemma 5.56 *For rational $q \in [0, 1) \cap \mathbb{Q}$ put $M(q) = \sum_l m_l$ as in Definition 5.51. The two paths $L_{q,\infty}$ and $\bigcup_l L_{m_l^{-1}0, m_l^{-1}\infty}$ have the same initial and end point.*

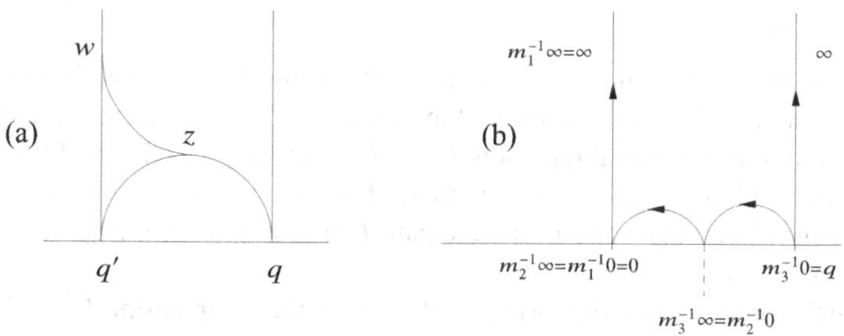

Fig. 5.2 Diagram (a) illustrates the paths used in the proof of Lemma 5.58. Diagram (b) illustrates the paths $L_{q,\infty}$ and $\bigcup_l L_{m_l^{-1}0, m_l^{-1}\infty}$ for, e.g., $q = \frac{2}{3}$.

Proof Let $M(q) = \bigcup_{l=1}^{L} m_l$ be defined as in 5.46. By construction $m_1^{-1} = I$ is the identity matrix and

$$m_l^{-1}0 = \begin{pmatrix} -a_{l-1} & a_l \\ -b_{l-1} & b_l \end{pmatrix} 0 = \frac{a_l}{b_l} \quad \text{and} \quad m_l^{-1}\infty = \begin{pmatrix} -a_{l-1} & a_l \\ -b_{l-1} & b_l \end{pmatrix} \infty = \frac{a_{l-1}}{b_{l-1}}$$

for all $l = 1, \ldots, L$. In particular $\frac{a_L}{b_L} = q$ and $\frac{a_0}{b_0} = -\infty$. Hence, we find

$$\begin{aligned} m_L^{-1}0 &= q, \\ m_l^{-1}\infty &= m_{l-1}^{-1}0 \text{ for all } l = 1, \ldots, L \quad \text{and} \\ m_1^{-1}\infty &= \infty. \end{aligned} \tag{5.48}$$

Moreover $\bigcup_{l=1}^{L} L_{m_l^{-1}0, m_l^{-1}\infty}$ is a path as union of simple paths and has the same initial and end point as $L_{q,\infty}$. □

Figure 5.2b gives an illustration of the paths $\bigcup_{l=1}^{3} L_{m_l^{-1}0, m_l^{-1}\infty}$ and $L_{q,\infty}$.

Lemma 5.57 *Let q, $L_{q,\infty}$, and $M(q)$ be as in Lemma 5.56. Both paths $L_{q,\infty}$ and $\bigcup_{l=1}^{L} L_{m_l^{-1}0, m_l^{-1}\infty}$ are in the first quadrant.*

Proof Obviously $L_{0,\infty} = \{it; t \geq 0\}$ and $L_{q,\infty} = \{q + it; t \geq 0\}$ lie in the first quadrant since $q \geq 0$.

The construction of $M(q)$ gives $m_l^{-1}0 = m_{l+1}^{-1}\infty \geq 0$, $m_1^{-1}\infty = \infty$, and $m_L^{-1}0 = q \geq 0$. Hence the paths $L_{m_l^{-1}0, m_l^{-1}\infty}$ lie in the first quadrant. The lemma follows by composition of these paths. □

Lemma 5.58 *For θ a closed 1-form on $\mathbb{H}$ such that $\int_L \theta$ exists for all simple paths L in $\mathbb{H}^\star$ we have*

$$\int_{L_{q,\infty}} \theta = \int_{L_{q,q'}} \theta + \int_{L_{q',\infty}} \theta$$

for all $q, q' \in \mathbb{Q}$.

Proof Choose points z and $w \in \mathbb{H}$ on the simple paths $L_{q,q'}$ (respectively $L_{q',\infty}$) as illustrated in Figure 5.2a. We consider the path $L^1 = L_{q',z} \cup L_{z,w} \cup L_{w,q'}$ constructed from the union of the two three paths $L_{q',z}$, $L_{z,w}$ and $L_{w,q'}$. Obviously, L^1 has the same initial and end point q'. The path L^2 has initial point q, end point ∞, and goes through the points z and w and the path L^3 has initial point z and end point ∞ through the point w.

Using the property that the form θ is closed and the composition $L^2 = L_{q,z} \cup L^3$, we find

$$\int_{L_{q,\infty}} \theta = \int_{L^2} \theta = \int_{L_{q,z}\cup L^3} \theta = \int_{L_{q,z}} \theta + \int_{L^3} \theta.$$

Moreover $\int_{L^1} \theta$ vanishes since the path L^1 is a closed path and θ is closed. Hence we have

$$\begin{aligned}\int_{L_{q,\infty}} \theta &= \int_{L_{q,z}} \theta + \int_{L^1} \theta + \int_{L^3} \theta \\ &= \int_{L_{q,z}} \theta + \int_{L_{z,w}} \theta + \int_{L_{w,q'}} \theta + \int_{L_{q',z}} \theta + \int_{L^3} \theta \\ &= \int_{L_{q,q'}} \theta + \int_{L_{q',\infty}} \theta.\end{aligned}$$

This concluses the proof. □

5.3.4 Hecke Operators for Period Functions for $\Gamma(1)$

Recall the Hecke operator T_n of level $n \in \mathbb{N}$, as introduced in (3.81) in §3.6. In particular, we see that T_n can be described by the set of representatives $\mathcal{T}(n)$ given in (1.139).

We will show below how the Hecke operators T_n induce operators $\tilde{H}_n$ on period functions.

Lemma 5.59 *Let $u \in S_{0,1}^{Maass}\big(\Gamma(1), \nu\big)$ be a Maass cusp form for the full modular group $\Gamma(1)$ of weight 0 and trivial multiplier 1. Let $P_{0,\nu}$ be the period function given by (5.16) in Definition 5.13 associated with the Maass cusp form u. For $A = \begin{pmatrix} * & * \\ 0 & d \end{pmatrix} \in \mathcal{T}(n)$ and $M(A0) = \sum_{l=1}^{L} m_l \in \mathcal{R}_1$, we have*

$$\begin{aligned} m^s d^{-2s} \int_{L_{A0,A\infty}} \eta_0\big(R_{0,\nu}(\cdot, A\zeta), u\big)(z) &= \sum_{l=1}^{L} \big(P_{0,\nu}\big|_{1-2\nu} m_l A\big)(\zeta) \\ &= \Big(P_{0,\nu}\big|_{1-2\nu} M(A0)A\Big)(\zeta) \end{aligned} \tag{5.49}$$

for all $\zeta > 0$.

Proof For $l \in \{1, \ldots, L\}$, we find

$$\int_{L_{m_l^{-1}0,m_l^{-1}\infty}} \eta_0\big(R_{0,\nu}(\cdot, A\zeta), u\big) = \int_{L_{0,\infty}} \eta_0\big(R_{0,\nu}(\cdot, A\zeta)\big\|_0 m_l^{-1}, u\big\|_0 m_l^{-1}\big)$$

using the second and third part of Lemma 4.86. Recall that the matrix $m_l^{-1} \in \Gamma(1)$ is in the full modular group and that u is a Maass cusp form on the full modular group. Hence, one has $u\big|_0 m_l^{-1} = u$. Using (4.168), we find for $m_l = \begin{pmatrix} * & * \\ c_l & d_l \end{pmatrix}$

$$\begin{aligned} R_{0,\nu}(\cdot, A\zeta)\big|_0 m_l^{-1} &= (c_l A\zeta + d_l)^{1-2\nu}\, R_{0,\nu}(m_l\cdot, m_l A\zeta)\big|_0 m_l m_l^{-1} \\ &= (c_l A\zeta + d_l)^{1-2\nu}\, R_{0,\nu}(m_l\cdot, m_l A\zeta). \end{aligned}$$

The expression $(c_l A\zeta + d_l)^{1-2\nu}$ is well defined since $c_l A\zeta + d_l > 0$ (see Lemma 5.53).

Recall that $A = \begin{pmatrix} * & * \\ 0 & d \end{pmatrix} \in \mathcal{T}(n)$ is a triangle matrix with $\det A = n$. We have

$$\begin{aligned} & m^{\nu-\frac{1}{2}} d^{1-2\nu} \int_{L_{m_l^{-1}0, m_l^{-1}\infty}} \eta_0\big(R_{0,\nu}(\cdot, A\zeta), u\big) \\ &\quad = m^{\nu-\frac{1}{2}} d^{1-2\nu} (c_l A\zeta + d_l)^{1-2\nu} \int_{L_{0,\infty}} \eta_0\big(R_{0,\nu}(m_l\cdot, m_l A\zeta), u\big) \\ &\quad = \big(P_{0,\nu}\big|_{2\nu-1} m_l A\big)(\zeta). \end{aligned}$$

Since Lemma 5.54 shows that $m_l A \in \mathcal{R}_n$ with $m_l A$ having only nonnegative entries, we see that $j(m_l A, \zeta) > 0$ for all $\zeta > 0$. Hence, the slash action above is well defined.

On the other hand, consider the term $\int_{L_{m_L^{-1}0,\infty}} \eta_0\big(R_{0,\nu}(\cdot, A\zeta)\big)$. Lemma 5.58 implies that

$$\begin{aligned} & \int_{L_{m_L^{-1}0,\infty}} \eta_0\big(R_{0,\nu}(\cdot, A\zeta), u\big) \\ &\quad = \int_{L_{m_L^{-1}0, m_L^{-1}\infty}} \eta_0\big(R_{0,\nu}(\cdot, A\zeta), u\big) + \int_{L_{m_{L-1}^{-1}0,\infty}} \eta_0\big(R_{0,\nu}(\cdot, A\zeta), u\big) \end{aligned}$$

since the matrices m_l satisfy (5.48). Iterating the argument gives

$$\int_{L_{m_L^{-1}0,\infty}} \eta_0\big(R_{0,\nu}(\cdot, A\zeta), u\big) = \sum_{l=1}^{L} \int_{L_{m_l^{-1}0, m_l^{-1}\infty}} \eta_0\big(R_{0,\nu}(\cdot, A\zeta), u\big).$$

Since $L_{A0,A\infty} = L_{m_L^{-1}0,\infty}$, we find

$$m^{\nu-\frac{1}{2}} d^{1-2\nu} \int_{L_{A0,A\infty}} \eta_0\big(R_{0,\nu}(\cdot, A\zeta), u\big) = \sum_{l=1}^{L} \big(P_{0,\nu}|_{1-2\nu} m_l A\big)(\zeta)$$

for $\zeta > 0$. □

Now, we can define linear operators $\tilde{H}_n$ on period functions.

Definition 5.60 For $n \in \mathbb{N}$, define $\mathcal{H}(n) \in \mathcal{R}_n$ (with only nonnegative entries) as

$$\begin{aligned} \mathcal{H}(n) :&= \sum_{\substack{d|n \\ 0 \le b < d}} M\big(\tfrac{b}{d}\big) \begin{pmatrix} \frac{n}{d} & b \\ 0 & d \end{pmatrix} \\ &= \sum_{A \in \mathcal{T}(n)} M\big(A0\big)\, A. \end{aligned} \tag{5.50}$$

For $s \in \mathbb{C}$, the formal sum $\mathcal{H}(n)$ induces an operator $\tilde{H}_n$ on the space of holomorphic functions on $(0, \infty)$ through

$$\tilde{H}_n f = n^{-1} f|_{1-2\nu} \mathcal{H}(n) \tag{5.51}$$

for all f a holomorphic function on $\mathbb{R}_{>0}$.

Remark 5.61 The matrices in the formal sum $\mathcal{H}(n)$ in (5.50) have only nonnegative integer entries as shown in Lemma 5.54. Hence $\big(n^{-1} f|_{1-2\nu}\mathcal{H}(n)\big)(\zeta)$ in (5.51) is well defined for all $\zeta > 0$.

Remark 5.62 Lemma 5.55 implies that the set

$$\left\{ m_l A;\ A \in \mathcal{T}(n),\ M(A0) = \sum_{l=1}^{L} m_l \right\},$$

that contains all matrices appearing in the formal sum $\mathcal{H}(n)$, is a subset of

$$\begin{aligned} S_n &= \left\{ \begin{pmatrix} a & b \\ c & d \end{pmatrix};\ a > c \ge 0, d > b \ge 0, ad - bc = n \right\} \\ &\subset \left\{ \begin{pmatrix} a & b \\ c & d \end{pmatrix} \in \mathrm{Mat}_2\big(\mathbb{Z}_{\ge 0}\big);\ ad - bc = n \right\}. \end{aligned}$$

In [96], it is shown that both sets are indeed equal. (The authors in [96] assume that $\gcd(a, b, c, d) = 1$ but this restriction is not necessary.) Hence $\mathcal{H}(n) \in \mathcal{R}_n$ with only nonnegative entries in the matrices is given by $\mathcal{H}(n) = \sum_{B \in S_n} B$.

Lemma 5.63 *Let $P_{0,\nu}$ be the period function associated with the Maass cusp form $u \in S_{0,1}^{Maass}\big(\Gamma(1), \nu\big)$ by* (5.16)*. For every $n \in \mathbb{N}$ satisfies the operator $\tilde{H}_n$ the identity*

$$\Big(\tilde{H}_n P_{0,\nu}\Big)(\zeta) = \int_0^{i\infty} \eta_0\big(R_{0,\nu}(\cdot, \zeta), T_n u\big) \tag{5.52}$$

for all $\zeta > 0$, where T_n is the Hecke operator given in (3.81).

Proof Let $P_{0,\nu}$ be the period function associated with the Maass cusp form $u \in S_{0,1}^{\mathrm{Maass}}\big(\Gamma(1), \nu\big)$ by (5.16), and let T_n be the Hecke operator given in (3.81).

For $A = \begin{pmatrix} a & b \\ 0 & d \end{pmatrix} \in \mathcal{T}(n)$, we have

$$\begin{aligned} \int_{L_{0,\infty}} \eta_0\big(R_{0,\nu}(\cdot, \zeta), u\big\|_0 A\big) &= \int_{L_{A0,A\infty}} \eta_0\big(R_{0,\nu}(\cdot, \zeta)\big\|_0 A^{-1}, u\big) \\ &= \int_{L_{A0,A\infty}} \eta_0\big(R_{0,\nu}(\cdot, A\zeta), u\big) \end{aligned}$$

using Lemma 4.86 and Equation (4.168). Applying Lemma 5.59, we find

$$\int_{L_{0,\infty}} \eta_0\big(R_{0,\nu}(\cdot, \zeta), u\big\|_0 A\big) = \big(P_{0,\nu}\big|_{1-2\nu} M(A0)A\big)(\zeta).$$

Next, we calculate $\int_{L_{0,\infty}} \eta_0\big(R_{0,\nu}(\cdot, \zeta), T_n u\big)$ for $T_n u$ in (3.81). We have

$$\begin{aligned} \int_{L_{0,\infty}} \eta_0\big(R_{0,\nu}(\cdot, \zeta), T_n u\big) &= \sum_{A \in \mathcal{T}(n)} \Big(P_{0\nu}\big|_{1-2\nu} M\,(A0)\,A\Big)(\zeta) \\ &= P_{0,\nu}\big|_{1-2\nu} \mathcal{H}(n)(\zeta) \end{aligned}$$

since $\mathcal{H}(n) = \sum_{A \in \mathcal{T}(n)} M(A0)A$. □

Proposition 5.64 *Let $P_{0,\nu}$ be the period function associated with the Maass cusp form $u \in S_{0,1}^{Maass}\big(\Gamma(1), \nu\big)$ by* (5.16)*. We denote by Q the period function associated with the Maass cusp form $T_n u$ with T_n given in* (3.81)*, $n \in \mathbb{N}$. Then, we have the identity*

$$n^{-1}\, P_{0,\nu}\big|_{1-2\nu} \mathcal{H}(n) = Q. \tag{5.53}$$

Proof This follows immediately from Lemma 5.63. □

Remark 5.65 Proposition 5.64 shows how the Hecke operators T_n on Maass cusp forms for $\Gamma(1)$ induce Hecke operators on period functions. These Hecke operators are the same as the operators presented in [96]. The authors of [96] derived the operators using only period functions, and respectively transfer operators for the groups $\Gamma_0(N)$; they construct period functions for the congruence subgroups $\Gamma_0(nm)$ for fixed n and arbitrary $m \in \mathbb{N}$ based on period functions for the group $\Gamma_0(n)$.

Another derivation of $\tilde{H}_n$ is given also in [148], using a criterion found by Choie and Zagier in [57]. (We refer to our discussion of Hecke operators on period polynomials in §2.4 where we state Choie and Zagier's result.) A similar representation of the Hecke operators for modular symbols has been given by Merel in [141].

Hecke operators for period functions on congruence subgroups are also discussed in the mathematical literature. We refer to the results in [149], [58], and also [96] that we already mentioned above.

5.4 Concluding Remarks

Historically, the interesting paper [122] by Lewis and Zagier serves essentially as the starting point of the concept of period functions. Both authors presented a lot of techniques that we also use in Section 5.1 to introduce and calculate period functions associated with (families of) Maass cusp forms. These period functions are essentially an analogue to period polynomials for modular cusp forms.

Another new development is the connection with the elements of the mixed parabolic cohomology group $H^1_{\mathrm{par}}\big(\Gamma(1), \mathcal{V}^{\omega}_{\nu}, \mathcal{V}^{\omega^*,\infty}_{\nu}\big)$, see [36]. The approach through cohomological arguments was opened with Zagier's cohomology theorem in [36], see also §5.2.3 for a short introduction. It would be interesting to see if one could transport arguments and results for the Eichler cohomology group and the associated modular cusp forms to the new setting of the mixed parabolic cohomology group and Maass cusp forms.

The development of (representations of) Hecke operators acting on period functions in Section 5.3 concludes this chapter. This section calculates the analogue of Hecke operators on period polynomials in Section 2.4 for period polynomials.

Chapter 6
Continued Fractions and the Transfer Operator Approach

The transfer operator approach is a concept first introduced by David Ruelle, a Belgian-French mathematical physicist, who worked on statistical physics and dynamical systems. He introduced the transfer operator a means to study dynamical zeta functions. A good introduction to chaotic systems, motivating and introducing dynamical zeta functions (in a general setting) is the web book [62]. Our goal in this chapter is more specific. We focus solely on variants of the Artin billiard and the Mayer's transfer operator approach. Surprisingly, the period functions, as defined in §5.1 of the previous chapter, play a significant role in Mayer's transfer operator theory.

The chapter is organized as follows: We first discuss continued fractions and their use in the coding of geodesics on the half-plane $\mathbb{H}$ in Section 6.1. This gives us the basic tools to describe closed geodesics on the quotient surface $\Gamma(1)\backslash\mathbb{H}$. Next, in Section 6.2, we introduce Ruelle's transfer operator approach for the Gauss map and its variant and the Mayer's transfer operator $\mathcal{L}_s$. In Section 6.3, we show how period functions fit into the whole discussion about transfer operators. We show that eigenfunctions of the transfer operator $\mathcal{L}_s$ are indeed related to period functions. In Section 6.4, we discuss another interesting aspect of the transfer operator $\mathcal{L}_s$: its relation to the Selberg zeta function. In the last Section 6.5, we conclude with few remarks on the recent developments in the context of transfer operators associated with Maass cusp forms.

6.1 The Artin Billiard and Continued Fractions

The Artin billiard describes the free motion of a point particle on a surface of constant negative curvature, a simple non-compact Riemann surface with one cusp. Basically the "billiard table" is given by half of the standard fundamental domain $\mathcal{F}_{\Gamma(1)}$ given in Definition 1.18. Reflections on the boundary of the fundamental

T. Mühlenbruch, W. Raji, *On the Theory of Maass Wave Forms*, Universitext,
https://doi.org/10.1007/978-3-030-40475-8_6

domain can be encoded in certain Möbius transformations. The whole dynamical billiard was first studied by Emil Artin [13] in 1924 using continued fraction expansions to describe the free motion of such particle.

6.1.1 Continued Fractions and the Gauss Map

An infinite *continued fraction* is a fraction of the form

$$[a_0; a_1, a_2, a_3, \dots] := a_0 + \cfrac{1}{a_1 + \cfrac{1}{a_2 + \cfrac{1}{a_3 + \cfrac{1}{\ddots}}}} \tag{6.1}$$

with $a_0 \in \mathbb{Z}$ and $a_i \in \mathbb{N}, i \in \mathbb{N}$.

We call (6.1) a *finite* continued fraction if the representation of the continued fraction in (6.1) stops after finitely many steps. Conversely, we call (6.1) an *infinite* continued fraction if the representation of the continued fraction in (6.1) does not stop.

For $[a_0; a_1, a_2, a_3, \dots]$, we call the finite continued fraction

$$\frac{P_n}{Q_n} := [a_0; a_1, \dots, a_n] \tag{6.2}$$

its nth *convergent*.

A *periodic continued fraction* is indicated by over-lining the periodic part; for example

$$\begin{aligned}[a_0; a_1, \dots, a_n, \underbrace{\underbrace{a_{n+1}, \dots, a_{n+l}}, \underbrace{a_{n+1}, \dots, a_{n+l}}, \dots}_{\text{periodic part}}] \\ = [a_0; a_1, \dots, a_n, \overline{a_{n+1}, \dots, a_{n+l}}]\end{aligned} \tag{6.3}$$

for some $n \in \mathbb{N}$ and $l \in \mathbb{Z}_{\geq 0}$. Here, l is the *period length* or *length of the periodic part*. We call l *prime period length* and the continued fraction *prime periodic* if $l \in \mathbb{N}$ is the smallest period length of the expansion. We call a continued fraction expansion *purely periodic* if the continued fraction expansion contains only a periodic part, for example

$$[0; \underbrace{\underbrace{a_1, \dots, a_l}, \underbrace{a_{l+1}, \dots, a_{2l}}, \dots}_{\text{periodic part}}] = [0; \overline{a_1, \dots, a_l}]. \tag{6.4}$$

We refer to Appendix A.15 for more details about continued fractions

Exercise 6.1 Put

$$C := \begin{pmatrix} 0 & 1 \\ 1 & 0 \end{pmatrix} \quad \text{and} \quad S \overset{(1.20)}{=} \begin{pmatrix} 1 & 1 \\ 0 & 1 \end{pmatrix}. \tag{6.5}$$

Use the Möbius transformation in (1.16) to write

$$[a_0; a_1, a_2, a_3, \ldots] = S^{a_0} \, C S^{a_1} \, C S^{a_2} \, C S^{a_3} \, \cdots \, 0. \tag{6.6}$$

In the case of an infinite continued fraction expansion, we understand the above identity as the limit

$$\begin{aligned} [a_0; a_1, a_2, a_3, \ldots] &= \lim_{n\to\infty} [a_0; a_1, a_2, a_3, \ldots, a_n] \\ &= \lim_{n\to\infty} \underbrace{S^{a_0} \, C S^{a_1} \, C S^{a_2} \, C S^{a_3} \, \cdots \, C S^{a_n}}_{\in \mathrm{GL}_2(\mathbb{R})} \, 0 \\ &= S^{a_0} \, C S^{a_1} \, C S^{a_2} \, C S^{a_3} \, \cdots \, 0. \end{aligned}$$

We call the map

$$\lfloor \cdot \rfloor : \mathbb{T} \to \mathbb{Z}; \quad x \mapsto \lfloor x \rfloor := \max\{n \in \mathbb{Z} \colon \; n \le x\} \tag{6.7}$$

the *floor function*.

Remark 6.2 The floor function associates with a real number x the largest integer which is smaller or equal to x. Usually this function is called the *Gauss bracket* $[\cdot]$. However, we prefer the name floor function, which is predominately used in more programming oriented literature, to avoid confusion with the continued fraction brackets $[a_0; a_1, \ldots]$.

Example 6.3 We have for example $\lfloor \pi \rfloor = 3$ and $\lfloor \sqrt{2} \rfloor = 1$.

Definition 6.4 The *Gauss function* $f_{\mathrm{G}} : \mathbb{R} \to [0, 1)$ is given by

$$f_{\mathrm{G}}(x) := \begin{cases} 0, & \text{if } x = 0, \\ x - \lfloor x \rfloor, & \text{if } x \notin [0, 1), \text{ and} \\ \frac{1}{x} - \left\lfloor \frac{1}{x} \right\rfloor, & \text{if } x \in (0, 1). \end{cases} \tag{6.8}$$

Example 6.5 We have $f_{\mathrm{G}}(0) = 0$, $f_{\mathrm{G}}(\pi) = \pi - 3 \approx 0.141592653589793\ldots$ and $f_{\mathrm{G}}\left(\frac{2}{3}\right) = \frac{3}{2} - 1 = \frac{1}{2}$.

Exercise 6.6 Write the Gauss function f_{G} in terms of the matrices C and S introduced in (6.5) and show that

$$f_G(x) = \begin{cases} 0, & \text{if } x = 0, \\ S^{-\lfloor x \rfloor} x, & \text{if } x \notin [0, 1), \text{ and} \\ S^{-\lfloor C x \rfloor} C x, & \text{if } x \in (0, 1) \end{cases} \tag{6.9}$$

for all $x \in \mathbb{R}$.

Remark 6.7 Some sources define the Gauss map as the function

$$[0, 1) \to [0, 1); \quad x \mapsto \begin{cases} \frac{1}{x} \bmod 1 & \text{if } x \neq 0 \text{ and} \\ 0 & \text{if } x = 0 \end{cases}$$

on the unit interval. Of course, the above map is identical to f_G restricted to the interval $[0, 1)$.

The Gauss map and the continued fractions are related by the following algorithm.

Lemma 6.8 *Let $x \in \mathbb{R}$ be a real number. Calculate the (finite or infinite) sequence of coefficients $(a_i)_i$ using the following algorithm:*

Step 0 *Put $a_0 := \lfloor x \rfloor$ and $x_1 := x - a_0$.*
Step 1 *Put $a_1 := \left\lfloor \frac{1}{x_1} \right\rfloor$ and $x_2 := \frac{1}{x_1} - a_1 = f_G(x_1)$.*
Step i Put $a_i := \left\lfloor \frac{1}{x_i} \right\rfloor$ and $x_{i+1} := \frac{1}{x_i} - a_i = f_G(x_i)$.
Stop criterium We stop or terminate the algorithm when $x_{i+1} = 0$ happens for the first time.

Then x admits the (regular) continued fraction

$$x = [a_0; a_1, a_2, \ldots].$$

Proof We see immediately that the algorithm admits only integer coefficient $a_0 \in \mathbb{Z}$ as first coefficient. All subsequent constructed x_i satisfy $x_i \in [0, 1)$. Hence, either we have $x_i = 0$ and the algorithm terminates or we have $0 < x_i < 1$ and hence $\frac{1}{x_i} > 1$. In the later case, we proceed with the algorithm and calculate $a_i = \left\lfloor \frac{1}{x_i} \right\rfloor \in \mathbb{N}$ with a strictly positive integer by construction. The next value x_{i+1} satisfies

$$x_{i+1} := \frac{1}{x_i} - a_i = \frac{1}{x_i} - \left\lfloor \frac{1}{x_i} \right\rfloor \overset{(6.8)}{=} f_G(x_i).$$

□

Exercise 6.9 Let $x \in \mathbb{Q}$ be a rational number. Show that x has a finite continued fraction expansion.

Lemma 6.10 *For $x \in \mathbb{R}$, the following two statements are equivalent:*

1. *x is a rational number.*
2. *x admits a finite continued fraction expansion.*

Proof One direction was shown in Exercise 6.9. The other one is obvious since a finite (regular) continued fraction expansion gives a rational number by construction. □

Exercise 6.11 Show that the Gauss function f_{G} acts as a "left shift" on the continued fraction expansions in the following sense:

$$f_{\mathrm{G}}\big([0; a_1, a_2, \ldots]\big) = [0; a_2, \ldots]. \tag{6.10}$$

Exercise 6.12 Assume that $x \in [0, 1)$ has the continued fraction expansion $x = [0; a_1, \ldots]$. Show the following statement for every $y \in [0, 1)$:

$$\begin{aligned} &f_{\mathrm{G}}(y) = x \\ \iff &y = [0; a, a_1, \ldots] \quad \text{for some } a \in \mathbb{N}. \end{aligned} \tag{6.11}$$

6.1.2 $\mathrm{GL}_2(\mathbb{Z})$ and the Artin Billiard

Recall the full modular group $\mathrm{SL}_2(\mathbb{Z})$ defined in (1.5). Theorem 1.14 shows that each element of $\Gamma(1)$ can be written as a product of the matrices

$$S = \begin{pmatrix} 1 & 1 \\ 0 & 1 \end{pmatrix} \quad \text{and} \quad T = \begin{pmatrix} 0 & -1 \\ 1 & 0 \end{pmatrix}, \tag{1.20}$$

up to a sign.

We would like to compare such products with continued fractions. Recalling Exercise 6.1, we see that we can write each (regular) continued fraction as a product of the matrices

$$S = \begin{pmatrix} 1 & 1 \\ 0 & 1 \end{pmatrix} \quad \text{and} \quad C = \begin{pmatrix} 0 & 1 \\ 1 & 0 \end{pmatrix}, \tag{6.5}$$

applied as Möbius transform to 0.

The matrix S appears in both lines. However, the other elements C and T differ slightly in the sign of the lower left matrix entry. This slight difference leads, for example, to different determinants ($\det C = -1$ and $\det T = 1$).

We already observed that we have to extend our matrix group $\mathrm{SL}_2(\mathbb{Z}) = \Gamma(1)$, the full modular group, slightly. We have to include matrices with integral entries of determinant -1. This leads to the matrix group $\mathrm{GL}_2(\mathbb{Z})$ that contains the full modular group $\mathrm{SL}_2(\mathbb{Z})$ as a subgroup of index 2.

6.1.2.1 The Group $\mathrm{GL}_2(\mathbb{Z})$

Definition 6.13 We define the group $\mathrm{GL}_2(\mathbb{Z}) \subset \mathrm{GL}_2(\mathbb{R})$ as

$$\mathrm{GL}_2(\mathbb{Z}) := \left\{ \begin{pmatrix} a & b \\ c & d \end{pmatrix} : \ a, b, c, d \in \mathbb{Z} \text{ and } ad - bc = \pm 1 \right\}. \tag{6.12}$$

Define

$$A := \begin{pmatrix} -1 & 0 \\ 0 & 1 \end{pmatrix}, \quad B := \begin{pmatrix} -1 & 1 \\ 0 & 1 \end{pmatrix}, \quad \text{and} \quad C \overset{(6.5)}{=} \begin{pmatrix} 0 & 1 \\ 1 & 0 \end{pmatrix}. \tag{6.13}$$

We see immediately that

1. $A, B, C \in \mathrm{GL}_2(\mathbb{Z})$ with $\det A = \det B = \det C = -1$ and
2. $A^2 = B^2 = C^2 = \mathbf{1}$.

Definition 6.14 Let $\mathcal{A}, \mathcal{B} \subset \mathrm{GL}_2(\mathbb{R})$ be two sets of matrices.
We define the *dot-multiplication* of two sets $\mathcal{A} \cdot \mathcal{B}$ by

$$\mathcal{A} \cdot \mathcal{B} := \{AB : \ A \in \mathcal{A}, B \in \mathcal{B}\}. \tag{6.14}$$

Lemma 6.15 *We have*

$$\begin{aligned} \mathrm{GL}_2(\mathbb{Z}) &= \mathrm{SL}_2(\mathbb{Z}) \cup \{A\} \cdot \mathrm{SL}_2(\mathbb{Z}) \\ &= \mathrm{SL}_2(\mathbb{Z}) \cup \mathrm{SL}_2(\mathbb{Z}) \cdot \{A\} \end{aligned} \tag{6.15}$$

as union of disjoint sets, using the dot-multiplication of sets introduced in Definition 6.14 above.

Remark 6.16 Lemma 6.15 shows in particular that the full modular group $\mathrm{SL}_2(\mathbb{Z})$ is a subgroup of $\mathrm{GL}_2(\mathbb{Z})$ with index 2.

Proof (of Lemma 6.15) We show both set of inclusions of the first identity $\mathrm{GL}_2(\mathbb{Z}) = \mathrm{SL}_2(\mathbb{Z}) \cup \{A\} \cdot \mathrm{SL}_2(\mathbb{Z})$ separately.

"$\Longrightarrow$" Take an element $M \in \mathrm{GL}_2(\mathbb{Z})$. We have either $\det M = +1$ or $\det M = -1$. If $\det M = +1$, then $M \in \mathrm{SL}_2(\mathbb{Z})$.
If $\det M = -1$, then we have

$$\det AM = \det A \cdot \det M = 1$$

using (6.13). Hence, $AM \in \mathrm{SL}_2(\mathbb{Z})$ by (1.5). Multiplying both sides by A from the left, we get

$$M \in \{A\}\, \mathrm{SL}_2(\mathbb{Z}),$$

using $A^2 = \mathbf{1}$.

"$\Longleftarrow$" We consider again two cases: either $M \in \mathrm{SL}_2(\mathbb{Z})$ or $M \in \{A\} \cdot \mathrm{SL}_2(\mathbb{Z})$.
If $M \in \mathrm{SL}_2(\mathbb{Z})$, then we immediately have $M \in \mathrm{GL}_2(\mathbb{Z})$.
If $M \in \{A\} \cdot \mathrm{SL}_2(\mathbb{Z})$, then M is a 2×2 matrices with integer entries with determinant -1, again by (1.5) and $\det A = -1$.

A similar argument proves the second set identity, see Exercise 6.17 below. □

Exercise 6.17 Prove the set identity

$$\mathrm{GL}_2(\mathbb{Z}) = \mathrm{SL}_2(\mathbb{Z}) \cup \mathrm{SL}_2(\mathbb{Z}) \cdot \{A\}$$

in Lemma 6.15.

Lemma 6.18 $\mathrm{GL}_2(\mathbb{Z})$ *is generated by* A, B, *and* C *in the sense that each element* M *in* $\mathrm{GL}_2(\mathbb{Z})$ *can be written as a finite product of* A, B, *and* C *defined in* (6.13).

Proof Let $M \in \mathrm{GL}_2(\mathbb{Z})$. Using Lemma 6.15, we may assume that either $M \in \mathrm{SL}_2(\mathbb{Z})$ or $AM \in \mathrm{SL}_2(\mathbb{Z})$.

Assume $M \in \mathrm{SL}_2(\mathbb{Z})$. Theorem 1.14 shows that we can write M as a product in S and T (recalling that a possible common sign $-$ can be written as left-multiplication by T^2). Direct calculation shows the identities

$$S = (AB)^{-1} \quad \text{and} \quad T = AC. \tag{6.16}$$

Hence, we can write M as a product in S and T into a product containing only the matrices A, B, and C.

Assume that $M \in \{A\}\,\mathrm{SL}_2(\mathbb{Z})$ which means $AM \in \mathrm{SL}_2(\mathbb{Z})$. We can apply the above reasoning to write AM as a product in the matrices A, B, and C. Multiplying the expression by $A^{-1} = A$ from the left shows that M can be written as a product in the matrices A, B, and C too. □

To motivate a connection between continued fractions, matrices in $\mathrm{GL}_2(\mathbb{Z})$, and Möbius transformations, consider the following exercise.

Exercise 6.19 Let $x = [0; \overline{a_1, \ldots, a_l}]$ be a purely periodic continued fraction in the sense of (6.4) with period length $l \in \mathbb{N}$. Then the matrix

$$M := CS^{a_1} \cdots CS^{a_l} \tag{6.17}$$

is an element of $\mathrm{GL}_2(\mathbb{Z})$ and satisfies

$$Mx = x, \tag{6.18}$$

using the Möbius transformation defined in (1.16).

6.1.2.2 $\mathrm{GL}_2(\mathbb{Z})$ and Möbius Transformations

In (1.16), we already have the Möbius transformation defined for matrices in $\mathrm{GL}_2(\mathbb{C})$. Restricting to the subgroup $\mathrm{GL}_2(\mathbb{Z})$ of $\mathrm{GL}_2(\mathbb{C})$, we realize that the Möbius transformation, as defined in (1.16), keeps the set $\mathbb{R} \cup \{\infty\}$ and also keeps the set $\mathbb{H} \cup \mathbb{H}^-$ invariant.

For example, the Möbius transform of $A = \begin{pmatrix} -1 & 0 \\ 0 & 1 \end{pmatrix}$ defined in (6.13) reads as

$$A\,z \mapsto -z.$$

This map interchanges the upper and lower half-plane $\mathbb{H}$ and $\mathbb{H}^-$ but keeps the set $\mathbb{R} \cup \{\infty\}$ invariant. This is true in general in the following sense:

Lemma 6.20 *Consider $M \in \mathrm{GL}_2(\mathbb{Z})$. Then the Möbius transform $z \mapsto M\,z$ defined in* (1.16) *keeps the sets $\mathbb{R} \cup \{\infty\}$ and $\mathbb{H} \cup \mathbb{H}^-$ invariant. If* $\det M = +1$, *then $z \mapsto M\,z$ also keeps $\mathbb{H}$ and separately $\mathbb{H}^-$ invariant. If* $\det M = -1$, *then $z \mapsto M\,z$ interchanges $\mathbb{H}$ and $\mathbb{H}^-$.*

Proof Using

$$\Im(M\,z) = \det M \,\frac{\Im(z)}{|j(M,z)|^2},$$

the proof follows directly. □

However, we would like that the Möbius transformation of $\mathrm{GL}_2(\mathbb{Z})$ acts on the upper half-plane $\mathbb{H}$, instead on $\mathbb{H} \cup \mathbb{H}^-$. We do this by modifying the Möbius transformation as follows.

Definition 6.21 For $z \in \mathbb{H} \cup \mathbb{R} \cup \{\infty\}$, we define the action of the group $\mathrm{GL}_2(\mathbb{Z})$ on $\mathbb{H} \cup \mathbb{R} \cup \{\infty\}$ by

$$\left(\begin{pmatrix} a & b \\ c & d \end{pmatrix}, z\right) \mapsto \begin{pmatrix} a & b \\ c & d \end{pmatrix} z := \begin{cases} \frac{az+b}{cz+d} & \text{if } z \in \mathbb{H} \text{ and } ad - bc = +1, \\ \frac{a\bar{z}+b}{c\bar{z}+d} & \text{if } z \in \mathbb{H} \text{ and } ad - bc = -1, \\ \frac{a}{c} & \text{if } z = \infty \text{ and} \\ \infty & \text{if } z = -\frac{d}{c}. \end{cases} \tag{6.19}$$

We call this group action also the *linear fractional transformation* or *Möbius transformation.*

Lemma 1.6 applies to this modified Möbius transformation as well.

Remark 6.22 Instead of modifying the Möbius transformation as above, we could have also identified $\mathbb{H}$ with its complex conjugated set $\mathbb{H}^-$ (by identifying z with $\bar{z}$). This would have the same effect.

Doing the calculation in (1.18) for this new modified Möbius transformation, we find the two cases

1. $M \in \mathrm{GL}_2(\mathbb{Z})$ with $\det M = 1$: Then the calculation in (1.18) goes through since Mz defined in (1.16) and in (6.19) coincide in this case.
2. $M \in \mathrm{GL}_2(\mathbb{Z})$ with $\det M = -1$: Write $M = \begin{pmatrix} a & b \\ c & d \end{pmatrix}$ in this case. We have

$$\begin{aligned} \Im(Mz) &\overset{(6.19)}{=} \Im\left(\frac{a\bar{z}+b}{c\bar{z}+d}\right) = \Im\left(\frac{(a\bar{z}+b)\overline{(c\bar{z}+d)}}{|c\bar{z}+d|^2}\right) \\ &= -\det M \, \frac{\Im(z)}{|cz+d|^2}, \end{aligned} \tag{6.20}$$

analog to (1.18). Hence, $\Im(Mz) > 0$ since $\det M = -1$ by assumption.

Example 6.23 The matrices A, B, and C defined in (6.13) induce the maps

$$z \mapsto Az = -\bar{z}, \quad z \mapsto Bz = 1 - \bar{z}, \quad \text{and} \quad z \mapsto Cz = \frac{1}{\bar{z}} = \frac{z}{|z|^2} \tag{6.21}$$

by the Möbius transformation (6.19). In Cartesian coordinates $z = x + iy$, we have

$$\begin{aligned} x + iy &\mapsto A(x+iy) = -x + iy, \\ x + iy &\mapsto B(x+iy) = 1 - x + iy, \quad \text{and} \\ x + iy &\mapsto C(x+iy) = \frac{x+iy}{\sqrt{x^2+y^2}}. \end{aligned} \tag{6.22}$$

Exercise 6.24 Calculate the sets in $\mathbb{H}$ that are invariant under the Möbius transforms induced by A, B, and C, respectively.

6.1.2.3 The Artin Billiard

We saw above that the group $\mathrm{GL}_2(\mathbb{Z})$ acts on $\mathbb{H}$ via the modified Möbius transformation given in Definition 6.21. Since the group is discrete, abusing the sense of the word in Definition 1.2.1, we may expect that a fundamental domain similar to $\mathcal{F}_{\Gamma(1)}$ in (1.22) in §1.2.2 exists.

We have to modify Definition 1.16 a bit.

Definition 6.25 Consider $\mathrm{GL}_2(\mathbb{Z})$. A *fundamental domain* $\mathcal{F} = \mathcal{F}_{\mathrm{GL}_2(\mathbb{Z})}$ of $\mathrm{GL}_2(\mathbb{Z})$ is an open subset $\mathcal{F} = \mathcal{F}_{\mathrm{GL}_2(\mathbb{Z})} \subset \mathbb{H}$ such that:

1. No two distinct points of $\mathcal{F}$ are equivalent under $\mathrm{GL}_2(\mathbb{Z})$: for two distinct points $z_1, z_2 \in \mathcal{F}$, $z_1 \neq z_2$, there is no matrix $V \in \mathrm{GL}_2(\mathbb{Z})$ with $V z_1 = z_2$.

2. Every point in $\mathbb{H}$ is equivalent to some point in the closure $\overline{\mathcal{F}}$ of $\mathcal{F}$ under $\mathrm{GL}_2(\mathbb{Z})$. This means that for every point $z \in \mathbb{H}$ there exists a $V \in \mathrm{GL}_2(\mathbb{Z})$ such that $Vz \in \overline{\mathcal{F}}$.

Theorem 6.26 *The open set*

$$\mathcal{F}_{\mathrm{GL}_2(\mathbb{Z})} = \left\{ z \in \mathbb{H}\colon \; |z| > 1,\; 0 < \Re(z) < \frac{1}{2} \right\} \tag{6.23}$$

is a fundamental domain for $\mathrm{GL}_2(\mathbb{Z})$.

Corollary 6.27 *Comparing* $\mathcal{F}_{\mathrm{GL}_2(\mathbb{Z})}$ *in* (6.23) *and* $\mathcal{F}_{\Gamma(1)}$ *given in* (1.22), *we have*

$$\overline{\mathcal{F}_{\Gamma(1)}} = \overline{\mathcal{F}_{\mathrm{GL}_2(\mathbb{Z})}} \cup A\,\overline{\mathcal{F}_{\mathrm{GL}_2(\mathbb{Z})}}. \tag{6.24}$$

Proof This follows directly from (1.22) and (6.23). We have

$$\begin{aligned}
\overline{\mathcal{F}_{\Gamma(1)}} &\overset{(1.22)}{=} \left\{ z \in \mathbb{H}\colon \; |z| \geq 1,\; |\Re(z)| \leq \frac{1}{2} \right\} \\
&= \left\{ z\in\mathbb{H}\colon \; |z| \geq 1,\; 0 \leq \Re(z) \leq \frac{1}{2} \right\} \cup \left\{ z \in \mathbb{H}\colon \; |z| \geq 1,\; -\frac{1}{2} \leq \Re(z) \leq 0 \right\} \\
&= \left\{ z \in \mathbb{H}\colon \; |z| \geq 1,\; 0 \leq \Re(z) \leq \frac{1}{2} \right\} \\
&\qquad \cup \left\{ -x+iy\colon \; x+iy = z \in \mathbb{H}, |z| \geq 1,\; 0 \leq \Re(z) \leq \frac{1}{2} \right\} \\
&\overset{(6.21)}{=} \left\{ z \in \mathbb{H}\colon \; |z| \geq 1,\; 0 \leq \Re(z) \leq \frac{1}{2} \right\} \\
&\qquad \cup A \left\{ z \in \mathbb{H}\colon \; |z| \geq 1,\; 0 \leq \Re(z) \leq \frac{1}{2} \right\} \\
&\overset{(6.23)}{=} \overline{\mathcal{F}_{\mathrm{GL}_2(\mathbb{Z})}} \cup A\,\overline{\mathcal{F}_{\mathrm{GL}_2(\mathbb{Z})}}.
\end{aligned}$$

□

Proof *(of Theorem 6.26)* Recall Lemma 6.15 which shows that the full modular group $\mathrm{SL}_2(\mathbb{Z})$ is a subgroup of $\mathrm{GL}_2(\mathbb{Z})$ with index 2, see also Remark 6.16. Its right representatives of $\mathrm{SL}_2(\mathbb{Z})$ in $\mathrm{GL}_2(\mathbb{Z})$ are given by $\mathbf{1}$ and A.

We already know that

$$\mathcal{F}_{\Gamma(1)} = \left\{ z \in \mathbb{H}\colon \; |z| > 1,\; |\Re(z)| < \frac{1}{2} \right\} \tag{1.22}$$

is a fundamental domain for the full modular group $\mathrm{SL}_2(\mathbb{Z})$. Hence, no two distinct points of $\mathcal{F}_{\Gamma(1)}$ are equivalent under $\mathrm{SL}_2(\mathbb{Z})$ and every point in $\mathbb{H}$ is equivalent to some point in the closure $\overline{\mathcal{F}_{\Gamma(1)}}$ of $\mathcal{F}_{\Gamma(1)}$ under $\mathrm{SL}_2(\mathbb{Z})$, see Definition 1.16.

Consider the proper subset

$$\mathcal{F}_{\mathrm{GL}_2(\mathbb{Z})} = \left\{ z \in \mathbb{H}: \ |z| > 1, \ 0 < \Re(z) < \frac{1}{2} \right\} \tag{6.23}$$

of $\mathcal{F}_{\Gamma(1)}$. To show that $\mathcal{F}_{\mathrm{GL}_2(\mathbb{Z})}$ is indeed a fundamental domain according to Definition 6.25, we have to check that the conditions also work for the additional matrix element A, since $GL_2(\mathbb{Z})$ and $\mathrm{SL}_2(\mathbb{Z})$ are related by (6.15).

No two distinct points of $\mathcal{F}_{\mathrm{GL}_2(\mathbb{Z})}$ are equivalent under $\mathrm{GL}_2(\mathbb{Z})$

Assume that the two points $z_1, z_2 \in \mathcal{F}_{\mathrm{GL}_2(\mathbb{Z})}$, $z_1 \neq z_2$, are distinct. Since $\mathcal{F}_{\mathrm{GL}_2(\mathbb{Z})} \subset \mathcal{F}_{\mathrm{SL}_2(\mathbb{Z})}$, we find no element $V \in \mathrm{SL}_2(\mathbb{Z})$ satisfying $V z_1 = z_2$. Now, assume that there exists an element $V \in \mathrm{SL}_2(\mathbb{Z})$ such that $AV z_1 = z_2$ holds. Then

$$\begin{aligned} V z_1 &= A z_2 \\ &\overset{(6.21)}{=} -\overline{z}_2. \end{aligned}$$

However, if $z_2 \in \mathcal{F}_{\Gamma(1)}$, then $-\overline{z}_2 \in \mathcal{F}_{\Gamma(1)}$ since the standard fundamental domain $\mathcal{F}_{\Gamma(1)}$ of the full modular group is symmetric along the imaginary axis, see, e.g., Figure 1.1. This implies that we can find a matrix $M \in \mathrm{SL}_2(\mathbb{Z})$ with $M z_1 = z_2$, contradicting the assumption.

Every point in $\mathbb{H}$ is $\mathrm{GL}_2(\mathbb{Z})$-equivalent to some point in $\overline{\mathcal{F}_{\mathrm{GL}_2(\mathbb{Z})}}$

For $z \in \mathbb{H}$, take a matrix $V \in \mathrm{SL}_2(\mathbb{Z})$ such that $V z \in \overline{\mathcal{F}_{\Gamma(1)}}$. We have either $V z \in \overline{\mathcal{F}_{\mathrm{GL}_2(\mathbb{Z})}}$ and we are done, or we have $V z \notin \overline{\mathcal{F}_{\mathrm{GL}_2(\mathbb{Z})}}$. By symmetry of the set $\mathcal{F}_{\Gamma(1)}$ along the imaginary axis, see, e.g., Figure 1.1, we immediately see that $A(V z)$ lies in the set $\overline{\mathcal{F}_{\mathrm{GL}_2(\mathbb{Z})}}$, completing this case.

□

Definition 6.28 We call the fundamental domain $\mathcal{F}_{\mathrm{GL}_2(\mathbb{Z})}$ in (6.23) the *Artin billiard.*

Remark 6.29 The Artin billiard $\mathcal{F}_{\mathrm{GL}_2(\mathbb{Z})}$ is an example of a chaotic billiard. This means that the dynamic induced by the free movement of a particle along the geodesics on $\mathrm{GL}_2(\mathbb{Z}) \backslash \mathbb{H}$ forms a chaotic system. A good reference for chaotic billiard systems in general is [56].

Lemma 6.30 *The boundary of the Artin billiard $\mathcal{F}_{\mathrm{GL}_2(\mathbb{Z})}$ is formed by segments of the sets in $\mathbb{H}$ that are invariant under the Möbius transforms of A, B, and C, respectively, given in Example 6.23.*

Proof We calculated in Exercise 6.24 the invariant sets

$$\{z \in \mathbb{H}:\ A\,z = z\} = \{z \in \mathbb{H}:\ \Re(z) = 0\},$$
$$\{z \in \mathbb{H}:\ B\,z = z\} = \left\{z \in \mathbb{H}:\ \Re(z) = \frac{1}{2}\right\} \quad \text{and} \tag{6.25}$$
$$\{z \in \mathbb{H}:\ C\,z = z\} = \{z \in \mathbb{H}:\ |z| = 1\}$$

of the Möbius transforms $z \mapsto A\,z$, $z \mapsto B\,z$, and $z \mapsto C\,z$, respectively, see Example 6.23. Comparing the union of the above three sets to the boundary of $\mathcal{F}_{\mathrm{GL}_2(\mathbb{Z})}$ given in (6.23), we immediately conclude the statement of Lemma 6.30.

□

6.1.3 Geodesics on the Artin Billiard

Recall that geodesics on the upper half-plane $\mathbb{H}$ in the Poincaré model are given by (oriented) vertical lines and (oriented) semi-circles. The (oriented) geodesic is uniquely determined by its two base points γ_- and $\gamma_+ \in \mathbb{R} \cup \{i\infty\}$. We refer to Appendix A.20 for a brief introduction. Usually, the base points of a geodesic γ are denoted by the tuple (γ_-, γ_+). Both notations are often identified, i.e., we write $\gamma = (\gamma_-, \gamma_+)$. Also, the Möbius transforms of $\mathrm{GL}_2(\mathbb{Z})$ act on geodesics in the usual way, as explained in Remark A.7 of Appendix A.20. We have $M\,\gamma = (M\,\gamma_-, M\,\gamma_+)$ for all $M \in \mathrm{GL}_2(\mathbb{Z})$.

Example 6.31

1. The matrix $T = \begin{pmatrix} 0 & -1 \\ 1 & 0 \end{pmatrix}$ given in (1.20) maps the geodesic $\gamma = (0, i\infty)$ onto itself (as set $i\mathbb{R}_{>0}$) and reserves the orientation: if the original geodesic runs from 0 towards $i\infty$ along the geodesic line $i\mathbb{R}_{>0}$, then $T\,\gamma = (T\,0, T\,i\infty) = (i\infty, 0)$ runs from $i\infty$ towards 0.
2. The matrix $C = \begin{pmatrix} 0 & 1 \\ 1 & 0 \end{pmatrix}$ given in (6.5) maps also the geodesic $\gamma = (0, i\infty)$ onto itself and reserves the orientation. We have $C\,\gamma = (C\,0, C\,i\infty) = (i\infty, 0)$.
3. The matrix $A = \begin{pmatrix} -1 & 0 \\ 0 & 1 \end{pmatrix}$ given in (6.5) maps the geodesic $\gamma = (\gamma_-, \gamma_+)$ to the geodesic $A\gamma = (A\gamma_-, A\,\gamma_+) = (-\gamma_-, -\gamma_+)$.

Now, consider the elements A, B, and C in (6.13). We have seen in Lemma 6.30 that the boundary of the fundamental domain $\mathcal{F}_{\mathrm{GL}_2(\mathbb{Z})}$ in (6.23), the Artin billiard, is formed by the parts invariant sets of the Möbius transformations of those three elements respectively. Moreover, we now recognize those invariant sets as the geodesics $(0, \infty)$, $\left(\frac{1}{2}, \infty\right)$, and $(-1, 1)$. But can we interpret the maps in (6.21) as some geometric property of the Artin billiard, too?

Consider a geodesic $\gamma = (\gamma_-, \gamma_+)$. Its image under the Möbius transformation $z \mapsto A\,z$ in (6.21) satisfies

$$A\,\gamma = (A\gamma_-, A\,\gamma_+) \overset{(6.22)}{=} (-\gamma_-, -\gamma_+).$$

Similarly we find

$$B\,\gamma = (B\gamma_-, B\,\gamma_+) \overset{(6.22)}{=} (1-\gamma_-, 1-\gamma_+).$$

and

$$C\,\gamma = (C\gamma_-, C\,\gamma_+) \overset{(6.22)}{=} \left(\frac{1}{\gamma_-}, \frac{1}{\gamma_+}\right).$$

These maps describe reflections along the geodesics $(0, \infty)$, $\left(\frac{1}{2}, \infty\right)$, and $(-1, 1)$, respectively. Hence, we can view the Artin billiard as a real billiard. If a force-free moving particle hits the boundary of the Artin billiard $\mathcal{F}_{\mathrm{GL}_2(\mathbb{Z})}$, then it gets reflected by the map $z \mapsto A\,z$, $z \mapsto B\,z$, or $z \mapsto C\,z$, respectively, depending which part of the geodesics $(0, \infty)$, $\left(\frac{1}{2}, \infty\right)$ and $(-1, 1)$ we hit.

Since A, B, and C generate all of $\mathrm{GL}_2(\mathbb{Z})$, we may view such a billiard as the projection of all $\mathrm{GL}_2(\mathbb{Z})$-equivalent geodesics of some initial geodesic γ. This leads to the notion of $\mathrm{GL}_2(\mathbb{Z})$-equivalent geodesics.

6.1.3.1 $\mathrm{GL}_2(\mathbb{Z})$-Equivalent Geodesics

Definition 6.32 Let $\gamma = (\gamma_-, \gamma_+)$ be a (oriented) geodesic on $\mathbb{H}$ with base points γ_- and γ_+. We call another geodesic $\delta = (\delta_-, \delta_+)$ $\mathrm{GL}_2(\mathbb{Z})$-equivalent if there exists a $M \in \mathrm{GL}_2(\mathbb{Z})$ with

$$\gamma = M\,\delta \iff (\gamma_-, \gamma_+) = (M\,\delta_-, M\,\delta_+).$$

We call the set of all geodesics $\mathrm{GL}_2(\mathbb{Z})$-equivalent to γ the $\mathrm{GL}_2(\mathbb{Z})$-equivalence class of γ.

Using the notion of $\mathrm{GL}_2(\mathbb{Z})$-equivalence of geodesics, we can prove the following result which is a result from Emil Artin [13] (Figure 6.1).

Theorem 6.33 ([13]) *Let $\gamma' = (\gamma'_-, \gamma'_+)$ be an oriented geodesic on $\mathbb{H}$. Then there exists a γ' $\mathrm{GL}_2(\mathbb{Z})$-equivalent geodesic to $\gamma = (\gamma_-, \gamma_+)$ with base-points satisfying*

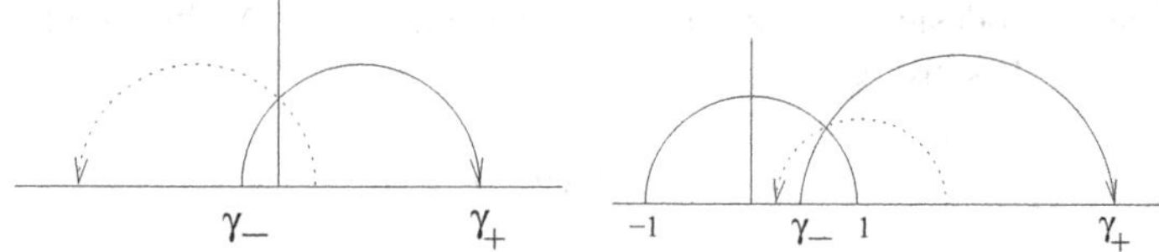

Fig. 6.1 An illustration of the reflections of an oriented geodesic γ and its base points (γ_-, γ_+) at the geodesic $(0, i\infty)$ or at the geodesic $(-1, 1)$.

$$-1 < \gamma_- \leq 0 \quad \textit{and} \quad \gamma_+ \in \mathbb{R}_{>1} \cup \{\infty\}. \tag{6.26}$$

Proof Define the *ceiling function* $\lceil \cdot \rceil : \mathbb{R} \to \mathbb{Z}$

$$\lceil \cdot \rceil := \min\{n \in \mathbb{Z} : \ x < n\}. \tag{6.27}$$

We consider several cases:

1. $\gamma'_- \in \mathbb{R}$ and $\gamma'_+ = \infty$: Recall $S = (AB)^{-1} \in \mathrm{GL}_2(\mathbb{Z})$, see (6.16). Since $\gamma'_- \in \mathbb{R}$, we find trivially that $\gamma'_- - \lceil \gamma'_- \rceil \in (-1, 0]$. Now apply $S^{-\lceil \gamma'_- \rceil}$ to the geodesic γ'. We get

$$S^{-\lceil \gamma'_- \rceil} \gamma' =: \gamma = (\gamma_-, \gamma_+)$$

with

$$\gamma_- := S^{-\lceil \gamma'_- \rceil} \gamma'_- \in (-1, 0]$$

and

$$\gamma_+ := S^{-\lceil \gamma'_- \rceil} \gamma'_+ = S^{-\lceil \gamma'_- \rceil} \infty = \infty.$$

Hence, the γ' $\mathrm{GL}_2(\mathbb{Z})$-equivalent geodesic to γ satisfies (6.26).

2. $\gamma'_- = \infty$ and $\gamma'_+ \in \mathbb{R}$: We consider the $\mathrm{GL}_2(\mathbb{Z})$-equivalent geodesic $S^{-\lfloor \gamma'_+ \rfloor} \gamma'$ which has the base-points

$$S^{-\lfloor \gamma'_+ \rfloor} \gamma' = \left(S^{-\lfloor \gamma'_+ \rfloor} \infty, S^{-\lfloor \gamma'_+ \rfloor} \gamma'_+ \right) = \left(\infty, S^{-\lfloor \gamma'_+ \rfloor} \gamma'_+ \right)$$

with $S^{-\lfloor \gamma'_+ \rfloor} \gamma'_+ \in [0, 1)$ using Example 6.31, see also Appendix A.20.

Next, consider the $\mathrm{GL}_2(\mathbb{Z})$-equivalent geodesic $C \ S^{-\lfloor \gamma'_+ \rfloor} \gamma'$ which has the base-points

$$C\, S^{-\lfloor \gamma'_+ \rfloor} \gamma' = \left(C\, \infty, C\, S^{-\lfloor \gamma'_+ \rfloor} \gamma'_+ \right) = \left(0, C\, S^{-\lfloor \gamma'_+ \rfloor} \gamma'_+ \right)$$

with $C\,S^{-\lfloor\gamma'_+\rfloor}\,\gamma'_+ > 0$, using Example 6.31, see also Appendix A.20, and that $z \mapsto C\,z$ acts as reflection at the geodesic $(-1, 1) = \{z \in \mathbb{H}\colon\ |z| = 1\}$. Hence, the γ' $\mathrm{GL}_2(\mathbb{Z})$-equivalent geodesic to $\gamma := C\,S^{-\lfloor\gamma'_+\rfloor}\,\gamma'$ satisfies (6.26).

3. $\gamma'_- \in (-1, 0]$ and $\gamma'_+ \in [0, 1)$: Consider the geodesic $C\,\gamma'$. Its new base-points satisfy

$$C\,\gamma'_- < -1 \quad \text{and} \quad C\,\gamma'_+ > 1,$$

since $z \mapsto C\,z$ acts as reflection at the geodesic $(-1, 1) = \{z \in \mathbb{H}\colon\ |z| = 1\}$.

If $C\,\gamma'_- = \infty$, we are in the setting of the second case. If $C\,\gamma'_+ = \infty$, we are in the setting of the first case.

Now, consider the remaining case $C\,\gamma'_- < -1$ and $C\,\gamma'_+ > 1$. Similar to the first case, we apply the translation by $S^{-\lfloor\gamma'_+\rfloor}$ to $C\,\gamma'$. We get

$$S^{-\lfloor\gamma'_+\rfloor}C\,\gamma' = \left(S^{-\lfloor\gamma'_+\rfloor}C\,\gamma'_-,\, S^{-\lfloor\gamma'_+\rfloor}C\,\gamma'_+\right)$$

with base points

$$S^{-\lfloor\gamma'_+\rfloor}C\,\gamma'_- \in (-1, 0] \quad \text{and}$$
$$S^{-\lfloor\gamma'_+\rfloor}C\,\gamma'_+ > 1 + \lfloor\gamma'_+\rfloor > 1.$$

Hence $S^{-\lfloor\gamma'_+\rfloor}C\,\gamma'$ satisfies (6.26).

4. $-1 < \gamma'_- < \gamma'_+ < 0$: We apply first the matrix A to the geodesic γ'. Since $z \mapsto A\,z$ acts as a reflection along $(0, \infty) = \{z \in \mathbb{H}\colon\ \Re(z) = 0\}$, we find for the base-points of the new geodesic $A\,\gamma'$ the relation

$$0 < A\,\gamma'_+ < A\,\gamma'_- < 1.$$

Now, apply Exercise 6.34 below. There exists a matrix $M \in \mathrm{GL}_2(\mathbb{Z})$ with

$$0 \le MA\,\gamma'_+ < 1 \le MA\,\gamma'_-,$$

where we might have $MA\,\gamma'_- = \infty$.

Assume $1 \le MA\,\gamma'_- < \infty$. Applying A gives

$$-\infty < AMA\,\gamma'_- \le -1 < AMA\,\gamma'_+ < 0.$$

Finally, a translation by $S^{-\lceil AMA\,\gamma'_-\rceil}$ gives

$$-1 < S^{-\lceil AMA\,\gamma'_-\rceil}AMA\,\gamma'_- \le 0 < S^{-\lceil AMA\,\gamma'_-\rceil}AMA\,\gamma'_+.$$

Hence, the geodesic $S^{-\lceil AMA\,\gamma'_-\rceil}AMA\,\gamma'$ satisfies either the third case above or it satisfies directly (6.26).

Now assume $MA\,\gamma'_- = \infty$. Applying C gives $CMA\,\gamma'$ with base points

$$CMA\,\gamma'_- = 0 \quad \text{and} \quad CMA\,\gamma'_+ \in \mathbb{R}_{>1} \cup \{\infty\}.$$

Hence, the geodesic $CMA\,\gamma'$ satisfies (6.26).

5. $\gamma'_- \in \mathbb{R} \smallsetminus (-1, 0]$ and $\gamma'_+ \in \mathbb{R} \smallsetminus [0, 1)$: If $\gamma'_- > \gamma'_-$, consider the $\mathrm{GL}_2(\mathbb{Z})$-equivalent geodesic $A\,\gamma'$ with base-points $A\,\gamma'_- = -\gamma'_- < A\,\gamma'_+ = \gamma'_+$, see Example 6.31, see also Appendix A.20. The base-points of the new geodesic reverse their ordering.

   We may assume that the base points of γ' satisfy the ordering $\gamma_- < \gamma_+$. Using the same arguments as in the first case, we may translate the geodesic γ' to the $\mathrm{GL}_2(\mathbb{Z})$-equivalent $S^{-\lceil \gamma'_- \rceil}\,\gamma'$, that satisfies

$$S^{-\lceil \gamma'_- \rceil}\,\gamma'_- \in (-1, 0] \quad \text{and} \quad S^{-\lceil \gamma'_- \rceil}\,\gamma'_- > S^{-\lceil \gamma'_- \rceil}\,\gamma'_-,$$

   since the translation $S^{-\lceil \gamma'_- \rceil}$ does not change the order of the base points.

   If $S^{-\lceil \gamma'_- \rceil}\,\gamma'_+ \geq 1$, we are done since the new geodesic satisfies (6.26). In the case of $0 \leq S^{-\lceil \gamma'_- \rceil}\,\gamma'_+ < 1$, we may apply the third case above.

   Hence, we assume that we have

$$-1 < S^{-\lceil \gamma'_- \rceil}\,\gamma'_- < S^{-\lceil \gamma'_- \rceil}\,\gamma'_+ < 0,$$

   which was handled in the fourth case above.

□

Exercise 6.34 Let $x, y \in \mathbb{R}$, $x \neq y$ two distinct real numbers satisfying

$$|x - y| < 1.$$

Show that there exists a matrix $M \in \mathrm{GL}_2(\mathbb{Z})$ such that

$$M\,x \in [0, 1) \quad \text{and} \quad M\,y \in \mathbb{R}_{>1} \cup \{\infty\}.$$

6.1.3.2 $\mathrm{GL}_2(\mathbb{Z})$-Equivalent Geodesics and Continued Fractions

In view of Theorem 6.33, we might consider the continued fraction expansions of the base points of geodesics γ satisfying (6.26).

Let $\gamma = (\gamma_-, \gamma_+)$ be a geodesic satisfying the slightly stronger condition

$$-\gamma_- \in [0, 1) \quad \text{and} \quad \gamma_+ \in \mathbb{R}_{>1}. \tag{6.28}$$

We can easily calculate the continued fraction expansions of $-\gamma_-$ and γ_+. Do these continued fraction expansions contain useful information in the view of the Artin billiard?

Remark 6.35 The only geodesics $\gamma = (\gamma_-, \gamma_+)$ which satisfy (6.26) but not (6.28) are geodesics with base-point $\gamma_+ = \infty$, which is the cusp $i\infty$ for the full modular group $\mathrm{SL}_2(\mathbb{Z})$.

Lemma 6.36 *Let $\gamma = (\gamma_-, \gamma_+)$ be a geodesic satisfying* (6.28) *and assume that the base points of γ admit the continued fraction expansions*

$$-\gamma_- = [0; a_1, a_2, \ldots] \quad \text{and} \quad \frac{1}{\gamma_+} = [0; b_0, b_1, \ldots], \tag{6.29}$$

generated by Lemma 6.8.

Applying the matrix CS^{-b_0} to the geodesic γ gives the $\mathrm{GL}_2(\mathbb{Z})$*-equivalent geodesic $\gamma' := CS^{-b_0}\gamma$ with the following properties:*

1. The base points of $\gamma' := CS^{-b_0}\gamma = \left(CS^{-b_0}\gamma_-, CS^{-b_0}\gamma_+\right)$ satisfy

$$\begin{aligned} -\gamma'_- &= -\left(CS^{-b_0}\gamma_-\right) = [0; b_0, a_1, a_2, \ldots] \quad \text{and} \\ \frac{1}{\gamma'_+} &= \frac{1}{CS^{-b_0}\gamma_+} = [0; b_1, \ldots]. \end{aligned} \tag{6.30}$$

2. *We have*

$$-\gamma'_- = CS^{-b_0}\left(-\gamma_-\right) \quad \text{and} \quad \frac{1}{\gamma'_+} = f_G\left(\frac{1}{\gamma_+}\right). \tag{6.31}$$

Proof Using Lemma 6.8 together with (6.9), we see immediately that the first and second statements of the lemma are equivalent. Therefore, we only have to show the first statement.

Let γ be a geodesic satisfying (6.29) and consider the $\mathrm{GL}_2(\mathbb{Z})$-equivalent geodesic $\gamma' := CS^{-b_0}\gamma$. We have

$$\begin{aligned} -\gamma'_- &= -CS^{-b_0}\gamma_- \\ &\overset{(6.29)}{=} -CS^{-b_0}\left(-[0; a_1, a_2, \ldots]\right) \\ &\overset{\substack{(6.6)\\(6.13)}}{=} ACS^{-b_0}ACS^{a_1}CS^{a_2}\cdots 0 \\ &= A\begin{pmatrix}1 & 0\\ 0 & -1\end{pmatrix} CS^{b_0}CS^{a_1}CS^{a_2}\cdots 0 \\ &\overset{(6.6)}{=} [0; b_0, a_1, a_2, \ldots], \end{aligned}$$

using $CS^{-b_0}A = \begin{pmatrix}1 & 0\\ 0 & -1\end{pmatrix} CS^{b_0}$ and that

$$A\begin{pmatrix}1 & 0\\ 0 & -1\end{pmatrix} = \begin{pmatrix}-1 & 0\\ 0 & -1\end{pmatrix} \overset{(1.13)}{=} -\mathbf{1}$$

acts trivially under the Möbius transformation, see Lemma 1.7.

The other base-point is calculated as

$$\begin{aligned}
\frac{1}{\gamma'_+} &\overset{(6.5)}{=} C\,\gamma'_+ \\
&= C\,C\,S^{-b_0}\,\gamma_- \\
&\overset{(6.29)}{=} S^{-b_0}\left(\frac{1}{[0;b_0,b_1,\ldots]}\right) \\
&\overset{(6.5)}{=} S^{-b_0}\,C\,[0;b_0,b_1,\ldots] \\
&\overset{(6.6)}{=} S^{-b_0}\,C\,C S^{b_0}\,C S^{b_1}\cdots 0 \\
&= C S^{b_1}\cdots 0 \\
&\overset{(6.6)}{=} [0;b_1,\ldots].
\end{aligned}$$

Hence, we find

$$\begin{aligned}
\frac{1}{\gamma'_+} &= [0;b_1,\ldots] \\
&\overset{(6.10)}{=} f_G\big([0;b_0,b_1,\ldots]\big) \\
&\overset{(6.29)}{=} f_G\left(\frac{1}{\gamma_+}\right),
\end{aligned}$$

using Exercise 6.11. This proves (6.30). □

We have also some related statement, where we move digits from the continued fraction expansion of $-\gamma_-$ to the one of $\frac{1}{\gamma_+}$.

Lemma 6.37 *Let $\gamma = (\gamma_-, \gamma_+)$ be a geodesic satisfying* (6.28) *and assume that the base-points of γ admit the continued fraction expansions*

$$-\gamma_- = [0;a_0,a_1,\ldots] \quad \text{and} \quad \frac{1}{\gamma_+} = [0;b_1,b_2,\ldots], \tag{6.32}$$

generated by Lemma 6.8.

Applying the matrix $S^{a_0}C$ to the geodesic γ gives the $\mathrm{GL}_2(\mathbb{Z})$*-equivalent geodesic $\gamma' := S^{a_0}C\,\gamma$ with the following properties:*

1. The base points of $\gamma' := S^{a_0}C\,\gamma = \big(S^{a_0}C\,\gamma_-, S^{a_0}C\,\gamma_+\big)$ satisfy

$$-\gamma'_- = -\left(S^{a_0} C\,\gamma_-\right) = [0; a_2, \ldots] \quad \textit{and}$$
$$\frac{1}{\gamma'_+} = \frac{1}{S^{a_0} C\,\gamma_+} = [0; a_0, b_1, b_2, \ldots]. \tag{6.33}$$

2. *We have*

$$-\gamma'_- = f_G(-\gamma_-) \quad \textit{and} \quad \frac{1}{\gamma'_+} = C\,S^{a_0}\,\frac{1}{\gamma_+}. \tag{6.34}$$

Proof Using Lemma 6.8 together with (6.9), we see immediately that the first and second statements of the lemma are equivalent. Therefore we only have to show the first statement.

Let γ be a geodesic satisfying (6.32) and consider the $\mathrm{GL}_2(\mathbb{Z})$-equivalent geodesic $\gamma' := S^{a_0} C\,\gamma$. We have

$$\begin{aligned}
\frac{1}{\gamma'_+} &\overset{(6.5)}{=} C\,\gamma'_+ \\
&= C\,S^{a_0} C\,\gamma_+ \\
&\overset{(6.5)}{=} C\,S^{a_0}\,\frac{1}{\gamma_+} \\
&\overset{(6.32)}{=} C\,S^{a_0}\,[0; b_1, b_2, \ldots] \\
&\overset{(6.6)}{=} CS^{a_0}\,CS^{b_1}\,\cdots\,0 \\
&\overset{(6.6)}{=} [0; a_0, b_1, b_2, \ldots].
\end{aligned}$$

The other base-point is calculated as

$$\begin{aligned}
-\gamma'_- &\overset{(6.13)}{=} A\,\gamma'_- \\
&= A\,S^{a_0} C\,\gamma_- \\
&\overset{(6.13)}{=} A\,S^{a_0} C\,AA\,\gamma_- \\
&= A\,S^{a_0} C\,A\,(-\gamma_-) \\
&\overset{(6.32)}{=} AS^{a_0} CA\,[0; a_0, a_1, \ldots] \\
&= \begin{pmatrix} -1 & 0 \\ 0 & -1 \end{pmatrix} S^{-a_0}\,C\,[0; a_0, a_1, \ldots] \\
&\overset{(6.6)}{=} \begin{pmatrix} -1 & 0 \\ 0 & -1 \end{pmatrix} S^{-a_0}\,C\,CS^{a_0}\,CS^{a_1}\,\cdots\,0
\end{aligned}$$

$$
\begin{aligned}
&= \begin{pmatrix} -1 & 0 \\ 0 & -1 \end{pmatrix} CS^{a_1} \cdots 0 \\
&\overset{(6.6)}{=} \begin{pmatrix} -1 & 0 \\ 0 & -1 \end{pmatrix} [0; a_1, \ldots] \\
&\overset{L\,1.7}{=} [0; a_1, \ldots],
\end{aligned}
$$

where we used

$$
AS^{a_0}CA = \begin{pmatrix} a_0 & -1 \\ -1 & 0 \end{pmatrix} = \begin{pmatrix} -1 & 0 \\ 0 & -1 \end{pmatrix} \begin{pmatrix} a_0 & 1 \\ 1 & 0 \end{pmatrix} = \begin{pmatrix} -1 & 0 \\ 0 & -1 \end{pmatrix} S^{-a_0}\, C
$$

and the fact that $\begin{pmatrix} -1 & 0 \\ 0 & -1 \end{pmatrix} = \mathbf{-1}$ acts trivially under the Möbius transformation, see Lemma 1.7. Recalling Exercise 6.11, we get

$$
\begin{aligned}
-\gamma'_- &= [0; a_1, \ldots] \\
&\overset{(6.10)}{=} f_G\big([0; a_0, a_1, \ldots]\big) \\
&\overset{(6.32)}{=} f_G\big(-\gamma_-\big).
\end{aligned}
$$

This proves (6.33). □

How can we relate properties of the continued fraction expansions in (6.29) to properties of the geodesics γ and vice versa? For example, assume that $[0; b_0, b_1, \ldots, b_l]$ is a finite continued fraction expansion of length $l + 1$. It is quite obvious that the associated geodesic γ in (6.29) is $\mathrm{GL}_2(\mathbb{Z})$-equivalent to a geodesic γ' with base-point $\gamma'_+ = \infty$.

To be more specific, we have the following characterization.

Proposition 6.38 *Let $\gamma = (\gamma_-, \gamma_+)$ be a geodesic satisfying* (6.28) *and assume that at least one of the base-points admits a finite continued fraction expansion in either* (6.29) *or* (6.32). *Then γ is $\mathrm{GL}_2(\mathbb{Z})$-equivalent to a geodesic $\gamma' = (\gamma'_-, \gamma'_+)$ with one base-point being ∞.*

Proof We assume that the base-point γ_+ of $\gamma = (\gamma_-, \gamma_+)$ has the finite continued fraction expansion

$$
\frac{1}{\gamma_+} = [0; b_0, b_1, \ldots, b_l]
$$

in (6.29). Applying Lemma 6.36 $l + 1$ times, we get a $\mathrm{GL}_2(\mathbb{Z})$-equivalent geodesic γ'' given by

$$
\gamma'' := CS^{-b_l} \cdots CS^{-b_1}\, CS^{-b_0}\, \gamma.
$$

The associated base-points γ''_-, γ''_+ of γ'' satisfy

$$-\gamma''_- = [0; b_l, \ldots, b_1, b_0, \ldots] \quad \text{and}$$
$$\frac{1}{\gamma''_+} = [0;].$$

Now, applying the matric C on γ'' gives the geodesic

$$\gamma' := C\,\gamma''$$

with base-points

$$\gamma'_- = C\,[0; b_l, \ldots, b_1, b_0, \ldots] = [b_l; b_{l-1}, \ldots, b_1, b_0, \ldots] \qquad \text{and}$$
$$\gamma'_+ = C\,0 = \infty.$$

We get the second case of γ_- of $\gamma = (\gamma_-, \gamma_+)$ having a finite continued fraction expansion by applying Lemma 6.8. The details are left to the reader. □

Exercise 6.39 Show the missing case in the proof of Proposition 6.38.

Another case would be if the continued fractions are periodic. This leads to the following result:

Lemma 6.40 *Let $\gamma = (\gamma_-, \gamma_+)$ be a geodesic satisfying* (6.28). *Assume that the base-points of γ admit periodic continued fraction expansions in* (6.29) *or* (6.32) *of the form*

$$-\gamma_- = [0; \overline{a_1, \ldots, a_l}] \quad \textit{and} \quad \frac{1}{\gamma_+} = [0; \overline{a_l, \ldots, a_1}]. \tag{6.35}$$

Then there exists an element $M \in \mathrm{GL}_2(\mathbb{Z})$ with $M^2 \neq \pm\mathbf{1}$ such that

$$\begin{aligned} &M\,\gamma = \gamma \\ \Longleftrightarrow\ &M\,\gamma_- = \gamma_- \quad \textit{and} \quad M\,\gamma_+ = \gamma_+. \end{aligned} \tag{6.36}$$

Remark 6.41 We see later in §6.2 that (6.36) with a hyperbolic matrix $M \in \mathrm{SL}_2(\mathbb{Z})$ in the sense of Definition 1.2.4 corresponds to periodic continued fraction expansions of the form in (6.35). This will give a partial inverse statement to Lemma 6.40.

Proof *(of Lemma 6.40)* Assume that the base-points of γ admit periodic continued fraction expansions in (6.29) or (6.32) of the form (6.36) with period length l. Applying Lemma 6.37 l-times gives us the $\mathrm{GL}_2(\mathbb{Z})$-equivalent geodesic γ' defined by

$$\gamma' := \underbrace{S^{a_l}C \cdots S^{a_1}C}_{=:M} \gamma$$

with base-points

$$-\gamma'_- = [0; \overline{a_1, \ldots, a_l}] \quad \text{and} \quad \frac{1}{\gamma'_+} = [0; \overline{a_l, \ldots, a_1}].$$

Hence, γ' and γ denote the same geodesic, since the base-points coincide, see for example Lemma A.10 in Appendix A.20. The matrix $M = S^{a_l}C \cdots S^{a_1}C$ is in $\mathrm{GL}_2(\mathbb{Z})$. Since the digits $a_1, \ldots, a_l \in \mathbb{N}$ are all positive, we have in particular that

$$M^2 = S^{a_l}C \cdots S^{a_1}C\, S^{a_l}C \cdots S^{a_1}C$$

contains only nonnegative integers and is not identical to $\pm\mathbf{1}$. □

6.1.4 Continued Fractions, Geodesics, and the Full Modular Group

We discussed the relation between continued fraction expansions, geodesics, and the group $\mathrm{GL}_2(\mathbb{Z})$ in §6.1.1–6.1.3. In particular, one main reason to introduce the group $\mathrm{GL}_2(\mathbb{Z})$ in §6.1.2 was to find a suitable matrix group that contains the elements S and C which are used in the continued fraction expansion (see (6.5) and (6.6)).

However, the main group appearing in the previous chapters is the full modular group $\mathrm{SL}_2(\mathbb{Z})$ defined in (1.5). Can we relate the results concerning continued fractions, geodesics, and $\mathrm{GL}_2(\mathbb{Z})$ to the full modular group $\mathrm{SL}_2(\mathbb{Z})$? In particular, can we formulate something similar to Theorem 6.33?

Fortunately, $\mathrm{GL}_2(\mathbb{Z})$ and $\mathrm{SL}_2(\mathbb{Z})$ are related by (6.15) in Lemma 6.15.

Let us introduce the concept of $\mathrm{SL}_2(\mathbb{Z})$-equivalent geodesics, similar to Definition 6.32.

Definition 6.42 Let $\gamma = (\gamma_-, \gamma_+)$ be a (oriented) geodesic on $\mathbb{H}$ with base points γ_- and γ_+. We call another geodesic $\delta = (\delta_0, \delta_+)$ $\mathrm{SL}_2(\mathbb{Z})$-equivalent if there exists a $M \in \mathrm{SL}_2(\mathbb{Z})$ with

$$\gamma = M\,\delta \iff \big(\gamma_-, \gamma_+\big) = \big(M\,\delta_-, M\,\delta_+\big).$$

We call the set of all geodesics $\mathrm{SL}_2(\mathbb{Z})$-equivalent to γ the $\mathrm{SL}_2(\mathbb{Z})$-equivalence class of γ.

Remark 6.43 We introduced the above notion of $\mathrm{SL}_2(\mathbb{Z})$-equivalent geodesics to highlight our interest in equivalence classes under $\mathrm{SL}_2(\mathbb{Z})$ instead of using $\mathrm{GL}_2(\mathbb{Z})$-

equivalences. Since Lemma 6.15 implies $\mathrm{SL}_2(\mathbb{Z}) \subset \mathrm{GL}_2(\mathbb{Z})$, we actually see that a $\mathrm{GL}_2(\mathbb{Z})$-equivalence implies the $\mathrm{SL}_2(\mathbb{Z})$-equivalence.

Use Lemma 6.15 and Theorem 6.33 to show the following result.

Exercise 6.44 Let $\gamma' = (\gamma'_-, \gamma'_+)$ be an oriented geodesic on $\mathbb{H}$. Show that there exists a γ' $\mathrm{SL}_2(\mathbb{Z})$-equivalent geodesic to $\gamma = (\gamma_-, \gamma_+)$ with base points satisfying

$$\begin{cases} -1 < \gamma_- \leq 0 \quad \text{and} \quad \gamma_+ \in \mathbb{R}_{>1} \cup \{\infty\}, \quad \text{or} \\ 0 \leq \gamma_- < 1 \quad \text{and} \quad \gamma_+ \in \mathbb{R}_{<-1} \cup \{\infty\}. \end{cases} \tag{6.37}$$

Moreover, the base-points of γ admit continued fraction expansions of the form

$$-\varepsilon\,\gamma_- = [0; a_1, a_2, \ldots] \quad \text{and} \quad \frac{\varepsilon}{\gamma_+} = [0; b_0, b_1, \ldots], \tag{6.38}$$

where $\varepsilon \in \{\pm 1\}$ encodes the suitable sign coinciding whether we have the first or second case in (6.37).

Remark 6.45 Caroline Series studied the coding of geodesics on $\mathbb{H}$. Her work [181] contains some more details about coding of geodesics in continued fractions, including slightly from the result in Exercise 6.44. The representation of γ in terms of continued fractions expansions of the base points as in (6.38) is sometimes called the *cutting sequence* of γ, see for example [181].

We would like to continue exploring the relation between oriented geodesics and continued fraction expansions. To do so, we first give the geodesics satisfying (6.37) a name.

Definition 6.46 Inspired by the description (6.37) of oriented geodesics up to $\mathrm{SL}_2(\mathbb{Z})$-equivalency, we define the set of oriented geodesics

$$\begin{aligned} \mathcal{A} := & \left\{\gamma = (\gamma_-, \gamma_+) \text{ geodesics on } \mathbb{H} \colon\ 0 \leq |\gamma_-| < 1 < |\gamma_+| \leq \infty,\ \gamma_-\,\gamma_+ \leq 0\right\} \\ = & \left\{\gamma = (\gamma_-, \gamma_+) \text{ geodesics on } \mathbb{H} \colon\ \gamma \text{ satisfies (6.37)}\right\}. \end{aligned} \tag{6.39}$$

Exercise 6.44 shows directly that the base points of each oriented geodesic $\gamma = (\gamma_-, \gamma_+) \in \mathcal{A}$ admits continued fraction expansions of its base points of the form (6.38).

We discuss the relation of the geodesics in $\mathcal{A}$ with the Gauss map below in §6.4.1.

6.1.5 *Closed Geodesics Under* $\mathrm{SL}_2(\mathbb{Z})$ *and Periodic Continued Fraction Expansions*

Consider a geodesic γ on $\mathbb{H}$, given by its base points $\gamma = (\gamma_-, \gamma_+)$.

Definition 6.47 We call a geodesic $\gamma = (\gamma_-, \gamma_+)$ *closed* for $\mathrm{SL}_2(\mathbb{Z})$ if there exists a matrix $M \in \mathrm{SL}_2(\mathbb{Z}) \setminus \{\pm\mathbf{1}\}$ such that the geodesic γ as set is invariant under the Möbius transformation under M:

$$z \in \gamma \quad \Longrightarrow \quad M\,z \in \gamma \tag{6.40}$$

or equivalently

$$M\,\gamma := \{M\,z\colon\ z \in \gamma\} \stackrel{!}{=} \gamma. \tag{6.41}$$

A simple characterization of closed geodesics is given by the following.

Lemma 6.48 *The following statements are equivalent:*

1. *γ is a closed geodesic.*
2. *$M\,\gamma = \gamma$ for some $M \in \mathrm{SL}_2(\mathbb{Z}) \setminus \{\pm\mathbf{1}\}$.*
3. *$\gamma_- = M\,\gamma_-$ and $\gamma_+ = M\,\gamma_+$ for some $M \in \mathrm{SL}_2(\mathbb{Z}) \setminus \{\pm\mathbf{1}\}$.*
4. *M is hyperbolic with the two distinct fixed points $\gamma_-, \gamma_+ \in \mathbb{R} \cup \{\infty\}$.*

Proof We have the following equivalences:

(1) $\Longleftrightarrow$ (2) This follows directly from Definition 6.47, using the defining relation (6.41).
(2) $\Longleftrightarrow$ (3) This follows from the characterization of geodesics as tuples of base points, see for example Appendix A.20 in particular Lemma A.6.
(3) $\Longleftrightarrow$ (4) This follows directly from a Proposition in Appendix A.18.

□

Next, we would like to combine closed geodesics discussed above with the result presented in Exercise 6.44: Let δ be a closed geodesic. Can we find a δ $\mathrm{SL}_2(\mathbb{Z})$-equivalent closed geodesic to γ whose base points satisfy (6.37)? And what effect does it have on the continued fraction expansions of the base points in (6.38)?
We discuss these questions in the next lemma.

Lemma 6.49 *Let γ and δ be two $\mathrm{SL}_2(\mathbb{Z})$ equivalent geodesics with $\gamma \in \mathcal{A}$. (See* (6.39) *for the definition of the set $\mathcal{A}$.)*

The following three statements are equivalent:

1. *δ is a closed geodesic with associated hyperbolic matrix N.*
2. *γ is a closed geodesic with associated hyperbolic matrix M conjugated to N: there exists a $V \in \mathrm{SL}_2(\mathbb{Z})$ with $M = V N V^{-1}$.*
3. *The base points of $\gamma = (\gamma_-, \gamma_+)$ satisfy purely periodic expansions of the form*

$$-\,\varepsilon\,\gamma_- = [0; \overline{a_{2n}, a_{2n-1}, \ldots, a_2, a_1}] \quad \text{and} \quad \frac{\varepsilon}{\gamma_+} = [0; \overline{a_1, a_2, \ldots, a_{2n}}] \tag{6.42}$$

for some number $n \in \mathbb{N}$ and suitable digits $a_i \in \mathbb{N}$.

Proof

(1) $\Longrightarrow$ (2) Let δ be a closed geodesic with associated hyperbolic matrix N. Exercise 6.44 shows that there exists a $\mathrm{SL}_2(\mathbb{Z})$-equivalent geodesic $\gamma \in \mathcal{A}$: there exists a $V \in \mathrm{SL}_2(\mathbb{Z})$ with $\gamma = V\,\delta$. Since we have

$$\begin{aligned}\gamma &= V\,\delta \\ &= VN\delta \\ &= VNV^{-1}\,\gamma.\end{aligned}$$

Hence $M := VNV^{-1}$ satisfies

$$\gamma = M\,\gamma.$$

Since the trace is invariant under conjugation,

$$\operatorname{trace}(M) = \operatorname{trace}\left(VNV^{-1}\right) = \operatorname{trace}(N),$$

we conclude with Definition 1.2.4 that M is hyperbolic iff N is hyperbolic.

(2) $\Longrightarrow$ (1) As in the argument above, we conclude that δ is a closed geodesic with associated hyperbolic matrix $N = V^{-1}MV$.

(2) $\Longleftrightarrow$ (3) For this direction we have to refer to a result by Caroline Series. Series showed in [181, §3.3.4] that a number $\alpha > 1$ has a purely periodic continued fraction expansion (of an even number digits) if and only if α is a reduced quadratic surd with α and its conjugate root $\bar{\alpha}$ satisfy

$$\alpha = [\overline{b_{2n}; b_{2n-1}, \ldots, b_1, b_{2n}}] \quad \text{and} \quad \frac{-1}{\bar{\alpha}} = [\overline{b_1; b_2, \ldots, b_{2n}, b_1}].$$

α being a reduced quadratic surd means that α and its conjugate root $\bar{\alpha}$ satisfy $\alpha = \tilde{M}\alpha$ and $\bar{\alpha} = \tilde{M}\bar{\alpha}$ for some matrix $\tilde{M} \in \mathrm{SL}_2(\mathbb{Z})$.

Using this result with $M = \tilde{M}$, $\alpha = \varepsilon\gamma_+$, and $\bar{\alpha} = \varepsilon\gamma_-$ we derive the desired equivalence.

□

6.2 Ruelle's Transfer Operator Approach for the Gauss Map

6.2.1 Motivation of Dynamical Zeta Functions and the Associated Transfer Operator

Ruelle introduced the transfer operator approach to study the dynamical zeta functions. We informally recall a few aspects of the approach, following an introductory article [174].

According to the introductory article [174], consider a discrete dynamical system with weight function (M, f, g). Here M is the base space on which the map $f : M \to M$ acts. f does not need to be invertible. However, we assume that the set

$$\mathrm{Fix}\big(f^{(m)}\big) := \Big\{x \in M: \underbrace{f\Big(f\big(\dots f(\cdot)\dots\big)\Big)}_{m \text{ times}} = x\Big\}$$

of fixed points under the mth iterated map

$$f^{(m)} := \underbrace{f\Big(f\big(\dots f(\cdot)\dots\big)\Big)}_{m \text{ times}}$$

is a finite set. The function $g : M \to \mathbb{C}$ is a weight function, usually expected to be positive.

The prototype of dynamical zeta function is

$$\zeta(z) = e^{\sum_{m=1}^{\infty} \frac{z^m}{m} \sum_{x\in \mathrm{Fix}\left(f^{(m)}\right)} \prod_{k=0}^{m-1} g\left(f^{(k)}(x)\right)}.$$

Ruelle introduced the transfer operator approach to study such dynamical zeta functions. Usually, the transfer operator $\mathcal{L}$ associated with the above discrete dynamical system (M, f, g) is formally defined by

$$\big(\mathcal{L}\phi\big)(x) := \sum_{y\in f^{-1}(x)} g(y)\,\phi(y)$$

acting on a suitable function space $\{\phi\}$. $f^{-1}(x) \subset M$ denotes the pre-image of x under the shift function f of the dynamical system. We refer to the nicely written introductory paper [174] for more details. One important fact is that, assuming nice notion of determinant of an operator, we get a relation between the determinant $\det\big(1 - z\mathcal{L}\big)$ and the dynamical zeta function $\zeta(z)$. In the case of the determinant $\det^c$ defined via the counting trace, we get

$$\zeta(z) = \frac{1}{\det^c\left(1 - z\lambda\right)}.$$

We refer to [174] for more details.

6.2.2 *The Transfer Operator for the Gauss Map*

We follow the approach presented in [135]. Recall the Gauss map f_{G} defined in (6.8) which acts as "left shift" operator on the continued fraction expansions, see (6.10) in Exercise 6.11. We define the associated transfer operator $\mathcal{L}_\beta$ formally by

$$\begin{aligned}\left(\mathcal{L}_\beta \phi\right)(x) :&= \sum_{y \in f_{\mathrm{G}}^{-1}} e^{-\beta r(y)} \phi(y) \\ &= \sum_{y \in f_{\mathrm{G}}^{-1}} e^{\beta \log\left|f_{\mathrm{G}}'(y)\right|} \phi(y)\end{aligned} \tag{6.43}$$

for $\phi : [0, 1) \to [0, 1)$ a suitably nice function and $\beta \in \mathbb{C}$ a parameter. (We will define later what "suitably nice" means.) Of course, we need to determine what the derivative $f_{\mathrm{G}}'(\cdot)$ and what the pre-image $f_{\mathrm{G}}^{-1}(x)$ is. We call the function

$$r : [0, 1) \to \mathbb{R}; \quad y \mapsto r(y) := \log\left|f_{\mathrm{G}}'(y)\right| \tag{6.44}$$

appearing in (6.43) the *recurrence time* function for the Gauss map f_{G}.

Lemma 6.50 *For $x \in (0, 1)$, the pre-image of the Gauss map f_G is given by*

$$f_G^{-1}(x) = \left\{\frac{1}{n + x} : \ n \in \mathbb{N}\right\}. \tag{6.45}$$

Proof Let $x \in (0, 1)$. We assume that x has the continued fraction expansion $x = [0; a_1, \ldots]$. Using Exercise 6.12, we see that the pre-image of $f_{\mathrm{G}}^{-1}(x)$ is given by

$$\begin{aligned}f_{\mathrm{G}}^{-1}(x) &\overset{(6.11)}{=} \left\{[0; n, a_1, a_2, \ldots] : \ n \in \mathbb{N}\right\} \\ &\overset{(6.1)}{=} \left\{\frac{1}{[0; a_1, a_2, \ldots]} : \ n \in \mathbb{N}\right\} \\ &\overset{(6.1)}{=} \left\{\frac{1}{n + x} : \ n \in \mathbb{N}\right\}.\end{aligned}$$

This proves (6.45) for $x \in (0, 1)$.

The definition (6.8) of the Gauss map shows

$$y \in [0, 1) \quad \text{with} \quad f_{\mathrm{G}}(y) = 0$$

is equivalent with

$$y \in \left\{0, \frac{1}{2}, \frac{1}{3}, \dots\right\}.$$

Since $f_{\mathrm{G}}(0) = f_{\mathrm{G}}(0)$, see (6.8), we replace the 0 in the above set with $1 = \frac{1}{1}$. Hence the pre-image of $f_{\mathrm{G}}^{-1}(x)$ is given by

$$f_{\mathrm{G}}^{-1}(0) = \left\{\frac{1}{n} : \ n \in \mathbb{N}\right\}.$$

This proves (6.45) for $x = 0$. □

Lemma 6.51 *The derivative of the Gauss map f_{G} restricted to* $[0, 1)$ *is piecewise continuous and satisfies*

$$f_{G}'(x) = \frac{-1}{x^2} \tag{6.46}$$

for almost all $x \in [0, 1)$.

Proof Let $n \in \mathbb{N}$ be a natural number. According to the definition (6.8), the Gauss map f_{G} restricted to $\left(\frac{1}{n+1}, \frac{1}{n}\right)$ is given by the map

$$\left(\frac{1}{n+1}, \frac{1}{n}\right) \to (0, 1); \quad x \mapsto \frac{1}{x} - n.$$

Its derivative is given by

$$\left(\frac{1}{n+1}, \frac{1}{n}\right) \to \mathbb{R}; \quad x \mapsto \frac{-1}{x^2}.$$

Hence, we have

$$f_{G}'(x) = \frac{-1}{x^2}$$

for all $x \in (0, 1) \setminus \{m^{-1};\ m \in \mathbb{N}\}$. Extending the map smoothly to the exceptional points, we find (6.46). □

Remark 6.52 Comparing to the model of a transfer operator, that we mentioned in the motivational section §6.2.1, we see that the weight function g is given by $\frac{-1}{f_G'}$. This choice corresponds to the preferred choice of a positive function.

Using Lemmas 6.50 and 6.51 allow us to concretize the definition of the transfer operator a bit. However, we still do not specify the function space on which the operator acts.

Definition 6.53 Let $s \in \mathbb{C}$ be a complex number. We define the *transfer operator* $\mathcal{L}_s$ formally by

$$(\mathcal{L}_s\phi)(x) := \sum_{n\in\mathbb{N}} \left(\frac{1}{n+x}\right)^{2s} \phi\left(\frac{1}{x+n}\right) \tag{6.47}$$

for $\phi : [0, 1) \to \mathbb{C}$ a suitably nice function.

Exercise 6.54 Show that $\mathcal{L}_1$ in (6.47) is equal to $\mathcal{L}$ in (6.43) in the sense that its formal action on a function ϕ is the same.

Let us move toward a complete definition of the transfer operator $\mathcal{L}_s$. Define the open disc $D \subset \mathbb{C}$ of radius $\frac{3}{2}$ around 1:

$$D := \left\{z \in \mathbb{C}:\ |z-1| < \frac{3}{2}\right\}. \tag{6.48}$$

For $n \in \mathbb{N}$, define the maps

$$\theta_n \to \mathbb{C} \quad z \mapsto \frac{1}{z+n}, \tag{6.49}$$

which appear in the formal definition (6.47) of $\mathcal{L}_s$. These maps are well defined and contractions on D, as the lemma below shows.

Lemma 6.55 *Let $n \in \mathbb{N}$.*

1. *The maps θ_n are well defined.*
2. *θ_n is a Möbius transformation in the sense that*

$$\theta_n(z) = CS^n\, z \tag{6.50}$$

for all $z \in D$.

3. *On D, they act as contractions in the sense that*

$$\theta_n(D) \subset \left\{z \in \mathbb{C}:\ |z-1| < \frac{3}{2} - \varepsilon\right\} \tag{6.51}$$

for some small $\varepsilon > 0$ independent of n.

Proof We show the statements separately.

1. The denominator $z+n$ of $\theta_n(z) \overset{(6.49)}{=} \frac{1}{z+n}$ vanishes for $z = -n$. Since all negative integers do not lie in the set D, i.e., $-m \notin D$ for all $m \in \mathbb{N}$, we see that the map θ_n in (6.49) is well defined.
2. Recalling C and S in (6.5), we see immediately that the Möbius transformation of $CS^n = \begin{pmatrix} 0 & 1 \\ 1 & n \end{pmatrix}$ can be written as

$$CS^n\, z \overset{(6.5)}{=} \begin{pmatrix} 0 & 1 \\ 1 & n \end{pmatrix} z \overset{(1.16)}{=} \frac{1}{z+n} \overset{(6.49)}{=} \theta_n(z)$$

   for all $z \in D$.
3. We first consider the image of the interval

$$I := \left(-\frac{1}{2}, \frac{5}{2}\right) = D \cap \mathbb{R},$$

   which appears if we consider $D \cap \mathbb{R}$. We have

$$\begin{aligned} & x \in I \\ \iff & -\frac{1}{2} < x < \frac{5}{2} \\ \iff & n - \frac{1}{2} < x + n < \frac{5}{2} + n \\ \iff & \frac{2}{2n-1} = \frac{1}{n-\frac{1}{2}} > \frac{1}{x+n} > \frac{1}{\frac{5}{2}+n} = \frac{2}{2n+5} \\ \Longrightarrow & 2 > \theta_n(x) > 0 \\ \Longrightarrow & \theta_n(x) \in (0,2) \subset \left(\varepsilon - \frac{1}{2}, \frac{5}{2} - \varepsilon\right) \subset I \end{aligned} \tag{6.52}$$

   for every $0 < \varepsilon < \frac{1}{2}$. Hence, θ_n restricted to $I = D \cap \mathbb{R}$ satisfies (6.51).

   The maps θ_n are conformal since they are given by Möbius transformations. This means in particular that we can extend the relation (6.51) restricted to $I = D \cap \mathbb{R}$ to the disc D. This proves (6.51).

□

Definition 6.56 Let $B(D)$ denote the Banach-space of continuous functions on the closed disc $\overline{D}$ that are holomorphic on the open disc D. The norm on $B(D)$ is given by the usual sup-norm $\|\cdot\|_\infty$ given by

$$\|\phi\|_\infty := \sup_{z\in\overline{D}} |\phi(z)|$$

for all $\phi \in B(D)$.

We are finally able to formulate the complete definition of the transfer operator.

Definition 6.57 Let $s \in \mathbb{C}$ with $\Re(s) > \frac{1}{2}$. We define the *transfer operator* $\mathcal{L}_s : B(D) \to B(D)$ by

$$\big(\mathcal{L}_s\phi\big)(z) := \sum_{n\in\mathbb{N}} \left(\frac{1}{n+z}\right)^{2s} \phi\left(\frac{1}{z+n}\right) \tag{6.53}$$

for $\phi \in B(D)$.

Exercise 6.58 Show that $\mathcal{L}_s$ in (6.53) satisfies

$$\big(\mathcal{L}_s\phi\big)(z) = \sum_{n\in\mathbb{N}} \big(-\theta_n'(z)\big)^s \,\phi\big(\theta_n(z)\big) \tag{6.54}$$

for $\phi \in B(D)$.

Exercise 6.59 Show that $\mathcal{L}_s$ in (6.53) is a linear operator, i.e.

$$\big(\mathcal{L}_s\phi + \alpha\psi\big)(z) = \big(\mathcal{L}_s\phi\big)(z) + \alpha\big(\mathcal{L}_s\psi\big)(z) \tag{6.55}$$

for $\phi, \psi \in B(D)$ and $\alpha \in \mathbb{C}$.

Some important analytic properties are given in the next theorem which is based on results by Mayer for the transfer operators $\mathcal{L}_1$ in [134, §1, Lemma] and $\mathcal{L}_s$ in [135, Proposition 1]. One ingredient of the next theorem is the aspect that the transfer operator is a nuclear operator. The fact that the operator is nuclear (of order 0) allows us to define and calculate the trace of the operator. We refer to Appendix A.19 and references therein for a brief introduction and a definition of nuclear operators.

Theorem 6.60 ([135, Proposition 1]) *Let $s \in \mathbb{C}$ with $\Re(s) > \frac{1}{2}$. The transfer operator $\mathcal{L}_s$ well defined on $B(D)$. Moreover, it is a nuclear operator of order zero.*

Proof Let $\phi \in B(D)$ be a function. We see in particular that f is a bounded function since it is continuous on the compact disc $\overline{D}$. This shows

$$\begin{aligned}
\big\|\big(\mathcal{L}_s\phi\big)(z)\big\|_\infty &\le \sum_{n\in\mathbb{N}} \|z+n\|^{-2\Re(s)} \,\big\|\phi\big(\theta_n(\cdot)\big)\big\|_\infty \\
&= \big\|\phi\big\|_\infty \sum_{n\in\mathbb{N}} \|z+n\|^{-2\Re(s)} \\
&< \infty
\end{aligned}$$

using that the series $\sum_{n\in\mathbb{N}} \left|z+n\right|^{-2\Re(s)}$ converges absolutely for all $z \in \overline{D}$. This shows that (6.53) defined an operator on $B(D)$. Hence $\mathcal{L}_s$ is well defined for $\Re(s) > \frac{1}{2}$.

Following the arguments in the proof of the [131, p. 198, Lemma], we have to show that $\mathcal{L}_s$ is a bounded and a linear operator on $B(D)$ and that there exists a neighborhood of the zero function in $B(D)$ which is mapped onto a bounded set in $B(D)$. Exercise 6.59 shows that $\mathcal{L}_s$ is linear.

The boundedness of $\mathcal{L}_s$ remains to be shown. Let

$$K := \left\{ z \in \mathbb{C}: \ |z-1| \leq \frac{3}{2} - \varepsilon \right\},$$

for some $0 < \varepsilon < \frac{1}{2}$ which is the closure of the set in (6.51) and a proper subset of D, and define

$$V_M(0) := \left\{ \phi \in B(D): \ \sup_{z\in K} |\phi(z)| \leq M \right\}$$

for some $M > 0$. Since $\phi_n(z) \in K$, see (6.51) of Lemma 6.55, we find that ϕ_n is a bounded function. We have in particular

$$\left|\phi(z)\right| \leq \frac{5}{2}$$

for all $z \in \overline{D}$.

For every $\phi \in V_M(0)$, we now find the estimate

$$\begin{aligned}
\left\|\mathcal{L}_s\phi\right\|_\infty &\overset{(6.53)}{=} \left\| \sum_{n\in\mathbb{N}} \left(\frac{1}{n+\cdot}\right)^{2s} \phi\left(\frac{1}{\cdot+n}\right) \right\|_\infty \\
&\leq \sum_{n\in\mathbb{N}} \left|\phi(\cdot)\right|^{2\Re(s)} \left| \phi\left(\underbrace{\frac{1}{\cdot+n}}_{\in K}\right) \right| \\
&\overset{(6.52)}{\leq} \sum_{n\in\mathbb{N}} \left(\frac{1}{n-\frac{1}{2}}\right)^{2\Re(s)} \left|\phi\left(\frac{5}{2}\right)\right| \\
&= \left|\phi\left(\frac{5}{2}\right)\right| \sum_{n\in\mathbb{N}} \left(\frac{1}{n-\frac{1}{2}}\right)^{2\Re(s)}
\end{aligned}$$

since $\sum_{n\in\mathbb{N}} \left(\frac{1}{n-\frac{1}{2}}\right)^{2\Re(s)}$ converges absolutely for every s with $\Re(s) > \frac{1}{2}$.

Hence, $\mathcal{L}_s$ is a bounded linear operator on $B(D)$. Applying arguments in [171] and [91] shows that $\mathcal{L}_s$ is nuclear of order 0. □

Then Mayer in [134, pp. 98–99] (for $s = 1$) and [135] (for arbitrary s) proceeds to show that the trace $(\mathcal{L}_s)$ and the Fredholm determinant of $\det\big(1 - z\mathcal{L}_s\big)$ with $z \in \mathbb{C}$ are well defined, using results of Alexandre Grothendieck's theory of nuclear operators in [91]. Without showing the proof, we have the following result:

Proposition 6.61 ([134, 135]) *Let $s \in \mathbb{C}$ with $\Re(s) > \frac{1}{2}$. The operator $\mathcal{L}_s$ is a linear operator which is nuclear of order* 0. *It admits the notion of a trace given by*

$$\operatorname{trace}(\mathcal{L}_s) = \sum_i \lambda_i$$

and the notion of a Fredholm determinant

$$\det(1 - z\mathcal{L}_s) = \prod_i \big(1 - z\lambda_i\big),$$

where λ_i denote the eigenvalues of $\mathcal{L}_s$. The eigenvalues are at most countable many.

Next, we try to extend the range of values $s \in \mathbb{C}$ for which $\mathcal{L}_s$ can be defined. We see already by the argument in the proof of Theorem 6.60 that we have to overcome the problem of convergence of the series $\sum_{n\in\mathbb{N}} \big(z + n\big)^{-2s}$. In particular, we see immediately that $s = \frac{1}{2}$ will be a problem, because we hit here the harmonic series that does not converge.

To facilitate the extension of $\mathcal{L}_s$ to other values of s, we need to recall the Hurwitz ζ function, see Appendix A.16.

Definition 6.62 For $s, q \in \mathbb{C}$ with $\Re(s) > 1$ and $\Re(q) > 0$, the *Hurwitz ζ-function* $\zeta(s, q)$ is given by

$$\zeta(s, q) := \sum_{n=0}^{\infty} (n + q)^{-s}. \tag{6.56}$$

This zeta function extends meromorphically to $s \in \mathbb{C}$ with one simple pole in $s = 1$, see Appendix A.16. The meromorphic extension is also denoted by $\zeta(s, q)$.

Exercise 6.63 Let $m \in \mathbb{Z}_{\geq 0}$ a nonnegative integer and $-m \neq s \in \mathbb{C}$. Calculate formally the image of the monomial

$$p_m(z) := z^m$$

under $\mathcal{L}_s$ and show

$$\big(\mathcal{L}_s p_m\big)(z) = \zeta(2s + m, z + 1) \tag{6.57}$$

for all $z \in \overline{D}$.

Recall that the elements of $B(D)$ are holomorphic functions on D which extend continuously to $\overline{D}$. This means in particular that such functions admit converging Taylor expansions centered around 0

$$\phi(z) = \sum_{m=0}^{\infty} a_m\, z^m.$$

Let $l \in \mathbb{Z}_{\geq 0}$. We define the projections of holomorphic functions onto parts of its Taylor expansion in the following way:

$$\begin{aligned} P_{\leq l}\left(\sum_{m=0}^{\infty} a_m\, z^m\right) &:= \sum_{m=0}^{l} a_m\, z^m = a_0 + a_1^z + a_2 z^2 + \ldots + a_l z^l \quad \text{and} \\ P_{>l}\left(\sum_{m=0}^{\infty} a_m\, z^m\right) &:= \sum_{m=l+1}^{\infty} a_m\, z^m \end{aligned} \tag{6.58}$$

for every $z \in \mathbb{C}$.

We use these projections to define two subspaces $B_{\leq l}(D)$ and $B_{>l}(D)$ of $B(D)$.

$$B(D)_{\leq l} := \left\{P_{\leq l}\phi\colon\ \phi \in B(D)\right\} \quad \text{and} \quad B(D)_{>l} := \left\{P_{>l}\phi\colon\ \phi \in B(D)\right\}. \tag{6.59}$$

Since we have

$$\phi = P_{\leq l}\phi + P_{>l}\phi$$

and

$$0 = P_{\leq l}\big(P_{>l}\phi\big) = P_{>l}\big(P_{\leq l}\phi\big),$$

we see that the space $B(D)$ has the direct composition

$$B(D) = B(D)_{\leq l} \otimes B(D)_{>l}. \tag{6.60}$$

How does $\mathcal{L}_s$ operate on these subspaces? We give the answer in the following two lemmas.

Lemma 6.64 *Let $l \in \mathbb{Z}_{\geq 0}$ be a nonnegative integer and consider $s \in \mathbb{C} \setminus \left\{\frac{1}{2}, 0, \frac{-1}{2}, \ldots, \frac{1-l}{2}\right\}$.*

The transfer operator $\mathcal{L}_s$ is well defined on the subspace $B_{\leq l}(D)$. We have

$$\big(\mathcal{L}_s\phi\big)(z) = \sum_{m=0}^{l} a_m\, \zeta(2s+m, z+1) \tag{6.61}$$

for every $\phi \in B_{\leq l}(D)$ *with*

$$\phi(z) = \sum_{m=0}^{l} a_m \, z^m.$$

Proof Let $\phi \in B_{\leq l}(D)$. According to (6.58)and (6.59), we see that ϕ is a polynomial of degree l, say

$$\phi(z) = \sum_{m=0}^{l} a_m \, z^m.$$

Using the linearity of $\mathcal{L}_s$, see Exercise 6.59, and the calculation in Exercise 6.63, we find

$$\begin{aligned}
\big(\mathcal{L}_s\phi\big)(z) &= \mathcal{L}_s\left(\sum_{m=0}^{l} a_m \, z^m\right) \\
&\overset{(6.55)}{=} \sum_{m=0}^{l} a_m \left(\mathcal{L}_s z^m\right) \\
&\overset{(6.57)}{=} \sum_{m=0}^{l} a_m \, \zeta(2s+m, z+1)
\end{aligned}$$

for all $z \in \overline{D}$. (The Hurwitz ζ function $\zeta(2s+m, z+1)$ is well defined since $\Re\,(2s+m) > 1$ and $\Re\,(z+1) \geq \frac{1}{2}$ for $z \in D$, see (6.48).)

Since $2s+m \neq 1$ for all $m \in \{0, 1, \ldots, l\}$, we see that the expression

$$\sum_{m=0}^{l} a_m \, \zeta(2s+m, z+1)$$

is well defined. This shows that the operator $\mathcal{L}_s$ restricted to the function space $B_{\leq l}(D)$ is well defined. □

Lemma 6.65 *Let* $l \in \mathbb{Z}_{\geq 0}$ *be a nonnegative integer. The transfer operator* $\mathcal{L}_s$ *is well defined on the subspace* $B_{>l}(D)$ *for all* $s \in \mathbb{C}$ *with* $\Re\,(s) > -\frac{l}{2}$.

Proof By construction of $B_{>l}(D)$ in (6.59), we see that every function $\phi \in B_{>l}(D)$ can be written as

$$\phi(z) = z^{l+1}\,\psi(z) \tag{6.62}$$

with $\psi \in B(D)$: since ϕ admits a Taylor expansion of the form $\sum_{m=l+1}^{\infty} a_m z^m$, we can factor out the power z^{l+1} and get

$$\phi(z) = z^{l+1} \sum_{m=l+1}^{\infty} a_m z^{m-l-1} = z^{l+1} \psi(z)$$

with $\psi(z) := \sum_{m=l+1}^{\infty} a_m z^{m-l-1}$.

Following the steps in the proof of Theorem 6.60, we find

$$\begin{aligned}
\left\|(\mathcal{L}_s\phi)(z)\right\|_\infty &\le \sum_{n\in\mathbb{N}} \left\|(z+n)^{-2s}\,\phi(\theta_n(z))\right\|_\infty \\
&\overset{(6.62)}{=} \sum_{n\in\mathbb{N}} \left\|(z+n)^{-2s-l-1}\,\psi(\theta_n(z))\right\|_\infty \\
&\le \sum_{n\in\mathbb{N}} \left\|z+n\right\|^{-2\Re(s)-l-1} \left\|\psi(\theta_n(\cdot))\right\|_\infty \\
&= \left\|\psi\right\|_\infty \sum_{n\in\mathbb{N}} \left|z+n\right|^{-2\Re(s)-l-1} \\
&< \infty
\end{aligned}$$

for all $2\Re(s) > -l$. Hence, $\mathcal{L}_s$ is well defined on the subspace $B_{>l}(D)$ for all $\Re(s) > -\frac{l}{2}$. □

Combining the above two lemmas gives the following.

Theorem 6.66 *Pick a nonnegative integer $l \in \mathbb{Z}_{\ge 0}$.*

The transfer operator $\mathcal{L}_s$ extends meromorphically in s from $\Re(s) > \frac{1}{2}$ to all $s \in \mathbb{C} \setminus \left\{\frac{1}{2}, 0, \frac{-1}{2}, \ldots, \frac{1-l}{2}\right\}$ with $\Re(s) > -\frac{l}{2}$ by

$$\mathcal{L}_s\phi := \left(\mathcal{L}_s P_{\le l}\phi\right) + \left(\mathcal{L}_s P_{>l}\phi\right), \tag{6.63}$$

where the first term $\mathcal{L}_s P_{\le l}\phi$ is defined in the sense of (6.61) *in Lemma 6.64 and the second term $\mathcal{L}_s P_{>l}\phi$ is defined in the sense of* (6.62) *in Lemma 6.65.*

Exercise 6.67 Prove Theorem 6.66 above.

6.2.3 Mayer's Transfer Operator

Mayer extended the transfer operator $\mathcal{L}_s$ in (6.53) towards an operator related to the base points of geodesics in $\mathcal{A}$, where the set $\mathcal{A}$ is given in Definition 6.46 of §6.1.4. One important step was the extension of the base interval $[0, 1)$ of the Gauss map f_G and the associated continued fractions used in the construction of $\mathcal{L}_s$ in §6.2.2 to the direct product $[0, 1) \times [0, 1)$.

Following the construction outlined in [55, §4.4], we extend the transfer operator $\mathcal{L}_s$ in (6.53) as follows:

Definition 6.68 Let $s \in \mathbb{C} \setminus \frac{1}{2}\mathbb{Z}_{\leq 1}$ be a complex number. We define *Mayer's transfer operator* $\tilde{\mathcal{L}}_s : B(D) \times B(D) \to \mathbb{C} \times \mathbb{C}$ by

$$\begin{aligned}\left(\tilde{\mathcal{L}}_s \begin{pmatrix}\phi_1\\ \phi_2\end{pmatrix}\right)(z) :&= \begin{pmatrix}(\mathcal{L}_s\phi_2)(z)\\ (\mathcal{L}_s\phi_1)(z)\end{pmatrix}\\ &= \begin{pmatrix}0 & \mathcal{L}_s\\ \mathcal{L}_s & 0\end{pmatrix}\begin{pmatrix}\phi_1\\ \phi_2\end{pmatrix}(z)\end{aligned} \tag{6.64}$$

with $\phi_1, \phi_2 \in B(D)$.

Exercise 6.69 Let $s \in \mathbb{C} \setminus \frac{1}{2}\mathbb{Z}_{\leq 1}$ be a complex number.

1. Show that $\tilde{\mathcal{L}}_s$ is a well-defined linear operator.
2. If $\Re(s) > \frac{1}{2}$, show that $\tilde{\mathcal{L}}_s$ satisfies

$$\left(\tilde{\mathcal{L}}_s \begin{pmatrix}\phi_1\\ \phi_2\end{pmatrix}\right)(z) = \sum_{n\in\mathbb{N}} \left(\frac{1}{n+z}\right)^{2s} \begin{pmatrix}\phi_1\\ \phi_2\end{pmatrix}\left(\frac{1}{z+n}\right) \tag{6.65}$$

for all $\phi_1, \phi_2 \in B(D)$ with $z \in \overline{D}$.

Exercise 6.70 Let $\phi \in B(D)$ be an eigenfunction of $\mathcal{L}_s$ with eigenvalue λ. Show that $\tilde{\phi}$ defined by

$$\tilde{\phi}(z) := \begin{pmatrix}\phi(z)\\ \pm\phi(z)\end{pmatrix} \qquad (z \in \overline{D}) \tag{6.66}$$

is an eigenfunction of $\tilde{\mathcal{L}}_s$ with eigenvalue $\pm\lambda$.

Exercise 6.71 Let $\lambda \in \mathbb{C}$ and $\phi_1, \phi_x \in B(D)$. The following two statements are equivalent:

1. $\tilde{\mathcal{L}}_s \begin{pmatrix}\phi_1\\ \phi_2\end{pmatrix} = \lambda \begin{pmatrix}\phi_1\\ \phi_2\end{pmatrix}$.
2. $\mathcal{L}_s\phi_2 = \lambda\phi_1$ and $\mathcal{L}_s\phi_1 = \lambda\phi_2$.

Exercise 6.72 Let $\begin{pmatrix}\phi_1\\ \phi_2\end{pmatrix} \in B(D) \times B(D)$ be an eigenfunction of $\tilde{\mathcal{L}}_s$ with eigenvalue λ. Show that $\begin{pmatrix}\phi_1\\ -\phi_2\end{pmatrix}$ is also an eigenfunction of $\tilde{\mathcal{L}}_s$ with eigenvalue $-\lambda$.

Using Proposition 6.61 and the above exercises, we can prove the following result.

Proposition 6.73 ([55, (4.39)–(4.40)]) *Let $s \in \mathbb{C}$ with $\Re(s) > \frac{1}{2}$. The operator $\tilde{\mathcal{L}}_s$ is a linear operator which is nuclear of order 0. It admits the notion of a trace given by*

$$\operatorname{trace}\left(\tilde{\mathcal{L}}_s\right) = \sum_i \lambda_i$$

and the notion of a Fredholm determinant

$$\det\left(1 - \tilde{\mathcal{L}}_s\right) = \prod_i \left(1 - \lambda_i\right),$$

where λ_i denote at most countable eigenvalues of $\tilde{\mathcal{L}}_s$.

The determinant $\det\left(1 - \tilde{\mathcal{L}}_s\right)$ *can be calculated by* $\det\left(1 \pm \mathcal{L}_s\right)$ *using the relation*

$$\begin{aligned}\det\left(1 - \tilde{\mathcal{L}}_s\right) &= \det\left(1 - \left(\mathcal{L}_s\right)^2\right)\\ &= \det\left(1 + \mathcal{L}_s\right)\,\det\left(1 - \mathcal{L}_s\right),\end{aligned} \tag{6.67}$$

where $\mathcal{L}_s$ denotes the transfer operator given in Definition 6.57.

Proof We see that $\tilde{\mathcal{L}}_s$ is a nuclear operator of order 0 since it is a matrix composition of the nuclear operator $\mathcal{L}_s$ of order 0, see Definition 6.68 and Theorem 6.60. As in Proposition 6.61, Grothendiek's theory of nuclear operators [91] shows that we can define the trace and the determinant. (We refer also to the arguments presented in [55] and the references therein.)

Hence, the expression of $\det\left(1 - \tilde{\mathcal{L}}_s\right)$ as Fredholm determinant is well defined. Using Exercise 6.71, we can write $\det\left(1 - \tilde{\mathcal{L}}_s\right)$ as

$$\det\left(1 - \tilde{\mathcal{L}}_s\right) = \det\left(1 - \left(\mathcal{L}_s\right)^2\right)$$

Then, using the multiplicative property of determinants and the third binomial formula $\left(a^2 - b^2\right) = (a + b)(a - b)$, we find

$$\det\left(1 - \left(\mathcal{L}_s\right)^2\right) = \det\left(1 + \mathcal{L}_s\right)\,\det\left(1 - \mathcal{L}_s\right).$$

This shows (6.67) and completes the proof. □

Remark 6.74 The above proposition shows in particular the following: If we are interested in eigenfunctions of Mayer's transfer operator $\tilde{\mathcal{L}}_s$ with eigenvalue 1, we can equivalently look at eigenfunctions of the transfer operator $\mathcal{L}_s$ with eigenvalues ± 1. This will be very useful, since we relate later eigenfunction of $\mathcal{L}_s$ with eigenvalue ± 1 to period functions, see §6.3. In §6.4, we will relate part of the values of s where $\det\left(1 - \tilde{\mathcal{L}}_s\right)$ vanishes indirectly to the existence of Maass cusp forms of weight 0 and the trivial multiplier for the full modular group $\mathrm{SL}_2(\mathbb{Z})$.

6.3 The Transfer Operator and Period Functions on $\Gamma(1)$

Assume now that we have $s \in \mathbb{C}$. Recall the defining equation (6.53) of the transfer operator $\mathcal{L}_s$. Since we are interested in eigenfunctions $\phi \in B(D)$, we assume that there exists a $\lambda \in \mathbb{C}_{\neq 0}$ with

$$\mathcal{L}_s \phi = \lambda \phi. \tag{6.68}$$

Using (6.53), the above eigenvalue equation can be written as

$$\sum_{n \in \mathbb{N}} \left(\frac{1}{n+z}\right)^{2s} \phi\left(\frac{1}{z+n}\right) = \lambda \phi(z) \tag{6.69}$$

for all $z \in \overline{D}$ and D being the disc given in (6.48). Now, some reindexing of the left-hand side gives

$$\begin{aligned}
\lambda \phi(z) &\overset{(6.69)}{=} \sum_{n \in \mathbb{N}} \left(\frac{1}{n+z}\right)^{2s} \phi\left(\frac{1}{z+n}\right) \\
&= \left(\frac{1}{1+z}\right)^{2s} \phi\left(\frac{1}{z+1}\right) + \sum_{n=2}^{\infty} \left(\frac{1}{n+z}\right)^{2s} \phi\left(\frac{1}{z+n}\right) \\
&= \left(\frac{1}{1+z}\right)^{2s} \phi\left(\frac{1}{z+1}\right) + \sum_{n=1}^{\infty} \left(\frac{1}{n+z+1}\right)^{2s} \phi\left(\frac{1}{z+n+1}\right) \\
&\overset{(6.53)}{=} \left(\frac{1}{1+z}\right)^{2s} \phi\left(\frac{1}{z+1}\right) + (\mathcal{L}_s \phi)(z+1) \\
&\overset{(6.68)}{=} \left(\frac{1}{1+z}\right)^{2s} \phi\left(\frac{1}{z+1}\right) + \lambda \phi(z+1)
\end{aligned}$$

for all $z \in \overline{D}$ satisfying $z + 1 \in \overline{D}$. Dividing by λ gives

$$\phi(z) = \phi(z+1) + \lambda^{-1} \left(\frac{1}{1+z}\right)^{2s} \phi\left(\frac{1}{z+1}\right) \tag{6.70}$$

for all $z \in \overline{D}$ satisfying $z+1 \in \overline{D}$.

Hence we have the following lemma.

Lemma 6.75 *Let $s \in \mathbb{C}$ with $\Re(s) > \frac{1}{2}$. Assume that $\phi \in B(D)$ is an eigenfunction of $\mathcal{L}_s$ with nonvanishing eigenvalue $\lambda \neq 0$, i.e., ϕ satisfies* (6.68). *Then ϕ satisfies the* three-term equation (6.70).

A converse statement can also be easily obtained with the additional assumption $\Re(s) > \frac{1}{2}$ on s. (We can relax the restriction $\Re(s) > \frac{1}{2}$ to $\Re(s) > -\frac{l}{2}$ for some $l \in \mathbb{Z}_{\geq 0}$ if we restrict our ϕ to be an element in the subspace $P_{>l}B(D)$, see Theorem 6.66.)

Lemma 6.76 *Let $s \in \mathbb{C}$ with $\Re(s) > \frac{1}{2}$ and $\lambda \in \mathbb{C}_{\neq 0}$. Assume that a set W with $\overline{D} \subset W \subset \{z \in \mathbb{C}\colon\ \Re(z) > -1\}$ exists such that for all $z \in W$, we have $z+1$ and $\frac{1}{1+z} \in W$. Let $\phi : W \to \mathbb{C}$ be a function which is holomorphic on the interior of W and satisfies the three-term equation* (6.70). *Additionally, assume that ϕ satisfies the limit*

$$\lim_{N\to\infty} \phi(z+N) = 0.$$

Then the $\overline{D}$ restricted function $\tilde{\phi} := \phi|_{\overline{D}}$ is an element of $B(D)$ and satisfies

$$\left(\mathcal{L}_s\tilde{\phi}\right)(z) = \lambda^{-1}\, \tilde{\phi}(z) \tag{6.71}$$

for all $z \in \overline{D}$.

Proof First, we show that $\phi(z+N)$ is well defined for every $z \in W$ and all $N \in \mathbb{N}$. This follows from the assumption on W: if $z \in W$, then $z+1$ and $\frac{1}{z+1}$ are in W. Applying the same assumption to $z+1 \in W$ shows that $z+2$ and $\frac{1}{z+2}$ are in W. Iterating this argument shows that $z+N$ and $\frac{1}{z+N}$ are in W for every $N \in \mathbb{N}$. A consequence of this argument is that the assumption on the limit, $\lim_{N\to\infty} \psi(z+N)$ in the lemma above, is well posed.

Now, consider the restriction

$$\tilde{\phi} := \phi|_{\overline{D}}.$$

The assumptions on ϕ show that $\tilde{\phi}$ is holomorphic on D and continuous on $\overline{D}$. Hence, Definition 6.56 shows $\tilde{\phi} \in B(D)$.

For $z \in W$ consider the three-term equation (6.70) repeatedly. We have

$$\begin{aligned}
\phi(z) &\overset{(6.70)}{=} \lambda^{-1}\left(\frac{1}{1+z}\right)^{2s}\phi\left(\frac{1}{z+1}\right)+\phi(z+1)\\
&\overset{(6.70)}{=} \lambda^{-1}\left(\frac{1}{1+z}\right)^{2s}\phi\left(\frac{1}{z+1}\right)+\lambda^{-1}\left(\frac{1}{2+z}\right)^{2s}\phi\left(\frac{1}{z+2}\right)+\phi(z+2)\\
&= \dots\\
&\overset{(6.70)}{=} \lambda^{-1}\left(\frac{1}{1+z}\right)^{2s}\phi\left(\frac{1}{z+1}\right)+\lambda^{-1}\left(\frac{1}{2+z}\right)^{2s}\phi\left(\frac{1}{z+2}\right)+\dots\\
&\qquad \dots+\lambda^{-1}\left(\frac{1}{N+z}\right)^{2s}\phi\left(\frac{1}{z+N}\right)+\phi(z+N+1)
\end{aligned}$$

for every $N \in \mathbb{N}$. The appearing $z+1,\dots,z+N$ and $\frac{1}{z+1},\dots,\frac{1}{z+N}$ are all in W, since we assume that for $z+n$ also $z+n+1$ and $\frac{1}{z+n+1}$ are contained in W. Taking the limit $N \to \infty$, we see that ϕ satisfies

$$\phi(z)=\lambda^{-1}\left(\sum_{n=1}^{\infty}\left(\frac{1}{n+z}\right)^{2s}\phi\left(\frac{1}{z+n}\right)\right) \quad + \lim_{N\to\infty}\psi(z+N).$$

The limit $\lim_{N\to\infty}\psi(z+N)$ on the right-hand side vanished by assumption. We see that ψ satisfies

$$\phi(z)=\lambda^{-1}\left(\sum_{n=1}^{\infty}\left(\frac{1}{n+z}\right)^{2s}\phi\left(\frac{1}{z+n}\right)\right).$$

We note that the series on the right-hand side is well defined. Since ψ is continuous in $0 \in \overline{D} \subset W$ by assumption and $\lim_{N\to\infty}\frac{1}{1+N} = 0$, we see that we can bound the terms $\phi\left(\frac{1}{z+n}\right)$ by some $M > 0$ for all $n \in \mathbb{N}$ uniformly in n. The absolute convergence now follows from $\Re(s) > \frac{1}{2}$:

$$\begin{aligned}
\sum_{n=1}^{\infty}\left|\left(\frac{1}{n+z}\right)^{2s}\phi\left(\frac{1}{z+n}\right)\right| &\le M\sum_{n=1}^{\infty}\left|\left(\frac{1}{n+z}\right)^{2s}\right|\\
&= M\sum_{n=1}^{\infty}\left|\frac{1}{n+z}\right|^{2\Re(s)}\\
&< \infty.
\end{aligned}$$

Summarizing, we have shown that ψ satisfies

$$\lambda\,\phi(z) = \left(\sum_{n=1}^{\infty} \left(\frac{1}{n+z}\right)^{2s} \phi\left(\frac{1}{z+n}\right)\right).$$

Using (6.53), we find

$$\big(\mathcal{L}_s\phi\big)(z) = \lambda\,\phi(z)$$

for all $z \in \overline{D} \subset W$. This proves (6.71). □

In the following subsection §6.3.1, we discuss the relation between eigenfunctions of the transfer operator $\mathcal{L}_s$ and solutions of the three-term equation (6.70) in a more explicit way.

Also, the three-term equation in (6.70) and the three-term equation of period functions in (5.1) are related, as the naming of these functional equations suggests. We will work out this relation in §6.3.2.

6.3.1 Eigenfunctions of the Transfer Operator and Solutions of a Three-Term Equation

We extend our assumption on the parameter s to $s \in \mathbb{C}$, $\Re(s) > 0$ throughout this section.

We can extend solutions of the three-term equation (6.70) to functions on the *cut-plane*

$$\tilde{\mathbb{C}}' = \mathbb{C} \smallsetminus \mathbb{R}_{\le -1}. \tag{6.72}$$

Proposition 6.77 *Let $s \in \mathbb{C}$ with $\Re(s) > 0$ and $\lambda \in \mathbb{C}_{\neq 0}$. The following statements are equivalent:*

1. $\mathcal{L}_s\phi = \pm\phi$,
2. *$\phi \in B(D)$ satisfies the three-term equation* (6.70), *and*
3. *ϕ extends to a holomorphic function on the cut-plane $\tilde{\mathbb{C}}'$ which satisfies the three-term equation* (6.70).

We split the proof of Proposition 6.77 into several steps. Lemma 6.78 establishes that every solution of the three-term equation 6.70 extends to a set which embeds the positive real line. Then, Lemma 6.79 establishes that we can move between solutions of the three-term equation and eigenfunctions of the transfer operator $\mathcal{L}_s$, where Lemma 6.81 deals with one of the additional assumptions used in Lemma 6.79. Next, we extend holomorphic solutions of the three-term equation from the set containing the positive real line to $\tilde{\mathbb{C}}'$ using the three steps given in the Lemmas 6.82–6.84.

Lemma 6.78 *Let $s \in \mathbb{C}$, $\lambda \in \mathbb{C}_{\neq 0}$ and assume that $\phi \in B(D)$ satisfies the three-term equation* (6.70). *The function ϕ extends holomorphically to the set*

$$\tilde{D} := \bigcup_{n=0}^{\infty} n + \overline{D}$$
$$= D \cup \{z+1\colon\ z \in \overline{D}\} \cup \{z+2\colon\ z \in \overline{D}\} \cup \{z+3\colon\ z \in \overline{D}\} \cup \ldots \tag{6.73}$$

and satisfies the three-term equation (6.70) *on $\tilde{D}$.*

Proof We can write the set $\tilde{D}$ as

$$\tilde{D} \overset{(6.73)}{=} \bigcup_{n=0}^{\infty} n + \overline{D}$$
$$= D \cup \{z+1\colon\ z \in \overline{D}\} \cup \{z+2\colon\ z \in \overline{D}\} \cup \{z+3\colon\ z \in \overline{D}\} \cup \ldots$$
$$\overset{(6.48)}{=} \bigcup_{n=0}^{\infty} \left\{ z \in \mathbb{C}\colon\ |z-n-1| \leq \frac{3}{2} \right\}.$$

Let $s \in \mathbb{C}$ and $\lambda \in \mathbb{C}_{\neq 0}$. We assume that $\phi \in B(D)$ satisfies the three-term equation (6.70) and extend ϕ holomorphically by induction.

$N = 0$ The function ϕ is well defined on $\overline{D} \overset{(6.48)}{=} \left\{ z \in \mathbb{C}\colon\ |z-n-1| \leq \frac{3}{2} \right\}$ with $n = 0$ by assumption and satisfies the three-term equation (6.70), see Lemma 6.75.

$N \to N+1$ Assume that ϕ extends already holomorphically to $\bigcup_{n=0}^{N} n + \overline{D}$ and satisfies the three-term equation (6.70) there. Consider the rewritten three-term equation

$$\phi(z) \overset{(6.70)}{=} \phi(z-1) - \lambda^{-1} \left(\frac{1}{z}\right)^{2s} \phi\left(\frac{1}{z}\right) \tag{6.74}$$

derived by (6.70) and the substitution $z \mapsto z-1$. Let $z \in N+1+\overline{D}$. We first notice that $z \neq 0$ by construction since $|z| \geq \frac{1}{2}$ by the construction of the set $N+1+\overline{D}$. Hence $\left|\frac{1}{z}\right| \leq 2$ follows. Moreover, we have

$$\frac{1}{z} \in D \subset \bigcup_{n=0}^{N} n + \overline{D} \qquad \text{and} \qquad z-1 \in \bigcup_{n=0}^{N} n + \overline{D}.$$

We see in particular that the right side of (6.74) can be calculated since ϕ is well defined in $\frac{1}{z}$ and $z-1$. Next, we extend ϕ by defining $\phi(z)$ by the right side of (6.74): Put

$$\phi(z) \overset{(6.70)}{:=} \phi(z-1) - \lambda^{-1}\left(\frac{1}{z}\right)^{2s}\phi\left(\frac{1}{z}\right).$$

ϕ being holomorphic on $\bigcup_{n=0}^{N} n+\overline{D}$ implies that the right side of (6.74) is holomorphic. Hence the $\bigcup_{n=0}^{N+1} n+\overline{D}$ extended to ϕ is holomorphic.

This induction proves the extension of ϕ to $\bigcup_{n=0}^{\infty} n+\overline{D}$ and that ϕ is holomorphic on the interior of $\bigcup_{n=0}^{\infty} n+\overline{D}$. □

Next, we extend the connection between eigenfunctions of $\mathcal{L}_s$ and solutions to the three-term equation (6.70), extending the two Lemmas 6.75 and 6.76 to $\Re(s)>0$.

Lemma 6.79 *Let $s \in \mathbb{C}$ with $\Re(s) > 0$ and $\lambda \in \mathbb{C}_{\neq 0}$.*

1. *Assume that $\phi \in B(D)$ is an eigenfunction of $\mathcal{L}_s$ with eigenvalue $\lambda \neq 0$, i.e., ϕ satisfies* (6.68). *Then ϕ satisfies the* three-term equation (6.70).
2. *Assume that a set W with $\overline{D} \subset W \subset \{z \in \mathbb{C}\colon\ \Re(z) > -1\}$ exists such that for all $z \in W$, we have $z+1$ and $\frac{1}{1+z} \in W$:*

$$\begin{cases} \overline{D} \subset W \subset \{z \in \mathbb{C}\colon\ \Re(z) > -1\} \\ z \in W \quad \Longrightarrow \quad z+1,\ \frac{1}{z+1} \in W. \end{cases} \tag{6.75}$$

Let $\phi : W \to \mathbb{C}$ be a function which is holomorphic on the interior of W and satisfies the three-term equation (6.70). *Additionally, assume that ϕ satisfies the limit*

$$\lim_{N\to\infty} \phi(z+N) = 0$$

and that it admits the estimate

$$\phi(z) = O(z)$$

near $z = 0$.

Then the $\overline{D}$ restricted function $\tilde{\phi} := \phi|_{\overline{D}}$ is an element of $B(D)$ and satisfies (6.71) *for all $z \in \overline{D}$.*

Remark 6.80 The condition $\phi(z) = O(z)$ near $z = 0$ implies that the holomorphic ϕ admits a Taylor expansion centered around $z = 0$ with no constant term. Using the operators $P_{\leq l}$ and $P_{>0}$ in (6.58), this means

$$\phi(z) = O(z) \quad \Longleftrightarrow \quad \Big(P_{\leq 0}\phi = 0 \quad \text{and} \quad P_{>0}\phi = \phi\Big). \tag{6.76}$$

Proof *(of Lemma 6.79)* The first part is just a rephrasing of Lemma 6.75.

Now consider the second part of the lemma. Following the steps in the proof of Lemma 6.76, we see that ϕ satisfies

$$\phi(z) = \lambda^{-1}\left(\frac{1}{1+z}\right)^{2s}\phi\left(\frac{1}{z+1}\right) + \lambda^{-1}\left(\frac{1}{2+z}\right)^{2s}\phi\left(\frac{1}{z+2}\right) + \dots$$
$$\dots + \lambda^{-1}\left(\frac{1}{N+z}\right)^{2s}\phi\left(\frac{1}{z+N}\right) + \phi(z+N+1)$$

for every $N \in \mathbb{N}$ and $z \in W$.

Next, we would like to take the limit $N \to \infty$. Using the additional assumption $\phi(z) = O(z)$ near $z = 0$ implies that the limit

$$\lim_{N\to\infty}\sum_{n=1}^{N}\left(\frac{1}{n+z}\right)^{2s}\phi\left(\frac{1}{z+n}\right) = \sum_{n=1}^{\infty}\left(\frac{1}{n+z}\right)^{2s}\phi\left(\frac{1}{z+n}\right)$$

exists since

$$\begin{aligned}\sum_{n=1}^{\infty}\left|\left(\frac{1}{n+z}\right)^{2s}\phi\left(\frac{1}{z+n}\right)\right| &= \sum_{n=1}^{\infty}\left|\frac{1}{n+z}\right|^{2\Re(s)} O\left(\frac{1}{z+n}\right)\\ &\le \sum_{n=1}^{\infty} O\left(\left|\frac{1}{n+z}\right|^{2\Re(s)+1}\right)\\ &< \infty.\end{aligned}$$

This shows that the series converges absolutely for all $\Re(s) > 0$.

Using the assumption that $\lim_{N\to\infty}\phi(z+N) = 0$ allows us to split the limit of

$$\lambda^{-1}\lim_{N\to\infty}\left[\sum_{n=1}^{N}\left(\frac{1}{n+z}\right)^{2s}\phi\left(\frac{1}{z+n}\right)\right] + \phi(z+N+1)$$

into the two existing limits $\sum_{n=1}^{\infty}\left(\frac{1}{n+z}\right)^{2s}\phi\left(\frac{1}{z+n}\right)$ and $\lim_{N\to\infty}\phi(z+N) = 0$. Hence, we have shown that ϕ satisfies

$$\phi(z) = \lambda^{-1}\sum_{n=1}^{\infty}\left(\frac{1}{n+z}\right)^{2s}\phi\left(\frac{1}{z+n}\right)$$

or, equivalently,

$$\lambda\phi = \mathcal{L}_s\phi.$$

□

Lemma 6.81 *Let W denote a set satisfying* (6.75) *in Lemma 6.79. Assume that $\psi : W \to \mathbb{C}$ is holomorphic, satisfies the three-term equation* (6.70) *on W, and satisfies the growth condition*

$$\lim_{N\to\infty} \psi(z+N) = 0$$

for all $z \in W$. Then ψ satisfies the estimate

$$\psi(z) = O(z)$$

near $z = 0$, i.e., ψ admits a locally convergent Taylor expansion around $z = 0$ without constant term.

Proof We exploit the three-term equation (6.70) together with the assumption on the limit behavior of ψ. Using the three-term equation, we find that ψ satisfies

$$\phi(z) - \phi(z+1) \overset{(6.70)}{=} \lambda^{-1} \left(\frac{1}{1+z}\right)^{2s} \phi\left(\frac{1}{z+1}\right).$$

Taking the limit $z \to \infty$ shows that the left side converges to 0:

$$\begin{aligned} \lim_{z\to\infty} \phi(z) - \phi(z+1) &= \lim_{z\to\infty} \phi(z) - \lim_{z\to\infty} \phi(z+1) \\ &= 0 - 0 = 0. \end{aligned}$$

Hence the limit of the right side for $z \to \infty$ vanishes. Since $\frac{1}{z+1} \longrightarrow 0$ for $z \to \infty$, we find

$$0 = \lim_{z\to\infty} \phi\left(\frac{1}{z+1}\right) = \lim_{z\to 0} \phi(z).$$

Using that ψ is holomorphic implies that $\psi(z)$ admits a locally convergent Taylor series near $z = 0$, we see that ψ satisfies the estimate

$$\psi(z) = O(z)$$

near $z = 0$ and that the constant term in the Taylor expansion vanishes. This completes the proof of the lemma. □

At this point, we discussed the relation between eigenfunction of the transfer operator $\mathcal{L}_s$ and solutions ϕ of the three-term equation (6.70) on a domain $\tilde{D}$ in (6.73). This domain $\tilde{D}$ is one example of the domain W assumed in (6.75) of Lemma 6.79.

Next, we would like to extend the domain $\tilde{D}$ to $\tilde{\mathbb{C}}'$ given in (6.72). We do this in several steps:

1. Extend the domain $\tilde{D}$ to a "wedge-shaped" domain $W_\delta \cup \overline{D}$ with $W_\delta = \{z \in \mathbb{C}\colon\ \Re(z) > 0$ and $|\arg(z)| \le \delta\}$ for some explicitly given $0 < \delta < \frac{\pi}{2}$.
2. Extend the "wedge-shaped" domain $W_\delta \cup \overline{D}$ to $\overline{D} \cup \{z \in \mathbb{C}\colon\ \Re(z) > 0\}$.
3. Extend $D \cup \{z \in \mathbb{C}\colon\ \Re(z) > 0\}$ iteratively to $\tilde{\mathbb{C}}'$.

Lemma 6.82 *Let $\Re(s) > 0$ and let ϕ be a solution of the three-term equation* (6.70) *on the domain $\tilde{D}$ given in* (6.73) *with the additional condition*

$$\lim_{N\to\infty} \phi(z+N) = 0 \qquad \textit{for all} \quad z \in \tilde{D}.$$

Then, there exists a $0 < \delta < \frac{\pi}{2}$ such that ϕ extends to a solution of the three-term equation on $\overline{D} \cup W_\delta$ (holomorphic on the interior and continuous on the closure). The wedge W_δ is given by

$$W_\delta = \left\{z \in \mathbb{C};\ |\arg(z)| \le \delta\right\}. \tag{6.77}$$

Proof Consider the point $\zeta \in \mathbb{C}$ with $\Im(\zeta) > 0$ and $\Re(\zeta) > 0$ that lies in the intersection between the boundaries of the two circles D and $D + 1$ where D is given in (6.48):

$$\zeta \in \left\{z \in \mathbb{C}\colon\ |z-1| = \frac{3}{2}\right\} \cap \left\{z \in \mathbb{C}\colon\ |z-2| = \frac{3}{2}\right\}.$$

A simple geometric calculation implies that ζ is given by

$$\zeta = \frac{3}{2} + i\sqrt{2}.$$

Put $\delta := \arg(\zeta) \in \left(0, \frac{\pi}{2}\right)$ and consider the wedge W_δ given in (6.77).

Let $z \in \overline{D} \cup W_\delta$. If $\Re(z) \le \frac{3}{2}$, then surely $z \in \overline{D}$ and $\phi(z)$ is well defined by assumption. Assume $z \in \overline{D} \cup W_\delta$ with $\Re(z) > \frac{3}{2}$. Using the second part of Lemma 6.79 shows that $\phi|_{\overline{D}}$ is an eigenfunction of $\mathcal{L}_s$ with eigenvalue λ^{-1}:

$$\phi(\tilde{z}) = \lambda^{-1} \sum_{n\in\mathbb{N}} \left(\frac{1}{\tilde{z}+n}\right)^{2s} \phi\left(\frac{1}{\tilde{z}+n}\right).$$

Since $|z+n| \ge \frac{3}{2} + 1 = \frac{5}{2}$, we have

$$\left|\frac{1}{z+n}\right| \le \frac{2}{5} < \frac{1}{2}$$

which implies $\frac{1}{z+n} \in \overline{D}$. Hence, we can use the above eigenvalue equation to define $\phi(z)$ by the right side

$$\phi(z) := \lambda^{-1} \sum_{n\in\mathbb{N}} \left(\frac{1}{z+n}\right)^{2s} \phi\left(\frac{1}{z+n}\right).$$

This shows that ϕ can be extended to $\left\{z \in W_\delta : \Re(z) \geq \frac{3}{2}\right\}$. Hence, we have extended ϕ to $W_\delta \cup \overline{D}$, continuously on the whole domain and holomorphic on the interior. □

Lemma 6.83 *Let ϕ be a solution of the three-term equation* (6.70)*, given on the domain $\overline{D} \cup W_\delta$ with W_δ derived in Lemma 6.82. ϕ is holomorphic in the interior and continuous on the closure of $\overline{D} \cup W_\delta$. Then ϕ extends holomorphically to $\overline{D} \cup \{z \in \mathbb{C} : \Re(z) > 0\}$.*

Proof Take $z \in \{z \in \mathbb{C} : \Re(z) > 0\}$. If $z \in \overline{D} \cup W_\delta$ we are done. Therefore, we assume $z \notin \overline{D} \cup W_\delta$. This means in particular that $|\Im(z)| \geq \sqrt{2}$, since the boundary of D intersects the imaginary axis in $\pm i\sqrt{2}$. Hence, $z + n$ satisfies

$$\begin{aligned} |z+n| &\geq |z+1| \\ &\geq \sqrt{(\Re(z)+1)^2 + \Im(z)^2} \\ &\geq \sqrt{1+2} \\ &= \sqrt{3} \end{aligned}$$

for every $n \in \mathbb{N}$. This shows

$$\left|\frac{1}{z+n}\right| \leq \frac{1}{\sqrt{3}} \leq \frac{3}{5}.$$

We conclude that $\frac{1}{z+n} \in D$ for every $n \in \mathbb{N}$.

Now, we use the three-term equation (6.70) iteratively (as in the proof of Lemma 6.76) and find

$$\begin{aligned} \phi(z) &\overset{(6.70)}{=} \lambda^{-1} \left(\frac{1}{1+z}\right)^{2s} \phi\left(\frac{1}{z+1}\right) + \phi(z+1) \\ &\overset{(6.70)}{=} \lambda^{-1} \left(\frac{1}{1+z}\right)^{2s} \phi\left(\frac{1}{z+1}\right) + \lambda^{-1} \left(\frac{1}{2+z}\right)^{2s} \phi\left(\frac{1}{z+2}\right) + \phi(z+2) \\ &\overset{(6.70)}{=} \lambda^{-1} \left(\frac{1}{1+z}\right)^{2s} \phi\left(\frac{1}{z+1}\right) + \lambda^{-1} \left(\frac{1}{2+z}\right)^{2s} \phi\left(\frac{1}{z+2}\right) + \ldots \end{aligned}$$

$$\dots + \lambda^{-1}\left(\frac{1}{N+z}\right)^{2s} \phi\left(\frac{1}{z+N}\right) + \phi(z+N+1)$$

for every $N \in \mathbb{N}$. We stop if $z+N+1 \in W_\delta$ which has to happen eventually, since W_δ is a wedge-shaped domain. This allows us to write $\psi(z)$ in terms of the form $\lambda^{-1}\left(\frac{1}{z+n}\right)^{2s} \phi\left(\frac{1}{z+n}\right)$ and $\phi(z+N+1)$ that are well defined since all points $\frac{1}{z+n}$ and the point $z+N+1$ are in $\overline{D} \cup W_\delta$. We also see that this extension of ψ is done holomorphically. □

Lemma 6.84 *Let ϕ be a solution of the three-term equation* (6.70)*, given on the domain $\overline{D} \cup \{z \in \mathbb{C}\colon \Re(z) > 0\}$ where ϕ is holomorphic in the interior. Then ϕ extends holomorphically to $\tilde{\mathbb{C}}' = \mathbb{C} \setminus \mathbb{R}_{\leq -1}$.*

Proof Assume that ϕ satisfies the three-term equation (6.70) on $\overline{D} \cup \{z \in \mathbb{C}\colon \Re(z) > 0\}$. As a first step, we extend the domain of ϕ to $\{z \in \mathbb{C}\colon \Re(z) > -1\}$. Consider $z \in \mathbb{C}$ with $\Re(z) > -1$. Hence $\Re(z+1) > 0$ and $\Re\left(\frac{1}{z+1}\right) > 0$. We have

$$z+1, \frac{1}{z+1} \in \overline{D} \cup \{z \in \mathbb{C}\colon \Re(z) > 0\}.$$

This allows us to use the three-term equation to define the extension of ϕ

$$\phi(z) \overset{(6.70)}{:=} \lambda^{-1}\left(\frac{1}{z+1}\right)^{2s} \phi\left(\frac{1}{z+1}\right) + \phi(z+1).$$

Hence, we have extended (holomorphically) ϕ to $\{z \in \mathbb{C}\colon \Re(z) > -1\}$.

As a next step, assume that $z \in \tilde{\mathbb{C}}'$ with $-n-1 < \Re(z) \leq -n$ for some $n \in \mathbb{N}$. Again, we see trivially, that $z+1$ satisfies $\Re(z+1) > -n$. Also $\Re\left(\frac{1}{z+1}\right) > -1$ since $\left|\frac{1}{z+1}\right| < 1$. Assuming that we already extended ϕ to the domain $\{z \in \tilde{\mathbb{C}}'\colon \Re(z) > -n\}$, then we extend ϕ to $\{z \in \tilde{\mathbb{C}}'\colon \Re(z) > -n-1\}$ by the three-term equation as stated above. Iterating the last step shows that ϕ can be extended holomorphically to $\tilde{\mathbb{C}}$. □

6.3.2 Solutions of a Three-Term Equation and Period Functions

Let $\nu \in \mathbb{C}$ and consider a period function P with trivial multiplier $v = \mathbf{1}$ in the sense of Definition 5.1: P is a holomorphic function on the cut-plane $\mathbb{C}' = \mathbb{C} \setminus (-\infty, 0]$, solving the three-term equation

$$P(\zeta) = P(\zeta+1) + (\zeta+1)^{2\nu-1}\, P\left(\frac{\zeta}{\zeta+1}\right) \tag{5.1}$$

for all $\zeta \in \mathbb{C}'$, and satisfying the growth condition

$$P(\zeta) = \begin{cases} O\left(z^{\max\{0,2\Re(\nu)-1\}}\right) & \text{as } \Im(z)=0,\ \zeta \searrow 0 \text{ and} \\ O\left(z^{\min\{0,2\Re(\nu)-1\}}\right) & \text{as } \Im(z)=0,\ \zeta \to \infty. \end{cases} \tag{5.2}$$

Exercise 6.85 Let $P : \mathbb{C}' \to \mathbb{C}$ be a period function in the sense of Definition 5.1 with parameter $\nu \in \mathbb{C}$ and trivial multiplier $\upsilon = \mathbf{1}$.

If, additionally, P satisfies

$$P\big|_{1-2\nu} C = \pm P \tag{6.78}$$

using the modified Möbius transformation given in (6.19) and $C = \begin{pmatrix} 0 & 1 \\ 1 & 0 \end{pmatrix}$ in (6.5), then P satisfies the functional equation

$$P(\zeta) = P(\zeta+1) \pm \zeta^{2\nu-1}\, P\left(1+\frac{1}{\zeta}\right) \tag{6.79}$$

for all $\zeta \in \mathbb{C}'$.

Exercise 6.86 Let $P : \mathbb{C}' \to \mathbb{C}$ be a function satisfying the three-term equation (5.1) with trivial multiplier $\upsilon = \mathbf{1}$ and some parameter $\nu \in \mathbb{C}$.

Show that the function $P\big|_{1-2\nu} C$ also satisfies the three-term equation (5.1) with trivial multiplier $\upsilon = \mathbf{1}$ and the same parameter $\nu \in \mathbb{C}$.

Exercise 6.87 Let $P : \mathbb{C}' \to \mathbb{C}$ be a function satisfying the three-term equation 6.79 with given sign $\pm$ and some parameter $\nu \in \mathbb{C}$.

Show that P is an eigenfunction under the map $P \mapsto P\big|_{1-2\nu} C$ with the same sign $\pm$.

The above exercise shows that we can split the solution space of the three-term equation (5.1).

Lemma 6.88 *Let FE_s denote the space of functions solving the three-term equation* (5.1) *and let $\widetilde{FE}_s^{\pm}$ denote the space of functions solving* (6.79). *We have*

$$FE_s = \widetilde{FE}_s^{+} \oplus \widetilde{FE}_s^{-}. \tag{6.80}$$

Proof Recall the matrix identity $C^2 = \mathbf{1}$ in (6.13). This implies that the map

$$P \mapsto P\big|_{1-2\nu} C \tag{6.81}$$

is an involution, i.e., applying the map twice gives the identity. This can also be calculated directly:

$$\begin{aligned}\left(\left(P\big|_{1-2\nu}C\right)\big|_{1-2\nu}C\right)(\zeta) &= \zeta^{2\nu-1}\left(P\big|_{1-2\nu}C\right)\left(\frac{1}{\zeta}\right)\\ &= \zeta^{2\nu-1}\left(\frac{1}{\zeta}\right)^{2\nu-1} P\left(\frac{1}{\frac{1}{\zeta}}\right)\\ &= P(\zeta).\end{aligned}$$

Hence, we can decompose every function P into its $+1$-eigenfunction part and its -1-eigenfunction part under (6.81) by considering

$$P \mapsto \frac{1}{2}\left(P + P\big|_{1-2\nu}C\right)$$

and

$$P \mapsto \frac{1}{2}\left(P - P\big|_{1-2\nu}C\right).$$

Exercise 6.86 shows that the above images solve also the three-term equation (5.1) and are eigenfunctions under the map (6.81) with eigenvalue $+1$ or -1 respectively. Exercise 6.85 implies that these functions satisfy the functional equation (6.79) with sign $+$ respectively $-$.

For the reverse direction, apply Exercise 6.87 to a solution of the functional equation (6.79). We see that each solution of (6.79) is an ± 1-eigenfunction to the map (6.81). The calculation in the proof of Exercise 6.85 shows that the above solution is also a solution of the three-term equation 5.1. □

Now, we can formulate our main result connecting period functions and solutions of (6.70) with $\lambda \in \{\pm 1\}$.

Proposition 6.89 *Let $\nu \in \mathbb{C}$. We have the following two statements.*

1. *If ϕ satisfies* (6.70) *on the cut-plane $\tilde{\mathbb{C}}'$, see* (6.72) *for the definition of the set $\tilde{\mathbb{C}}'$, with $\lambda = \pm 1$ and parameter $s \in \mathbb{C}$, then $P(\zeta) := \phi(\zeta - 1)$, $\zeta \in \mathbb{C}'$, satisfies the three-term equation* (5.1) *of period functions on $\mathbb{C}'$ with $\nu = \frac{1}{2} - s$.*
2. *If P satisfies the three-term equation* (5.1) *of a period function, we decompose P into two functions $P^{\pm}$ given by*

$$P^{+} := \frac{1}{2}\left(P + P\big|_{1-2\nu}C\right) \quad \textit{and} \quad P^{-} := \frac{1}{2}\left(P - P\big|_{1-2\nu}C\right) \tag{6.82}$$

satisfying $P = P^{+} + P^{-}$ and $P^{\pm} = \pm\left(P\big|_{1-2\nu}C\right)$. Then

$$\phi^{\pm}(z) := P^{\pm}(z+1), \tag{6.83}$$

for $z \in \tilde{\mathbb{C}}'$ *satisfies the three-term equation* (6.70) *on* $\tilde{\mathbb{C}}'$ *with* $\lambda = \pm 1$.

Proof We consider both statements separately.

1. Assume ϕ on $\tilde{\mathbb{C}}'$ satisfies (6.70) with $\lambda = \pm 1$. This implies that P defined by $P(\zeta) := \phi(\zeta - 1)$, $\zeta \in \mathbb{C}'$ satisfies

$$\begin{aligned} P(\zeta) &= \phi(\zeta - 1) \\ &\overset{(6.70)}{=} \phi(\zeta) \pm \left(\frac{1}{\zeta}\right)^{2s} \phi\left(\frac{1}{\zeta}\right) \\ &= \phi(\zeta) \pm \zeta^{-2s}\, \phi\left(\frac{1}{\zeta}\right) \\ &= P(\zeta + 1) \pm \zeta^{-2s}\, P\left(1 + \frac{1}{\zeta}\right). \end{aligned}$$

Hence, P satisfies the three-term equation (6.79). Applying Lemma 6.88 (and its proof) shows that P satisfies the three-term equation (5.1) on $\mathbb{C}'$ with

$$2\nu - 1 = -2s \iff \nu = \frac{1}{2} - s,$$

too.

2. Assume P satisfies the three-term equation (5.1) on $\mathbb{C}'$ with given parameter $\nu \in \mathbb{C}$. Then $P^{\pm}$ defined in (6.82) satisfies the three-term equation (6.79) with sign $\pm$. Defining ψ by

$$\psi^{\pm}(z) := P^{\pm}(z + 1)$$

for all $z \in \tilde{\mathbb{C}}'$, we see that ψ satisfies

$$\begin{aligned} \psi^{\pm}(z) &= P^{\pm}(z + 1) \\ &\overset{(6.79)}{=} P^{\pm}(z + 2) + (z + 1)^{2\nu - 1}\, P^{\pm}\left(1 + \frac{1}{z + 1}\right) \\ &= \psi^{\pm}(z + 1) + (z + 1)^{2\nu - 1}\, \psi^{\pm}\left(\frac{1}{z + 1}\right) \\ &= \psi^{\pm}(z + 1) + \left(\frac{1}{z + 1}\right)^{1 - 2\nu}\, \psi^{\pm}\left(\frac{1}{z + 1}\right). \end{aligned}$$

Hence $\psi^{\pm}$ satisfies the three-term equation (6.70) with parameter

$$2s = 1 - 2\nu \iff \nu = \frac{1}{2} - s$$

and $\lambda = \pm 1$.

□

Remark 6.90 Proposition 6.89 shows that period functions for the parameter ν and trivial multiplier $v = \mathbf{1}$ in the sense of Definition 5.1 are directly related to solutions of the three-term equation (6.79) and with §6.3.1 to eigenfunctions of the transfer operator $\mathcal{L}_{\frac{1}{2}-\nu}$ with eigenvalue ± 1.

Using for example Theorem 5.26 now gives us a direct relation between eigenfunction of the transfer operator $\mathcal{L}_{\frac{1}{2}-\nu}$ with eigenvalue ± 1 and families of Maass cusp forms on the full modular group with vanishing weight $k = 0$ and trivial multiplier $v = \mathbf{1}$.

6.4 The Transfer Operator and the Selberg Zeta Function

We will relate the transfer operator developed in §6.2 to the dynamical zeta function introduced by Ruelle. Then, we describe the connection between the dynamical zeta function and the Selberg zeta function on the full modular group $\Gamma(1)$.

6.4.1 Geodesics on $\mathbb{H}$*,* $\mathrm{SL}_2(\mathbb{Z})$ *Periodic Orbits, and Continued Fractions*

We follow the construction of the Poincaré map described in [55] which itself is based on the L-R-coding constructed in [181]. However, we restrict ourselves to considering only closed geodesics, since these are the objects of interest in §6.4.2 and in particular in Definition 6.98 of the Selberg zeta function.

Consider the upper half-plane $\mathbb{H}$ and its tessellation derived by copying the standard fundamental domain $\mathcal{F}_{\Gamma(1)}$ by $\mathrm{SL}_2(\mathbb{Z})$-actions, see, e.g., (1.22) for $\mathcal{F}_{\Gamma(1)}$ and Definition 1.16 of a fundamental domain. Now, consider a geodesic $\gamma = (\gamma_-, \gamma_+)$ on $\mathbb{H}$ that intersects the fundamental domain $\mathcal{F}_{\Gamma(1)}$. Continuing such a geodesic line outside of the fundamental domain, we see that it enters an $\mathrm{SL}_2(\mathbb{Z})$ equivalent copy $M\mathcal{F}_{\Gamma(1)}$ of the fundamental domain $\mathcal{F}_{\Gamma(1)}$ (for some $M \in \mathrm{SL}_2(\mathbb{Z})$). Hence, we can copy the piece of γ in $M\mathcal{F}_{\Gamma(1)}$ to the standard fundamental domain $\mathcal{F}_{\Gamma(1)}$ by applying M^{-1} on γ and taking the intersection of the geodesic line γ with $\mathcal{F}_{\Gamma(1)}$. This way, we can study the whole geodesic γ inside the standard fundamental domain $\mathcal{F}_{\Gamma(1)}$. (The whole idea is analog to studying lines in the Euclidian plane $\mathbb{R}^2$ modulo $\mathbb{Z}^2$ as a fundamental domain of the quotient $\mathbb{R}^2/\mathbb{Z}^2$ is the unit square $[0, 1] \times [0, 1]$ with suitably identified boundaries.)

Remark 6.91 If we start with a closed geodesic in the sense of Definition 6.47, we see that the above description gives the geodesic image in $\mathcal{F}_{\Gamma(1)}$ which means that we only have finitely many line segments. We illustrate this in Figure 6.2 with

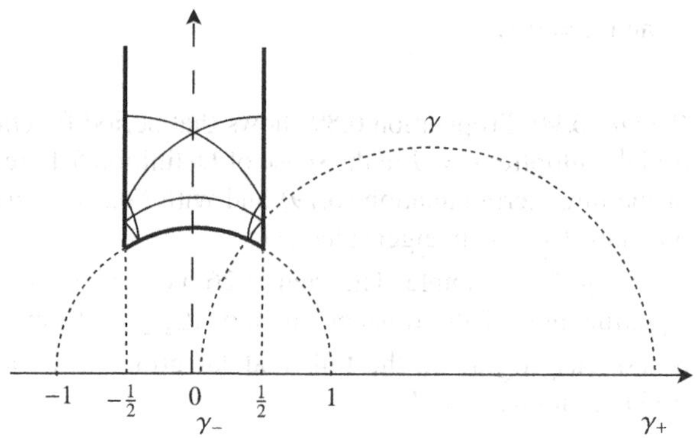

Fig. 6.2 An illustration of the $\mathrm{SL}_2(\mathbb{Z})$ copies of the (closed) geodesic $\gamma = (\gamma_-, \gamma_+)$ inside the fundamental domain $\mathcal{F}_{\Gamma(1)}$. The copies inside $\mathcal{F}_{\Gamma(1)}$ are denoted by γ^*.

γ^* denoting the $\mathrm{SL}_2(\mathbb{Z})$ copies of the closed geodesic $\gamma = (\gamma_-, \gamma_+)$ inside the fundamental domain $\mathcal{F}_{\Gamma(1)}$.

Now, start with a geodesic $\gamma \in \mathcal{A}$ as defined in (6.39). We assume that the base points of $\gamma = (\gamma_-, \gamma_+)$ admit infinite continued fraction expansions of the form (6.38):

$$\gamma_- = -\varepsilon[0; a_1, a_2, \ldots] \quad \text{and} \quad \gamma_+ = \frac{\varepsilon}{[0; b_0, b_1, \ldots]}.$$

Here, $\varepsilon = +1$ means that the geodesic γ crosses the upper imaginary axis $i\mathbb{R}_{>0}$ from left to right and $\varepsilon = -1$ means that the geodesics crosses the axis from right to left.

Let us consider the case $\varepsilon = +1$. Since the matrix $CS^{-b_0} \in \mathrm{GL}_2(\mathbb{Z})$ maps γ to $\gamma' = CS^{-b_0}\gamma$ with base points

$$\begin{aligned} -\gamma'_- &= -\left(CS^{-b_0}\,\gamma_-\right) = [0; b_0, a_1, a_2, \ldots] \quad \text{and} \\ \frac{1}{\gamma'_+} &= \frac{1}{CS^{-b_0}\,\gamma_+} = [0; b_1, \ldots], \end{aligned} \tag{6.30}$$

see Lemma 6.36, we see that the matrix $TS^{-b_0} \in \mathrm{SL}_2(\mathbb{Z})$ to a geodesic γ'' with base points

$$\begin{aligned} +\gamma''_- &= +\left(TS^{-b_0}\,\gamma_-\right) = [0; b_0, a_1, a_2, \ldots] \quad \text{and} \\ \frac{-1}{\gamma'_+} &= \frac{-1}{TS^{-b_0}\,\gamma_+} = [0; b_1, \ldots]. \end{aligned} \tag{6.84}$$

Similarly, we find for $\varepsilon = -1$ that the matrix $TS^{-b_0} \in \mathrm{SL}_2(\mathbb{Z})$ maps γ to a geodesic γ'' with base points

$$-\gamma_-'' = -\left(TS^{-b_0}\gamma_-\right) = [0; b_0, a_1, a_2, \ldots] \quad \text{and}$$
$$\frac{+1}{\gamma_+'} = \frac{+1}{TS^{-b_0}\gamma_+} = [0; b_1, \ldots]. \tag{6.85}$$

This shows that the map induced by the matrix $TS^{-b_0} \in \mathrm{SL}_2(\mathbb{Z})$ shifts a digit in the continued fraction expansions of the base points and, additionally, changes the sign of ε used in (6.38). Since the left shift on the first entry of

$$\frac{\varepsilon}{\gamma_+} = [0; b_0, b_1, \ldots] \longmapsto \frac{-\varepsilon}{TS^{-b_0}\gamma_+} = [0; b_1, \ldots]$$

is just the application of the Gauss map f_{G} on the continued fraction expansion $[0; b_0, b_1, \ldots]$ and a change of sign $\varepsilon \mapsto -\varepsilon$.

We just showed that applying TS^{-b_0} to a geodesic γ with base points given by (6.38) maps one element of the set $\mathcal{A}$ to another element of $\mathcal{A}$. Here, b_0 is the first digit in the continued fraction expansion of $\frac{\varepsilon}{\gamma_+}$ in (6.38).

If we now assume a closed geodesic δ on $\mathbb{H}$ in the sense of Definition 6.47. Taking the $\mathrm{SL}_2(\mathbb{Z})$-equivalent geodesic in $\gamma \in \mathcal{A}$, see Exercise 6.44, we see that γ is also closed in the sense of Definition 6.47.

Exercise 6.92 Let δ be a closed geodesic δ on $\mathbb{H}$ in the sense of Definition 6.47. Show that the $\mathrm{SL}_2(\mathbb{Z})$-equivalent geodesic $\gamma \in \mathcal{A}$, see Exercise 6.44, is also closed in the sense of Definition 6.47.

Combining this exercise with the characterization of closed geodesics in $\mathcal{A}$ Lemma 6.49 we get the following

Lemma 6.93 *A set of representatives of all* $\mathrm{SL}_2(\mathbb{Z})$*-equivalent closed geodesics is the set of geodesics in* $\mathcal{A}$ *with purely periodic continued fraction expansions of the form* (6.42).

Remark 6.94 According to, e.g., [132, §3], we can define a distant function $d(\cdot, \cdot)$ on the upper half-plane, which measure the hyperbolic distance between two points $z, w \in \mathbb{H}$. The function $d : \mathbb{H} \times \mathbb{H} \to \mathbb{R}_{\geq 0}$ can be characterized by

$$\cosh d(z, w) = +\frac{|z - w|^2}{2\Im(z)\,\Im(w)}. \tag{6.86}$$

Now, consider a closed geodesic γ. According to Definition 6.47 and Lemma 6.48, there exists an associated hyperbolic matrix $M \in \mathrm{SL}_2(\mathbb{Z})$ that fixes γ in the sense $M\gamma = \gamma$. Marklof stated in [132, (39)] that the length $l(\gamma)$ of the geodesic γ can be calculated by taking the infimum of all hyperbolic distances between $z \in \mathbb{H}$ and its image Mz under the matrix M. We have

$$l(\gamma) = \inf_{z \in \mathbb{H}} d\left(z, Mz\right). \tag{6.87}$$

Hejhal showed in [94] that in the above setting the length $l(\gamma)$ of a closed geodesic γ can be calculated in terms of the trace of M. We have the relation

$$2\cosh\left(\frac{l(\gamma)}{2}\right) = \left(e^{l(\gamma)}\right)^{\frac{1}{2}} + \left(e^{l(\gamma)}\right)^{-\frac{1}{2}} = |\mathrm{trace}\,(M)|\,, \tag{6.88}$$

see [55, (3.18)] or alternatively [132, (41)].

6.4.2 *The Dynamical Zeta Function by Ruelle and the Selberg Zeta Function*

Consider closed geodesics (which correspond to the hyperbolic matrices in $\mathrm{SL}_2(\mathbb{Z})$ in Lemma 6.48). We define the set of all such closed geodesics by H.

We call a closed geodesic $\gamma \in H$ *primitive*, if the associated purely periodic continued fraction expansion in (6.42) of the base points, see Lemma 6.49, is purely *prime* periodic. Here a periodic continued fraction expansion

$$[a_0; a_1, \ldots, a_n, \overline{a_{n+1}, \ldots, a_{n+l}}]$$

is called *prime* periodic if the length of the periodic part (i.e., the number l of digits) is minimal, see also Appendix A.15.

The set of all primitive closed geodesics is denoted by $H_* \subset H$.

Remark 6.95 The notation H_* for primitive closed geodesics is borrowed from [132, §9, between(166) and (167)].

Remark 6.96 Lemma 6.49 and the above definition of primitive closed geodesics imply that we have a natural isomorphism between the set H_* of primitive closed geodesics and the prime purely periodic continued fractions.

Following [172] and [173], the dynamical zeta function ζ_{R} on the full modular group $\Gamma(1)$ is given by the following.

Definition 6.97 ([173]) The *dynamical zeta function* ζ_{R} *on* $\Gamma(1)$ is given by

$$\zeta_{\mathrm{R}}(s) := \prod_{\gamma\in H_*}\left[1 - e^{-s\,l(\gamma)}\right], \tag{6.89}$$

where the product runs over all primitive closed geodesics $\gamma \in H_*$. periodic continued fractions. The term $l(\gamma)$ denotes the length of the (primitive) closed geodesic γ in the sense of Remark 6.94.

We follow [199] for the definition of the Selberg zeta function $Z_{\mathrm{S}}(s)$:

Definition 6.98 ([199, Chapter I]) The *Selberg zeta function* Z_S *on* $\Gamma(1)$ is given by

$$Z_S(s) := \prod_{\gamma \in H_*} \prod_{k=0}^{\infty} \left[1 - e^{-(s+k)\, l(\gamma)} \right], \tag{6.90}$$

where the outer product runs over all primitive closed geodesics. The term $l(\gamma)$ describes the length of the (primitive) closed geodesic γ in the sense of Remark 6.94.

Comparing (6.89) with (6.90), we immediately find the relation

$$Z_S(s) = \prod_{k=0}^{\infty} \zeta_R(s+k)^{-1}. \tag{6.91}$$

Remark 6.99 No brief presentation of the Selberg zeta function would be adequate; the importance of the Selberg zeta function and the related topic of the Selberg trace formula deserves a much better presentation. Moreover, it appears that we do not have any immediate use of the Selberg zeta function to our applications, except for the few lines below. Thus readers are referred to other sources, for example Marklof's introductory article [132] or the rich but hard to understand treatment in Hejhal's book [95].

Chang and Mayer relate the dynamical zeta function ζ_R to the Fredholm determinant of the transfer operator $\tilde{\mathcal{L}}_s$ using periodic points of the Gauss function f_G. They show in [55, (5.42)] that

$$\zeta_R(s) = \frac{\det\left(1 - \tilde{\mathcal{L}}_{s+1}\right)}{\det\left(1 - \tilde{\mathcal{L}}_s\right)}. \tag{6.92}$$

Using (6.91), together with the above relation (6.92), gives

$$\begin{aligned} Z_S(s) &= \det\left(1 - \tilde{\mathcal{L}}_s\right) \\ &\overset{(6.67)}{=} \det\left(1 + \mathcal{L}_s\right) \det\left(1 - \mathcal{L}_s\right). \end{aligned} \tag{6.93}$$

Remark 6.100 The above relation was already shown in [136, 137] by Mayer.

We combine the result in (6.93) with Propositions 6.77 and 6.89. Using known properties of the zeros of the Selberg zeta function Z_S, we find the following.

Proposition 6.101 ([122, p. 255]) *Let* $s \neq \frac{1}{2}$ *be a complex number with* $\Re(s) > 0$. *Then:*

1. *there exists a nonzero function* $h \in B(D)$ *with* $\mathcal{L}_s h = -h$ *if and only if* s *is the spectral parameter corresponding to an odd Maass wave form on* $\Gamma(1)$;

2. *there exists a nonzero function $h \in B(D)$ with $\mathcal{L}_s h = h$ if and only if s is either the spectral parameter corresponding to an even Maass form, or $2s$ is a zero of the Riemann zeta function, or $s = 1$.*

Remark 6.102 The above result gives an interesting relation between the spectral parameters of Maass cusp forms and eigenfunctions of the transfer operator $\mathcal{L}_s$ with eigenvalue ± 1. This connection is in particular independent of Selberg trace formula, discussed for example in [95].

6.5 Concluding Remarks

We presented in §6.1–6.4 the transfer operator approach by Mayer. This method attracted interest when Zagier and Lewis related ± 1-eigenfunctions of $\mathcal{L}_s$ to period functions associated with Maass wave forms in [122]. Since then there has been a few developments:

1. Mayer's approach using continued fraction expansions got extended to cover Hecke triangle groups. These are groups generated by the matrices $T = \begin{pmatrix} 0 & -1 \\ 1 & 0 \end{pmatrix}$ and $S_q = \begin{pmatrix} 1 & 2\cos\left(\frac{\pi}{q}\right) \\ 0 & 1 \end{pmatrix}$, $q \in \{3, 4, 5, \ldots\}$, satisfying the relation $T^2 = (T S_q)^q = -\mathbf{1}$. Details can be found in [138–140].
2. Mayer's approach can also be exploited to calculate the Selberg zeta function numerically. Details can be found in [193] for the full modular group and in [34, 81] for the congruence subgroup $\Gamma_0(4)$.
3. Pohl starts with groups and then constructs transfer operators from these groups. The advantage here is that no continued fractions as intermediate step are required. Details can be found in [143, 158–162].
4. Bruggeman, Lewis, and Zagier introduced a cohomology theorem in [35] where they connect the space of Maass wave forms with certain cohomology groups. We presented a short introduction to Zagier's cohomology theorem in §5.2 of Chapter 4. However, the proof of this result relies on certain averaging operators, see [36, §4], which are in structure somewhat similar to the transfer operators. It would be nice to know if this is just a coincidence or if there is some underlying connecting principle.
5. There is another connection relating elements of the mixed parabolic cohomology group $H^1_{\text{par}}(\Gamma(1), \mathcal{V}^{\omega}_{\nu}, \mathcal{V}^{\omega^*,\infty}_{\nu})$ in Zagier's cohomology theorem (Theorem 5.32 in §5.2.3) to transfer operators.

One can consider $\Gamma(1)$ as a Hecke triangle group. Then, constructing the transfer operator for Hecke triangle groups — this technique was developed in the papers [138–140] — shows that Maass cusp forms correspond to certain solutions of a four-term equation. (This is in contrast to the three-term equation (5.1) appearing in Definition 5.1 of period functions.) The authors of [33] showed that solutions of this four-term equation can be related explicitly to solutions of the three-term equation. It would be interesting to have a similar relationship for general Fuchsian groups.

Chapter 7
Weak Harmonic Maass Wave Forms

In this chapter, we introduce the newly emerging concept of weakly harmonic Maass wave forms. The development of the theory started just after the turn of the century with the PhD Thesis of Zwegers [211] and with the work done by Bruinier and Funke [40]. The basic idea behind weakly harmonic Maass cusp forms is the following: we extend the hyperbolic Laplace operator $\Delta_0 = -y^2 \left(\frac{\partial^2}{\partial x^2} + \frac{\partial^2}{\partial y^2} \right)$ for weight 0 (see (3.8) in Section 3.2) with a weight k dependent component. In Section 3.2, we "added" the weight dependent differential $iky \frac{\partial}{\partial x}$ to Δ_0, we constructed the second-order differential operator Δ_k in (3.9), the hyperbolic Laplace operator Δ_k of weight k, and we introduced Maass wave forms of real weight k in the sense of Definition 3.31 that are eigenfunctions of Δ_k.
Now, we "add" the weight dependent differential

$$iky \left(\frac{\partial}{\partial x} + i \frac{\partial}{\partial y} \right) = 2iky \frac{\partial}{\partial \overline{z}}$$

to Δ_0 for the setting of weakly harmonic Maass wave forms.

The chapter is organized as follows: In Section 7.1, we introduce weak harmonic Maass wave forms, following the approach by Brunier and Funke [40, §3] and Bruinier, Ono, and Rhoades [48]. A short sidestep to mock modular forms is also presented in Section 7.2. In Section 7.3, we discuss the associated L-series, following ideas from Bringmann, Fricke, Guerzhoy, Kent, and Ono in [22, 24, 25]. The three papers mentioned above also developed the Eichler-Shimura theory for weakly harmonic Maass forms. We give in Section 7.4 a brief motivation towards these results by introducing period polynomials and period functions associated with weak harmonic Maass wave forms.

T. Mühlenbruch, W. Raji, *On the Theory of Maass Wave Forms*, Universitext,
https://doi.org/10.1007/978-3-030-40475-8_7

7.1 Weak Harmonic Maass Wave Forms

Recall the discussion of Maass wave forms in Chapter 3. Two important aspects about Maass wave forms are that these forms are eigenfunctions of the hyperbolic Laplace-operator Δ_k given in (3.9) and that they satisfy the transformation property in (3.6). (We refer to Definition 3.31 for the complete description of the defining properties of a Maass wave form.) The transformation property under the double-slash operator in (3.5), uses a unitary weight $j_u(M, z)$ as can be seen in (3.2)and (3.3), and a unitary multiplier system $v(M)$ as given in Definition 1.44.

Bruinier and Funke in [40] took another approach, which led to the notion of weak Maass forms. Instead of focusing on unitary multipliers and weight actions, they wanted to preserve the holomorphic automorphic factor $j(M, z)$ defined in (1.29). This led to the use of the slash operator in (1.131). However, in order to preserve that the hyperbolic Laplacian and the slash operator in (1.131) commute, they had to adapt the Laplacian into a suitable version. To simplify our discussion, we restrict ourselves again to the full modular group $\Gamma(1) = \mathrm{SL}_2(\mathbb{Z})$ with trivial multiplier $v \equiv 1$.

Let $k \in \mathbb{Z}$ and denote again $\mathrm{SL}_2(\mathbb{Z})$ by $\Gamma(1)$. In the context of weak harmonic Maass wave forms, we define the *hyperbolic Laplace operator* of *weight* k by the differential

$$\begin{aligned}\tilde{\Delta}_k := &-y^2\left(\frac{\partial^2}{\partial x^2}+\frac{\partial^2}{\partial y^2}\right)+iky\left(\frac{\partial}{\partial x}+i\frac{\partial}{\partial y}\right)\\ &= -\left(R_{k-2}\,L_k\right) = -\left(L_{k+2}\,R_k+k\right),\end{aligned} \tag{7.1}$$

where R_k and L_k are the differential operators

$$R_k = 2i\frac{\partial}{\partial z}+\frac{k}{y} \qquad \text{and} \qquad L_k = -2iy^2\frac{\partial}{\partial \overline{z}} \tag{7.2}$$

with $z = x + iy \in \mathbb{H}$ and $x = \Re(z)$, the real part and $y = \Im(z)$, the imaginary part of z. These are called the *raising* and *lowering operators* or *Maass operators* of non-holomorphic modular forms of weight k.

Remark 7.1 Relation (7.1) satisfied by the raising and lowering operators R_k and L_k and the hyperbolic Laplace operator $\tilde{\Delta}_k$ is similar to the Relation (3.12) between the Maass operators $\mathrm{E}_k^{\pm}$ and the hyperbolic Laplace operator Δ_k.

However, one observed difference is that $\mathrm{E}_k^{\pm}$ are quite similar and just exchange z with $\overline{z}$, see (3.11). On the other hand, the raising and lowering operators R_k and L_k favor the anti-holomorphic partial derivative in L_k as can be seen in (7.2); the holomorphic partial derivative in the raising operator R_k has an additional weight component.

Exercise 7.2 Let g be an eigenfunction of $\tilde{\Delta}_k$ with eigenvalue λ: $\tilde{\Delta}_k g = \lambda g$. Show that $R_k g$ satisfies $\tilde{\Delta}_{k+2} R_k g = (\lambda + k) R_k g$.

Similar to the definition of Maass wave forms in Section 3.3, we introduce weak harmonic Maass wave forms as follows.

Definition 7.3 Let $k \in 2\mathbb{Z}$. A twice continuously differentiable function $f : \mathbb{H} \to \mathbb{C}$ is called a *weak harmonic Maass wave form* of *weight k* and trivial multiplier on the full modular group $\Gamma(1)$, if it satisfies the following three conditions:

1. f satisfies (1.28): $f(Mz) = j(M,z)^k f(z)$ for all $z \in \mathbb{H}$ and $M \in \Gamma(1)$,
2. $f(z) = O\left(e^{Cy}\right)$ as $\Im(z) = y \to \infty$, and
3. $\tilde{\Delta}_k f = 0$.

We denote the space of weak harmonic Maass wave forms by $H_k = H_k\big(\Gamma(1)\big)$.

For simplicity, we denote the space of real analytic functions f on $\mathbb{H}$ satisfying the transformation property (1.28) (which is the first condition above) by $A_k = A_k\big(\Gamma(1)\big)$.

Remark 7.4 Comparing the above Definition 7.3 of weakly harmonic Maass wave forms with Definition 1.59 of weakly holomorphic modular forms, we see that the main difference is the assumption that F is holomorphic, i.e., $\frac{\partial}{\partial \bar{z}} F(z) = 0$, is replaced with the condition $\tilde{\Delta}_k f = 0$. This explains the word "harmonic" in the name "weakly harmonic Maass wave forms" since functions vanishing under a Laplace operator are usually called harmonic functions.

Definition 7.5 Let $k \in 2\mathbb{Z}$. We define the subset $H_k^+ = H_k^+\big(\Gamma(1)\big)$ of $H_k\big(\Gamma(1)\big)$ by: $f \in H_k^+$ if and only if $f \in H_k$ and f satisfies the condition

(4) there exists a (nonpositive) number $n_+ \in \mathbb{Z}$ and a polynomial

$$P_f = \sum_{n_+ \le n \le 0} c_f^+(n)\, q^n \in \mathbb{C}\left[q^{-1}\right]$$

such that

$$f(z) - P_f\left(e^{2\pi i z}\right) = O\left(e^{-c\,\Im(z)}\right) \tag{7.3}$$

as $\Im(z) \to \infty$ for some $c > 0$.

We call the polynomial $P_f(q)$ or sometimes the function $P_f\left(e^{2\pi i z}\right)$, the *principal part* of f.

Remark 7.6 Note that P_f defined in Definition 7.5 is indeed a polynomial in q^{-1} since the negative summation index n (satisfying $n_+ \le n \le 0$) appears as positive exponent $-n$ in the power of the variable q^{-1}.

Similar to our discussion in §3.2.2, we can expand a weakly harmonic Maass wave form f into an expansion of Fourier-type. The following result was shown in [40, §3, p. 55].

Lemma 7.7 ([40]) *Let f be a weak harmonic Maass wave form of weight $k \in 2\mathbb{Z}$. Then f has a unique decomposition into two functions f_+ and f_-,*

$$f(z) = f_+(z) + f_-(z) \tag{7.4}$$

for all $z \in \mathbb{H}$, where f_+ and f_- admit expansions of the following form: There exists suitable $n_+, n_- \in \mathbb{Z}$ such that

$$\begin{aligned} f_+(z) &= \sum_{\substack{n\in\mathbb{Z}\\ n\geq n_+}} a_+(n)\, e^{2\pi i n z} \qquad \text{and} \\ f_-(z) &= a_-(0)\, \Im(z)^{1-k} + \sum_{\substack{n\in\mathbb{Z}_{\neq 0}\\ n\leq n_-}} a_-(n)\, \Gamma\big(1-k, -4\pi n \Im(z)\big)\, e^{2\pi i n z}, \end{aligned} \tag{7.5}$$

where $\Gamma\big(\alpha, x\big) = \int_x^\infty t^{\alpha-1} e^{-t}\, \mathrm{d}t$ denotes the incomplete Γ-function. (We refer to [153, (8.2.2)] and Appendix A.17 for more details on the incomplete Γ-function.)

Remark 7.8 In the words of [40, p. 56], the second condition in Definition 7.3 implies that all but finitely many coefficients $a_+(n)$ (respectively $a_-(n)$) with negative (respectively positive) index n vanish.

Remark 7.9 If $f \in H_k^+$, then the decomposition of f into f_+ and f_- in (7.5) satisfies: There exists an $n_+ \in \mathbb{Z}$ with

$$\begin{aligned} f_+(z) &= \sum_{n\in\mathbb{Z}_{\geq n_+}} a_+(n)\, e^{2\pi i n z} \qquad \text{and} \\ f_-(z) &= \sum_{n\in\mathbb{Z}_{<0}} a_-(n)\, \Gamma\big(1-k, 4\pi |n| \Im(z)\big)\, e^{2\pi i n z}, \end{aligned} \tag{7.6}$$

for all $z \in \mathbb{H}$. Note that the expansion of f_- only contains terms with negative index n.

Remark 7.10 The holomorphic part f_+ of a weakly harmonic Maass wave form $f = f_+ + f_- \in H_k\big(\Gamma(1)\big)$ is also known as a *mock modular form*. This name goes back to the 17 examples of "mock theta functions" in a letter sent by Ramanujan to Hardy in 1920, see, e.g., [166] and his lost notebook [165]. Ramanujan mentioned that these functions have an asymptotic expansion at the cusps, which is similar to that of modular forms of weight $\frac{1}{2}$, but these cannot be expressed in terms of "normal" theta functions. He called these functions "mock theta functions." Zwegers discovered the connection of the mock theta functions with modular forms in his PhD Thesis [211]. He realized that the mock theta functions could be written as a sum of two functions, where one is a real-analytic modular form of weight $\frac{1}{2}$ and is an eigenfunction of an appropriate Laplace operator. The other function is a theta series associated with a negative definite unitary quadratic form. We refer to [67] for a slightly longer overview.

The raising and lowering operators R_k and L_k defined in (7.2) raise and lower the weight k of real analytic functions f on $\mathbb{H}$ that satisfy the transformation property (1.28) (for weight k) to $k+2$ and $k-2$ respectively. This is, in some sense, an analogous result to Corollary 3.41 for Maass forms in the sense of Definition 3.31.

Lemma 7.11 ([40]) *For $k \in 2\mathbb{Z}$, we have*

$$L_k : A_k\big(\Gamma(1)\big) \to A_{k-2}\big(\Gamma(1)\big) \quad \text{and} \quad R_k : A_k\big(\Gamma(1)\big) \to A_{k+2}\big(\Gamma(1)\big). \tag{7.7}$$

Additionally, to the raising and lowering of the weight k, we see that the lowering operator L_k admits the following result.

Lemma 7.12 *Let $f \in H_k$ be a weakly harmonic Maass wave form that admits the decomposition* (7.4) *with respective expansions* (7.5). *Then $L_k f$ admits the expansion*

$$\begin{aligned}\big(L_k f\big)(z) &= \big(L_k f_-\big)(z)\\ &= -2\Im(z)^{2-k}\left((k-1)a_-(0) + \sum_{\substack{n\in\mathbb{Z}_{\neq 0}\\ n\leq n_-}} a_-(n)\big(-4\pi n\big)^{1-k} e^{2\pi i n\bar{z}}\right).\end{aligned} \tag{7.8}$$

Exercise 7.13 Let $g : \mathbb{H} \to \mathbb{C}$ be a holomorphic function. The image of g under the lowering operator vanishes:

$$L_k\, g = 0. \tag{7.9}$$

Using the above exercise, we can embed the weakly holomorphic modular form of weight k and trivial multiplier on $\Gamma(1)$, see Definition 1.59, into H_k.

Lemma 7.14 *Let $k \in 2\mathbb{Z}$. A weakly holomorphic modular form g of weight k and trivial multiplier system on $\Gamma(1)$ is a weakly harmonic Maass wave form. We have the set inclusion*

$$M^!_{k,1}\big(\Gamma(1)\big) \subset H_k\big(\Gamma(1)\big). \tag{7.10}$$

Proof Let $g \in M^!_{k,1}\big(\Gamma(1)\big)$ denote a weakly holomorphic modular form in the sense of Definition 1.59. Exercise 7.13 implies $L_k\, g = 0$. Hence, g satisfies

$$\begin{aligned}\tilde{\Delta}_k\, g &\overset{(7.1)}{=} -R_{k-2}\, L_k\, g\\ &\overset{(7.9)}{=} -R_{k-2}\, 0\\ &= 0.\end{aligned}$$

Moreover, g as weakly holomorphic modular form satisfies the transformation property (1.28) and is meromorphic at the cusp ∞, see Definition 1.59. This implies trivially that g satisfies the required growth condition and the transformation property (1.28) required by Definition 7.3. Hence g is a weakly harmonic Maass wave form of weight k. □

There are more differential operators acting on the space of weakly harmonic Maass forms.

Proposition 7.15 *Let $k \in 2\mathbb{Z}$.. The differential operator ξ_k given by*

$$\begin{aligned}(\xi_k f)(z) :&= \Im(z)^{k-2}\overline{L_k f(z)} \\ &= R_{-k}\,\Im(z)^k\overline{f(z)}\end{aligned} \tag{7.11}$$

defines an anti-linear mapping

$$\xi_k : H_k\big(\Gamma(1)\big) \to M^!_{2-k,1}\big(\Gamma(1)\big) \tag{7.12}$$

with kernel

$$ker(\xi_k) = M^!_{k,1}\big(\Gamma(1)\big). \tag{7.13}$$

Restricting ξ_k to the space $H_k^+\big(\Gamma(1)\big)$, we have the map

$$\xi_k : H_k^+\big(\Gamma(1)\big) \to S_{2-k,1}\big(\Gamma(1)\big). \tag{7.14}$$

Proof To show the identity in (7.11), consider

$$\begin{aligned}
R_{-k}\left(y^k\overline{f(z)}\right) &\overset{(7.2)}{=} 2i\frac{\partial}{\partial z}\left(y^k\overline{f(z)}\right) + \frac{-k}{y}y^k\overline{f(z)} \\
&= -ky^{k-1}\overline{f(z)} + 2i\frac{\partial}{\partial z}\left(\left(\frac{z-\overline{z}}{2i}\right)^k\overline{f(z)}\right) \\
&= \underbrace{-ky^{k-1}\overline{f(z)} + k\left(\frac{z-\overline{z}}{2i}\right)^{k-1}\overline{f(z)}}_{=0} + 2i\left(\left(\frac{z-\overline{z}}{2i}\right)^k\frac{\partial}{\partial z}\overline{f(z)}\right) \\
&= y^{k-2}\left(2iy^2\frac{\partial}{\partial z}\overline{f(z)}\right) \\
&= y^{k-2}\overline{\left(-2iy^2\frac{\partial}{\partial \overline{z}}f(z)\right)} \\
&\overset{(7.2)}{=} \Im(z)^{k-2}\overline{L_k f(z)},
\end{aligned}$$

using the usual $z = x + iy$ and that we can write $y = \frac{z-\overline{z}}{2i}$.

Next, we use Lemma 7.12. The expansion in (7.8) shows that $L_k f$ is a holomorphic function on $\mathbb{H}$ since there appears no dependency on $\overline{z}$ in the expansion anymore. Moreover, relation (7.8) shows $L_k f = 0$ if and only if $f_- = 0$ vanishes. This implies

$$\xi_k f = 0 \iff f_- = 0.$$

Hence, $\ker(\xi_k)$ contains only weakly harmonic Maass forms f with $f = f_+$ and $f_- = 0$ in the sense of Lemma 7.7. We conclude that $f = f_+$ is holomorphic on $\mathbb{H}$ and meromorphic at ∞, which we conclude from the expansion of f_+ in (7.5). Moreover, f satisfies (1.28). This shows that $\xi_k f = 0$ is equivalent to $f \in M_{k,1}^!(\Gamma(1))$ in the sense of Definition 1.59. We now have proven (7.13).

Since $\xi_k f$ has the Fourier expansion

$$\begin{aligned}
(\xi_k f)(z) &\overset{(7.11)}{=} \Im(z)^{k-2}\overline{L_k f(z)} \\
&\overset{(7.8)}{=} \overline{-2\left((k-1)a_-(0) + \sum_{\substack{n\in\mathbb{Z}_{\neq 0}\\ n\le n_{-\infty}}} a_-(n)\left(-4\pi n\right)^{1-k} e^{2\pi i n\overline{z}}\right)} \\
&= -2(k-1)a_-(0) - 2\sum_{\substack{n\in\mathbb{Z}_{\neq 0}\\ n\le n_{-\infty}}} a_-(n)\left(-4\pi n\right)^{1-k} e^{2\pi i|n|z}, \qquad (7.15)
\end{aligned}$$

we conclude that $\xi_k f$ is holomorphic on $\mathbb{H}$ and meromorphic at the cusp ∞, i.e., $\xi_k f$ satisfies the second and third conditions in Definition 1.59.

It remains to show that $\xi_k f$ satisfies the transformation property (1.28) for weight $2-k$. For $M \in \Gamma(1)$ and $z \in \mathbb{H}$, consider $(\xi_k f)(Mz)$. We find

$$\begin{aligned}
(\xi_k f)(Mz) &\overset{(7.11)}{=} \Im(Mz)^{k-2}\overline{L_k f(Mz)} \\
&\overset{(7.7)}{=} \Im(Mz)^{k-2}\overline{j(M,z)^{k-2}\left(L_k f\right)(z)} \\
&\overset{(1.18)}{=} \left(\frac{\Im(z)}{j(M,z)\overline{j(M,z)}}\right)^{k-2}\overline{j(M,z)^{k-2}\left(L_k f\right)(z)} \\
&= j(M,z)^{2-k}\,\Im(z)^{k-2}\overline{\left(L_k f\right)(z)}.
\end{aligned}$$

Hence, $\xi_k f$ satisfies the transformation property (1.28) for weight $2-k$. Summarizing, we conclude $\xi_k f \in H_{2-k}(\Gamma(1))$.

For the remaining relation (7.14), recall the definition of the space $H_k^+(\Gamma(1))$ in Definition 7.5 and the expansions of its elements described in (7.6) of Remark 7.9. We see in particular that the Fourier expansion of f_- only runs through negative indices $n \in \mathbb{Z}_{<0}$. This means that $(\xi_k f)(z)$ in (7.15) admits the expansion

$$\begin{aligned}(\xi_k f)(z) &= -2 \sum_{n\in\mathbb{Z}_{<0}} a_-(n)\big(-4\pi n\big)^{1-k} e^{-2\pi i n z}\\ &= -2\sum_{n\in\mathbb{N}} a_-(-n)\big(4\pi n\big)^{1-k} e^{2\pi i n z}.\end{aligned}$$

Hence, we have $\xi_k f \in S_{k,1}\big(\Gamma(1)\big)$, concluding the proof. □

Remark 7.16 It was shown in [40, Theorem 3.7] that ξ_k in (7.12) is a surjective map.

Now, consider weight $k \in 2\mathbb{N}$. Next to ξ_k, there exists another distinguished differential operator D^{k-1}. Consider the differential operator

$$\begin{aligned}D &:= \frac{1}{2\pi i}\frac{\mathrm{d}}{\mathrm{d}z}\\ &\overset{(7.2)}{=} -4\pi R_k + \frac{k}{2},\end{aligned} \tag{7.16}$$

where we used the defining relation (7.2) for the raising operator R_k. Using the recursive defined notation

$$\begin{cases} R_k^0 := \mathbf{1},\\ R_k^1 := R_k,\\ R_k^2 := R_{k+2}\circ R_k = R_{k+2}\, R_k,\\ \vdots\\ R_k^n := R_{k+2(n-1)}\circ R_k^{n-1} = R_{k+2(n-1)}\circ\ldots\circ R_k \end{cases} \tag{7.17}$$

for every $n \in \mathbb{Z}_{\geq 0}$.

Exercise 7.17 Let g be an eigenfunction of $\tilde{\Delta}_{2-k}$ with eigenvalue λ: $\tilde{\Delta}_{2-k} g = \lambda\, g$. Show that $R_{2-k}^{k-2} g$ satisfies $\tilde{\Delta}_{k-2} R_{2-k}^{k-2} g = (2-k+\lambda) R_{2-k}^{k-2} g$.

With the above notation in (7.17), we can state the following relation given for example in [48].

Lemma 7.18 ([48, Lemma 2.1]) *Let $k \in 2\mathbb{N}$. We have*

$$D^{k-1} = \frac{1}{(-4\pi)^{k-1}} R_{2-k}^{k-1}. \tag{7.18}$$

Remark 7.19

1. The above identity (7.18) is also known as *Bol's identity*.
2. Sometimes, Bol's identity is given such that it contains the action of a slash operator, for example in the form of (2.13) in Exercise 2.18, [98, p. 566], or [122, (4.14)].

3. Bol's identity in the form of (7.19) below implies a linear map between the weakly holomorphic modular forms of weight $2-k$ and k:

$$\frac{\mathrm{d}^{k+1}}{\mathrm{d}z^{k+1}} : M^!_{2-k,v}\big(\Gamma\big) \to M^!_{k,v}\big(\Gamma\big). \tag{7.19}$$

The differential operator D^{k-1} acts on weakly harmonic Maass wave forms, extending the linear map in (7.19). This implies the result derived in [48].

Proposition 7.20 ([48, Theorem 1.1]) *Let $k \in 2\mathbb{N}$ and $f \in H_{2-k}\big(\Gamma(1)\big)$ a weak harmonic Maass wave form. Then $D^{k-1} f$ is a weakly holomorphic modular form of weight k and trivial multiplier on $\Gamma(1)$:*

$$D^{k-1} f \in M^!_{k,1}\big(\Gamma(1)\big). \tag{7.20}$$

Assume additionally that $f \in H^+_{2-k}\big(\Gamma(1)\big)$. If $f = f_+ + f_-$ admits the expansion given in (7.6), we have $D^{k-1} f \in S^!_{k,1}\big(\Gamma(1)\big)$ with

$$\begin{aligned} D^{k-1} f &= D^{k-1} f_+ \\ &= \sum_{n\in\mathbb{Z}_{\geq n_+}} a_+(n)\, n^{k-1} e^{2\pi i n z} \end{aligned} \tag{7.21}$$

for $z \in \mathbb{H}$.

Proof For $k \in 2\mathbb{N}$, let $f \in H_{2-k}\big(\Gamma(1)\big)$ be a weak harmonic Maass wave form of weight $2-k$. Lemma 7.11 together with Lemma 7.18 show that

$$\begin{aligned} D^{k-1} f &\overset{(7.18)}{=} \frac{1}{(-4\pi)^{k-1}} R^{k-1}_{2-k} f \\ &\overset{(7.17)}{=} \frac{1}{(-4\pi)^{k-1}} R_k \cdots R_{2-k} f \end{aligned}$$

satisfies the transformation property (1.28) for weight k, since f satisfies (1.28) for weight $2-k$.

To prove (7.20), we have to show that $D^{k-1} f$ is holomorphic. Recall the lowering operator L_{2-k} in (7.2). If a function $g : \mathbb{H} \to \mathbb{C}$ is holomorphic, then $\frac{\partial}{\partial \overline{z}} g(z) = 0$. Hence, $L_{2-k}\, g = 0$ implies that g is holomorphic by (7.2).

Since f is a weakly harmonic Maass wave form of weight $2-k$, we have

$$\tilde{\Delta}_{2-k} f = 0$$

by Definition 7.3. Using Exercise 7.17, we find

$$\tilde{\Delta}_{k-2}\left(R_{2-k}^{k-2} f\right) \overset{E\,7.17}{=} (2-k)\, R_{2-k}^{k-1} f.$$

With (7.1), we find

$$\begin{aligned} L_k R_{2-k}^{k-1} f &\overset{7.17}{=} L_k R_{k-2}\, R_{2-k}^{k-2} f \\ &\overset{7.1}{=} \big(-\tilde{\Delta}_{k-2} - (k-2)\big)\, R_{2-k}^{k-2} f \\ &\overset{E\,7.17}{=} \big(-(2-k) - (k-2)\big)\, R_{2-k}^{k-2} f \\ &= 0. \end{aligned}$$

This shows the first part of the proposition.

To show the second part, assume that $f \in H_{2-k}^{+}$ admits the expansion

$$\begin{aligned} f(z) &\overset{(7.4)}{=} f_{-}(z) + f_{-}(z) \\ &\overset{(7.6)}{=} \sum_{n \in \mathbb{Z}_{\geq n_{+}}} a_{+}(n)\, e^{2\pi i n z} + \\ &\quad + \sum_{n \in \mathbb{Z}_{<0}} a_{-}(n)\, \Gamma\big(k-1, 4\pi\, |n| \Im(z)\big)\, e^{2\pi i n z}. \end{aligned}$$

Calculating $R_{2-k}^{k-2} f$ term by term, we find

$$\begin{aligned} R_{2-k}^{k-2} f(z) = &\sum_{n \in \mathbb{Z}_{\geq n_{+}}} a_{+}(n)\, \Gamma\big(k-1, 4\pi\, |n| \Im(z)\big)\, \big(-\Im(z)\big)^{2-k}\, e^{2\pi i n \bar{z}} \\ &+ \big((k-2)!\big)^{2} \sum_{n \in \mathbb{Z}_{<0}} a_{-}(n)\, \big(\Im(z)\big)^{2-k}\, e^{2\pi i n \bar{z}}. \end{aligned}$$

Applying R_{k-2} to $R_{2-k}^{k-2} f$, we have

$$\begin{aligned} R_{2-k}^{k-1} f(z) &\overset{(7.17)}{=} R_{k-2}\, R_{2-k}^{k-2} f(z) \\ &= \sum_{n \in \mathbb{Z}_{\geq n_{+}}} a_{+}(n)\, (-4\pi n)^{k-1}\, e^{2\pi i n z}. \end{aligned}$$

Now, Lemma 7.18 implies

$$D^{k-1} f = \sum_{n \in \mathbb{Z}_{\geq n_{+}}} a_{+}(n)\, n^{k-1}\, e^{2\pi i n z}.$$

□

7.2 Mock Modular Forms

This section expands on Remark 7.10 where we briefly mention mock modular forms. We restrict our discussion to half-integral weight for this part. The whole section is based on the discussion of mock modular forms in [63, §7].

7.2.1 *Pure Mock Modular Forms*

Definition 7.21 Let $k \in \frac{1}{2}\mathbb{Z}$. We consider $g \in M_{2-k}$ which is a modular form of weight $2-k$ in the sense of Definition 1.59 and Remark 1.61. Assume that g admits a Fourier expansion of the form

$$g(z) = \sum_{n=0}^{\infty} a_n \, e^{2\pi i n z}. \tag{7.22}$$

We define the *non-holomorphic Eichler integral* g^* of g by

$$g^*(z) := \begin{cases} \overline{a_0} \, \frac{(4\pi \Im(z))^{1-k}}{k-1} + \sum_{n=1}^{\infty} n^{k-1} \, \overline{a_n} \, \Gamma\big(1-k, 4\pi n \Im(z)\big) \, e^{-2\pi i n z} \\ \qquad \text{for} k \neq 1 \text{and} \\ -\overline{a_0} \, \log(4\pi \Im(z)) + \sum_{n=1}^{\infty} n^{k-1} \, \overline{a_n} \, \Gamma\big(1-k, 4\pi n \Im(z)\big) \, e^{-2\pi i n z} \\ \qquad \text{for } k = 1. \end{cases} \tag{7.23}$$

The expression $\Gamma(s, x) = \int_x^{\infty} e^{-t} t^{s-1} \, dt$ denotes the incomplete Γ-function, see Appendix A.17 or [153, §8] for more details.

Remark 7.22 The series expansion of g^* in (7.23) converges. This follows from the incomplete Γ-function which satisfies the estimate $\Gamma(1-k, 4\pi n y) = O\big(x^{-k} \, e^{-4\pi n y}\big)$, see Appendix A.17. This estimate compensates sufficiently for the exponential growth of the term $e^{2\pi n y}$, so that we have exponential decay in n overall.

Exercise 7.23 Let $k \in \frac{1}{2}\mathbb{Z}$. Show that $g \in M_{2-k}$ given in (7.22) and g^* given in (7.23) satisfy the first-order partial differential equation

$$\big(4\pi \Im(z)\big)^k \, \frac{\partial g^*(z)}{\partial \overline{z}} = -2\pi i \, \overline{g(z)}. \tag{7.24}$$

Exercise 7.24 Let $k \in \frac{1}{2}\mathbb{Z}$ and assume that $g \in M_{2-k}$ admits a Fourier expansion (7.22) without the constant term. In other words, we assume $a_0 = 0$ and $g \in S_{2-k}$ is a cusp form. Show that the function

$$\tilde{g}^*(z) := \left(\frac{i}{2\pi}\right)^{k-1} \int_{-\bar{z}}^{i\infty} (z+w)^{-k}\, \overline{g(-\bar{z})}\, \mathrm{d}w \tag{7.25}$$

satisfies the first-order partial differential equation (7.24).

Remark 7.25 Let $k \in \frac{1}{2}\mathbb{Z}$. For $g \in S_{2-k}$, consider g^* and $\tilde{g}^*$ defined in (7.23) and (7.25). Since both functions solve the same first-order differential equation (7.24), we see that both functions are identical up to a multiplicative constant. In fact, one can show that both functions are equal, see, e.g., [87, (104) and (107)].

The integral representation $\tilde{g}^*$ in (7.25) is sometimes known as a Niebur integral, see, e.g., [87, (104)].

Now, we can define mock modular forms.

Definition 7.26 Let $k \in \frac{1}{2}\mathbb{Z}$ and (h, g) be a pair of functions on the upper complex half-plane $\mathbb{H}$ satisfying the following conditions:

1. h is a holomorphic function on $\mathbb{H}$ that is meromorphic at all cusps in the sense of Definition 1.58.
2. g is a holomorphic modular form of weight $2-k$, i.e., $g \in M_{2-k}$ in the sense of Definition 1.59 and Remark 1.61.
3. $\widehat{h} := h + g^*$ transforms like a holomorphic modular form of weight k for some congruence subgroup, i.e., $\widehat{h}$ satisfies

$$\widehat{h}\big(M z\big) = j(M, z)^k\, \widehat{h}(z) \tag{7.26}$$

for all $z \in \mathbb{H}$ and $M \in \Gamma$ where Γ is a congruence subgroup in the sense of Definition 1.2.

We call h the weakly holomorphic *pure mock modular form* of weight k, g its *shadow*, and $\widehat{h}$ the *completion* of h. The space of weakly holomorphic pure mock modular forms of weight k is denoted by $M^!_{k|0}$.

Remark 7.27 The 0 in the notation $M^!_{k|0}$ of the space of weakly holomorphic pure mock modular forms of weight k indicates the setting of *pure* mock modular forms. We replace the 0 with an arbitrary integer or half-integer in §7.2.3 where we discuss a more general situation.

Example 7.28 Let $k \in \frac{1}{2}\mathbb{Z}$ and let $h \in M^!_k$ be a weakly holomorphic modular form in the sense of Definition 1.59. Then h is a mock modular form in the sense of Definition 7.26 with associated shadow $g = 0$. The completion $\widehat{h} = h + g^* = h + 0$ transforms trivially like a holomorphic modular form of weight k for the full modular group.

Lemma 7.29 *Let $k \in \frac{1}{2}\mathbb{Z}$ and (f, g) be a pair of functions as in Definition 7.26. The completion $\widehat{h}$ of h satisfies the differential equation*

$$\big(4\pi \Im(z)\big)^k \frac{\partial \widehat{h}(z)}{\partial \bar{z}} = -2\pi i\, \overline{g(z)}. \tag{7.27}$$

Proof The mock modular form h is a holomorphic function by definition. This implies that the derivative $\frac{\partial}{\partial \overline{z}} h(z) = 0$ vanishes. Hence

$$\frac{\partial \widehat{h}(z)}{\partial \overline{z}} = \frac{\partial \big(h(z) + g^*(z)\big)}{\partial \overline{z}} = \frac{\partial g^*(z)}{\partial \overline{z}}.$$

This shows that the completion $\widehat{h}$ of the mock modular form h satisfies (7.27). □

Definition 7.30 Let $k \in \frac{1}{2}\mathbb{Z}$ and (h, g) a mock modular form with its shadow in the sense of Definition 7.26. The map

$$\mathcal{S} : M^!_{k|0} \to \overline{M_{2-k}}; \quad h \mapsto \overline{g} \tag{7.28}$$

is called the *shadow map*.

Remark 7.31 The space $\overline{M_{2-k}}$ denotes the space of complex conjugated functions $\overline{g}$ with $g \in M_{2-k}$ a modular forms of weight $2 - k$. In other words, we have

$$\overline{M_{2-k}} := \{\overline{g} : \ g \in M_{2-k}\}. \tag{7.29}$$

Proposition 7.32 *Let $k \in \frac{1}{2}\mathbb{Z}$. The sequence*

$$0 \longrightarrow M^!_k \longrightarrow M^!_{k|0} \xrightarrow{\mathcal{S}} \overline{M_{2-k}} \tag{7.30}$$

is exact.

Proof The map $M^!_k \longrightarrow M^!_{k|0}$ is obviously well defined by Example 7.28. We see that the associated shadows vanish.

On the other hand, assume that $h \in M^!_{k|0}$ has a vanishing shadow $g = 0$. Then Definition 7.26 implies that $h = \widehat{h}$ and h transforms trivially like a holomorphic modular form of weight k on the full modular group. Hence, h is a weakly holomorphic modular form of weight k, i.e., $h \in M^!_k$.

This shows that the sequence in (7.30) is exact. □

Remark 7.33 For convenience, we state an example by Zagier in [63, p. 42, §7.1, Example 2] and that goes back to Zagier in [209].

The example is the generating function $\mathbf{H} : \mathbb{H} \to \mathbb{C}$ of the Hurwitz class numbers $H(N)$. The *Hurwitz class numbers* is a function $H : \mathbb{Z} \to \mathbb{R}$ defined by the following properties [116, p. 47]: For $N < 0$, put $H(N) = 0$ and for $N = 0$, put $H(0) = -\frac{1}{12}$. For $N > 0$, denote by $H(N)$ the number of $\mathrm{PSL}_2(\mathbb{Z})$-equivalence classes of integral binary quadratic forms of discriminant $-N$, weighted by the reciprocal of the number of their automorphisms (if $-N$ is the discriminant of an imaginary quadratic field K other than $\mathbb{Q}(i)$ or $\mathbb{Q}(\sqrt{-3})$, this is just the class number of K). Zagier shows in [209] that the function

$$\mathbf{H}(z) := \sum_{n=0}^{\infty} H(N)\, e^{2\pi i n z}$$

is a mock modular form of weight $\frac{3}{2}$ on $\Gamma_0(4)$. Its shadow is the classical theta function

$$\theta(z) := \sum_{n=0}^{\infty} e^{2\pi i\, n^2}.$$

Notice that $H(z)$ is regular at the cusps in the sense of Definition 1.58.

The first 74 entries of the Hurwitz class number, i.e., the entries of the sequence $\big(H(0), \ldots, H(74)\big)$, are listed in the On-Line Encyclopedia of Integer Sequences under reference [196, A259825].

7.2.2 Mock Jacobi Forms

Following [63, §7.2] a mock Jacobi form $\phi : \mathbb{C} \times \mathbb{H} \to \mathbb{C}$ of integer weight k and index $m \in \mathbb{N}$ is "nearly a Jacobi form" in the sense of Remark 1.133 with the requirement that the coefficients $h_l(z)$ in the theta series expansion of ϕ in Remark 1.133 are mock modular forms of weight $k - \frac{1}{2}$ and that the modular transformation property holds for the completed function $\hat{\phi}$. We give the definition below.

Definition 7.34 A *mock Jacobi form* of *level* 1, *weight* k and *index* m is a function $\phi : \mathbb{C} \times \mathbb{H} \to \mathbb{C}$ satisfying

1. $\phi(w + \lambda z + \mu, z) = e^{-2\pi i m(\lambda^2 z + 2\lambda w)}\, \phi(w, z)$ for all integers $\lambda, \mu \in \mathbb{Z}$,
2. ϕ has a Fourier expansion of the form

$$\phi(w, z) = \sum_{n \in \mathbb{Z}_{\geq 0}} \sum_{r^2 \leq 4mn} c(n, r)\, e^{2\pi i (nz + rw)}$$

   with suitable coefficients $c(n, r)$ and
3. The coefficients $c(n, r)$ in the Fourier expansion above satisfy $c(n, r) = 0$ for all $4mn < r^2$.

Moreover, ϕ admits a theta series expansion of the form

$$\phi(w, z) = \sum_{l \in \{0, \ldots, 2m-1\}} h_l(z)\, \Theta_{m,l}(w, z), \tag{7.31}$$

where the coefficients $h_l(z)$ are mock modular forms of weight $k - \frac{1}{2}$ in the sense of Definition 7.26.

The completion $\widehat{\phi}$ of ϕ given by

$$\widehat{\phi}(w, z) = \sum_{l \in \{0, 2m-1\}} \widehat{h}_l(z)\, \Theta_{m,l}(w, z) \tag{7.32}$$

and satisfies

$$\phi\left(\frac{w}{cz+d}, \frac{az+b}{cz+d}\right) = (cz+d)^k\, e^{\frac{2\pi imcw^2}{cz+d}}\, \phi(w, z) \tag{7.33}$$

for all $\begin{pmatrix} a & b \\ c & d \end{pmatrix} \in \mathrm{SL}_2(\mathbb{Z})$, $w \in \mathbb{C}$ and $z \in \mathbb{H}$.

Lemma 7.35 ([63, §7.2]) *Let ϕ be a mock Jacobi form of weight k and index m in the sense of Definition 7.34 and let g_l denote the shadows of the mock modular forms h_l in the theta series expansion* (7.32).

The mock Jacobi form satisfies

$$\widehat{\phi}(w, z) = \phi(w, z) + \sum_{l \in \{0, 2m-1\}} g_l^*(z)\, \Theta_{m,l}(w, z) \tag{7.34}$$

with g_l^* given as in Definition 7.34. Using (7.24), *we have*

$$\psi(w, z) := \Im(z)^{k-\frac{1}{2}} \frac{\partial}{\partial \bar{z}} \widehat{\phi}(w, z) = C \sum_{l \in \{0, 2m-1\}} \overline{g_l(z)}\, \Theta_{m,l}(w, z). \tag{7.35}$$

Here, C denotes some constant not depending on z and w.

More details can be found in [63, §7.2].

7.2.3 General Mock Modular Forms

The previous discussion about pure mock modular forms in §7.2.1 allowed the mock modular form h to be meromorphic at the cusps. This means in particular that these functions fulfill an exponential growth condition at the cusps; for example at $i\infty$ of the form $O\left(e^{m\Im(z)}\right)$ as $z \to i\infty$ with bounded $\Re(z)$ for some positive $m > 0$.

Replacing this condition by requiring that the function is regular at the cusps would lead to the class of strongly holomorphic mock modular forms.

Definition 7.36 Let $k \in \frac{1}{2}\mathbb{Z}$ and (h, g) be a pair of functions on the upper complex half-plane $\mathbb{H}$ satisfying the following conditions:

1. h is a holomorphic function on $\mathbb{H}$ that is regular at all cusps in the sense of Definition 1.58.
2. g is a holomorphic modular form of weight $2-k$, i.e., $g \in M_{2-k}$ in the sense of Definition 1.59 and Remark 1.61.
3. $\widehat{h} := h + g^*$ transforms like a holomorphic modular form of weight k for some congruence subgroup, i.e., $\widehat{h}$ satisfies

$$\widehat{h}(M\,z) = j(M,z)^k\,\widehat{h}(z) \tag{7.36}$$

for all $z \in \mathbb{H}$ and $M \in \Gamma$ where Γ is a congruence subgroup in the sense of Definition 1.2.

We call h the strongly holomorphic *pure mock modular form* of weight k, g its *shadow*, and $\widehat{h}$ the *completion* of h.

The space of strongly holomorphic pure mock modular forms of weight k is denoted by $\mathbb{M}_{k|0}$.

Remark 7.37 In analogy with the discussion in §7.2.1, we can derive the exact sequence

$$0 \longrightarrow M_k \longrightarrow \mathbb{M}_{k|0} \stackrel{S}{\longrightarrow} \overline{M_{2-k}}. \tag{7.37}$$

However, the concept of strongly holomorphic pure mock modular forms is not very useful. As Zagier points out in [63, §7.3], there are almost no examples of strongly holomorphic pure mock modular forms. Essentially, the only examples are the function $\mathbf{H}$ mentioned in Remark 7.33 and its variants.

The notion of strongly holomorphic mock modular forms becomes useful if we generalize to mixed mock modular forms of mixed weight $k|l$ with $k, l \in \frac{1}{2}\mathbb{Z}$

Definition 7.38 Let $k, l \in \frac{1}{2}\mathbb{Z}$ and let h and f_i, g_i, $i \in \{1, \ldots, i_0\}$ (a finite number of) functions on the upper complex half-plane $\mathbb{H}$ satisfying the following conditions:

1. h is a holomorphic function on $\mathbb{H}$ that is regular at all cusps in the sense of Definition 1.58.
2. $f_i \in M_l$ and $g_i \in M_{2-k+l}$ for all $i \in \{1, \ldots, i_0\}$ in the sense of Definition 1.59 and Remark 1.61.
3. $\widehat{h} := h + \sum_i f_i\, g_i^*$ transforms like a holomorphic modular form of weight k for some congruence subgroup, i.e., $\widehat{h}$ satisfies

$$\widehat{h}(M\,z) = j(M,z)^k\,\widehat{h}(z) \tag{7.38}$$

for all $z \in \mathbb{H}$ and $M \in \Gamma$ where Γ is a congruence subgroup in the sense of Definition 1.2.

We call h the strongly holomorphic *mixed mock modular form* of weight k, g_i are its *shadows*, and $\widehat{h}$ the *completion* of h. The space of strongly holomorphic mixed mock modular forms of weight k is denoted by $M_{k|l}$.

The above definition implies that we have to adapt the shadow map to the strongly holomorphic mixed modular forms.

Definition 7.39 Let $k, l \in \frac{1}{2}\mathbb{Z}$ and let h a strongly holomorphic mixed mock modular form with its shadows in the sense of Definition 7.38. The map

$$\mathcal{S} : M_{k|l} \to M_l \otimes \overline{M_{2-k+l}}; \quad h \mapsto \sum_i f_i \, \overline{g_i} \tag{7.39}$$

is called the (extended) *shadow map*.

Exercise 7.40 Let $k \in \frac{1}{2}\mathbb{Z}$ be a half-integral weight and $h \in M_{k|0}$ be a strongly holomorphic mixed mock modular form with its shadows in the sense of Definition 7.38. Show that h is also a pure mock modular form in the sense of Definition 7.26.

Similar to the proof of Proposition 7.32, we show the following.

Proposition 7.41 *Let $k, l \in \frac{1}{2}\mathbb{Z}$. The sequence*

$$0 \longrightarrow M_k \longrightarrow M_{k|l} \xrightarrow{\mathcal{S}} M_l \otimes \overline{M_{2-k+l}} \tag{7.40}$$

is exact.

We extend the space of mixed mock modular forms $M_{k|l}$ by considering also functions whose shadows are finite sum of products $f_i \, \overline{g_i}$ with $f_i \in M_{l_i}$ and $g_i \; in M_{2-k+l_i}$ allowing different weights l_i. This makes the mixed mock modular form "even more mixed" in some sense.

Definition 7.42 Let $k \in \frac{1}{2}\mathbb{Z}$ and let h and $f_i, g_i, i \in \{1, \ldots, i_0\}$ (a finite number of) functions on the upper complex half-plane $\mathbb{H}$ satisfying the following conditions:

1. h is a holomorphic function on $\mathbb{H}$ that is regular at all cusps in the sense of Definition 1.58.
2. There exist weights $l_i \in \frac{1}{2}\mathbb{Z}$ such that $f_i \in M_{l_i}$ and $g_i \in M_{2-k+l_i}$ for all $i \in \{1, \ldots, i_0\}$ in the sense of Definition 1.59 and Remark 1.61.
3. $\widehat{h} := h + \sum_i f_i \, g_i^*$ transforms like a holomorphic modular form of weight k for some congruence subgroup, i.e., $\widehat{h}$ satisfies

$$\widehat{h}\big(M\, z\big) = j(M, z)^k \, \widehat{h}(z) \tag{7.41}$$

for all $z \in \mathbb{H}$ and $M \in \Gamma$ where Γ is a congruence subgroup in the sense of Definition 1.2.

We call h a strongly holomorphic *mock modular form* of weight k, g_i are its *shadows*, and $\widehat{h}$ the *completion* of h. The space of strongly holomorphic mixed mock modular forms of weight k is denoted by $\mathbb{M}_k$.

Similar to the proof of Proposition 7.32, we show the following.

Proposition 7.43 *Let $k \in \frac{1}{2}\mathbb{Z}$. The sequence*

$$0 \longrightarrow M_k \longrightarrow \mathbb{M}_k \xrightarrow{S} \bigoplus_i M_{l_i} \otimes \overline{M_{2-k+l_i}} \tag{7.42}$$

is exact.

Remark 7.44 According to [63, §7.3], the functions in the space $\mathbb{M}_k$ are the most natural objects.

To justify this claim, the authors of [63] present the following argument: Consider the space $\mathfrak{M}_{k,l}$ of real-analytic functions on the upper half-plane $\mathbb{H}$ with at most polynomial growth – this corresponds to being regular in the sense of Definition 1.58 in the setting of holomorphic functions – that transform according to

$$F\left(\begin{pmatrix} a & b \\ c & d \end{pmatrix} z\right) = (cz+d)^k \, (c\overline{z}+d)^l \, F(z)$$

for all $z \in \mathbb{H}$ and for all matrices $\begin{pmatrix} a & b \\ c & d \end{pmatrix}$ belonging to some subgroup of finite index of the full modular group $\mathrm{SL}_2(\mathbb{Z})$. Since the imaginary part $\Im(z)$ transforms like (1.18), the spaces $\mathfrak{M}_{k,l}$ and $\Im(\cdot)^r \, \mathfrak{M}_{k+r,l+r}$ coincide for all $k, l, r \in \frac{1}{2}\mathbb{Z}$.

Obviously, the space $\mathfrak{M}_{k,l}$ contains the space $M_k \otimes \overline{M_l}$. Also the above argument shows that it also contains the direct sum

$$\bigoplus_{r \in \mathbb{Z}} \Im(\cdot)^r \, \mathfrak{M}_{k+r,l+r}.$$

The authors of [63] call elements of this subspace *decomposable*.

Assume that $F \in \mathfrak{M}_{k,0}$ has the property that its $\overline{z}$-derivative (which automatically belongs to $\mathfrak{M}_{k,2}$) is decomposable:

$$\frac{\partial}{\partial \overline{z}} F = \sum_i \Im(\cdot)^{r_i} \, f_i \, \overline{g_i}.$$

Then F is the sum of a holomorphic function h and a multiple of $\sum_i f_i \, g_i^*$ and h is a mock modular form in the sense of Definition 7.42.

Conversely, a holomorphic function h is a mock modular form of weight k if and only if there exist numbers r_i and modular forms $f_i \in M_{k+r_i}$ and $g_i \in M_{2+r_i}$ such that the sum

$$\widehat{h} = h + \sum_i f_i\, g_i^*$$

belongs to $\mathbb{M}_k$.

Remark 7.45 The above argument, presented in Remark 7.44, leads the authors of [63] to the following description:

A mock modular form is the holomorphic part of a real-analytic modular form whose $\overline{z}$-derivative is decomposable.

7.3 L-Functions of Weakly Harmonic Maass Wave Forms

Let $F \in S_{k,1}\big(\Gamma(1)\big)$ be a holomorphic cusp form in the sense of Definition 1.59. Recall that F admits a Fourier expansion of the form

$$F(z) = \sum_{n=1}^{\infty} a(n)\, e^{2\pi i n z} \tag{1.157}$$

for all $z \in \mathbb{H}$. Then, we continue in §1.7 and associate the L-series

$$L(s) = \sum_{n=1}^{\infty} a_n\, n^{-s} \tag{1.153}$$

to F and discuss its properties. In §4.2, we discuss L-series associated with admissible families of Maass cusp forms. This raises the following question: Can we associate an L-series to weakly harmonic Maass wave forms? We follow an approach introduced in [22], [24], and [25].

7.3.1 *Regularized L-Series for* $S^!_{k,1}\big(\Gamma(1)\big)$

Let $g \in \mathbb{H} \to \mathbb{C}$ be a continuous function that satisfies the following growth condition: There exists an $c > 0$ such that

$$g(z) = O\left(e^{c\Im(z)}\right) \tag{7.43}$$

uniformly in $\Re(z)$ as $\Im(z) \to \infty$.

Exercise 7.46 Let $g : \mathbb{H} \to \mathbb{C}$ be continuous and satisfies the growth condition (7.43). Show that the integral

$$\int_{z_0}^{i\infty} e^{uiw}\, g(w)\, \mathrm{d}w \tag{7.44}$$

converges for $u \in \mathbb{C}$ with $\Re(u) > c$. Here, c is the constant appearing in (7.43). The path of integration lies within a vertical strip, and $z_0 \in \mathbb{H}$ is an arbitrary but fixed reference point.

Definition 7.47 Let $g : \mathbb{H} \to \mathbb{C}$ be continuous and satisfies the growth condition (7.43). If the integral in (7.44) has an analytic continuation in the variable u to $u = 0$, we define the *regularized integral* of f by

$$R.\int_{z_0}^{i\infty} g(w)\, \mathrm{d}w := \left[\int_{z_0}^{i\infty} e^{uiw}\, g(w)\, \mathrm{d}w\right]_{u=0}, \tag{7.45}$$

where the right-hand side means that we take the value $u = 0$ of the analytic continuation of the integral as function in u.

We extend the above definition to parabolic cusps other than $i\infty$ in the following way: Let A be the scaling matrix of the cusp $q = A\, i\infty$ in the sense of Remark 1.29. Then, we define the *regularized integral* of f at the cusp q by

$$R.\int_{z_0}^{q} g(w)\, \mathrm{d}w := R.\int_{A^{-1}z_0}^{i\infty} g\big|_2 A(w)\, \mathrm{d}w. \tag{7.46}$$

For two cusps $q_1 = A_1\, i\infty$ and $q_2 = A_2\, i\infty$ with scaling matrices A_1 and A_2, we set

$$R.\int_{q_1}^{q_2} g(w)\, \mathrm{d}w := R.\int_{z_0}^{q_2} g(w)\, \mathrm{d}w - R.\int_{z_0}^{q_1} g(w)\, \mathrm{d}w. \tag{7.47}$$

Remark 7.48 We are mainly interested in the regularized integral

$$R.\int_{0}^{i\infty} g(w)\, w^{s-1}\, \mathrm{d}w$$

for some $s \in \mathbb{C}$. To shorten the notation, we write

$$R.\int_{0}^{\infty} g(iy)\, y^{s-1}\, \mathrm{d}y. = i^{-s}\, R.\int_{0}^{i\infty} g(w)\, w^{s-1}\, \mathrm{d}w,$$

which can be seen by substituting $w = iy$.

Now, we define the L-series associated with weakly holomorphic modular forms $f \in S^{!}_{k,1}\big(\Gamma(1)\big)$, following the approach in [24].

Definition 7.49 For $k \in 2\mathbb{N}$, let $f \in S^{!}_{k,1}\big(\Gamma(1)\big)$ be a weakly holomorphic modular form in the sense of Definition 1.59. We assume that f admits a Fourier expansion of the form

$$f(z) = \sum_{\substack{n=-\nu \\ n\neq 0}}^{\infty} a(n)\, e^{2\pi i n z} \tag{7.48}$$

with vanishing constant $a(0) = 0$.

The *L-series* associated with f is defined by

$$L_f(s) := \frac{(2\pi)^s}{\Gamma(s)}\, L_f^\star(s) \qquad \text{with}$$

$$L_f^\star(s) := \sum_{n\geq -\nu} \frac{a(n)\,\Gamma(s, 2\pi n t_0)}{(2\pi n)^s} + i^k \sum_{n\geq -\nu} \frac{a(n)\,\Gamma\left(k-s, \frac{2\pi n}{t_0}\right)}{(2\pi n)^{k-s}} \tag{7.49}$$

for some reference value $t_0 > 0$. Here, $\Gamma\big(\alpha, x\big) = \int_x^\infty t^{\alpha-1} e^{-t}\, dt$ denotes the incomplete Γ-function, see [153, (8.2.2)] and Appendix A.17.

Remark 7.50 The authors of [24] show that $L_f^\star(s)$ is well defined for weakly holomorphic modular forms $f \in S^!_{k,1}\big(\Gamma(1)\big)$ and independent of the choice of t_0.

Theorem 7.51 ([24, Theorem 2.2]) *For $k \in 2\mathbb{N}$, let $f \in S^!_{k,1}\big(\Gamma(1)\big)$ be a weakly holomorphic modular form in the sense of Definition 1.59. We have*

$$L_f^\star(s) = R.\int_0^\infty f(iy)\, y^{s-1}\, \mathrm{d}y. \tag{7.50}$$

Furthermore, $L_f^\star(s)$ satisfies the functional equation

$$L_f^\star(k-s) = L_f^\star(s). \tag{7.51}$$

Proof Consider the right-hand side of (7.50). Splitting the integral in the sense of (7.47) at some $t_0 \geq 0$, we find

$$R.\int_0^\infty f(iy)\, y^{s-1}\, \mathrm{d}y = R.\int_{t_0}^\infty f(iy)\, y^{s-1}\, \mathrm{d}y - R.\int_{t_0}^0 f(iy)\, y^{s-1}\, \mathrm{d}y.$$

Plugging in the Fourier expansion (7.48) of f into the first term of the right-hand side, we get

$$\begin{aligned} R.\int_{t_0}^\infty f(iy)\, y^{s-1}\, \mathrm{d}y &\overset{(7.45)}{=} \left[\int_{t_0}^\infty e^{-uy}\, f(iy)\, y^{s-1}\, \mathrm{d}y\right]_{u=0} \\ &\overset{(7.48)}{=} \left[\int_{t_0}^\infty \sum_{\substack{n=-\nu \\ n\neq 0}}^{\infty} a(n)\, e^{-2\pi n y - uy}\, y^{s-1}\, \mathrm{d}y\right]_{u=0} \end{aligned}$$

$$= \left[\sum_{\substack{n=-\nu \\ n \neq 0}}^{\infty} a(n) \int_{t_0}^{\infty} e^{-2\pi n y - u y} \, y^{s-1} \, \mathrm{d}y \right]_{u=0}$$

$$= \left[\sum_{\substack{n=-\nu \\ n \neq 0}}^{\infty} \frac{a(n)}{(u + 2\pi n)^s} \int_{2\pi n t_0 + u t_0}^{\infty} e^{-\xi} \, \xi^{s-1} \, \mathrm{d}\xi \right]_{u=0}$$

$$= \sum_{\substack{n=-\nu \\ n \neq 0}}^{\infty} \frac{a(n)}{(2\pi n)^s} \int_{2\pi n t_0}^{\infty} e^{-y} \, y^{s-1} \, \mathrm{d}y$$

$$= \sum_{\substack{n=-\nu \\ n \neq 0}}^{\infty} \frac{a(n)}{(2\pi n)^s} \Gamma\big(s, 2\pi n t_0\big),$$

using the substitution $-2\pi n y - u y = \xi$ and the integral representation of the incomplete Γ-function, see Appendix A.17. For the second term, we similarly get

$$-R. \int_{t_0}^{0} f(iy) \, y^{s-1} \, \mathrm{d}y = i^{-k} \sum_{\substack{n=-\nu \\ n \neq 0}}^{\infty} \frac{a(n)}{(2\pi n)^{k-s}} \Gamma\left(k - s, \frac{2\pi n}{t_0}\right).$$

This shows (7.50).

Using the fact that the definition of $L_f^\star(s)$ is independent of the choice of $t_0 > 0$, we may choose $\frac{1}{t_0}$ instead and find

$$L_f^\star(k-s) \overset{(7.49)}{=} \sum_{n \geq -\nu} \frac{a(n) \left(k - s, \frac{2\pi n}{t_0}\right)}{(2\pi n)^{k-s}} + i^k \sum_{n \geq -\nu} \frac{a(n) \, \Gamma\big(s, 2\pi n t_0\big)}{(2\pi n)^s}$$

$$\overset{(7.49)}{=} i^k \, L_f^\star(s).$$

□

Corollary 7.52 *For $f \in S_{k,1}\big(\Gamma(1)\big)$, the two L-series associated by Definition 1.205 and by Definition 7.49 coincide.*

Proof Let $f \in S_{k,1}\big(\Gamma(1)\big)$. For $s \in \mathbb{C}$ with $\Re(s) > k + 1$ large enough, we have

$$L_f^\star(s) \overset{(7.50)}{=} R. \int_0^{\infty} f(iy) \, y^{s-1} \, \mathrm{d}y$$

$$= \left[\int_{t_0}^{\infty} e^{-uy} \, f(iy) \, y^{s-1} \, \mathrm{d}y \right]_{u=0} - \left[\int_0^{t_0} e^{-uy} \, f(iy) \, y^{s-1} \, \mathrm{d}y \right]_{u=0}$$

$$
\begin{aligned}
&= \int_{t_0}^{\infty} f(iy)\, y^{s-1}\, dy - \int_0^{t_0} f(iy)\, y^{s-1}\, dy \\
&= \int_0^{\infty} f(iy)\, y^{s-1}\, dy \\
&\overset{(1.160)}{=} L^\star(s).
\end{aligned}
$$

The interchange of the limit $u \to 0$ and the integration is valid since we can apply the dominated convergence theorem using the same arguments as in the proof of Lemma 1.207. □

7.3.2 Regularized L-Series for $H_k(\Gamma(1))$

For $k \in 2\mathbb{N}$, consider a weakly harmonic Maass wave form $f \in H_k(\Gamma(1))$ with a decomposition $f = f_+ + f_-$ as in Lemma 7.7. We assume additionally that the 0th Fourier coefficient $a_-(0)$ of f_-, see the expansion in (7.5), vanishes: $a_-(0) = 0$. Recall Propositions 7.15 and 7.20 where we establish the maps

$$
\begin{aligned}
\xi_{2-k} &: H_{2-k}(\Gamma(1)) \longrightarrow M^!_{k,1}(\Gamma(1)) \quad \text{and} \\
D^{k-1} &: H_{2-k}(\Gamma(1)) \longrightarrow M^!_{k,1}(\Gamma(1)).
\end{aligned}
$$

Using these maps, we may associate with a weakly harmonic Maass wave form $f \in H^+_{2-k}(\Gamma(1))$ two functions $\xi_{2-k} f \in S_{k,1}(\Gamma(1))$ and $D^{k-1} f \in S^!_{k,1}(\Gamma(1))$:

$$
\begin{aligned}
\xi_{2-k} &: H^+_{2-k}(\Gamma(1)) \longrightarrow S_{k,1}(\Gamma(1)) \quad \text{and} \\
D^{k-1} &: H^+_{2-k}(\Gamma(1)) \longrightarrow S^!_{k,1}(\Gamma(1)).
\end{aligned}
\tag{7.52}
$$

This way, we may use Definition 7.49 to associate two L-series with f:

$$
L^\star_{\xi_{2-k} f}(s) \quad \text{and} \quad L^\star_{D^{k-1} f}(s).
$$

An interesting relation between these two L-series' is as follows.

Proposition 7.53 ([24, Corollary 3.2]) *For $4 \le k \in 2\mathbb{N}$, consider the function $f \in H^+_{2-k}(\Gamma(1))$. We have*

$$
L_{\xi_{2-k} f}(n+1) = (-1)^n \frac{(4\pi)^{k-1}}{(k-2)!} \overline{L_{D^{k-1} f}(n+1)} \tag{7.53}
$$

for all integers $0 < n < k-2$.

We will give the proof of this proposition at the end of §7.4.2.

7.4 Periods and Weakly Harmonic Maass Wave Forms

7.4.1 The Mock Modular Period Function

Recall that any weakly harmonic Maass wave form $f \in H_k(\gamma(1))$ can be decomposed into f_+ and f_- in the sense of Lemma 7.7. In particular, the holomorphic part f_+ admits the Fourier expansion in (7.5). The authors of [22] associate a period function with f_+, mimicking the idea of period polynomials associated with Eichler integrals, see §2.1.2.

Definition 7.54 For $k \in 2\mathbb{N}$, let $f = f_+ + f_- \in H_{2-k}(\gamma(1))$ denote a weakly harmonic Maass form with holomorphic part f_+ in the sense of Lemma 7.7.

For each $M \in \Gamma(1)$, define the *M-mock modular period function* $\mathbb{P}(f_+, M; \cdot) : \mathbb{H} \to \mathbb{C}$ by

$$\mathbb{P}(f_+, M; z) := \frac{(4\pi)^{k-1}}{\Gamma(k-1)} \left(f_+ - f_+\big|_{2-k}M\right)(z). \tag{7.54}$$

Remark 7.55 Comparing the above definition with Definition 2.14, we see that f_+ is an Eichler integral if $\mathbb{P}(f_+, M; \cdot)$ is a polynomial of degree at most $k-2$. Moreover, following the proof of Proposition 2.16, we conclude that $\mathbb{P}(f_+, T; \cdot)$ satisfies the relations in (2.4) of a period polynomial:

$$\mathbb{P}(f_+, T; \cdot)\big|_{2-k}(1+T) = 0 \quad \text{and} \quad \mathbb{P}(f_+, T; \cdot)\big|_{2-k}(1+TS+TSTS) = 0. \tag{7.55}$$

A first result, generalizing Proposition 2.9 to M-mock modular period functions, was obtained in [22].

Proposition 7.56 ([22, Theorem 1.1]) *For $4 \leq k \in 2\mathbb{N}$, let $f \in H^+_{2-k}(\Gamma(1))$ be a weakly harmonic Maass wave form.*

We have the relation

$$\overline{\mathbb{P}(f_+, T; \overline{z})} = \sum_{l=0}^{k-2} \frac{L(l+1)}{(k-2-l)!} (2\pi i z)^{k-2-l} \tag{7.56}$$

for all $z \in \mathbb{H}$. Here, $L(s)$ denotes the usual L-series in (1.153) *associated with the cusp form $\xi_{2-k} f \in S_k(\Gamma(1))$ of weight k.*

Exercise 7.57 For $4 \leq k \in 2\mathbb{N}$, let $f \in H^+_{2-k}(\Gamma(1))$ be a weakly harmonic Maass wave form and denote $g := \xi_{2-k} f$. Since $g \in S_k$ is a cusp form, see (7.14) of Lemma 7.15, show that the right side of $\sum_{l=0}^{k-2} \frac{L(l+1)}{(k-2-l)!} (2\pi i z)^{k-2-l}$ indeed corresponds to the right-hand side of (2.6) in Proposition 2.9 for the L-series $L(s)$ associated with the cusp form g up to a constant. In other words, show that

$$\sum_{l=0}^{k-2} \binom{k-2}{l} (-z)^{k-l-2}\, i^{l+1}\, L^\star(l+1)$$
$$= \frac{(k-2)!}{(-2\pi i)^{k-1}} \sum_{l=0}^{k-2} \frac{L(l+1)}{(k-l-2)!} (2\pi i z)^{k-l-2}. \tag{7.57}$$

Corollary 7.58 *For $4 \le k \in 2\mathbb{N}$, let $f \in H^+_{2-k}\big(\Gamma(1)\big)$ be a weakly harmonic Maass wave form and $P_{\xi_{2-k}f}(X)$ denotes the period polynomial in* (2.1) *associated with the cusp form $\xi_{2-k}f \in S_k\big(\Gamma(1)\big)$ of weight k. We have*

$$-\frac{(k-2)!}{(2\pi i)^{k-1}} \overline{\mathbb{P}\big(f_+, T; \overline{z}\big)} = P_{\xi_{2-k}f}(z) \tag{7.58}$$

for all $t \in \mathbb{H}$.

Proof Combining Propositions 2.9 and 7.56 with Exercise 7.57 gives

$$\begin{aligned} P_{\xi_{2-k}f}(z) &\overset{(2.6)}{=} \sum_{l=0}^{k-2} \binom{k-2}{l} (-X)^{k-l-2}\, i^{l+1}\, L^\star(l+1) \\ &\overset{(7.57)}{=} \frac{(k-2)!}{(-2\pi i)^{k-1}} \sum_{l=0}^{k-2} \frac{(L(l+1)}{(k-l-2)!} (2\pi i z)^{k-l-2} \\ &\overset{(7.56)}{=} \frac{(k-2)!}{(-2\pi i)^{k-1}} \overline{\mathbb{P}\big(f_+, T; \overline{z}\big)} \end{aligned}$$

for all $z \in \mathbb{H}$. Since k is even, we conclude $(-2\pi i)^{k-1} = -(2\pi i)^{k-1}$. This proves (7.58). □

Remark 7.59 A version of the Eichler-Shimura theorem for mock modular forms also exists. Bingmann, Diamantis, and Raum presented this "mock Eichler-Shimura theorem" in [21].

7.4.2 Period Functions via Formal Eichler Integrals

Let $k \in 2\mathbb{N}$. Recall the formal Eichler integral $\mathcal{E}_f$ introduced in Remark 2.21 for weakly modular forms $f \in M^!_{k,1}\big(\Gamma(1)\big)$. Following the spirit of Definition 2.14 and Proposition 2.16 in §2.1.2, we define the period function associated with a weakly holomorphic modular form as follows.

Definition 7.60 For $k \in 2\mathbb{N}$, let $f \in M^!_{k,1}\big(\Gamma(1)\big)$ be a weakly holomorphic modular form of weight k. The associated *period function* $r(f;\cdot) : \mathbb{H} \to \mathbb{C}$ is given by

$$r(f;z) := -\frac{\Gamma(k-1)}{(2\pi i)^{k-1}}\big(\mathcal{E}_f - \mathcal{E}_f\big|_{2-k}T\big)(z) \tag{7.59}$$

for all $z \in \mathbb{H}$ where $\mathcal{E}_f$ denotes the formal Eichler integral of f given in (2.16) of Remark 2.21.

Exercise 7.61 For $k \in 2\mathbb{N}$, let $f \in S_{k,1}\big(\Gamma(1)\big)$ be a cusp form of weight k.

Using the embedding $S_{k,1}\big(\Gamma(1)\big) \subset M^!_{k,1}\big(\Gamma(1)\big)$, show that the associated period function $r(f,z)$ to f in (7.59) and the period polynomial $P_f(z)$ in (2.1) coincide:

$$r(f;z) = P_f(z) \tag{7.60}$$

for all $z \in \mathbb{H}$.

We see that the formal Eichler integral $\mathcal{E}_f$ appears in the definition of period functions of weakly holomorphic modular forms, see Definition 7.60. It would be nice to have another expression for $\mathcal{E}_f$ in terms of a regularized integral. This would generalize the observation made in (2.21) of Remark 2.25 for cusp forms.

Lemma 7.62 *For $k \in 2\mathbb{N}$, let $f \in S^!_{k,1}\big(\Gamma(1)\big)$. We have*

$$\mathcal{E}_f(z) = -\frac{(2\pi i)^{k-1}}{\Gamma(k-1)}\, R.\int_z^{i\infty} f(w)\,(w-z)^{k-2}\,\mathrm{d}w \tag{7.61}$$

for $z \in \mathbb{H}$.

Proof Recall that f admits a Fourier expansion of the form

$$f(z) = \sum_{0\neq n\in\mathbb{Z}_{\geq n_+}} a(n)\, e^{2\pi i n z}$$

for $z \in \mathbb{H}$ and given $n_+ \in \mathbb{Z}$.

The integrand $f(z)\,(w-z)^{k-2}$ has a possible pole at the cusp $i\infty$. Hence, we calculate the regularized integral as follows:

$$\begin{aligned} & R.\int_z^{i\infty} f(w)\,(w-z)^{k-2}\,\mathrm{d}w \\ &\overset{(7.45)}{=} \left[\int_z^{i\infty} e^{uiw}\, f(w)\,(w-z)^{k-2}\,\mathrm{d}w\right]_{u=0} \\ &= \sum_{n\neq\in\mathbb{Z}_{\geq n_+}} a(n) \left[\int_z^{i\infty} e^{uiw+2\pi i n w}\,(w-z)^{k-2}\,\mathrm{d}w\right]_{u=0}. \end{aligned}$$

Substituting $w - z = \tau$ gives

$$\begin{aligned} R.\int_z^{i\infty} f(w)\,(w-z)^{k-2}\,\mathrm{d}w &= \sum_{n\neq\in\mathbb{Z}_{\geq n_+}} a(n)\left[\int_z^{i\infty} e^{uiw+2\pi inw}\,(w-z)^{k-2}\,\mathrm{d}w\right]_{u=0} \\ &= \sum_{n\neq\in\mathbb{Z}_{\geq n_+}} a(n)\left[\int_0^{i\infty} e^{ui\tau+2\pi in\tau}e^{uiz+2\pi inz}\,\tau^{k-2}\,\mathrm{d}\tau\right]_{u=0} \\ &= \sum_{n\neq\in\mathbb{Z}_{\geq n_+}} a(n)\,e^{2\pi inz}\left[\int_0^{i\infty} e^{ui\tau+2\pi in\tau}e^{uiz}\,\tau^{k-2}\,\mathrm{d}\tau\right]_{u=0}. \end{aligned}$$

Let us consider the integral $\int_0^{i\infty} e^{ui\tau+2\pi in\tau}e^{uiz}\,\tau^{k-2}\,\mathrm{d}\tau$, separately. Assuming $\Re(u) > -n$, we have

$$\begin{aligned} \int_0^{i\infty} e^{ui\tau+2\pi in\tau}e^{uiz}\,\tau^{k-2}\,\mathrm{d}\tau &= \frac{1}{(ui+2\pi in)^{k-1}}\int_0^{\infty} e^{-y}\,y^{k-2}\,\mathrm{d}y \\ &\overset{(1.155)}{=} \frac{\Gamma(k-1)}{(-ui-2\pi in)^{k-1}}, \end{aligned}$$

using the integral representation of the Γ-function in (1.155) and the substitution $-i(u+2\pi n)\tau = y$. Since k is even, we have $(-u-2\pi n)^{k-1} = -(u+2\pi n)^{k-1}$. Now taking $u\to 0$, we find

$$\left[\int_0^{i\infty} e^{ui\tau+2\pi in\tau}e^{uiz}\,\tau^{k-2}\,\mathrm{d}\tau\right]_{u=0} = -\frac{\Gamma(k-1)}{(2\pi in)^{k-1}}.$$

Hence the regularized integral $R.\int_z^{i\infty} f(w)\,(w-z)^{k-2}\,\mathrm{d}w$ is given by

$$\begin{aligned} R.\int_z^{i\infty} f(w)\,(w-z)^{k-2}\,\mathrm{d}w &= -\frac{\Gamma(k-1)}{(2\pi i)^{k-1}}\sum_{n\neq\in\mathbb{Z}_{\geq n_+}} a(n)\,n^{1-k}e^{2\pi inz} \\ &\overset{(2.16)}{=} -\frac{\Gamma(k-1)}{(2\pi i)^{k-1}}\,\mathcal{E}_f(z). \end{aligned}$$

□

The above Lemma allows us to write $r(f;\cdot)$ as a regularized integral.

Proposition 7.63 *For $k\in 2\mathbb{N}$, let $f\in S^!_{k,1}(\Gamma(1))$ then we have*

$$r(f;z) = R.\int_0^{i\infty} f(w)\,(w-z)^{k-2}\,\mathrm{d}w. \qquad (7.62)$$

Proof We have

$$\begin{aligned} r(f;z) &\overset{(7.59)}{=} -\frac{\Gamma(k-1)}{(2\pi i)^{k-1}}\big(\mathcal{E}_f - \mathcal{E}_f\big|_{2-k}T\big)(z) \\ &\overset{(7.61)}{=} R.\int_z^{i\infty} f(w)\,(w-z)^{k-2}\,\mathrm{d}w - \\ &\quad - z^{k-2}\,R.\int_{\frac{-1}{z}}^{i\infty} f(w)\left(w - \frac{-1}{z}\right)^{k-2}\mathrm{d}w. \end{aligned}$$

With the substitution $w \mapsto \frac{-1}{w}$ in the second regularized integral, we find

$$\begin{aligned} r(f;z) &= R.\int_z^{i\infty} f(w)\,(w-z)^{k-2}\,\mathrm{d}w \;-\; z^{k-2}\,R.\int_{\frac{-1}{z}}^{i\infty} f(w)\left(w-\frac{-1}{z}\right)^{k-2}\mathrm{d}w \\ &= R.\int_z^{i\infty} f(w)\,(w-z)^{k-2}\,\mathrm{d}w - \\ &\quad - z^{k-2}\,R.\int_z^{0} f\left(\frac{-1}{w}\right)\left(\frac{-1}{w}-\frac{-1}{z}\right)^{k-2}\frac{1}{w^2}\mathrm{d}w. \end{aligned}$$

Using the transformation property (1.28) of the weakly holomorphic form $f \in S^!_{k,1}\big(\Gamma(1)\big)$, we get

$$f\left(\frac{-1}{w}\right) = w^k\, f(w).$$

Hence, the integrand in the second regularized integral reads

$$f\left(\frac{-1}{w}\right)\left(\frac{-1}{w}-\frac{-1}{z}\right)^{k-2}\frac{1}{w^2} = f(w)\left(-1+\frac{w}{z}\right)^{k-2}.$$

This gives for the second regularized integral

$$\begin{aligned} & z^{k-2}\,R.\int_z^{0} f\left(\frac{-1}{w}\right)\left(\frac{-1}{w}-z\right)^{k-2}\frac{1}{w^2}\mathrm{d}w \\ &= z^{k-2}\,R.\int_z^{0} f(w)\left(-1+\frac{w}{z}\right)^{k-2}\mathrm{d}w \\ &= R.\int_z^{0} f(w)\,(w-z)^{k-2}\,\mathrm{d}w. \end{aligned}$$

Hence, we have

$$\begin{aligned} r(f;z) &= R.\int_z^{i\infty} f(w)\,(w-z)^{k-2}\,\mathrm{d}w - \\ &\quad - z^{k-2}\,R.\int_z^0 f\left(\frac{-1}{w}\right)\left(\frac{-1}{w}-\frac{-1}{z}\right)^{k-2}\frac{1}{w^2}\mathrm{d}w \\ &= R.\int_z^{i\infty} f(w)\,(w-z)^{k-2}\,\mathrm{d}w \;-\; R.\int_z^0 f(w)\,(w-z)^{k-2}\,\mathrm{d}w \\ &\overset{(7.47)}{=} R.\int_0^{i\infty} f(w)\,(w-z)^{k-2}\,\mathrm{d}w. \end{aligned}$$

This proves the proposition. □

We can also express $r(f;\cdot)$ in terms of the L-series associated with f in (7.49), similar to Proposition 2.9 for cusp forms.

Proposition 7.64 *For $k \in 2\mathbb{N}$, let $f \in S^!_{k,1}(\Gamma(1))$ we have*

$$\begin{aligned} r(f;z) &= \sum_{l=0}^{k-2}\binom{k-2}{l}(-z)^{k-l-2}\,i^{l+1}\,L^\star_f(l+1) \\ &= -\frac{(k-2)!}{(2\pi i)^{k-1}}\sum_{l=0}^{k-2}\frac{L_f(l+1)}{(k-l-2)!}(2\pi i z)^{k-l-2}. \end{aligned} \tag{7.63}$$

Proof We copy the proof of Proposition 2.9.

Using the binomial formula

$$(w-z)^{k-2} = \sum_{l=0}^{k-2}\binom{k-2}{l}w^l\,(-z)^{k-l-2},$$

we have

$$\begin{aligned} r(f;z) &\overset{(7.62)}{=} R.\int_0^{i\infty}(w-z)^{k-2}\,f(w)\,\mathrm{d}w \\ &= R.\int_0^{i\infty}\sum_{l=0}^{k-2}\binom{k-2}{l}w^l(-z)^{k-l-2}\,f(w)\,\mathrm{d}w \\ &= \sum_{l=0}^{k-2}\binom{k-2}{l}(-z)^{k-l-2}\,R.\int_0^{i\infty} w^l\,f(w)\,\mathrm{d}w \qquad . \end{aligned}$$

$$\overset{w\mapsto iy}{=} \sum_{l=0}^{k-2} \binom{k-2}{l} i^{l+1} (-z)^{k-l-2} \int_0^\infty w^l\, F(iy)\, dw$$

$$\overset{(7.50)}{=} \sum_{l=0}^{k-2} \binom{k-2}{l} i^{l+1} (-z)^{k-l-2}\, L_f^\star(l+1).$$

This shows the first relation.

Using $L_f^\star(s) = (2\pi)^{-s}\,\Gamma(s)\,L_f(s)$, see (7.49), we can perform the same calculation leading to the identity (7.57) in Exercise 7.57. This gives us

$$r(f;z) = \sum_{l=0}^{k-2} \binom{k-2}{l} i^{l+1} (-z)^{k-l-2}\, L_f^\star(l+1)$$

$$\overset{(7.57)}{\underset{(7.49)}{=}} \frac{(k-2)!}{(-2\pi i)^{k-1}} \sum_{l=0}^{k-2} \frac{L_f(l+1)}{(k-l-2)!} (2\pi i z)^{k-l-2}.$$

This shows the second identity using that k is even. □

Now, consider a weakly harmonic Maass wave form $f \in H_{2-k}^+\big(\Gamma(1)\big)$. Using the maps ξ_{2-k} and D^{k-1} on f, see (7.52), we end up with a cusp form $\xi_{2-k} f \in S_{k,1}\big(\Gamma(1)\big)$ and a weakly holomorphic form $D^{k-1} f \in S_{k,1}^!\big(\Gamma(1)\big)$. This shows that the associated period polynomial $P_{\xi_{2-k} f}(\cdot)$ and period function $r(D^{k-1} f;\cdot)$ are well defined. However, it immediately raises the following question: How are the associated period polynomial $P_{\xi_{2-k} f}(\cdot)$, see (2.1), and period function $r(D^{k-1} f;\cdot)$, see (7.59), related?

This question was answered by Kathrin Bringmann et al. in [22]. To state the result, we need the following notation.

Definition 7.65 We say two functions $f(z)$ and $g(z)$ are equivalent modulo $z^{k-2}-1$ and write

$$f(z) \equiv g(z) \pmod{z^{k-2}-1}$$

if

$$f(z) - g(z) = c\big(z^{k-2}-1\big)$$

for some $c \in \mathbb{C}$. In other words, the difference $f(z)$ and $g(z)$ is a constant multiple of $z^{k-2}-1$.

Proposition 7.66 ([22, Theorem 1.4]) *For $k \in 2\mathbb{N}$, consider a weakly harmonic Maass wave form $f \in H_{2-k}^+\big(\Gamma(1)\big)$, its associated period polynomial $P_{\xi_{2-k} f}(\cdot)$, and period function $r(D^{k-1} f;\cdot)$ defined in* (2.1) *and* (7.59). *We have*

$$P_{\xi_{2-k}f}(z) \equiv -\frac{(4\pi)^{k-1}}{\Gamma(k-1)}\overline{r(D^{k-1}f;\overline{z})} \pmod{z^{k-2}-1}. \tag{7.64}$$

Moreover, there exists a weakly harmonic Maass wave form $g \in H^+_{2-k}(\Gamma(1))$ *satisfying*

$$P_{\xi_{2-k}f}(z) = -\frac{(4\pi)^{k-1}}{\Gamma(k-1)}\overline{r(D^{k-1}g;\overline{z})}. \tag{7.65}$$

Remark 7.67 Already Fray mentions the relation between harmonic automorphic forms, automorphic integrals, and cocycles for the shadow in [77, p. 145].

Comparing the right-hand side of (7.64) in Proposition 7.66 and (7.58) in Corollary 7.58, we immediately derive the following.

Corollary 7.68 *For* $4 \le k \in 2\mathbb{N}$, *let* $f \in H^+_{2-k}(\Gamma(1))$ *be a weakly harmonic Maass wave form. We have*

$$\begin{aligned} P_{\xi_{2-k}f}(z) &= -\frac{(k-2)!}{(2\pi i)^{k-1}}\overline{\mathbb{P}(f_+,T;\overline{z})} \\ &\equiv -\frac{(4\pi)^{k-1}}{\Gamma(k-1)}\overline{r(D^{k-1}f;\overline{z})} \pmod{z^{k-2}-1}. \end{aligned} \tag{7.66}$$

Combining Corollary 7.68 with Proposition 7.56, we also derive a proof of Proposition 7.53 in §7.3.2.

Proof *(of Proposition 7.53.)* For $4 \le k \in 2\mathbb{N}$, consider $f = f_+ + f_- \in H^+_{2-k}(\Gamma(1))$ and denote by $L(s)$ the L-series associated with the cusp form $\xi_{2-k}f \in S_k(\Gamma(1))$ of weight k in (1.153).

Using Proposition 7.56, we have

$$\overline{\mathbb{P}(f_+,T;\overline{z})} \overset{(7.52)}{\underset{(7.56)}{=}} \sum_{l=0}^{k-2} \frac{L_{\xi_{2-k}f}(l+1)}{(k-2-l)!}(2\pi i z)^{k-2-l}.$$

Combining this with Corollary 7.68 gives

$$\begin{aligned} P_{\xi_{2-k}f}(z) &\overset{(7.66)}{=} -\frac{(k-2)!}{(2\pi i)^{k-1}}\overline{\mathbb{P}(f_+,T;\overline{z})} \\ &\overset{(7.52)}{\underset{(7.56)}{=}} -\frac{(k-2)!}{(2\pi i)^{k-1}}\sum_{l=0}^{k-2}\frac{L_{\xi_{2-k}f}(l+1)}{(k-2-l)!}(2\pi i z)^{k-2-l}. \end{aligned}$$

Using Corollary 7.68 again, we have

$$\overline{\mathbb{P}\big(f_+, T; \overline{z}\big)} \overset{(7.66)}{\equiv} -\frac{(4\pi)^{k-1}}{\Gamma(k-1)} \overline{r(D^{k-1}f; \overline{z})} \pmod{z^{k-2}-1}.$$

Applying Proposition 7.64, we find

$$\begin{aligned}
P_{\xi_{2-k}f}(z) &\overset{(7.66)}{\equiv} -\frac{(4\pi)^{k-1}}{\Gamma(k-1)} \overline{r(D^{k-1}f; \overline{z})} \qquad \pmod{z^{k-2}-1} \\
&\overset{(7.63)}{\equiv} \frac{(4\pi)^{k-1}}{\Gamma(k-1)} \overline{\frac{(k-2)!}{(2\pi i)^{k-1}} \sum_{l=0}^{k-2} \frac{L_{D^{k-1}f}(l+1)}{(k-l-2)!} (2\pi i \overline{z})^{k-l-2}} \\
&\qquad \pmod{z^{k-2}-1} \\
&\equiv \frac{(4\pi)^{k-1}}{\Gamma(k-1)} \frac{(k-2)!}{(-2\pi i)^{k-1}} \sum_{l=0}^{k-2} \frac{\overline{L_{D^{k-1}f}(l+1)}}{(k-l-2)!} (-2\pi i z)^{k-l-2} \\
&\qquad \pmod{z^{k-2}-1}.
\end{aligned}$$

Combining the above two identities, we get

$$\begin{aligned}
&\frac{(4\pi)^{k-1}}{\Gamma(k-1)} \frac{(k-2)!}{(-2\pi i)^{k-1}} \sum_{l=0}^{k-2} \frac{\overline{L_{D^{k-1}f}(l+1)}}{(k-l-2)!} (-2\pi i z)^{k-l-2} \\
&\quad \equiv P_{\xi_{2-k}f}(z) \qquad \pmod{z^{k-2}-1} \qquad (7.67) \\
&\quad \equiv -\frac{(k-2)!}{(2\pi i)^{k-1}} \sum_{l=0}^{k-2} \frac{L_{\xi_{2-k}f}(l+1)}{(k-2-l)!} (2\pi i z)^{k-2-l}
\end{aligned}$$

for all z. Hence we can compare coefficients and get

$$\begin{aligned}
&\frac{(4\pi)^{k-1}}{(-2\pi i)^{k-1}} \overline{L_{D^{k-1}f}(l+1)} (-1)^{k-l-2} = -\frac{(k-2)!}{(2\pi i)^{k-1}} L_{\xi_{2-k}f}(l+1) \\
\iff\ & L_{\xi_{2-k}f}(l+1) = -\frac{(2\pi i)^{k-1}}{(k-2)!} \frac{(4\pi)^{k-1}}{(-2\pi i)^{k-1}} \overline{L_{D^{k-1}f}(l+1)} (-1)^{k-l-2} \\
\iff\ & L_{\xi_{2-k}f}(l+1) = (-1)^l \frac{(4\pi)^{k-1}}{(k-2)!} \overline{L_{D^{k-1}f}(l+1)}
\end{aligned}$$

for all $0 < n < k-2$, since we only have the identity (7.67) modulo $z^{k-2}-1$. □

7.5 Concluding Remarks

Weak harmonic Maass forms and mock modular forms are active topics of research. In this chapter, we only presented a brief introduction to the topic.

We mention few interesting connections to illustrate the research done on mock modular forms. We have for example the following:

1. Lawrence and Zagier showed a connection between mock theta functions and quantum invariants of 3-manifolds in [119].
2. A relation between mock theta functions, infinite-dimensional Lie superalgebras, and two-dimensional conformal field theory was shown in [180].
3. Troost [198] showed that the modular completions of mock modular forms arise as elliptic genera of conformal field theories with continuous spectrum.
4. Mock theta functions were used in the proof of the umbral moonshine conjecture [72].
5. The authors of [63] presented a connection between mock modular forms and some degeneracies of quantum black holes in certain string theories.
6. Duke, Imamoḡlu, and Tóth discuss the connection between weakly holomorphic modular forms and rational period functions in [70].

A good starting point for a more detailed look at some essential features of the theory of harmonic Maass forms and mock modular forms, together with a wide variety of applications to algebraic number theory, combinatorics, elliptic curves, mathematical physics, quantum modular forms, and representation theory is the recent textbook [26]. The book also includes an introduction of the results obtained by Zwegers in his PhD Thesis [211].

Another informal introduction to the world of weak harmonic Maass forms and mock modular forms are the slides of Amanda Folsom's plenary lecture "Mock and Quantum Modular Forms" [78] at the Connecticut Summer School in Number Theory 2016.

An interesting (historic) overview on harmonic Maass forms is also Ono's article [155].

Appendix A
Background Material

Nobody knows everything

A.1 Chinese Remainder Theorem – Special Case

[76, Chinese remainder theorem] Suppose $n_1, n_2 \in \mathbb{N}$ are coprime. Then, for every given pair of integers a_1, a_2, there exists $x \in \mathbb{Z}$ solving the system of simultaneous congruences

$$x \equiv a_1 \pmod{n_1} \quad \text{and} \quad x \equiv a_2 \pmod{n_2}.$$

Furthermore, all solutions x of this system are congruent modulo of the product $N = n_1 n_2$. The solution is given by

$$\begin{aligned} x &= a_1 b_1 \frac{N}{n_1} + a_2 b_2 \frac{N}{n_2} \mod N \\ &= a_1 b_1 n_2 + a_2 b_2 n_1 \mod n_1 n_2 \end{aligned}$$

for suitable $b_1, b_2 \in \mathbb{Z}$.

A.2 Phragmén-Lindelöf principle

[114, Section XIII.5] Let $\phi(s)$ be a holomorphic function on the upper part of a vertical strip

$$\left\{ s \in \mathbb{C} : \ \sigma_1 \leq \Re(s) \leq \sigma_2, \ \Im(s) > c \right\}$$

for some $\sigma_1 < \sigma_2$ and $c \in \mathbb{R}$ such that

T. Mühlenbruch, W. Raji, *On the Theory of Maass Wave Forms*, Universitext,
https://doi.org/10.1007/978-3-030-40475-8

$$\phi(s) = O\left(e^{\Im(s)^\alpha}\right)$$

for some $\alpha > 0$ on the strip. If the function ϕ satisfies

$$\phi(s) = O\left(\Im(s)^M\right) \quad \text{for all } s \text{ in the strip with} \Re(s) \in \{\sigma_1, \sigma_2\}$$

on the vertical boundary of the strip for some $M \in \mathbb{R}$, then

$$\phi(s) = O\left(\Im(s)^M\right) \quad \text{for all } s \text{ in the strip.}$$

A.3 Stirling's Estimate of the Γ-Function

Let $\Gamma(s)$ denote the Γ-function defined in (1.155). Uniformly for $|\sigma| \leq 2$ and $|t| \geq 2$, we have

$$\log \Gamma(\sigma + it) = \frac{-\pi|t|}{2} + \left(\sigma - \frac{1}{2}\right) \log|t| + it \log|t| - it + O(1),$$

see [153, (5.11.1)]. In particular, in this range there exist positive constants $0 < C_1 < C_2$ such that

$$C_1 e^{-\frac{\pi|t|}{2}} |t|^{\sigma - \frac{1}{2}} \leq |\Gamma(s)| \leq C_2 e^{-\frac{\pi|t|}{2}} |t|^{\sigma - \frac{1}{2}}.$$

Moreover, we have the asymptotic estimate

$$\Gamma(s) \sim \sqrt{2\pi}\, e^{(s - \frac{1}{2}) \log s - s}$$

uniformly for $s \to \infty$ and $|\arg(s)| < \pi - \delta$ for some $\delta > 0$, see [153, (5.11.3)]. This is also known as *Stirling's formula.*

A.4 Cauchy-Riemann Equations

[76, Cauchy-Riemann equations] Let $f : U \to \mathbb{C}$ with $U \subset \mathbb{C}$ open be a holomorphic function and write the real and imaginary parts of f as functions of x and y:

$$z = x + iy \quad \text{and} \quad f(x + iy) =: u(x, y) + iv(x, y)$$

with real valued functions $u(x, y)$ and $v(x, y)$ of two real variables x and y. Then u and v satisfy

$$\frac{\partial u}{\partial x} = \frac{\partial v}{\partial y} \quad \text{and} \quad \frac{\partial u}{\partial y} = -\frac{\partial v}{\partial x}.$$

Using the complex coordinates $z = x + iy$ and $\overline{z} = x - iy$ the above Cauchy-Riemann equation can be written as

$$\frac{\partial}{\partial \overline{z}} f(z) = 0.$$

A.5 Cauchy's Integral Theorem

[76, Cauchy integral theorem] If $D \subset \mathbb{C}$ is a simply connected open set and $f : D \to \mathbb{C}$ is a holomorphic function, then the integral of $f(z)$ along every closed rectifiable curve $\gamma \subset D$ vanishes, i.e.,

$$\int_{\gamma} f(z)\,\mathrm{d}z = 0.$$

An equivalent version states that (under the same assumptions as above), given a (rectifiable) path $\eta : [0, 1] \to D$, the integral

$$\int_{\eta} f(z)\,\mathrm{d}z$$

depends only upon the two endpoints $\eta(0)$ and $\eta(1)$, and hence it is independent of the choice of the path of integration η.

A.6 Rouché's Theorem

[76, Rouché theorem] Let $f(z)$ and $g(z)$ be holomorphic functions of a complex variable z in a domain D. Let Γ be a simple closed piecewise-smooth curve that is a boundary of a domain G in D. Assume that on Γ the inequality $|f(z)| > |g(z)|$ is valid. Then in the domain G, $f(z) + g(z)$ has the same number of zeros as $f(z)$.

A.7 The Beta Function and Γ-Function

[153, §5.2–§5.5] The Γ-function $\Gamma(s)$ is defined for $\Re(s) > 0$ by

$$\Gamma(s) = \int_0^\infty e^{-t} t^{s-1}\,\mathrm{d}t$$

and for $\Re(s) \leq 0$ by analytic continuation. It is a function with no zeros and with simple poles at $s \in \mathbb{Z}_{\leq 0}$. The Γ-function generalized the factorial symbol in the following sense:

$$\Gamma(n) = (n-1)! \qquad \text{for all} n \in \mathbb{N}.$$

This function satisfies the functional equations

$$\Gamma(s+1) = s\,\Gamma(s) \quad \text{and} \quad \Gamma(s)\Gamma(1-s) = \frac{\pi}{\sin(\pi s)}.$$

A variant of the Γ-function is the beta function defined by

$$B(x,y) = \int_0^1 t^{x-1}(1-t)^{y-1},$$

see [89, 90, (8.380.1)]. It is related to the Γ-function by

$$B(x,y) = \frac{\Gamma(x)\,\Gamma(y)}{\Gamma(x+y)}.$$

Moreover, $B(x,y)$ admits the integral representation

$$B(x,y) = 2\int_0^\infty \frac{t^{2x-1}}{(1+t^2)^{x+y}}\,\mathrm{d}t,$$

see [89, 90, (8.380.3)].

A.8 Unique Continuation Principle For Holomorphic Functions

[76, Analytic function] Let $f, g : D \to \mathbb{C}$ be two analytic functions in a domain $D \subset \mathbb{C}$. If the two functions agree on some set with an accumulation point in D, then they are identical everywhere.

An important consequence is the following corollary: if D is a connected open set and $f : D \to \mathbb{C}$ a holomorphic function which is not identically zero, then each zero z_0 of f is isolated. In addition, for some neighborhood U of z_0, there is a holomorphic function $g : U \to \mathbb{C}$ that never vanishes and a natural number n (called the multiplicity of the zero z_0, or order of vanishing of f at z_0) such that $f(z) = (z - z_0)^n g(z)$ for every $z \in U$.

A.9 Whittaker Functions

[153, Chapter XIII Confluent Hypergeometric Functions] We collect a few facts from the Digital Library of Mathematical Functions [153] and from some other references.

Whittaker's normalized differential equation is the second-order ordinary differential equation

$$\frac{\partial^2}{\partial y^2} G(y) + \left(-\frac{1}{4} + \frac{k}{y} + \frac{\frac{1}{4} - \nu^2}{y^2} \right) G(y) = 0$$

for smooth functions $G : (0, \infty) \to \mathbb{C}$ and $\nu \notin -\frac{1}{2}\mathbb{N}$. According to [153, (13.14.2), (13.14.3)], see also [130, Chapter VII], we have two solutions $M_{k,\nu}(y)$ and $W_{k,\nu}(y)$ with different behavior as $y \to \infty$:

$$M_{k,\nu}(y) \sim \frac{\Gamma(1 + 2\nu)}{\Gamma\left(\frac{1}{2} + \nu + k\right)} e^{\frac{1}{2}y} y^{-k} \qquad \text{and}$$

$$W_{k,\nu}(y) \sim e^{-\frac{1}{2}y} y^k.$$

The asymptotic behavior is valid for $k - \nu \notin \left\{\frac{1}{2}, \frac{3}{2}, \frac{5}{2}, \dots\right\}$, see [153, (13.14.20), (13.14.21)]. These functions also satisfy some recurrence relations [153, (13.15.1), (13.15.11)] and differentiation relations [153, (13.15.17), (13.15.20) and (13.15.23), (13.15.26)] and [130, p. 302, §7.2.1].

We have the following identities and relations with other functions:

1. [153, (13.18.2)]

$$W_{k,k-\frac{1}{2}}(y) = W_{k,\frac{1}{2}-k}(y) = e^{-\frac{y}{2}} y^k \qquad (k \in \mathbb{R}, y > 0),$$

2. [153, (13.18.9)]

$$W_{0,\nu}(2y) = \sqrt{\frac{2y}{\pi}} K_\nu(y) \qquad (\nu \in \mathbb{C} \setminus -\frac{1}{2}\mathbb{N}, y > 0), \text{ and}$$

3. [186, (3.5.12)] if $\Re\left(\frac{1}{2} - \nu - \frac{k}{2}\right) > 0$, we have

$$\Gamma\left(\frac{1}{2}-\nu-\frac{k}{2}\right) W_{\frac{k}{2},-\nu}(y) = y^{\frac{1}{2}-\nu} 2^{2\nu} \int_1^\infty e^{-\frac{1}{2}yt}(t-1)^{-\frac{1}{2}-\nu-\frac{k}{2}}(t+1)^{-\frac{1}{2}-\nu+\frac{k}{2}}\, dt$$

for $y > 0$.

(Slater states the formula only under the condition $\Re\left(\frac{1}{2}-\nu\pm k\right) > 0$, but it is obvious that the integral converges also under the slightly relaxed condition given here.)

A.10 Generalized Hypergeometric Functions

[153, Chapter XVI], [186, §1.1.1] We collect a few facts from the Digital Library of Mathematical Functions [153] and from some other references about the generalized hypergeometric function ${}_2F_1$.

For $p, q \in \mathbb{Z}_{\geq 0}$, the *generalized hypergeometric function* ${}_pF_q$ is defined by the power series expansion

$${}_pF_q\left[\begin{matrix} a_1, \ldots, a_p \\ b_1, \ldots, b_q \end{matrix}\middle| z\right] = \sum_{n=0}^{\infty} \frac{(a_1)_n \cdots (a_p)_n}{(b_1)_n \cdots (b_q)_n} \frac{z^n}{n!},$$

where the *Pochammer symbol* $(a)_n$ is defined by

$$(a)_0 := 1, \quad (a)_1 := a, \quad \text{and} \quad (a)_n := a(a+1)(a+2)\cdots(a+n-1)$$

for $a \in \mathbb{C}$ and $n \in \mathbb{Z}_{\geq 0}$.

Several families of special functions can be derived by generalized hypergeometric functions if we choose the parameters accordingly. For example, we have the following connection to the Whittaker $W_{k,\nu}$-function:

$$M_{k,\nu}(y) = {}_1F_1\left[\begin{matrix} \frac{1}{2}-k+\nu \\ 1+2\nu \end{matrix}\middle| y\right] e^{-\frac{y}{2}}\, y^{\frac{1}{2}+\nu}$$

and

$$W_{k,\nu}(y) = \Gamma(-2\nu)\frac{{}_1F_1\left[\begin{matrix} \frac{1}{2}-k+\nu \\ 1+2\nu \end{matrix}\middle| y\right]}{\Gamma(\frac{1}{2}-k-\nu)} e^{-\frac{y}{2}}\, y^{\frac{1}{2}+\nu} + \\ + \Gamma(2\nu)\frac{{}_1F_1\left[\begin{matrix} \frac{1}{2}-k-\nu \\ 1-2\nu \end{matrix}\middle| y\right]}{\Gamma(\frac{1}{2}-k+\nu)} e^{-\frac{y}{2}}\, y^{\frac{1}{2}-\nu}$$

for $k \in \mathbb{R}$, $\nu \in \mathbb{C}$ and $y > 0$, see [186, (1.9.7), (1.9.10)].

A.11 Poisson Summation Formula

[190, Chapter IV, §1 and Theorem 2.4] For each $a > 0$ denote by $\mathfrak{F}_a$, the class of all functions f that satisfy the conditions

1. the function f is holomorphic in the horizontal strip

$$S_a = \{z \in \mathbb{C}\colon\ |\Im(z)| < a\},\ \text{and}$$

2. there exists a constant $A > 0$ such that

$$|f(x+iy)| \le \frac{A}{1+x^2} \qquad \text{for all } x+iy \in S_a.$$

We denote by $\mathfrak{F}$ the class of all functions that belong to $\mathfrak{F}_a$ for some a.

For $f \in \mathfrak{F}$ the *Poisson summation formula* is the relation

$$\sum_{n\in\mathbb{Z}} f(n) = \sum_{n\in\mathbb{Z}} \widehat{f}(n),$$

where $\widehat{f}$ denotes the *Fourier transformation*

$$\widehat{f}(k) = \int_{-\infty}^{\infty} f(x)\, e^{-2\pi ikx}\, \mathrm{d}x$$

of f.

A.12 Liouville's Theorem

[76, Liouville theorems] Liouville's theorem states that every bounded entire function must be constant.

In other words, let $f : \mathbb{C} \to \mathbb{C}$ be a holomorphic function and $M > 0$ a positive constant such that

$$|f(z)| \le M$$

for all $z \in \mathbb{C}$. Then f is constant.

A.13 Residue Theorem

[76, Residue of an analytic function] The theory of residues is based on the *Cauchy integral formula*.

A simple form of the Cauchy integral formula is the following: Suppose $U \subset \mathbb{C}$ is open and simply connected, $f : U \to \mathbb{C}$ is holomorphic, and γ is a positively oriented simple closed curve completely contained in U with winding number 1 around a. Then

$$f(a) = \frac{1}{2\pi i} \oint_\gamma \frac{f(z)}{z-a}\, \mathrm{d}z,$$

where the contour integral is taken counter-clockwise.

The *residue theorem* states: suppose U is a simply connected open subset of the complex plane, and $a_1, \ldots, a_n$ are finitely many points in U and let f be a meromorphic function on U with poles at $a_1, \ldots, a_n$. If γ is a closed rectifiable curve in U which does not meet any of the a_k, we have

$$\oint_\gamma f(z)\, \mathrm{d}z = 2\pi i \sum_{k=1}^{n} \mathrm{I}(\gamma, a_k)\, \mathrm{res}_{z=a_k}(f),$$

where $I(\gamma, a_k)$ is the winding number of γ around a_k.

Its slightly simplified form reads as follows: if γ is a positively oriented simple closed curve and $I(\gamma, a_k) = \begin{cases} 1 & \text{if } a_k \text{ is in the interior of } \gamma, \\ 0 & \text{elsewhere,} \end{cases}$ then we have

$$\oint_\gamma f(z)\, \mathrm{d}z = 2\pi i \sum \mathrm{res}_{z=a_k}(f),$$

where the sum runs over those k for which a_k is inside γ.

A.14 Geometric Series

[76, Geometric progression], [11] The *geometric series* is given by

$$\sum_{n=0}^{\infty} q^n.$$

This series converges absolutely for all $|q| < 1$ and has the value

$$\sum_{n=0}^{\infty} q^n = \frac{1}{1-q}.$$

Its variants

$$\sum_{n=0}^{\infty} n^k q^n$$

converge also absolutely for $|q| < 1$. After interchanging differentiation and summation, we can calculate the series and get

$$\begin{aligned}\sum_{n=0}^{\infty} n\, q^n &= q \frac{\mathrm{d}}{\mathrm{d}q} \sum_{n=0}^{\infty} q^n \\ &= q \frac{\mathrm{d}}{\mathrm{d}q} \frac{1}{1-q} \\ &= \frac{q}{(1-q)^2}\end{aligned}$$

and

$$\begin{aligned}\sum_{n=0}^{\infty} n^2 q^n &= q \frac{\mathrm{d}}{\mathrm{d}q} q \frac{\mathrm{d}}{\mathrm{d}q} \sum_{n=0}^{\infty} q^n \\ &= q \frac{\mathrm{d}}{\mathrm{d}q} q \frac{\mathrm{d}}{\mathrm{d}q} \frac{1}{1-q} \\ &= q \frac{\mathrm{d}}{\mathrm{d}q} \frac{q}{(1-q)^2} \\ &= \frac{q(1+q)}{(1-q)^3}.\end{aligned}$$

Similar arguments show the growth estimate

$$\sum_{n=0}^{\infty} n^k q^n = O\left((1-q)^{-C}\right) \qquad \text{as } q \to 1$$

for some $C > 0$.

A.15 Continued Fractions

[76, Continued fraction] A *continued fraction* is a finite or infinite expression of the form

$$a_0 + \cfrac{b_1}{a_1 + \cfrac{b_2}{a_2 + \cdots + \cfrac{b_n}{a_n + \cdots}}}$$

where $(a_n)_n$ and $(b_n)_n$ are finite or infinite sequences of complex numbers.

The *regular Gauss continued fraction* is defined as the expression

$$a_0 + \cfrac{1}{a_1 + \cfrac{1}{a_2 + \cdots + \cfrac{1}{a_n + \cdots}}}$$

with $a_0 \in \mathbb{Z}$ and all following $a_n \in \mathbb{N}$ (if they exist). The above continued fraction is called *finite* if the expansion terminates after finitely many fractions (i.e., $a_n = 0$ for all $n \geq n_0$ for some n_0). Otherwise, the continued fraction is called *infinite*.

We denote a *periodic continued fraction* by over-lining the periodic part. For example, we write

$$[a_0; a_1, \ldots, a_n, \underbrace{a_{n+1}, \ldots, a_{n+l}}, \underbrace{a_{n+1}, \ldots, a_{n+l}}_{\text{periodic part}}, \ldots]$$

$$= [a_0; a_1, \ldots, a_n, \overline{a_{n+1}, \ldots, a_{n+l}}]$$

for some $n \in \mathbb{N}$ and $l \in \mathbb{Z}_{\geq 0}$. The integer l is called the *period length* or *length of the periodic part*. We call l the *prime period length* and the associated periodic part of the continued fraction *prime periodic* if $l \in \mathbb{N}$ is the smallest period length of the expansion. We call a continued fraction expansion *purely periodic* if the continued fraction expansion contains only a periodic part, i.e., it is of the form

$$[0; \underbrace{a_1, \ldots, a_l}, \underbrace{a_{l+1}, \ldots, a_{2l}}_{\text{periodic part}}, \ldots] = [0; \overline{a_1, \ldots, a_l}].$$

As an example, consider the expansions of π and e. The first few digits of the continued fraction expansion of π and the Euler constant e are respectively given by

$$\pi = 3 + \cfrac{1}{7 + \cfrac{1}{15 + \cfrac{1}{1 + \cfrac{1}{292 + \cfrac{1}{1 + \cfrac{1}{1 + \cfrac{1}{1 + \cfrac{1}{2 + \cfrac{1}{1 + \cfrac{1}{3 + \cfrac{1}{1 + \cfrac{1}{14 + \cfrac{1}{2 + \dots}}}}}}}}}}}}}$$

$$= [3; 7, 15, 1, 292, 1, 1, 1, 2, 1, 3, 1, 14, 2, \dots],$$

see [196, A001203], and

$$e = 2 + \cfrac{1}{1 + \cfrac{1}{2 + \cfrac{1}{1 + \cfrac{1}{1 + \cfrac{1}{4 + \cfrac{1}{1 + \cfrac{1}{1 + \cfrac{1}{6 + \cfrac{1}{1 + \cfrac{1}{1 + \cfrac{1}{8 + \cfrac{1}{1 + \cfrac{1}{1 + \dots}}}}}}}}}}}}}$$

$$= [2; 1, 2, 1, 1, 4, 1, 1, 6, 1, 1, 8, 1, 1, \dots],$$

see [196, A003417].

Consider either a finite or infinite Gauss continued fraction $a = [a_0; a_1, a_2, \dots, a_{n_0}]$ or $a = [a_0; a_1, a_2, \dots]$ with positive a_n for all $n \in \{1, \dots, n_0\}$ or $n \in \mathbb{N}$ respectively. We denote the index set of the digits a_n appearing in the Gauss continued fraction by M. (Either $M = \{0, 1, \dots n_0\}$ or $M = \mathbb{Z}_{\leq 0}$.) The recurrence equations

$$P_n = a_n P_{n-1} + P_{n-2} \quad \text{and}$$

$$Q_n = a_n Q_{n-1} + Q_{n-2},$$

for all $n \in M$ with the initial conditions

$$P_{-2} = 0, \quad P_{-1} = 1, \quad Q_{-2} = 1, \quad \text{and} \quad Q_{-1} = 0$$

determine two sequences $(P_n)_n$ and $(Q_n)_n$ of complex numbers for all n. As a rule, it is assumed that these sequences are such that $Q_n \neq 0$ for all $n \in \mathbb{Z}_{\geq 0}$ in the sequence. The above recurrence relation terminates if $Q_{n+1} = 0$.

The fraction $\frac{P_n}{Q_n}$ is called the nth *convergent* of the continued fraction. We have

$$\frac{P_0}{Q_0} = a_0, \quad \frac{P_1}{Q_1} = a_0 + \frac{b_1}{a_1}, \quad \frac{P_2}{Q_2} = a_0 + \frac{b_1}{a_1 + \frac{b_2}{a_2}}, \dots.$$

Moreover,

$$\frac{P_n}{Q_n} - \frac{P_{n-1}}{Q_{n-1}} = \frac{(-1)^{n-1}\, b_1 \cdots b_n}{Q_n\, Q_{n-1}}.$$

It is convenient to denote the nth convergent of the regular Gauss continued fraction by

$$[a_0; a_1, \dots, a_n] := a_0 + \cfrac{1}{a_1 + \cfrac{1}{a_2 + \dots + \cfrac{1}{a_n}}}.$$

Its convergents satisfy the equalities

$$[a_n; \dots, a_1] = \frac{Q_n}{Q_{n-1}} \qquad \text{for } n \geq 1$$

and

$$[a_n; \dots, a_0] = \frac{P_n}{P_{n-1}} \qquad \text{for } a_0 \neq 0 \text{ and } n \geq 0.$$

If the continued fraction is an infinite expression and the sequence of convergents $\frac{P_n}{Q_n}$ converges to some limit l, then the continued fraction is called *convergent* and the number l is its value. If the continued fraction is a finite expression, then its value is defined as the last of its convergents.

The convergents of a regular Gauss continued fraction satisfy

$$a_0 = \frac{P_0}{Q_0} < \frac{P_2}{Q_2} < \frac{P_4}{Q_4} < \dots < l < \dots < \frac{P_5}{Q_5} < \frac{P_3}{Q_3} < \frac{P_1}{Q_1} = a_0 + \frac{1}{a_1}.$$

Regular continued fractions are very convenient tools for the approximating real numbers by rational numbers. The following propositions hold.

1. If $\frac{P_n}{Q_n}$ and $\frac{P_{n+1}}{Q_{n+1}}$ are neighboring convergents of the expansion of a number r in a regular Gauss continued fraction, then

$$\left|r - \frac{P_n}{Q_n}\right| \geq \left|r - \frac{P_{n+1}}{Q_{n+1}}\right| \quad \text{and} \quad \left|r - \frac{P_n}{Q_n}\right| \leq \frac{1}{Q_n\, Q_{n+1}}.$$

   Equality holds in the estimate on the right side only if $r = \frac{P_{n+1}}{Q_{n+1}}$.
2. For two neighboring convergents of the expansion of a number r in a regular Gauss continued fraction, at least one satisfies the inequality

$$\left|r - \frac{P_n}{Q_n}\right| = \frac{1}{2Q_n^2}.$$

3. Let a and b be integers, $b \geq 1$, and let r is a real number. If

$$\left|r - \frac{a}{b}\right| \leq \frac{1}{2b^2},$$

   then $\frac{a}{b}$ is a convergent of the expansion of r in a regular continued fraction.
4. Let $\frac{P_n}{Q_n}$ be a convergent of the expansion of a number r into a regular Gauss continued fraction. If the integers a and b with $b > 0$, $\frac{P_n}{Q_n} \neq \frac{a}{b}$ satisfy

$$\left|r - \frac{a}{b}\right| \leq \left|r - \frac{P_n}{Q - n}\right|,$$

   then we have $b \geq Q_n$.

Example One classical application of continued fractions outside of the scope of this book is the construction of a fundamental solution to Pell's equation.

Let $n \in \mathbb{N}$ be a non-square integer (i.e., n is not the square of another integer). We call the Diophantine equation

$$x^2 - n\, y^2 = 1 \tag{A.1}$$

Pell's equation. The solution set $\{(x, y) \in \mathbb{R}^2 : \text{ x and y satisfy (A.1)}\}$ forms a hyperbola as illustrated in Figure A.1. We are interested in integer solutions x, y of (A.1).

Joseph Louis Lagrange (1736–1813) proved that, as long as n is not a perfect square, Pell's equation (A.1) has infinitely many distinct integer solutions. These solutions may be used to accurately approximate the square root $\sqrt{n}$ by rational numbers of the form $\frac{x}{y}$. In particular, one of the first methods to calculate the

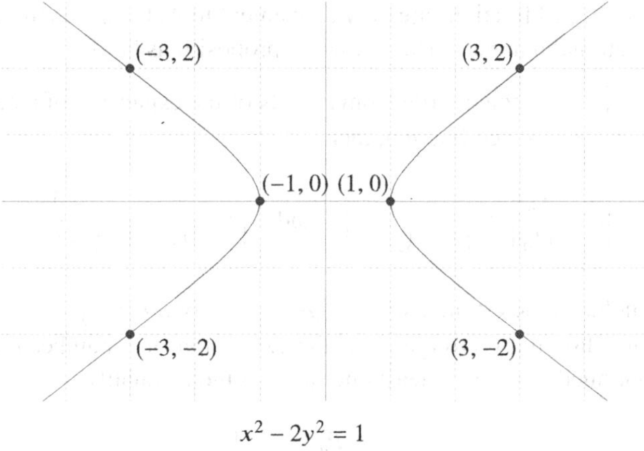

Fig. A.1 Pell's equation for $n = 2$ and some of its integer solutions.

fundamental solution of (A.1), that is, the smallest strictly positive solution, uses continued fractions.

We refer to the review article [120] for a more in-depth review of Pell's equation.

A.16 Hurwitz Zeta Function

[76, Hurwitz zeta function], [130] The Hurwitz zeta function $\zeta(s, q)$ is defined for $\Re(q) > 0$ and $\Re(s) > 1$ as

$$\zeta(s, q) = \sum_{n=0}^{\infty} (n + q)^{-s}.$$

The series converges absolutely and defines, for $\Re(s) > 1$, an analytic function. The function possesses an analytic continuation to the whole s-plane except for a simple pole of residue 1 at $s = 1$.

$\zeta(s, q)$ admits the integral representation

$$\zeta(s, q) = \frac{1}{\Gamma(s)} \int_0^{\infty} \frac{t^{s-1} e^{-qt}}{1 - e^{-t}} \, dt$$

for $\Re(s) > 1$ and $\Re(q) > 0$.

We have

$$\lim_{s \to 1} \left[\zeta(s, q) - \frac{1}{s - 1} \right] = -\frac{\Gamma'(q)}{\Gamma(q)},$$

where $\Gamma(q)$ denotes the Γ-function, see Appendix A.7, and $\Gamma'(q)$ its derivative.

A.17 Incomplete Γ-Function

[153, §8] The incomplete Γ-function $\Gamma(s, x)$ is defined for $\Re(s) > 0$ and $x \geq 0$ by

$$\Gamma(s, x) = \int_x^{\infty} e^{-t} t^{s-1}\, \mathrm{d}t$$

and for $\Re(s) \leq 0$ by analytic continuation.

If $x \neq 0$, then the function $s \mapsto \Gamma(s, x)$ is analytic, see [153, §8.2(ii)].

It satisfies the growth estimate

$$\Gamma(s, x) = O\left(x^{s-1}\, e^{-x}\right)$$

as $x \to \infty$, see [153, (8.11.2)].

The incomplete Γ-function satisfies

$$\frac{\mathrm{d}}{\mathrm{d}x}\left(e^x\, \Gamma(s, x)\right) = -(1-s)\, e^x\, \Gamma(s-1, x)$$

and

$$\Gamma(s+1, x) = a\, \Gamma(s, x) + x^s\, e^{-x}$$

see [153, (8.8.2) and (8.8.19)].

A.18 Properties of Linear Fractional Transformations

The linear fractional transformations given by $V = \begin{pmatrix} a & b \\ c & d \end{pmatrix} \in \mathrm{SL}_2(\mathbb{R})$ can be separated, according to their action on the upper half-plane $\mathbb{H}$, into three distinct classes with different properties.

Definition Let $V \in \mathrm{SL}_2(\mathbb{R})$ be such that it acts nontrivially on $\mathbb{H}$, i.e., $V \notin \{\mathbf{1}, -\mathbf{1}\}$. We say:

1. V is *elliptic* if $|\mathrm{trace}\,(V)| < 2$.
2. V is *hyperbolic* if $|\mathrm{trace}\,(V)| > 2$.
3. V is *parabolic* if $|\mathrm{trace}\,(V)| = 2$.

Definition Let $V \in \mathrm{GL}_2(\mathbb{R})$ be a matrix. A point $z \in \mathbb{S} = \mathbb{C} \cup \{\infty\}$ is called a *fixed point* of V, if $V z = z$.

For nontrivial elements of $\mathrm{SL}_2(\mathbb{R})$, the notion of a fixed point is related to the above classification.

Proposition *Let* $V \in \mathrm{SL}_2(\mathbb{R})$, $V \notin \{\mathbf{1}, -\mathbf{1}\}$ *be a matrix. We have the following equivalences:*

1. *V is elliptic* $\Longleftrightarrow$ *V has two distinct fixed points in* $\mathbb{H} \cup \mathbb{H}^-$ *that are conjugates of each other.*
2. *V is hyperbolic* $\Longleftrightarrow$ *V has two distinct fixed points in* $\mathbb{R} \cup \{\infty\}$.
3. *V is parabolic* $\Longleftrightarrow$ *V has one fixed point in* $\mathbb{R} \cup \{\infty\}$.

Different types of matrices in $\mathrm{SL}_2(\mathbb{R})$ have names that we introduce below.

Definition For obvious reasons, we call a matrix of the form $\begin{pmatrix} \pm 1 & b \\ 0 & \pm 1 \end{pmatrix}$ with $b \neq 0$ a *translation.*

Proposition

1. *Let* $V \in \mathrm{SL}_2(\mathbb{R})$, $V \neq \pm\mathbf{1}$. *A parabolic transformation V fixes* ∞ *if and only if it is a translation.*
2. *A transformation* $V \in \mathrm{GL}_2(\mathbb{C})$ *that fixes both* 0 *and* ∞ *acts as*

$$V\, z = \kappa\, z$$

   for some $\kappa \neq 0$.
3. *Let* $V \in \mathrm{SL}_2(\mathbb{R})$. *If there are two distinct points* $\alpha, \beta \in \mathbb{S}$ *satisfying*

$$V\alpha = \beta \quad and \quad V\beta = \alpha,$$

   then V is elliptic.

It is known that a necessary and sufficient condition for a set of real fractional transformations to commute is that they fix the same set of points.

Proposition *Suppose* V_1 *and* V_2 *are nontrivial real linear fractional transformations (i.e.,* $V_1, V_2 \in \mathrm{SL}_2(\mathbb{R})$ *and* $V_1, V_2 \notin \{\mathbf{1}, -\mathbf{1}\}$*), then* $V_1 V_2 = V_2 V_1$ *if and only if* V_1 *and* V_2 *have the same set of fixed points.*

We can denote the commuting property in the proposition above also as a vanishing commutator relation:

$$V_1 V_2 = V_2 V_1 \Longleftrightarrow [V_1, V_2] = 0.$$

As usual, the *commutator brackets* are defined as

$$[V_1, V_2] := V_1 V_2 - V_2 V_1 .$$

A.19 Nuclear Operator

[76, Nuclear operator] We recall briefly a characterization of compact operators on Hilbert spaces. Let H be a Hilbert space with inner product $\langle \cdot, \cdot \rangle$. An operator $L : H \to H$ is *compact* if L can be written in the form

$$L(h) = \sum_{i=1}^{\infty} \lambda_i \langle f_i, h \rangle g_i,$$

where $\{f_1, f_2, \dots\}$ and $\{g_1, g_2, \dots\}$ are orthonormal sets and λ_i are real numbers satisfying $\lim_{i \to \infty} \lambda_i = 0$. The above series representation converges in the inner product induced norm.

We call the operator L *nuclear* if the values λ_i additionally satisfy

$$\sum_{i=1}^{\infty} |\lambda_i| < \infty.$$

We call the operator L *nuclear of order* 0 if the values λ_i additionally satisfy

$$\sum_{i=1}^{\infty} |\lambda_i|^p < \infty$$

for every $p > 0$.

A nuclear operator on a Hilbert space has the important property that its trace is well defined, finite, and independent of the choice of basis. Such operators are also often called "trace class" operators.

The above definition of "trace class" operators was extended to Banach spaces by Alexander Grothendieck [91] in 1955.

Let A be a Banach space. We denote its *dual Banach* space by A', which is the set of all continuous or, equivalently, bounded linear functionals on A with the usual norm.

Now, let A and B be Banach spaces. An operator

$$L : A \to B$$

is *nuclear* if there exist sequences of vectors $\{g_i\}$, $g_i \in B$ with $\|g_i\|_2 \leq 1$, functionals $\{f_i'\}$, $f_i' \in A'$ with $\|f_i'\|_2 \leq 1$, and complex numbers λ_i with $\sum_{i=1}^{\infty} |\lambda_1| < \infty$ such that L can be represented by

$$L(h) = \sum_{i=1}^{\infty} \lambda_i \, f_i'(h) \, g_i$$

with the sum converging in the operator norm.

We call the operator L *nuclear of order* 0 if the values λ_i additionally satisfy

$$\sum_{i=1}^{\infty} |\lambda_i|^p < \infty$$

for every $p > 0$.

For a nuclear operator L acting on an arbitrary Banach space, it is possible to give conditions under which traces of the operator L exist and that they are equal. In particular, if L is a nuclear operator of order 0, then trace (L) exists.

A.20 Geodesics on the Upper Half-Plane (Poincaré Model of a Hyperbolic Plane)

[187] [100, §2.1] We briefly introduce the concept of geodesics on the Riemannian surface $\mathbb{H}$ equipped with the hyperbolic metric $\mathrm{d}s$, as introduced in §1.3.6 and refer to [187] for general background information about the geodesics on the hyperbolic surface $\mathbb{H}$.

Recall the hyperbolic distance function $d : \mathbb{H} \times \mathbb{H} \to \mathbb{R}_{\geq 0}$ introduced in (1.60). Following [100, §2.1], we have an alternative variational characterization of the hyperbolic distance function.

Definition A.1 Let $z, z' \in \mathbb{H}$ and $L : [0, 1] \to \mathbb{H}$ be a smooth map with $L(0) = z$ and $L(1) = z'$. We call L a *smooth curve* in $\mathbb{H}$ *from* z *to* z'.

Sometimes, we denote its coordinate projections by $x : [0, 1] \to \mathbb{R}$ and $y : [0, 1] \to \mathbb{R}_{>0}$ where

$$[0, 1] \to \mathbb{H}; \quad t \mapsto L(t) = x(t) + iy(t).$$

Lemma A.2 ([100, p. 25]) *Let $z, w \in \mathbb{H}$ be two distinct points in the upper half-plane. The hyperbolic distance function d in* (1.60) *satisfies*

$$d\big(z, w\big) = \min_{L} \int_0^1 \sqrt{\big(x'(t)\big)^2 + \big(y'(t)\big)^2}\, \frac{\mathrm{d}t}{y(t)}, \tag{A.2}$$

where the minimum is taken over the range of smooth curves L, with $L(t) = x(t) + iy(t)$, from z to w.

Definition A.3 Let $z, w \in \mathbb{H}$ be two distinct points in the upper half-plane. A curve L which admits the minimum in (A.2) is called a *geodesic segment* connecting z and w.

A *geodesic* G is a smooth map $G : \mathbb{R} \to \mathbb{H}$ such that each restriction $G\big|_{[t_0,t_1]}$ is a geodesic segment from $G(t_0)$ to $G(t_1)$, up to re-parametrization.

We may view geodesics as *oriented* if we fix the orientation, e.g., $G\big|_{[t_0,t_1]}$.

Fig. A.2 An illustration of different geodesics in $\mathbb{H}$.

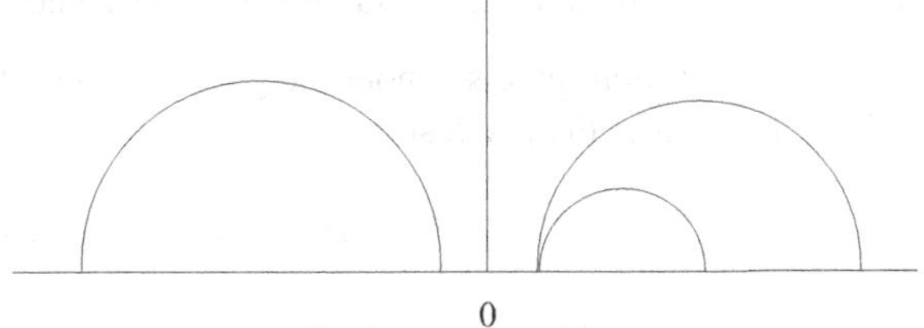

Remark A.4 A physical interpretation is that geodesics are those paths on which a particle or atom would travel on the Riemannian surface $\mathbb{H}$ with the hyperbolic metric $\mathrm{d}s$ if it is not accelerated or influenced by an external force. In this interpretation, we have that the orientation induces a sense of future and past: the particle or atom will travel in one direction and return from the other side (Figure A.2).

We can describe geodesics in the usual presentation of $\mathbb{H}$ as a subset of the usual Euclidean complex plane as follows:

Lemma A.5 ([100, p. 26]) *The set of (non-oriented) geodesics on the Riemannian surface $\mathbb{H}$ with the hyperbolic metric $\mathrm{d}s$ is given by the set of Euclidean semi-circles and vertical lines on $\mathbb{H}$ as a subset of the Euclidean complex plane.*

This means that the set

$$\bigcup_{x\in\mathbb{R}} \big\{z\colon \Re(z) = x\big\} \cup \bigcup_{\substack{x_0\in\mathbb{R}\\ r\in\mathbb{R}_{>0}}} \big\{z \in \mathbb{H}\colon\ |z - x_0| = r\big\}, \tag{A.3}$$

i.e., the union of all vertical lines and all semi-circles with origin on $\mathbb{R}$, describes all geodesics on $\mathbb{H}$.

Lemma A.6 *Let γ denote a geodesic on $\mathbb{H}$. Then γ can be characterized by a tuple (γ_-, γ_+) of distinct points $\gamma_-, \gamma_+ \in \mathbb{R}\cup\{\infty\}$, $\gamma_- \neq \gamma_+$.*

We use the notation $\gamma = (\gamma_-, \gamma_+)$.

Proof (of Lemma A.6) We consider two cases: either $\gamma = \big\{z; \Re(z) = x\big\}$ for some $x \in \mathbb{R}$ is a vertical line or $\gamma = \big\{z \in \mathbb{H}\colon\ |z - x_0| = r\big\}$ for some $x_0 \in \mathbb{R}$ and $r \in \mathbb{R}_{>0}$ is a semi-circle.

$\gamma = \big\{z; \Re(z) = x\big\}$ Putting either $\gamma_- = x$ and $\gamma_+ = \infty$ or $\gamma_- = \infty$ and $\gamma_+ = x$ describes the line in $\mathbb{H}$ between the real point x and the cusp $i\infty$. The choice is the orientation: if we orient the geodesic from x to $i\infty$ or orient the geodesic from $i\infty$ towards x.

$\gamma = \big\{z \in \mathbb{H}\colon\ |z - x_0| = r\big\}$ Its points intersecting with the boundary $\mathbb{R}\cup\{\infty\}$ of $\mathbb{H}$ are $x_0 - r$ and $x_0 + r$. Putting either $\gamma_- = x_0 - r$ and $\gamma_+ = x_0 + r$ or $\gamma_- = x_0 + r$ and $\gamma_+ = x_0 - r$ describes the semi-circle in $\mathbb{H}$ between the real points $x_0 - r$ and $x_0 + r$. The choice is the orientation: if we orient the geodesic from $x_0 - r$ to $x_0 + r$ or orient the geodesic from $x_0 + r$ towards $x_0 - r$.

In any case, the tuple (γ_-, γ_+) identifies the oriented geodesic. □

To be a bit more precise about the geodesics and their connection to $\mathrm{SL}_2(\mathbb{R}) \cup \mathrm{GL}_2(\mathbb{Z})$, consider the geodesic

$$i\mathbb{R}_{>0} = \{z \in \mathbb{H}\colon \Re(z) = 0\} \tag{A.4}$$

as a reference geodesic. We can obtain every other geodesic γ on the upper half-plane $\mathbb{H}$ in the following way. There exists a matrix $\begin{pmatrix} a & b \\ c & d \end{pmatrix} \in \mathrm{SL}_2(\mathbb{R}) \cup \mathrm{GL}_2(\mathbb{Z})$ such that the map

$$i\mathbb{R}_{>0} \to \gamma\colon \quad z \mapsto \begin{pmatrix} a & b \\ c & d \end{pmatrix} z,$$

using the Möbius transformation defined in (6.19), has the geodesic γ as its image. This motivates the definition of the bijection

$$i\mathbb{R}_{>0} \to \gamma\colon \quad z \mapsto \begin{pmatrix} a & b \\ c & d \end{pmatrix} z \tag{A.5}$$

between the reference geodesic $i\mathbb{R}_{>0}$ in (A.4) and the geodesic γ.

Remark A.7 The map (A.5) motivates the definition of *Möbius transforms on geodesics*. For a given geodesic γ and an element $\begin{pmatrix} a & b \\ c & d \end{pmatrix} \in \mathrm{SL}_2(\mathbb{R}) \cup \mathrm{GL}_2(\mathbb{Z})$, we define $\begin{pmatrix} a & b \\ c & d \end{pmatrix} \gamma$ as

$$\begin{pmatrix} a & b \\ c & d \end{pmatrix} \gamma := \left\{\begin{pmatrix} a & b \\ c & d \end{pmatrix} z\colon \ z \in \gamma\right\}. \tag{A.6}$$

This set is again a geodesic on the upper half-plane $\mathbb{H}$ since (A.5) maps the reference geodesic $i\mathbb{R}_{>0}$ in (A.4) to another geodesic.

Starting with the reference geodesic $i\mathbb{R}_{>0}$, we consider the map

$$\mathbb{R} \to i\mathbb{R}_{>0}; \quad t \mapsto i\,e^t. \tag{A.7}$$

This way, we run through the geodesic $i\mathbb{R}_{>0}$ upwards from 0 toward $i\infty$ if we run through the set $\mathbb{R}$ from $-\infty$ to ∞. This gives us an orientation on the reference geodesic $i\mathbb{R}_{>0}$ (Figure A.3).

A geodesic $\gamma = \begin{pmatrix} a & b \\ c & d \end{pmatrix} i\mathbb{R}_{>0}$ can be orientated similarly by taking

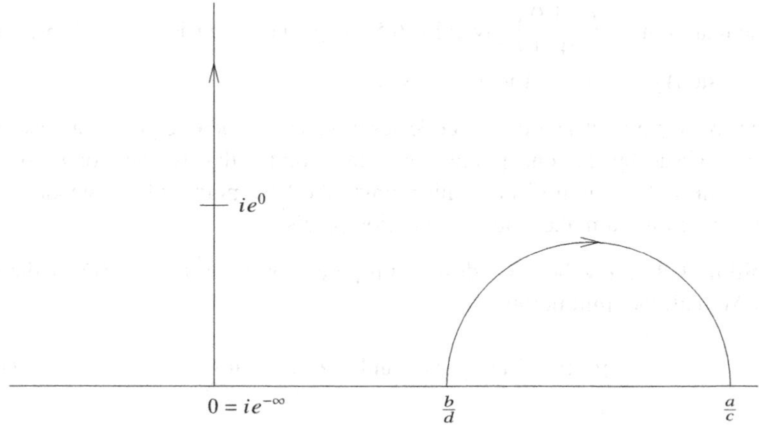

Fig. A.3 An illustration of the reference geodesic $i\mathbb{R}_{>0}$ in (A.7) and the geodesic $\begin{pmatrix} a & b \\ c & d \end{pmatrix} i\mathbb{R}_{>0}$ in (A.8).

$$\mathbb{R} \to \gamma: \quad t \mapsto \gamma(t) := \begin{pmatrix} a & b \\ c & d \end{pmatrix} i\, e^t = \frac{iae^t + b}{ice^t + d} \tag{A.8}$$

for some suitable matrix $\begin{pmatrix} a & b \\ c & d \end{pmatrix} \in \mathrm{SL}_2(\mathbb{R}) \cup \mathrm{GL}_2(\mathbb{Z})$.

Example A.8

1. The matrix $T = \begin{pmatrix} 0 & -1 \\ 1 & 0 \end{pmatrix}$ given in (1.20) maps the geodesic $i\mathbb{R}_{>0}$ onto itself and reserves the orientation given in (A.7):

$$\mathbb{R} \to i\mathbb{R}_{>0}; \quad t \mapsto T\, ie^t = \frac{-1}{ie^t} = ie^{-t}. \tag{A.9}$$

   Here $T\, ie^t$ runs from $i\infty$ towards 0 along the geodesic line $i\mathbb{R}_{>0}$ if t runs from $-\infty$ to ∞ in $\mathbb{R}$.

2. The matrix $C = \begin{pmatrix} 0 & 1 \\ 1 & 0 \end{pmatrix}$ given in (6.5) maps also the geodesic $i\mathbb{R}_{>0}$ onto itself and reverses the orientation given in (A.7):

$$\mathbb{R} \to i\mathbb{R}_{>0}; \quad t \mapsto C\, ie^t = \frac{1}{\overline{ie^t}} = ie^{-t}. \tag{A.10}$$

   Here $C\, ie^t$ runs from $i\infty$ towards 0 along the geodesic line $i\mathbb{R}_{>0}$ if t runs from $-\infty$ to ∞ in $\mathbb{R}$.

3. The matrix $A = \begin{pmatrix} -1 & 0 \\ 0 & 1 \end{pmatrix}$ given in (6.5) maps the geodesic $\gamma = (\gamma_-, \gamma_+)$ to the geodesic $A\gamma = (A\gamma_-, A\,\gamma_+) = (-\gamma_-, -\gamma_+)$.

The above parametrization of geodesics motivates another equivalent description of the geodesic by its "end points" or "limit points" that we get for $t \to \pm\infty$. We call these "end points" or "limit points" the base points of the geodesic. The following definition makes the above notion precise.

Definition A.9 Let γ be a geodesic with parametrization $t \mapsto \gamma(t)$ of the form (A.8). We call the limit points

$$\gamma_- := \lim_{t\to-\infty} \gamma(t) \quad \text{and} \quad \gamma_+ := \lim_{t\to+\infty} \gamma(t) \tag{A.11}$$

the *base points* of the geodesic γ and denote $\gamma = (\gamma_-, \gamma_+)$ (Figure A.4).

Lemma A.10 *Let γ be a geodesic on $\mathbb{H}$ with base points (γ_-, γ_+). The base points satisfy:*

1. $\gamma_- \neq \gamma_+$ *and*
2. $\gamma_-, \gamma_+ \in \mathbb{R} \cup \{\infty\}$.
3. *We can recover the geodesic γ from its base points (γ_-, γ_+).*

Proof Consider the reference geodesic $i\mathbb{R}_{>0}$ in (A.4) and its parametrization in (A.7). We also obviously have

$$\gamma_- = \lim_{t\to-\infty} i\,e^t = 0 \quad \text{and} \quad \gamma_+ = \lim_{t\to+\infty} i\,e^t = i\infty.$$

We have obviously $\gamma_- \neq \gamma_+$ and $\gamma_-, \gamma_+ \in \mathbb{R} \cup \{\infty\}$ for the reference geodesic $\gamma = i\mathbb{R}_{>0}$. Moreover, we can recover the reference geodesic $i\mathbb{R}_{>0}$ from its base points 0 and $i\infty$ by connecting the boundary points with the line $i\mathbb{R}_{>0}$.

Now, consider an arbitrary geodesic γ parameterized as in (A.8). We have

$$\gamma_- = \lim_{t\to-\infty} \begin{pmatrix} a & b \\ c & d \end{pmatrix} i\,e^t = \begin{pmatrix} a & b \\ c & d \end{pmatrix} 0 = \frac{c}{d}$$

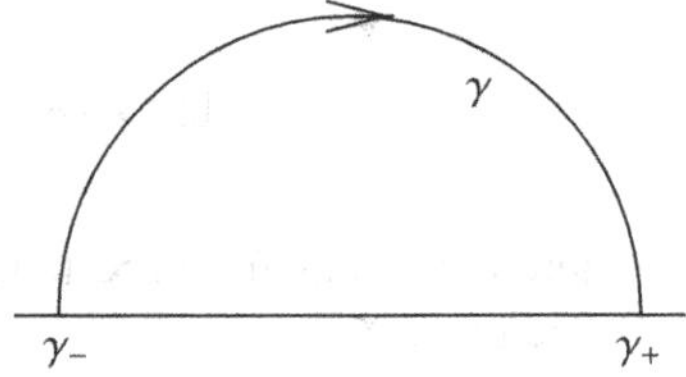

Fig. A.4 An illustration of an oriented geodesic γ and its base points (γ_-, γ_+).

and

$$\gamma_+ = \lim_{t \to +\infty} \begin{pmatrix} a & b \\ c & d \end{pmatrix} i\, e^t = \begin{pmatrix} a & b \\ c & d \end{pmatrix} i\infty = \frac{a}{c}.$$

The condition $\begin{pmatrix} a & b \\ c & d \end{pmatrix} \in \mathrm{SL}_2(\mathbb{R})$ implies $\frac{a}{c} \neq \frac{b}{d}$ which shows the first part. Since $a, b, c, d \in \mathbb{R}$, we also see that $\frac{a}{c}, \frac{b}{d} \in \mathbb{R} \cup \{\infty\}$, where we interpret $\frac{\cdot}{0} = \infty$. This proves the second part of the lemma.

Given the base points $\gamma_{\pm}$, we can write them as fractions $\gamma_+ = \frac{a}{c}$ and $\gamma_- = \frac{b}{d}$ such that $a, b, c, d \in \mathbb{R}$, and $ad - bc = 1$. Hence, we can recover γ by writing it as image of the reference geodesic $i\mathbb{R}_{>0}$ using the matrix $\begin{pmatrix} a & b \\ c & d \end{pmatrix} \in \mathrm{SL}_2(\mathbb{R})$. This shows the last part of the lemma. □

References

1. S. Ahlgren, Distribution of the partition function modulo composite integers M. Math. Ann. **318**(4), 795–803 (2000)
2. S. Ahlgren, N. Andersen, Kloosterman sums and Maass cusp forms of half-integral weight for the modular group. Int. Math. Res. Not. 2, 492–570 (2018)
3. S. Ahlgren, B. Kim, Mock modular grids and Hecke relations for mock modular forms. Forum Math. **26**(4), 1261–1287 (2014)
4. N. Andersen, Classification of congruences for mock theta functions and weakly holomorphic modular forms. Q. J. Math. **65**(3), 781–805 (2014)
5. N. Andersen, Periods of the j-function along infinite geodesics and mock modular forms. Bull. Lond. Math. Soc. **47**(3), 407–417 (2015)
6. N. Andersen, Arithmetic of Maass forms of half-integral weight. Ph.D. thesis, University of Illinois, 2016
7. N. Andersen, Vector-valued modular forms and the mock theta conjectures. Res. Number Theory **2**, 14 (2016). Art. 32
8. G.E. Andrews, On the theorems of Watson and Dragonette for Ramanujan's mock theta functions Am. J. Math. **88**, 454–490 (1966)
9. G.E. Andrews, *Partitions: Yesterday and Today* (New Zealand Mathematical Society, Wellington, 1979)
10. G.E. Andrews, Ramanujan's "lost" notebook. I. Partial θ-functions. Adv. Math. **41**(2), 137–172 (1981)
11. G.E. Andrews, The geometric series in calculus. Am. Math. Mon. **105**(1), 36–40 (1998). https://doi.org/10.2307/2589524
12. T.M. Apostol, *Modular Functions and Dirichlet Series in Number Theory*. Graduate Texts in Mathematics, vol. 41, 2nd edn. (Springer, New York, 1990). ISBN 978-0-387-97127-8. https://doi.org/10.1007/978-1-4612-0999-7
13. E. Artin, *Ein mechanisches System mit quasiergodischen Bahnen*. Abh. Math. Sem. Univ. Hamburg **3**(1), 170–175 (1924). MR3069425. https://doi.org/10.1007/BF02954622; http://www.ams.org/mathscinet-getitem?mr=3069425
14. L. Bers, F. John, M. Schechter, *Partial Differential Equations*. Lectures in Applied Mathematics, vol. 3 (Interscience Publishers, Geneva, 1964). ISBN 978-0-8218-0049-2
15. G. Bol, Invarianten linearer Differentialgleichungen. Abh. Math. Sem. Univ. Hamburg **16**(3–4), 1–28 (1949). MR33411. http://www.ams.org/mathscinet-getitem?mr=33411; https://doi.org/10.1007/BF03343515
16. A.R. Booker, A. Strömbergsson, A. Venkatesh, Effective computation of Maass cusp forms. Int. Math. Res. Not. **2006**, 71281. https://doi.org/10.1155/IMRN/2006/71281

T. Mühlenbruch, W. Raji, *On the Theory of Maass Wave Forms*, Universitext,
https://doi.org/10.1007/978-3-030-40475-8

17. A. Borel, *Automorphic Forms on $SL_2(\mathbb{R})$* (Cambridge University Press, Cambridge, 1997). ISBN 978-0511896064. https://doi.org/10.1017/CBO9780511896064
18. K. Bringmann, K. Ono, The $f(q)$ mock theta function conjecture and partition ranks. Invent. Math. **165**(2), 243–266 (2006). https://doi.org/10.1007/s00222-005-0493-5
19. K. Bringmann, K. Ono, Lifting cusp forms to Maass forms with an application to partitions. Proc. Natl. Acad. Sci. U. S. A. **104**(10), 3725–3731 (2007). https://doi.org/10.1073/pnas.0611414104
20. K. Bringmann, K. Ono, Coefficients of harmonic Maass forms. in *Partitions, q-Series, and Modular Forms*. Developments in Mathematics, vol. 23 (Springer, New York, 2012), pp. 23–38. https://doi.org/10.1007/978-1-4614-0028-8_3
21. K. Bringmann, N. Diamantis, M. Raum, Mock period functions, sesquiharmonic Maass forms, and non-critical values of L-functions. Adv. Math. **233**, 115–134 (2013). https://doi.org/10.1016/j.aim.2012.09.025
22. K. Bringmann, P. Guerzhoy, Z. Kent, K. Ono, Eichler-Shimura theory for mock modular forms. Math. Ann. **355**(3), 1085–1121 (2013). https://doi.org/10.1007/s00208-012-0816-y
23. K. Bringmann, B. Kane, R.C. Rhoades, Duality and differential operators for harmonic Maass forms, in *From Fourier Analysis and Number Theory to Radon Transforms and Geometry*. Developments in Mathematics, vol. 28 (Springer, New York, 2013), pp. 85–106. ISBN 978-1-4614-4074-1. https://doi.org/10.1007/978-1-4614-4075-8_6
24. K. Bringmann, K.-H. Fricke, Z. Kent, Special L-values and periods of weakly holomorphic modular forms. Proc. Am. Math. Soc. **142**(10), 3425–3439 (2014). MR3238419. http://www.ams.org/mathscinet-getitem?mr=3238419; https://doi.org/10.1090/S0002-9939-2014-12092-2
25. K. Bringmann, P. Guerzhoy, B. Kane, On cycle integrals of weakly holomorphic modular forms. Math. Proc. Cambridge Philos. Soc. **158**(3), 439–449 (2015). MR3335420. http://www.ams.org/mathscinet-getitem?mr=3335420; https://doi.org/10.1017/S0305004115000055
26. K. Bringmann, A. Folsom, K. Ono, L. Rolen, *Harmonic Maass Forms and Mock Modular Forms: Theory and Applications*. Colloquium Publications, vol. 64 (American Mathematical Society, Providence, RI, 2017). ISBN 978-1-4704-1944-8, ISBN 978-1-4704-4313-9
27. R.W. Bruggeman, Fourier coefficients of cusp forms. Invent. Math. **45**, 1–18 (1978). MR0472701. http://www.ams.org/mathscinet-getitem?mr=0472701
28. R.W. Bruggeman, Modular forms of varying weight. I. Math. Z. **190**(4), 477–495 (1985). MR808915. http://www.ams.org/mathscinet-getitem?mr=808915; https://doi.org/10.1007/BF01214747
29. R.W. Bruggeman, Modular forms of varying weight. II. Math. Z. **192**(2), 297–328 (1986). https://doi.org/10.1007/BF01179430
30. R.W. Bruggeman, Modular forms of varying weight. III. J. Reine Angew. Math. **371**, 144–190 (1986). https://doi.org/10.1515/crll.1986.371.144
31. R.W. Bruggeman, *Families of Automorphic Forms*. Monographs in Mathematics (Birkhäuser Boston Inc., Boston, MA, 1994). ISBN 3-7643-5046-6
32. R.W. Bruggeman, Y. Choie, Modular cocycles and cup product. Adv. Math. **351**, 296–342 (2019). arXiv:1811.10359. https://arxiv.org/abs/1811.10359; https://doi.org/10.1016/j.aim.2019.05.008
33. R.W. Bruggeman, T. Mühlenbruch, Eigenfunctions of transfer operators and cohomology. J. Number Theory **129**(1), 158–181 (2009). https://doi.org/10.1016/j.jnt.2008.08.003
34. R.W. Bruggeman, M. Fraczek, D. Mayer, Perturbation of zeros of the Selberg zeta function for $\Gamma_0(4)$. Exp. Math. **22**(3), 217–242 (2013). https://doi.org/10.1080/10586458.2013.776381
35. R.W. Bruggeman, J. Lewis, D. Zagier, Function theory related to the group $PSL_2(\mathbb{R})$, in *From Fourier Analysis and Number Theory to Radon Transforms and Geometry*. Developments in Mathematics, vol. 28 (Springer, New York, 2013), pp. 107–201. ISBN 978-1-4614-4074-1. https://doi.org/10.1007/978-1-4614-4075-8_7

36. R.W. Bruggeman, J. Lewis, D. Zagier, Period functions for Maass wave forms and cohomology. Mem. Am. Math. Soc. **237**(1118) (2015). ISBN 978-1-4704-1407-8; 978-1-4704-2503-6. https://doi.org/10.1090/memo/1118
37. R.W. Bruggeman, Y. Choie, N. Diamantis, Holomorphic automorphic forms and cohomology. Mem. Am. Math. Soc. **253**, (1212) (2018). ISBN 978-1-4704-2855-6, ISBN 978-1-4704-4419-8. https://doi.org/10.1090/memo/1212
38. J.H. Bruinier, Borcherds products and Chern classes of Hirzebruch-Zagier divisors. Invent. Math. **138**(1), 51–83 (1999)
39. J.H. Bruinier, Harmonic Maass forms and periods. Math. Ann. **357**(4), 1363–1387 (2013)
40. J.H. Bruinier, J. Funke, On two geometric theta lifts. Duke Math J. **125**(1), 45–90 (2004)
41. J.H. Bruinier, J. Funke, Traces of CM values of modular functions. J. Reine Angew. Math. **594**, 1–33 (2006)
42. J.H. Bruinier, Y. Li, Heegner divisors in generalized Jacobians and traces of singular moduli. Algebra Number Theory **10**(6), 1277–1300 (2016)
43. J.H. Bruinier, K. Ono, Heegner divisors, L-functions and harmonic weak Maass forms. Ann. Math. (2) **172**(3), 2135–2181 (2010)
44. J.H. Bruinier, K. Ono, Algebraic formulas for the coefficients of half-integral weight harmonic weak Maass forms. Adv. Math. **246**, 198–219 (2013)
45. J.H. Bruinier, M. Schwagenscheidt, Algebraic formulas for the coefficients of mock theta functions and Weyl vectors of Borcherds products. J. Algebra **478**, 38–57 (2017)
46. J.H. Bruinier, F. Strömberg, Computation of harmonic weak Maass forms. Exp. Math. **21**(2), 117–131 (2012)
47. J.H. Bruinier, W. Kohnen, K. Ono, The arithmetic of the values of modular functions and the divisors of modular forms. Compos. Math. **140**(3), 552–566 (2004)
48. J.H. Bruinier, K. Ono, R. Rhoades, Differential operators and harmonic weak Maass forms. Math. Ann. **342**, 673–693 (2008). https://doi.org/10.1007/s00208-008-0252-1. Errata: Math. Ann. **345**, 31 (2009). https://doi.org/10.1007/s00208-009-0338-4
49. J.H. Bruinier, J. Funke, Ö. Imamoḡlu, Regularized theta liftings and periods of modular functions. J. Reine Angew. Math. **703**, 43–93 (2015)
50. J.H. Bruinier, B. Howard, T. Yang, Heights of Kudla-Rapoport divisors and derivatives of L-functions. Invent. Math. **201**(1), 1–95 (2015)
51. J. Bolte, F. Steiner, *Hyperbolic Geometry and Applications in Quantum Chaos and Cosmology*. London Mathematical Society, Lecture Note Series, vol. 397. (Cambridge University Press, Cambridge, 2012). MR2885180. http://www.ams.org/mathscinet-getitem?mr=2885180
52. D. Bump, *Automorphic Forms and Representations*. Cambridge Studies in Advanced Mathematics, vol. 55 (Cambridge University Press, Cambridge, 1997). ISBN 0-521-55098-X
53. D. Bump, J.W. Cogdell, E. de Shalit, D. Gaitsgory, E. Kowalski, S.S. Kudla, *An Introduction to the Langlands Program* (Birkhäuser Boston, Cambridge, 2004). ISBN 978-0-8176-3211-3. https://doi.org/10.1007/978-0-8176-8226-2
54. V.A. Bykovskii, Functional equations for Hecke-Maass series. Funct. Anal. Appl. **34**, 23–32, 95 (2000)
55. C.-H. Chang, D. Mayer, Thermodynamic formalism and Selberg's zeta function for modular groups. Regul. Chaotic Dyn. **5**(3), 281–312 (2000). MR1789478. http://www.ams.org/mathscinet-getitem?mr=1789478; https://doi.org/10.1070/rd2000v005n03ABEH000150
56. N. Chernov, R. Markarian, *Chaotic Billiards*. Mathematical Surveys and Monographs, vol. 127 (American Mathematical Society, Providence, RI, 2006). ISBN 0-8218-4096-7. https://doi.org/10.1090/surv/127
57. Y.J. Choie, D. Zagier, Rational period functions for PSL(2, $\mathbb{Z}$), in *A Tribute to Emil Grosswald: Number Theory and related Analysis*, ed. by M. Knopp, M. Sheingorn. Contemporary Mathematics, vol. 143 (American Mathematical Society, Providence, RI, 1993), pp. 89–108

58. D. Choi, S. Lim, T. Mühlenbruch, W. Raji, Series expansion of the period function and representations of Hecke operators. J. Number Theory **171**, 301–340 (2017). https://doi.org/10.1016/j.jnt.2016.07.020
59. A. Cohn, Über die Anzahl der Wurzeln einer algebraischen Gleichung in einem Kreise. Math. Z. **14**, 110–148 (1922). https://doi.org/10.1007/BF01215894
60. J.B. Conrey, D.W. Farmer, Ö. Imamoglu, The nontrivial zeros of period polynomials of modular forms lie on the unit circle. Int. Math. Res. Not. **2013**, 4758–4771 (2013). https://doi.org/10.1093/imrn/rns183
61. R. Courant, D. Hilbert, *Methods of Mathematical Physics. Vol. I* (Interscience Publishers, Geneva, 1953). MR65391. http://www.ams.org/mathscinet-getitem?mr=65391
62. P. Cvitanović, R. Artuso, R. Mainieri, G. Tanner, G. Vattay, *Chaos: Classical and Quantum*. Edition 16.0 from 2018-01-22 (Niels Bohr Institute, Copenhagen, 2018). http://chaosbook.org/
63. A. Dabholkar, S. Murthy, D. Zagier, *Quantum Black Holes, Wall Crossing, and Mock Modular Forms* (2012). Preprint. arXiv:1208.4074v2[hep-th]. https://arxiv.org/abs/1208.4074
64. P. Deligne, Formes modulaires et représentations l-adiques. Séminaire Bourbaki **355** (février 1969); Lect. Notes Math. **179**, 139–172 (1971)
65. N. Diamantis, L. Rolen, Period polynomials, derivatives of L-functions, and zeros of polynomials. Res. Math. Sci. **5**(9) (2018). https://doi.org/10.1007/s40687-018-0126-4
66. F. Diamond, J. Shurman, *A First Course in Modular Forms*. Graduate Texts in Mathematics (Springer, New York, 2005). ISBN 978-0-387-23229-4. https://doi.org/10.1007/978-0-387-27226-9
67. W. Duke, Almost a century of answering the question: what is a mock theta function? Not. Am. Math. Soc. **61**(11), 1314–1320 (2014). https://doi.org/10.1090/noti1185
68. W. Duke, P. Jenkins, Integral traces of singular values of weak Maass forms. Algebra Number Theory **2**(5), 573–593 (2008)
69. W. Duke, Y. Li, Harmonic Maass forms of weight 1. Duke Math. J. **64**(1), 39–113 (2015)
70. W. Duke, Ö. Imamoḡlu, Á. Tóth, Rational period functions and cycle integrals. Abh. Math. Semin. Hambg. **80**, 255–264 (2010)
71. W. Duke, Ö. Imamoḡlu, Á. Tóth, Regularized inner products of modular functions. Ramanujan J. **41**, 13–29 (2016). https://doi.org/10.1007/s11139-013-9544-5
72. J.F.R. Duncan, M.J. Griffin, K. Ono, Proof of the umbral moonshine conjecture. Res. Math. Sci. 2–26 (2015). https://doi.org/10.1186/s40687-015-0044-7
73. M. Eichler, Eine Verallgemeinerung der Abelschen Integrale. Math. Z. **67**, 267–298 (1957). https://doi.org/10.1007/BF01258863
74. M. Eichler, D. Zagier, *The Theory of Jacobi Forms*. Progress in Mathematics, vol. 55 (Birkhäuser Boston, Cambridge, 1985). ISBN 978-0-8176-3180-2
75. A. El-Guindy, W. Raji, Unimodularity of roots of period polynomials of Hecke eigenforms. Bull. Lond. Math. Soc. **46**, 528–536 (2014). https://doi.org/10.1112/blms/bdu007
76. Encyclopedia of Mathematics (2013). http://www.encyclopediaofmath.org/
77. J.D. Fay, Fourier Coefficients of the Resolvent for a Fuchsian Group. J. Reine Angew. Math. **293/294**, 143–203 (1977). https://doi.org/10.1515/crll.1977.293-294.143
78. A. Folsom, Mock and quantum modular forms. Plenary Lecture at the Connecticut Summer School in Number Theory 2016. https://ctnt-summer.math.uconn.edu/schedules-and-abstracts/. Video: https://www.youtube.com/playlist?list=PLJUSzeW191Qzs7obSgtg7mK5Zn1qNVvno. PDF: http://ctnt-summer.math.uconn.edu/wp-content/uploads/sites/1632/2016/02/CTNT2016talk-Amanda-Folsom.pdf
79. A. Folsom, K. Ono, Duality involving the mock theta function $f(q)$. J. Lond. Math. Soc. (2) **77**, 320–334 (2008). https://doi.org/10.1112/jlms/jdm119

80. M. Fraczek, D. Mayer, A realization of the Hecke algebra on the space of period functions for $\Gamma_0(n)$. J. Reine Angew. Math. **603**, 133–163 (2007). https://doi.org/10.1515/CRELLE.2007.014
81. M. Fraczek, D. Mayer, Symmetries of the transfer operator for $\Gamma_0(N)$ and a character deformation of the Selberg zeta function for $\Gamma_0(4)$. Algebra Number Theory **6**(3), 587–610 (2012). https://doi.org/10.2140/ant.2012.6.587
82. E. Freitag, *Siegelsche Modulfunktionen*. Grundlehren der Mathematischen Wissenschaften, vol. 254 (Springer, Berlin, Heidelberg, 1983). ISBN 978-3-642-68650-4. https://doi.org/10.1007/978-3-642-68649-8
83. E. Freitag, *Hilbert Modular Forms* (Springer, Berlin, Heidelberg, 1990). ISBN 978-3-540-50586-0. https://doi.org/10.1007/978-3-662-02638-0
84. P.B. Garrett, *Holomorphic Hilbert Modular Forms* (Wadsworth & Brooks/Cole Advanced Books & Software, Pacific Grove, CA, 1990). ISBN 0-534-10344-8
85. S.S. Gelbart, *Automorphic Forms on Adele Groups*. Annals of Mathematics Studies, vol. 83 (Princeton University Press, Princeton, 1975). ISBN 978-0-691-08156-4
86. S.S. Gelbart, An elementary introduction to the Langlands program. Bull. Am. Math. Soc. (N.S.) **10**(2), 177–219 (1984)
87. J. Gimenez, T. Mühlenbruch, W. Raji, Construction of vector-valued modular integrals and vector-valued mock modular forms. Ramanujan J. 1–64 (2014). https://doi.org/10.1007/s11139-014-9606-3
88. A. Good, Cusp forms and eigenfunctions of the Laplacian. Math. Ann. **255**(4), 523–548 (1984). https://doi.org/10.1007/BF01451932
89. I.S. Gradshteyn, I.M. Ryzhik, *Table of Integrals, Series, and Products*, 4th edn. (Academic, New York/London, 1965). MR197789. http://www.ams.org/mathscinet-getitem?mr=197789
90. I.S. Gradshteyn, I.M. Ryzhik, *Table of Integrals, Series, and Products*. (Academic, New York, 1980). ISBN 0-12-294760-6
91. A. Grothendieck, Produits tensoriels topologiques et espaces nucléaires. Mem. Am. Math. Soc. **16** (1955). MR0075539. http://www.ams.org/mathscinet-getitem?mr=75539
92. G.H. Hardy, E.M. Wright, *An Introduction to the Theory of Numbers*, 4th edn. (Oxford University Press, Oxford, 1975)
93. E. Hecke, Über die Bestimmung Dirichletscher Reihen durch ihre Funktionalgleichung. Math. Ann. **112**, 664–699 (1936)
94. D.A. Hejhal, The Selberg trace formula and the Riemann zeta function. Duke Math. J. **43**(3) (1976), 441–482. https://doi.org/10.1215/S0012-7094-76-04338-6
95. D.A. Hejhal, *The Selberg Trace Formula for* PSL(2, $\mathbf{R}$), *Vol. 2*. Lecture Notes in Mathematics, vol. 1001 (Springer, Berlin, 1983). ISBN 3-540-12323-7
96. J. Hilgert, D. Mayer, H. Movasati, Transfer operators for $\Gamma_0(n)$ and the Hecke operators for period functions of PSL(2, $\mathbb{Z}$). Math. Proc. Camb. Philos. Soc. **139**, 81–116 (2005). https://doi.org/10.1017/S0305004105008480
97. A. Hurwitz, Ueber die angenäherte Darstellung der Zahlen durch rational Brüche. Math. Ann. **44**, 417–436 (1894). https://doi.org/10.1007/BF01446417
98. S.Y. Husseini, M. Knopp, Eichler cohomology and automorphic forms. Ill. J. Math. **15**, 565–577 (1971). http://projecteuclid.org/euclid.ijm/1256052512
99. M.N. Huxley, Scattering matrices for congruence subgroups, in *Modular Forms (Durham, 1983)*. Ellis Horwood Series in Mathematics and its Applications: Statistics and Operational Research (Horwood, Chichester, 1984), pp. 141–156. MR803366. http://www.ams.org/mathscinet-getitem?mr=803366
100. H. Iwaniec, *Topics in Classical Automorphic Forms*. Graduate Studies in Mathematics, vol. 17 (American Mathematical Society, Providence, RI, 1997). ISBN 0-8218-0777-3. https://doi.org/10.1090/gsm/017
101. H. Iwaniec, *Spectral Methods of Automorphic Forms*. Graduate Studies in Mathematics, vol. 53, 2nd edn. (American Mathematical Society, Providence, RI); Revista Matemática Iberoamericana, Madrid, 2002. ISBN 0-8218-3160-7. https://doi.org/10.1090/gsm/053

102. H. Iwanieca, P. Sarnak, The non-vanishing of central values of automorphic L-functions and Landau-Siegel zeros. Israel J. Math. **120**, 155–177 (2000)
103. T. Kato, *Perturbation Theory for Linear Operators*. Reprint of the 1980 edition. Classics in Mathematics (Springer, Berlin, 1995). ISBN 3-540-58661-X
104. H. Klingen, *Introductory Lectures on Siegel Modular Forms* (Cambridge University Press, Cambridge, 1990). ISBN 978-0-511-61987-8. https://doi.org/10.1017/CBO9780511619878
105. H.D. Kloosterman, On the representation of numbers in the form $ax^2 + by^2 + cz^2 + dt^2$. Acta Math. **49**(3–4), 407–464 (1927). https://doi.org/10.1007/BF02564120
106. M. Knopp, *Modular Functions in Analytic Number Theory* (Markham Publishing Co., Chicago, IL, 1970). ISBN 9780841010000. MR265287. http://www.ams.org/mathscinet-getitem?mr=265287
107. M. Knopp, W. Raji, Eichler cohomology for generalized modular forms II. Int. J. Number Theory **6**(5), 1083–1090 (2010). https://doi.org/10.1142/S179304211000340X
108. M. Knopp, J. Lehner, W. Raji, Eichler cohomology for generalized modular forms. Int. J. Number Theory **5**(6), 1049–1059 (2009). https://doi.org/10.1142/S1793042109002547
109. N. Koblitz, *Introduction to Elliptic Curves and Modular Forms*, 2nd edn. Graduate Texts in Mathematics, vol. 97 (Springer, New York, 1993). ISBN 0-387-97966-2. https://doi.org/10.1007/978-1-4612-0909-6
110. J. Kramer, Leben und Werk von Martin Eichler. Elemente der Mathematik **49**, 45–60 (1994). Überarbeitete PDF-Fassung
111. T. Kubota *Elementary Theory of Eisenstein Series* (Kodansha Ltd., Tokyo/Wiley, New York, 1973)
112. C. Kurth, L. Long, Computations with finite index subgroups of $\mathrm{PSL}_2(\mathbb{Z})$ using Farey symbols, in *Advances in Algebra and Combinatorics* (World Scientific Publishing, Hackensack, NJ, 2008), pp. 225–242. https://doi.org/10.1142/9789812790019_0015
113. M. Lalín, C. Smyth, Unimodularity of zeros of self-inversive polynomials. Acta Math. Hungar. **138**, 85–101 (2013). https://doi.org/10.1007/s10474-012-0225-4
114. S. Lang, *Algebraic Number Theory*, 2nd edn. Graduate Texts in Mathematics, vol. 110 (Springer, New York, 1994). ISBN 0-387-94225-4
115. S. Lang, $\mathrm{SL}_2(\mathbf{R})$ (Springer, New York, 1985). Corrected second printing, 1998. ISBN 0-387-96198-4
116. S. Lang, *Introduction to Modular Forms*, 3rd edn. Grundlehren der Mathematischen Wissenschaften, vol. 222 (Springer, Berlin, 2001). ISBN 978-3-540-07833-3. https://doi.org/10.1007/978-3-642-51447-0
117. R.P. Langlands, Problems in the theory of automorphic forms, in *Lectures in Modern Analysis and Applications, III*. Lecture Notes in Mathematics, vol. 170 (Springer, New York, 1970), pp. 18–61. ISBN 978-3-540-05284-5. https://doi.org/10.1007/BFb0079065
118. P.D. Lax, R.S. Phillips, *Scattering Theory for Automorphic Functions*. Annals of Mathematics Studies, vol. 87 (Princeton University Press, Princeton, NJ, 1976). ISBN 0-691-08179-4
119. R. Lawrence, D. Zagier, Modular forms and quantum invariants of 3-manifolds, Sir Michael Atiyah: a great mathematician of the twentieth century, Asian J. Math. **3**, 93–107 (1999)
120. H.W. Lenstra Jr., Solving the Pell equation. Not. Am. Math. Soc. **49**(2), 182–192 (2002). MR1875156 http://www.ams.org/mathscinet-getitem?mr=1875156
121. J. Lewis, Spaces of holomorphic functions equivalent to the even Maass cusp forms. Invent. Math. **127**, 271–306 (1997). https://doi.org/10.1007/s002220050120
122. J. Lewis, D. Zagier, Period functions for Maass wave forms. I. Ann. Math. **153**, 191–258 (2001)
123. S. Liu, Nonvanishing of central L-values of Maass forms. Adv. Math. **332**, 403–437 (2018). arXiv:1702.07084v2. https://arxiv.org/abs/1702.07084v2; https://doi.org/10.1016/j.aim.2018.05.017
124. The LMFDB Collaboration, *The L-Functions and Modular Forms Database* (2019). http://www.lmfdb.org. Accessed 4 Apr 2020
125. L. Long, Finite index subgroups of the modular group and their modular forms (2007). Preprint. arXiv:0707.3315v1. https://arxiv.org/abs/0707.3315v1

126. A. Lozano-Robledo, *Elliptic Curves, Modular Forms, and Their L-Functions* (American Mathematical Society, Providence, RI, 2015). ISBN 978-0-8218-5242-2
127. W. Luo, Z. Rudnick, P. Sarnak, On Selberg's eigenvalue conjecture. Geom. Funct. Anal. **5**(2), 387–401 (1995). https://doi.org/10.1007/BF01895672
128. H. Maass, *Über eine neue Art von nichtanalytischen automorphen Funktionen und die Bestimmung Dirichletscher Reihen durch Funktionalgleichungen*. Math. Ann. **121**, 141–183 (1949). https://doi.org/10.1007/BF01329622
129. H. Maass, *Lectures on Modular Functions of One Complex Variable*, 2nd edn. (Tata Institute of Fundamental Research, Bombay, 1983). http://www.math.tifr.res.in/~publ/ln/tifr29.pdf
130. W. Magnus, F. Oberhettinger, R.P. Soni, *Formulas and Theorems for the Special Functions of Mathematical Physics*. 3rd edn. Die Grundlehren der mathematischen Wissenschaften, vol. 52 (Springer, New York, 1966). MR0232968. http://www.ams.org/mathscinet-getitem?mr=232968
131. Ju.I. Manin, Periods of cusp forms, and p-adic Hecke series. Mat. Sb. (N. Ser.) **92**(134), 378–401 (1973). English translation appeared in: Math. USSR-Sb. **21**(3), 371–393 (1973). MR345909. http://www.ams.org/mathscinet-getitem?mr=345909
132. J. Marklof, Selberg's trace formula: an introduction. in *Hyperbolic Geometry and Applications in Quantum Chaos and Cosmology*, ed. by J. Bolte, F. Steiner (Cambridge University Press, Cambridge, 2011), pp. 83–119
133. B.N. Maurizi, Construction of an ordinary Dirichlet series with convergence beyond the Bohr strip. Missouri J. Math. Sci. **25**, 110–133 (2013). MR3161629. http://www.ams.org/mathscinet-getitem?mr=3161629. Zbl 06242000: http://zbmath.org/?q=an:06242000
134. D. Mayer, On a ζ function related to the continued fraction transformation. Bulletin de la S. M. F. **104**, 195–203 (1976)
135. D. Mayer, On the thermodynamic formalism for the Gauss map. Commun. Math. Phys. **130**(2), 311–333 (1990). MR1059321. http://www.ams.org/mathscinet-getitem?mr=1059321
136. D. Mayer, The thermodynamic formalism approach to Selberg's zeta function for PSL(2,Z). Bull. A.M.S. **25**, 55–60 (1991)
137. D. Mayer, Continued fractions and related transformations, in *Ergodic Theory, Symbolic Dynamics and Hyperbolic Spaces*, ed. by T. Bedford, M. Keane, C. Series (Oxford University Press, New York, 1991), pp. 175–222
138. D. Mayer, T. Mühlenbruch, From the transfer operator for geodesic flows on modular surfaces to the Hecke operators on period functions of $\Gamma_0(n)$, in *Algebraic and Topological Dynamics*. Contemporary Mathematics, vol. 385 (American Mathematical Society, Providence, RI, 2005), pp. 137–161. https://doi.org/10.1090/conm/385/07195
139. D. Mayer, F. Strömberg, Symbolic dynamics for the geodesic flow on Hecke surfaces. J. Mod. Dyn. **2**(4), 581–627 (2008). https://doi.org/10.3934/jmd.2008.2.581
140. D. Mayer, T. Mühlenbruch, F. Strömberg, The transfer operator for the Hecke triangle groups. Discrete Contin. Dyn. Syst. **32**(7), 2453–2484 (2012). https://doi.org/10.3934/dcds.2012.32.2453
141. L. Merel, Universal Fourier expansions of modular forms. in *On Artkins Conjecture for Odd 2-Dimensional Representations*, ed. by G. Frey. Lecture Notes in Mathematics, vol. 1585 (Springer, Berlin, 1994)
142. T. Miyake, *Modular Forms* (Springer, Berlin, 1989). ISBN 0-540-50268-8
143. M. Möller, A. Pohl, Period functions for Hecke triangle groups, and the Selberg zeta function as a Fredholm determinant. Ergodic Theory Dyn. Syst. **33**(1), 247–283 (2013). https://doi.org/10.1017/S0143385711000794
144. H.L. Montgomery, R.C. Vaughan, *Multiplicative Number Theory. I. Classical Theory*. Cambridge Studies in Advanced Mathematics, vol. 97 (Cambridge University Press, Cambridge, 2007). ISBN 978-0-521-84903-6
145. Y. Motohashi, *Spectral Theory of the Riemann Zeta-Function* (Cambridge University Press, Cambridge, 1997). ISBN 978-0-511-98339-9. https://doi.org/10.1017/CBO9780511983399
146. Y. Motohashi, A note on the mean value of the zeta and L-functions. XII. Proc. Jpn. Acad. Ser. A Math. Sci. **78**(3), 36–41 (2002)

147. T. Mühlenbruch, Systems of automorphic forms and period functions. Ph.D. Thesis, Utrecht University, 2003
148. T. Mühlenbruch, Hecke operators on period functions for the full modular group. Int. Math. Res. Not. 77, 4127–4145 (2004). https://doi.org/10.1155/S1073792804143365
149. T. Mühlenbruch, Hecke operators on period functions for $\Gamma_0(n)$. J. Number Theory **118**, 208–235 (2006). https://doi.org/10.1016/j.jnt.2005.09.003
150. T. Mühlenbruch, W. Raji, Generalized Maass wave forms. Proc. Am. Math. Soc. **141**, 1143–1158 (2013)
151. T. Mühlenbruch, W. Raji, Eichler integrals for Maass cusp forms of half-integral weight. Ill. J. Math. **57**(2), 445–475 (2013)
152. R. Narasimhan, Y. Nievergelt, *Complex Analysis in One Variable*, 2nd edn. (Birkhäuser Boston Inc., Boston, 2001). ISBN 0-8176-4164-5
153. NIST Digital Library of Mathematical Functions. Version 1.0.8. Release date 25 April 2014. https://dlmf.nist.gov/
154. J.J. O'Connor, E.F. Robertson, *Goro Shimura*. MacTutor History of Mathematics archive, University of St Andrews. http://www-history.mcs.st-andrews.ac.uk/Biographies/Shimura.html
155. K. Ono, Unearthing the visions of a master: harmonic Maass forms and number theory. Curr. Dev. Math. **2008**, 347–454 (2009). https://projecteuclid.org/euclid.cdm/1254748659
156. PARI/GP, version 2.3.4, Bordeaux (2008). http://pari.math.u-bordeaux.fr/
157. P.C. Pasles, Multiplier systems. Acta Arith. **108**, 235–243 (2003). MR1980509. http://www.ams.org/mathscinet-getitem?mr=1980509; https://doi.org/10.4067/aa108-3-3
158. A. Pohl, A dynamical approach to Maass cusp forms. J. Mod. Dyn. **6**(4), 563–596 (2012). https://doi.org/10.3934/jmd.2012.6.563
159. A. Pohl, Period functions for Maass cusp forms for $\Gamma_0(p)$: a transfer operator approach. Int. Math. Res. Not. **14**, 3250–3273 (2013)
160. A. Pohl, Symbolic dynamics for the geodesic flow on two-dimensional hyperbolic good orbifolds. Discrete Contin. Dyn. Syst. **34**(5), 2173–2241 (2014). https://doi.org/10.3934/dcds.2014.34.2173
161. A. Pohl, A thermodynamic formalism approach to the Selberg zeta function for Hecke triangle surfaces of infinite area. Commun. Math. Phys. **337**(1), 103–126 (2015). https://doi.org/10.1007/s00220-015-2304-1
162. A. Pohl, Odd and even Maass cusp forms for Hecke triangle groups, and the billiard flow. Ergodic Theory Dyn. Syst. **36**(1), 142–172 (2016). https://doi.org/10.1017/etds.2014.64
163. L. Pukánszky, The Plancherel formula for the universal covering group of SL$(R,\ 2)$. Math. Ann. **156**, 96–143 (1964)
164. H. Rademacher, *Topics in Analytic Number Theory* (Springer, New York, 1973)
165. S. Ramanujan, *The Lost Notebook and Other Unpublished Papers* (Springer, New York, 1988). ISBN 978-3-540-18726-4. MR947735. http://www.ams.org/mathscinet-getitem?mr=947735
166. S. Ramanujan, *Collected Papers of Srinivasa Ramanujan* (AMS Chelsea Publishing, Providence, RI, 2000). ISBN 978-0-8218-2076-6. MR2280843. http://www.ams.org/mathscinet-getitem?mr=2280843
167. R.A. Rankin, On horocyclic groups. Proc. Lond. Math. Soc. (3) **4**, 219–234 (1954). MR0061676. http://www.ams.org/mathscinet-getitem?mr=61676
168. K. Ribet, W.A. Stein, *Lectures on Modular Forms and Hecke Operators*. https://wstein.org/books/ribet-stein/
169. W. Roelcke, Automorphe formen in der hyperbolishce Ebene, I. Math. Ann. **167**, 292–337 (1966)
170. W. Rudin, *Real and Complex Analysis* (McGraw-Hill, New York, 1966). ISBN 0-070-54234-1
171. D. Ruelle, Zeta-functions for expanding maps and Anosov flows. Invent. Math. **34**(3), 231–242 (1976). MR0420720. http://www.ams.org/mathscinet-getitem?mr=420720; https://doi.org/10.1007/BF01403069

172. D. Ruelle, Dynamical zeta functions for maps of the interval. Bull. Am. Math. Soc. **3**, 212–214 (1994). MR1246470. http://www.ams.org/mathscinet-getitem?mr=1246470; https://doi.org/10.1090/S0273-0979-1994-00489-6
173. D. Ruelle, *Dynamical Zeta Functions for Piecewise Monotone Maps of the Interval.* CRM Monograph Series, vol. 4 (American Mathematical Society, Providence, RI, 1994). ISBN 0-8218-6991-4. MR1274046. http://www.ams.org/mathscinet-getitem?mr=1274046
174. D. Ruelle, Dynamical Zeta Functions and Transfer Operators. Not. Am. Math. Soc. **49**(8), 887–895 (2002)
175. P. Sarnak, Selberg's eigenvalue conjecture. Not. Am. Math. Soc. **42**, 1272–1277 (1995)
176. B. Schoenberg, J.R. Smart, E. Schwandt, *Elliptic Modular Functions: An Introduction.* Grundlehren Der Mathematischen Wissenchaften Series, vol. 203 (Springer, New York, Heidelberg, Berlin, 1974)
177. A.J. Scholl, Modular forms on noncongruence subgroups, in *Séminairede Theorie des Nombres, Paris 1985–86.* Progress in Mathematics, vol. 71 (Birkhauser Boston, Boston, MA, 1987), pp. 199–206. https://doi.org/10.1007/978-1-4757-4267-1_14
178. R. Schulze-Pillot, A. Yenirce, Petersson products of bases of spaces of cusp forms and estimates for Fourier coefficients. Int. J. Number Theory **14**, 2277–2290 (2018). https://doi.org/10.1142/S1793042118501385
179. A. Selberg, On the estimation of Fourier coefficients of modular forms, in *Theory of Numbers*, ed. by A.L. Whiteman. Proceedings of Symposia in Pure Mathematics VIII (American Mathematical Society, Providence, RI, 1965), pp. 1–15. ISBN 978-0-8218-1408-6. MR0182610. http://www.ams.org/mathscinet-getitem?mr=0182610
180. A.M. Semikhatov, A. Taormina, I.Yu. Tipunin, Higher-level Appell functions, modular transformations, and characters. Commun. Math. Phys. **255**, 469–512 (2005). https://doi.org/10.1007/s00220-004-1280-7
181. C. Series, The modular surface and continued fractions. J. Lond. Math. Soc. (2) **31**(1), 69–80 (1985). MR0810563. http://www.ams.org/mathscinet-getitem?mr=810563; https://doi.org/10.1112/jlms/s2-31.1.69
182. J.P. Serre, *A Course in Arithmetic* (Springer, New York, 1973). ISBN 978-0-387-90040-7. https://doi.org/10.1007/978-1-4684-9884-4
183. G. Shimura, Sur les intégrales attachées aux formes automorphes. J. Math. Soc. Jpn. **11**, 291–311 (1959). https://doi.org/10.4099/jmath.11.291
184. G. Shimura, *Introduction to the Arithmetic Theory of Automorphic Functions* (Princeton University Press, Princeton, NJ, 1971)
185. N.P. Skoruppa, D. Zagier, Jacobi forms and a certain space of modular forms. Invent. Math. **94** (1988). https://doi.org/10.1007/BF01394347
186. L.J. Slater, *Confluent Hypergeometric Functions* (Cambridge University Press, New York, 1960). MR107026. http://www.ams.org/mathscinet-getitem?mr=107026
187. S. Stahl, *The Poincaré Half-Plane* (Jones and Bartlett, Burlington, MA, 1993). ISBN 0-86720-298-X
188. W.A. Stein, *Explicitly Computing Modular Forms*. Electronic notes, 2006. https://wstein.org/msri06/refs/stein-book-on-modular-forms.pdf
189. W.A. Stein, *Modular Forms, A Computational Approach*. Graduate Studies in Mathematics, vol. 79 (American Mathematical Society, Providence, RI, 2007). ISBN 0-8218-3960-8. http://wstein.org/books/modform/
190. E.M. Stein, R. Shakarchi, *Complex Analysis*. Princeton Lectures in Analysis II (Princeton University Press, Princeton, 2003). ISBN 978-0-691-1138-2
191. F. Strömberg, Computational aspects of Maass waveforms. PhD-thesis, Uppsala Universitet, 2005. ISBN 91-506-1794-X. URN: urn:nbn:se:uu:diva-4778: http://urn.kb.se/resolve?urn=urn%3Anbn%3Ase%3Auu%3Adiva-4778. DiVA:diva2:165722: http://uu.diva-portal.org/smash/record.jsf?pid=diva2%3A165722
192. F. Strömberg, Hecke operators for Maass waveforms on $PSL(2, \mathbb{Z})$ with integer weight and eta multiplier. Int. Math. Res. Not. **2007** (2007). https://doi.org/10.1093/imrn/rnm062

193. F. Strömberg, Computation of Selberg's zeta functions on Hecke triangle groups (2008). Preprint. arXiv:0804.4837. http://arxiv.org/abs/0804.4837
194. A. Terras, *Harmonic Analysis on Symmetric Spaces and Applications, I* (Springer, New York, 1985)
195. A. Terras, *Harmonic Analysis and Symmetric Spaces and Applications, II* (Springer, Berlin, 1988)
196. The On-Line Encyclopedia of Integer Sequences (2016). Published electronically at http://oeis.org
197. H. Then, *Maass cusp forms for large eigenvalues.* Math. Comp. **74**, 363–381 (2005). https://doi.org/10.1090/S0025-5718-04-01658-8
198. J. Troost, The Non-Compact Elliptic Genus: Mock or Modular. J. High Energy Phys. **2010**(6), 104 (2010). https://doi.org/10.1007/JHEP06(2010)104
199. A. Venkov, *Spectral Theory of Automorphic Functions and Its Applications.* Mathematics and its Applications (Soviet Series), vol. 51 (Kluwer Academic Publishers Group, Dordrecht, 1990). ISBN 0-7923-0487-X. https://doi.org/10.1007/978-94-009-1892-4
200. J. Voight, Computing fundamental domains for Fuchsian groups. J. Théor. Nombres Bordeaux **21**, 469–491 (2009). https://doi.org/10.5802/jtnb.683
201. I. Wagner, Harmonic Maass form eigencurves. Res. Math. Sci. (2018) **5**, 24. https://doi.org/10.1007/s40687-018-0141-5
202. W. Walter, *Analysis I.* Grundwissen Mathematik, vol. 3 (Springer, New York, 1985)
203. Whittaker, Watson, *A Course in Modern Analysis*, 4th edn. (Cambridge University Press, Cambridge, 1927)
204. A.J. Wiles, Modular elliptic curves and Fermat's last theorem. Ann. Math. (2) **141**(3), 443–551 (1995). MR1333035. http://www.ams.org/mathscinet-getitem?mr=1333035
205. B. Williams, Remarks on the theta decomposition of vector-valued Jacobi forms. J. Number Theory **197**, 250–267 (2019). https://doi.org/10.1016/j.jnt.2018.08.013
206. W. Winnie Li, L. Long, Z. Yang, Modular forms for noncongruence subgroups. Quart. J. Pure Appl. Math. **1**(1), 205–221 (2005). MR2155139. http://www.ams.org/mathscinet-getitem?mr=2155139
207. E. Wirtinger, Zur formalen Theorie der Funktionen von mehr komplexen Veränderlichen. Math. Ann. **97**, 357–375 (1926, in German). https://doi.org/10.1007/BF01447872
208. K. Wohlfahrt, An extension of F. Klein's level concept. Ill. J. Math. **8**, 529–535 (1964). https://doi.org/10.1215/ijm/1256059574
209. D. Zagier, Nombres de classes et formes modulaires de poids 3/2. C. R. Acad. Sc. Paris **281**, 883–886 (1975).
210. D. Zagier, Ramanujan's mock theta functions and their applications [d'aprés Zwegers and Bringmann-Ono]. Séminaire Bourbaki, 60ème année, 2006–2007, n° 986, Astérisque **326** (2009), Soc. Math. de France, pp. 143–164. MR2605321. http://www.ams.org/mathscinet-getitem?mr=2605321
211. S.P. Zwegers, Mock theta functions. PhD thesis, Utrecht University, 2002. ISBN 90-393-3155-3. http://dspace.library.uu.nl/handle/1874/878

Symbol Index

T. Mühlenbruch, W. Raji, *On the Theory of Maass Wave Forms*, Universitext,
https://doi.org/10.1007/978-3-030-40475-8

Index

T. Mühlenbruch, W. Raji, *On the Theory of Maass Wave Forms*, Universitext,
https://doi.org/10.1007/978-3-030-40475-8

N

O

P

9783030404772